Hormonal Carcinogenesis V

International Symposium on Hormonal Carcinogenesis • 1991 •

Proceedings of the
Fifth International Symposium

ADVANCES IN EXPERIMENTAL MEDICINE AND BIOLOGY

Recent Volumes in this Series

Volume 610
TARGET THERAPIES IN CANCER
Edited by Francesco Colotta and Alberto Mantovani

Volume 611
PETIDES FOR YOUTH
Edited by Susan Del Valle, Emanuel Escher, and William D. Lubell

Volume 612
RELAXIN AND RELATED PETIDES
Edited by Alexander I. Agoulnik

Volume 613
RECENT ADVANCES IN RETINAL DEGENERATION
Edited by Joe G. Hollyfield, Matthew M. LaVail, and Robert E. Anderson

Volume 614
OXYGEN TRANSPORT TO TISSUE XXIX
Edited by Kyung A. Kang, David K. Harrison, and Duane F. Bruley

Volume 615
PROGRAMMED CELL DEATH IN CANCER PROGRESSION AND THERAPY
Edited by Roya Khosravi-Far, and Eileen White

Volume 616
TRANSGENIC MICROALGAE AS GREEN CELL FACTORIES
Edited by Rosa León, Aurora Graván, and Emilio Fernández

Volume 617
HORMONAL CARCINOGENESIS V
Edited by Jonathan J. Li

Jonathan J. Li • Sara A. Li • Suresh Mohla
Henri Rochefort • Thierry Maudelonde
Editors

Hormonal Carcinogenesis V

Springer

Jonathan J. Li, PhD
University of Kansas Medical Center
Department of Pharmacology
Kansas City, KS, USA
jli1@kumc.edu

Sara A. Li, PhD
University of Kansas Medical Center
Department of Pharmacology
Kansas City, KS, USA
sli@kumc.edu

Suresh Mohla, PhD
Division of Cancer Biology
National Cancer Institute
Rockville, MD, USA
mohlas@mail.nih.gov

Henri Rochefort, MD/PhD
Montpellier University and INSERM
Hormone and Cancer Unit, U540
Montpellier, France
henri.rochefort@valdorel.fnclcc.fr

Thierry Maudelonde, MD/PhD
Department of Cell Biology
Montpellier University 1-INSERM, U540
Hospital Arnaud de Villeneuve
Montpellier, France
t-maudelonde@chu-montpellier.fr

ISBN: 978-1-4419-2399-8 e-ISBN: 978-0-387-69080-3
DOI: 10.1007/978-0387-69078-3

Printed on acid-free paper.

9 8 7 6 5 4 3 2 1

springer.com

This volume is dedicated to all young investigators who are inspired to understand both fundamental and translational aspects of hormonal oncogenesis. We challenge them to emulate the example of Theodor Boveri (1862–1915), whose dedication, keen observation, and originality have laid the foundation of our renewed understanding of neoplasia.

Acknowledgements

Patrons

USA

National Cancer Institutes of Health (NIH):
Office of Research on Women's Health
National Cancer Institute

France

Institut National du Cancer
INSERM and IFR3
La Région Languedoc Roussillon
Université Montpellier 1

Pharmaceutical Companies

Astra Zeneca
Pfizer
Wyeth
Novartis Pharma
Eli Lilly
Pierre Fabre
Sanofi Aventis
Servier
GTx

Sponsors

National Institute of Environmental Health Sciences
Canceropole GSO

Contributors

Beckman Coulter
Roche
Ligue Régionale contre le Cancer

Preface

The chapters presented in this volume show that impressive advances have been made in our understanding of hormone-associated neoplasms with respect to their causation, progression, and new approaches for their prevention and treatment. However, there is much more to accomplish in order to successfully prevent, treat, and eradicate this prevalent group of hormonal-related neoplasms; particularly in the breast, prostate, ovary, and endometrium. The incidence of these cancers continues to increase in most countries, and their etiology is not fully understood. There are a number of emerging as well as ongoing fields that are having a major impact on studies of hormonal oncogenesis. These include adult and tumor stem cells, mitotic kinases, centrosome amplification and genomic instability, biomarkers for early detection and progression, and recent advances in steroid receptor coactivation and repression, which finely regulate hormone action. Advances in proteomics and genomics will continue to be important in hormone dependency, independency, and resistance. New to this Symposium was attention to the role of hormones in colorectal and lung cancer. We are pleased that the Hormonal Carcinogenesis Symposia have facilitated communication and collaboration among investigators in this field. It is further heartening to witness the increased response to these Symposia over the years. The current meeting, our largest to date, brought together 155 attendants from 23 countries, and stands as a tribute to all involved in its organization as well as to the excellence of the science presented in this volume.

The Executive Board is especially grateful to the Scientific Advisory Board for assisting in the assembly of an outstanding program of speakers, all leaders in their respective fields. We warmly acknowledge the host country, France, both the local committee from INSERM and Alpha Visa Congres for their excellent and gracious hospitality and partnership. Their time, effort, and energy contributed to an outstanding Symposium. We also appreciate the careful examination of the manuscripts by many reviewers: Albert Brinkmann, Robert Clarke, Janet Daling, Stephen Hillier, Elwood Jensen, Scott Kahn, Jonathan Li, Sara Antonia Li, Gordon Mills, Gerald Mueller, Polly Newcomb, Ross Prentice, Henri Rochefort, Gilbert Smith, and Donald Tindall. We offer special thanks to Ms. Tandria Price and Ms. Stephanie Schock for their dedication and professionalism in their respective involvement in this Symposium and in the preparation of this volume. Finally, we especially thank the National Institutes of Health, Institut National du Cancer, other institutions, and the pharmaceutical companies for their generous support for this Symposium.

Kansas City, KS, USA	Jonathan J. Li
Kansas City, KS, USA	Sara Antonia Li
Bethesda, MD, USA	Suresh Mohla
Montpellier, France	Henri Rochefort
Montpellier, France	Thierry Maudelonde

Executive and Scientific Advisory Boards

L to R: J. Li*, E. Baulieu, S. Mohla*, J. Pickar†, S.A. Li†, E. Jensen†, M. Varella-Garcia†, H. Rochefort*, T. Maudelonde*, and S. Liao†

Absent SAB members: M. Beato, W. Brinkley, A. Brinkmann, Z. Culig, G. Mueller, J. Salisbury, D. Tindall, B. Vonderhaar

*Executive Board
†Scientific Advisory Board

Participants of the Fifth International Symposium on Hormonal Carcinogenesis

Contents

Liver Cancer

Hormone Metabolism

Contributors

Alvarez, Silvia
Instituto de Investigaciones Biomedicas, Madrid, Spain

Anose, Bynthia
Bethel University, St. Paul, MN, USA

Applanat, Martine Inserm
Unit 553 Hospital Saint Louis, Paris, France

Asadi, Farrokh
Harold Washington College Department of Biology, Chicago, IL, USA

Azzam, Diana
Department of Pathology Royal Perth Hospital, Perth, Western Australia

Barbachano, Antonio
Instituto de Investigaciones Biomedicas, Madrid, Spain

Baulieu, Etienee
French Academy of Sciences College de France,
Le Kremlin-Bicetre, France

Beato, Miguel
Centre for Genomic Regulation, Barcelona, SPAIN

Bhat, Hari K.
Dept. of Environmental Health Science Mailman School
of Public Health Columbia University, New York, NY, USA

Boizet-Bonhoure, Brigitte
Institut de Genetique Humaine, Montpellier, France

Borges, Sahra
Physiologie Integrative Cell. Et Mol. Universite Claude Bernard, Villeurbanne,
France

Boulle, Nathalie
Lab. De Biologie Cell. et Hormonale Hopital Aranud
de Villeneuve, Montpellier, France

Bret, Caroline
Inserm Unit 540, Montpellier, France

Brinkley, Willam R.
Baylor College of Medicine, Houston, TX, USA

Brinkmann, Albert O.
Department of Reproduction and Dev. Erasmus University Center,
Rotterdam, Netherlands

Busson, Muriel
Inserm Unit 540, Montpellier, France

Cance, Jean-Yves
Inserm Unit 540, Montpellier, France

Cavailles, Vincent
Inserm Unit 540, Montpellier, France

Cavarretta, Ilaria
Department of Urology Innsbruck Medical University,
Anichstrasse, Innsbruck, Austria

Chalbos, Dany
Endocrinologie Moleculaire et Cellulaire des Cancers, Montpellier, France

Chambon, Monique
Inserm Unit 540, Montpellier, France

Chauchereau, Anne
Business Unit Prostate lab de Recherche Translationnelle
Institut Gustave Roussy, Villejuif, France

Chung, Chuhan
Section of Digestive Diseases Yale University School of Medicine,
West Haven, CT, USA

Clarke, Michael F.
Deputy Director of the Stem and Regenerative Medicine
Inst. Stanford University, Palo Alto, CA, USA

Clarke, Robert
Breast Biology Group Division of Cancer Studies University of Manchester
Christie Hospital (NHS) Trust, Manchester, United Kingdom

Cornwell, Mona
Northwestern University Feinberg School of Medicine Department of Pathology,
Chicago, IL, USA

Coulomb, Helene Oncology Department Ipsen Research Laboratories Institut Henri Beaufour, Les Ulis, France

Crawford, Susan
Northwestern University Feinberg School of Medicine Department of Pathology, Chicago, IL, USA

Culig, Zoran
Department of Urology University of Innsbruck, Anichstrasse, Innsbruck, Austria

Culine, Stephanie
Clinical Research Montpellier Cancer Center, Montpellier, France

Cuzick, Jack
Cancer Research UK Centre for Epidemiol., Mathematics, and Statistics, Wolfson Institute of Preventive Medicine, London, United Kingdom

Daling, Janet R.
Fred Hutchinson Cancer Research Center, Seattle, WA, USA

Davies, Richard
Department of Surgery Hackensack University Medical Center, Hackensack, NJ, USA

Diaz-Prado, Silvia
PINAR 6 BIS 4G, Madrid, SPAIN

Drai, Laurent
Market Development Manager Immunodiagnostics and Nucleic Acid Testing Business Center, Marseille, France

Dromard, Mathilde
Inserm Unit 540 Mol. and Cell. Endocrin. of Cancers, Montpellier, France

Dubois, Jean Bernard
Montpellier Cancer Center, Montpellier, France

Duensing, Stefan M.
Molecular Virology Program University of Pittsburgh Cancer Institute, Pittsburgh, PA. USA

Dumay, Anne
Institut Gustave Roussy, Villejuif, France

Duong, Vanessa
Inserm Unit 540, Montpellier, France

Edery, Marc
Museum National D'Histoire Naturelle, Paris, France

Eeckhoute, Jerome
Dana Farber Cancer Institute, Boston, MA, USA

Escande, Aurelie
Inserm Unit 540, Montpellier, France

Esslimani-Sahla Majida
Department of Pathologie CRLC Val D'Aurelle ParcEuromedecine, Montpellier, France

Fabre, Alban
Inserm ERI-20 Institut Gustave Roussy, Villejuif, France

Fabris, Victoria
Lab de Carcinogenesis Hormonal Instituto de Biologia Y Med. Exp., Vuelta de Obligado, Capital Federal, Argentina

Fajas, Lluis
Inserm Unit 540, Montpellier, France

Fournier, Agnes
Inserm ERI-20 Institut Gustave Roussy, Villejuif, France

Gance, Christine IMS
Health, London, United Kingdom

Garcia, Marcel
Inserm Unit 540, Montpellier, France

Ghosh, Paramita
Research Services VA Northern California HCS, Mather, CA, USA

Giton, Frank
Faculte de Medecine Centre de Recherches Chirurgicales, Creteil, France

Gulyaeva, Lyudmila F.
Insitute of Molecular Biology and Biophysics Academy of Medical Sciences, Novosibirsk, RUSSIA

Gunnarsson, Cecilia
IBK/Oncology Linkoping University, Linkoping, Sweden

Guzman, Raphael C.
491 Life Science Addition University of California, Berkeley, CA, USA

Hankinson, Susan E.
Harvard Medical School Channing Laboratory, Boston, MA, USA

Heemers, Hannelore V.
Mayo Clinic Department of Urology Research, Rochester, MN, USA

Hillier, Stephen G.
University of Edinburgh Center for Reproductive Biology The Queen's Medical Research Institute, Edinburgh, Scotland

Hontz, Adrianne (Beth)
Department of Pharmacology University of Kansas University Medical Center, Kansas City, KS, USA

Howell, Anthony
CRC Department Medical Oncology University of Manchester Christie Hospital and Holt Radium Inst., Manchester, united kingdom

Jacobson, Herbert
Albany Medical College, Albany, NY, USA

Jalagier, Stephan
Inserm Unit 540, Montpellier, France

Jansson, Agneta
IBK/Oncology Linkoping University, Linkoping, Sweden

Jensen, Elwood
Dept. of Cell Biology Vontz Center for Molecular Studies University of Cincinnati, Cincinnati, OH, USA

John, Placheril J.
Department of Zoology University of Rajasthan, Jaipur, India

Junco Barranco, Jesús A
Center for Genetic Engineering and Biotechnology of Camaguey, Camaguey, CUBA

Kahn, Scott
Columbia University/Roosevelt Hospital Department of Urology, New York, NY, USA

Karlsson, Sandra
University of Skövde School of Life Sciences – Biomedicine, Skövde, Sweden

Kerr, Alan
Hormone Receptor Laboratory Department of Biochemistry and Molecular Biology University of Louisville – HSC, Louisville, KY, USA

Kim, Sung-Hoon
Lab of Angiogenesis and Chemoprevention Oriental Medical College Kyunghee University, Seoul, Korea

Kobiela, Jarek
Department of Medical Chemistry Medical University of Gdansk, Gdansk, Poland

Lafanechere, Sylviane
Beckman Coulter, Chaska, MN, USA

Lapouge, Gaëlle
Laboratoire de Cancerologie Experimentale et de Radiobiologie, Strasbourg, France

Larsson, Dennis
University of Skövde School of Life Sciences – Biomedicine, Skövde, Sweden

Lazennec, Gwendal
Inserm Unit 540, Montpellier, France

Lehmann, Leane
Institute of Applied Biosciences University of Karlsruhe, Karlsruhe, Germany

Lemamy, Guy Joseph
Faculte de Medecine Universite des Sciences de la Sante, Libreville (Central Africa), Gabon

Li , Jonathan J.
Department of Pharmacology University of Kansas University Medical Center, Kansas City, KS, USA

Li, Sara Antonia
Department of Pharmacology University of Kansas University Medical Center, Kansas City, KS, USA

Liao, Shutsung
Ben May Institute for Cancer Research Center of Integrative Sciences University of Chicago, Chicago, IL, USA

Liaudet-Coopman,
Emmanuelle Inserm Unit 540, Montpellier, France

Lingle, Wilma L.
Departmental Laboratory of Medicine and Pathology Mayo Clinic Foundation, Rochester, MN, USA

Logothetis, Christopher J.
Department of GU Medical Oncology UT MD Anderson Cancer Center, Houston, TX, USA

Loius, Sherif
Manitoba Institute of Cell Biology, Winnipeg, MB, Canada

Mai, Sabine
Manitota Institute of Cell Biology, Winnipeg, MB, Canada

Markicevic, Milan
Laboratory for Receptors Department of Experimental Oncology Institute of Oncology and Radiology of Serbia, Serbia, Yugoslavia

Márquez-Garbán, Diana C.
University of California – LA Division of Hematology-Oncology,
Los Angeles, CA, USA

Maskarinec, Gertraud
Cancer Research Center of Hawaii, Honolulu, HI, USA

Maudelonde, Thierry
Department of Cell Biology University of Montpellier 1-INSERM Unit 540
Hospital Arnaud de Villeneuve, Montpellier, France

Menendez, Javier A.
Fundación Privada de Investigación Biomédica de Girona Dr. Joseph Trueta
Instituto Catalá de Oncología de Girona, Catalonia, Spain

Mertani, Hichem-Claude
Physiologie Moleculaire Universite Claude Bernard Lyon1, Villeurbanne, France

Metivier, Raphaël
Campus de Beaulieu Batiment 13, Rennes, France

Mignone, Joseph
Ponte Vedra Beach, FL, USA

Mills, Gordon
Department of Molecular Therapeutics MD Anderson Cancer Center, Houston,
TX, USA

Mohla, Suresh
Division of Cancer Biology National Cancer Institute, Rockville, MD, USA

Mohler, James
Urologic Oncology University of North Carolina, Chapel Hill, NC, USA

Mostaghel, Elahe
Fred Hutchinson Cancer Research Center, Seattle, WA, USA

Munoz, Alberto
Instituto de Investigaciones Biomedicas, Madrid, Spain

Musgrove, Elizabeth
Cancer Research Program Garvan Institute, Darlinghurst, Sydney, Australia

Nandi, Satyabrata
Cancer Research Laboratory University of California, Berkeley, CA, USA

Newcomb, Polly A.
Cancer Prevention Public Health Science Fred Hutchinson Cancer Research
Center, Seattle, WA, USA

Nguyen, Annelise
Department of Diagnostic Medicine/Pathology Kansas State University,
Manhattan, KS, USA

Nicolas, Jean-Claude
Inserm Unit 540, Montpellier, France

Nirde, Philippe
Inserm Unit 540, Montpellier, France

Nordlander, Carola
Department of Cellular and Molecular Biology-Genetics Lundberg Institute, Göteborg, Sweden

Notarnicola, Cecile
Institut de Genetique Humaine, Montpellier, France

Olsson, Birgit
IBK/Oncology Linköping University, Linköping, Sweden

Ostadrahimi, Alireza
Department of Nutrition Health and Nutrition Faculty Tabriz University of Medicinal Science, Tabriz, Iran

Ostrander, Elaine A.
National Human Genome Research Institute National Institutes of Health, Bethesda, MD, USA

Ostrowski, Michael C.
Department of Molecular and Cellular Biochemistry Ohio State University, Columbus, OH, USA

Pfeffer, Ulrich
Functional Genomics National Cancer Research Institute, Genova, Italy

Pickar, James H.
Clinical Research and Development Wyeth Research, Collegeville, PA, USA

Pinn, Vivian W.
Office of Research on Women's Health, Bethesda, MD, USA

Prentice, Ross L.
Public Health Sciences Fred Hutchinson Cancer Research Center, Seattle, WA, USA

Prigent, Claude
Universite Rennes Fac. de Medecine, Rennes, France

Raccurt, Mireille
Physiologie Integrative Cell. Et Mol. Universite Claude Bernard Lyon1-43, Villeurbanne, France

Rochefort, Henri
Montpellier University and INSERM Unit 540 Hormone and Cancer Unit, Montpellier, France

Roger, Pascal
Lab. D'Anatomie et Cytologie Pathol. Hospital Lapeyronie, Montpellier, France

Roy-Burman, Pradip
Department of Pathology University of Southern California Medical School, Los Angeles, CA, USA

Salisbury, Jeffrey L.
Tumor Biology Program Mayo Clinic, Rochester, MN, USA

Samuelson, Emma
Göteborg University Sahlgrenska Academy Department of Clinical Genetics, Göteborg, Sweden

Sancho-Garnier, Helene
Montpellier Cancer Center Parc Euromedecine, Montpellier, France

Seera, Sreenivasa Rao
Department of Pathology Kidwai Memorial Institute of Oncology, Bangalore, INDIA

Seewaldt, Victoria
Duke University Cancer Center, Durham, NC, USA

Sen, Subrata Mol. Pathology MD
Anderson Cancer Center, Houston, TX, USA

Sierralta Lundin, Walter Daniel
INTA-Universidad de Chile, Santiago, Chile

Singer, Dinah S.
National Cancer Institute, Rockville, MD, USA

Solassol, Jerome
INSERM Unit 540, Montpellier, France

Stål, Olle
IBK/Oncology Linkoping University, Linkoping, Sweden

Tasanen, Merja
Orion Pharma, Turku, Finland

Tawfik, Ossama
Department of Pathology and Laboratory Medicine University of Kansas University Medical Center, Kansas City, KS, USA

Terry, Stephane
Gauguin, Creteil, France

Thatcher, Greg
Department of Medicinal Chemistry and Pharmacology College of Pharmacy University of Illinois, Chicago, IL, USA

Tindall, Donald
Department of Urology Mayo Clinic, Rochester, MN, USA

Trossö, Fredrik
Högskolan i Skövde, Skövde, Sweden

Valle, Noelia
Instituto de Investigaciones Biomedicas, Madrid, Spain

Vanacker, Jean-Marc
INSERM EMI 229 Centre de Recherche en Cancerologie, Montpellier, France

Varella-Garcia, Marileila
Department of Medicine/Medicinal Oncology University of Colorado Health Science Center, Aurora, CO, USA

Vlaeminck-Guillem, Virginie
LBMC LR5, Lyon, France

Vonderhaar, Barbara
Mammary Biology And Tumorigenesis Laboratory Center for Cancer Research NCI, Bethesda, MD, USA

Vouyovitch, Cecile
Physiologie Intefrative Cell. Mol. Universite Claude Bernard, Villeurbanne, France

Vuillaume, Marie-Laure
Laboratory Diagnostic Genetique et Molecular Centre Jean Perrin, Clermont-Ferrand, France

Wagner, Jörg
Insitote of Applied Biosciences Section of Food Chemistry and Toxicology University of Karlsruhe, Karlsruhe, Germany

Warren, Rachel
Springer, New York, NY, USA

Weigel, Nancy
Dept. of Molecular and Cellular Biology Baylor College of Medicne, Houston, TX, USA

Weinstein, Bernard
Columbia University, New York, NY, USA

Wittliff, James L.
Hormone Receptor Laboratory Deptartment of Biochemistry and Molecular Biology University of Louisville, Louisville, KY, USA

Wong, Y.C.
Cancer Biology Lab Department of Anatomy Li Ka Shing Faculty of Medicine University of Hong Kong, Hong Kong, China

Wu, Hong
Dept. of Molecular and Medicinal Pharmacology UCLA School of Medicine Howard Hughes Medical Inst., Los Angeles, CA, USA

Yang, Shen K.
Cancer Etiology Branch Division of Cancer Biology National Cancer Institute, Bethesda, MD, USA

Zarghami, Nosratolla
Department of Clinical Biochemistry and RIA Drug Applied Research Center Tabriz University of Medical Science, Tabriz, Iran

Zurawa-Janicka, Dorota
Department of Biochemistry Institute of Biology University of Gdansk, Gdansk, Poland

Abbreviations

βERKO	Estrogen receptor β null
ΔMEK1	Mutant MEK1
Δraf	Mutant Raf
√	Ratio1/ratio2
μF	Microfarads
Σ	Depression angle
1,25D	1, 25-Dihydroxyvitamin D_3
1,25-MARRS	1,25D-Membrane associated rapid response steroid binding protein
2-HO-E_2	2-Hydroxyestradiol
3-D rECM	3-Dimensional prepared extracellular matrix culture
4-HO-E_2	4-Hydroxyestradiol
4-MeOEN	4-Methoxyequilenin
4-OHEN	4-Hydroxyequilenin
11βHSD	11β-Hydroxysteroid dehydrogenase
17-4-MeOEN	17β-Dihydro-4-methoxy-equilenin
AA	African American
ADH	Atypical ductal hyperplasia
ADI	Androgen depletion independent
ADT	Androgen deprivation therapy
AE	Antiestrogens
AF	Amniotic fluid
AF1,2	Activation function
AhpC	Alkyl hydroperoxide reductase
AI	Allelic imbalance
AI	Androgen independent
AIPC	Androgen-independent PC
AIs	Aromatase inhibitors
ANOVA	Analysis of variance
AP	Alkaline phosphatase
AP-1	Activated protein-1
APC/C	Anaphase promoting complex/cyclosome
APF	Aminophenylfluorescein

AR	Androgen receptor
ARE	Androgen response element
ASD	Androstenedione
ATCC	American Type Culture Collection
AU	Arbitrary units
Aur	Aurora kinase
BBD	Benign breast diseases
BC	Breast cancer
BCCL	Breast cancer derived cell lines
BdRU	Bromodeoxyuridine
BFGF	Basic fibroblast growth factor
BHC	Benzene hexa chlororide
BIRCWH	Building Interdisciplinary Research Careers in Women's Health
BM	Bone marrow
BMI	Body mass index
BN	Brown Norway
BPH	Benign prostate hyperplasia
C	Capacitance
CARM1	Coactivator-associated arginine methyltransferase
CAT	Catalase
CatD	Cathepsin D
CBI	Center for Biotechnology Information
CBP	CREB-binding protein
CBP	p300/cAMP-response element-binding protein
CD	Cathepsin D
CD	Charcoal/dextran
CD-FCS	Medium containing steroid-depleted FCS
CDSS	Computerized decision support system
c-erb-B2	HER-2
CFTRI	Central Food Technological Research Institute
CGH	Comparative genomic hybridization
CHD	Coronary heart disease
ChIP	Chromatin immunoprecipitation
CHO	Chinese hamster ovary cells
Chr	Chromosome location
CHRT	Combined hormone replacement therapy
CHU	Montpellier University Hospital
CI	Confidence interval
CIMP	CpG island methylator phenotype
CIN	Chromosomal instabilities
CIS	Carcinoma *in-situ*
CIs	Confidence intervals
CIS	Cytokine-inducible SH2 containing
CKI	Casein kinase I

CL	Cell lines
CM	Conditioned medium
COMT	Catechol-*O*-methyltransferase
CORE	Continued outcomes relevant to evista
COX2	Cyclooxygenase-2 isozyme
cP	Cyclized 9-mer peptide
CRC	Colorectal cancer
CREB	cAMP response element binding protein
CSCs	Cancer stem cells
CT	Clinical trial
C_T	Threshold-crossing
CTE	C terminal end
CV	Cardiovascular
CYP19	Aromatase enzyme
CYPs	Cytochrom P450-dependent monooxygenases
CYPc	Cytochrome c
D	Hinge domain
DAI	Daidzein
DBD	DNA-binding domain
DCIS	Ductal carcinoma *in-situ*
DDT	Dichloro diphenyl trichloro ethane
DEIS	Ductal epithelial impedance spectroscopy
DES	Diethylstilbestrol
DFS	Disease-free survival
DGGE	Denaturing gradient gel electrophoresis
DHEA	Dehydroepiandrosterone
DHEAS	Dehydroepiandrosterone sulfate
DHT	5α-Dihydrotestosterone
DIA	Digital image analysis
DM	Dietary modification
DM	Double minutes
DMBA	12-Dimethylbenz(a)anthracene
DNMT	DNA methyltransferases
HLOD	Heterogeneity logarithm of the odds
DRIP/TRAP	VDR-interacting proteins complex
DSMB	Data and Safety Monitoring Committee
dT	Oligo
E	Estrogen
E_2	17β-estradiol
E_3	Estriol
EAC	Endometrial adenocarcinoma
E-alone	Estrogen alone
EC	Endometrial cancer
ECD	Electron capture detector
ECM	Extracellular matrix

ED	Estrogen deprivation
EGCG	Epigallocatechin gallate
EGF	Epidermal growth factor
EGFR	Epidermal growth factor receptor
ELISA	Enzyme-linked immuno-sorbent assay
EMSA	Electrophoretic mobility shift assay
EMU4	E- and c-*myc*-upregulated
EOC	Epithelial ovarian cancer
EP	Estrogen and progestin
EPA	Environmental Protection Agency
EPIC	European Prospective Investigation into Nutrition and Cancer
ER	Estrogen receptor
ERβcx	Estrogen receptor β2
ERE	Estrogen response element
ERE-BP	Estrogen receptor element-binding proteins
ERGDB	Databases of E-responsive genes
ERK/MAPK	Extracellular signal-regulated mitogen activated protein kinase
EROD	Ethoxyresorufin-*O*-deethylase
ERR	Estrogen-receptor-related receptor
ERRE	Estrogen-receptor related specific response elements
ERT	Estrogen-replacement therapy
ESCC	Endometrial squamous cell carcinoma
ESI	Electrospray ionization
ESR1	Estrogen receptor α
EST	Expressed sequence tag
EUTK	Estrogen-induced ectopic uterine stem cell tumors
FAP	Familial adenomatous polyposis
FBS	Fetal bovine serum
f_c	Characteristic frequency
FCS	Fetal calf serum
FFQ	Food frequency questionnaire
FHCRC	Fred Hutchinson Cancer Research Center
FISH	Fluorescence *in-situ* hybridization
FISFRAP	Fluorescence recovery after photo-bleaching
Ful	Fulvestrant
GEN	Genistein
GFR	Growth factor receptor
GH	Growth hormone
GHR	Growth hormone receptor
GnRH	Gonadotrophin releasing hormone
GO	Gene ontology
GPR30	G protein-coupled receptor
GPx	Glutathione peroxidase

GR	Glucocorticoid receptor
GR	Glutathione reductase
GRIP-1	Glucocorticoid receptor interacting protein-1
GRO	Growth-related oncogene
GSK	Glycogen synthase kinase
GST	Glutathione-*S*-transferase
GTC	Green tea catechins
H	Hinge region
H&E	Hematoxylin and eosine
HAT	Histone acetyl transferase
HBD	Hormone-binding domain
HCC	Hepatocellular carcinoma
hCG	Human chorionic gonadotropin
HDACs	Histone deacetylase
HELU	Hyperplastic enlarged lobular units
hERα	Human estrogen receptor alpha
hGH	Human growth hormone
HL	Liver
HLH	Helix-loop-helix
hMAM	Human mammaglobin
HMECs	Human mammary epithelial cells
HMG	Normal human mammary gland
HMTs	Histone methyl transferases
HNPCC	Hereditary nonpolyposis colorectal cancer
HOC	Human ovarian cancer
HOSE	Human OSE
HP	Hewlett Packard
HPC	Hereditary prostate cancer
HPRT	Hypoxanthine-guanine phosphoribosyltransferase
hPSA-luc	Human PSA promoter tagged to a luciferase construct
HPV	Human papilloma virus
HR	Hazard ratio
HRE	Hypoxia response element
HRPC	Hormone refractive prostate cancer
HRT	Hormone replacement therapy
HSP	Heat shock protein
HSR	Homogeneously stained regions
HT	Hormone therapy
Htsg	Hypothetical tumors suppressor gene
Hz	Hertz
IBIS	International Breast Intervention Study
ICI	Internet Computer Integration Program
ICPCG	International Consortium of Prostate Cancer Genetics
Id-1	Inhibitor of differentiation
IFN-β	Interferon beta

IFN-γ	Type II interferon
IGF	Insulin-like growth factor
IGFBP	Insulin-like growth factor-binding proteins
IHC	Immunohistochemistry
IL	Interleukin
ILR	Interleukin receptor
INR	Internally normalized ratio
Iressa	International educational resource AstraZeneca
IRF-1	Interferon regulatory factor-1
ISGs	Interferon-stimulated genes
IVT	*In vitro* transcription
JAK	Janus kinases
JAMA	Journal of the American Medical Association
JHU	Johns Hopkins University
JNK	c-Jun N-terminal kinase
K18	Keratin 18
kDa	kilo Daltons
KGF	Keratin growth factor
KO	Knock-out
kΩ	kiloOhms
LBD	Ligand-binding domain
LC	Liquid chromatography
LC	Lung cancer
LC-ESI-Q-TOF	Quadrupole-time of flight mass spectrometer
LCM	Laser capture microdissection
LDH	Lactate dehydrogenase
LH	Luteinizing hormone
LCA	Lithocholic acid
LNCaP	Lymph node prostate cancer cell line
LOD	Limit of detection
LOH	Loss of heterozygosity
LOX	Lysyl oxidase
LRCs	Label-retaining cells
m/z	Mass/charge
MALDI-TOF-MS	Matrix-assisted laser desorption ionization time-of-flight mass spectrometry
MAO	Monoamine oxidase
MAP	Mitogen-activated protein
MAPK	Mitogen-activated protein kinase
MBC	Metastatic BC
MCAD	Medium chain acyl-dehydrogenase
Mcl-1	Myeloid cell leukemia 1
MC-LR	MC-leucine-arginine
MCs	Microcystins
MG	Mammary gland

MIB-I	Mindbomb homolog 1
MIN	Microsatellite
MIP	Molecular inversion probe
MISS	Membrane initiated steroid signaling
MMPs	Matrixmetalloproteinases
MMR	Mismatch repair
MMTV	Mammary tumor virus
MNU	Methylnitrosourea
MO	Month
MOD	Mean optical density
MORE	Multiple Outcomes of Raloxifene Evaluation
MPM	Malignant peritoneal mesothelioma
MPT	Mitochondrial permeability transition
MR	Mineralocorticoid
MS	Mass spectrometry
MSI	Microsatellite-unstable/instability
MSS	Microsatellite-stable
MTD	Maximum tolerated does
MTOC	Microtubule organizing center
MXC	Mixed carcinomas
myr	Myristoylated
N	Nuclear
NCIC	National Cancer Institute of Canada
NEAA	Nonessential amino acids
NFAT	Nuclear factor of activated T cells
NFκB	Nuclear factor-κB
NHS	Nurses' Health Study
NIH	National Institutes of Health
NISS	Nuclear initiated signaling
NLS	Nuclear localization signal
NPL	Nonparametric linkage
NR3	NR family three
NRs	Nuclear receptors
NSABP	National Surgical Adjuvant Breast Project
NSCLC	Nonsmall cell lung cancers
NT	Untreated rats
nVDR	Nuclear/cytosolic vitamin D receptor
OC	Ovarian cancer
OHT	4-OH-tamoxifen
ORWH	Office of Research on Women's Health
OS	Observational study
OSE	Ovarian surface epithelial
P	Progesterone
p300-HAT	p300 that lacks HAT activity
PAH	Phenylalanine hydroxylase

Pax	Paired-box
pAkt	Phosphorylated Akt
PBS	Phosphate buffered saline
PC	Prostate cancer
PCA	Principal component analysis
PCM	Pericentriolar material
PCNA	Proliferating cell nuclear antigen
PCR	Polymerase chain reaction
PCR-RFLP	Polymerase chain reaction-restriction fragment length polymorphism
PDGFR	Platelet-derived growth factor receptor
PDGFr	Platelet-derived growth factor receptor
PEBP	Phosphatidylethanolamine-binding protein
PEDF	Pigment epithelium-derived factor
pERK1/2	Phosphorylated ERK1/2
PI3K	Phosphatidylinositol 3-kinase
PI-MEC	Parity-induced mammary epithelial cells
PIN	Prostate intraepithelial neoplasia
PIN	Prostatic intraepithelial neoplasia
PKG	Specific protein kinase G
PM	Peritoneal mesothelioma
PMF	Peptide mass fingerprinting
PMH	Postmenopausal hormone
pN+	Positive lymph node status
POD	Guajacol peroxidase
Pol II	RNA polymerase II complex
PolyQ	Polyglutamine
PonA	Ponasterone A
PP1	Protein phosphatase 1
PPRE	PPAR-response elements
PR	Progesterone receptor
PS	Penicillin/streptomycin
PSA	Prostate-specific antigen
PSMA	Prostate-specific membrane antigen
PTEN	Phosphatase and tensin
QIC	Quantitative IHC
QR	Quinone-oxidoreductase
Ral	Raloxifene
RARα	Retinoid receptor α
Rb	Retinoblastoma
RC	Resistor–capacitor
Re	Epithelial resistance
Re	Resistance
REs	Response elements
RIA	Radioimmunoassay

RIN	RNA integrity number
RKIP	raf kinase inhibitor protein
RNO4	Rat chromosome 4
ROS	Reactive oxygen species
RPPA	Reverse phase tissue lysate array
RR	Relative risk
RT-PCR	Real-time polymerase chain reaction
RT-PCR	Reverse-transcriptase polymerase-chain reaction
RUTH	Raloxifene Use and the Heart
RXR	Retinoid X receptor
SAGE	Serial analysis of gene expression
SAGEmap	SAGE tag to gene mapping
SBEM	Small breast epithelial mucin
SBMA	Spinal bulbar muscular atrophy
SCCs	Squamous cell carcinomas
SCORs	Specialized Centers of Research
SD rats	Sprague-Dawley rats
SD	Standard deviation
SDS-PAGE	Sodium dodecyl sulphate-polyacrylamide gel electrophoresis
SeBP	Selenium-binding protein
SELECT	The National Clinical Selinium, Vitamin E, Chemoprevention Trial
SEM	Standard error of measurement or mean
SERD	Selective estrogen receptor downregulator
SERMs	Selective estrogen receptor modulators
SF	Serum free
SHBG	Specific membrane receptor, R_{SHBG}
SHBG	Sex hormone-binding globulin
SIR	Standardized incidence ratio
siRNA	Small/silencing interfering RNA
siRNA	Small RNA interfering
SMS	Sawai Man Singh
SNPs	Single nucleotide polymorphisms
SOCS	Suppressor of cytokine signaling
SPRDcu	Sprague-Dawley-curly3
SR	Steroid receptors
SRC-1	Steroid receptor coactivator
SRO	Smallest region of overlap
SSCP	Single-strand conformational polymorphism
STAR	Study of Tamoxifen and Raloxifene
STAT	Signal transducer and activator of transcription
SULT	Sulfotransferase
T	Testosterone
TACC	Transforming acidic coiled coil

TAM	Tamoxifen
TAQMAN	A rapid fluorophore-based real-time PCR method
Tarceva	Erlotinib
tblastn program	Translating basic local alignment search tool for nucleotide
TBP	TATA-binding protein
TCC	Transitional cell carcinoma
TCDD	Tetrachloro[p]dibenzodioxine
TCF/LEF	T cell factor/ lymphoid-enhancing factor
TCGA	The Cancer Genome Atlas
TDLU	Terminal duct lobular unit
TEP	Transepithelial potential
TFF1	pS2
TGFβ	Tumor-derived growth factor beta
TGF-β	Transforming growth factor beta
Th	T helper
TKI	Tyrosine kinase inhibitors
TLR	Toll-like receptor
Tpm	Tag/million
TPX2	Targeting protein for Xklp2
TSA	Thiol specific antioxidant
Tsc	Tubersclerosis
TSGs	Tumor suppessor genes
TT	Tetanic toxoid
UGT	UDP-glucuronosyltransferase
U gene	Undetermined gene
uPA	Urokinase type plasminogen activator
VDR	Vitamin D receptor
VDRE	Vitamin D response elements
VEGF	Vascular endothelial growth factor
v-erb B	Erythroblastosis virus
W/O	Water in oil
WBC	White blood cells
WHI	Women's Health Initiative
WSF	Without specific findings
Wt	Wild-type
XB	Extraction buffer
Yrs	Years
Z	Reactance
ZEB	Zinc finger E-box binding factor
αFP	α-Fetoprotein

SYMPOSIUM ADDRESS

Relevance of the Concept of Oncogene Addiction to Hormonal Carcinogenesis and Molecular Targeting in Cancer Prevention and Therapy

Bernard Weinstein

Prevalence of Hormonal Carcinogenesis in Humans

The topic of this Symposium "Hormonal Carcinogenesis" is highly relevant to human cancer since in the USA and Western Europe, breast cancer (BC) is the most common cancer in women and prostate cancer (PC) is the most common cancer in men. Indeed, BC and PC account for about one-third of the total cancer incidence in women and men, respectively (1). Furthermore, although a few decades ago the incidence and mortality of BC and PC were relatively low in Asian countries, in recent years both of these cancers are becoming more frequent in Japan, Korea, Singapore, and urban China (2–4). The latter findings coupled with migration studies demonstrating an increase in BC and PC in Japanese living in Hawaii and the United States clearly indicate that exogenous factors play a major role in the causation of BC and PC. Therefore, in principle, it should be possible to prevent a major fraction of these cancers by identifying and modifying the various exogenous causative factors. Presumably, the recent increased incidence of BC and PC in Asian countries reflects "westernization", but the specific factors have not been identified. This is an important and challenging task if we want to interrupt this literal epidemic of breast and prostate cancer in Asian countries.

BC Causation

Several factors have been implicated in the causation of BC (5). Although highly penetrant single gene inherited factors, like BRCA1 and BRCA2, are important in a subset of BCs, they appear to account for ~5% of all BC. Excessive exposure to ionizing radiation, particularly during puberty, is an established risk factor and this is of interest with respect to mechanisms of breast carcinogenesis, but it is not a prevalent causative factor. Reproductive factors have been frequently implicated in BC causation. These include an early age of menarche and late age of menopause, nulliparity or low parity, late age of first pregnancy, and prolonged use of postmenopausal hormone supplements. Dietary factors, including postmenopausal obesity and excess alcohol consumption are considered risk factors, while fruits, vegetables, and Vitamin D consumption may be protective. Exercise/physical activity may also be protective. Some of these risk factors, such as early menarche and late menopause, appear to act by increasing the exposure to estrogenic factors, but other growth factors like those related to the IGF1 axis may also play an important role, thus broadening the role of "hormonal factors" in breast oncogenesis. Aside from radiation, specific genotoxic carcinogens have not been established as causative factors of human BC. Personally, I believe that these aspects, and the possible role of infectious agents, are important and challenging areas for future studies on BC causation and prevention, especially since reproductive and dietary factors do not appear to totally explain variations in BC incidence in diverse populations.

PC Causation

The causative factors of human PC are less defined than those for BC (6). Familial factors appear to play a role in some patients, but, thus far, single highly penetrant mutant genes have not been identified, although several candidate chromosomal loci are under investigation. Ageing itself is a risk factor. Other possible risk factors include excess calories and fat. Fruits, vegetables, selenium, lycopene, vitamin E, and exercise may be protective but have not been well-established. As with BC, abnormalities in the IGF axis may also play a role as well as chronic prostate inflammation. The recent association of PC with the novel retrovirus XMRV (7) is a potentially important lead to PC causation in humans.

Ethnic Variations and Subtypes

Generally, it is well recognized that African American (AA) women with BC and AA men with PC have a poorer prognosis than Causation BC and PC patients (1, 5, 6). In addition, the age of onset of PC tends to be earlier in AA than in Caucasian men. At the present time, it is unclear to what extent these ethnic differences are due to disparities in cancer screening and treatment, differences in dietary or lifestyle factors, or true biologic differences related to variations in specific hereditary risk factors.

The increasing ability to subclassify subtypes of BC and PC on the basis of molecular profiling may lead to the identification of risk factors that are specific for precise subtypes of these diseases, thus clarifying their etiologies. It is already known that postmenopausal BCs are more likely to be estrogen receptor $(ER\alpha)^{+}$ than premenopausal BCs, perhaps reflecting a greater role of estrogenic factors in the development of the former disease (5).

Multistage Carcinogenesis and the Concept of Oncogene Addiction

An axiom in oncology is that human cancers often evolve through a multistage process that extends over a period of decades. The explosion of molecular biology studies within the past three decades has revealed that this process is driven by the progressive accumulation of mutations, and epigenetic abnormalities resulting in the expression of multiple genes that have diverse biochemical functions. Thus, a carcinoma of the breast or prostate often displays mutations in multiple oncogenes and tumor suppressor genes, epigenetic abnormalities that result in the increase/decrease in the expression of hundreds of genes, and chromosomal abnormalities, including aneuploidy and loss of heterozygosity at numerous loci (8, 9). Therefore, it is surprising that despite this virtual chaos in cancer cells genomes, there are several examples in both experimental systems and in cancer patients in which the reversal of only one or few of these abnormalities can profoundly inhibit cancer cell growth, and in some cases lead to improved survival rates. A few years ago, we described this phenomenon as "oncogene addiction," to emphasize the apparent dependency of some cancers on one or few genes to maintain the malignant phenotype (8). The field of hormonal carcinogenesis provided the earliest examples of the exquisite dependence of specific malignant types of cancer on specific growth factors and pathways for the maintenance of their growth and eventual tumorigenesis. Thus, in 1886 Beatson (10) demonstrated that oophorectomy in women caused the regression of BC. Subsequent experimental studies, in the early twentieth century, demonstrated that oophorectomy also decreased mammary tumors in mice (11, 12). These BC findings were paralleled to Huggins demonstration, in 1941 (13), that castration caused PC regression in men. In recent decades, the estrogen dependence of $ER\alpha^+$ human BC growth has been exploited through the use of specific ER modulators, tamoxifen and raloxifene, and aromatase inhibitors; and the dependence of PC on androgens has been exploited through the use of luteinizing hormone-releasing hormone antagonists, anti-androgens, and 5α-reductase inhibitors.

The dependence of specific types of human BC or PCs on estrogen- or androgen-related signaling pathways, respectively, despite the presence in these tumors of numerous genetic and epigenetic abnormalities, is paralleled by the exquisite dependence of various types of cancer on the activity of specific oncogenes. We have previously reviewed such examples in mouse models of human cancer, in human cancer cell lines, and in clinical studies with molecular targeting agents (8, 9) These examples are directly relevant to hormonal cancers, especially BC. Thus, when mammary tumors were induced in mice with either the *her2/neu* (14, 15), c-*myc* (16), or w*nt1* (17) oncogenes, subsequent "switching-off" of the respective oncogene caused tumor regression. Furthermore, inhibition of expression of the *her2* oncogene with an antisense oligonucleotide in a *her2*$^+$ human BC cell line inhibited the growth of these cells (18), and inhibition of the function of *her2* with the antibody Trastuzumab, combined with treatment with cytotoxic agents, is a well-established procedure for treating patients with *her2*$^+$ BC (19, 20).

It is of interest that in the above-described c-*myc* BC mice model, when the c-*myc* oncogene was switched off, although 50% of the tumors regressed, the remaining 50% showed partial regression. Furthermore, breast tumors that recurred were c-*myc* independent and some of them displayed an activated K-*ras* oncogene (16). Similarly, in the *her2/neu* breast tumor model (14), tumors that recurred were *her-2/neu* independent, and were associated with increased expression of the transcription factor Snail (15). In the *wnt-1* murine model (17), even though down regulation of *wnt-1* resulted in rapid and extensive regression of aneuploid and invasive breast tumors and pulmonary metastases, a number of breast tumors recurred that were *wnt-1* independent. Apparently, this recurrence was caused by acquisition of mutations in the *p53* tumor suppressor gene (17). These examples of "escape from oncogene addiction" may reflect the genomic instability of cancer cells, and suggest that exploiting the phenomenon of oncogene addiction in cancer therapy will require an appropriate combination of drugs to prevent this type of emergence of resistant cancer cells (9). A major challenge is to design the optimum combination of drugs.

Similar examples of targeting specific oncogenes in the therapy of PC are not yet available. On the other hand, there is evidence that even so-called "hormone independent" PC remains dependent on the continued function of the androgen receptor (AR) for tumor growth (21). In this sense they remain "addicted to" the function of the AR, which therefore remains a potential therapeutic target. This aspect is further discussed below.

The Cell Cycle Control Gene *cdk6* can Bind to and Activate the Function of the Androgen Receptor

As discussed earlier, most PCs are initially dependent on androgens for their growth. However, a major barrier to therapy is the eventual progression of the disease to what appears to be an "androgen-independent" stage. This can occur by several mechanisms including: (1) point mutations in the AR that allow it to be activated by ligands other than 5α-dihydrotestosterone (DHT); (2) AR gene amplification and increased expression, which increases ligand sensitivity; (3) increase in AR coregulators; (4) increased tumor synthesis of active androgens; and (5) activation of other signaling molecules, i.e., MAPKS, PTEN, c-*myc*, EGFR, IGF-IR, etc. that enhance or bypass the AR pathway (21, 22).

We recently discovered that, in addition to its known role of enhancing the progression of cells through the G1 phase of the cell cycle, in human PC cells, cdk6 can bind to and markedly enhance the transcriptional activity of the AR in a ligand-dependent fashion (22). Although cyclin D1 binds to and activates the serine/threonine kinase activity of cdk6, we found that cyclin D1 actually inhibited the ability of cdk6 to stimulate the activity of the AR. This finding may explain why cyclin D1 is only rarely overexpressed in human PC. Furthermore, our studies with mutant forms of cdk6 indicated that its kinase activity is not required for its ability to stimulate the AR. Presumably, cdk6 exerts its effect on the AR by associating with it and other transcriptional control proteins. Indeed, we obtained evidence that in LNCaP cells, cdk6 is associated with a transcriptional complex that contains the AR and the promoter sequence for the androgen responsive gene PSA. Several studies suggest that AA men, who are at higher risk for PC than Caucasian or Asian men, more frequently display shorter CAG repeats in exon 1 of the AR. We found that a decrease in CAG repeats in the AR enhances stimulation of its transcriptional activity by cdk6. This finding suggests that overexpression of cdk6 may be especially effective in stimulating the activity of the AR in men whose AR has shorter CAG repeats, thereby further enhancing the development of PC in these individuals. Our data indicating that cdk6 can stimulate the transcriptional activity of the AR is not confined to transient transfection reporter assays, since derivatives of LNCaP cells that stably overexpress cdk6 display increased expression of both prostate specific antigen (PSA) mRNA and protein and increased secretion of the PSA protein. In addition, the cells that overexpress cdk6 display enhanced growth in the presence of DHT. We found that a point mutated AR originally identified in human PCs, designated AR T877A, displayed a dramatic stimulation of transcriptional activity by cdk6 in the presence of DHT, 17β-estradiol, or progesterone.

Taken together, these findings suggest that in some cases of "androgen-independent" PCs, AR-mediated pathways of gene expression that enhance growth are maintained, even at low DHT levels or other steroids, by the stimulatory effects of cdk6 on mutant forms of the AR (22). Indeed, a preliminary study indicates that cdk6 is overexpressed in 44% of a series of 34 cases of PC. It will be of interest to examine a larger series of PC cases for possible correlations between overexpression of cdk6 and specific clinical and pathologic parameters. Hopefully, these clinical studies and further mechanistic studies will indicate whether cdk6 may provide a useful prognostic marker or a target for the PC therapy.

Novel Agents for the Prevention and Treatment of Breast and Prostate Cancer: Inhibitory Effects of Sulindac Derivatives on Human Prostate and Breast Cancer

Sulindac is a nonsteroidal antiinflammatory drug that has been used predominantly for its analgesic, antipyretic, and platelet inhibitory activities to treat patients with chronic inflammatory diseases. This drug has also been used for the treatment of patients with familial adenomatous polyposis (FAP) of the colon. Sulindac sulfone (exisulind), a normal oxidative metabolite of sulindac, has been shown to also induce polyp regression and prevent polyp recurrence in FAP patients, although it does not inhibit COX-2 (23). We found that the metabolites sulindac sulfide (SULS) and exisulind (EXSUL), as well as two more potent analogs of EXSUL, CP248 and CP461, induce growth inhibition and apoptosis in several human PC cell lines, irrespective of COX-1/-2 expression, bcl-2 overexpression, or androgen-dependence (24). Although the growth of normal prostate epithelial cells was also inhibited, these cells did not display apoptosis, even at high doses of these compounds. Related studies showed that oral administration of EXSUL inhibited growth and induced apoptosis in xenografts of LNCaP cell in nude mice (25). Furthermore, oral administration of EXSUL to postprostatectomy patients with PC lowered their serum PSA levels (26). In subsequent cell culture studies, we found that doses of these compounds that induce growth inhibition and apoptosis in LNCaP cells also decreased PSA secretion and cellular levels of PSA in these cells (27). The PSA effects appear to be a result of inhibition of the AR signaling pathway since the treated cells also displayed decreases in the level of the AR protein and mRNA, and inhibition of transcription of an AR promoter luciferase reporter in transient transfection assays. The same compounds also induced apoptosis and caused growth inhibition in androgen-independent PC-3 cell lines. Taken together, these data suggest that these compounds can exert growth inhibition in human PC cells by two independent mechanisms: (1) inhibition of the AR signal transduction pathway and (2) induction of apoptosis by an independent mechanism (27). The latter mechanism appears to be due to activation of the enzyme protein kinase G. Because of these dual effects, these compounds provide a novel approach to the prevention and treatment of human PC.

We found that SULS and EXSUL also cause growth inhibition, arrest cells at G_1, and induce apoptosis in human BC cell lines. These effects were associated with decreased expression of cyclin D1 (28). In a recent study, we focused on the effects of SULS and EXSUL on hormone signaling components in BC cells (29). We found that $ER\alpha^+$ and progesterone receptor $(PR)^+$ T47D BC cells were somewhat more sensitive to growth inhibition by SULS or EXSUL than ER^-/PR^- MB-MDA-468 BC cells. Further studies indicated that SULS and EXSUL caused marked down regulation of expression of the ERα and PR-A and PR-B in T47D cells, both at the mRNA and protein levels. However, neither compound caused a major change in expression of the retinoid receptor α (RARα), RARβ, or RXRα in T47D cells. SULS and EXSUL also caused a decrease in expression of the ERα in estrogen-responsive MCF-7 BC cells. Both compounds also markedly inhibited estrogen-stimulated activation of an estrogen-responsive promoter in transient transfection reporter assays. Treatment of T47D cells with specific protein kinase G (PKG) activators did not cause a decrease in ERα or PR expression. Therefore, although SULS and EXSUL can cause PKG activation, the inhibitory effects of these compounds on ERα and PR expression do not appear to be mediated by PKG. Our data suggest that the growth inhibition by SULS and EXSUL in $ER\alpha^+/PR^+$ human BC cells may be mediated, in part, by inhibition of ERα/ PR signaling. Thus, these and related compounds may also provide a novel approach to the prevention and treatment of human BC, especially those that are $ER\alpha^+$ (29).

Inhibitory Effects of EGCG and Related Catechins Present in Green Tea on Breast, Prostate, Colon, and Liver Cancer Cells

Epidemiologic and experimental studies provide evidence that epigallocatechin gallate (EGCG) and related catechins present in green tea can exert anticancer effects in various types of cancer cells, including BC and PC (30). Some of the evidence related to BC is summarized in Table 1. This evidence includes epidemiologic studies in Asian–American women (31, 32), inhibitory effects on DMBA-induced BCs in rats (33), and our studies (34) indicating that EGCG inhibits *her2* activation and signaling in human BC cells. The latter effect has also been seen by other investigators in murine BC cells (35). Furthermore, we found that EGCG acted synergistically with taxol in inhibiting the growth of human BC cells (34).

Several studies demonstrating inhibitory effects of green tea catechins (GTC) on PC are summarized in Table 2. They include epidemiologic studies (36), inhibition of 5α reductase (37), experimental studies in the TRAMP mouse model of PC (38, 39), and a recent clinical study demonstrating that oral administration of a mixture of GTC to patients inhibited the progression of high grade PIN lesions to PC (40).

In mechanistic studies utilizing human colon cancer and hepatoma cells, we found that EGCG inhibits activation of several receptor tyrosine kinases including EGFR, HER2, HER3, and IGF-1R (30, 34, 41, 42). These effects were associated with inhibition of the activation of several downstream signaling molecules, including ERK and AKT, and inhibition of several transcription factors. It is of interest that the inhibition of activation of IGF-1R was associated with decreased expression of IGF1, which activates the IGF-R, and increased expression of IGFBP-3, which inhibits the activity of IGF1 (42). Thus, EGCG can "normalize" abnormalities in the IGF1 axis in cancer cells. The latter effects are consistent with those seen with GTC by other investigators in the TRAMP mouse model of PC (38).

Table 1 Green tea catechins and breast cancer

1	Green tea intake is associated with reduced risk of BC in Asian–American women (31). This finding is linked to a catechol-*O*-methyltransferase (COMT) deficient genotype, an enzyme that metabolizes catechins (32).
2	Green tea catechins inhibit DMBA-induced development of BCs in rats, and inhibits the in vitro growth of human BC cells (33).
3	EGCG inhibits *her2* activation and signaling in human (34) and murine BC cell lines (35).

Table 2 Green tea catechins and prostate cancer

1	Epidemiologic studies suggest protective effects of green tea consumption on PC in Asian countries (36).
2	GTCs inhibit 5-α-reductase (37).
3	Oral administration of GTC (600 mg per day) to patients with high grade PIN lesions inhibited progression of these lesions to PC (40).

In view of the above diverse effects of EGCG and mixtures of GTC on cancer cells, we believe that these compounds merit further clinical trials for the prevention and treatment of BC and PC, and possibly other hormone-related cancers. In some of these studies it may be efficacious to combine these compounds with other agents (30, 34, 48, 49). Low concentrations of EGCG can exert synergistic growth inhibitory effects on human cancer cell lines when combined with other natural products or chemotherapy agents.

Future Directions in Hormonal Carcinogenesis

Finally, I want to present a brief overview of what I think are current critical issues and future important directions in the field of hormonal carcinogenesis. As emphasized earlier in this chapter, at the present time we do not have a complete understanding of the exogenous and endogenous factors that cause BC and PC and the mechanisms by which these factors enhance carcinogenesis. This information is critical to mounting more effective efforts to prevent the occurrence of these two major diseases. The possible roles of abnormalities in energy metabolism, inflammation, and infectious agents need to be further explored. With respect to the role of disturbances in energy metabolism further studies on the role of the IGF1 axis are warranted since several studies have implicated perturbation in the IGF1 axis in breast and prostate cancer causation (43, 44). As discussed earlier, it is of interest that GTC can inhibit this pathway in cancer cells. Recent studies have highlighted the role of Sirtuins (a family of NAD^+-dependent deacetylases) in metabolism, energy restriction, and ageing (45) Therefore, it may be fruitful to explore the potential roles of Sirtuins in hormonal carcinogenesis. Because of the association of obesity with BC and PC, the possible roles of adipokines, hormones secreted by fat cells, in hormonal carcinogenesis also merit investigation. The recently developed subclassification of human BC into four subtypes (46), Luminal A ($ER\alpha^+$), Luminal B ($ER\alpha^+$), Basal (ER^-/*her2*$^+$), and Basal ($ER\alpha^-$/*her2*$^-$), has profound implications with respect to both etiology, prognosis, and molecularly targeted therapy. Thus, a recent study suggests that there is a higher prevalence of basal-like breast tumors in young AA women with BC and that this might contribute to the poorer prognosis in these patients (47). It remains to be determined whether gene expression analyses will also lead to the subclassification of PCs and other hormonal cancers, and thereby enhance our understanding of the etiology and the development of more effective targeted therapies for these cancers. With respect to further exploiting the concept of oncogene addiction in the therapy of hormonal cancers, network theory and systems biology approaches may be required to comprehend the regulatory circuitry of subtypes of hormonal cancers and thereby identify the most critical oncogenes, i.e., the “Achilles heel” with respect to molecular-targeted therapy (9). Further studies on the tumor stem cell subpopulation in human hormonal cancers are likely to provide important insights into the origin of these tumors, as well as tumor latency and recurrence. Because of likely differences in profiles of gene expression between tumor stem cells and their more differentiated progeny cells, including differences in the expression of specific nuclear receptors, the state of oncogene addiction may differ between these two subpopulations (9). If this is true, then new types of drugs may be required that specifically target the tumor stem cell population.

My final future perspective is to emphasize the need for increased research funding for young investigators in the field of hormonal cancers and related disciplines. This is crucial for bringing to fruition the exciting progress in research that has already been made, and developing more effective approaches for the prevention and treatment of breast, prostate, and other hormonal cancers.

Acknowledgments The author is indebted to numerous students and collaborators who have greatly contributed to the research in his laboratory. I also thank Ms. Barbara Castro for her expertise in preparing this manuscript. This research was supported by funds from the NCI, the EIF-NCCR, NFCR, and T.J. Martell Foundations.

References

1. Jemal A, Siegel R, Ward E, et al. (2007) Cancer statistics, 2007. CA Cancer J Clin 57: 43–66.
2. Bray F, McCarron P, Parkin DM (2004) The changing global patterns of female breast cancer incidence and mortality. Breast Cancer Res 6: 229–239.
3. Sim HG, Cheng CW (2005) Changing demography of prostate cancer in Asia. Eur J Cancer 41: 834–845.
4. Park SK, Sakoda LC, Kang D, et al. (2006) Rising prostate cancer rates in South Korea. Prostate 66: 1285–1291.
5. Willett, WR, B, Hankinson SE. Non-Genetic factors in the causation of breast cancer, Personal communication.
6. Giovannaccu E (2006) in Prostate Cancer: Principles and Practice (ed. Kirby R, P. A., Feneley M, Parsons JK eds.) (Taylor & Francis, 2006).
7. Dong B, Kim S, Hong S, et al. (2007) An infectious retrovirus susceptible to an IFN antiviral pathway from human prostate tumors. Proc Natl Acad Sci USA 104(5): 1655–1660.
8. Weinstein IB (2002) Cancer. Addiction to oncogenes – the Achilles heal of cancer. Science 297: 63–64.
9. Weinstein IB, Joe AK (2006) Mechanisms of disease: Oncogene addiction – a rationale for molecular targeting in cancer therapy. Nat Clin Pract Oncol 3: 448–457.
10. Beatson G (1896) On the treatment of inoperable cases of carcinoma of the mamma: suggestions for a new method of treatment with illustrative cases. Lancet ii: 104–107 and 162–165.
11. Lathrop AE, Loeb L (1916) Further investigations on the origin of tumors in mice. III. On the part played by internal secretion in the spontaneous development of tumors. J Cancer Res 1: 1–19.
12. Murray W (1928) Ovarian secretion and tumor incidence. Cancer Res 12: 18–25.
13. Huggins CB, SR, Jr, Hodges CV (1941) Studies in prostatic cancer. II. The effect of castration on advanced carcinoma of the prostate gland. Arch Surg 209–223.
14. Moody SE, Sarkisian CJ, Hahn KT, et al. (2002) Conditional activation of Neu in the mammary epithelium of transgenic mice results in reversible pulmonary metastasis. Cancer Cell 2: 451–461.

15. Moody SE, Perez D, Pan TC, et al. (2005) The transcriptional repressor Snail promotes mammary tumor recurrence. Cancer Cell 8: 197–209.
16. D'Cruz CM, Gunther EJ, Boxer RB, et al. (2001) c-MYC induces mammary tumorigenesis by means of a preferred pathway involving spontaneous Kras2 mutations. Nat Med 7: 235–239.
17. Gunther EJ, Moody SE, Belka GK, et al. (2003) Impact of p53 loss on reversal and recurrence of conditional Wnt-induced tumorigenesis. Genes Dev 17: 488–501.
18. Colomer R, Lupu R, Bacus SS, et al. (1994) erbB-2 antisense oligonucleotides inhibit the proliferation of breast carcinoma cells with erbB-2 oncogene amplification. Br J Cancer 70: 819–825.
19. Slamon DJ, Leyland-Jones B, Shak S, et al. (2001) Use of chemotherapy plus a monoclonal antibody against HER2 for metastatic breast cancer that overexpresses HER2. N Engl J Med 344: 783–792.
20. Piccart-Gebhart MJ, Procter M, Leyland-Jones B, et al. (2005) Trastuzumab after adjuvant chemotherapy in HER2-positive breast cancer. N Engl J Med 353: 1659–1672.
21. Pienta KJ, Bradley D (2006) Mechanisms underlying the development of androgen-independent prostate cancer. Clin Cancer Res 12: 1665–1671.
22. Lim JT, Mansukhani M, Weinstein IB (2005) Cyclin-dependent kinase 6 associates with the androgen receptor and enhances its transcriptional activity in prostate cancer cells. Proc Natl Acad Sci USA 102: 5156–5161.
23. Deguchi A, Thompson WJ, Weinstein IB (2004) Activation of protein kinase G is sufficient to induce apoptosis and inhibit cell migration in colon cancer cells. Cancer Res 64: 3966–3973.
24. Lim JT, Piazza GA, Han EK, et al. (1999) Sulindac derivatives inhibit growth and induce apoptosis in human prostate cancer cell lines. Biochem Pharmacol 58: 1097–1107.
25. Goluboff ET, Shabsigh A, Saidi JA, et al. (1999) Exisulind (sulindac sulfone) suppresses growth of human prostate cancer in a nude mouse xenograft model by increasing apoptosis. Urology 53: 440–445.
26. Goluboff ET, Prager D, Rukstalis D, et al. (2001) Safety and efficacy of exisulind for treatment of recurrent prostate cancer after radical prostatectomy. J Urol 166: 882–886.
27. Lim JT, Piazza GA, Pamukcu R, et al. (2003) Exisulind and related compounds inhibit expression and function of the androgen receptor in human prostate cancer cells. Clin Cancer Res 9: 4972–4982.
28. Han EK, Arber N, Yamamoto H, et al. (1998) Effects of sulindac and its metabolites on growth and apoptosis in human mammary epithelial and breast carcinoma cell lines. Breast Cancer Res Treat 48: 195–203.
29. Lim JT, Joe AK, Suzui M, et al. (2006) Sulindac sulfide and exisulind inhibit expression of the estrogen and progesterone receptors in human breast cancer cells. Clin Cancer Res 12: 3478–3484.
30. Shimizu M, Deguchi A, Lim JT, et al. (2005) (-)-Epigallocatechin gallate and polyphenon E inhibit growth and activation of the epidermal growth factor receptor and human epidermal growth factor receptor-2 signaling pathways in human colon cancer cells. Clin Cancer Res 11: 2735–2746.
31. Sun CL, Yuan JM, Koh WP, et al. (2006) Green tea, black tea and breast cancer risk: a meta-analysis of epidemiological studies. Carcinogenesis 27: 1310–1315.
32. Wu AH, Tseng CC, Van Den Berg D, et al. (2003) Tea intake, COMT genotype, and breast cancer in Asian–American women. Cancer Res 63: 7526–7529.
33. Kavanagh KT, Hafer LJ, Kim DW, et al. (2001) Green tea extracts decrease carcinogen-induced mammary tumor burden in rats and rate of breast cancer cell proliferation in culture. J Cell Biochem 82: 387–398.
34. Masuda M, Suzui M, Lim JT, et al. (2003) Epigallocatechin-3-gallate inhibits activation of HER-2/neu and downstream signaling pathways in human head and neck and breast carcinoma cells. Clin Cancer Res 9: 3486–3491.
35. Pianetti S, Guo S, Kavanagh KT, et al. (2002) Green tea polyphenol epigallocatechin-3 gallate inhibits Her-2/neu signaling, proliferation, and transformed phenotype of breast cancer cells. Cancer Res 62: 652–655.

36. Jian L, Xie LP, Lee AH, et al. (2004) Protective effect of green tea against prostate cancer: a case-control study in southeast China. Int J Cancer 108: 130–135.
37. Hiipakka RA, Zhang HZ, Dai W, et al. (2002) Structure-activity relationships for inhibition of human 5alpha-reductases by polyphenols. Biochem Pharmacol 63: 1165–1176.
38. Adhami VM, Siddiqui IA, Ahmad N, et al. (2004) Oral consumption of reen tea polyphenols inhibits insulin-like growth factor-I-induced signaling in an autochthonous mouse model of prostate cancer. Cancer Res 64: 8715–8722.
39. Caporali A, Davalli P, Astancolle S, et al. (2004) The chemopreventive action of catechins in the TRAMP mouse model of prostate carcinogenesis is accompanied by clusterin overexpression. Carcinogenesis 25: 2217–2224.
40. Bettuzzi S, Brausi M, Rizzi F, et al. (2006) Chemoprevention of human prostate cancer by oral administration of green tea catechins in volunteers with high-grade prostate intraepithelial neoplasia: a preliminary report from a one-year proof-of-principle study. Cancer Res 66: 1234–1240.
41. Shimizu M, Deguchi A, Joe AK, et al. (2005) EGCG inhibits activation of HER3 and expression of cyclooxygenase-2 in human colon cancer cells. J Exp Ther Oncol 5: 69–78.
42. Shimizu M, Deguchi A, Hara Y, et al. (2005) EGCG inhibits activation of the insulin-like growth factor-1 receptor in human colon cancer cells. Biochem Biophys Res Commun 334: 947–953.
43. Renehan AG, Zwahlen M, Minder C, et al. (2004) Insulin-like growth factor (IGF)-I, IGF binding protein-3, and cancer risk: systematic review and meta-regression analysis. Lancet 363: 1346–1353.
44. Blakely CM, Stoddard AJ, Belka GK, et al. (2006) Hormone-induced protection against mammary tumorigenesis is conserved in multiple rat strains and identifies a core gene expression signature induced by pregnancy. Cancer Res 66: 6421–6431.
45. Guarente L (2006) Sirtuins as potential targets for metabolic syndrome. Nature 444: 868–874.
46. Hu Z, Fan C, Oh DS, et al. (2006) The molecular portraits of breast tumors are conserved across microarray platforms. BMC Genomics 7: 96
47. Carey LA, Perou CM, Livasy CA, et al. (2006) Race, breast cancer subtypes, and survival in the Carolina Breast Cancer Study. JAMA 295: 2492–2502.
48. Suganuma M, Okabe S, Kai Y, et al. (1999) Synergistic effects of (–)-epigallocatechin gallate with (–)-epicatechin, sulindac, or tamoxifen on cancer-preventive activity in the human lung cancer cell line PC-9. Cancer Res 59: 44–47.
49. Khafif A, Schantz SP, Chou TC, et al. (1998) Quantitation of chemopreventive synergism between (-)-epigallocatechin-3-gallate and curcumin in normal, premalignant and malignant human oral epithelial cells. Carcinogenesis 19: 419–424.

STATE OF THE ART SPEAKERS

1
What Can We Learn about Breast Cancer from Stem Cells?

Michael F. Clarke

To survive into adulthood, long-lived organisms such as man need to maintain the many diverse organs and tissues that are necessary for the myriads of essential functions such as absorption of nutrients, protection from infection, and replacement of cells damaged by insults including toxins, radiation and trauma. This need to constantly replenish the mature cells of a tissue presents a particularly vexing problem for complex multicellular animals. Cells must be able to replicate in order to replace the old or damaged cells, but this replication must be tightly regulated to prevent the accumulation of errors that result in the development of tumors and eventually cancer (1, 2).

Thus, animals have developed a remarkable system to cope with the dilemma of tissue maintenance. In organs such as the skin, blood, and gut the cells are constantly exposed to conditions that put them at risk for suffering cellular damage and mutations. Other tissues such as the breast need to expand a duct system for milk production that is transiently needed during nursing. The fully developed milk ducts are lined by luminal epithelial cells that are surrounded by myoepithelial cells, which can contract to expel the milk produced by alveolar cells (3). After weaning, the milk producing system undergoes involution to its original nonpregnant steady state until such time that it is needed again.

To accomplish this constant tissue regeneration, there is a cellular hierarchy in which a minority population of stem cells gives rise to a progenitor, or transiently amplifying, cell (4–15). The progenitor cells in turn undergo a series of mitoses accompanied by differentiation to generate the mature cells of the organ. Although both the stem cells and the progenitor cells both have the ability to replicate, they differ in a very significant way. Stem cells have the ability to regenerate an exact replica of themselves via a process called self renewal. In a renewing cell division, either one or both daughter cells have the same differentiation capacity and an extensive ability to replicate. On the other hand, with each cell division a progenitor cell becomes progressively more differentiated until it finally loses the ability to undergo mitoses (1). Because of their ability to self-renew, a single mammary stem cell can reconstitute a mouse breast, but progenitor cells can only do so transiently (16).

Cancer cells and stem cells share the ability to undergo large numbers of mitoses. However, they differ in one fundamental way. The expansion of normal stem cells is under strict genetic regulation resulting in a steady-state level above

J.J. Li et al. (eds.), *Hormonal Carcinogenesis V.*

which normal stem cells do not expand (17). Some cancer cells have lost these constraints and thus continuously expand forming lethal tumors. Multiple genes limit this expansion of normal stem cells, making it impossible for a single simple mutation to result in a malignant tumor (17).

Since aberrant self renewal is central to tumor development, it is important to understand the molecular regulation of this process. Although our understanding of these renewal pathways is still sketchy, some progress has been made in recent years. In adult tissues, the blood system is best understood (18–37). In the blood, a series of elegant studies have allowed the identification of the hematopoietic stem cell (HSC) as well as each of the progenitor cell compartments. Since the HSC, which comprises only 1 in 20,000 bone marrow cells, can be isolated this has opened the door to molecular and cellular analyses. Using a microarray analysis, we found that the polycomb gene *Bmi1* is absolutely required for the maintenance of adult HSCs (33). In the absence of Bmi1, the stem cells cannot self renew. Subsequently, Bmi1 has also been found to be required for the self renewal of neural stem cells (29).

Polycomb functions to repress transcription of target genes. Genes that appear to be repressed by Bmi1 include genes that have been implicated in senescence, apoptosis, and differentiation. Notably, expression of both $p16^{Ink4a}$ and $p19^{Arf}$ are increased in Bmi1 mutant mice (38). Each of these genes is known to be a tumor suppressor and has been implicated in multiple types of human cancers. Recent evidence suggests that $p16^{Ink4a}$ plays a role in stem cell senescence in multiple tissues including the blood, brain, and pancreas islet cells (39–41).

Recent evidence has implicated polycomb in the self renewal of embryonic stem (ES) cells. Several transcription factors including Oct4, Nanog, and Sox2 are needed for ES cell renewal (21, 23, 27, 35). These transcription factors bind to genes needed for self renewal. In addition, they also target genes such as members of the Hox complex that direct tissue specification fate decisions. However, in ES cells, polycomb binds to and represses transcription of differentiation-inducing genes and allows the ES cell renewal (20, 42). Thus, similar to HSCs, polycomb appears to repress genes involved in differentiation decisions of ES cells.

Not surprisingly, the genes thus far implicated in regulating self renewal of normal stem cells have also been implicated in cancer. *Bmi1*, $p16^{Ink4a}$, and $p19^{Arf}$ have been shown to be involved several types of adult cancers. Oct4 and Nanog are expressed at very high levels in germ cell tumors. In addition, other genes implicated in both the self renewal of normal stem cells and cancer include members of the Wnt/β-catenin pathway, the Notch family, sonic hedgehog pathway, and multiple genes mutated in human leukemia (43).

Clearly in cancer, some oncogenic mutations must result in the expansion of abnormally self renewing cancer cells. However, the mechanisms by which self renewal is deregulated in cancer cells are not well understood. As stated previously, the isolation of normal blood and brain stem cells was central to the identification of genes involved in their renewal. Purification of the stem cells enabled molecular and functional analysis of these stem cells (33). Thus, to understand how oncogenic mutations disrupt the normal constraints on the expansion of self renewing cells, one must first identify the self renewing cancer cells in a tumor. Until recently, it was assumed that all of the cancer cells in a solid tumor were capable of driving the growth and spread of the tumor. However, it has long been known that there is marked heterogeneity of the cancer cells in a tumor. In breast tumors, the differentiation state of the cancer cells within a tumor shows considerable variability. Many tumors consist of a mixture of very immature tumors cells and more differentiated cells. The latter cells can even produce milk proteins.

Such differences in cell maturity and marker expression in tumors could be a result of influences of the microenvironment (44). Alternatively, these differences could be a result of a cellular hierarchy in which cancer stem cells give rise to aberrantly differentiated nontumorigenic progeny (44). Using the tools of stem cell biology, we found that this latter explanation appears to be true. Using an immunodeficient mouse model in which human tumors could be established using dissociated tumor cells, in eight of nine tumors tested, we found that the $CD44^{+}CD24^{-/low}$ cancer cells were highly enriched for the ability to establish new tumors. As few as 200 of these cancer cells could form a tumor when injected into the breast of a mouse (45). By contrast, tens of thousands of the other cancer cells failed to form tumors when injected into a mouse breast. Importantly, the $CD44^{+}CD24^{-/low}$ cancer cells could be serially transplanted and in each tumor they regenerated the cellular heterogeneity seen in the original tumor. These cells, therefore, had the two cardinal features of stem cells, they self renewed and they produced differentiated progeny. Thus, these cells can be called cancer stem cells. Since the identification of the breast cancer stem cell, we and others have identified a stem cell population in many other types of human and mouse solid tumors (46–48).

The identification and isolation of cancer stem cells has implications for the prognosis, treatment, study, and diagnosis of cancer. We have recently identified a "cancer stem cell" gene signature. Remarkably, this gene signature can be used to predict survival of patients with multiple types of cancer (49). The isolation of cancer stem cells also will aid in the investigation of the cancer. In collaboration with the laboratory of A. Thomas Look, we investigated del (5Q) leukemias. Using markers identified by John Dick et al., and refined by Craig Jordan et al., we isolated leukemic stem cells from patients with these leukemias. We then asked whether there was a difference in the genes expressed by the leukemic stem cells compared with their normal HSC counterparts. Remarkably, in many patients we found that expression of α-catenin was markedly reduced in the AML stem cells. The α-catenin protein forms an adhesion complex with E-caderin and β-catenin. Disruption of this complex in *Drosophila* germ cells and mouse skin induces expansion of the stem cell pool. These findings suggest that this adhesion pathway's role in stem cell biology is conserved from fruit fly to man and its disruption plays a role in human cancer. In many patients, this difference was only seen when the AML stem cell compartment was examined (49). Thus, without first being able to isolate the leukemia stem cells the role of α-catenin in del (5Q) AML would not have been discovered.

In summary, the genetic changes that lead to cancer perturb the normal developmental hierarchy of a tissue in which a cancer arises. Remarkably, in many tumors there remains a vestige of this hierarchy in which only a subset of the cancer cells can self renew. These self renewing cells, called cancer stem cells, appear to be responsible for driving the growth and spread of the tumor. Thus, these cancer stem cells appear to be the ultimate cause of morbidity in patients' with cancer. The ability to isolate cancer stem cells should help us to develop new ways to target these cells and this will hopefully improve outcome in patients with cancer.

References

1. Clarke MF, Fuller M (2006) Stem cells and cancer: two faces of eve. Cell 124:1111–5.
2. Fearon ER, Vogelstein B (1990) A genetic model for colorectal tumorigenesis. Cell 61:759–767.
3. Smith GH, Chepko G (2001) Mammary epithelial stem cells. Microsc Res Tech 52:190–203.
4. Blanpain C, Lowry WE, Geoghegan A, et al. (2004) Self-renewal, multipotency, and the existence of two cell populations within an epithelial stem cell niche. Cell 118:635–48.
5. Nilsson E, Parrott JA, Skinner MK (2001) Basic fibroblast growth factor induces primordial follicle development and initiates folliculogenesis. Mol Cell Endocrinol 175:123–30.
6. Vasioukhin V, Bauer C, Degenstein L, et al. (2001) Hyperproliferation and defects in epithelial polarity upon conditional ablation of alpha-catenin in skin. Cell 104:605–17.
7. Spangrude GJ, Heimfeld S, Weissman IL (1988) Purification and characterization of mouse hematopoietic stem cells. Science 241:58–62.
8. Blyszczuk P (2004) Embryonic stem cells differentiate into insulin-producing cells without selection of nestin-expressing cells. Int J Dev Biol 48:1095–104.
9. Bouwens L, De Blay E (1996) Islet morphogenesis and stem cell markers in rat pancreas. J Histochem Cytochem 44:947–51.
10. Doetsch F, Petreanu L, Caille I, et al. (2002) EGF converts transit-amplifying neurogenic precursors in the adult brain into multipotent stem cells. Neuron 36:1021–34.
11. Chepko G, Smith GH (1999) Mammary epithelial stem cells: our current understanding. J Mammary Gland Biol Neoplasia 4:35–52.
12. Easterday MC (2003) Neural progenitor genes. Germinal zone expression and analysis of genetic overlap in stem cell populations. Dev Biol 264:309–22.
13. Taipale J, Beachy PA (2001) The Hedgehog and Wnt signalling pathways in cancer. Nature 411:349–54.
14. van Es JH (2005) Notch/gamma-secretase inhibition turns proliferative cells in intestinal crypts and adenomas into goblet cells. Nature 435:959–63.
15. Chan EF, Gat U, McNiff JM, Fuchs EA (1999) Common human skin tumour is caused by activating mutations in beta-catenin. Nat Genet 21:410–3.
16. Shackleton M (2006) Generation of a functional mammary gland from a single stem cell. Nature 439:84–8.
17. Morrison SJ, et al. (2002) A genetic determinant that specifically regulates the frequency of hematopoietic stem cells. J Immunol 168:635–42.
18. Allman D, Aster JC, Pear WS (2002) Notch signaling in hematopoiesis and early lymphocyte development. Immunol Rev 187:75–86.
19. Beachy PA, Karhadkar SS, Berman DM (2004) Tissue repair and stem cell renewal in carcinogenesis. Nature 432:324–31.
20. Boyer LA, et al. (2005) Core transcriptional regulatory circuitry in human embryonic stem cells. Cell 122:947–56.

21. Cavaleri F, Scholer HR (2003) Nanog: a new recruit to the embryonic stem cell orchestra. Cell 113:551–2.
22. Ema H, Takano H, Sudo K, Nakauchi H (2000) In vitro self-renewal division of hematopoietic stem cells. J Exp Med 192:1281–8.
23. He XC, et al. (2004) BMP signaling inhibits intestinal stem cell self-renewal through suppression of Wnt-beta-catenin signaling. Nat Genet 36:1117–21.
24. Hitoshi S, et al. (2002) Notch pathway molecules are essential for the maintenance, but not the generation, of mammalian neural stem cells. Genes Dev 16:846–58.
25. Huntly BJ, et al. (2004) MOZ-TIF2, but not BCR-ABL, confers properties of leukemic stem cells to committed murine hematopoietic progenitors. Cancer Cell 6:587–96.
26. Kyba M, Perlingeiro RC, Daley GQ (2002) HoxB4 confers definitive lymphoid-myeloid engraftment potential on embryonic stem cell and yolk sac hematopoietic progenitors. Cell 109, 29–37.
27. Loh YH, et al. (2006) The Oct4 and Nanog transcription network regulates pluripotency in mouse embryonic stem cells. Nat Genet 38:431–40.
28. Mitsui K, et al. (2003) The homeoprotein Nanog is required for maintenance of pluripotency in mouse epiblast and ES cells. Cell 113:631–42.
29. Molofsky AV, et al. (2003) Bmi-1 dependence distinguishes neural stem cell self-renewal from progenitor proliferation. Nature 425:962–7.
30. Morrison SJ, et al. (2000) Transient notch activation initiates an irreversible switch from neurogenesis to gliogenesis by neural crest stem cells. Cell 101:499–510.
31. Nakamura Y, et al. (2000) The bHLH gene *hes1* as a repressor of the neuronal commitment of CNS stem cells. J Neurosci 20:283–93.
32. Ohta H, et al. (2002) Polycomb group gene *rae28* is required for sustaining activity of hematopoietic stem cells. J Exp Med 195:759–70.
33. Park IK, et al. (2003) Bmi-1 is required for maintenance of adult self-renewing haematopoietic stem cells. Nature 423:302–5.
34. Reya T, et al. (2003) A role for Wnt signalling in self-renewal of haematopoietic stem cells. Nature 423:409–14.
35. Xu RH, et al. (2005) Basic FGF and suppression of BMP signaling sustain undifferentiated proliferation of human ES cells. Nat Meth 2:185–90.
36. Ying QL, Nichols J, Chambers I, Smith A (2003) BMP induction of Id proteins suppresses differentiation and sustains embryonic stem cell self-renewal in collaboration with STAT3. Cell 115:281–92.
37. Zaehres H, et al. (2005) High-efficiency RNA interference in human embryonic stem cells. Stem Cells 23:299–305.
38. Liang SH, Clarke MF (2001) Regulation of p53 localization. Eur J Biochem 268:2779–83.
39. Krishnamurthy J, et al. (2006) p16INK4a induces an age-dependent decline in islet regenerative potential. Nature 443:453–7.
40. Molofsky AV, et al. (2006) Increasing p16INK4a expression decreases forebrain progenitors and neurogenesis during ageing. Nature 443:448–52.
41. Janzen V, et al. (2006) Stem-cell ageing modified by the cyclin-dependent kinase inhibitor p16INK4a. Nature 443:421–6.
42. Lee TI, et al. (2006) Control of developmental regulators by Polycomb in human embryonic stem cells. Cell 125:301–13.
43. Akala OO, Clarke MF (2006) Hematopoietic stem cell self-renewal. Curr Opin Genet Dev 16:496–501.
44. Reya T, Morrison SJ, Clarke MF, Weissman IL (2001) Stem cells, cancer, and cancer stem cells. Nature 414:105–11.
45. Al-Hajj M, Wicha MS, Benito-Hernandez A, Morrison SJ, Clarke MF (2003) Prospective identification of tumorigenic breast cancer cells. Proc Natl Acad Sci USA 100:3983–8.
46. Clarke MF (2004) Neurobiology: at the root of brain cancer. Nature 432:281–2.
47. Fang D, et al. (2005) A tumorigenic subpopulation with stem cell properties in melanomas. Cancer Res 65:9328–37.

48. Singh SK, et al. (2003) Identification of a cancer stem cell in human brain tumors. Cancer Res 63:5821–8.
49. Liu R, et al. (2007) The prognastic role of a gene signature from tumorigenic breast-cancers cells. N Engl J Med 356:217–26.

2
Ovarian Cancer: Linking Genomics to New Target Discovery and Molecular Markers – The Way Ahead

Bryan T. Hennessy, Mandi Murph, Meera Nanjundan, Mark Carey, Nelly Auersperg, Jonas Almeida, Kevin R. Coombes, Jinsong Liu, Yiling Lu, Joe W. Gray, and Gordon B. Mills

Introduction

Epithelial ovarian cancer (EOC) is the 4th leading cause of cancer deaths among USA women and the most common cause of death from gynecologic cancers (1, 2). It arises from ovarian surface epithelial (OSE) cells, and is currently classified by surgical and histological appearance (see Table 1), although the predictive value of this morphologic classification is limited (3).

However, while the exact cell of origin and the reasons underlying morphological differences in ovarian tumors are unclear, the histological differentiation of specific ovarian tumors appears to be due to the expression and activity of homeobox genes that regulate differentiation and proliferation of the Mullerian duct (4). The dismal outcome of EOCs is due to the majority of patients presenting with advanced stage III/IV disease. Despite improved median survival in patients with paclitaxel/carboplatin chemotherapy after surgical debulking, relapse occurs in most patients with advanced disease, and only 20% are alive and disease-free at 5 years. Although current efforts to improve outcomes focus on both earlier diagnosis and the incorporation of additional cytotoxic chemotherapy drugs and novel-targeted therapies into treatment, these efforts have not provided significant improvements in patient outcomes in comparison to paclitaxel/carboplatin therapy alone. The results of the ICON5 trial, for example, demonstrate that adding a third cytotoxic agent (gemcitabine, topotecan, or liposomal doxorubicin) to paclitaxel/carboplatin does not prolong progression-free survival in first-line therapy of EOC (5). Current targeted approaches include utilization of epidermal growth factor receptor (EGFR) family inhibitors, or inhibitors of Kit and platelet-derived growth factor receptor (PDGFR), although EGFR and Kit mutations, associated with targeted therapy responsiveness in other tumor types, are rare in EOC.

Like many solid tumors, EOC has a high degree of chromosomal instability, and both total and regional instability are associated with altered patient outcomes (6). Unpublished studies from the Gray group and the Ovarian Cancer SPORE support this contention (personal communication). EOC evolves through the multistep acquisition of genomic and epigenetic aberrations that initially deregulate normal cell growth control, followed by autonomous proliferation, and eventually other

J.J. Li et al. (eds.), *Hormonal Carcinogenesis V.*

Table 1 Current concepts regarding the molecular pathology of specific histologic forms of epithelial ovarian carcinoma[a]

Histology	Likely precursor	Molecular features
High-grade serous carcinoma	"De novo" in epithelial inclusion cysts	p53 mutation and BRCA1 dysfunction (promoter methylation)
High-grade endometrioid carcinoma	Epithelial inclusion glands/cysts	p53 mutation and BRCA1 dysfunction (promoter methylation) PIK3CA mutation
Low-grade serous carcinoma	Cystadenoma-borderline tumor-carcinoma sequence	Mutations in K-ras and/or b-raf
Mucinous carcinoma	Cystadenoma-borderline tumor-carcinoma sequence	Mutations in K-ras; ? p53 mutation associated with transition from borderline tumor to carcinoma
Low-grade endometrioid carcinoma	Endometriosis and endometrial-like hyperplasia[b]	Mutations in CTNNB1 (B-catenin gene) and PTEN with microsatellite instability
Clear cell carcinoma	? endometriosis	? PTEN mutation/loss of heterozygosity PIK3CA mutation

[a] *PIK3CA* is the gene at 3q26 that encodes the p110α subunit of phosphatidylinositol-3-kinase (reproduced from reference 3 with permission)
[b] Endometriosis and adjacent low-grade endometrioid carcinoma share common genetic events such as loss of heterozygosity involving the same allele (i.e., PTEN). In contrast, high-grade and poorly differentiated endometrioid carcinomas are similar to high-grade serous carcinomas

"hallmark" features of malignancy (7). Although there are only few "hallmarks", many genomic aberrations contribute to the underlying process, and many of these events act in concert to generate the tumorigenic phenotype. Indeed, aberrations in single oncogenes frequently result in senescence or oncogene-induced death. Combined with the evidence for multiple different nonoverlapping defects in cancer cells, this suggests a major degree of complexity. An improved understanding of the underlying genomic aberrations driving EOC, and how they disrupt protein function leading to oncogenesis is greatly needed. Thus, we have advocated a comprehensive characterization of genomic anomalies in EOC as a common first step in studies that attempt to advance our understanding of ovarian oncogenesis and to discover promising new markers and targets for its therapy.

The selection of EOC as one of three tumor types for extensive characterization as part of The Cancer Genome Atlas (TCGA) project offers an excellent opportunity to extend our understanding of this disease. Although genomic mutations commonly affect only a limited number of genes in EOC, chromosomal copy number aberrations typically involve relatively large numbers of genes and other innovative techniques are needed in an attempt to determine a more limited list of genes that act as "key drivers" of oncogenesis. Using breast cancer (BC) as an example, the *her2/neu* oncogene is considered a "key driver" of the *her2/neu* amplicon on chromosome 17q, although multiple other nearby genes are frequently coamplified with *her2/neu*. Whether these coamplified genes are "innocent" passengers or act cooperatively, *her2/neu* remains an unanswered question. Theoretically, genes selected for mutation as well as "key drivers" of chromosomal copy number alterations that play a key role in the oncogenesis process will manifest significant aberrations at the transcription level and/or protein expression/function than other aberrant gene "passengers" at copy number changes. "Key drivers" represent potential markers for early diagnosis and targets for therapy, their identification are critical. Therefore, study of the transcriptional and proteomic effects of genomically aberrant genes in EOC is essential to determine their importance in the oncogenic process and to explore the antitumor efficacy and proteomic effects of therapies designed to specifically target the protein products of these aberrant genes. Alternatively, a systematic down regulation of each gene in an amplicon may expose phenotypic responses in cells where the genes are amplified. However, such approaches may miss critical aberrations with context-dependent effects such as those that manifest only in vivo. Moreover, as there are too many candidates for systematic analysis of all potential drivers at genomic aberrations, we have concentrated our efforts specifically on genomic changes that correlate with patient outcomes. This imperfect filter focuses on genes that are most likely to provide novel targets or molecular markers.

Mutations in Ovarian Cancer

Genomic mutations play a key role in the pathogenesis of multiple forms of cancer. High prevalence mutations (>5%) have been identified only in a limited number of genes in EOC. Genes known to develop mutations at high rate in certain EOC histologies and their frequency are shown in Table 1 (3–10). Current evidence, in particular the subtype-restricted expression patterns, suggests that specific mutations may be important in the pathogenesis of specific EOC subtypes. Moreover, germline mutations of BRCA1, BRCA2, and the hereditary non-polyposis coli (Lynch syndrome II) genes are all associated with DNA repair processes, a markedly increased risk of EOC, and account for 5–10% of all EOCs (11). Hereditary BRCA1/2-related EOCs tend to occur at an earlier age than sporadic tumors, possibly by their predisposition to genomic instability, and are usually high-grade serous tumors with p53 dysfunction.

Recent high-throughput approaches such as sequencing of specific functional groups of genes (i.e., the kinome) have not revealed other commonly mutated genes in EOC, although only limited sections of the genome have been probed (12). A large number of "rare" events have been identified that may act coordinately during oncogenesis. Whole genome sequencing has now been reported preliminarily for breast and colorectal cancers but has similarly not uncovered novel unknown mutations occurring at high frequency; rather, it has demonstrated that infrequent mutations affect a significantly higher number of genes in individual tumors than previously suspected (13). Moreover, there is a great diversity in the genes targeted, with few mutations in common. Thus, high-throughput sequencing by the Sanger Center (www.sanger.ac.uk), as proposed by TCGA in EOC, will likely identify a large number of low frequency mutations, although pathway or network analyses as part of a systems biology approach may demonstrate that these "rare" mutations result in coordinate aberrations in common critical processes. Therefore, it will be a challenge to determine specific mutations that play an important role in oncogenesis vs. gene mutations that result as a consequence of genomic instability.

By ectopically manipulating genes in mice OSE cells, it has been possible to induce a very poorly differentiated form of EOC by combining p53 loss with activation/overexpression of two of three oncogenes (*myc, K-ras,* and *AKT1*) (14, 15). While the relevance of this model to human EOC remains disputable, it has the potential to allow further confirmation for the role of specific oncogenes in EOC, as well as allowing potential evaluation of targeted therapeutics. Similar studies in human epithelium have demonstrated that introduction of SV40 large and small T antigen (TAg and tag) increase the number of cell divisions that normal OSE undergoes before entering senescence (16). Similarly, introduction of telomerase (hTERT) into these cells results in an extended life span in the absence of a transformed phenotype (17). Disrupting retinoblastoma function replaces TAg in cellular immortalization. Immortalized OSE can be transformed by the introduction of activated *ras* but not other oncogenes. Whether this represents a specific role in the transformation of OSE or complementation of TAg and hTERT is unknown. OSE cells provide potent reagents to determine mechanisms underlying immortalization and transformation (18). However, these studies, while supportive, do not definitively establish OSE as the EOC cell of origin.

Although the transcriptional and proteomic effects of mutations on protein expression and signaling in EOC as well as the roles played by these proteomic effects in oncogenesis remain only partly understood, mutations do allow identification of a number of logical potential markers and targets to facilitate earlier diagnosis and better treatment. On the other hand, successful targeting of the consequences of mutations in human EOC treatment has not been achieved. In an early example, despite promising preclinical and clinical data leading to the initiation of an international randomized phase II/III trial of *p53* gene-therapy using replication-deficient adenoviral vectors in combination with first-line chemotherapy, a therapeutic benefit was not shown (19). However, there were multiple problems including some that potentially affected the successful intratumoral delivery of p53. Only a limited number of targeted therapeutics relevant to the genomic aberrations present in EOC has been assessed. Importantly, a number of therapeutics targeting aberrations relevant to EOC such as mutations in PIK3CA will enter clinical trials over the next year. The successful application of genomic mutations in EOC therapy will require not only effective delivery of the agent to the tumor cells, but a comprehensive understanding of the transcriptional and functional proteomic effects of these mutations, changes induced in response to in vitro and in vivo targeting of the protein product of the mutant gene, and its interaction with other genomic changes. After all, there are multiple genetic changes in EOC and it is possible that these may bypass the tumor dependence on a single mutation/change. Functional studies of the protein signaling and related pathways associated with or affected by mutant genes will facilitate "real time" monitoring of mechanisms associated with the ability of a cell to survive therapy targeted specifically toward an important mutational event.

Comprehensive Genomic Copy Number Screening

Genomic instability is very frequent in EOC and it represents at least two theoretically different processes in ovarian oncogenesis. Of less importance are those genomic areas that are destabilized as a result of the oncogenic process per se. More important are particular areas of the genome that may be targeted and selected specifically due to their functional importance for tumor initiation and progression. Although such anomalous areas of DNA may carry multiple genes, only a limited number of these are "key drivers" of the process. These "key drivers" are perceived as the most critical markers and potential treatment targets. An identification of the "key drivers" of genomic copy number aberrations could greatly improve our understanding of EOC and contribute to improved patient management. It is not clear, however, whether each genomic aberration harbors a single "key driver" or whether mulitple genes within a genomically aberrant region may contribute to "cooperative oncogenesis". These two alternatives may be reflected in narrow alterations areas in contrast to large regional aberrations. Alternatively, the underlying mechanisms of fragile areas in the genome may determine the aberration type.

A number of relatively new high throughput technologies are now capable of profiling copy number changes across the entire genome in cancer. Comparative genomic hybridization (CGH) is one such technology and has recently become a conventional first step in the identification of potentially important markers and targets in EOC. The resolution of single nucleotide polymorphism (SNP)-based CGH approaches, and that of other recent methodologies developed to identify and map genome wide copy number changes in EOC, such as digital karyotyping (Fig. 1) or ROMA, has improved rapidly and dramatically (20, 21). A novel emerging molecular inversion probe (MIP) has the potential to supplant previous low resolution methods. Conventionally, strategies to distinguish candidate "drivers" from other less important coamplified or deleted genes in areas of chromosomal copy number change have focused on finding aberrations that induce significant changes at both the RNA and protein levels resulting in alterations in the phenotypic behavior of manipulated cells in vitro, and in significant changes to patient outcome (22). Important induced changes at the transcriptional level generally entail a significant increase (amplified genes) or decrease (deleted genes) in the amount of mRNA, and changes in expression and, more importantly, function at the protein level. For example, *her2/neu* gene amplification in BC dramatically increases not only HER2 protein expression but, importantly, activation of downstream proteins such as protein kinase B (AKT).

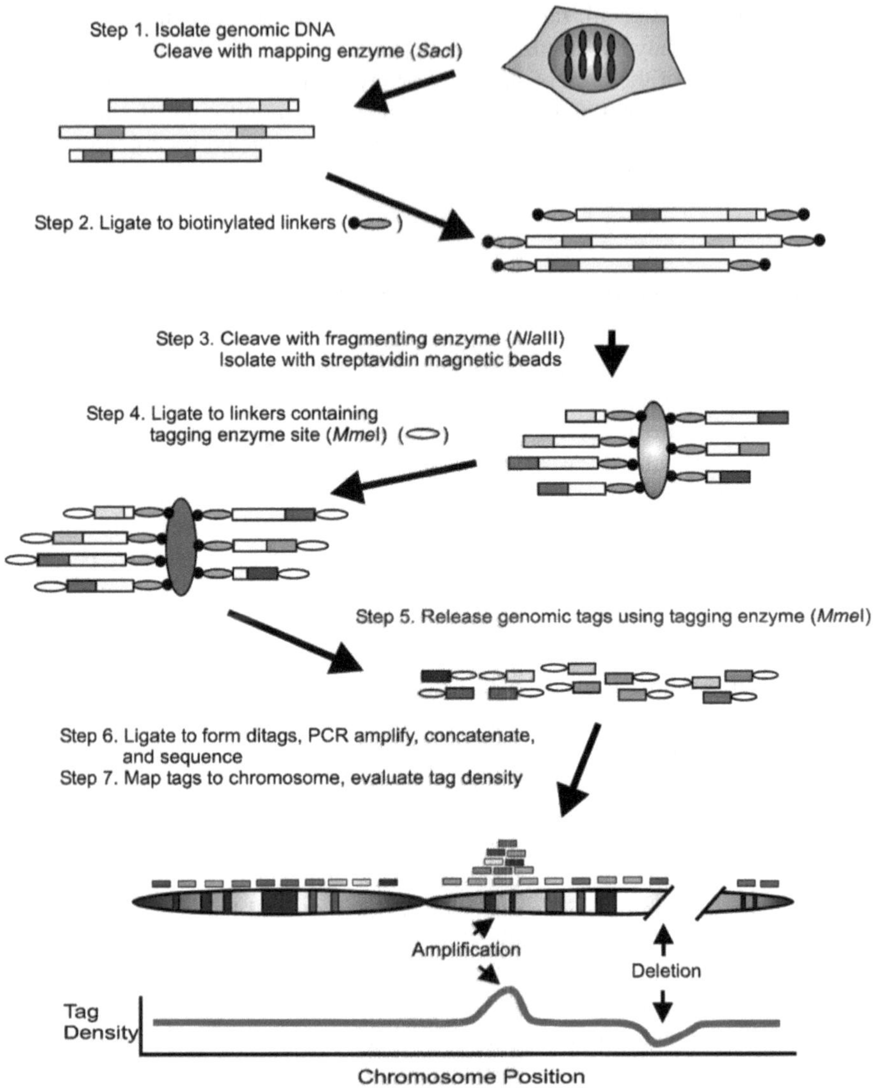

Fig. 1 Digital karyotyping approach: *Dark boxes* represent genomic tags; *Small ovals* represent linkers; *Larger ovals* represent streptavidin-coated magnetic beads (From (20) with permission)

These "drivers" may be applicable through small interfering (si)RNA technology or overexpression of candidate genes. The challenge is to determine which strategies and assays will reveal the spectrum of functional candidates within a genomic aberration. For example, knockdown and expression approaches may result in false positive data when a gene may be important for normal cellular function but is not specifically involved in oncogenesis.

In EOC, frequent chromosomal losses have been observed at 4p11-14, 9p, 11p14-15, 13q22-34, and 18q22-23, while common areas of gain include 1q22, 3q26, 5p15, 7q32-36, 8q23-24, 12p11-12, 17q12-23, 20p, and 20q12-13 (23). Multiple studies have identified candidate "drivers" by exploring genes that impact transcription, protein expression and function, cell behavior in vitro, and patient outcome. Since inhibition of protein function is more likely to occur than restoration of function, a perception reinforced by the clinical failure in EOC to exploit p53 loss, and by the success of the anti-HER2 monoclonal antibody trastuzumab in BC, most studies have focused on areas of chromosomal gain (amplicons) rather than on areas of deletion. Candidate "drivers" at areas of copy number anomaly have been identified. They include the small G protein *RAB25* at 1q22, evi1, protein kinase C iota (*PKCi*), *SnoN, BCL6*, and *PIK3CA* [the gene that encodes the p110α protein subunit of phosphatidylinositol-3-kinase (PI3K)] at 3q26.2, *myc* and *PVT1* at 8q24.2, remodeling and spacing factor 1 (rsf-1, HBXAP) and *PAK1* at 11q13, *her-neu* at 17q12, *AKT2* at 19q13.2 and *ZNF217* and *EEF1A2*, both at 20q13.2 (21, 24–31).

In addition to mutations and DNA copy number changes, rearrangements, epigenetic changes, and imprinting also affect cellular function potentially identifying important markers and therapy targets. Several imprinted genes including *ARH1* (*NOEY2*) and *LOT1/ZAC* function as tumor suppressor genes, and these and other potential tumor suppressors (i.e., *WWOX, HIC1, OVCA1*, and *OVCA2*) may be lost in EOC, either by epigenetic mechanisms and/or deletions (32–33). New comprehensive profiling approaches to detect methylation patterns and to explore imprinted genes are likely to identify new candidates (34). To date, however, none of the aforementioned potential targets have had an impact on patient care in the clinic; in fact, clinical studies have not yet even begun to assess therapies that are targeted to the majority of these aberratations. Most current clinical studies in EOC still focus on alternative cytotoxic chemotherapy regimens, in addition to treatments that target EGFR, despite the lack of evidence of genomic aberrations in the gene that encodes EGFR, and vascular endothelial growth factor (VEGF) or its receptors; indeed, it may be VEGF-targeted therapies that will prove most effective of this particular list of treatments (35).

Transcriptional Profiling

Transcriptional profiling approaches have identified expression patterns associated with the different histological types of EOC (36). The similar expression patterns in EOCs and Mullerian tissues and the comparable histological features strengthen concerns regarding the tissue of origin and differentiation patterns in EOC. A number of transcriptional profiles that correlate with patient outcomes have been identified. However, none of these parameters have sufficient power or been confirmed to warrant alterations in patient management. A combination of a platin and taxane remain the primary therapy for EOC. However, since only 70% of patients respond to this chemotheapy and fewer than 20% are cured by it, approaches to identify patients likely to respond and, in particular, unresponsive are desperately needed to allow triage to alternative therapies including clinical trials.

Transcriptional profiling approaches have also identified a number of candidate genes relevant to EOC. The availability of public databases such as www.oncomine.org, and the prerequisite of journals and granting agencies that transcriptional profiling data be deposited in MIAME-compliant databases such as the Gene Expression Ominibus (GEO http://www.ncbi.nlm.nih.gov/geo) will provide a robust resource. This will be greatly strengthened by the efforts of TCGA. Indeed, it behoves the research community to rapidly and efficiently exploit the outcomes of these efforts.

However, transcriptional profiling has major limitations. The absence of the clear identity of the precursor cell of each EOC histotype and the limited population of "relevant stem cells" make interpretation of transcriptional profiles difficult. Is the "pattern" observed in a tumor representative of the underlying defects driving oncogenesis, or is simply a reflection of the tissue of origin of the tumor? Moreover, if there is a "tumor stem cell" population in EOC, does the observed transcriptional profile have relevance to the stem cell population, or does it reflect the irrelevant but predominant differentiated cell population? A potentially critical problem regarding current transcriptional profling efforts is the alternate and aberrant splicing that occurs in cancer and specifically in EOC. Indeed, our recent data indicate that EOC may be a disease of "aberrant" splicing, since a wide variety of splice aberrations are present in multiple critically important genes (37). DNA copy number errors may contribute to EOC by causing aberrant splicing. We have identified functionally relevant splicing aberrations in multiple genes implicated in the inititation and progression of EOC. Therefore, new splicing arrays such as those being incorporated into the TCGA may reveal a whole new level of complexity.

Functional Proteomics

As discussed earlier, novel technologies such as whole genome sequencing and SNP-based CGH have revolutionized our ability to visualize genomic changes in cancer cells. Quantitative polymerase chain reaction (QPCR) and transcriptional profiling are both well established approaches to measure changes in mRNA levels in cancer cells. DNA copy number, mutation, rearrangement, and methylation alterations act in coordination to alter critical cellular functions through changes at the protein level; however, because of posttranslational modification and other events, protein levels and function offer a limited correlation to DNA and RNA changes. Thus, it is necessary to develop and implement functional proteomics approaches that can assay posttranslational and other protein modifications. In terms of our ability to globally assess cancer cells across DNA, RNA, and protein platforms, functional proteomics is significantly less advanced than genomic technologies. An efficient and clinical applicable functional proteomics strategy is needed.

Mass spectroscopy-based approaches represent the emerging most exciting technology, although they are not applicable to patient management. Currently, they are low throughput, in depth analytical strategies for a limited number of specimens. New mass spectroscopy platforms and the efforts of the Clinical Proteome Analysis Technology Consortium provide new strategies and opportunities. ELISA, Western blotting, and immunohistochemistry are still the most robust and, therefore, the standard approaches to assess protein expression and activation. However, these conventional proteomic technologies are semiquantitative and lack high-throughput capability and capacity to comprehensively assess changes in the activation of the multiple signaling pathways and events that may be perturbed by genomic aberrations. Thus, efforts currently focus on mass spectrometry and particularly on protein array platforms in an effort to address these shortcomings.

We have focused our efforts on the development of a novel moderate-throughput and quantitative proteomic technology known as reverse phase tissue lysate array (RPPA) (38–40). Using this approach, cell/tumor protein lysates are arrayed in serial dilutions on nitrocellulose (or other affinity reagent)-coated glass slides and probed with monospecific antibodies that quantify protein expression or a specific posttranslational modification (e.g., phosphorylation), followed by signal amplification using a colorimetric or similar methodology. This approach can be multiplexed with fluorescent, near infrared, or quantum dot methologies; however, we have restricted our approaches to single reagents to decrease the possibility of interference by multiple probes. Quantification involves software-based construction of serial dilution-signal intensity curves for individual samples with logistic fit models that use logarithmic values representative of the intensity of individual spots in serial dilutions of each sample. A representative value of each sample curve on the slide is then generated and used as a relative quantification of the amount of the particular protein in each sample. Absolute quantification can be achieved by comparison to samples with known amounts of protein or of the immunizing peptide. Protein loading is corrected among samples using loading controls of relatively constant protein concentrations or using an average of all proteins assayed. Each slide is stained with a single antibody under optimal conditions to allow comparison and quantification. Slides can be replicated robotically, allowing analysis of many samples (currently up to 1 000 per slide) with multiple antibodies targeting functionally relevant antigens in signaling and other pathways (we have validated almost 100 different antibodies). As the assay is "solid phase" it requires little starting material (nanograms) and is extremely sensitive (femtograms of target). Thus, it is ideal for valuable patient samples and is highly applicable to laser-aided microdissection or needle biopsies. A major drawback of the RPPA approach is the lack of validated monospecific antibodies able to assess expression and activation of the majority of proteins that make up the proteome. New technologies for the generation of affinity reagents as well as improved validation approaches are ameliorating this challenge. Further, RPPA is primarily a candidate gene approach based on known proteins with high quality antibodies, and, therefore, it is not ideal for discovery. It may, however, be useful for the discovery of proteomic "patterns" indicative of prognosis or predictive of response to therapy.

A major advantage of RPPA is that it can assess activity of many relevant signaling pathways at multiple levels. This allows an integration of information across several pathways. Our initial data suggest that many of the "standard" signaling diagrams may not be applicable to patient samples. Further, genomic aberrations may impact signaling pathways at different levels. The ability to comprehensively analyze particular signaling cascades will provide evidence to identify which pathways are important in tumor progression. RPPA is thus ideally suited to determine a "global" proteomic view of signaling changes induced by genomic aberrations and occurring in response to the targeting of proteins encoded by these aberrations. Its major limitation is that it does not assay spatial organization; thus it is best combined with immunohistochemistry.

RPPA can also play a major role in drug development and implementation. The process from target identification, to drug evaluation and implementation, into patient care is slow and fraught with a high failure rate. RPPA can be used early in drug development as a secondary screen to identify potential "hit" targets from high-throughput screens. It can also be used in structure/function analysis to decrease the likelihood of off-target activity. RPPA may also identify critical feedback loops that lead to unexpected drug effects and rationale therapy combinations. Indeed, we have utilized RPPA to comprehensively map feedback loops driven by inhibitors of mTOR and AKT. Such loops may play a role in resistance to targeted therapeutics or lead to unexpected deleterious effects such as toxicity or increased tumor growth. RPPA, particularly when used early in drug development, may also be capable of identifying patterns of protein expression representative of likelihood of response to targeted therapeutics. It also allows determination of appropriate dosing and scheduling of drugs (Hennessy, et al., in preparation). Despite the difficulty in obtaining serial biopsies for tailoring dose/response in human tumors, RPPA may be applicable to accessible surrogate tissues such as white blood cells and, more importantly, may prove critical to the identification and validation of targets for molecular imaging.

Early Detection of Ovarian Cancer

The great majority of EOC patients are diagnosed with advanced disease (stage III/IV). Early detection implies the diagnosis of EOC at an early stage in its development with resultant expected improvements in patient outcome using conventional treatment strategies. For early diagnosis, technologies must be available that are capable of detecting small tumors at an early stage; in addition, since screening of the entire population with such technologies is often not practical, guidelines as to patient risk assessment are often necessary to appropriately target technologies for the purposes of early cancer diagnosis and prevention. Age, for example, is a major factor used to delineate risk in current population screening approaches to breast and colorectal cancers. As with BC, BRCA1 and BRCA2 mutations are associated with a particularly high risk of EOC in many cases.

Serum or urine markers have clear potential utility in identifying patients with early stage EOC as well as those at high risk who will likely benefit from chemoprevention. It has become the "holy grail" of most studies aimed at early detection of EOC. Such markers also offer foreseeable means to better facilitate treatment planning and even individualization of care than is currently possible. The most well-known serum marker associated with EOC is CA125. Its routine use in screening is impeded by a lack of sufficient sensitivity and specificity (41). Since EOC prevalence is low and secondary screening approaches such as CAT scans are expensive and inaccurate, an extremely high specificity is necessary to prevent frequent false positive data with resultant morbidity and high financial costs. Combinations of multiple markers will likely be necessary to provide sufficient sensitivity and specificity for early diagnosis. Thus, it has been theorized that there are proteomic panels that can accurately identify tumors at an early stage of development. Indeed, several groups have applied proteomic approaches such as surface-enhanced laser desorption/ionization-time of flight (SELDI-TOF) mass spectrometry (MS) for identification of the theorized serum biomarker panels that will allow early detection of EOC. For example, there is evidence that serum transthyretin (TTR), β-hemoglobin (Hb), apolipoprotein (Apo), AI, and transferrin (TF), when combined with CA125, significantly improve the detection of early stage EOC (42). However, although potentially exciting, these and similar studies suffer from potential drawbacks including impracticality, expense, difficulty in identifying specific candidate proteins at recognized peaks, and lack of prospective validation in large studies. Nevertheless, they have important potential to significantly impact EOC mortality.

RPPA may have a role in the analysis and evaluation of serum markers. With appropriate high quality validated antibodies, RPPA can be developed with single antibodies rather than sandwich approaches. This provides a potential advantage over ELISA or Luminex methodologies. Indeed, with the sensitivity of RPPA, it may also be possible to quantify low levels of markers that cannot be assessed effectively by other methods. Further, the solid phase approach may be less sensitive to "interference" by contaminating compounds. We have successfully quantitated potential markers in as little as one nanoliter of serum when compared with the 100–500 μl required for most ELISA assays. The ability to analyze up to 1000 samples at a time greatly decreases cost when compared with ELISA and Luminex.

Although data regarding the potential usefulness of circulating tumor cells in EOC are currently very limited, these as well as ascitic fluid cells can theoretically provide an easily accessible and abundant source of tumor genome and proteome that may be useful in early cancer detection and in novel molecular studies of markers and therapy targets. Circulating and ascitic tumor cell DNA in EOC patients is thus currently being studied for a potential role in early diagnosis, as a prognostic factor, and in monitoring response to treatment (43). One early study found peripheral blood circulating tumor-specific p53 sequences in only a small proportion (15.3%) of FIGO stage III/IV EOC patients, demonstrating that this approach, as it stands, will not be of benefit for early EOC detection (44). However, current methods for enrichment of circulating tumor cells including immunomagnetic enrichment technologies such as CellSearch™ (Veridex, LLC) or enhanced density gradient systems may increase yield (45, 46). The application of sensitive mass spectrometry approaches such as those used by Sequenom (San Diego, CA) to assess fetal DNA in perinatal diagnosis may also have major advantages. A major challenge remains the ability to detect and amplify, without error, DNA from as little as 1–5 cancer cells per each 10 ml of blood. Expansion of the list of probed genomic changes to more than a limited number of specific p53 sequences may also increase yield. As the sensitivity of these assays improves, the usefulness of circulating tumor cells in guiding early diagnosis and treatment may increase.

Novel high-throughput molecular technologies are adding multiple new possibilities for early EOC detection, particularly with integration of data across DNA, RNA, and protein platforms. Ideally, such markers will not only allow early detection of cancer but will also facilitate prognosis and prediction/treatment planning. In other words, they should ideally provide information about the presence of both current and potentially exploitable molecular targets. These studies may also offer a route to preferentially target treatments guided by specific genomic or other cancer cell changes. However, the application of such a global approach to early detection is still very much in its infancy.

Challenges

Before studies that link genomics to the discovery of new targets and molecular markers are ready for routine clinical implementation, many challenges must be addressed and overcome. As mentioned earlier, potential targets discovered by the approaches described in this chapter have yet to be tested in EOC clinical trials. The "accessibility" of the target itself may be partly responsible in some cases. HER2, for example, is a membrane protein with a large and easily accessed extracellular domain, while many of the potentially important targets already uncovered in EOC are intracellular and more difficult to attain. Indeed, the success of targeting HER2 in BC led to a number of clinical trials in EOC. However, based on those studies, it appears that less than 10% of patients express elevated levels of HER2. This makes the successful completion of targeted therapeutic trials against HER2 technically impractical in a relatively uncommon disease such as EOC. Moreover, while targeted monoclonal antibodies such as trastuzumab are a major part of current targeted therapy armamentarium, they are generally not useful for intracellular targets. More interaction between basic, translational, and clinical investigators along with an improved general understanding of the processes involved, and the events that drive each of these stages will also facilitate more logical and targeted translation of the results of preclinical studies into human clinical trials. At present, while marker and target discovery using high throughput genomic approaches with initial validation using transcriptional and proteomic technologies is becoming easier to implement in preclinical studies, many challenges lie in the way of their practical application to human clinical trials. Therefore, each stage of research must be planned in a fashion that better considers how best to integrate any derive findings and data into subsequent stages of research. Concurrent development of targeted therapeutics and molecular markers capable to direct clinical evaluation (theragnostics) is critically important.

Studies utilizing human tumor cell lines have been criticized for their inability to adequately represent the molecular heterogeneity of human cancers in vivo. In some cases, this may contribute to failure to replicate findings from in vitro studies in human clinical trials. Newly developed targeted treatments have frequently shown promising activity in cell line studies but fail to subsequently demonstrate efficacy in human cancer patients. Multiple reasons have been postulated to explain the difficulty in translating early studies to improved patient outcomes. Most of the reasons remain purely speculative. However, it is quite likely that the use of small numbers of cell lines and the failure to match molecular aberrations (i.e., underlying genetic changes) in cell lines to genomic and proteomic changes in specific human tumors are some of the most important underlying problems. Therefore, it is probable that using baseline characterization of molecular abnormalities in a large number of cell lines to link particular cell lines to populations of patient tumors will allow better in vitro modeling of the molecular heterogeneity of human tumors. This will require development of a large set of EOC cell lines that are comprehensively characterized for genomic, transcriptional, and proteomic aberrations. We have implemented this process with academic/industrial partnerships and philanthropic support to embark on a major genomic, transcriptional, and functional proteomic evaluation of over 40 EOC cell lines and isogenic pairs that reflect known genomic aberrations in EOC patients. These lines will also be characterized for responsiveness to therapeutics with a major emphasis on targeted drugs. The resulting database will be made available to the research community to facilitate improved patient management. Indeed, we request that investigators provide additional cell lines to aid in their characterization. The data and lines will be shared with the community to fulfill our covenant with patients who donated the material and with the public that provides support for EOC research.

Likewise, the development of more appropriate animal tumor models driven by specific genetic aberrations will be necessary. Although current therapies are mainly developed based on the "average" patient and not embracing the concept of patient and tumor heterogeneity, overcoming these aforementioned difficulties may allow identification of effective targeted therapies in specific patient tumors that may otherwise be able to be dismissed as inactive in clinical trials. It will be a challenge, however, to introduce personalized molecular medicine in a relatively uncommon disease such as EOC, particularly because of difficulty in implementing clinical trials. We may need to "borrow" information and strategies derived from more common tumors. Thus, as discussed in the next section, as high throughput approaches become more important to study the genome and the proteome in preclinical and clinical research, the creation of widely accessible and user-friendly databases will be essential to maximize their impact and avoid effort duplication. With such resources, characterization of the genomic, transcriptional, and proteomic aberrations of human tumor and cell line panels, along with the responsiveness of each to multiple therapies, will advance our understanding of the heterogeneity of EOC and facilitate logical studies that utilize more targeted cell lines, along with predictive drug activity models in specific populations of EOC patients.

Following the identification of high-quality target tools, including chemical genomics and siRNA, it will be necessary to apply them concurrently with high-throughput genomic and proteomic studies to allow determination of on- and off-target effects of inhibition and pathway crosstalk. Although siRNA validation offers specific and effective knockdown, it may not mimic the effects of drugs that potentially have other, including off-target, functions. Further, because of the slow action of siRNA, the cells studied may represent those that survived the stress of siRNA knockdown of critical targets. These cells may have extensive "pathogenic" rewiring of critical pathways, making interpretation of the data challenging. Early in development, methods that combine novel targeted treatments with standard radiation or chemotherapy should also be sought as this may better facilitate translation of novel targets to clinical trials.

Clinical trials will also need to advance in a fashion that allows clinical drug development to keep pace with preclinical target development (47). Mandated tumor biopsies with well-designed correlative studies in molecular-marker-driven trials will be necessary for the efficient evaluation of novel targeted therapeutics. It will be crucial to distinguish on- from off-target activity to prevent the elimination of a good target because of an incorrect drug. We must begin to employ the concept of "biologically relevant dose" rather than maximum tolerated dose (MTD) to determine drug dosing by correlating biologic effects with clinical efficacy; this open the possibility to merge the three clinical trial phases into one continuous clinical/biological assessment of drug efficacy in patients. The goals of Phase I trials in particular will need to be addressed if the initial drug testing is not driven by MTD. In preclinical studies when no validated mechanism(s) have been discovered, large, randomized Phase III trials are particularly necessary to assess drugs that benefit a small proportion of cancer patients to separate likely responders from nonresponders. After drug approval, studies must continue to improve the identification of likely responders and associated tumor molecular aberrations, to facilitate early identification of nonresponders and to monitor off-target effects of each drug. When emergent resistance in tumors possessing the target is observed, it is important to initiate biological studies to determine molecular mechanism(s) likely to be predictable secondary events in many cases. Indeed, well-designed preclinical studies should uncover potential mechanisms of resistance to targeted therapies before resistance becomes a clinical problem in patients, particularly as functional proteomic technologies improve. Comprehensive characterization of human ovarian tumor genomic and proteomic aberrations, and their similarity to particular groups of EOC cell lines, may ultimately make the preclinical to clinical translation of targets and specific therapies a more logical and predictable process than it is at the present.

Data Integration

As studies characterize EOC cell lines and human EOCs using high-throughput genomic, transcriptional, and ultimately proteomic approaches, it is critical that adequate and centralized computational infrastructure, in addition to bioinformatics and biostatistical support, be developed to allow storage and utilization/integration of the vast and highly heterogeneous data derived from "omics" technologies (48). Such a computational resource should be made available to all investigators in a manner that is easy to use but also protects patient information and confidentiality. This will avoid duplication of efforts. In addition, these resources should facilitate data mining, retrieval, and automatic analysis with statistical software packages such as R or Matlab, thus allowing automatic high-throughput data integration across genomic, transcriptional, and proteomic platforms, between datasets (i.e., to model specific groups of cell lines to human ovarian tumors with similar patterns of molecular aberrations), and analysis of the association of specific aberrations and changes with clinical endpoints, and with the characterized responsiveness of cell lines to multiple targeted and chemotherapies. Repeated updating should be facilitated as novel "omics" technologies are continually introduced and upgraded (49). Such access would also allow and encourage novel bioinformatics and biostatistical approaches that further our ability to work with and integrate large amounts of data across multiple platforms to advance our understanding of the pathogenesis and molecular biology of EOC.

Potential Pitfalls

Overall, we are still limited by poor understanding of the molecular mechanisms underlying the development and selection of genomic aberrations in EOC. Moreover, the complexity of molecular changes raises the very daunting possibility that the "key driver" hypothesis is overly simple. If each of the large number of molecular anomalies plays an individual minor role in most EOCs, indicating that they are truly molecularly heterogeneous, then identification and targeted therapy planning will be considerably more difficult and complicated than is presumed today.

Conclusion

EOC is characterized by a large number of complicated genomic and proteomic changes with functional consequences that converge to create the relatively few "hallmarks" of cancer. Novel high-throughput technologies enable us to comprehensively profile these changes and identify "key drivers" in oncogenesis. Although these "key drivers" are potentially important targets for patient treatment, currently, their clinical application is considerably more difficult than their identification.

Thus, basic, translational, and clinical studies must evolve in a reciprocal fashion as part of an integrated and overlapping continuum. As our capacity to investigate EOC molecular changes increases exponentially, more organized and accessible data collection as well as more formal bioinformatics and biostatistical support is becoming necessary. The current focus of the TCGA on EOC offers a major opportunity to the research community that will hopefully be rapidly translated into improved patient outcomes. However, the complexity of genomic changes may preclude a comprehensive understanding of the many interactions of EOC genetic and proteomic aberrations. Indeed, this complexity will likely bypass the ability of current reductionist approaches to deal with the amount and complexity of data. New computational and systems biology approaches may allow generation of usable models that result in testable predictions.

Finally, data from high-throughput approaches must be integrated with the results of other more conventional research studies to allow a comprehensive and real understanding of EOC. EOC pathogenesis involves a number of cellular functions such as DNA repair and intracellular kinase signaling that are not only influenced by genomic aberrations, but potentially, by other processes, as well, including autocrine and membrane-initiated signaling loops involving molecules such as VEGF and phospholipids including lysophosphatidic acid (LPA) (3, 50). Knowledge of these and other processes will expand our knowledge of potential markers and targets, allowing the formulation of approaches that best assimilate all of these data. After a recent important study in EOC pathogenesis identifying a specific role of changes on homeobox gene expression, researchers are now attempting to integrate this novel finding with the presence of EOC histological-specific mutations and other genomic changes. These studies may ultimately uncover the sequence of events underlying the initiation of EOC oncogenesis (50).

References

1. Jemal A, et al. (2003) Cancer statistics, 2003. CA Cancer J Clin 53:5–26.
2. Cannistra SA (2004) Cancer of the ovary. N Engl J Med 351:2519–2529.
3. Hennessy BT, Mills GB, et al. (2006) Ovarian cancer: Homeobox genes, autocrine/paracrine growth, and kinase signaling. Int J Biochem Cell Biol 38:1450–1456.
4. Taylor HS, Vanden Heuvel GB, Igarashi P (1997) A conserved Hox axis in the mouse and human female reproductive system: late establishment and persistent adult expression of the Hoxa cluster genes. Biol Reprod 57:1338–1345.
5. International Collaborative Ovarian Neoplasm Group (2002) Paclitaxel plus carboplatin versus standard chemotherapy with either single-agent carboplatin or cyclophosphamide, doxorubicin, and cisplatin in women with ovarian cancer: the ICON3 randomised trial. Lancet 360:505–515.
6. Suzuki S, Moore III DH, Ginzinger DG, et al. (2000) An approach to analysis of large-scale correlations between genome changes and clinical endpoints in ovarian cancer. Cancer Res 60:5382–5385.
7. Hanahan D, Weinberg RA (2000) The hallmarks of cancer. Cell 100:57–70
8. Levine DA, Bogomolniy F, Yee CJ, et al. (2005) Frequent mutation of the PIK3CA gene in ovarian and breast cancers. Clin Cancer Res 11:2875–2878.
9. Campbell IG, Russell SE, Choong DY, et al. (2004) Mutation of the PIK3CA gene in ovarian and breast cancer. Cancer Res 64:7678–7681.

10. Shih IeM, Kurman RJ (2004) Ovarian tumorigenesis: a proposed model based on morphological and molecular genetic analysis. Am J Pathol 164:1511–1518.
11. Narod SA, Boyd J (2002) Current understanding of the epidemiology and clinical implications of BRCA1 and BRCA2 mutations for ovarian cancer. Curr Opin Obstet Gynecol 14:19–26.
12. Stephens P, Edkins S, Davies H, et al. (2005) A screen of the complete protein kinase gene family identifies diverse patterns of somatic mutations in human breast cancer. Nat Genet 37:590–592.
13. Sjoblom T, Jones S, Wood LD, et al. (2006) The Consensus Coding Sequences of Human Breast and Colorectal Cancers. Science 314:268–274.
14. Orsulic S, Li Y, Soslow RA, et al. (2002). Induction of ovarian cancer by defined multiple genetic changes in a mouse model system. Cancer Cell 1:53–62.
15. Dinulescu DM, Ince TA, Quade BJ, et al. (2005) Role of K-ras and Pten in the development of mouse models of endometriosis and endometrioid ovarian cancer. Nat Med 11:63–70.
16. Yang G, Rosen DG, Mercado-Uribe I, et al. (2007) Knockdown of p53 combined with expression of the catalytic subunit of telomerase is sufficient to immortalize primary human ovarian surface epithelial cells. Carcinogenesis 28:174–182.
17. Alvero AB, Fishman DA, Qumsiyeh MB, et al. (2004) Telomerase prolongs the lifespan of normal human ovarian surface epithelial cells without inducing neoplastic phenotype. J Soc Gynecol Investig 11:553–561.
18. Auersperg N, Wong AS, Choi KC, et al. (2001) Ovarian surface epithelium: biology, endocrinology and pathology. Endocr Rev 22:255–288.
19. Zeimet AG, Marth C (2003) Why did *p53* gene therapy fail in ovarian cancer? Lancet Oncol 4:415–422.
20. Wang TL, Maierhofer C, Speicher MR, et al. (2002) Digital karyotyping. Proc Natl Acad Sci USA 99:16156–16161.
21. Shih IeM, Sheu JJ, Santillan A, et al. (2005) Amplification of a chromatin remodeling gene, Rsf-1/HBXAP, in ovarian carcinoma. Proc Natl Acad Sci USA 102:14004–14009.
22. Hennessy BT, Nanjunden M, Cheng KW, Nolden L, Mills GB (2006) Ovarian cancer: identification of remodeling and spacing factor 1 (rsf-1, HBXAP) at chromosome 11q13 as a putative oncogene in ovarian cancer (News and Commentary). Eur J Hum Genet 14:381–383.
23. Watanabe T, Imoto I, Kosugi Y, et al. (2001) A novel amplification at 17q21-23 in ovarian cancer cell lines detected by comparative genomic hybridization. Gynecol Oncol 81:172–177.
24. Shayesteh L, Lu Y, Kuo WL, et al. (1999) *PIK3CA* is implicated as an oncogene in ovarian cancer. Nat Genet 21:99–102.
25. Cheng KW, Lahad JP, Kuo WL, et al. (2004) The Rab 25 small GTPase determines aggressiveness of ovarian and breast cancers. Nat Med 10:1251–1256.
26. Eder A, Sui X, Rosen D, et al. (2005) Atypical PKCiota contributes to poor prognosis through loss of apical-basal polarity and cyclin E overexpression in ovarian cancer. Proc Natl Acad Sci USA 102:12519–12524.
27. Brown LA, Irving J, Parker R, et al. (2006) Amplification of EMSY, a novel oncogene on 11q13, in high grade ovarian surface epithelial carcinomas. Gynecol Oncol 100:264–70.
28. Hughes-Davies L, Huntsman D, Ruas M, et al. (2003) EMSY links the BRCA2 pathway to sporadic breast and ovarian cancer. Cell 115:523–535.
29. Anand N, Murthy S, Amann G, et al. (2002) Protein elongation factor EEF1A2 is a putative oncogene in ovarian cancer. Nat Genet 31:301–305.
30. Schraml P, Schwerdtfeger G, Burkhalter F, et al. (2003) Combined array comparative genomic hybridization and tissue microarray analysis suggest PAK1 at 11q13.5-q14 as a critical oncogene target in ovarian carcinoma. Am J Pathol 163:985–992.
31. Li P, Maines-Bandiera S, Kuo WL, et al. (2007) Multiple roles of the candidate oncogene *znf217* in ovarian epithelial neoplastic progression. Int J Cancer 120:1863–73.
32. Kamikihara T, Arima T, Kato K, et al. (2005) Epigenetic silencing of the imprinted gene *ZAC* by DNA methylation is an early event in the progression of human ovarian cancer. Int J Cancer 115:690–700.
33. Strathdee G, Appleton K, Illand M, et al. (2001) Primary ovarian carcinomas display multiple methylator phenotypes involving known tumor suppressor genes. Am J Pathol 158:1121–1127.

34. Wei SH, Balch C, Paik HH, et al. (2006) Prognostic DNA methylation biomarkers in ovarian cancer. Clin Cancer Res 12:2788–2794.
35. Wright JD, Hagemann A, Rader JS, et al. (2006) Bevacizumab combination therapy in recurrent, platinum-refractory, epithelial ovarian carcinoma: a retrospective analysis. Cancer 107:83–89.
36. Smith DI (2002) Transcriptional profiling develops molecular signatures for ovarian tumors. Cytometry 47:60–62.
37. Nanjundan M, Zhang F, Schmandt R, Smith-McCune K, Mills GB (2007) Identification of a novel splice variant of AML1b in ovarian cancer patients conferring loss of wild-type tumor suppressive functions. Oncogene 26:2574–84.
38. Charboneau L, Scott H, Chen T, et al. (2002) Utility of reverse phase protein arrays: applications to signalling pathways and human body arrays. Brief Funct Genomic Proteomic 1:305–315.
39. Sheehan KM, Calvert VS, Kay EW, et al. (2005) Use of reverse phase protein microarrays and reference standard development for molecular network analysis of metastatic ovarian carcinoma. Mol Cell Proteomics 4:346–355.
40. Tibes R, Qiu Y, Lu Y, et al. (2006) Reverse phase protein array (RPPA): validation of a novel proteomic technology and utility for analysis of primary leukemia specimens and hematopoetic stem cells (HSC). Mol Cancer Ther 5:2512–21.
41. Duffy MJ (2006) Serum tumor markers in breast cancer: are they of clinical value? Clin Chem 52:345–351.
42. Kozak KR, Su F, Whitelegge JP, et al. (2005) Characterization of serum biomarkers for detection of early stage ovarian cancer. Proteomics 5:4589–4596.
43. Judson PL, Geller MA, Bliss RL, et al. (2003) Preoperative detection of peripherally circulating cancer cells and its prognostic significance in ovarian cancer. Gynecol Oncol 91:389–394.
44. Swisher EM, Wollan M, Mahtani SM, et al. (2005) Specific p53 sequences in blood and peritoneal fluid of women with epithelial ovarian cancer. Am J Obstet Gynecol 193:662–667.
45. Cristofanilli M, Budd GT, Ellis MJ, et al. (2004) Circulating tumor cells, disease progression, and survival in metastatic breast cancer. New Engl J Med 351:781–791.
46. Muller V, Stahmann N, Riethdorf S, et al. (2005) Circulating tumor cells in breast cancer: correlation to bone marrow micrometastases, heterogeneous response to systemic therapy and low proliferative activity. Clin Can Res 11:3678–3685.
47. Hennessy BT, Smith DL, Ram PT, Lu Y, Mills GB (2005) Exploiting the PI3K/AKT pathway for cancer drug discovery. Nat Rev Drug Discov 4:988–1004.
48. Wang X, Gorlitsky R, Almeida JS (2005) From XML to RDF: how semantic web technologies will change the design of 'omic' standards. Nat Biotechnol 23:1099–1103.
49. Almeida JS, C Chen, R Gorlitsky, et al. (2006) Data integration gets 'Sloppy'. Nat Biotechnol 24:1070–1071.
50. Cheng W, Liu J, Yoshida H, et al. (2005) Lineage infidelity of epithelial ovarian cancers is controlled by *HOX* genes that specify regional identity in the reproductive tract. Nat Med 11:531–537.

3
Cdk1, Plks, Auroras, and Neks: The Mitotic Bodyguards

Patrick Salaun, Yoann Rannou, and Claude Prigent

Introduction

"*Omnis cellula e cellula,*" in 1858, an important dogma in cell biology was born, when Rudolf Virchow established that every cell must derive from a preexisting cell. And indeed cell division is the only way for life to expend, it is also the way for immortalization, and unfortunately when uncontrolled also the way for cancer. But unrevealing mechanisms leading to cell division took quite a while. How does a mother cell divide to give two daughters? This is known as the cell cycle, which describes a series of events that insures faithfully transition of the genetic information from one cell generation to the next. These dividing mechanisms have been conserved throughout evolution; they underlie growth and development in all living organisms and are central to their heredity and evolution.

In eukaryotic cells, the cell cycle was first described as two distinct phases: interphase and mitosis that just precedes cell division (Fig. 1). The interphase was later on divided into three phases, S-phase standing for DNA synthesis surrounded by two G-phases G1 and G2 standing for Gap-phases. Fully described by Walter Flemming in 1882, mitosis remains the most spectacular and sophisticated part of the cell cycle. In less than an hour, the mother cell organizes a complex machine aim to separate its genetic information and all its subcellular components into two identical sets that will be inherited by the two daughter cells. If mitosis proceeds without any error it eventually ends up with cytokinesis corresponding to the physical separation of the two daughter cells. Theodor Boveri predicted errors during mitosis to be at the origin of cancer in 1902. Hundred years later the scientific community is still debating on whether or not this might be true. The coordination of progression through mitosis is mainly orchestrated by protein phosphorylation insured by several serine/threonine kinases. In this short review we will focus on the four main mitotic kinase families: the cyclin-dependent kinase: Cdks, the polo-like kinases: Plks, the Aurora kinases, and the NIMA-related kinases: Neks.

"Cyclin-dependant kinases," Cdks that must associate to a cyclin to become active kinases are key regulators of cell cycle progression. There are now about 12 Cdks; the first one Cdk1 (or cdc2) has long been considered as THE cell cycle master kinase, thought to be responsible for all cell cycle transitions (1). This is true

J.J. Li et al. (eds.), *Hormonal Carcinogenesis V.*

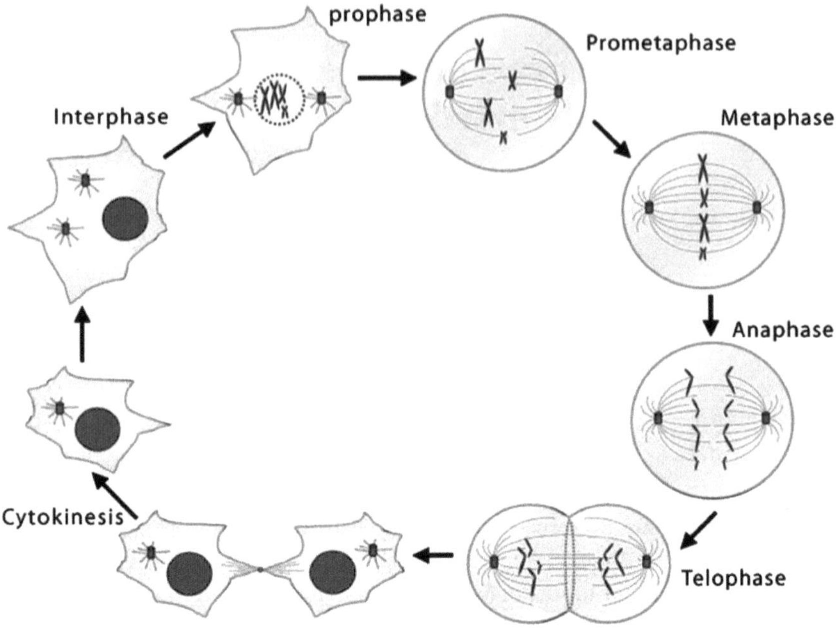

Fig. 1 The different phases of the mitosis

in yeast where Cdk1 kinase activity is required for the G1/S and the G2/M transition (2). In mammalian cell, however, Cdk1 activity is only required for the G2/M transition (3). Cdk1 binds to cyclin A, cyclin B, or Ringo to become an active kinase (4, 5).

The "Polo-like kinases," Plks form a family of four different proteins that regulates many aspects of the cell cycle progression. They all share small conserved domains named polo-box required for protein localization. Only Plk1 that is the most extensively studied is a true mitotic kinase homolog to the *Drosophila* polo kinase (6). Plk2, Plk3, and Plk4 are more likely involved only in interphase. However, Plk4 activity is required for centriole duplication, an event that must be achieved before entering mitosis, and necessary to assemble the bipolar mitotic spindle (7).

Aurora kinases were first identified in *S. cerevisae* and *Drosophila* (8, 9). Yeast cells possess only one Aurora-related kinase, invertebrates such *Drosophila* and *C. elegans* have two (A and B type) and mammals have three, named Aurora A, B, and C (10). From an evolution point of view, the A and B types have evolved from a common ancestor, while C type has evolved from the B type (11). Consequently, Aurora A has distinct functions while Aurora B and C share same functions, though all three kinases are involved in the control of many processes required for mitosis.

"NIMA-related kinases," Neks belong to a very large family of protein kinases with 13 different Nek proteins in human, from Nek1 to Nek11 (Nek2A and Nek2B, and Nek11L and Nek11S) (12). The belonging to the Nek family is defined by the sequence homology with the kinase NIMA (never in mitosis A), a true *Aspergillus nidulans* mitotic kinase (13). However, not all of the Nek kinases are involved in mitosis (14). Nek2 is the most studied of all; its activity is required for centrosome behavior and for cytokinesis (15, 16).

During the interphase, the cell's nucleus is well defined, with two pairs of centrioles adjacent to the nucleus. At the end of the interphase, the genome has been duplicated but the chromosomes are not distinguishable. When prophase starts, the nucleoli disappear and the chromatin starts to coil and fold into observable chromosomes, the spindle forms and the centrosomes move apart. During prometaphase, the nuclear membrane breaks down and some of spindle microtubules attach to sister chromatids at the kinetochores. The microtubules start to deplace the chromatid pairs to form a metaphase plate. At the metaphase, the chromosomes have moved to the center of the dividing cell along the metaphase plate. Identical chromatids are attached to kinetochore fibers radiating from opposite ends of the parent cell. The sister chromatids begin to separate at the anaphase when the spindle microtubules pull separating chromosomes to opposite poles. During telophase, daughter nuclei begin to assemble with nuclear envelopes appearing around chromosomes. Nucleoli reappear and chromosomes decondense. The last step of mitosis is the cytokinesis step. It occurs when a contractile ring of actin and myosin filaments constricts the plasma membrane at the equator, triggering the physical division of the two daughter cells.

To Get Ready for Mitosis. Mitosis comprises many complex events that must be accomplished in less than an hour. The length of a human full cell cycle is approximately 24 h during which a dividing cell is preparing itself to enter mitosis. First of all the cell must have replicated its DNA (S phase) and possess two full copies of its genome (G2 phase). Second, the cell must also have duplicated its centrosome and possess two centrosomes (four centrioles). These centrosomes then need to go through a maturation process, meaning that proteins involved in mitotic microtubule nucleation, such as γ-tubulin for instance, must have been recruited to the centrosome before cells may enter mitosis.

Prophase: Leaving the Starting Blocks. During the prophase stage, the chromatin start to condense to form well-defined chromosomes, each chromosome consists of two sister chromatids connected at the level of their centromeres. While centrosome maturation is continuing during prophase, duplicated centrosomes must have separated and started to migrate around the nucleus to reach opposite position (the two centrosomes are now separated by the nucleus). By the end of prophase, the nuclear membrane starts to breakdown.

Prometaphase: A Cell Without Nucleus. At this the stage, the nuclear membrane has been dissolved, the chromosomes have become thicker. Centrosomes nucleate asters of microtubule that search for chromosomes to attach to. Other microtubules nucleated by the chromosomes will help to assemble the bipolar spindle. The chromosome centromeres where the kinetochores are assembled are an important attachment point for the microtubules. This attachment is controlled by the metaphase spindle checkpoint.

Metaphase: Being Under Surveillance. The chromosomes have reached their maximum condensation state. One pair of sister chromatids linked together by cohesins forms each chromosome. Each pair of chromatid kinetochores must have one kinetochore attached to microtubules nucleated by a centrosome and the opposite kinetochore attached to microtubules emanating from the opposite centrosome. During all this process, the spindle formation is controlled by the dynamic instability of the microtubules. At the end of metaphase, the spindle must be under tension with all the chromosome kinetochores attached to both centrosomes and aligned at the metaphase plate. Cell will remain in metaphase until all the above conditions are fulfilled leading to the spindle checkpoint switch off.

Anaphase: Chromosome Segregation. This stage is triggered once the cell has controlled the spindle under tension and all the kinetochores have been captured by microtubules. When the spindle seems the most stable, the cohesins that maintain the sister chromatids are degraded and each sister chromatid is pulled toward each centrosome forming two identical set of chromosomes.

Telophase: Get Ready for Cell Division. During anaphase while chromosomes are moving, many kinetochore proteins get detached from chromosomes to remain at the center of the cell where a central spindle is assembled. A contractile actin ring forms under the surface of the plasma membrane, around the central spindle. All these events lead to a contraction of the plasma membrane at the middle of the cell that will form two cells attached by the midbody.

Cytokinesis and Abscission: Daughter Cells Separation. This is the less understood event of mitosis: the two daughter cells must separate. To do so, the midbody must be broken and one of the cells will inherit a flemming body (remaining of the midbody). But more importantly, the cell must repair the plasma membrane to avoid leaking of cell contents. This is achieved by recruiting membrane vesicles from the previously dissolved Golgi. These vesicles also carry proteins required for cytokinesis. The very last step of cytokenesis called abscission is the physical separation of the two daughter cells.

Control of Mitosis by Phosphorylation. The protein kinases described here are all involved in the regulation of multiple events during mitotic progression. Analyzing the function of a mitotic kinase is not easy since knock down of the protein expression by RNA interference usually generates a phenotype that corresponds to the first event controlled by the enzyme. For instance, eliminating CDK1 leads to a cell cycle arrest in G2 phase. The cell does not enter mitosis because CDK1 is required for the G2/M transition. But CDK1 is also required for progression through mitosis. The function of each kinase is also tightly linked to their localizations during progression through mitosis, "being at the right place at the right time" (Fig. 2). One can for instance rescue Aurora B knock down by an Aurora A kinase chimera containing Aurora B localization sequences (17).

CDK1/Cyclin B Activity Delimits Mitosis. Cdk1/cyclin B activity appears in late G2 and peaks at metaphase (the middle of M phase) and is inactivated upon exit from mitosis by cyclin B destruction, degraded first on the spindle at the chromosome level together with cohesins (18). Cdk1 kinase plays important roles in early stages

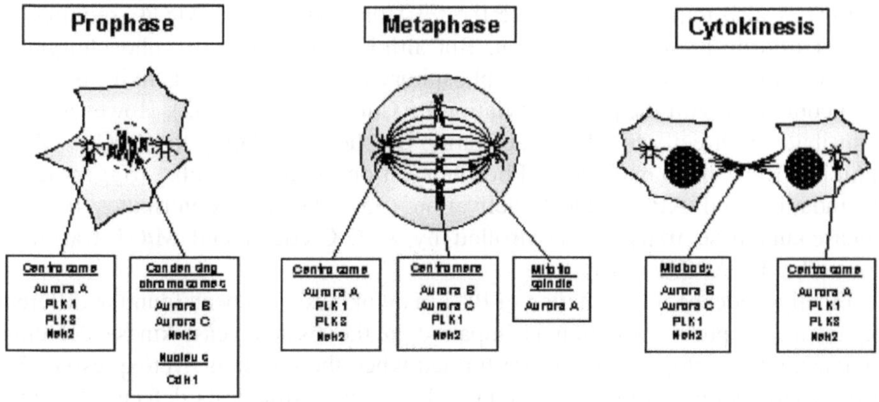

Fig. 2 Localization of the major mitotic kinase through the mitotic phase. One of the clues to succeed in mitosis for mitotic kinases is to be "at the right place at the right moment." The short summary of where the kinases have been found gives an idea of the complexity of the controls insured by mitotic protein kinases

that contribute to the G2/M transition. Cdk1 phosphorylates motor proteins involved in centrosomes separation required for bipolar spindle assembly (19). Cdk1 phosphorylates lamina inducing a destabilization of the nuclear structure leading to nuclear envelope breaks down (20). It also phosphorylates condensin contributing to chromosome condensation (21). When Cdk1 activity is maximum, it participates to the activation of the APC/C that insure the ubiquitination of the proteins targeted to be degraded at the metaphase/anaphase transition, including cyclin B and securin (22).

Plk1: A Very Busy Kinase. Plk1 kinase activity peaks in mitosis. The kinase is composed of a catalytic domain and a PBD (polo box domain) that must bind to a docking protein previously phosphorylated by a priming kinase to allow Plk1 activation (23). Also, Plk1 is activated by phosphorylation of its T-loop by an activated kinase (24). Plk1 localizes to the centrosomes, the kinetochores, and the midbody during mitosis. The kinase plays multiple roles during mitosis; it participates to the G2/M transition, its inhibition delays entry in mitosis. Among the Plk1 substrates one finds all the major players involved in the G2/M transition, CDC25, Myt1, and cyclin B1 (25–27). Plk1 would be involved in the feed back loop that controls the activation of Cdk1/cyclin B.

Plk1 activity is also required for centrosome maturation by recruiting protein necessary to nucleate the microtubules that will participate to bipolar spindle assembly; the kinase also interacts with and phosphorylates many proteins involved in microtubules dynamic (28, 29). In addition to be localized and active at the centrosome level, Plk1 also localizes to the chromosome kinetochores (30) where its activity participates to the localization of spindle checkpoint proteins. The exact function of Plk1 at the kinetochores and its participation to the spindle checkpoint remains to be clarified.

Plk1 is also required to activate the E3 Ubiquitine ligase APC/C required to trigger mitotic protein degradation. But although Plk1 directly phosphorylates APC/C subunits, the effect of this phosphorylation on APC/C activity is minor (31) compared with the phosphorylation by Cdk1/cyclin B1 (32). However, Plk1 contributes indirectly to APC/C activation by phosphorylating the APC/C-cdc20 inhibitor Emi1 in somatic cells. Phosphorylation of Emi1 by Plk1 triggers Emi1 degradation and APC/C-cdc20 activation (33, 34). This contributes to metaphase–anaphase transition controlled by APC/C-cdc20 and M/G1 transition controlled by APC/C-cdh1.

Finally evidence for a function of Plk1 in cytokinesis has been found in different organisms. Septum formation is impaired in the fission yeast kinase defective mutants, while ectopic septums are formed when the kinase is overexpressed (35, 36). In *Drosophila*, polo kinase mutant also shows cytokinesis defects at various stages of spermatogenesis (37). In vertebrate cells, the kinase localizes at the midbody (38). Plk1 also interact with and phophorylates kinesin proteins required in cytokinesis such as MKLP1 (39).

Aurora A: A Centrosome Protein. Aurora A is activated by binding to some of its substrates like TPX2, a mechanism that insures a local activation of the kinase (40). Aurora A is restricted to the centrosome area where it phosphorylates CDC25B contributing to G2/M transition (41). But unlike Cdk1, Aurora A is dispensable; its absence only delays entry into mitosis (42). Aurora A activity is required for centrosomes separation and maturation that consists in recruiting proteins involved in microtubule nucleation. The kinase phosphorylates motor proteins (43) and proteins required for astral microtubule nucleation (44). Aurora A might also be involved later in mitosis because its overexpression induced a bypass of the Taxol-induced mitotic checkpoint (45). The kinase is involved in cytokinesis since its overexpression induced polyploidy aggravated in the absence of p53 (46). However, these two last points need to be investigated further, in particular the relationship between Aurora A and p53. Upon exit from mitosis, Aurora A is degraded by the proteasome in a CDH1 dependant manner (47, 48).

Aurora B: A Chromosome Passenger Protein. Aurora B participates to at least two protein complexes with INCENP, and with INCENP/survivin/Borealin (49). Those proteins form the chromosome passenger protein family, they localize to the kinetochores until the metaphase–anaphase transition occurs then they relocalize to the midbody (50). Like Aurora A, Aurora B is activated by binding to some of its substrates. Aurora B clearly fulfills three distinct functions during mitosis. Aurora B is a histone kinase, it phosphorylates serines 10 and 28 on histone H3 and the serine 7 in the centromere histone variant CENP-A (51–53). The function of these phosphorylations is still debated: chromosome condensation? Loading of mitotic proteins on chromosome? Signaling mitosis?(54)

Aurora B also phosphorylates MCAK (mitotic centromere-associated kinesin) that results in the inactivation of its microtubule depolymerase catalytic activity and its targeting to the kinetochores. MCAK is involved in the spindle checkpoint by correcting the nonamphitelic attachments of microtubules to the kinetochores

(55). Aurora B RNA interference or inhibition mainly induces the formation of polyploid cells indicating that Aurora B activity is required for cytokinesis. And indeed, the kinase phosphorylates vimentin, the kinesin ZEN-4/MKLP1, and MgcRacGAP, a GTPase activating protein (GAP) all required for cytokinesis (56–58).

Aurora C: An Aurora B Substitute? Aurora C is expressed only in testis (59). However, overexpression of Aurora C has been observed in number of cancer cell lines and tumors (60, 61). Aurora C was first described as an anaphase centrosome protein (60). However, it turns out that Aurora C when overexpressed behaved just like Aurora A in interphase and like Aurora B in mitosis (62, 63). Aurora C like other Aurora is activated by some of its substrates, in particular by Aurora B substrate such as INCENP (63). Not only Aurora C mimics Aurora B in mitosis but also it rescues Aurora B depleted cells (63). Strikingly, nobody has yet localized the endogenous protein or analyzed whether the kinase is expressed in normal cells and what would be its function.

NIMA: A Kinase with Many Relatives. NIMA (never in mitosis A) is an *Aspergillus nidulans* protein kinase. Mutations that inactivate the kinase led to a late G2 arrest with cells harboring duplicated but unseparated centrosomes (64). Among the 13 mammalian Nek, Nek2 is the closest NIMA relatives, its activity is absolutely required for mitosis. Nek2 phosphorylates C-Nap1 (centrosomal Nek2-associated protein 1). Its phosphorylation is required for centrosome separation that is a prerequisite to bipolar spindle assembly (65). Nek2, at least in *Drosophila*, might also have a role late in mitosis since its overexpression leads to cytokinesis defects (16). These functions again are related to the localization of the kinase: the centrosomes and the kinetochores. Human cells express two isoforms of Nek2, Nek2A and Nek2B. Nek2A is degraded by the APC/C upon entry into mitosis whereas Nek2B remains stable during mitosis (15, 66). Nek2 is activated by trans-autophophorylation and inhibited by dephosphorylation by the phosphatase PP1 (67).

Other members of the Nek kinase family play roles during mitosis. Nek6 that is highly expressed during mitosis is required for mitosis progression because its inhibition provokes a metaphase arrest (68, 69). Nek9 that is a mitotic centrosome kinase phosphorylates and activates Nek6 (68). Inhibition of Nek9 impairs bipolar spindle assembly (70).

Mitotic Kinases and Cancer. In 1914, Boveri proposed aneuploidy (abnormal chromosome number) arising from mitotic defects as a mechanism that might lead to oncogenesis. Abnormal mitosis can indeed generate cells with multiple centrosomes and abnormal number of chromosomes frequently observed in cancer cells (71).

Boveri was right; chromosome instability and aneuploidy generate genetic defects that are hallmarks of tumorigenesis. They arise through defects during mitosis when chromosomes are unequally segregated between the two daughter cells. Neoplastic development is a multistep mechanism due to an accumulation of genetic defects that breaks the balance between growth-inhibitory signal and division-promoting signal. Tumors then would derive from one cell in which the growth-inhibitory signal is down regulated (loss of tumor suppressor genes) and the division-promoting signal is elevated (gain of proto-oncogenes). One of the best examples is the transformation of human cell line achieved by coexpression of the SV40 large-T oncoprotein, the *H-ras* oncogene, and the telomerase catalytic subunit (72).

Chromosome segregation is a finely regulated process insured by the mitotic spindle that is a highly dynamic microtubule-based structure. The mitotic spindle is composed of two centrosomes connected by microtubules to the chromosomes aligned at the metaphase plate in the centre of the structure. In every pair of sister chromatid, each chromatid is connected to opposite centrosomes forming amphitelic attachment. During segregation, each sister chromatid migrates to one pole of the cell leading to the formation of two identical groups of chromosome. This bipolarity is necessary to form two daughter cells with the same DNA content during cytokinesis. In mammalian cells, each spindle pole is organized around a centrosome. Remarkably, in cancer cells, the number, the structure, and the function of centrosomes are often abnormal and correlate with aneuploidy and chromosome instability.

So, because the mitotic spindle plays a central role in chromosome segregation, many proteins involved in its establishment and in its regulation are often misregulated in cancers. This is the case for the mitotic protein kinases Cdk1, Aurora A, B, and C, Plk1, and Nek2 protein kinases (Table 1).

Table 1 An overview of relationship between mitotic kinase expression and cancer

Protein kinase	Expression in cancer	Cancers associated
Aurora A	Amplification of gene locus (20q13) Over-expression	Breast, colon, pancreatic, bladder & ovarian cancer, prostate cancer & neuroblastoma
Aurora B	Over-expression	Colorectal cancer, thyroid carcinoma, non-small cell lung carcinoma & prostate cancer
Aurora C	Over-expression	Thyroid carcinoma
Plk1	Over-expression	Non-small cell lung carcinoma, head and neck cancer, oesophageal and gastric cancer, breast, ovarian & endometrial cancer, colorectal cancer, thyroid cancer, glioma & melanoma
Plk3	Down-regulation	Head/neck squamous carcinoma, lung carcinoma & colon tumour
	Over-expression	Epithelial ovarian tumour
Nek2	Over-expression	Ewing's tumour, non-hodgkin lymphoma, breast, cervical and prostate carcinoma
	Amplification of gene locus (1q32)	Breast & gastric cancer
Cdk1	Over-expression	Colon adenoma, colorectal carcinoma and esophageal adenocarcinoma

CDKs. Alterations of Cdks have rarely been observed in cancer, and overexpression of Cdk1 and Cdk2 has been reported in colon adenomas (73, 74). However, alterations of proteins that regulate Cdks such as cyclins, Cdk-activating enzymes, and CKI are frequently observed (75, 76). For instance, cyclin A is overexpressed in lung carcinoma and elevated expression correlated with shorter survival (77). But the best proves that Cdk hyperactivities are involved in cancer is the fact that drugs inhibiting Cdks are particularly effective to inhibit tumor progression. A large variety of compounds are actually on the market and their successes are due to their effects, inhibition of cell proliferation, activation of apoptosis, and in some cases they can even trigger differentiation (78–80).

Aurora A. The gene encoding Aurora A is located on chromosome 20q13. This chromosome region is frequently found amplified in human cancers, and the amplification is associated with overexpression of the protein kinase (81). This Aurora A amplification/overexpression is detected in cancers like breast, colon, pancreatic, bladder, ovarian, prostate cancer, and neuroblastoma (82–85). Furthermore, the *Aurora A* gene copy number correlates with chromosomal instability and aneuploidy in human bladder tumor. Overexpression of the kinase also correlates with clinical aggressiveness of the tumor (84, 86). Aurora A has also been found mutated in several cancers, the mutation that is proposed to be due to polymorphism designs Aurora A as a candidate susceptibility gene (87).

In vitro Aurora A kinase overexpression induces aneuploidy and abnormal centrosome numbers leading to cell tumorigenic transformation (82, 83). Long-term overexpression of Aurora A is also sufficient to induce tumor formation in mice, after a long period of genomic instability (88). However, it is not clear how the kinase induces tumorigenesis, it has been proposed that Aurora A overexpression would be sufficient to escape negative regulation by tumor suppressor pathway. Aurora A kinase interacts with and phosphorylates the tumor suppressor protein p53. Phosphorylation of p53 induced its degradation through mdm2 (89) and reduced it transactivation activity (90). Because p53 plays a major role in carcinogenesis, its interaction with Aurora A might be important in the kinase oncogenic activity. Moreover, Aurora A kinase is a RasGap Src homology 3 domain binding protein and forms a complex with RasGap and Survivin proteins. This interaction inhibits Aurora A activity (91). Because RasGap is also a negative regulator of Ras pathway, it has been suggested that in cell overexpressing Ras, there would not be enough RasGap to inactivate both Ras and Aurora A leading to Aurora A hyperactivity participating to oncogenesis.

As mitosis regulators, the expression levels of the major mitotic kinases are crucial for cell division. In many cancers, up and down regulation of their expression have been observed, underlying the importance to follow the expression level of mitotic kinases for developing new targeted therapy.

Aurora B. Aurora B overexpression has been found in many cancers like colorectal cancer (92, 93) or thyroid carcinoma (94). In colorectal and prostate cancer, Aurora B overexpression increases in correlation with the tumor malignancy (92, 95). However, unlike Aurora A, *Aurora B* gene has never been found amplified and the origin of Aurora B overexpression is actually unknown. Also unlike Aurora A, Aurora B is not an oncogene, but its overexpression induces metastasis. Aurora B overexpression results in hyperphosphorylation of histone H3 on serine 10 (96). This increase in serine10 phosphorylation is observed on lagging chromosomes during mitosis (96). Does hyperphosphorylation of histone H3 induce chromosome instability and aneuploidy? It seems so since, lagging chromosomes have been observed in cells transfected with a Ser10 phospho-mimetic form of histone H3 (96). Whether hyperphosphorylation of H3 participates to segregation defect is obvious, whether it participates to metastases apparition is not clear. Although Aurora B overexpression in cancer cells correlates with genetic instability (97), how Aurora B expression is linked to cancer remains to be determined. However, inhibition of Aurora kinases (especially with anti-Aurora B drugs) efficiently reduced tumor growth in mice (98).

Aurora C. Little is know about this third member of Aurora kinase family. In normal physiological conditions, Aurora C is expressed only in testis, (59). However, cancer cell lines expressed the kinase (60). Aurora C is highly expressed in human thyroid carcinoma cell lines and tissues, where its expression correlates with the aggressiveness of the tumor (61). Overexpression of Aurora C gives rise to polyploid cells. Like for the other Aurora kinases, the phenotype is aggravated in the absence of p53 (62). Because Aurora C is very close to Aurora B, one would expect overexpression of Aurora C to have the same consequences than over-expression of Aurora B.

Plk1. Overexpression of Plk1 in rodent cells is sufficient to confer a transform phenotype indicating that Plk1 is a potential oncogene (99). In agreement with this data, Plk1 has been found overexpressed in a large variety of cancers (100 for review). And high level of Plk1 is a sign of bad prognosis in several cancers (101–105). Beside overexpression, mutation in Plk1 has been observed in cancers; some of the mutations inhibit the interaction of Plk1 with Hsp90 and stabilize the kinase leading to a hyperactivity of Plk1 (106). Like for Aurora kinases, overexpression of Plk1 generates genomic instability by triggering the formation of polyploid cells (107) frequently observed in cancer cells (108). Taking together these data designed Plk1 as a good target for inhibitors used as anticancer drugs. And indeed inhibition of Plk1 was reported to have different effects in cancer cells vs. normal cells (28). Inhibition of Plk1 arrests tumor cells in culture as well as it reduces tumor growth in mice indicating that the kinase is absolutely required for cells that highly proliferate (109–110).

Nek2. Ewing's tumor cell line derived (a paediatric osteosarcoma) and non-Hodgkin lymphoma show elevated level of Nek2 mRNA, and the transcript level increase correlates with aggressiveness (111). Nek2 is also overexpressed in cervical and prostate carcinoma as well as in gastric and breast in which the chromosomal region 1q32 corresponding to the human *Nek2* gene locus is amplified (six times in breast cancer) (112, 113). Nek2 is not a proto-oncogene; however, its overexpression provokes defects in centrosome organization and function. In HBL100 cells, over-expression of human Nek2 induces the formation of aneuploid cells containing abnormal numbers of centrosomes a hallmark of cancer cells (71).

Conclusion

This short review only describes the most known mitotic protein kinases, we voluntarily omitted checkpoint kinases and others to make this review comprehensive for people that are not familiar with the protein kinase world. It is important to notice that many other mitotic protein kinases remain to be discovered and studied as demonstrated by recently performed screens in search for novel kinases (114).

Acknowledgments I express my apologies to all the authors who have published work related to this review but have not been cited here owing to space constraints. Research in the Claude Prigent laboratory is financed by the CNRS, the INCa and the «Canceropole grand Ouest». Patrick Salaun is a fellow of the LNCC and Yoann Rannou a fellow of the French Minister.

References

1. Fisher DL, Nurse P (1996) A single fission yeast mitotic cyclin B p34cdc2 kinase promotes both S-phase and mitosis in the absence of G1 cyclins. EMBO J 15(4):850–60.
2. Durkacz B, Carr A, Nurse P (1986) Transcription of the *cdc2* cell cycle control gene of the fission yeast *Schizosaccharomyces pombe*. EMBO J 5(2):369–373.
3. Draetta G, Beach D (1988) Activation of cdc2 protein kinase during mitosis in human cells: cell cycle-dependent phosphorylation and subunit rearrangement. Cell 54(1):17–26.
4. Draetta G, Luca F, Westendorf J, et al. (1989) Cdc2 protein kinase is complexed with both cyclin A and B: evidence for proteolytic inactivation of MPF. Cell 56(5):829–38.
5. Karaiskou A, Perez LH, Ferby I, et al. (2001) Differential regulation of Cdc2 and Cdk2 by RINGO and cyclins. J Biol Chem 276(38):36028–34.
6. Nigg EA (1998) Polo-like kinases: positive regulators of cell division from start to finish. Curr Opin Cell Biol 10:776–783.
7. Habedanck R, Stierhof YD, Wilkinson CJ, Nigg EA (2005) The Polo kinase Plk4 functions in centriole duplication. Nat Cell Biol 7(11):1140–6.
8. Chan CS, Botstein D (1993) Isolation and characterization of chromosome-gain and increase-in-ploidy mutants in yeast. Genetics 135(3):677–91.
9. Glover DM, Leibowitz MH, McLean DA, et al. (1995) Mutations in aurora prevent centrosome separation leading to the formation of monopolar spindles. Cell 81(1):95–105.
10. Nigg EA (2001) Mitotic kinases as regulators of cell division and its checkpoints. Nat Rev Mol Cell Biol 2(1):21–32.
11. Brown JR, Koretke KK, Birkeland ML, et al. (2004) Evolutionary relationships of Aurora kinases: implications for model organism studies and the development of anti-cancer drugs. BMC Evol Biol 4(1):39.
12. O'Connell MJ, Krien MJ, Hunter T (2003) Never say never. The NIMA-related protein kinases in mitotic control. Trends Cell Biol 13(5):221–8.
13. Osmani SA, May GS, Morris NR (1987) Regulation of the mRNA levels of *nimA*, a gene required for the G2-M transition in *Aspergillus nidulans*. J Cell Biol 104(6):1495–504.
14. Belham C, Roig J, Caldwell JA, et al. (2003) Mitotic cascade of NIMA family kinases. Nercc1/Nek9 activates the Nek6 and Nek7 kinases. J Biol Chem 278(37):34897–909.
15. Fry AM (2002) The Nek2 protein kinase: a novel regulator of centrosome structure. Oncogene 21:6184–6194.
16. Prigent C, Glover DM, Giet R (2005) *Drosophila* Nek2 protein kinase knockdown leads to centrosome maturation defects while overexpression causes centrosome fragmentation and cytokinesis failure. Exp Cell Res 303(1):1–13.

17. Scrittori L, Skoufias DA, Hans F, et al. (2005) A small C-terminal sequence of Aurora B is responsible for localization and function. Mol Biol Cell 16(1):292–305.
18. Clute P, Pines J (1999) Temporal and spatial control of cyclin B1 destruction in metaphase. Nat Cell Biol 1(2):82–7.
19. Blangy A, Lane HA, d'Herin P, et al. (1995) Phosphorylation by p34cdc2 regulates spindle association of human Eg5, a kinesin-related motor essential for bipolar spindle formation in vivo. Cell 83(7):1159–69.
20. Peter M, Nakagawa J, Doree M, et al. (1990) In vitro disassembly of the nuclear lamina and M phase-specific phosphorylation of lamins by cdc2 kinase. Cell 61(4):591–602.
21. Kimura K, Hirano M, Kobayashi R, et al. (1998) Phosphorylation and activation of 13S condensin by Cdc2 in vitro. Science 282(5388):487–90.
22. Rudner AD, Murray AW (2000) Phosphorylation by Cdc28 activates the Cdc20-dependent activity of the anaphase-promoting complex. J Cell Biol 149(7):1377–90.
23. Elia AE, Rellos P, Haire LF, et al. (2003) The molecular basis for phosphodependent substrate targeting and regulation of Plks by the Polo-box domain. Cell 115(1):83–95.
24. Qian YW, Erikson E, Maller JL (1998) Purification and cloning of a protein kinase that phosphorylates and activates the polo-like kinase Plx1. Science 282(5394):1701–4.
25. Kumagai, A, Dunphy WG (1996) Purification and molecular cloning of Plx1, a Cdc25-regulatory kinase from Xenopus egg extracts. Science 273:1377–1380.
26. Nakajima H, Toyoshima-Morimoto F, Taniguchi E, et al. (2003) Identification of a consensus motif for Plk (Polo-like kinase) phosphorylation reveals Myt1 as a Plk1 substrate. J Biol Chem 278(28):25277–80.
27. Nakajima Toyoshima-Morimoto F, Taniguchi E, Shinya N, et al. (2001) Polo-like kinase 1 phosphorylates cyclin B1 and targets it to the nucleus during prophase. Nature 410(6825):215–20.
28. Lane, Ham, Nigg, EA (1996) Antibody microinjection reveals an essential role for human polo-like kinase 1 (Plk1) in the functional maturation of mitotic centrosomes. J. Cell Biol 135:1701–1713.
29. do Carmo Avides M, Tavares A, Glover DM (2001) Polo kinase and Asp are needed to promote the mitotic organizing activity of centrosomes. Nature Cell Biol 3:421–424.
30. Arnaud L, Pines J, Nigg EA (1998) GFP tagging reveals human Polo-like kinase 1 at the kinetochore/centromere region of mitotic chromosomes. Chromosoma 107(6–7):424–9.
31. Golan A, Yudkovsky Y, Hershko A (2002) The cyclin-ubiquitin ligase activity of cyclosome/APC is jointly activated by protein kinases Cdk1-cyclin B and Plk. J Biol Chem 277(18):15552–7.
32. Kraft C, Herzog F, Gieffers C, et al. (2003) Mitotic regulation of the human anaphase-promoting complex by phosphorylation. EMBO J 22(24):6598–609.
33. Hansen DV, Loktev AV, Ban KH, et al. (2004) Plk1 regulates activation of the anaphase promoting complex by phosphorylating and triggering SCFbetaTrCP-dependent destruction of the APC Inhibitor Emi1. Mol Biol Cell (12):5623–34.
34. Moshe Y, Boulaire J, Pagano M, et al. (2004) Role of Polo-like kinase in the degradation of early mitotic inhibitor 1, a regulator of the anaphase promoting complex/cyclosome. Proc Natl Acad Sci USA 101(21):7937–42.
35. Ohkura H, Hagan IM, Glover DM (1995) The conserved *Schizosaccharomyces pombe* kinase plo1, required to form a bipolar spindle, the actin ring, and septum, can drive septum formation in G1 and G2 cells. Genes Dev 9(9):1059–73.
36. Bahler J, Steever AB, Wheatley S, et al. (1998) Role of polo kinase and Mid1p in determining the site of cell division in fission yeast. J Cell Biol 143(6):1603–16.
37. Carmena M, Riparbelli MG, Minestrini G, et al. (1998) Drosophila polo kinase is required for cytokinesis. J Cell Biol 143(3):659–71.
38. Qian YW, Erikson E, Maller JL (1999) Mitotic effects of a constitutively active mutant of the Xenopus polo-like kinase Plx1. Mol Cell Biol (12):8625–32.
39. Neef R, Preisinger C, Stucliffe J, et al. (2003) Phosphorylation of mitotic kinesin-like protein 2 by polo-like kinase 1 is required for cytokinesis. J Cell Biol 162:863–875.

40. Tsai MY, Wiese C, Cao K, et al. (2003) A Ran signalling pathway mediated by the mitotic kinase Aurora A in spindle assembly. Nat Cell Biol 5(3):242–8.
41. Dutertre S, Cazales M, Quaranta M, et al. (2004) Phosphorylation of CDC25B by Aurora A at the centrosome contributes to the G2-M transition. J Cell Sci 117(Pt 12):2523–31.
42. Marumoto T, Hirota T, Morisaki T, et al. (2002) Roles of Aurora A kinase in mitotic entry and G2 checkpoint in mammalian cells. Genes Cells 7(11):1173–82.
43. Giet R, Uzbekov R, Cubizolles F, et al. (1999) The Xenopus laevis aurora-related protein kinase pEg2 associates with and phosphorylates the kinesin-related protein XlEg5. J Biol Chem 274(21):15005–13.
44. Giet R, McLean D, Descamps S, et al. (2002) Drosophila Aurora A kinase is required to localize D-TACC to centrosomes and to regulate astral microtubules. J Cell Biol 156(3):437–51.
45. Anand S, Penrhyn-Lowe S, Venkitaraman AR (2003) AURORA A amplification overrides the mitotic spindle assembly checkpoint, inducing resistance to Taxol. Cancer Cell 3(1):51–62.
46. Meraldi P, Honda R, Nigg EA (2002) Aurora A overexpression reveals tetraploidization as a major route to centrosome amplification in p53–/– cells. EMBO J 21(4):483–92.
47. Castro A, Arlot-Bonnemains Y, Vigneron S, et al. (2002a) APC/Fizzy-Related targets Aurora A kinase for proteolysis. EMBO Rep 3(5):457–62.
48. Castro A, Vigneron S, Bernis C, et al. (2002b) The D-Box-activating domain (DAD) is a new proteolysis signal that stimulates the silent D-Box sequence of Aurora A. EMBO Rep 3(12):1209–14.
49. Giet R, Petretti C, Prigent C (2006) Aurora kinases, aneuploidy and cancer, a coincidence or a real link? Trends in Cell Biol 15(5):241–50.
50. Adams RR, Wheatley SP, Gouldsworthy AM, et al. (2000) INCENP binds the Aurora-related kinase AIRK2 and is required to target it to chromosomes, the central spindle and cleavage furrow. Curr Biol 10(17):1075–8.
51. Hsu JY, Sun ZW, Li X, et al. (2000) Mitotic phosphorylation of histone H3 is governed by Ipl1/aurora kinase and Glc7/PP1 phosphatase in budding yeast and nematodes. Cell 102(3):279–91.
52. Giet R, Glover DM (2001) *Drosophila* aurora B kinase is required for histone H3 phosphorylation and condensin recruitment during chromosome condensation and to organize the central spindle during cytokinesis. J Cell Biol 152(4):669–82.
53. Zeitlin SG, Shelby RD, Sullivan KF (2001) CENP-A is phosphorylated by Aurora B kinase and plays an unexpected role in completion of cytokinesis. J Cell Biol 155(7):1147–57.
54. Prigent C, Dimitrov S (2003) Phosphorylation of serine 10 in histone H3, what for? J Cell Sci 116(Pt 18):3677–85.
55. Ohi R, Sapra T, Howard J, et al. (2004) Differentiation of cytoplasmic and meiotic spindle assembly MCAK functions by Aurora B-dependent phosphorylation. Mol Biol Cell 15(6):2895–906.
56. Goto H, Yasui Y, Kawajiri A, et al. (2003) Aurora B regulates the cleavage furrow-specific vimentin phosphorylation in the cytokinetic process. J Biol Chem 278(10):8526–30.
57. Minoshima Y, Kawashima T, Hirose K, et al. (2003) Phosphorylation by aurora B converts MgcRacGAP to a RhoGAP during cytokinesis. Dev Cell 4(4):549–60.
58. Guse A, Mishima M, Glotzer M (2005) Phosphorylation of ZEN-4/MKLP1 by aurora B regulates completion of cytokinesis. Curr Biol 15(8):778–86.
59. Hu HM, Chuang CK, Lee MJ, et al. (2000) Genomic organization, expression, and chromosome localization of a third aurora-related kinase gene, Aie1. DNA Cell Biol 19(11):679–88.
60. Kimura M, Matsuda Y, Yoshioka T, et al. (1999) Cell cycle-dependent expression and centrosome localization of a third human aurora/Ipl1-related protein kinase, AIK3. J Biol Chem 274(11):7334–40.
61. Ulisse S, Delcros JG, Baldini E, et al. (2006) Expression of Aurora kinases in human thyroid carcinoma cell lines and tissues. Int J Cancer 119(2):275–82.

62. Dutertre S, Hamard-Peron E, Cremet JY, et al. (2005) The absence of p53 aggravates polyploidy and centrosome number abnormality induced by Aurora C overexpression. Cell Cycle 4(12):1783–7.
63. Sasai K, Katayama H, Stenoien DL, et al. (2004) Aurora C kinase is a novel chromosomal passenger protein that can complement Aurora B kinase function in mitotic cells. Cell Motil Cytoskeleton 59(4):249–63.
64. Osmani AH, McGuire SL, Osmani SA (1991) Parallel activation of the NIMA and p34cdc2 cell cycle-regulated protein kinases is required to initiate mitosis in *A. nidulans*. Cell 67(2):283–91.
65. Fry AM, Mayor T, Meraldi P, et al. (1998) C-Nap1, a novel centrosomal coiled-coil protein and candidate substrate of the cell cycle-regulated protein kinase Nek2. J Cell Biol 141(7):1563–74.
66. Fry AM, Arnaud L, Nigg EA (1999) Activity of the human centrosomal kinase, Nek2, depends on an unusual leucine zipper dimerization motif. J Biol Chem 274:16304–10.
67. Helps NR, Luo X, Barker HM, et al. (2000) NIMA-related kinase 2 (Nek2), a cell-cycle-regulated protein kinase localized to centrosomes, is complexed to protein phosphatase 1. Biochem J 349(Pt 2):509–18.
68. Belham C, Roig J, Caldwell JA, et al. (2003) A mitotic cascade of NIMA family kinases. Nercc1/Nek9 activates the Nek6 and Nek7 kinases. J Biol Chem 278:34897–909.
69. Yin MJ, Shao L, Voehringer D, et al. (2003). The serine/threonine kinase Nek6 is required for cell cycle progression through mitosis. J Biol Chem 278:52454–60.
70. Roig J, Groen A, Caldwell J, et al. (2005). Active Nercc1 protein kinase concentrates at centrosomes early in mitosis, and is necessary for proper spindle assembly. Mol Biol Cell 16:4827–40.
71. Lingle WL, Barrett SL, Negron VC, et al. (2002) Centrosome amplification drives chromosomal instability in breast tumor development. Proc Natl Acad Sci USA 99(4):1978–83.
72. Hahn WC, Counter CM, Lundberg AS, et al. (1999) Creation of human tumour cells with defined genetic elements. Nature 400(6743):464–8.
73. Yamamoto H, Monden T, Miyoshi H, et al. (1998) Cdk2/cdc2 expression in colon carcinogenesis and effects of cdk2/cdc2 inhibitor in colon cancer cells. Int J Oncol 13(2):233–9.
74. Kim JH, Kang MJ, Park CU, et al. (1999) Amplified CDK2 and cdc2 activities in primary colorectal carcinoma. Cancer 85(3):546–53.
75. Sherr CJ (1996) Cancer cell cycles. Science 274(5293):1672–7.
76. McDonald ER 3rd, El-Deiry WS (2000) Cell cycle control as a basis for cancer drug development. Int J Oncol 16(5):871–86.
77. Dobashi Y, Shoji M, Jiang SX, et al. (1998) Active cyclin A-CDK2 complex, a possible critical factor for cell proliferation in human primary lung carcinomas. Am J Pathol 153(3):963–72.
78. Matushansky I, Radparvar F, Skoultchi AI (2000) Reprogramming leukemic cells to terminal differentiation by inhibiting specific cyclin-dependent kinases in G1. Proc Natl Acad Sci USA 97:14317–22.
79. Damiens E, Baratte B, Marie D, et al. (2001) Anti-mitotic properties of indirubin-3-monoxime, a CDK/GSK-3 inhibitor: induction of endoreplication following prophase arrest. Oncogene 20:3786–97.
80. Edamatsu H, Gau CL, Nemoto T, et al. (2000) Cdk inhibitors, roscovitine and olomoucine, synergize with farnesyl transferase inhibitor (FTI) to induce efficient apoptosis of human cancer cell lines. Oncogene 19:3059–68.
81. Sen S, Zhou H, Zhang RD, et al. (2002). Amplification/overexpression of a mitotic kinase gene in human bladder cancer. J Natl Cancer Inst 94:1320–9.
82. Bischoff JR, Anderson L, Zhu Y, et al. (1998) A homologue of Drosophila aurora kinase is oncogenic and amplified in human colorectal cancers. Embo J 17:3052–65.
83. Zhou H, Kuang J, Zhong L, et al. (1998). Tumour amplified kinase STK15/BTAK induces centrosome amplification, aneuploidy and transformation. Nat Genet 20:189–93.
84. Tanaka T, Kimura M, Matsunaga K, et al. (1999) Centrosomal kinase AIK1 is overexpressed in invasive ductal carcinoma of the breast. Cancer Res 59:2041–4.

85. Han H, Bearss DJ, Browne LW, et al. (2002). Identification of differentially expressed genes in pancreatic cancer cells using cDNA microarray. Cancer Res 62:2890–6.
86. Miyoshi Y, Iwao K, Egawa C, et al. (2001) Association of centrosomal kinase STK15/BTAK mRNA expression with chromosomal instability in human breast cancers. Int J Cancer 92:370–3.
87. Ewart-Toland A, Briassouli P, de Koning JP, et al. (2003) Identification of Stk6/STK15 as a candidate low-penetrance tumor-susceptibility gene in mouse and human. Nat Genet 34(4):403–12.
88. Wang X, Zhou YX, Qiao W, et al. (2006) Overexpression of aurora kinase A in mouse mammary epithelium induces genetic instability preceding mammary tumor formation. Oncogene 25:7148–58.
89. Katayama H, Sasai K, Kawai H, et al. (2004) Phosphorylation by aurora kinase A induces Mdm2-mediated destabilization and inhibition of p53. Nat Genet 36:55–62.
90. Liu Q, KaNeko S, Yang I, et al. (2004) Aurora A abrogation of p53 dna binding and transactivation activity by phosphorylation of serine 215. J Biol Chem 279:52175–82.
91. Gigoux V, L'Hoste S, Raynaud F, et al. (2002) Identification of Aurora kinases as RasGAP Src homology 3 domain-binding proteins. J Biol Chem 277:23742–6.
92. Katayama H, Ota T, Jisaki F, et al. (1999) Mitotic kinase expression and colorectal cancer progression. J Natl Cancer Inst 91:1160–2.
93. Takahashi T, Futamura M, Yoshimi N, et al. (2000) Centrosomal kinases, HsAIRK1 and HsAIRK3, are overexpressed in primary colorectal cancers. Jpn J Cancer Res 91:1007–14.
94. Adams RR, Eckley DM, Vagnarelli P, et al. (2001) Human INCENP colocalizes with the Aurora-B/AIRK2 kinase on chromosomes and is overexpressed in tumour cells. Chromosoma 110:65–74.
95. Chieffi P, Cozzolino L, Kisslinger A, et al. (2006) Aurora B expression directly correlates with prostate cancer malignancy and influence prostate cell proliferation. Prostate 66(3):326–33.
96. Ota T, Suto S, Katayama H, et al. (2002) Increased mitotic phosphorylation of histone H3 attributable to AIM-1/Aurora-B overexpression contributes to chromosome number instability. Cancer Res 62:5168–77.
97. Smith SL, Bowers NL, Betticher DC, et al. (2005) Overexpression of aurora B kinase (AURKB) in primary non-small cell lung carcinoma is frequent, generally driven from one allele, and correlates with the level of genetic instability. Br J Cancer 93(6):719–29.
98. Harrington EA, Bebbington D, Moore J, et al. (2004) VX-680, a potent and selective small-molecule inhibitor of the Aurora kinases, suppresses tumor growth in vivo. Nat Med 10(3):262–7.
99. Smith MR, Wilson ML, Hamanaka R, et al. (1997) Malignant transformation of mammalian cells initiated by constitutive expression of the polo-like kinase. Biochem Biophys Res Commun 234:397–405.
100. Eckerdt F, Yuan J, Strebhardt K (2005) Polo-like kinases and oncogenesis. Oncogene 24(2):267–76.
101. Weichert W, Denkert C, Schmidt M, et al. (2004). Polo-like kinase isoform expression is a prognostic factor in ovarian carcinoma. Br J Cancer 90:815–21.
102. Tokumitsu Y, Mori M, Tanaka S, et al. (1999) Prognostic significance of polo-like kinase expression in esophageal carcinoma. Int J Oncol 15:687–92.
103. Knecht R, Oberhauser C, Strebhardt K (2000) PLK (polo-like kinase), a new prognostic marker for oropharyngeal carcinomas. Int J Cancer 89:535–6.
104. Kneisel L, Strebhardt K, Bernd A, et al. (2002) Expression of polo-like kinase (PLK1) in thin melanomas: a novel marker of metastatic disease. J Cutan Pathol 29(6):354–8.
105. Yamada S, Ohira M, Horie H, et al. (2004) Expression profiling and differential screening between hepatoblastomas and the corresponding normal livers: identification of high expression of the PLK1 oncogene as a poor-prognostic indicator of hepatoblastomas. Oncogene 23(35):5901–11.

106. Simizu, S, Osada H (2000) Mutations in the Plk gene lead to instability of Plk protein in human tumour cell lines. Nat Cell Biol 2:852–4.
107. Mundt KE, Golsteyn RM, Lane HA, et al. (1997) On the regulation and function of human polo-like kinase 1 (PLK1): effects of overexpression on cell cycle progression. Biochem Biophys Res Commun 239(2):377–85.
108. Yamamoto Y, Matsuyama H, Kawauchi S, et al. (2006) Overexpression of polo-like kinase 1 (PLK1) and chromosomal instability in bladder cancer. Oncology 70(3):231–7.
109. Spankuch-Schmitt B, Wolf G, Solbach C, et al. (2002) Downregulation of human polo-like kinase activity by antisense oligonucleotides induces growth inhibition in cancer cells. Oncogene 21(20):3162–71.
110. Spankuch B, Matthess Y, Knecht R, et al. (2004) Cancer inhibition in nude mice after systemic application of U6 promoter-driven short hairpin RNAs against PLK1. J Natl Cancer Inst 96(11):862–72.
111. Weiss MM, Kuipers EJ, Postma C, et al. (2004) Genomic alterations in primary gastric adenocarcinomas correlate with clinicopathological characteristics and survival. Cell Oncol 26(5–6):307–17.
112. Schultz SJ, Fry AM, Sutterlin C, et al. (1994) Cell cycle-dependent expression of Nek2, a novel human protein kinase related to the NIMA mitotic regulator of Aspergillus nidulans. Cell Growth Differ 5(6):625–35.
113. Loo LW, Grove DI, Williams EM, et al. (2004) Array comparative genomic hybridization analysis of genomic alterations in breast cancer subtypes. Cancer Res 64(23):8541–9.
114. Bettencourt-Dias M, Giet R, Sinka R, et al. 2004 Genome-wide survey of protein kinases required for cell cycle progression. Nature 432:23–30.

4
Women's Health Research: Perspectives from the National Institutes of Health

Vivian W. Pinn

Introduction

During the past 15 years, much progress has been made in understanding women's health. A notable recent comment about advancements in the science of women's health was found in the March 22, 2006, women's health theme issue of the Journal of the American Medical Association (JAMA). An editorial in that issue indicates that in the 5 years since JAMA's previous theme issue on women's health, interest, knowledge, and understanding about medical problems and issues related to women have increased importantly: "The keen interest in the topic of women's health is evidenced by the 412 manuscripts, a record for JAMA, submitted for consideration in this issue."

That explosion of new information about women's health has been driven by the recognition of several key understandings: (1) diseases may affect women differently than men, (2) some women may be affected differently than other women, and (3) women may be affected differently based on an array of interacting variables.

Credit for these advances is well-deserved by the basic scientists and clinical researchers whose knowledge and creativity have produced the hypotheses and study results, but we also must give credit to the dedicated women's health advocates who understood the need to question the bases for assumptions about women's health and who advocated for policies that would ensure that the appropriate attention and funding would accrue to those research concepts. Underlying the Women's Health Movement of the 1990s were such questions as: *Shouldn't women be used as the norm in evaluating how medical interventions would affect them rather than assuming that studies conducted primarily on men would have the same clinical application for women? Why isn't there more research on women's conditions such as breast cancer (BC), menopause, and HIV/AIDS in women?*

Credit also belongs to the policy makers in Congress, at federal agencies, and at foundations and organizations with health research missions who responded to the need to guide attention to women's health research through policies that required consideration of innate and environmental differences that would affect their health and health care differently. Among those entities establishing policies and funding mechanisms for research on women's health are the National Institutes of Health (NIH) and the NIH Office of Research on Women's Health (ORWH).

J.J. Li et al. (eds.), *Hormonal Carcinogenesis V.*

The Roles of NIH and ORWH NIH Policy on Inclusion

The NIH, one of many entities in the USA Department of Health and Human Services, is the major funding agency of biomedical research for the USA federal government. NIH consists of 27 separate institutes and centers located in Bethesda, Maryland, and numerous other satellite locations.

In the early 1990s, NIH strengthened its inclusion guidelines for clinical research to ensure that norms for health, disease, treatments, and other medical interventions would be appropriately applicable to all populations, women and men, and diverse racial/ethnic groups, because they would be based on scientific evidence derived from studies of those populations. These guidelines conform to the NIH Revitalization Act of 1993, which require NIH to:

- Ensure that women and members of minority groups and their subpopulations are included in all human subject research.
- Ensure that women and minorities and their subpopulations must be included in phase III clinical trials, such that valid analysis of differences in intervention effect can be accomplished.
- Prohibit cost as an acceptable reason for excluding these groups.
- Initiate programs and support for outreach efforts to recruit these groups into clinical studies.

Any exception requires a clear and compelling rationale and justification that inclusion would be inappropriate with respect to the health of the subjects or the purposes of the research.

ORWH Function and Initiatives

The ORWH is administratively located in the Office of the Director of NIH and was established in 1990 to serve as the focal point for women's health across the NIH. The USA Congress mandated establishment of ORWH. The mission of ORWH includes a mandate to:

- Set a national agenda for future directions in women's health research
- Enhance and increase research projects on women's health and related sex/gender factors
- Ensure that women are appropriately represented in biomedical clinical research studies
- Develop opportunities for the recruitment, retention, reentry, and advancement of girls and women in biomedical careers, and support career development for men and women in women's health and sex/gender research

As part of the Office of the Director, ORWH does not have direct funding authority, but rather works with and through the institutes and centers to either fund or cofund research projects or to bring a focus on women's health to their funding initiatives. For example, the funding or cofunding may support research grants and contracts, requests for applications and program announcements for research, career development programs, conferences and scientific workshops, or educational outreach.

In addition, women's health research at the NIH is a partnership that involves professional organizations, scientists and practitioners, women in clinical trials, and advocates. For example, the NIH research agenda for women's health was developed and later revised with input and direct participation of these groups and individuals nationwide.

Women's Health and the Science of Sex and Gender Differences

The perspective of 15 years, since the founding of ORWH, reveals important benchmarks in the continuum of women's health research, care, and leadership. As women's health research has developed, and continues to do so, there are clear advances in the understanding of the very concept of what constitutes women's health, of differences and similarities between men and women resulting from sex and gender comparisons, of health disparities and differences among various populations, of the necessity for interdisciplinary research and health care, of the importance of women's roles in health research and policy, and of the need for women and men to be women's health researchers.

The current concept of and approach to women's health take a lifespan perspective of health status, health outcomes, and environment. The disparities and differences among populations of women, and men, and between women and men, are a result of complex, interacting influences and factors: genes, biology, behavior, environment, socioeconomic status, and access to health care, etc. To clarify discussion and investigation of these factors, the Institute of Medicine Report Exploring the Biological Contributions to Human Health: Does Sex Matter? recommends that definitions for the terms "sex" and "gender" be used consistently by all medical journals. The report recommended that the term "sex" be used when differences are primarily biological in origin and may be genetic or phenotypic (genetic or physiological characteristics of being male or female), and that "gender" be used when referring to responses to social and cultural influences based on sex.

Discovering and distinguishing these differences through basic and clinical research will ultimately lead to improved clinical care for individuals. Indeed, some implications for clinical care based on sex and gender have already become apparent. For example, several studies that did not include men were directed to learning more about how to provide the best prevention and care for women. For example, the Women's Ischemia Syndrome Evaluation Study explored the distinctive ways in which women present with myocardial infarction or chest pain and developed diagnostic algorithms specifically for use in women.

The Women's Health Study evaluated use of low-dose aspirin for primary prevention of cardiovascular (CV) disease in women and reported different outcomes for women than those that had been shown in earlier studies on men. In this large, primary-prevention trial among women, aspirin lowered the risk of stroke without affecting the risk of myocardial infarction for women under the age of 65, whereas there was a significant reduction of major CV events, including myocardial infarction, among women 65 or older. Similar findings were reported from a randomized clinical trial of the Women's Health Study on the use of vitamin E. Benefit for the prevention of CV events was demonstrated only for older women.

Differences with clinical implications for women and men also are now known for the manifestations and progression of HIV/AIDS, for the response to pain and pain therapies, the expression and management of depression, the prevalence and care of Type 2 diabetes, and the incidence and effects of musculoskeletal diseases such as osteoarthritis and osteoporosis. Some generalizations can be made about the clinical implications of pharmacodynamics: (1) Sex differences are clinically important when determining the initial dosage of drugs with a narrow therapeutic range; (2) Pharmacodynamic sex differences potentially could be clinically significant with drugs with either wide or narrow therapeutic ranges; (3) These sex differences may be related to the higher incidence of adverse drug reactions in women, and similar sex differences in response to other interventions need to be studied.

To help scientists, clinicians, and members of academia gain a basic scientific understanding of the major physiological differences between the sexes and the importance of their implications for policy, medical research and health care, ORWH and the Office of Women's Health of the Food and Drug Administration developed, as a collaborative effort, an online course titled "The Science of Sex and Gender in Human Health." The initial course offerings are six lessons:

- Understanding the importance of sex and gender in biomedical research
- Federal requirements for the inclusion of human subjects in clinical research
- Cell physiology
- Developmental biology, including psychological and social development
- Pharmacokinetics and pharmacodynamics
- Clinical applications of pharmacogenomics

Women's Health and Cancer

The second leading cause of death for all women of all races in the USA is cancer, 22% in 2002. The leading cause is heart disease, 29% in 2002. The recently released Annual Report to the Nation on the Status of Cancer, 1975–2003 reports that the long-term decline in overall cancer death rates in the USA continued through 2003 for all races and both sexes combined. However, the declines were greater for men (1.6% per year 1993–2003) than for women (0.8% per year 1992–2003). The death rates decreased for 11 of the 15 most common cancers in men, and for 10 of the most common cancers in women.

For men, the incidence rates decreased for colon, rectal, lung, oral, and stomach cancers. The rates increased for prostate, myeloma, leukemia, liver, kidney, and esophageal cancers.

For women, the incidence rates decreased for breast, colon, rectal, uterine, ovarian, oral, cervical, and stomach cancers. However, the rates increased for women for non-Hodgkin's lymphoma, melanoma, leukemia, lung, bladder, kidney, and thyroid cancers.

Among the cancers, BC is the most common nonskin cancer in women, but lung cancer is the leading cause of cancer death for women in the USA. So, it is particularly alarming that lung cancer rates continue to increase for American women. Both of these cancers are among priorities in the NIH research agenda for women's health and for the NCI. Recognizing that lung cancer is an example of the interaction of biological and behavioral factors, and that compared with men, women develop lung cancer after smoking fewer cigarettes over a shorter time period, efforts continue to be strengthened to develop more effective smoking cessation methods for women, reliable biomarkers and imaging technologies, chemopreventive agents, and advances in cancer treatment that can contribute to the decline of this deadly disease. The decrease in deaths from BC are thought to be related to increased detection and better treatment, although some have questioned whether the decline in use of menopausal hormonal therapy, since the results of the WHI were reported, may also be a contributing factor.

Advances in genetics have improved our understanding of a number of cancers in women, especially those of the breast, endometrium, and ovary. NIH priorities include building on that knowledge as well as our understanding of hormonal and environmental contributors to women's cancers. And, developing better imaging modalities for earlier detection of cancers such as those of the breast remains a concern for investigative efforts. The discovery of the BRCA1 and BRCA2 mutations is one of the most striking breakthroughs in our understanding of genetic risks for cancer and has resulted in many new clinical preventive approaches for women who are found through genetic testing to be predisposed to BC. Yet, only a small proportion of BCs are associated with these mutations, and, although decreasing, death rates from BC remain high. Thus, continued exploration of the entire spectrum of detection, pathogenesis, prevention, and treatment of BC must remain among the priorities for women's health research. As research is clarifying more about the interaction of environment and hormones on genetic susceptibility to a number of cancers that affect women, death rates and the impact of associated morbidity on women and their families can be reduced. Other priorities for research on malignancies in women are to address sex differences in lung cancer, the leading cause of cancer deaths in women in the USA, as well as gender differences in techniques for smoking prevention and cessation. Successful development of biomarkers for ovarian cancer and efficacious vaccines to prevent the transmission of the human papilloma virus (HPV) will result in a decrease in the incidence and mortality of cervical cancer. Many studies have been reported on cancer disparities among women of racial and ethnic minorities, but more work is needed to address the biological characteristics and the medical, behavioral, and societal interventions that can reduce these disparities.

Other priorities for research are cervical, ovarian, uterine, endometrial, and colorectal cancers. Table 1 presents examples of recent studies on cancer and tumors that ORWH has cofunded with other various NIH institutes.

Table 1 Recent studies on cancer and tumors sponsored by ORWH/NIH

• Pharmacogenetics of the endocrine treatment of breast cancer (National Institute of General Medical Sciences)
• Phytoestrogens and aging: Dose, time, and tissue and breast cancer progression (National Institute on Aging)
• Tumorigenic subversion of mural cells in breast cancer (NCI)
• Clinical trials of two HPV-like particle vaccines (NCI)
• Molecular etiology of leiomyoma uteri (National Institute on Child Health and Human Development)
• Mediators and moderators of exercise behavior change (NCI)
• Social cognitive theory and physical activity (NCI)
• Patient-centered communication during chemotherapy (NCI)
• Caregivers' strengths/skills: managing older cancer patients (NCI)
• Impact of domestic violence on cancer treatment (NCI)
• Ongoing effects on fatigue and cognitive function after treatment for breast cancer (National Center for Complementary and Alternative Medicine)
• Iyengar yoga for breast cancer survivors with persistent fatigue (National Center for Complementary and Alternative Medicine)

Worldwide, cervical cancer annually accounts for over 400,000 cases, resulting in approximately 200,000 deaths. The impact is particularly devastating in developing countries where women are medically underserved and access to Pap smear screening is not readily available. NCI, with support from the ORWH, is continuing a large, double blinded, randomized clinical trial to evaluate the efficacy of an HPV 16/18 vaccine for the ultimate prevention of HPV-induced cervical cancer. The promise of a vaccine that would not only prevent the transmission of a sexually transmitted disease but that could also eradicate a malignancy that is a leading cause of cancer death in women globally represents one of the most important women's health advances from research of recent years.

Women's Health and Interdisciplinary Research and Career Development Programs

Accompanying the advances in understanding the clinical implications for sex/gender research has been the understanding of a theme of the early advocates for women's health research: the need to decrease the fragmentation of health care for women. In the research arena, this perception has led to a new focus on interdisciplinary research for women's health and sex/gender studies. The broad concept of what constitutes women's health has led to the recognition that research priorities in women's health must be comprehensive and interdisciplinary and should include not only clinical studies but also the full spectrum of research, from molecular and genetic studies to those of prevention, behavior, outcomes of interventions, and clinical translation of newly proven hypotheses. Interdisciplinary research can facilitate the integration and synergy of basic science, clinical research, translational research, population studies, behavioral and social research, and outcomes research.

An additional focus on bioengineering and biomedical informatics, genomics, proteomics, imaging, and metabolomics is increasingly relevant to research on women's health and sex/gender factors.

The concept of interdisciplinary research has been reinforced and enhanced through NIH Roadmap initiatives for novel interdisciplinary training and clinical approaches. With avenues of interdisciplinary communication and collaboration established through research efforts, women's interdisciplinary clinical collaboration may also benefit. Support also comes from major recommendations from the Institute of Medicine Report on Sex and Gender in Basic Biological Research that encourages and support interdisciplinary research on sex differences to achieve:

- Synergy between and among basic scientists, epidemiologists, social scientists, and clinical researchers
- Enhanced collaboration across medical specialties
- Improved translation research and interlevel integration of data (cellular, to animal, to human)

In collaboration with many of the NIH institutes, and the FDA and Agency for Health Care Research and Policy (AHRQ), ORWH initiated two large-scale programs to promote interdisciplinary research: Building Interdisciplinary Research Careers in Women's Health (BIRCWH), and the Specialized Centers of Research (SCORs) on Sex and Gender Factors Affecting Women's Health.

The BIRCWH program pairs senior investigators in an interdisciplinary, mentoring environment with junior researchers who have completed clinical training or postdoctoral fellowships and who are beginning basic, translational, clinical, or health services research relevant to women's health. To date, ORWH has funded 35 BIRCWH centers, and a Request for Applications has been issued for more awards to be made in September of 2007. Since 2000, a total of 265 BIRCWH scholars have thus far been trained in interdisciplinary women's health research, of which 74% are women, 26% men. Scholars have recorded over 882 publications and 872 abstracts based upon their BIRWCH funded interdisciplinary research.

Eleven SCORs have been funded and, as for the BIRCWHs, more awards will be made in September 2007. A SCOR is envisioned as a national resource associated with one or more major medical complexes and dedicated to furthering the interdisciplinary research effort on women's health or sex/gender factors to translate basic research into clinical applications. Examples of research in the SCORs are presented in Table 2.

ORWH has sponsored/collaborated with other NIH components in a number of other research initiatives or conferences such as the state-of-the-science conference on Management of Menopause-Related Symptoms (http://consensus.nih.gov/2005/2005MenopausalSymptomsSOS025html.htm), a symposium on Family Hormonal Health: Pituitary Disorders, the Second International Congress on Advances in Uterine Leiomyoma Research (http://orwh.od.nih.gov/health/fibroidsrevisedmarch2006.pdf), a scientific workshop on Neuroimmune Mechanisms and Chronic Fatigue Syndrome, and a scientific workshop on the Regulation of Inflammatory Responses:

Table 2 Research sponsored by SCOR

• Role of sex and gender differences in substance abuse relapse
• Genes, androgens, and intrauterine environment in polycystic ovarian syndrome
• Sex and gender factors in the pathophysiology of irritable bowel syndrome and interstitial cystitis
• Mechanisms underlying female urinary incontinence
• Sex differences in pain sensitivity

Influence of Sex and Gender. In March 2005, ORWH sponsored a wrap-up conference on the Legacy to Future Generations of Women of the Women's Health Initiative (WHI). This symposium presented the comprehensive results of the WHI (http://orwh.od.nih.gov/WHIConference.htm). It also brought together women from each of the WHI clinical trial sites who participated in the meeting and spoke to their personal perspectives about being a part of a research study of postmenopausal women.

Women in Biomedical Careers

The ORWH mandate includes developing opportunities for the recruitment, retention, reentry, and advancement of women in biomedical careers. One of ORWH's earliest initiatives was a nationwide outreach to determine the issues and barriers that women face in having a biomedical career. Certainly, progress has been made in the USA in recruiting more women into the biomedical sciences; for example, about 50% of students entering and graduating from medical schools are now women, and there continue to be increases in the numbers of women obtaining advanced degrees in the life sciences. Nevertheless, most of the barriers identified 15 years ago still remain as described in Table 3.

Now, women constitute a larger percentage of faculties in medical school and university science departments; however, they are not present in equitable numbers at the higher ranks of tenured faculty or organizational leadership positions. More difficult to identify and quantify are the attitudes women encounter in biomedical career paths, attitudes that assume they are incompetent or not committed, attitudes that do not encourage or facilitate their retention or advancement in research or scientific careers, or attitudes that keep them from opportunities and rewards they merit.

ORWH has directed its efforts to increase opportunities and support for girls and women in biomedical careers through programs directed to women themselves, intended toward the roles of institutions, professional and scientific societies, and government sponsored programs can have. Working collaboratively with professional societies, scientific organizations, institutions, and other government agencies, ORWH has implemented numerous other initiatives to facilitate careers in women's health research for women and men, as well as to promote networking and sharing of successful interventions and activities that assist women in securing effective mentoring and other information that can help them to overcome barriers to successful scientific careers.

Table 3 Barriers to/factors for biomedical careers for women

• Recruiting women and girls into scientific careers
• Lack of female role models and mentors
• Equity of career paths/rewards (salaries, promotions, etc.)
• Family responsibilities/dual roles
• Need for reentry into biomedical careers
• Sex discrimination and sexual harassment
• Gender sensitivity
• Racial bias/special needs of minority women
• Research initiatives on women's health

The National Academies of Science in 2006 launched a new study, Maximizing the Potential of Women in Academic Science and Engineering, chaired by Donna Shalala, President of the University of Miami and former Secretary of the USA Department of Health and Human Services. The study committee was charged to produce a comprehensive interdisciplinary guide to effective policies and practices for recruiting, hiring, and retaining women faculty, considering implicit and explicit barriers to career advancement, and the intersectionaly of race and gender across all fields of science and engineering. The results of this study will guide faculty, department chairs and deans, academic leaders, funding organizations, and government officials on how to maximize the potential of women science and engineering researchers. The committee will focus on academe, but will examine other research sectors to determine effective practices and develop findings and recommendations for recruiting, hiring, promotion, and retention of women scientists and engineers (Beyond Bias and Barriers: Fulfilling the Potential of Women in Academic Science and Engineering, available at http://books.nap.edu/catalog/11741.html.) The NIH will consider the recommendations from this committee and determine new or expanded existing programs and initiatives that can respond to the challenges identified.

Conclusion

There has been an expansion of science-driven initiatives on women's health, and progress in understanding the multiple factors that influence the differences between women and men, and among different women and racial/ethnic groups. Fifteen years ago, policy at NIH began to require the inclusion of women and minorities in clinical research, and there was some initial resistance. With the science-based findings that have come to the fore, however, determining sex and gender differences or similarities is now recognized by most of the biomedical research and health care communities as scientifically valid and appropriate. Further, to reduce health disparities between racial and ethnic minority women, more work is necessary to address biological characteristics or medical, behavioral and societal interventions needed to improve the health of all women and men As research is providing information about women's health, how to prevent their diseases, and when health care strategies or disease treatment should differ for women and men, this information becomes imperative for addressing malignancies in women, and especially when considering factors related to hormonal carcinogenesis.

Suggested Reading

1. DeAngelis CD, Glass RM (2006) Women's health – Advances in knowledge and understanding. JAMA 295:1448–1450.
2 National Institutes of Health (1994). NIH Guidelines on the Inclusion of Women and Minorities as Subjects in Clinical Research. NIH Guide http://grants2.nih.gov/grants/funding/women_min/guidelines_amended_10_2001.htm.
3. NIH Revitalization Act of 1993, Public Law No. 103–143.
4. Office of Research on Women's Health (1999). Agenda for Research on Women's Health for the 21st Century: A Report of the Task Force on the NIH. http://orwh.od.nih.gov/research/resagenda.html.
5. Wizemann TM, Pardue M-L (2001) Institute of Medicine Committee on Understanding the Biology of Sex and Gender Differences. Exploring the Biological Contributions to Human Health: Does Sex Matter? Washington, DC: National Academy Press, p.17.
6. Merz NB, Johnson BD, Kelsey PSF, et al. (2001). Women's Ischemia Syndrome Evaluation. Diagnostic, prognostic, and cost assessment of coronary artery disease in women. Am J Manag Care 7:959–965.
7. Johnson BD, Shaw LJ, Buchthal SD, et al. (2004). Prognosis in women with myocardial ischemia in the absence of obstructive coronary disease: results from the National Institutes of Health-National Heart, Lung, and Blood Institute-Sponsored Women's Ischemia Syndrome Evaluation (WISE). Circulation109:2993–2999.
8. Ridker PM, Cook NR, Lee IM, et al. (2005). A randomized trial of low-dose aspirin in the primary prevention of cardiovascular disease in women. N Engl J Med 352:1293–1304.
9. Jacobs, EJ, Thun MJ (2005). Vitamin E in the primary prevention of cardiovascular disease and cancer. JAMA 294:56–65.
10. Redberg, RF (2005). Vitamin E and cardiovascular health: does sex matter? JAMA 294:107–109.
11. The Science of Sex and Gender in Human Health Online Course Web Site. http://sexandgendercourse.od.nih.gov/index.aspx.
12. Howe HL, Wu X, Ries LAG, et al. (2006) Annual Report to the Nation on the Status of Cancer, 1975–2003, Featuring Cancer among U.S. Hispanic/Latino Populations. Cancer.
13. Narod SA (2005). Clinical implications of the breast cancer susceptibility genes *BRCA1* and *BRCA2*. Women's Health1:27–34.
14. Cook JD, Davis BJ, Cai S-L, et al. (2005) Interaction between genetic susceptibility and early-life environmental exposure determines tumor-suppressor-gene penetrance. PNAS 102:8644–8649.
15. Glanz K, Croyle RT, Chollette VY, Pinn VW. Cancer-related health
16. NIH Roadmap. Available at: http://nihroadmap.nih.gov/.
17. Wizemann TM, Pardue M-L (2001). Institute of Medicine Committee on Understanding the Biology of Sex and Gender Differences. Exploring the Biological Contributions to Human Health: Does Sex Matter? Washington, DC: National Academy Press, p. 9.
18. http://orwh.od.nih.gov/interdisciplinary/bircwhmenu.html. http://www.orwh.od.nih.gov/interdisciplinary/SCORs.html.
19. Office of Research on Women's Health (2007). Women in Biomedical Careers: Dynamics of Change: Strategies for the 21st Century. Order from: http://orwh.od.nih.gov/pubs/pubs_list.html.
20. Glanz K, Croyle RT, Chollette VY, Pinn VW (2003). Cancer-related health disparities in women. Am J Public Health 93:292–298.

Part 1
Celluar Origins of Endocrine-related Cancers

5 Stem Cells, Hormones, and Mammary Cancer

Gilbert H. Smith

Introduction

The incidence of breast cancer (BC) is influenced by age, genetics, ethnicity, diet, socioeconomic status, and reproductive history. The latter is the strongest and most reliable risk factor besides age and genetic susceptibility (1). Reproductive factors have been associated with risk for BC since the seventeenth century, when the disease was noted to be more prevalent among Catholic nuns. It is now a well-established fact that a full-term pregnancy early in life is associated with a long-term risk reduction for developing BC. A woman who has her first child after the age of 35 has approximately twice the risk of developing BC as a woman who has a child before age 20 (see current NCI Cancer Fact Sheet on Pregnancy and BC Risk). Despite this long-term reduction in BC risk in parous women, epidemiologists agreed at a recent NCI-sponsored workshop on "Early Reproductive Events and Breast Cancer" (http://nci.nih.gov/cancerinfo/ere) that each gestation *increases temporarily* the likelihood for developing BC (2). This transient increase in BC risk lasts for a few years after a full-term pregnancy.

Pregnancy and Breast Cancer

Pregnancy has a very similar dual effect on the etiology of mammary cancer in animal models. Like humans, parous rats and mice have a greatly reduced susceptibility to chemically induced mammary tumorigenesis compared with their nulliparous siblings (3). Humans who carry germ line mutations in tumor susceptibility genes do not benefit from the protective effects of pregnancy, but have a significantly greater risk of developing the disease following one or multiple gestation cycles (4). There are, however, conflicting reports whether lactation influences the onset of BC in women with *BRCA1* mutations.

The current view on BC as a stem cell disease is founded on compelling evidence that many BCs may arise as clonal expansions from epithelial progenitors with an infinite lifespan (5). It has been hypothesized that unique properties of mammary stem cells, such as self-renewal, make this population a prime target for

J.J. Li et al. (eds.), *Hormonal Carcinogenesis V*.

transformation and tumorigenesis. Several experimental BC models support this hypothesis. The most venerable is the mammary tumor virus (MMTV) (6) model in mice, where MMTV proviral insertions produce mutated mammary cells, which attain immortality (escape from growth senescence) and produce clones of mammary cells with increased propensity to develop mammary cancer. Serial transplantations of these preneoplastic lesions result in the formation of hyperplastic/dysplastic ductal trees, suggesting that multipotent cells are affected by MMTV transformation and that they pass on their neoplastic properties to their descendants (7). Morphologically undifferentiated cells, reminiscent of stem/progenitor cells are present in both premalignant and malignant mammary populations (Fig. 1). Reproductive history has a profound impact on breast tumorigenesis, thus it is

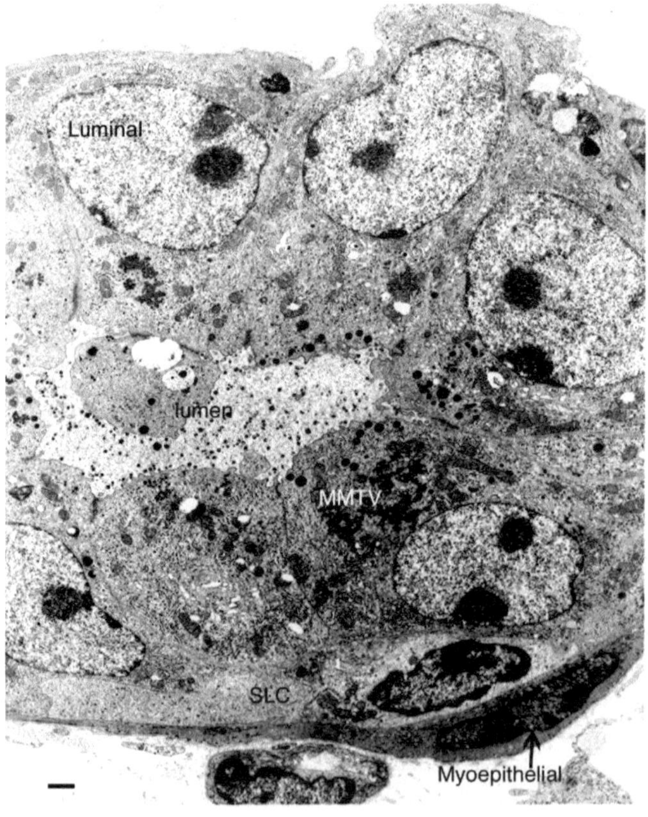

Fig. 1 This electron micrograph depicts an ultra thin section through one of the acini in an MMTV-induced alveolar hyperplasia. There is evidence of virus replication (MMTV) of secretory activity leading to secretory granule formation in the apical cytoplasm of the luminal cells and release into the lumen. An undifferentiated suprabasal cell (SLC) is present and proximal to it, a differentiated myoepithelial cell (*arrow*). Bar equals 1.0 μM

reasonable to assume that pregnancy and lactation have enduring effects on the cancer susceptibility of multipotent stem/progenitor cells.

Evidence that cancer stem cells sustain solid neoplasms has recently emerged (5, 8). Whether these "cancer stem cells" arise de novo or result from mutations within normal tissue stem/progenitor cells is presently unknown. A shift in the microenvironment of mammary epithelial cells as the result of pregnancy is a plausible mechanism by which to explain the greater refractivity of mammary tissue after early parity to cancer induction or progression. In a rat chemical carcinogenesis model, Nandi et al. have argued that there is no difference in the susceptibility of the mammary epithelium between nulliparous and parous females to initiation (malignant transformation) by NMU; rather, it is a reduction in the incidence of progression of the "initiated" cells to frank malignancy. This difference in "progression" is completely reversible when the parous rodents are subjected to various hormonal regimens or given growth factors such as IGF-1 (9). If this is correct, then epithelial (stem/progenitor) cell targets for carcinogenesis are the same but behave differently in their respective microenvironments (niches) during homeostatic tissue maintenance in the parous female.

Pregnancy Mediates Permanent Changes in Mammary Epithelial Cells

The basic principle for the dual phenomenon of pregnancy and BC is that a gestation cycle induces massive proliferation and an endpoint differentiation of epithelial subtypes. Either permanent systemic changes following a full-term pregnancy (such as a decrease in circulating levels of hormones) or the alteration of the mammary tissue itself could explain the difference in BC risk between nulliparous and parous women. Sivaraman et al. (10) suggested that the hormonal milieu of pregnancy affects the developmental state of a subset of mammary epithelial cells and their progeny.

Microarray evidence suggests that pregnancy mediates persistent changes in the gene expression profile in parous females (11, 12). These pregnancy-induced changes can be imitated through a transient administration of hormones, in particular estrogen and progesterone or human chorionic gonadotropin. Ginger et al. (12) used subtractive hybridization as a method to identify differentially expressed genes between hormone-treated Wistar–Furth rats and their untreated controls. Twenty-eight days after the last treatment, they identified approximately 100 differentially expressed loci. In a more comprehensive study, D'Cruz et al. (11) utilized oligonucleotide arrays to examine differences in the expression profile of approximately 5,500 genes between parous mice and their nulliparous controls. These initial results were verified by more laborious methods (northern blot analysis and in situ hybridization) and across several mouse strains as well as in two rat models.

The Origination of Parity-Induced Mammary Epithelial Cells During Late Pregnancy and Lactation

Using the Cre-lox technology, a mammary epithelial subtype, which is abundant in nonlactating and nonpregnant, parous mice, was recently described (13). These parity-induced mammary epithelial cells (PI-MEC) then permanently reside at the terminal ends of ducts (i.e., lobuloalveolar units) after postlactational remodeling.

Two lines of evidence exist that the presence of PI-MECs in the involuted mammary gland is not an artifact caused by a deregulated activation of the promoter of our randomly integrated WAP-Cre construct. First, the WAP-Cre transgenic expression closely follows the activation of the endogenous WAP locus and Ludwig et al. (14) have reported similar observations in genetically engineered mice that express Cre recombinase under the endogenous *Wap* gene promoter (i.e., WAP-Cre knock-in mutants). Second, limiting dilution transplantation assays with dispersed epithelial cells from nulliparous female mice demonstrate the existence of lobule-limited and duct-limited progenitors (15).

These studies were carried out with epithelial cells from WAP-LacZ transgenic mice where LacZ is expressed from the whey acidic protein promoter in late pregnant mice. Lobule-limited outgrowths positive for LacZ expression were observed in the implanted fat pads at parturition. Similar lobule-limited outgrowths were developed when PI-MEC were inoculated in limiting dilution into the cleared mammary fat pads of subsequently impregnated hosts (Fig. 2). These structures like those described earlier (15) comprised both secretory luminal cells and myoepithelial

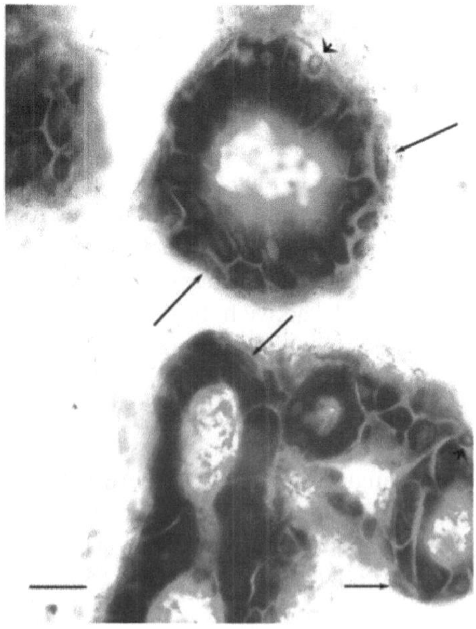

Fig. 2 The image shows a section through a lobule-limited LacZ-positive outgrowth in full term pregnant host composed entirely of progeny from PI-MEC. The growth comprises both luminal and myoepithelial (*long arrows*) cells and small-undifferentiated light cells (*short arrows*). Bar equals 20 μM

cells and were 100% positive for LacZ activity indicating that they were developed entirely from PI-MEC. Therefore, it is quite likely that PI-MEC arise from the lobule-limited progenitor population discovered by Smith (15) among the mammary epithelial cells present in nulliparous unbred females. In addition, to luminal and myoepithelial progeny, PI-MEC produced both small (SLC) and large undifferentiated light cells (ULLC) in the lobules. SLC and ULLC have essential roles in mammary stem/progenitor cell function (7). The existence of committed mammary alveolar precursors in mice and rats has been proposed earlier (16, 17).

PI-MEC are Self-Renewing and Pluripotent

When fragments of gland containing PI-MEC were transplanted to gland-free fat pads in nulliparous hosts, PI-MEC contributed to ductal elongation in a very significant manner. The vast majority of resulting outgrowths contained X-Gal positive cells, and in >75% of the transplants, PI-MEC-derived cells were present throughout the *entire* ductal tree. These results clearly demonstrated that PI-MEC exhibit two important features of multipotent stem cells: self-renewal and contribution to diverse epithelial populations in ducts and alveoli. We demonstrated, for the first time, that the progeny from cells previously expressing an *alveolar* differentiation marker (i.e., WAP) could contribute to the formation of primary and secondary *ducts*. When the transplanted hosts were impregnated, the self-renewed PI-MEC at the tips of duct side branches proliferated during early pregnancy to form the new secretory acini. The transplantation procedure itself had no effect on the activation of the WAP-Cre and Rosa-LacZ transgenes because mammary fragments from nulliparous double transgenic donors never produced outgrowths with uniformly distributed X-Gal-positive cells (13).

To establish an estimate of the self-renewing ability of PI-MECs, mammary fragments containing X-Gal positive cells were transferred through four transplant generations (13). Each successful transplant resulted in a 400.0-fold increase of the implanted epithelial population, which represents roughly an 8.0–9.0 (8.65)-fold doubling of the implanted cells.

To determine to what extent the presence of neighboring X-Gal-negative epithelial cells contributed to the self-renewing capacity of labeled PI-MEC, dispersed mammary epithelial cells from multiparous WAP-Cre/Rosa-LacZ females were inoculated at limiting dilutions into cleared fat pads, and the hosts were subsequently impregnated. All outgrowths contained LacZ-expressing cells, even though PI-MEC represented only 20% of the inoculated epithelial cells. Notably, no epithelial outgrowths were comprised entirely from unlabeled (LacZ-negative) cells. Both lobule-limited and duct-limited outgrowths were, however, entirely comprised from PI-MEC (and their LacZ-expressing descendents), as determined by serial sections through these structures. These results indicate that all luminal, myoepithelial, and cap cells of terminal buds may be derived from PI-MEC and their progeny. This conclusion was confirmed by demonstrating that the X-gal positive cells in these structures

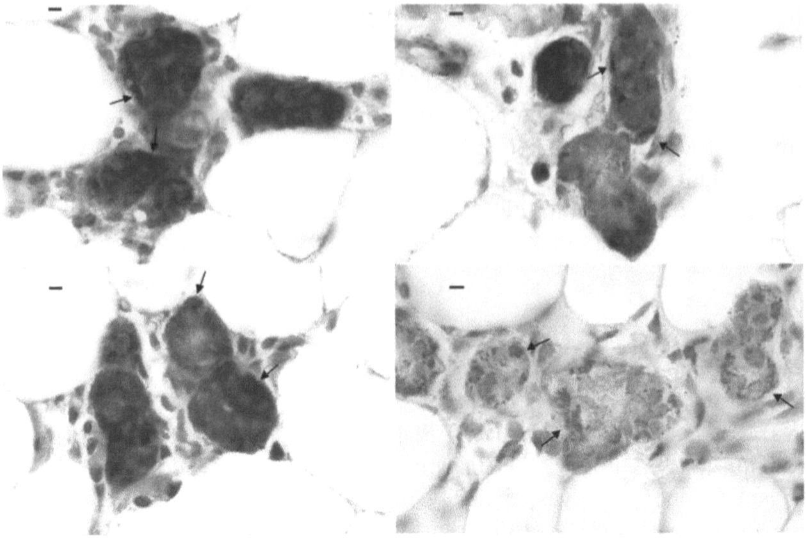

Fig. 3 This composite shows LacZ-negative acini stained for SMA in the *upper left panel* and LacZ-positive acini in the remaining *three panels*. The *arrows* indicate the myoepithelial cells demonstrated by positive SMA-staining. The LacZ-positive cells appear as *dark gray* in this gray-scale figure. Bars equal 5 μM

could be doubly stained for mammary cell lineage markers for myoepithelium (smooth muscle actin, Fig. 3), estrogen receptor alpha (ERα), or progesterone receptor (PR).

Thus, PI-MEC are not only self-renewing, but they are pluripotent as well, giving rise to progeny that differentiate along all the epithelial cell lineages of the mammary gland.

WAP-TGF-β1 Expression Aborts Self-Renewal of PI-MEC in Transplants

The reproductive capacity of the mammary epithelial stem cell is reduced coincident with the number of symmetric divisions it must perform. In a study using WAP-TGF-β1 transgenic mice, it was observed that mammary epithelial stem cells were prematurely aged because of ectopic expression of TGF-β1 under the regulation of the *WAP* gene promoter (18). To assess whether TGF-β1 expression in PI-MEC abolishes their capacity to self-renew, mammary epithelia from WAP-TGF-β1/WAP-Cre/Rosa-LacZ triple transgenic mice were transplanted into wild type recipients. It is important to note that the percentage of labeled cells in the triple transgenic glands after a single parity was indistinguishable from that observed in WAP-Cre/Rosa-LacZ double transgenic controls. As expected, mammary tissue implants and dispersed cells from the triple transgenic females, after either a single pregnancy or multiple gestation cycles, failed to produce full lobular development in full-term pregnant hosts. Perhaps more importantly, X-Gal positive cells were not observed in the ducts in these transplant outgrowths either in nulliparous or early pregnant hosts. LacZ-expressing cells did appear in the transplant population and were present in the lobular structures during late pregnancy in these transplants (after 15 days to parturition). In summary, the results of these studies demonstrate that the PI-MEC that develop during pregnancy and survive subsequent tissue remodeling in the absence of lactation in WAP-TGF-β1 females were incapable or severely limited in their ability to self-renew in transplants and could not contribute to ductal development in subsequent transplant outgrowths. Therefore, self-renewal (expansion outside of a stem cell niche) and proliferation competence (asymmetric divisions within a niche) appear to be properties independently affected by autocrine TGF-β1 expression in the PI-MEC.

By definition, the self-renewal of stem cells occurs by two different processes. In asymmetric divisions, the most common activity of stem cells residing in a niche, the stem cell is preserved and one daughter becomes committed to a particular cell fate. Alternatively, a stem cell may divide symmetrically and expand to produce two or more stem cell daughters that retain stem cell properties. This latter form of self-renewal is essential for expansion of the stem cell population during allometric growth of the tissue (i.e., during ductal growth and expansion in the postpubertal female or when the mammary epithelial implant is growing in the transplanted mammary fat pad). The negative effect of TGF-β1 on the expansive self-renewal of PI-MEC supports our earlier observation regarding protection from mouse mammary tumor virus (MMTV)-induced mammary tumorigenesis in WAP-TGF-β1 transgenic females (18). This might suggest that the cellular targets for MMTV-mediated neoplastic transformation are PI-MEC because multiple pregnancies accelerate MMTV-induced oncogenesis (7).

PI-MEC and Mammary Tumorigenesis

Pregnancy has a dual effect on human BC (protection or promotion), depending on the age of an individual, the period after a pregnancy, and the genetic predisposition. In genetically engineered strains that are highly susceptible to mammary tumorigenesis and exhibit accelerated tumor development in postpartum or parous females, one might expect that PI-MEC serve as targets for neoplastic transformation. The unique growth properties of PI-MEC (i.e., responsiveness to pregnancy hormones, survival during involution, and ability to self renew) make this epithelial subtype a potential target for pregnancy-associated tumorigenesis. Transgenic mice expressing the wild type *her2/neu* (*ErbB2*) oncogene under transcriptional regulation of the MMTV-LTR seem to be suitable for studying the involvement of PI-MEC in pregnancy-associated mammary tumorigenesis since this animal model exhibits a relatively long latency of tumorigenesis (T50 of 205 days). Using this animal model, we demonstrated that: (a) multiparous females consistently exhibited accelerated tumorigenesis compared with their nulliparous littermate controls in a mixed genetic background and (b) PI-MEC were, indeed, primary targets of neoplastic transformation in this model (19). The de novo generation and amplification of a large number of hormone-responsive and apoptosis-resistant epithelial cells (i.e., PI-MEC) during the first and subsequent reproductive cycles might, therefore, account for the significantly increased cancer susceptibility of parous MMTV-neu transgenic females.

To further substantiate that PI-MEC are primary targets for neoplastic transformation in MMTV-neu transgenic mice, we eliminated or greatly impaired the growth of PI-MEC by deleting the *Tsg101* gene in cells that transiently activated WAP-Cre (i.e., females that carry two transgenes, MMTV-neu and WAP-Cre, in a homozygous *Tsg101* conditional knockout background). The complete deletion of *Tsg101* can serve as an excellent negative "selection marker" for WAP-Cre expressing cells since this gene is indispensable for the survival of normal, immortalized, and fully transformed cells (20). In multiparous MMTV-neu females, impaired genesis or elimination of PI-MEC resulted in a significantly reduced tumor onset, suggesting that restraining the growth and survival of differentiating alveolar cells during pregnancy (and therefore PI-MEC in parous mice) eliminates the cellular basis for transformation in this model.

Some PI-MEC are Asymmetrically Dividing Long-Label Retaining Cells

It was proposed over 30 years ago that somatic stem cells avoid accumulation of genetic errors resulting from DNA synthesis prior to dividing by selectively retaining their template DNA strands and passing the newly synthesized strands to their committed daughters (21). Therefore, somatic epithelial stem/progenitor cells labeled by DNA analogs during their inception will become long-label retaining epithelial cells (LREC). Label retention has long been considered to be a characteristic of somatic stem cells and this propensity to retain DNA label has been explained by postulating that somatic stem cells seldom divide and are mainly proliferatively quiescent. Recent studies have, however, shown that in multiple tissues long-label-retaining cells are actively dividing and asymmetrically retain their labeled template DNA strands while passing the newly synthesized DNA to their differentiating progeny (22, 23). Interestingly, PI-MEC that have proliferated extensively in transplants give rise to LacZ-positive progeny, which retain the original DNA label for long periods and when pulsed with a second alternative DNA label prove to be actively traversing the cell cycle and thus become doubly labeled incorporating the second label into new DNA strands. Subsequent to a short chase period, the second label is transferred along with the new DNA strands to LacZ-positive progeny (Fig. 4). This evidence demonstrates that during self-renewal PI-MEC produce progeny (in addition to luminal and myoepithelial offspring) that behave as asymmetrically dividing stem/progenitor cells responsible for the steady-state

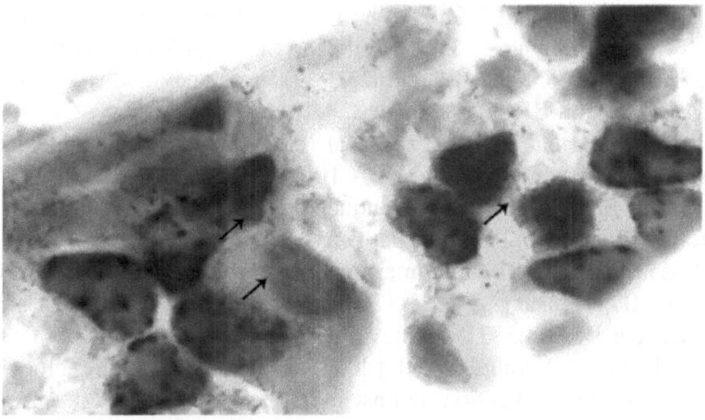

Fig. 4 PI-MEC long-label-retaining [^{3}H]-thymidine with autoradiographic grains were doubly labeled with 5BrdU (*darker nuclei*) in transplant labeled with thymidine 7 weeks earlier following a 2 day pulse with 5BrdU and produced 5BrdU-labeled-only (*arrows*) daughters after a 6 day chase

maintenance of the diverse LacZ-positive mammary epithelial population in the resulting outgrowth.

Conclusion

Stem cells are defined by how they act physiologically in the context of heterologous cells, i.e., the microenvironment or stem cell niche that balances protecting stem cells from exhaustion and protecting the host from unregulated stem cell growth. In addition to this complex model, it has been demonstrated recently that not only normal tissues, but also neoplastic lesions contain heterogeneous (hierarchical) types of stem cells (24). The discovery and genetic labeling of a parity-induced mammary epithelial cell population that is specific for parous females makes it possible to further examine the concept of stem cell hierarchy in the mammary gland and the homeostasis of mammary stem cells within the niche. In addition, our study of this progenitor population provides direct evidence for the proof of principle that a stem/progenitor cell may be the target of carcinogenic events and also that progression to frank malignancy is dependent upon the continued ability of the affected cell to expansively proliferate.

References

1. Kelsey JL, Gammon MD, John EM (1993) Reproductive factors and breast cancer. Epidemiol Rev 15:36–47.
2. Lambe M, Hsieh C, Trichopoulos D, et al. (1994) Transient increase in the risk of breast cancer after giving birth. N Engl J Med 331:5–9.

3. Guzman RC, Yang J, Rajkumar L, et al. (1999) Hormonal prevention of breast cancer: mimicking the protective effect of pregnancy. Proc Natl Acad Sci USA 96:2520–2525.
4. Jernstrom H, Lerman C, Ghadirian P, et al. (1999) Pregnancy and risk of early breast cancer in carriers of BRCA1 and BRCA2. Lancet 354:1846–1850.
5. Al-Hajj M, Wicha MS, Benito-Hernandez A, et al. (2003) Prospective identification of tumorigenic breast cancer cells. Proc Natl Acad Sci USA 100:3983–3988.
6. Callahan R, Smith GH (2000) MMTV-induced mammary tumorigenesis: gene discovery, progression to malignancy and cellular pathways. Oncogene 19:992–1001.
7. Smith GH (2005) Stem cells and mammary cancer in mice. Stem Cell Rev 1:215–223.
8. Singh SK, Clarke ID, Terasaki M, et al. (2003) Identification of a cancer stem cell in human brain tumors. Cancer Res 63:5821–5828.
9. Thordarson G, Slusher N, Leong H, et al. (2004) Insulin-like growth factor (IGF)-I obliterates the pregnancy-associated protection against mammary carcinogenesis in rats: evidence that IGF-I enhances cancer progression through estrogen receptor-alpha activation via the mitogen-activated protein kinase pathway. Breast Cancer Res 6:R423–R436.
10. Sivaraman L, Medina D (2002) Hormone-induced protection against breast cancer. J Mammary Gland Biol Neoplasia 7:77–92.
11. D'Cruz CM, Moody SE, Master SR, et al. (2002) Persistent parity-induced changes in growth factors, TGF-beta3, and differentiation in the rodent mammary gland. Mol Endocrinol 16:2034–2051.
12. Ginger MR, Gonzalez-Rimbau MF, Gay JP, et al. (2001) Persistent changes in gene expression induced by estrogen and progesterone in the rat mammary gland. Mol Endocrinol 15:1993–2009.
13. Boulanger CA, Ku W, Smith GH (2005) Parity-induced mammary epithelial cells are pluripotent, self-renewing and sensitive to TGF-β1 expression. Oncogene 24:552–560.
14. Ludwig T, Fisher P, Murty V, et al. (2001) Development of mammary adenocarcinomas by tissue-specific knockout of BRCA2 in mice. Oncogene 20:3937–3948.
15. Smith GH (1996) Experimental mammary epithelial morphogenesis in an in vivo model: evidence for distinct cellular progenitors of the ductal and lobular phenotype. Breast Cancer Res Treat 39:21–31.
16. Kordon EC, Smith GH (1998) An entire functional mammary gland may comprise the progeny from a single cell. Development 125:1921–1930.
17. Kamiya K, Gould MN, Clifton KH (1998) Quantitative studies of ductal versus alveolar differentiation from rat mammary clonogens. Proc Soc Exp Biol Med 219:217–225.
18. Boulanger CA, Smith GH (2001) Reducing mammary cancer risk through premature stem cell senescence. Oncogene 20:2264–2272.
19. Henry MD, Triplett AA, Oh KB, et al. (2004) Parity-induced mammary epithelial cells facilitate tumorigenesis in MMTV-neu transgenic mice. Oncogene 23:6980–6985.
20. Krempler A, Henry MD, Triplett AA, et al. (2002) Targeted deletion of the Tsg101 gene results in cell cycle arrest at G1/S and p53-independent cell death. J Biol Chem 277:43216–43223.
21. Cairns J (1975) Mutation selection and the natural history of cancer. Nature 255:197–200.
22. Potten CS, Owen G, Booth D (2002) Intestinal stem cells protect their genome by selective segregation of template DNA strands. J Cell Sci 115:2381–2388.
23. Smith GH (2005) Label-retaining epithelial cells in mouse mammary gland divide asymmetrically and retain their template DNA strands. Development 132:681–687.
24. Hope KJ, Jin L, Dick JE (2004) Acute myeloid leukemia originates from a hierarchy of leukemic stem cell classes that differ in self-renewal capacity. Nat Immunol 5:738–743.

6
The Origin of Estrogen Receptor α-Positive and α-Negative Breast Cancer

Robert B. Clarke, Andrew H. Sims, and Anthony Howell

Introduction

Recent advances in global gene expression analyses have led to a classification of breast tumours based on their intrinsic molecular signature rather than histological appearance, or presence/absence of one particular molecular marker. In a series of seminal papers, Perou and Sorlie working in Botstein's laboratory in Stanford delineated four basic molecular sub-types of breast cancer (BC) that have been generally confirmed and corroborated by subsequent studies (1–5). The four major sub-types, termed basal, HER2 and luminal types A and B, have been demonstrated to engender different prognostic outcomes (6). This is partly explained by their heterogeneous expression of the ERα, with absent or low levels in the first two sub-types and moderate or strong expression in the latter two sub-types. In this chapter, we review what is known about normal breast epithelial stem and progenitor cells, expression of ERα and how this informs us about the likely cellular origins of these cancer sub-types and their ERα status.

Breast Epithelial Stem Cells

Adult tissue stem cells are long-lived, generally quiescent cells defined by their ability both to self renew and to produce progeny that can differentiate into all the functional cell types of a particular tissue (7, 8). This may occur by symmetric or asymmetric cell division giving rise to either two new stem cells or a stem cell and an undifferentiated progenitor cell. The progenitor cells will then divide by transit amplification and generate the lineage-restricted progenitors that subsequently undergo terminal differentiation to form the functional cells of a tissue (9–11). Experimental and clinical data suggest that tissue-specific stem cells, because of their longevity, may represent the major target for mutations leading to cancer (7).

The adult mammary gland has a lobulo-alveolar structure, composed of two basic cell lineages: myoepithelial cells that form the basal layer of ducts and luminal epithelial cells that synthesize milk proteins (12). As is the case in other tissues,

J.J. Li et al. (eds.), *Hormonal Carcinogenesis V.*

the cellular repertoire of the human mammary gland is quite likely to be generated by a stem cell component. Evidence for the existence of mammary stem cells is suggested by the cyclic development, involution, and subsequent redevelopment of the mammary gland with each successive pregnancy and lactation. Seminal transplantation experiments in mice first demonstrated nearly half a century ago that isolated segments from any portion of the mammary gland are capable of regenerating a complete mammary ductal and alveolar network (13, 14). More recently, this transplantable, reconstitutive capacity was shown in the progeny of a single retrovirally marked mammary epithelial cell (15). In the past year, parallel experiments in two laboratories have confirmed that an entire mouse mammary gland can be regenerated by transplanting single cells with defined cell surface markers into cleared mammary fat pads (16, 17).

To study the functional properties of stem cells, one needs to identify and prospectively purify them, a task that has proved technically difficult because of the scarcity of stem cells in the tissue of origin, and the lack of universal morphologic traits for stem cells. Using cell surface markers, as in the studies described earlier, the purity of stem cells achieved by selection was never higher than approximately 5% of sorted cells. In the normal breast, by definition, stem cells should rarely divide and persist throughout reproductive life. They can be identified on the basis that they will retain label after the administration of labelled DNA precursors, such as [^{3}H]-thymidine or bromodeoxyuridine (BdUr). This approach has been used by a number of groups including our own demonstration using human mammary tissue implanted into athymic nude mice (18–21). In our study, 7–9% of label-retaining cells (LRCs) after two weeks chase expressed the putative stem cell markers p21$^{CIP1/WAF1}$ (cyclin-dependent kinase inhibitor) and Musashi-1 (Msi1; RNA-binding protein) (18). A characteristic also shared by a number of stem cells is their ability to exclude dyes like Hoechst or rhodamine as a result of increased expression of membrane transporter proteins, such as P-glycoproteins or BCRP (breast cancer resistance proteins) (22). This increased dye exclusion has been used to identify a sub-population of mouse mammary epithelial cells termed the "side population" (SP) similar to the SP containing haematopoietic stem cells in bone marrow (23). The multipotency of breast SP cells was demonstrated by their ability to regenerate the mouse mammary gland upon transplantation (24). A similar SP to that observed in the mouse mammary gland has also been identified by several groups in normal human breast tissue obtained from reduction mammoplasty and other non-cancer breast surgery (18, 24, 25). In the three groups who have performed human breast tissue SP analyses, the proportion of breast SP cells varied from ~0.2% to ~5%. Their stem cell nature has been analysed and compared with the non-SP cells using various in vitro cell culture methods. In support of their putative stem cell nature, only the cells within the SP possessed the ability to produce colonies comprising both myoepithelial and luminal epithelial cell types. In our own study, we identified a SP fraction of undifferentiated cells isolated from normal human mammary tissue, which lacked both the myoepithelial and luminal markers CALLA and MUC-1. In contrast to non-SP cells, a small fraction of SP cells, ~1/50, formed branching structures in matrigel, reminiscent of lobular structures in vivo, and which included cells of both luminal and myoepithelial lineages. The SP cells were also sixfold enriched for ERα$^+$ cells and expressed several fold higher levels of the putative stem cell genes, *p21*^{CIP1} and *Msi1*, than non-SP cells (18).

Estrogen Receptor α Expression and Stem Cells

On the basis of our results and those of other breast stem cell studies, we have concluded that there is both a basally located, primitive ERα$^-$ stem cell active early in development and a more luminal located ERα$^+$ stem cell active during adult tissue homeostasis. Studies on reconstitution of mouse mammary glands using single cells demonstrate that they express high levels of the basal epithelial markers CD29 or CD49f (26). However, our findings from human breast epithelium using SP analysis and putative stem cell markers indicate ERα expression in undifferentiated cells that are intermediate between basal and luminal-like (18). There is strong epidemiological evidence that the cumulative exposure to oestrogen significantly influences the lifetime risk of developing BC, including ERα$^-$ sub-types of BC (27). The use of human autopsy samples to study the ontogeny and expression of the *ERα* gene in foetal breast tissue demonstrated that no ERα$^+$ cells could be found in the rudimentary ductal tree of the human mammary gland before 30 weeks of gestation (28). After 30 weeks, ERα can be detected in breast epithelium and is markedly upregulated shortly after birth, suggesting the existence of ERα$^+$ progenitor cells that arise from stem cells late in foetal development. Our data and those of others demonstrate that the ERα$^+$ cells rarely proliferate in adult tissues during hormonal cycles and pregnancy, in all species of mammals examined (29–32). In fact, in the adult mammary gland, dividing cells are nearly all ERα$^-$. In contrast, ERα$^+$ cells are highly proliferative during experimental mammary gland development, for example, when stimulated with hormones after ovariectomy or after tissue transplantation, developmental periods when stem cells are likely to be actively dividing (33, 34). ERα$^+$ cells are also highly proliferative in premalignant lesions and invasive BC suggesting their close relationship to the ERα$^+$ stem cells present in normal breast development (30, 35).

In addition, two groups have reported attempts to define mammary gland stem cells in relation to their ERα status in the mouse using injection of DNA label and a pulse chase experiment to characterise LRCs. These studies produced different results based on the DNA label that was used, the developmental stage at which labelling was conducted and the length for which LRCs were followed. When the period of labelling occurred during puberty in the mouse, the frequency of association between ERα and LRCs decreased during the subsequent chase period of between 5 and 9 weeks, suggesting that during puberty the stem cells were ERα$^-$ (20). In contrast, when labelling was performed during oestrus in post-pubertal mice, 95% of the LRCs expressed ERα, suggesting that during estrus ERα$^+$ stem cells are active (21). These data on stem cells and ERα expression in normal breast epithelium suggest that differentiated lineages are derived from stem cells with

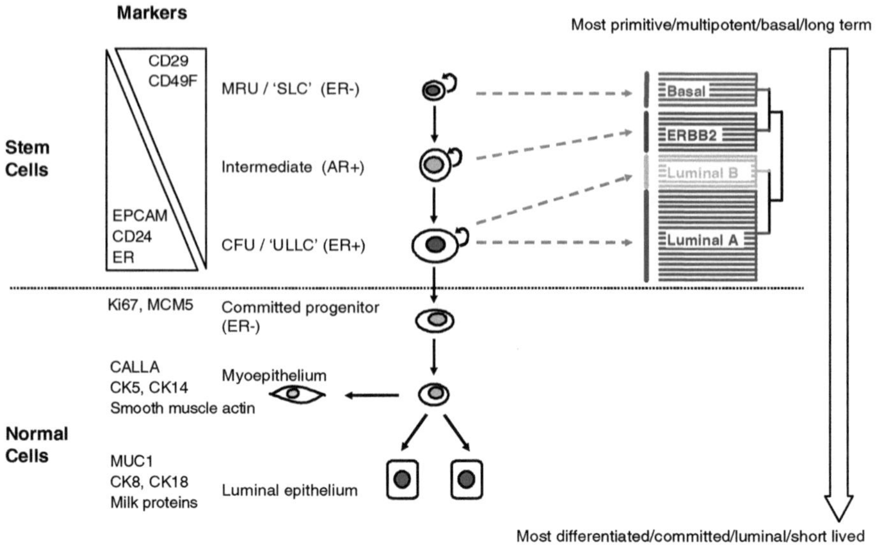

Fig. 1 A model of stem cell hierarchy and how it may account for the origins of molecular portraits of BC. *MRU* Mammary repopulating unit; *CFU* Colony forming unit (Stingl 2005); *SLC* small light cell; *ULLC* undifferentiated large light cell (Smith 2003); Cellular markers: *CALLA* common acute lymphoblastic leukaemia antigen; *CK* cytokeratin; *AR* androgen receptor; *CD24* small cell lung carcinoma cluster 4 antigen; *EPCAM* epithelial cell adhesion molecule; *CD29* integrin beta 1; *CD49f* integrin alpha chain alpha 6; *MUC1* mucin 1; cell surface associated. Proliferation markers: *Ki67* antigen identified by monoclonal antibody Ki-67; *MCM5* MCM5 minichromosome maintenance deficient 5

different potentials at different times in development; the long term ERα^- developmental stem cell being most primitive and capable of reconstituting a cleared mammary fat pad as a single cell and the ERα^+ short term stem cell capable of producing colonies in vitro and patches of epithelium in the adult during tissue homeostasis. These stem cell types are most likely a continuum of phenotypes in vivo that give rise to the dividing ERα^- transit amplifying cells that ultimately produce differentiated myo- and luminal-epithelial cells in the adult tissue (Fig. 1).

The Origins of Breast Cancer Sub-Types

Thus, the existence of a continuum of stem cells active at different points in development may provide an explanation for the existence of ERα^+ and ERα^- negative BC sub-types, and perhaps for the molecular sub-types recognised by their intrinsic gene expression (6). On the basis of the stem cell model of mammary carcinogenesis, one would predict that poorly differentiated ERα^- breast tumours would arise from the most primitive stem cells and the particular subset of mutations in these tumours must prevent differentiation into ERα$^+$ cells. The basal sub-type, which express basal cytokeratins, the basal marker p63 and are highly EGFR$^+$ and ERα$^-$, may be derived from transformation of these cells (Fig. 1). The HER2 sub-type has been described as baso-luminal (36), or molecular apocrine (37) may be derived from a stem cell midway along the continuum, while luminal sub-types A and B have no basal markers at all and have various expression levels of ERα. The well-differentiated ERα$^+$ luminal tumour sub-types would be predicted to arise from the transformation of ERα$^+$ stem cells (Fig. 1). Their further classification into sub-types A and B is reported to relate to ERα levels, expression of ERα-regulated genes and their prognosis; luminal B has lower ERα and a worse prognosis than A (1, 3, 6). Whether luminal sub-types A and B arise from phenotypically different ERα$^+$ stem cells in the normal epithelium remains unknown. The de facto origin of these distinct breast tumour phenotypes remains to be confirmed experimentally.

Breast Cancer Stem Cells

Using different systems, a number of investigators have demonstrated that only a minority of cells in human cancers, cancer stem cells (CSCs), are capable of self-renewal and reconstitution of the original tumour (7). This has been demonstrated in human breast tumours by examining the ability of sub-populations of tumour cells identified by cell surface markers (ESA$^+$/CD44$^+$/CD24$^-$/low) to form tumours when transplanted into immune-suppressed NOD-SCID mice (38). Most recently, putative breast CSCs have been isolated from three BC lesions and propagated in vitro and in vivo, and stem-like cells have been identified in BC cell lines (39, 40). These reports suggesting the existence of stem-like cells in breast tumours have important implications for tumour therapy. Most traditional cancer treatments target proliferating cells and while this might eliminate the mass of the tumour, relatively quiescent tumour stem cells could be spared. One of the most important issues is to understand the pathways that govern the self-renewal of normal stem cells since these same pathways may be active in CSCs. Inhibition of these pathways has recently been proposed to be a novel therapeutic modality that would aim to target stem-like cells within the tumour (41–43).

Conclusion

The recent description and confirmation of four major BC sub-types by gene expression analysis has been paralleled by advances in molecular and cellular biological advances in describing normal breast epithelial stem cell types. In this review, we speculate that ERα$^+$ and negative sub-types of BC, and probably the four molecular sub-types, can be explained by their origins in the different stem cells that operate at different points in mammary gland development.

In addition, there is increasing proof that an infrequent population of breast CSCs can recapitulate the entire tumour and these are likely candidates for the origin of cancer recurrence. Information on signalling pathways that regulate normal and cancer stem cell self renewal may lead to novel therapies. For example, knowledge of signalling pathways regulating stem cells could be used to induce differentiation of the CSC or promote their apoptosis. These could potentially be used in conjunction with conventional cancer treatments to eradicate proliferative cells and the quiescent CSCs, thus achieving improved cure rates for BC.

Acknowledgements This work was supported by Cancer Research UK and Christie Hospital Charitable Funds.

References

1. Fan C, Oh DS, Wessels L, et al. (2006) Concordance among gene-expression-based predictors for breast cancer. N Engl J Med 355(6):560–569.
2. Hu Z, Fan C, Oh DS, et al. (2006) The molecular portraits of breast tumors are conserved across microarray platforms. BMC Genom 7:96.
3. Oh DS, Troester MA, Usary J, et al. (2006) Estrogen-regulated genes predict survival in hormone receptor-positive breast cancers. J Clin Oncol 24(11):1656–1664.
4. Perou CM, Sorlie T, Eisen MB, et al. (2000) Molecular portraits of human breast tumours. Nature 406(6797):747–752.
5. Sorlie T, Perou CM, Tibshirani R, et al. (2001) Gene expression patterns of breast carcinomas distinguish tumor subclasses with clinical implications. Proc Natl Acad Sci USA 98(19):10869–10874.
6. Sorlie T, Tibshirani R, Parker J, et al. (2003) Repeated observation of breast tumor subtypes in independent gene expression data sets. Proc Natl Acad Sci USA 100(14):8418–8423.
7. Reya T, Morrison SJ, Clarke MF, et al. (2001) Stem cells, cancer, and cancer stem cells. Nature 414(6859):105–111.
8. Smalley M, Ashworth A (2003) Stem cells and breast cancer: a field in transit. Nat Rev Cancer 3(11):832–844.
9. Clarke RB, Anderson E, Howell A, et al. (2003) Regulation of human breast epithelial stem cells. Cell Prolif 36 (Suppl 1):45–58.
10. Dontu G, Al-Hajj M, Abdallah WM, et al. (2003) Stem cells in normal breast development and breast cancer. Cell Prolif 36 (Suppl 1):59–72.
11. Smith GH, Boulanger CA (2003) Mammary epithelial stem cells: transplantation and self-renewal analysis. Cell Prolif 36 (Suppl 1):3–15.
12. Daniel CW, Smith GH (1999) The mammary gland: a model for development. J Mammary Gland Biol Neoplasia 4(1):3–8.
13. Daniel CW, Deome KB (1965) Growth of mouse mammary glands in vivo after monolayer culture. Science 149:634–636.
14. Deome KB, Faulkin LJ Jr, Bern HA, et al. (1959) Development of mammary tumors from hyperplastic alveolar nodules transplanted into gland-free mammary fat pads of female C3H mice. Cancer Res 19(5):515–520.
15. Kordon EC, Smith GH (1998) An entire functional mammary gland may comprise the progeny from a single cell. Development 125(10):1921–1930.
16. Shackleton M, Vaillant F, Simpson KJ, et al. (2006) Generation of a functional mammary gland from a single stem cell. Nature 439(7072):84–88.

17. Stingl J, Eirew P, Ricketson I, et al. (2006) Purification and unique properties of mammary epithelial stem cells. Nature 439(7079):993–997.
18. Clarke RB, Spence K, Anderson E, et al. (2005) A putative human breast stem cell population is enriched for steroid receptor-positive cells. Dev Biol 277(2):443–456.
19. Smith GH (2005) Label-retaining epithelial cells in mouse mammary gland divide asymmetrically and retain their template DNA strands. Development 132(4):681–687.
20. Welm BE, Tepera SB, Venezia T, et al. (2002) Sca-1(pos) cells in the mouse mammary gland represent an enriched progenitor cell population. Dev Biol 245(1):42–56.
21. Zeps N, Bentel JM, Papadimitriou JM, et al. (1998) Estrogen receptor-negative epithelial cells in mouse mammary gland development and growth. Differentiation 62(5):221–226.
22. Goodell MA, Rosenzweig M, Kim H, et al. (1997) Dye efflux studies suggest that hematopoietic stem cells expressing low or undetectable levels of CD34 antigen exist in multiple species. Nat Med 3(12):1337–1345.
23. Smalley MJ, Clarke RB (2005) The mammary gland "side population": a putative stem/progenitor cell marker? J Mammary Gland Biol Neoplasia 10 (1):37–47.
24. Alvi AJ, Clayton H, Joshi C, et al. (2003) Functional and molecular characterisation of mammary side population cells. Breast Cancer Res 5(1):R1–R8.
25. Clayton H, Titley I, Vivanco M (2004) Growth and differentiation of progenitor/stem cells derived from the human mammary gland. Exp Cell Res 297(2):444–460.
26. Asselin-Labat ML, Shackleton M, Stingl J, et al. (2006) Steroid hormone receptor status of mouse mammary stem cells. J Natl Cancer Inst 98(14):1011–1014.
27. Anderson WF, Matsuno R (2006) Breast cancer heterogeneity: a mixture of at least two main types? J Natl Cancer Inst 98(14):948–951.
28. Keeling JW, Ozer E, King G, Walker F (2000) Oestrogen receptor alpha in female fetal, infant, and child mammary tissue. J Pathol 191(4):449–451.
29. Capuco AV, Ellis S, Wood DL, et al. (2002) Postnatal mammary ductal growth: three-dimensional imaging of cell proliferation, effects of estrogen treatment, and expression of steroid receptors in prepubertal calves. Tissue Cell 34(3):143–154.
30. Clarke RB, Howell A, Potten CS, et al. (1997) Dissociation between steroid receptor expression and cell proliferation in the human breast. Cancer Res 57(22):4987–4991.
31. Russo J, Ao X, Grill C, et al. (1999) Pattern of distribution of cells positive for estrogen receptor alpha and progesterone receptor in relation to proliferating cells in the mammary gland. Breast Cancer Res Treat 53(3):217–227.
32. Seagroves TN, Lydon JP, Hovey RC, et al. (2000) C/EBPbeta (CCAAT/enhancer binding protein) controls cell fate determination during mammary gland development. Mol Endocrinol 14(3):359–368.
33. Booth BW, Smith GH (2006) Estrogen receptor-alpha and progesterone receptor are expressed in label-retaining mammary epithelial cells that divide asymmetrically and retain their template DNA strands. Breast Cancer Res 8(4):R49.
34. Dimitrakakis C, Zhou J, Wang J, et al. (2006) Co-expression of estrogen receptor-alpha and targets of estrogen receptor action in proliferating monkey mammary epithelial cells. Breast Cancer Res 8(1):R10.
35. Shoker BS, Jarvis C, Clarke RB, et al. (1999) Estrogen receptor-positive proliferating cells in the normal and precancerous breast. Am J Pathol 155(6):1811–1815.
36. Laakso M, Tanner M, Nilsson J, et al. (2006) Basoluminal carcinoma: a new biologically and prognostically distinct entity between basal and luminal breast cancer. Clin Cancer Res 12(14 Pt 1):4185–4191.
37. Farmer P, Bonnefoi H, Becette V, et al. (2005) Identification of molecular apocrine breast tumours by microarray analysis. Oncogene 24(29):4660–4671.
38. Al-Hajj M, Wicha MS, Benito-Hernandez A, et al. (2003) Prospective identification of tumorigenic breast cancer cells. Proc Natl Acad Sci USA 100(7):3983–3988.
39. Locke M, Heywood M, Fawell S, et al. (2005) Retention of intrinsic stem cell hierarchies in carcinoma-derived cell lines. Cancer Res 65(19):8944–8950.

40. Ponti D, Costa A, Zaffaroni N, et al. (2005) Isolation and in vitro propagation f tumorigenic breast cancer cells with stem/progenitor cell properties. Cancer Res 65(13):5506–5511.
41. Behbod F, Rosen JM (2005) Will cancer stem cells provide new therapeutic targets? Carcinogenesis 26(4):703–711.
42. Kalirai H, Clarke RB (2006) Human breast epithelial stem cells and their regulation. J Pathol 208(1):7–16.
43. Liu S, Dontu G, Wicha MS (2005) Mammary stem cells, self-renewal pathways, and carcinogenesis. Breast Cancer Res 7(3):86–95.

7
Hormone Refractory Prostate Cancer: Lessons Learned from the PTEN Prostate Cancer Model

David J. Mulholland, Jing Jiao, and Hong Wu

Introduction

Progressive prostate cancer (PC) generally takes many years to clinically manifest and, therefore, is generally a disease associated with the elderly. Although PC patients live for many years, those that are diagnosed with advanced disease earlier in life are the ones most likely to require therapeutic intervention. For PC patients, the mainstay of treatment for advanced, extracapsular disease is hormone blockade or androgen ablation. Anti-androgens and castration reduce circulating 5α-dihydrotestosterone (DHT), thereby promoting apoptosis of androgen-dependent prostatic epithelium; however, this therapy is only palliative. Many patients after 12–16 months will undergo prostate specific antigen (PSA) failure, which is generally associated with hormone refractive prostate cancer (HRPC) and poor clinical outcome (1).

The ability to study mechanisms by which prostatic epithelium escape androgen dependency has been hampered by the ability to acquire tissue samples that accurately represent the full spectrum of progressive disease. Most surgical samples, in addition to being heterogeneous, are not hormone naive, and they may be biased by previous hormone blockade therapy. For these reasons, the development of appropriate model systems is important to study PC progression. Despite the ability of some cell lines to undergo androgen-independent (AI) tumorigenesis, there are distinct limitations that prevent consideration of important questions pertaining to HRPC. Perhaps the major disadvantage is that the starting point of many well-studied xenografts is cell lines isolated from a metastatic lesion, a scenario that does not reflect the natural progression of hormone regulated PC. For these reasons, the development of models, frequently generated in mice, that accurately represent human PC progression are of clinical significance.

J.J. Li et al. (eds.), *Hormonal Carcinogenesis V.*

The PTEN Prostate Model: A System to Study Progressive Prostate Cancer

Over the past 10 years, there has been overwhelming evidence indicating that perturbations in the PI3-kinase/Akt signaling pathway play a critical role in many human cancers. The predominant molecular lesions leading directly to pathway activation are loss of function mutations in the tumor suppressor gene phosphatase and tensin homolog (*PTEN*) or activating mutations in the catalytic subunit of PI3-kinase p110. Frequently, a range of tumor types are affected by these mutations including PC, breast cancer (BC), glioblastoma, and endometrial cancer, as well as many others. In addition, less direct mechanisms of PI3-kinase/Akt pathway activation occur through cell surface receptor signaling, as exemplified by tyrosine kinases, such as the insulin-like growth factor 1 (IGF-1) receptor, or her2/her3 heterodimers of the epidermal growth factor receptor (EGFR) (2).

The etiologic role of *PTEN* in cancer has been amply demonstrated through the genesis of transgenic or knockout mouse models in which perturbation of *PTEN* controlled signal pathway clearly leads to cancer in various tissue types. To accurately assess *PTEN*'s biological functions in the initiation, progression, and development of AI PC, we and others have generated a series mouse models carrying either conventional (3–6) or conditional (7–9) deletion of *PTEN*'s murine homolog, the *PTEN* gene. Deletion of both alleles of *PTEN* gene causes early embryonic lethality, suggesting that *PTEN* function is indispensable for normal embryonic development (3). Although *PTEN* heterozygous mice are prone to the development of tumors in multiple organs, only neoplasia (prostatic intraepithelial neoplasia, PIN) lesions are found in the prostate. Therefore, most valuable information, especially regarding the development of hormone refractory PC, are derived from analyzing the PTEN conditional knock-out mice model (7–9).

We employed probasin promoter driven Cre recombinase line to achieve exon 5-specific excision of the *PTEN* locus. In contrast to PTEN heterozygotes, conditional deletion of both PTEN alleles in the prostate significantly reduced the latency of PIN lesion development from 10 months to 6 weeks. Importantly, PTEN-null animals progress from PIN, to localized adenocarcinoma, to metastasis with precise kinetics (7), suggesting that the onset and progression of prostate cancer is *PTEN* dosage-dependent (Fig. 1).

Although the loss of *PTEN* appears sufficient to promote PC, not all human PCs are subject to biallelic *PTEN* loss (10). That is, single allele loss of *PTEN* or epigenetic events may dictate a partial reduction in *PTEN* expression (11, 12). Thus, while *PTEN* dosage is important for the initiation and progression of PC, *PTEN* can also synergize with other related genes, including *Nkx3.1*, *p27*, and *p53* to promote carcinogenesis (13, 14) and AI PC (15, 16).

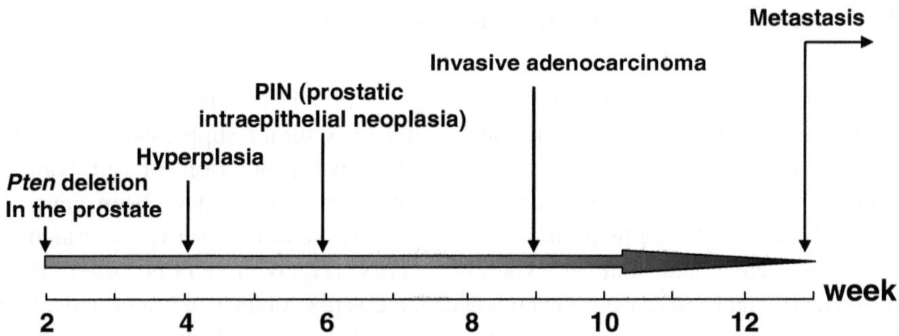

Fig. 1 Progression of the PTEN conditional prostate cancer model. Genetic deletion of the PTEN alleles is initiated by androgen-regulated expression of the Probasin promoter driving Cre recombinase expression. At 4 weeks of age, mice display hyperplastic expansion of luminal epithelium and neoplasia (PIN), at 6–8 weeks of age. Subsequently, mice demonstrate invasiveness (8–10 weeks) and eventually metastasis (>12 weeks)

The Onset of HRPC in the PTEN Prostate Cancer Model

The ability to assay for cellular and histological changes occurring from the initiation of tumorgenesis in homogenous epithelial environment serves as a powerful tool to study PC. Although probasin is regulated by androgen, PB-Cre-mediated PTEN deletion becomes irreversible after the excision of exon 5, thus AI. As a result, the PTEN conditional model can be used subsequent to castration for PC cellular and histological analysis or during the onset of HRPC. This is in contrast with other transgene-based PC models, in which the tumor initiation and further progression events are all driven by androgen-dependent promoters. Such models are generally not applicable for studies subsequent to castration or hormone ablation.

The PTEN PC model provides important information about the onset of HRPC. Although PTEN loss appears to be critical for the progression of HRPC (7), mutant mice that undergo hormone ablation (gonadal castration) still undergo massive apoptosis of the prostatic epithelium (7). Despite this, the proliferative index remains constant (15–20%), indicating that a select population of cells remain resistant to the stress of androgen withdrawal while the vast majority undergo apoptosis (7). Understanding the cell population that maintains proliferative capacity at castrate levels of androgens and the signaling cues that mobilize reconstitution of prostate epithelium, subsequent to androgen ablation, are important issues that are currently studied in our laboratory. It is known that, relative to wild type (wt) mice, PTEN mutants display considerable expansion of cells positive for markers of stemness including p63 and CK5 (17). Whether such basal cells are in fact those that repopulate either during AI or upon reconstitution with androgens is currently being investigated.

PTEN as a Dose-Dependent Tumor Suppressor

Increasing evidences support the critical nature of PTEN dosage in dictating its phenotypic output. The haploinsufficiency of PTEN tumor suppressor function is, therefore, extremely important for the progressive promotion of survival and enhanced proliferation. Several genetic loss-of-function tools have been generated that clearly demonstrate this argument. A hypomorphic mutant series generated in a PTEN dose-dependent manner whereby PTEN (Hy/WT)>PTEN (+/–)>PTEN (Hy/–) clearly manifests a dose-dependent cancerous phenotype (8). Importantly, while PTEN heterozygotes (+/–) develop hyperplasia and PIN through time, transcriptional interference (Hy/–) of the remaining PTEN allele results in high grade PIN and invasive carcinoma, representing a dramatic acceleration of the phenotype. These observations are recapitulated in the PTEN conditional PC model, in which PTEN (–/–) mice display significantly enhanced progression compared with PTEN (+/–) mice (7). PTEN dose-dependent effects are also relevant to human disease as exemplified by recent analysis using fluorescent in situ hybridization (FISH) in which PIN specimens and high grade PC samples displayed decreasing genomic PTEN expression (18, 10).

Although PTEN dosage is clearly critical to its phenotypic manifestations, single allelic PTEN loss may also cooperate with other growth factors and/or genetic events to accelerate its loss of function. This is clearly demonstrated in recent reports describing conditional crosses of fibroblast growth factor 8 (FGF8) expressing mice and PTEN (+/–). PTEN (+/–) mice display neoplasia, while compound FGF8b/ PTEN (+/–) mutants progress to adenocarcinoma (19). Despite the presence of this phenotype generated from the FGF/PTEN compound mutant, the authors did not observe a strong correlation with phosphor-AKT or p27. Such observations underscore the differences between conditional, biallelic loss of PTEN, and over-expression of Akt. The PTEN loss model can promote PC, while the MPAkt model only yields PIN (20). This indicates that other "non-Akt" oncogenic factors may be cooperating with the loss of a single PTEN allele to promote oncogenesis. These observations also indicate that other stromal or epithelial factors may function as "second hits" for PTEN loss of function, possibly including well-studied factors such as TGFβ, Wnts, EGF, IGF, and FGF isoforms.

PTEN has also been proposed to cooperate with the frequently mutated tumor suppressor, *p53,* in a manner whereby PTEN regulates *p53* function and *p53* can alter *PTEN* transcription. Recently, it has also been proposed that loss of *PTEN* promotes cellular senescence and that a subsequent genetic hit of *p53* loss is required to advance prostatic epithelium to a highly invasive and lethal phenotype at 7 months (15). Given that other genetic aberrations have been shown to potentiate the effects of *PTEN* to an invasive phenotype, and *PTEN* can regulate *p53* level and transcription activity (21, 22), it will be important to ascertain whether alteration in *p53* status is absolutely critical for a progressive phenotype. What these observations affirm is that biallelic *PTEN* loss can be potentiated by subsequent *p53* loss, a late stage genetic event observed in advanced human PC (23). An additional tumor suppressor gene that has been shown to potentiate the haploinsufficiency of *PTEN*

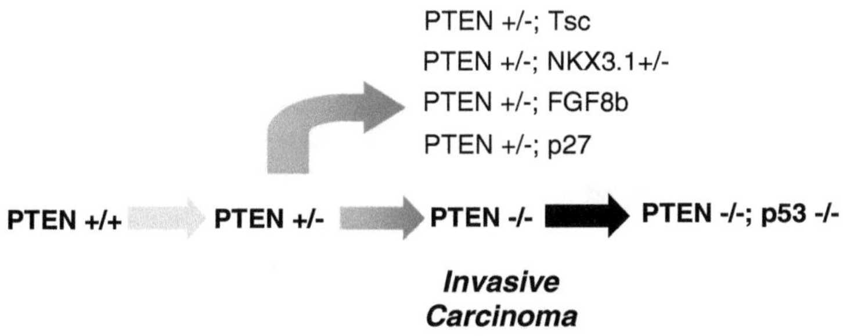

Fig. 2 PTEN as a dose-dependent tumor suppressor in PC models. The haploinsuffiency of PTEN is potentiated by other growth tumor suppressors, growth factors, and cell cycle regulators resulting in adenocarcinoma

is Tubersclerosis (*Tsc*), a tumor suppressor regulator of mTOR kinase function (24). Interestingly, this study revealed an asymmetrical relationship between *PTEN* and *Tsc*. Compound PTEN:Tsc heterozygote mice demonstrated reciprocal potentiation of single mutants, while only *PTEN* loss of function was enhanced by decreased *Tsc* expression. Conversely, *Tsc* loss of function was not potentiated by *PTEN* loss (25). Although the mechanisms governing these observations are not entirely clear, it appears that *PTEN* loss may function preferentially as an initial genetic hit.

The analysis of PC systems in which changes in *PTEN* expression can be controlled provide clinically relevant information when considering that the majority of human samples at clinical presentation show at least single *PTEN* allele loss (12), while biallelic loss is generally associated with advanced disease (18). These data also indicate that *PTEN* loss-of-function is not only important at late stage PC, but that its partial loss of expression may be sufficient for collaboration with other proliferative factors to promote oncogenesis (Fig. 2).

Factors known to regulate *PTEN* expression are, in general, poorly understood. However, using a genetic screen in *Drosophila* mutants, *DJ-1* was identified as a suppressor of *PTEN* function (26). Importantly, *DJ-1* both enhanced Akt phosphorylation, increased cell cycle regulators, correlated negatively with *PTEN* expression in BC samples (26). Such observations beg the question as to a potential role of *DJ-1* in reducing *PTEN* expression during progressive PC. Moreover, if histological PC progression is based upon threshold of *PTEN* expression, then repression of *DJ-1* function may represent an attractive therapeutic option.

NKX3.1: An Androgen-Regulated Target that Cooperates with PTEN-Loss to Promote Progressive Prostate Cancer

Using cDNA microarray analysis, we have assayed for genetic alterations occurring in the PTEN-null PC model. Among those identified include the androgen receptor (AR) target gene, *Nxk3.1*, a homeobox gene, and prostate specific tumor suppressor (7). Recently, our laboratory has made use of renal capsule reconstitution assays to

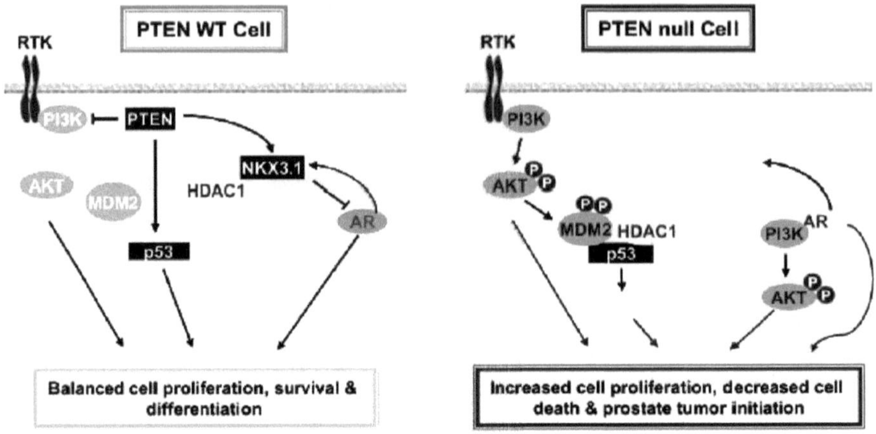

Fig. 3 Involvement of NKX3.1 in PTEN-controlled prostate-tumorigenesis. In PTEN wt prostatic epithelium (*left panel*), PTEN negatively regulates PI3K/AKT pathway but positively modulates *p53* level and activity. The net result is balanced cell proliferation, differentiation, and cell death, which prevents PC initiation. Upon PTEN loss (*right panel*), the balance is broken, and *Nek3.1* and its controlled signaling pathways are severely downregulated. AR, no longer under the control of *Nek3.1*, becomes overexpressed and activates its targets and downstream pathways, including PI3K/AKT pathway

further understand the dynamic interplay between AR and *Nkx3.1* in both normal prostate and in PTEN-null mediated PC (22). In normal environments, *PTEN* functions to maintain both proliferation and survival by positively regulating both *Nkx3.1* and *p53* function (22). Additionally, AR and *Nkx3.1* form a regulatory loop, likely involved in regulating the balance of proliferation and apoptotic index. Although AR positively regulates *Nkx3.1*, *Nkx3.1* can negatively regulate AR function at the transcriptional level. Using PTEN-null epithelium infected with exogenous *Nkx3.1*, decreased levels of AR were observed, in a PI3K/Akt-dependent manner (22). Thus, the balance of AR and *Nkx3.1* function appears to dictate the degree of proliferative vs. tumor suppressor function, respectively. However, during progressive PC and the loss of *PTEN,* the balanced function of *Nkx3.1* and AR may be disrupted. Moreover, the coordinated loss of both *PTEN* (27, 28) and *Nkx3.1*, both tumor suppressors and negative regulators of AR, provides a mechanism for potentiation of AR function during progression (Fig. 3). Collectively, *Nkx3.1* and AR form a feedback loop that is important both for development and for PC progression.

Novel Cell Lines Derived from the PTEN Model

Many commonly used PC cells lines, including LNCaP, PC3, and DU145, were isolated from advanced metastatic lesions from patients that had undergone and failed, therapeutic intervention (29, 30). Although these lines have been the mainstay for in vitro PC studies, each represents considerably different cytological traits that may not be representative of a primary prostate tumor. Specifically, in some instances these lines may harbor traits that have occurred as a result of adaptation to hormone blockade therapy. For example, the LNCaP line contains the T877A point mutation promoting promiscuity to non-androgen ligands, representing a mutational adaptation to castrate levels of androgens (31, 32). PC3 cells are AR negative, a phenotype that is inconsistent with the increased AR levels observed in most advancing PC (33). The isolation of cell lines from progressive PC systems is important for in vitro determination and clarification of events thought to be important for in vivo PC progression. In general, however, the development of cell lines from human hormone naïve, PC patients has been difficult. Nevertheless, isolation of lines from an increasing number of PC mouse models may provide powerful tools to study PC pathogenesis. Recently, our laboratory has isolated several novel lines from the PTEN-null PC model. The PTEN-CaP8 clone was isolated from a 10-month, intact PTEN-null mouse (34). Several features of the PTEN-CaP8, which is null or PTEN, make it amenable to in vitro studies because of this straightforward transfection and infection, expression of wt AR and high mitotic index. The PTEN-CaP8 line is also tumorgenic and capable of anchorage-independent growth. In addition, it is a valuable tool to study the effects of therapeutic intervention, including the effects of inhibitors targeting the PI3K/Akt and mTOR signaling axis. For example, recent studies from our laboratory indicate that PTEN-CaP8 cells are highly sensitive to inhibitors targeting the IGF1 receptor, Akt1/2, and mTOR, results that recapitulate in vivo dosing patterns with the PTEN PC model (Mulholland et al., unpublished observations). Thus, the PTEN-CaP8 cell line represents a valuable in vitro tool to study PTEN-null mediated, AI PC.

Conclusion and Perspectives

The PTEN PC model and related compound mutants have provided critical information regarding the molecular mechanisms contributing to PTEN-mediated progressive PC. Such models show clearly that *PTEN* dose is a critical factor in precipitating an invasive phenotype. It will be critical to determine which factors *PTEN* is capable of collaborating with in order to unveil its haploinsufficiency. Of clinical significance is the fact that partial reduction of *PTEN* expression may, upon activation by a “secondary hit,” manifest a phenotype that is equivalent to biallelic *PTEN* loss. Since this model can rapidly progress to HRPC, we now are in a position to rapidly test the efficacy of pharmacological interventions thereby providing valuable preclinical information for drugs with potential of clinical trials.

Given the dose-dependent manner in which PTEN is lost but permits gain-of-function for proliferation and survival pathways, it is tempting to contemplate therapeutic manipulation of *PTEN* expression, especially when *PTEN* silencing is accomplished via epigenic mechanisms. Although technically challenging, targeted pharmacological enhancement of *PTEN* function may be amenable to patients with low, medium, or even high grade PC. If feasible, PTEN mouse models, compound mutant variants, may provide a perfect preclinical testing ground. Collectively, as the PTEN conditional PC model and its variants are further studied, they are quite likely to provide lessons to those in the field of PC research.

References

1. Feldman BJ, Feldman D (2001) The development of androgen-independent prostate cancer. Nat Rev Cancer 1:34–45.
2. Vivanco I, Sawyers CL (2002) The phosphatidylinositol 3-Kinase AKT pathway in human cancer. Nat Rev Cancer 2:489–501.
3. Di Cristofano A, Pesce B, Cordon-Cardo C, et al. (1998) Pten is essential for embryonic development and tumour suppression. Nat Genet 19:348–355.
4. Suzuki A, de la Pompa JL, Stambolic V, et al. (1998) High cancer susceptibility and embryonic lethality associated with mutation of the PTEN tumor suppressor gene in mice. Curr Biol 8:1169–1178.
5. Freeman D, Lesche R, Kertesz N, et al. (2006) Genetic background controls tumor development in PTEN-deficient mice. Cancer Res 66:6492–6496.
6. Podsypanina K, Ellenson LH, Nemes A, et al. (1999) Mutation of Pten/Mmac1 in mice causes neoplasia in multiple organ systems. Proc Natl Acad Sci U S A 96:1563–1568.
7. Wang S, Gao J, Lei Q, et al. (2003) Prostate-specific deletion of the murine Pten tumor suppressor gene leads to metastatic prostate cancer. Cancer Cell 4:209–221.
8. Trotman LC, Niki M, Dotan ZA, et al. (2003) Pten dose dictates cancer progression in the prostate. PLoS Biol 1:E59.
9. Ma X, Ziel-van der Made AC, Autar B, et al. (2005) Targeted biallelic inactivation of Pten in the mouse prostate leads to prostate cancer accompanied by increased epithelial cell proliferation but not by reduced apoptosis. Cancer Res 65:5730–5739.
10. Yoshimoto M, Cutz JC, Nuin PA, et al. (2006) Interphase FISH analysis of PTEN in histologic sections shows genomic deletions in 68% of primary prostate cancer and 23% of high-grade prostatic intra-epithelial neoplasias. Cancer Genet Cytogenet 169:128–137.
11. Whang YE, Wu X, Suzuki H, et al. (1998) Inactivation of the tumor suppressor PTEN/MMAC1 in advanced human prostate cancer through loss of expression. Proc Natl Acad Sci U S A 95:5246–5250.
12. Gray IC, Stewart LM, Phillips SM, et al. (1998) Mutation and expression analysis of the putative prostate tumour-suppressor gene PTEN. Br J Cancer 78:1296–1300.
13. Kim MJ, Cardiff RD, Desai N, et al. (2002) Cooperativity of Nkx3.1 and Pten loss of function in a mouse model of prostate carcinogenesis. Proc Natl Acad Sci U S A 99:2884–2889.
14. Abate-Shen C, Banach-Petrosky WA, Sun X, et al. (2003) Nkx3.1; Pten mutant mice develop invasive prostate adenocarcinoma and lymph node metastases. Cancer Res 63:3886–3890.
15. Chen Z, Trotman LC, Shaffer D, et al. (2005) Crucial role of p53-dependent cellular senescence in suppression of Pten-deficient tumorigenesis. Nature 436:725–730.
16. Gao H, Ouyang X, Banach-Petrosky WA, et al. (2006) Emergence of androgen independence at early stages of prostate cancer progression in nkx3.1; pten mice. Cancer Res 66:7929–7933.
17. Wang S, Garcia AJ, Wu M, et al. 2006 Pten deletion leads to the expansion of a prostatic stem/progenitor cell subpopulation and tumor initiation. Proc Natl Acad Sci U S A 103:1480–1485.
18. Verhagen PC, van Duijn PW, Hermans KG, et al. (2006) The PTEN gene in locally progressive prostate cancer is preferentially inactivated by bi-allelic gene deletion. J Pathol 208:699–707.
19. Zhong C, Saribekyan G, Liao CP, et al. (2006) Cooperation between FGF8b overexpression and PTEN deficiency in prostate tumorigenesis. Cancer Res 66:2188–2194.

20. Majumder PK, Febbo PG, Bikoff R, et al. (2004) mTOR inhibition reverses Akt-dependent prostate intraepithelial neoplasia through regulation of apoptotic and HIF-1-dependent pathways. Nat Med 10:594–601.
21. Freeman DJ, Li AG, Wei G, et al. (2003) PTEN tumor suppressor regulates p53 protein levels and activity through phosphatase-dependent and -independent mechanisms. Cancer Cell 3:117–130.
22. Lei Q, Jiao J, Xin L, et al. (2006) NKX3.1 stabilizes p53, inhibits AKT activation, and blocks prostate cancer initiation caused by PTEN loss. Cancer Cell 9:367–378.
23. Heidenberg HB, Sesterhenn IA, Gaddipati JP, et al. (1995) Alteration of the tumor suppressor gene p53 in a high fraction of hormone refractory prostate cancer. J Urol 154:414–421.
24. Guertin DA, Sabatini DM (2005) An expanding role for mTOR in cancer. Trends Mol Med 11:353–361.
25. Ma L, Teruya-Feldstein J, Behrendt N, et al. (2005) Genetic analysis of Pten and Tsc2 functional interactions in the mouse reveals asymmetrical haploinsufficiency in tumor suppression. Genes Dev 19:1779–1786.
26. Kim RH, Peters M, Jang Y, et al. (2005) DJ-1, a novel regulator of the tumor suppressor PTEN. Cancer Cell 7:263–273.
27. Nan B, Snabboon T, Unni E, et al. 2003 The PTEN tumor suppressor is a negative modulator of androgen receptor transcriptional activity. J Mol Endocrinol 31:169–183.
28. Lin HK, Hu YC, Lee DK, et al. 2004 Regulation of androgen receptor signaling by PTEN (phosphatase and tensin homolog deleted on chromosome 10) tumor suppressor through distinct mechanisms in prostate cancer cells. Mol Endocrinol 18:2409–2423.
29. Sobel RE, Sadar MD 2005 Cell lines used in prostate cancer research: a compendium of old and new lines–part 1. J Urol 173:342–359.
30. Sobel RE, Sadar MD 2005 Cell lines used in prostate cancer research: a compendium of old and new lines–part 2. J Urol 173:360–372.
31. Veldscholte J, Berrevoets CA, Ris-Stalpers C, et al. 1992 The androgen receptor in LNCaP cells contains a mutation in the ligand binding domain which affects steroid binding characteristics and response to antiandrogens. J Steroid Biochem Mol Biol 41:665–669.
32. Klein KA, Reiter RE, Redula J, et al. 1997 Progression of metastatic human prostate cancer to androgen independence in immunodeficient SCID mice. Nat Med 3:402–408.
33. Chen CD, Welsbie DS, Tran C, et al. 2004 Molecular determinants of resistance to antiandrogen therapy. Nat Med 10:33–39.
34. Jiao J, Wang S, Qiao R, Vivanco I, Watson PA, Sawyers CL, Wu H. (2007) Murine cell lines derived from Pten null prostate cancer show the critical role of PTEN in hormone refractory prostate cancer development. Cancer Res 67:6083–9.

Part 2
Mitotic Kinases, Centrosome Amplification, and Genomic Instability

8
Functional Significance of Aurora Kinase A in Centrosome Amplification and Genomic Instability

Subrata Sen, Hiroshi Katayama, and Kaori Sasai

Introduction

Aurora kinase (Aur) A is the member of a Serine/Threonine protein kinase family that is represented by a single prototypic Ipl1 kinase in yeast and additional paralogues in metazoan organisms. Among mammals, this kinase family consists of three members, AurA, AurB, and AurC while in *Drosophila*, *C. elegans* and *Xenopus,* two members, AurA and B have been identified (1). Since its discovery about a decade ago, Aur family of kinases has received significant attention because of frequent over expression of all three kinases detected in human cancers and their roles as critical regulators of mitotic cell proliferation and chromosome segregation processes (2). Ectopic elevated expression of AurA in mammalian cells in vitro was reported to induce oncogenic transformation in cells along with centrosome amplification and chromosomal instability (3). Chromosomal ploidy alterations correlating with AurA over expression has since been detected in several human cancers (4–8), rat (9, 10), and mouse (11) in vivo mammary cancer model systems. These findings suggest that chromosomal instability is a genetically determined mutant phenotype induced due to anomalies in the molecular pathways critical to the development of malignant transformation in cells.

The three Aur are expressed at the highest levels during G2 through M phases of mitosis and show discretely predominant subcellular distribution patterns coinciding with their functional involvement in different cellular pathways at the respective cellular organelles of their localization. AurA is mainly localized on the centrosomes and along the mitotic microtubules, while AurB and AurC are detected primarily around the centromeres during prophase through metaphase and then at the spindle midzone followed by midbody during anaphase through cytokinesis (1). Interestingly, AurA also localizes in the nucleus (12) during G2-M phase and along the midbody during cytokinesis. Such discrete subcellular distributions of AurA correlate with functional interactions involving multiple different proteins implicated in the centrosome duplication–maturation–separation processes as well as in the maintenance of chromosomal stability. In this chapter, we first discuss the structural organization and regulation of AurA and then describe its functional interactions identified on centrosomes, and along microtubules as well as kinetochores in reference

J.J. Li et al. (eds.), *Hormonal Carcinogenesis V.*

to their possible roles in the induction of centrosome amplification and chromosomal instability.

Structural Organization and Regulation of Aurora Kinase A

Phylogenetic analyses show that the AurA and AurB are well defined in the vertebrates but the differences are not so obvious in invertebrates (1). The three mammalian Aurora kinase paralogues of various peptide lengths share extensive sequence similarities of more than 70% within the carboxy-terminal catalytic domain but their amino-terminal domains differ in length as well as sequence content. In human, AurA is a 46 kDa protein consisting of 403 amino acids (Fig. 1).

At the amino-terminal domain, three putative conserved Aur boxes can be identified. There is some suggestive evidence that these may be involved in subcellular localization or substrate recognition of the protein (13). Also, one of the serine residues in the A box II has been shown to be involved in the degradation of the protein (14). AurA content goes down at late mitosis through proteasome-mediated degradation (15) in presence of Cdh1/Fizzy related interacting anaphase promoting complex (APC^{Cdh1}). A putative destruction box sequence with the consensus motif, RxxLxxVxE near carboxyl terminal, mediates the proteolysis of AurA (16, 17). A signature motif for Aurora kinases in the catalytic domain spanning the activation loop between subdomains VII and VIII is DFGWSxxxxxxxRxTxCGTxDYLPPE. This sequence contains a conserved Threonine residue within the RxT motif that is phosphorylated for activation of AurA (18). Crystal structure of AurA catalytic

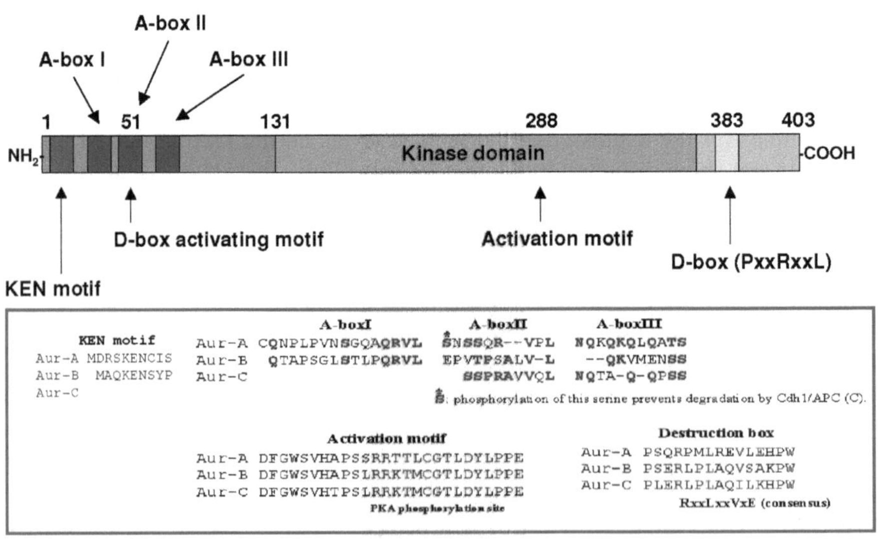

Fig. 1 Structure of human AurA and alignment of conserved domains in human Aur kinase family

domain has been resolved (19). This reveals the typical bilobal kinase fold comprising of an N-terminal β strand domain and a C-terminal α helical domain. These domains are linked together by a hinge region that plays an important role in the formation of the catalytic active site. The N-terminal domain is responsible for positioning ATP through an αC helix while the activation loop within the C-terminal domain harbor substrates (19, 20).

Aurora Kinase A and Centrosome Structure–Function

AurA in human cells localizes on the centrosomes during G1-S phase transition remaining at this location through the centrosome duplication process at the end of S phase and during mitotic progression through G2-M phases till the protein is degraded following mitotic exit in the early G1 phase. Detailed deletion mapping of the human AurA in cellular targeting experiments in human cells have revealed that 62 amino acids in the catalytic domain of the protein can localize to the centrosome (unpublished results). These data differ from a previous report in *Xenopus* AurA, where the noncatalytic protein domain was localized to the centrosome (21). The protein amount and the kinase activity peak during late G2-M phase. In addition, the protein maintains dynamic association with various subcellular sites during the cell cycle. Fluorescence recovery after photo-bleaching (FRAP) analyses have revealed that rapid redistribution of AurA is maintained between the centrosomes and the spindle microtubules during G2-M phases of mitosis (22). This localization profile suggests that AurA is functionally involved in various mitotic processes at multiple subcellular locations. Several upstream regulators as well as downstream AurA substrates have been identified at the intracellular sites of its localization, which play important roles at different stages of mitosis (Table 1).

The centrosome is a nonmembranous organelle often associated with the nuclear membrane and composed of a pair of centrioles connected with a protenaceous matrix, surrounded by aggregates of proteins referred to as the pericentriolar material (PCM). The centrioles are made up of nine triplets of microtubules arranged in the form of a cylinder while the PCM composition varies through the cell cycle with many proteins getting transiently associated with the structure. The PCM provides a scaffold for proteins important for the regulation of centrosome duplication and function, including the unit for microtubule nucleation, the γ-tubulin ring complex, a complex of six proteins, which in humans contains γ-tubulin and GCP2-6 (23). Estimates suggest that centrosomes comprise hundreds of proteins including large coil/coil scaffold proteins serving as the docking sites for regulatory and other activities (24).

Following completion of mitosis, a normal diploid cell inherits one centrosome with two centrioles that duplicate during the S phase, and after undergoing maturation, by recruiting additional proteins, separate during the G2-M phase transition, and subsequently, forming the microtubule organizing centers (MTOC) at the spindle poles during the M phase. The molecular details of the centrosome duplication

Table 1 Aurora-A binding partners and their nuclear localization

Aurora-A binding partner		Localization
CENP-A		Centromere
Ajuba	CPEB	
Bora	P160Rock	Centrosome/Mitotic spindle poles
HEF1	UBE2N	
PAK1	BRCA1	
Centrosomin	Lats2	
CDC25B	NM23-H1	
MAP215	p53	
TACC3		
MAP215		
TACC3		Mitotic spindles
HURP		
TPX2		
Eg5		
AIP		
CDC20		
Gadd45a		unknown
PP1		
RasGAP		
Survivin		

process have not yet been elucidated. It is known that cyclin E/A-cdk2 complexes are required for both centrosome duplication and S phase entry (25), while overexpression of cyclin E centrosome localization domain disrupts centrosome localization of the endogenous cyclin E preventing entry into the S phase (26). Ectopically expressed cyclin E accelerates entry into S phase even with a mutation that abolishes binding to cdk2 but not with a mutant localization signal. In addition to cyclin E/A-cdk2 complex, a number of other kinases including AurA (3, 27), polo-like kinases (28), and NIMA-related kinase Nek2 (29) have all been implicated in regulating centrosome structure, duplication, and function. As mentioned earlier, the molecular details of the pathways regulating centrosome structure, duplication, and function are not yet known, but phenomenological observations on the effects of kinases on centrosome structure and function together, with documented interactions with other centrosome associated proteins, some also affecting the centrosome duplication and function, provide compelling evidence of their direct roles in these processes. In this chapter, we will describe only the observations made with AurA and its interacting proteins on and around centrosomes to discuss their possible roles in the centrosome duplication, maturation, microtubule nucleation, and separation processes.

Although ectopic expression of AurA induces centrosome amplification in cells in vitro (3, 27) reflecting a possible role for this kinase in the duplication of centrosomes, the possible molecular pathways involved have not been established yet. In fact, the observation that a kinase inactive mutant of AurA can also induce centrosome amplification suggesting that AurA-induced centrosome amplification may be the indirect result of tetraploidization of cells (27). Our unpublished data in aphidicolin-treated S phase arrested human cells, however, reveal that AurA can affect extra rounds of centrosome duplication within a single cell cycle. Therefore, we favor the notion that AurA may regulate the cross talk between the centrosome cycle and the nuclear cycle. It is possible that AurA may, for example, control the "license" to duplicate by modifying the centrosomal substrates. Such potential licensing mechanisms for centrosome duplication and maturation involving other centrosomal kinases have also been suggested. For example, displacement of Ninein-like protein Nlp1 following Plk1-mediated phosphorylation (30) has been implicated in centrosome maturation and Nek2 phosphorylated C-Nap1 displacement, as well as its dephosphorylation by protein phosphatase 1 (PP1) are just some of the known important elements involved in splitting the centrosomes prior to the onset of mitosis (31, 32). It is relevant in this context that AurA and PP1 are involved in feedback regulation of their respective enzyme activities that influence the mitotic progression of cells (33). It is conceivable, therefore, that AurA regulation of PP1 could influence displacement of C-Nap1, and thus, separation of centrosomes.

In addition to phosphorylation-mediated modification of centrosomal proteins, activities of E3 ubiquitin ligases involved in bringing together the charged E2 ubiquitin conjugating enzyme with centrosomal substrate proteins have also been shown to be important in centrosome regulation. The E3 ubiquitin ligases implicated in such functions include the SCF complex (Skp1, Cullin, F-box) (34), the heterodimeric BRCA1/BARD1 complex (35), and the anaphase promoting complex/cyclosome (APC/C) (36). *BRCA1* inhibition in mammary tissue derived cells in culture cause centrosome amplification (35), and the APC/C substrates on the centrosome are the kinases, which regulate centrosome function. Ubiquitination of target proteins by APC/C is itself regulated both by phosphorylation of the core components and by the inhibitors of the adapter proteins such as Cdh1. An inhibitor of cdh1 protein complex Emi1 stabilizes cyclin A, an important kinase for initiating centrosome duplication in somatic cells (37). Considering that cdh1 is the adaptor protein required for APC/C mediated ubiquitination of AurA, it is expected that Emi1 also stabilizes AurA possibly facilitating duplication of centrosomes. Furthermore, in view of the fact that AurA phosphorylates BRCA1 that is important for mitotic progression of cells (38) and AurA is a target of APC/C (39), it may be suggested that AurA regulated pathways involving these two E3 ubiquitin ligases participate in controlling centrosome duplication, separation, maturation processes. Recent discovery on the role of protease separase along with APC/C that is required for activation of separase in centriole disengagement and licensing centriole duplication (40) raises the interesting possibility that AurA activity level at the onset of anaphase may serve as a signal for activating the license to duplicate the centrosomes.

The role of AurA in centrosome maturation has been suggested not only based on morphological assays, but also on compelling molecular evidence available. AurA depletion in human cells growing in culture results in inhibition of centrosome maturation (41). Several components of the PCM including the γ tubulin present in the centrosome become deficient, and the microtubule mass of spindles decreases by ~60%. Molecular mechanism underlying AurA-mediated centrosome maturation has been proposed based on the finding of AurA interaction with transforming acidic coiled coil (TACC) proteins (42), implicated in the centrosome maturation process. TACC phosphorylation by AurA facilitates recruitment of TACC to the centrosome early in mitosis leading to complex formation with Msps/XMAP215

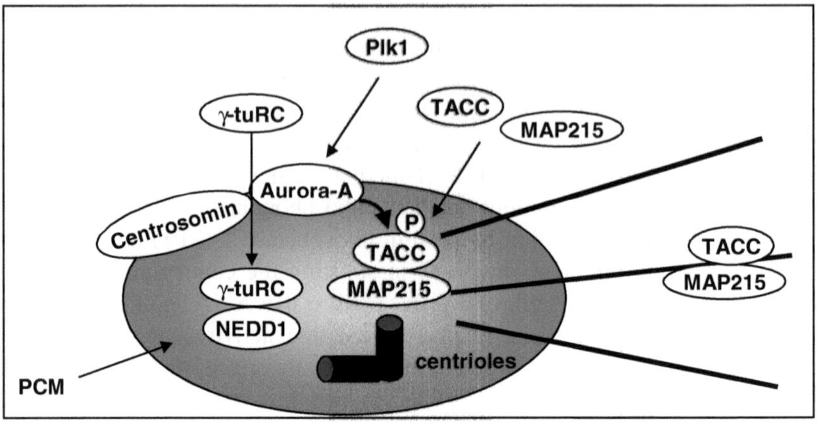

Fig. 2 Molecular interactions during centrosome maturation

microtubule associated proteins, which promote microtubule growth at both the minus and the plus ends (Fig. 2).

Centrosomes separate and migrate around the nuclear envelope to move to the poles and organize the bipolar spindle. This movement depends on microtubules and molecular motors. Defects in centrosome separation in the absence of AurA function have been detected in *C. elegans* (43), *Xenopus* (44), and *Drosophila* (45). In human HeLa cells, microinjection of affinity purified AurA antibodies at late G2 phase was also reported to inhibit separation of centriole pairs (46), and it was suggested that human AurA regulated centriole pair separation is mediated through phosphorylation of kinesin-like motor protein Eg5, known to be a substrate of the *Xenopus* AurA homologue Eg2.

Aurora Kinase A Interaction with Centrosome Associated Proteins in the Regulation of Mitosis and Spindle Assembly

Activated AurA is first detected on centrosomes in late G2 phase of HeLa cells. RNA interference studies in synchronized cells indicated that this activation is required for cdk1-cyclin B1 recruitment to the centrosome and commitment of the cells to the nucleus, with activated cdk1-cyclin B1 in turn required for full activation of AurA (41). Additionally, it has also been demonstrated that AurA phosphorylates Serine 353 of the phosphatase cdc25B, the dual specificity phosphatase required for the activation of cdks at the centrosomes during G2 phase, also facilitating the onset of mitosis (47). A number of other substrates that also activate AurA on the centrosome have been identified. These include the LIM protein Ajuba (41), a multidomain focal adhesion scaffolding protein HEF1 (48), the p21 activated protein kinase Paks (49), and, most recently, an evolutionarily conserved protein named Bora, originally identified in a protein screen involved in asymmetric cell division in *Drosophila* (50). Although the functional significance of Bora-mediated activation is not yet clear, it is significant that activation of AurA by Ajuba, HEF1, and Pak1 indicate cross talk between focal adhesion and the mitotic apparatus. It is possible that disassembly of focal adhesions releases the protein complexes leading to the activation of AurA, thereby connecting loss of adhesion to cell proliferation, and perhaps linking integrin signals to centrosome maturation.

One of the essential prerequisites for properly regulating mitotic cell division is to ensure that chromosomes are faithfully and equally segregated to the daughter cells. This is achieved by the mitotic spindle, a bipolar array of microtubules focused at each pole by the two centrosomes. This is accomplished with the help of microtubules attaching to the kinetochore region of each chromosome and the stabilization as well as bundling of microtubules to form kinetochore fibers (*k*-fibers). These fibers help biorientation of each chromosome to kinetochores from duplicated sister chromatids attached to microtubules from opposite poles of the spindle. When the fidelity of this process is compromised, abnormal numbers of chromosomes are segregated to the daughter cells giving rise to chromosomal instability. AurA together with Ran-GTP is now recognized to be a critical regulator of this process that ensures proper assembly and stabilization of spindle microtubules (51). Ran-GTP is a small GTPase, the active GTP bound configuration of which is generated by the chromatin associated guanine nucleotide exchange factor RCC1 and promotes the release of spindle assembly factors from inhibitory complexes with importins. The generation of Ran-GTP at the chromosomes possibly guide spindle assembly by creating a positional gradient signal. Among the factors released from Ran-GTP importin inhibition, the most important one is TPX2 that binds AurA at the centrosome and targets it to the spindle microtubules proximal to the pole (52). TPX2 regulates the kinase activity of AurA by counteracting the PP1 phosphatase activity and stimulating AurA autophopshorylation at the critical residue in the activation loop essential for kinase activity. More recently, it has been shown that another oncoprotein, hepatocarcinoma upregulated protein (HURP) is also under the regulation of Ran during mitosis (53, 54). HURP is a microtubule associated protein (MAP) and a substrate of AurA. In *Xenopus*, HURP forms part of a complex containing two additional MAPs, TPX2 and XMAP215, together with the plus end directed kinesin Eg5 and AurA (53, 54). AurA activity is essential for the formation and function of the complex, which appears to relate to microtubule polymerization and stabilization of the *k*-fibers and not to the astral microtubules. Since experimental HURP depletion results in misalignment of the chromosomes, it is reasonable to suggest that chromosomal instability detected in cells with deregulated expression of AurA also is due to abnormal functional organization of the HURP complex.

In addition to its role in the functioning of the HURP complex involving the MAPs, in spindle microtubule polymerization and stabilization, AurA has also been implicated in the regulation of proper chromosome alignment and segregation through phosphorylation of CENP-A, a variant of histone H3 in the nucleosome of the centromeric chromatin at the inner plate of the kinetochore (55).

On the basis of the number of AurA functional interactions identified with a host of proteins on the centrosomes, microtubules, and along the chromatin described earlier, it becomes abundantly clear that cells with abnormal AurA expression multiple pathways involving these interacting proteins get deranged leading to centrosome anomalies and chromosomal instability. However, it is also important to point out that since AurA overexpression can induce cell transformation, it is quite possible that AurA participates in other critical intracellular signaling pathways relevant to malignant transformation processes. These would include the recently described involvement of AurA in p53 tumor suppressor pathway (56) and the MAPK1/ERK2 pathway (57) aberrantly expressed in human cancers. It will be extremely interesting to find out how the different AurA-mediated intracellular signaling pathways cross talk in maintaining centrosomal and chromosomal stability in normal cells, as well as how these pathways respond in vivo when AurA expression is deregulated.

Although studies to elucidate the detailed pathways in experimentally manipulated in vitro grown cells are been conducted, more functional studies will have to be done with normally proliferating cells to understand the physiological relevance of the pathways in question. In the cancer context, the same principle will have to be applied to identify the genetic consequences of naturally AurA overexpression occurring in human malignancies. These data will be extremely helpful to eventually determinate the relevance of AurA pathway target proteins as biomarkers, and as possible future therapeutic targets.

Acknowledgments We acknowledge Dr. William Brinkley collaboration. This work was supported by grants from the National Institute of Health.

References

1. Carmena M, Earnshaw WC (2003) The cellular geography of Aurora kinases. Nat Rev of Mol Cell Biol 4:842–854.
2. Katayama H, Brinkley WR, Sen S (2003) The Aurora kinases: role in cell transformation and tumorigenesis. Cancer Metastasis Rev 22:451–464.
3. Zhou H, Kuang J, Zhong L, et al. (1998) Tumor amplified kinase STK 15/BTAK induces centrosome amplification, aneuploidy and transformation. Nat Genet 20:189–193.
4. Bischoff JR, Anderson L, Zhu Y, et al. (1988) A homologue of *Drosophilia* aurora kinase is oncogenic and amplified in human colorectal cancers. EMBO J 17:3052–3065.
5. Tanaka T, Kimura M, Matsunga K, et al. (1999) Centrosomal kinase AIK1 is overexpressed in invasive ductal carcinoma of the breast. Cancer Res 59:2041–2044.
6. Sen S, Zhou H, Zhang RD, et al. (2002) Amplification/overexpression of a mitotic kinase gene in human bladder cancer. J Natl Cancer Inst 94:1320–1329.
7. Li D, Zhu J, Sen S, et al. (2003) Overexpression of oncogenic STK15/BTAK/Aurora A kinase in human pancreatic cancer. Clin Cancer Res 9:991–997.
8. Tanaka E, Hashimoto Y, Ito T, et al. (2005) The clinical significance of Aurora-A/STK15/BTAK expression in human esophageal squamous cell carcinoma. Clin Cancer Res 11:1827–1834.
9. Goepfert TM, Adigun YE, Zhong L, et al. (2002) Centrosome amplification and overexpression of aurora A are early events in rat mammary carcinogenesis. Cancer Res 62:4115–4122.

10. Li JJ, Weroha SJ, Lingle WL, et al. (2004) Estrogen mediates Aurora-A overexpression, centrosome amplification, chromosomal instability, and breast cancer in female ACI rats. Proc Natl Acad Sci USA 101:18123–18128.
11. Wang X, Zhou YX, Qiao W, et al. (2006) Overexpression of aurora kinase A in mouse mammary epithelium induces genetic instability preceding mammary tumor formation. Oncogene Adv (online publication).
12. Crosio C, Fimia GM, Sen S, et al. (2002) Mitotic phosphorylation of histone H3: spatio-temporal regulation by mammalian Aurora kinases. Mol Cell Biol 22:874–885.
13. Dutertre S, Descamps S, Prigent C (2002) On the role of aurora-A in centrosome function. Oncogene 21:6175–6183.
14. Littlepage LE, Wu H, Anderson T, et al. (2002) Identification of phosphorylated residues that affect the activity of the mitotic kinase Aurora-A. Proc Natl Acad Sci USA 99:15440–15445.
15. Honda K, Mihara H, Kato Y, et al. (2000) Degradation of human Aurora2 protein kinase by the anaphase-promoting complex-ubiquitin-proteasome pathway. Oncogene 19:2812–2819.
16. Castro A, Vigneron S, Bennis C, et al. (2002) The D box activating domain (DAD) is a new proteolysis signal that stimulates the silent D-box sequence of Aurora A. EMBO Rep 3:1209–1214.
17. Crane R, Kloepfer A, Ruderman JV (2004) Requirements of the destruction of human Aurora-A. J Cell Sci 117:5975–5983.
18. Walter AO, Seqhezzi W, Korner W, et al. (2002) The mitotic serine/threonine kinase Aurora2/AIK is regulated by phosphorylation and degradation. Oncogene 19:4906–4916.
19. Cheetham GM, Knegtel RM, Coll JT, et al. (2002) Crystal structure of aurora-2, an oncogenic serine/threonine kinase. J Biol Chem 277:42419–42422.
20. Bayliss R, Sardon T, Vernos I, et al. (2003) Structural basis of Aurora-A activation by TPX2 at the mitotic spindle. Mol Cell 12:851–862.
21. Giet R, Prigent C (2001) The non-catalytic domain of the *Xenopus laevis* aurora A kinase localises the protein to the centrosome. J Cell Sci 114:2095–2104.
22. Stenoien DL, Sen S, Mancini MA, et al. (2003) Dynamic association of a tumor amplified kinase, aurora A, with the centrosome and mitotic spindle. Cell Motil Cytoskeletin 55:134–146.
23. Murphy SM, Preble AM, Patel UK, et al. (2001) GCP5 and GCP6: Two new members of the human gamma-tubulin complex. Mol Biol Cell 12:3340–3352.
24. Doxsey S, Zimmerman W, Mikule K (2005) Centrosome control of the cell cycle. Trends Cell Biol 15:303–311.
25. Lacey KR, Jackson PK, Stearns T (1999) Cyclin-dependent kinase control of centrosome duplication. Proc Natl Acad Sci USA 96:2817–2822.
26. Matsumoto Y, Malles JL (2004) A centrosomal localization signal in cyclin E required for Cdk2-independent S phase entry. Science 306:885–888.
27. Meraldi P, Honda R, Nigg EA (2002) Aurora-A overexpression reveals tetraploidization as a major route to centrosome amplification in p53–/– cells. EMBO J 21:483–492.
28. Habedanck R, Stierhof YD, Wilkinson CJ, et al. (2005) The Polo kinase Plk4 functions in centriole duplication. Nat Cell Biol 7:1140–1146.
29. Eto M, Elliott E, Prinkett TD, et al. (2002) Inhibitor-2 regulates protein phosphatase-1 complexed with NimA-related kinase to induce centrosome separation. J Biol Chem 277:44013–44020.
30. Casenghi M, Meraldi P, Weinhart U, et al. (2003) Polo-like kinase 1 regulates Nip, a centrosome protein involved in microtubule nucleation. Dev Cell 5:113–125.
31. Fry AM, Mayor T, Meraldi P, et al. (1998a) C-Nap1, a novel centrosomal coiled-coil protein and candidate substrate of the cell cycle-regulated protein kinase Nek2. J Cell Biol 141:1563–1574.
32. Fry AM, Meraldi P, Nigg EA (1998b) A centrosomal function for the human Nek2 protein kinase, a member of the NIMA family of cell cycle regulators. EMBO J 17:470–481.
33. Katayama H, Zhou H, Li Q, et al. (2001) Interaction and feedback regulation between STK15/BTAK/Aurora-A kinase and protein phosphatase 1 through mitotic cell division cycle. J Biol Chem 276:46219–46224.
34. Nakayama K, Nagahama H, Minamishima YA, et al. (2000) Targeted disruption of Skp2 results in accumulation of cyclin E and p27(Kip1), polyploidy and centrosome overduplication. EMBO J 19:2069–2081.

35. Sharita LM, Machida Y, Sankaran S, et al. (2004) BRCA1-dependent ubiquitination of gamma-tubulin regulates centrosome number. Mol Cell Biol 24:8457–8566.
36. Hsu JY, Reimann JD, Sorensen CS, et al. (2002) E2F-dependent accumulation of hEmi1 regulates S phase entry by inhibiting APC(Cdh1). Nat Cell Biol 4:358–366.
37. Meraldi P, Lukas J, Fry AM, et al. (1999) Centrosome duplication in mammalian somatic cells requires E2F and Cdk2-cyclin A. Nat Cell Biol 1:88–93.
38. Ouchi M, Fujiuchi N, Sasai K, et al. (2004) BRCA1 phosphorylation by Aurora-A in the regulation of G2 to M transition. J Biol Chem 279:19643–19648.
39. Littlepage LE, Rudeman JV (2002) Identification of a new APC/C recognition domain, the A box, which is required for the Cdh1-dependent destruction of the kinase Aurora-A during mitotic exit. Genes Dev 16:2272–2285.
40. Tsou MF, Stearns T (2006) Mechanism limiting centrosome duplication to once per cell cycle. Nature 442:947–951.
41. Marumoto T, Honda S, Hara T, et al. (2003) Aurora-A kinase maintains the fidelity of early and late mitotic events in HeLa cells. J Biol Chem 278:51786–51795.
42. Hirota T, Kunitoku N, Sasayama T, et al. (2003) Aurora-A and in interacting activator, the LIM protein Ajuba, are required in mitotic commitment in human cells. Cell 114:585–598.
43. Hannak E, Kirkham M, Hyman AA, et al. (2001) Aurora-A kinase is required for centrosome maturation in *Caenorhabditis elegans*. J Cell Biol 155:1109–1116.
44. Roghi C, Geit R, Uzebekov R, et al. (1998) The Xenopus protein kinase pEg2 associates with the centrosome in a cell cycle-dependent manner, binds to the spindle microtubules and is involoved in bipolar mitotic spindle assembly. J Cell Sci 111:557–572.
45. Glover DM, Leibowitz MH, McLean DA, et al. (1995) Mutations in aurora prevent centrosome separation leading to the formation of monopolar spindles. Cell 81:95–105.
46. Geit R, McLean D, Descamps S, et al. (2002) *Drosophila* Aurora A kinase is required to localize D-TACC to centrosomes and to regulate astral microtubules. J Cell Biol 156:437–451.
47. Dutertre S, Cazales M, Quaranta M, et al. (2004) Phosphorylation of CDC25B by Aurora-A at the centrosome contributes to the G2-M transition. J Cell Sci 117:2523–2531.
48. Pugacheva EN, Golemis EA (2005) The focal adhesion scaffolding protein HEF1 regulates activation of the Aurora-A and Nek2 kinsases at the centrosome. Nat Cell Biol 7:937–946.
49. Zhao ZS, Lim JP, Ng YW, et al. (2005) The GIT associated kinase PAK target to the centrosome and regulates Aurora-A. Mol Cell 20:237–249.
50. Hutterer A, Berdnik D, Weitz-Peitz F, et al. (2006) Mitotic activation of the kinase Aurora-A requires its binding partner Bora. Dev Cell 11:147–157.
51. Tsai MY, Wiese C, Cao K, et al. (2003) A Ran signaling pathway mediated by the mediated by the mitotic kinase Aurora A in spindle assembly. Nat Cell Biol 5:242–248.
52. Kufer TA, Sillje HH, Korner R, et al. (2002) Human TPX2 is required for targeting Aurora-A kinase to the spindle. J Cell Biol 158:617–623.
53. Koffa MD, Casanova CM, Santarella R, et al. (2006) HURP (Hepatocarcinoma-Upregulated) is part of a Ran-dependent complex involved in spindle formation. Curr Biol 16:743–754.
54. Sillje HH, Nagel S, Komer R, et al. (2006) HURP is a Ran-importin beta-regulated protein that stabilizes kinetochore microtubules in the vicinity of chromosomes. Curr Biol 16:731–742.
55. Kunitoku N, Sasayama T, Marumoto T, et al. (2003) CENP-A phosphorylation by Aurora-A in prophase is required for enrichment of Aurora-B at inner centromeres and for kinetochore function. Dev Cell 5:853–864.
56. Katayama H, Sasai K, Kawai H, et al. (2004) Phosphorylation by aurora kinase A induces Mdm2-mediated destabilization and inhibition of p53. Nat Genet 36:55–62.
57. Furukawa T, Kanai N, Shiwaker HO (2006) AURKA is one of the downstream targets of MAPK1/ERK2 in pancreatic cancer. Oncogene Adv (online publication).

9
Centrosome-Mediated Chromosomal Instability and Steroid Hormones as Co factors in Human Papillomavirus-Associated Cervical Carcinogenesis: Small Viruses Help to Answer Big Questions

Anette Duensing and Stefan Duensing

Human Papillomaviruses and Cervical Cancer

Human papillomaviruses (HPVs) are small DNA tumor viruses that contain a double-stranded genome of only approximately 8,000 base pairs. HPVs infect epithelial tissue and cause benign lesions such as common warts but also malignant and premalignant lesions of the anogenital tract and the oropharynx. Over 200 HPV types have been identified so far. The mucosa-associated high-risk HPV types such as HPV-16 or HPV-18 are intimately associated with squamous cell carcinomas (SCCs) of the uterine cervix, the second-most common cause of cancer-related deaths in women worldwide. High-risk HPVs are also involved in SCCs of the anus, penis, vulva, and a subset of oropharyngeal carcinomas (1). Despite the implementation of screening programs and the successful introduction of prophylactic vaccines (2), HPV-associated diseases remain a global health problem, mostly in medically underserved populations and patients with additional risk factors including immunosuppression or inherited cancer susceptibility syndromes.

Over 99% of cervical carcinomas contain high-risk HPV DNA integrated into host cell chromosomes (3). Viral integration is usually a late event after infection and leads to the deletion of large parts of the viral DNA. The remaining integrated viral DNA usually encodes the two viral oncoproteins, E6 and E7, which are consistently overexpressed in cervical carcinomas. High-risk HPV E6 and E7 fulfill crucial functions during the productive viral life cycle. HPVs gain access to basal or suprabasal cells of epithelial tissues through microtraumas. However, keratinocytes undergo a differentiation program that is accompanied by a permanent cell cycle arrest. High-copy viral genome amplification, capsid protein synthesis, and virion assembly occur in the upper, differentiated epithelial layers. Since HPVs do not encode DNA polymerases, the virus has evolved to disrupt critical host cell tumor suppressor pathways thereby creating a replication competent, S-phase like state in differentiated host keratinocytes (4). This strategy enables the replication of viral DNA by host cell replication enzymes. The viral oncoproteins E6 and E7 are pivotal in facilitating this process (5).

J.J. Li et al. (eds.), *Hormonal Carcinogenesis V.*

The High-Risk HPV E6 Oncoprotein

High-risk HPV E6 oncoproteins inactivate the p53 tumor suppressor protein by targeting *p53* for proteasomal degradation. This function is achieved by reprogramming a host cell HECT (homology to E6 C terminus) domain ubiquitin ligase, E6-AP (6). The C terminus of high-risk HPV E6 proteins has additional functions important for tumor formation (7) and contains a PDZ binding domain that mediates interactions with host cellular proteins involved in intercellular contact and cell signaling such as hDLG and others (5). The high-risk HPV E6 oncoprotein was also found to activate telomerase (8).

The High-Risk HPV E7 Oncoprotein

High-risk HPV E7 is a short-lived small, 98 amino acid phosphoprotein that has unknown enzymatic activity. Its functions are mediated by protein–protein interactions that include binding and degradation of the retinoblastoma tumor suppressor protein (pRB). In addition, high-risk HPV E7 oncoproteins bind and degrade the pRB family members p107 and p130 (5). These activities are mediated through a LXCXE motif as well as additional sequences that are conserved with the adenovirus E1A protein and SV40 T antigen (9). Besides inactivation of pRB family members, the high-risk HPV E7 oncoprotein targets additional cell cycle regulatory proteins including the cdk inhibitor $p21^{Cip1}$ (10, 11). High-risk HPV E7 directly interacts with cyclin A and indirectly with cyclin E through p107. High-risk HPV E7 can bind to histone deacetylases (HDACs) 1 and 2 independently of pRB through the Mi2 protein (4). High-risk HPV E7 transactivates the cdc25A promoter thereby accelerating S phase entry (12). The HPV E7 oncoprotein has been shown to cause abnormal cyclin B/cdk1 activity (13), and HPV oncoprotein expressing cells have an altered expression pattern of additional mitotic kinases such polo-like kinase 1 (Plk1), Aurora A and Mps1. These findings further underscore that both the G1/S and the G2/M transition are subverted in high-risk HPV E7-expressing cells. In addition to the aforementioned proteins, HPV E7 oncoproteins interact with several other host cellular proteins (14).

In contrast, E6 and E7 proteins encoded by low-risk HPVs such as HPV-6 or HPV-11 do not interact with the targets of high-risk HPV-encoded oncoproteins or do so much less efficiently.

Cofactors for Malignant Progression: Genomic Instability and Steroid Hormones

Continued expression of high-risk HPV E6 and E7 is required for the transformed phenotype and expression of the two oncoproteins in rodent cells leads to transformation (5). In primary human cells, high-risk HPV E6 and E7 cause an extension of the life span and occasionally immortalization. Malignant transformation, however, is generally a rare event, which has led to the concept that additional factors are necessary for malignant conversion (5).

Genomic Instability. The ability of HPV oncoproteins to destabilize the host cell genome and the subsequent accumulation of additional genetic alterations is quite likely to contribute importantly to malignant conversion and tumor progression (15). High-risk HPV-associated cervical carcinomas are characterized by numerical and structural chromosomal abnormalities. Aberrations in chromosome numbers (aneuploidy) are the most frequent cytogenetic alterations in solid tumors including cervical cancer and can be detected already in premalignant lesions. Aneuploidy is strongly associated with multipolar mitoses, a long-known hallmark of high-risk HPV-associated cervical neoplasms. Such mitotic aberrations are commonly caused by supernumerary numbers of mitotic spindle poles (16). Spindle poles are formed by centrosomes, small cytoplasmic organelles that serve as major microtubule organizing centers in most animal and human cells (17). In a normal cell, a single centrosome duplicates precisely once prior to mitosis in a process that is poorly understood. The HPV-16 E7 oncoprotein was found to rapidly stimulate the over-duplication of centrosomes (see later). In contrast, centrosomes were found to accumulate in HPV-16 E6-expressing cells, most likely through cytokinesis defects and not directly by disruption of the centrosome duplication cycle per se. Besides multipolar mitoses, several other aberrations are frequently detected in HPV oncoprotein-expressing cells including premature chromosome segregation and anaphase bridges (18). Furthermore, HPV oncoproteins have been found to disrupt mitotic checkpoint control (19, 20), and, as expected, tetra- or polyploidy are frequent findings in HPV oncoprotein-expressing cell populations (21).

Steroid Hormones. There is compelling evidence that steroid hormone signaling are additional important cofactors for HPV-associated cervical carcinogenesis (Fig. 1). The cervical transformation zone is not only the HPV predilection site for infection, but it is also uniquely sensitive to estrogen (22). Chronic estrogen exposure can lead to enhanced proliferation and epithelial hyperplasia, which may promote HPV-associated carcinogenesis through unscheduled cell cycle entry and replication stress (23). Importantly, estrogen has been shown to mediate upregulation of Aurora A, a centrosomal kinase that can provoke abnormal centrosome numbers and genomic instability (24–26). Epidemiological studies have implicated the use of oral contraceptives, in particular long-term use, as a risk factor for cervical SCCs as well as adenocarcinomas of the cervix (27). Cervical carcinoma cells express estrogen receptors (28) and is has been proposed that steroid hormones increase viral transcription through steroid response elements in the upstream regulatory region (URR) (29). In line with this notion, there are reports showing increased viral transcription following estrogen treatment (30). Conversely, HPV oncoproteins may also modulate steroid hormone signaling. A recent report by Baldwin et al. shows that HPV-16 E7 associates with steroid receptor coactivator 1 (SRC-1) and can relocalize SRC-1 to the cytoplasm thereby dysregulating SRC-1-mediated transcription (31). Persuasive evidence for cooperative effects between estrogen and HPV oncoproteins stems from a study by Arbeit et al. (32). In this

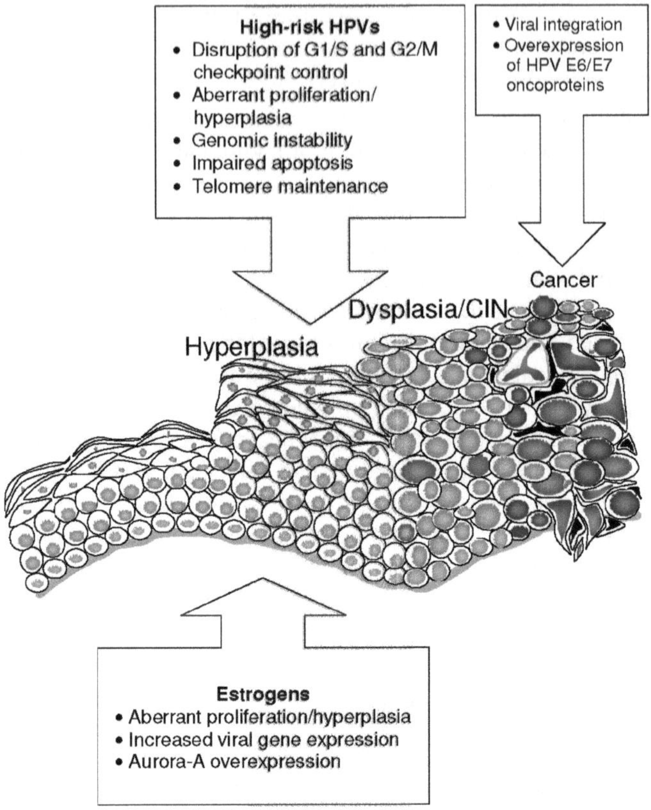

Fig. 1 Cervical carcinogenesis is a multistep process. Molecular, epidemiological and in vivo data strongly suggest that high-risk HPV types and estrogens cooperate to stimulate cancer development and progression

report, transgenic mice were generated that expressed the entire HPV-16 early region under control of the human keratin-14 (K14) promoter. These animals did not develop any reproductive tract tumors; however, chronic estrogen exposure resulted in multistage neoplastic progression in the squamous epithelia of cervix and vulva in 100% of the mice (32). In a follow-up study, it was shown that the carcinogenic functions of HPV-16 in low-dose estrogen-treated animals were mainly attributable to HPV-16 E7. Transgenic mice expressing HPV-16 E6 individually did not show any cervical carcinomas, whereas 80% of animals expressing HPV-16 E7 developed cancer (33). However, HPV-16 E6 cooperated with HPV-16 E7 and led to a further increase of invasive carcinomas.

When centrosome aberrations were determined in premalignant and malignant cervical lesions from these mice, it became obvious that besides a general increase of the level of abnormalities with severity of the lesions, tumors in HPV-16 E6- or HPV-16 E7-transgenic animals contained a comparable proportion of cells with centrosome aberrations (33). These data are in line with several other findings that support a role of HPV-16 E7 as a driving force for centrosome-mediated cell division errors, chromosomal instability, and carcinogenic progression.

The High-Risk HPV E7 Oncoprotein Stimulates Centriole Overduplication

As suggested by findings in transgenic mice and careful analysis of human tumors (34), abnormal centrosome numbers may differ substantially in their functional relevance for genomic instability (35, 36). In HPV-16 E7 oncoprotein-expressing cells, numerical centrosome aberrations developed rapidly (within 24–48 h) and before genomic instability became evident (37). Moreover, HPV-16 E7 was found to trigger the formation of abnormal numbers of centrioles, the core forming units of a centrosome in the presence of a single maternal centriole (38). In striking contrast, HPV-16 E6-associated centrosome abnormalities were not detected after transient expression and developed preferentially in cells that showed a severely altered nuclear morphology with multinucleation and micronuclei following prolonged expression of the HPV E6 oncoprotein (16). These phenotypical changes are typically present in cells that are genomically unstable and/or have failed to complete cytokinesis. It is, therefore, conceivable that centrosome aberrations in HPV-16 E6-expressing cells develop as a consequence of chromosomal instability and not as a direct cause (36). Clearly, simultaneous expression of HPV-16 E6 and E7 has cooperative effects (16), which is in agreement with animal studies discussed earlier (33).

The ability of HPV-16 E7 to trigger a bona fide centriole overduplication was dependent on the LXCXE core pRB-binding motif, which has also been implicated in the inactivation of the cdk inhibitor $p21^{Cip1}$ (39). In keeping with these data are findings showing that HPV-16 E7-induced centriole overduplication requires cdk2 activity whereas normal duplication does not (40). This does not rule out the possibility that cdk2 has functions during normal centriole duplication that are compensated by other kinases when cdk2 is absent. Nonetheless, these results show that HPV-16 E7-induced centriole overduplication differs from normal centriole duplication to an extent that allows to specifically target this process with small molecules such as the cdk inhibitor indirubin (41).

Is Centriole Overduplication Within a Single Cell Division Cycle Possible?

Centriole duplication is tightly regulated in normal cells and noncancerous tissues usually containing only a few cells with supernumerary centrosomes (our own unpublished results). Current models that have been proposed to explain this regulation focus mostly on cyclin/cdk2 complexes and their regulation by cdk inhibitors (42). Impaired G1/S checkpoint control promotes centriole amplification, for example, when cells are exposed to genotoxic stress (43). In addition, an intrinsic block to reduplication has been suggested that is established by the inability of centrioles to reproduce in an "engaged" state, i.e., when a daughter centriole is present at a maternal centriole (44). The central idea of all current models is that centriole reproduction occurs once and only once during each cell division cycle and that each duplication event gives rise to only one daughter centriole (45). These premises do not readily explain the rapid induction of abnormal centriole numbers by HPV-16 E7 within approximately a single cell division cycle. However, there are several possibilities that may help to reconcile these results. It might be possible that a maternal centriole gives rise to more than one daughter cell during a single round of duplication. Using a marker for mature centrioles, Cep170, it has been shown that HPV-16 E7 can induce multiple immature centrioles in the presence of a single mature centriole. The daughter centrioles had approximately the same size and were frequently found surrounding the maternal centriole (38). Although these findings are suggestive for a simultaneous formation of more than one daughter, more conclusive evidence is needed to corroborate this intriguing possibility. Centrioles in *Drosophila melanogaster* wing disc cells have been shown to give rise to more that one daughter centriole at the same time (46). In addition, daughter centrioles in *Drosophila* cells were found to nucleate additional daughter centrioles without evidence for prior maturation. Whether these mechanisms contribute to the formation of supernumerary centrioles in human tumor cells warrants further testing.

Lessons Learned from HPV Oncoproteins: Toward a Better Mechanistic Understanding of Genomic Instability in Cancer

Several conclusions can be drawn from studies using the HPV-16 E7 oncoprotein as a unique tool that disrupts G1/S checkpoint control at multiple levels: (1) Disruption of cell cycle checkpoint control at the G1/S boundary does not only lead to increased proliferation, but at the same time, instigates cellular alterations that can set the stage for genomic instability and malignant progression; (2) Genomic instability is quite likely to require a set of oncogenic insults including those that drive genomic instability and those that allow genomically unstable cells to survive and proliferate; (3) Abnormal centrosome aberrations per se should not be used as markers for genomic instability since they can occur without ongoing genomic instability (47) and in cells that have an uncertain proliferative potential. However, certain oncogenic stimuli including the HPV-16 E7 oncoprotein and G1/S checkpoint disruption (43) can drive centriole overduplication. The impact of centrosome anomalies generated under such conditions on genomic instability in tumors warrants further examination. Lastly, the molecular requirements for normal and abnormal centriole/centriole duplication differ, which may be exploited for cancer prevention and therapy using small molecule inhibitors that block aberrant centriole duplication but leave normal cell cycle and centriole duplication unaffected.

These findings raise important questions for future studies. How exactly does HPV-16 E7 stimulate a rapid centriole overduplication and which are the precise molecular mechanisms? How frequent is tumor centriole overduplication and what is the impact on cell division errors and genomic integrity? And finally: Does targeting aberrant centriole duplication in premalignant lesions prevent outgrowth of genomically unstable subclones, and ultimately malignant progression? Addressing these problems is quite likely to yield exiting insights into centriole duplication errors and cancer development in general.

Acknowledgments Work in the authors' laboratory is supported by the National Institutes of Health (National Cancer Institute grant R01CA112598) and the Susan G. Komen Breast Cancer Foundation.

References

1. Zur Hausen, H (2002) Papillomaviruses and cancer: from basic studies to clinical application. Nat Rev Cancer 2:342–50.
2. Koutsky LA, Ault KA, Wheeler CM, et al. (2002) A controlled trial of a human papillomavirus type 16 vaccine. N Engl J Med 347:1645–51.
3. Walboomers JM, Jacobs MV, Manos MM, et al. (1999) Human papillomavirus is a necessary cause of invasive cervical cancer worldwide. J Pathol 189:12–9.
4. Longworth MS, Laimins LA (2004) Pathogenesis of human papillomaviruses in differentiating epithelia. Microbiol Mol Biol Rev 68:362–72.
5. Munger K, Howley PM (2002) Human papillomavirus immortalization and transformation functions. Virus Res 89:213–28.
6. Scheffner M, Huibregtse JM, Vierstra RD, Howley PM (1993) The HPV-16 E6 and E6-AP complex functions as a ubiquitin-protein ligase in the ubiquitination of p53. Cell 75:495–505.
7. Lambert PF, Pan H, Pitot HC, et al. (1993) Epidermal cancer associated with expression of human papillomavirus type 16 E6 and E7 oncogenes in the skin of transgenic mice. Proc Natl Acad Sci USA 90:5583–7.
8. Veldman T, Horikawa I, Barrett JC, et al. (2001) Transcriptional activation of the telomerase *hTERT* gene by human papillomavirus type 16 E6 oncoprotein. J Virol 75:4467–72.
9. Phelps WC, Munger K, Yee CL, et al. (1992) Structure-function analysis of the human papillomavirus type 16 E7 oncoprotein. J Virol 66:2418–27.
10. Jones DL, Alani RM, Munger K (1997) The human papillomavirus E7 oncoprotein can uncouple cellular differentiation and proliferation in human keratinocytes by abrogating $p21^{Cip1}$-mediated inhibition of cdk2. Genes Dev 11:2101–11.
11. Funk JO, Waga S, Harry JB, et al. (1997) Inhibition of CDK activity and PCNA-dependent DNA replication by p21 is blocked by interaction with the HPV-16 E7 oncoprotein. Genes Dev 11:2090–100.
12. Nguyen DX, Westbrook TF, McCance DJ (2002) Human papillomavirus type 16 E7 maintains elevated levels of the cdc25A tyrosine phosphatase during deregulation of cell cycle arrest. J Virol 76:619–32.
13. Steinmann KE, Pei XF, Stoppler H, et al. (1994) Elevated expression and activity of mitotic regulatory proteins in human papillomavirus-immortalized keratinocytes. Oncogene 9:387–94.

14. Munger K, Baldwin A, Edwards KM, et al. (2004) Mechanisms of human papillomavirus-induced oncogenesis. J Virol 78:11451–60.
15. White AE, Livanos EM, Tlsty TD (1994) Differential disruption of genomic integrity and cell cycle regulation in normal human fibroblasts by the HPV oncoproteins. Genes Dev 8:666–77.
16. Duensing S, Lee LY, Duensing A, et al. (2000) The human papillomavirus type 16 E6 and E7 oncoproteins cooperate to induce mitotic defects and genomic instability by uncoupling centrosome duplication from the cell division cycle. Proc Natl Acad Sci USA 97:10002–7.
17. Nigg EA (2004) Centrosomes in Development and Disease. Wiley-VCH, Weinheim, Germany.
18. Plug-Demaggio AW, McDougall JK (2002) The human papillomavirus type 16 E6 oncogene induces premature mitotic chromosome segregation. Oncogene 21:7507–13.
19. Thomas JT, Laimins LA (1998) Human papillomavirus oncoproteins E6 and E7 independently abrogate the mitotic spindle checkpoint. J Virol 72:1131–7.
20. Thompson DA, Belinsky G, Chang TH, et al. (1997) The human papillomavirus-16 E6 oncoprotein decreases the vigilance of mitotic checkpoints. Oncogene 15:3025–35.
21. Southern SA, Noya F, Meyers C, et al. (2001) Tetrasomy is induced by human papillomavirus type 18 *E7* gene expression in keratinocyte raft cultures. Cancer Res 61:4858–63.
22. Elson DA, Riley RR, Lacey A, et al. (2000) Sensitivity of the cervical transformation zone to estrogen-induced squamous carcinogenesis. Cancer Res 60:1267–75.
23. Bartkova J, Horejsi Z, Koed K, et al. (2005) DNA damage response as a candidate anti-cancer barrier in early human tumorigenesis. Nature 434:864–70.
24. Li JJ, Weroha SJ, Lingle WL, et al. (2004) Estrogen mediates Aurora-A overexpression, centrosome amplification, chromosomal instability, and breast cancer in female ACI rats. Proc Natl Acad Sci USA 101:18123–8.
25. Goepfert TM, Adigun YE, Zhong L, et al. (2002) Centrosome amplification and overexpression of aurora A are early events in rat mammary carcinogenesis. Cancer Res 62:4115–22.
26. Zhou H, Kuang J, Zhong L, et al. (1998) Tumour amplified kinase STK15/BTAK induces centrosome amplification, aneuploidy and transformation. Nat Genet 20:189–93.
27. Moodley M, Moodley J, Chetty R, et al. (2003) The role of steroid contraceptive hormones in the pathogenesis of invasive cervical cancer: a review. Int J Gynecol Cancer 13:103–10.
28. Monsonego J, Magdelenat H, Catalan F, et al. (1991) Estrogen and progesterone receptors in cervical human papillomavirus related lesions. Int J Cancer 48:533–9.
29. von Knebel Doeberitz M, Bauknecht T, Bartsch D, et al. (1991) Influence of chromosomal integration on glucocorticoid-regulated transcription of growth-stimulating papillomavirus genes *E6* and *E7* in cervical carcinoma cells. Proc Natl Acad Sci USA 88:1411–15.
30. Mitrani-Rosenbaum S, Tsvieli R, Tur-Kaspa R (1989) Oestrogen stimulates differential transcription of human papillomavirus type 16 in SiHa cervical carcinoma cells. J Gen Virol 70:2227–32.
31. Baldwin A, Huh KW, Munger K (2006) Human papillomavirus E7 oncoprotein dysregulates steroid receptor coactivator 1 localization and function. J Virol 80:6669–77.
32. Arbeit JM, Howley PM, Hanahan D (1996) Chronic estrogen-induced cervical and vaginal squamous carcinogenesis in human papillomavirus type 16 transgenic mice. Proc Natl Acad Sci USA 93:2930–45.
33. Riley RR, Duensing S, Brake T, et al. (2003) Dissection of human papillomavirus E6 and E7 function in transgenic mouse models of cervical carcinogenesis. Cancer Res 63:4862–71.
34. Lingle WL, Barrett SL, Negron VC, et al. (2002) Centrosome amplification drives chromosomal instability in breast tumor development. Proc Natl Acad Sci USA 99:1978–83.
35. Duensing S (2005) A tentative classification of centrosome abnormalities in cancer. Cell Biol Int 29:352–9.
36. Nigg EA (2002) Centrosome aberrations: cause or consequence of cancer progression? Nature Rev Cancer 2:1–11.

37. Duensing S, Duensing A, Crum CP, et al. (2001) Human papillomavirus type 16 E7 oncoprotein-induced abnormal centrosome synthesis is an early event in the evolving malignant phenotype. Cancer Res 61:2356–60.
38. Guarguaglini G, Duncan PI, Stierhof YD et al. (2005) The Forkhead-associated domain protein Cep170 interacts with polo-like kinase 1 and serves as a marker for mature centrioles. Mol Biol Cell 16:1095–107.
39. Duensing S, Münger K (2003) Human papillomavirus type 16 E7 oncoprotein can induce abnormal centrosome duplication through a mechanism independent of inactivation of retinoblastoma protein family members. J Virol 77:12331–35.
40. Duensing A, Liu Y, Tseng M, et al. (2006) Cyclin-dependent kinase 2 is dispensable for normal centrosome duplication but required for oncogene-induced centrosome overduplication. Oncogene 25:2943–9.
41. Duensing S, Duensing A, Lee DC, et al. (2004) Cyclin-dependent kinase inhibitor indirubin-3'-oxime selectively inhibits human papillomavirus type 16 E7-induced numerical centrosome anomalies. Oncogene 23:8206–15.
42. Sluder G (2004) Centrosome duplication and its regulation in the higher animal cell. In: Nigg EA (Ed) Centrosomes in Development and Disease. Wiley-VCH, Weinheim, Germany, pp. 167–89.
43. D'Assoro AB, Busby R, Suino K, et al. (2004) Genotoxic stress leads to centrosome amplification in breast cancer cell lines that have an inactive G1/S cell cycle checkpoint. Oncogene 23:4068–75.
44. Tsou MF, Stearns T (2006) Mechanism limiting centrosome duplication to once per cell cycle. Nature 442:947–51.
45. Tsou MF, Stearns T (2006) Controlling centrosome number: licenses and blocks. Curr Opin Cell Biol 18:74–8.
46. Vidwans SJ, Wong ML, O'Farrell PH (2003) Anomalous centriole configurations are detected in *Drosophila* wing disc cells upon cdk1 inactivation. J Cell Sci 116:137–43.
47. Duensing S, Lee BH, Dal Cin P, et al. (2003) Excessive centrosome abnormalities without ongoing numerical chromosome instability in a Burkitt's lymphoma. Mol Cancer 2:30.

Part 3
New Developments in Steroid Receptor Interactions

10
Regulation of Hormone Signaling by Nuclear Receptor Interacting Proteins

Vanessa Duong, Patrick Augereau, Eric Badia, Stéphan Jalaguier, and Vincent Cavailles

Introduction

Control of Nuclear Receptor Transactivation. Nuclear receptors are ligand-activated transcription factors that subsequently bind to specific responsive elements located in the regulatory region of target gene promoters (1). They stimulate transcription using both a constitutive amino-terminal and a ligand-dependent carboxyl-terminal activation function (AF1 and AF2, respectively), the latter being associated with the ligand-binding domain. These activation functions act independently or synergistically depending on the cell type and promoter context, by recruiting a number of cofactors (2).

Role of Coregulators in Hormone-Related Cancers. Usually, transcription cofactors function as part of multimolecular complexes, acting either by stabilisation of the basal transcription machinery or by chromatin remodeling, which implicates various enzymatic activities such as histone acetyltransferases and deacetylases (3). Different types of posttranslational modifications (phosphorylation, acetylation, etc.) also regulate their activity, localization, or interaction with receptors (4). The physiological roles of these cofactors begin to be investigated by invalidation of the corresponding gene in mice. They are also implicated in several pathologies such as hormone insensitivity syndromes or hormone-dependent cancers (5). Clinical studies have investigated the potential interest of these molecules as prognosis or diagnosis markers. They have also described quantitative or qualitative alterations of these genes in tumors. Finally, several approaches are underway to define whether steroid receptor transcription cofactors may be valuable targets in cancer treatment (modulation of their binding to receptors, modification of their expression, or inhibition of their enzymatic activity).

Negative Regulation of Nuclear Receptor Activity by RIP140

RIP140 is a widely expressed protein of 1158 residues, which was isolated through its recruitment by estrogen receptor (ERα) AF2 in the presence of ligand (6). Subsequently, RIP140 was shown to interact with many nuclear receptors such as ERα, thyroid, retinoic acid (7), androgen (AR) (8), vitamin D (9), peroxisome

J.J. Li et al. (eds.), *Hormonal Carcinogenesis V.*

proliferator-activated receptor α and liver-X-receptor alpha (10), glucocorticoid (11), and orphan receptors SF1 and DAX-1 (12). More recently, gene knockout in mice indicated that it is an essential protein for female fertility and energy homeostasis (13, 14).

Effects on Nuclear Receptor Activity. Our recent work has highlighted both the interaction and transrepression activity of RIP140 on different members of the nuclear receptor superfamily, i.e., estrogen receptor-related receptors (ERRs) (15) and AR (16).

Estrogen-Receptor-Related Receptors. ERRα, ERRβ, and ERRγ exhibit strong sequence similarity to ERα (17, 18) and interfere with estrogen signaling (19) in part through binding to the same DNA-binding elements, namely the estrogen receptor related-element (ERRE) for ERR-response element or the classical estrogen response element (ERE) (20). As expected from results obtained with other nuclear receptors, we have shown that different regions of RIP140 interact in vitro with the three ERRs (15). Indeed, we have observed a significant binding of radiolabeled ERRα, ERRβ, and ERRγ with chimaeric GST-RIP140 proteins encompassing the amino-terminal (residues 27–439), the central (residues 429–582), or the carboxyl-terminal region of the molecule (residues 683–1158). In transient transfection experiments, RIP140 inhibits ERRα, ERRβ, and ERRγ activity on ERE- or ERRE-containing reporter constructs (artificial reporter plasmids and natural promoters harboring these binding sites such as pS2, ERRα, or osteopontin) (15).

Androgen Receptor. The AR is a cytoplasmic protein that undergoes nuclear translocation upon hormone binding (21). Recently, we have reported that the interaction between AR and RIP140, and delineated the interaction between the ligand-binding domain (LBD) of AR, and several domains of RIP140, which contain LxxLL motifs (16). In R1881-treated LNCaP prostate cancer cells, RIP140 is recruited to AR-target genes such as the prostate specific antigen (*PSA*) gene and completely relocalized from small nuclear foci to a diffuse pattern. Interestingly, the antagonist-bound AR, which also translocates to the nucleus, is not able to induce such a redistribution of RIP140. Our data indicate that RIP140 is a *bona fide* AR repressor since it inhibits AR-mediated transactivation and reverses the TIF2-induced overactivation of AR. Moreover, in mouse embryo fibroblasts lacking the *RIP140* gene, AR activity is significantly increased when compared with the wild type counterpart cells (22).

Mechanisms of Transrepression. The strong transcriptional repressive activity of RIP140, which was initially attributed to competition with coactivator binding to nuclear receptors (23), also involves an intrinsic inhibitory effect because of several domains that recruit various repressive effectors. We have identified two evolutionary conserved CtBP (carboxyl-terminal binding proteins) binding motifs (PIDLS and PINLS motifs located, respectively, between residues 440–444 and 565–569), which explain partly the repressive action of RIP140 (24). RIP140 also interacts with class I and II histone deacetylases (HDAC). We show that RIP140 directly binds HDACs through a sequence comprised between residues 115 and 199, and using fluorescent chimaeric proteins, we have evidenced its colocalization with HDAC5 in intact cells (24). However, our results indicate that (1) the repressive activity of RIP140 could be partially affected by inhibition of HDAC enzymatic activity and (2) two additional domains in the C-terminal region of the protein support strong repressive activity but do not require HDAC activity or CtBPs (24), thus suggesting that the global repressive activity of RIP140 may depend on the interplay among several negative regulatory modules.

Effect on Hypoxia-Related-Element (HRE)-Independent Transactivation. RIP140 also regulates HRE-independent transactivation by nuclear receptors that involves their indirect recruitment on target genes through protein–protein interaction in particular with AP1 or Sp1 factors (25). Previously, we reported the inhibition by RIP140 of 17β-estradiol (E_2)-induced AP-1-dependent transcription of ERα (26). RIP140 antagonizes the stimulatory effect of GRIP1 and competes for binding to c-jun and ERα both in vitro and in intact cells. More recently, we have demonstrated that overexpression of RIP140 strongly increases ERRα- and ERRγ-mediated transactivation via Sp1-response sites, on both isolated sites and natural promoters (15). This positive regulation exerted by RIP140 involves HDAC either directly or indirectly as suggested by overexpression of HDAC1 or treatment with TSA.

Regulatory Loops Involving RIP140. Cloning of the *RIP140* gene and analysis of transcriptional regulatory mechanisms revealed that it is involved in several feedback loops. RIP140 mRNA levels are rapidly and directly increased by E_2 in MCF-7 human breast cancer cells (27). This estrogenic regulation, which is preferentially mediated by ERα and not restricted to mammary cancer cells, involves a consensus ERE that allows efficient binding of ERα, both in vitro and in intact cells (28). Interestingly, the regulatory feed-back loop that we have demonstrated for ERα also exists for several other nuclear receptors, such as retinoid (29) and ARs (16), or for the dioxin receptor (AhR) (28, 30).

Effect of MDM2 on ER Signaling

Several studies have shown that binding of E_2 to ERα significantly decreases its stability. This shorter half-life in the presence of hormone appears to implicate the ubiquitin/proteasome pathway since ERα is ubiquitinated (31) and its ligand-dependent down-regulation is blocked by proteasome inhibitors (32–34).

Previous studies have suggested that ubiquitin-conjugating enzymes or ATPase subunits of the proteasome complex bind nuclear receptors and modulate their functions. In a recent study, we have shown that the MDM2 oncoprotein is also involved in the ligand-dependent decrease of ERα stability (35).

Regulation of ERα Expression and Activity. The *mdm2* oncogene is overexpressed in a wide variety of human cancers (36) and its role in tumorigenesis is linked to its ability to act as an E3 ubiquitin-ligase, which is required for the ubiquitination and proteasome-dependent degradation of several growth regulatory proteins including p53 (37–39).

Our data indicate that *mdm2* regulates ERα expression as a ternary complex with *p53* (35). Using a modified mammalian two-hybrid system and an in vitro protein–protein interaction assay, we have shown that *mdm2* coexists with ERα and p53 within the same protein complex in intact cells. Using transient transfection into p53/mdm2–/– cells, we have demonstrated that *p53* and *mdm2* are required for ligand-dependent ERα turn-over. By chase experiments using cycloheximide, we have found that *mdm2* overexpression decreases the apparent stability of the ERα protein, thus confirming its role in the posttranslational regulation of ERα expression. Finally, a mutant of *mdm2* (mdm2ΔRING) deleted in the carboxyl-terminal part of the protein that contains the RING domain required for its ubiquitin-ligase activity (40) still interacts with ERα in GST-pull down experiment but does not decrease ERα accumulation when compared with the effect of its wild-type counterpart suggesting that the E3 ubiquitin-ligase activity of *mdm2* is directly involved in ERα degradation.

Effect of Stress Inducing Agents. Since cellular stress results in an increased accumulation of *p53* mainly due to the inability of *mdm2* to degrade the protein, we have tested the effect of various stress-inducing agents (which stabilize *p53*) on ERα turn-over (35). We show that treatments that increase *p53* levels in MCF-7 human breast cancer (such as UV irradiation or treatment with RITA that inhibits the interaction of *p53* with *mdm2*) concomitantly suppress the hormone-dependent down-regulation of ERα. In the case of UV irradiation, our data demonstrate that the effect results from an increase in ERα stability.

Ligand-Dependent Turn-Over and Transactivation. Previous studies proposed that the E_2-dependent decrease of ERα accumulation was required for transcriptional activity of the receptor (41). However, our data provide several lines of evidence showing that the E_2-dependent turnover of the receptor is not necessary for ERE-mediated transactivation (35). Indeed, we have found (using p53/mdm2–/– cells or UV irradiation of MCF-7 cells) that ERα strongly activates transcription in cells where E_2 up-regulates receptor levels thus dissociating the effect of E_2 on ERα degradation and activity.

Conclusion

The data presented herein highlight the complexity of the regulatory mechanisms that control nuclear receptor expression and activity. RIP140 appears as an unconventional transcriptional regulator acting as an anticoactivator. Several negative regulatory modules are involved in transcriptional repression and posttranslational modifications might be key regulatory events controlling this activity. *RIP140* gene expression appears also finely tuned and the different regulatory loops and cross-talks that take place could be of importance in the regulation of breast cancer proliferation by hormones and environmental contaminants. Finally, our unpublished data suggest that RIP140 could be involved in the regulation of other transcription factors that control cell cycle progression and this obviously reinforces the relevance of RIP140 as a key factor in breast carcinogenesis.

ERα turn-over is also under complex regulation and, among other factors; the *mdm2* oncogene plays an important role. It regulates concomitantly the stability of two major proteins in breast cancer, i.e., ERα and *p53*, which are both stabilized upon cellular stress. Although it does not seem to be required for hormone-dependent transcriptional activity, ligand-induced turn-over of ERα could target ER-associated factors to the proteasome and as a consequence indirectly regulate cell proliferation, apoptosis, or invasion.

Acknowledgments We thank all the members of the "Hormone Signaling, Environment and Cancer Laboratory" for their contribution to the work presented in this chapter. We are particularly grateful to A. Castet, S. Carascossa, and P. Balaguer, and we also thank S. Bonnet, M. Fuentes, A. Lucas, and F. Rabenoelina for technical help. These studies were supported by the ''Institut National de la Santé et de la Recherche Médicale", the University of Montpellier I, the "Association pour la Recherche sur le Cancer" (Grant 3494), "The Ligue Régionale Contre le Cancer'' (Grant RAB05002FFA), the "Fondation Lejeune" and the National Research Agency.

References

1. Aranda A, Pascual A (2001) Nuclear hormone receptors and gene expression. Physiol Rev 81:1269–1304.
2. Hall JM, McDonnell DP (2005) Coregulators in nuclear estrogen receptor action: from concept to therapeutic targeting. Mol Interv 5:343–357.
3. Lonard DM, O'Malley BW (2005) Expanding functional diversity of the coactivators. Trends Biochem Sci 30:126–132.
4. Smith CL, O'Malley BW (2004) Coregulator function: a key to understanding tissue specificity of selective receptor modulators. Endocr Rev 25:45–71.
5. Kumar R, Gururaj AE, Vadlamudi RK, et al. (2005) The clinical relevance of steroid hormone receptor corepressors. Clin Cancer Res 11:2822–2831.
6. Cavailles V, Dauvois S, L'Horset F, Lopez G, et al. (1995) Nuclear factor RIP140 modulates transcriptional activation by the estrogen receptor. EMBO J 14:3741–3751.
7. L'Horset F, Dauvois S, Heery DM, et al. (1996) RIP-140 interacts with multiple nuclear receptors by means of two distinct sites. Mol Cell Biol 16:6029–6036.
8. Ikonen T, Palvimo JJ, Janne OA (1997) Interaction between the amino- and carboxyl-terminal regions of the rat androgen receptor modulates transcriptional activity and is influenced by nuclear receptor coactivators. J Biol Chem 272:29821–29828.
9. Masuyama H, Brownfield CM, St-Arnaud R, et al. (1997) Evidence for ligand-dependent intramolecular folding of the AF-2 domain in vitamin D receptor-activated transcription and coactivator interaction. Mol Endocrinol 11:1507–1517.
10. Miyata KS, McCaw SE, Meertens LM, et al. (1998) Receptor-interacting protein 140 interacts with and inhibits transactivation by, peroxisome proliferator-activated receptor alpha and liver-X-receptor alpha [In Process Citation]. Mol Cell Endocrinol 146:69–76.
11. Subramaniam N, Treuter E, Okret S (1999) Receptor interacting protein RIP140 inhibits both positive and negative gene regulation by glucocorticoids. J Biol Chem 274:18121–18127.
12. Sugawara T, Abe S, Sakuragi N, et al. (2001) RIP 140 modulates transcription of the steroidogenic acute regulatory protein gene through interactions with both SF-1 and DAX-1. Endocrinology 142:3570–3577.

13. Leonardsson G, Steel JH, Christian M, et al. (2004) Nuclear receptor corepressor RIP140 regulates fat accumulation. Proc Natl Acad Sci USA 101:8437–8442.
14. White R, Leonardsson G, Rosewell I, et al. (2000) The nuclear receptor co-repressor nrip1 (RIP140) is essential for female fertility. Nat Med 6:1368–1374.
15. Castet A, Herledan A, Bonnet S, et al. (2006) Receptor-interacting protein 140 differentially regulates estrogen receptor-related receptor transactivation depending on target genes. Mol Endocrinol 20:1035–1047.
16. Carascossa S, Gobinet J, Georget V, et al. (2006) Receptor-interacting protein 140 is a repressor of the androgen receptor activity. Mol Endocrinol 20:1506–1518.
17. Giguere V, Yang N, Segui P, et al. (1988) Identification of a new class of steroid hormone receptors. Nature 331:91–94.
18. Horard B, Vanacker JM (2003) Estrogen receptor-related receptors: orphan receptors desperately seeking a ligand. J Mol Endocrinol 31: 349–357.
19. Giguere V (2002) To ERR in the estrogen pathway. Trends Endocrinol Metab 13:220–225.
20. Vanacker JM, Pettersson K, Gustafsson JA, et al. (1999) Transcriptional targets shared by estrogen receptor-related receptors (ERRs) and estrogen receptor (ER) alpha, but not by ERbeta. EMBO J 18:4270–4279.
21. Pratt WB, Toft DO (1997) Steroid receptor interactions with heat shock protein and immunophilin chaperones. Endocr Rev 18:306–360.
22. Christian M, Kiskinis E, Debevec D, et al. (2005) RIP140-targeted repression of gene expression in adipocytes. Mol Cell Biol 25:9383–9391.
23. Treuter E, Albrektsen T, Johansson L, et al. (1998) A regulatory role for RIP140 in nuclear receptor activation. Mol Endocrinol 12:864–881.
24. Castet A, Boulahtouf A, Versini G, et al. (2004) Multiple domains of the receptor-interacting protein 140 contribute to transcription inhibition. Nucleic Acids Res 32:1957–1966.
25. Safe S, Kim K (2004) Nuclear receptor-mediated transactivation through interaction with Sp proteins. Prog Nucleic Acid Res Mol Biol 77:1–36.
26. Teyssier C, Belguise K, Galtier F, et al. (2003) Receptor-interacting protein 140 binds c-jun and inhibits estradiol-induced activator protein-1 activity by reversing glucocorticoid receptor-interacting protein 1 effect. Mol Endocrinol 17:287–299.
27. Thenot S, Charpin M, Bonnet S, et al. (1999) Estrogen receptor cofactors expression in breast and endometrial human cancer cells. Mol Cell Endocrinol 156:85–93.
28. Augereau P, Badia E, Fuentes M, et al. (2006) Transcriptional regulation of the human *NRIP1/RIP140* gene by estrogen is modulated by dioxin signalling. Mol Pharmacol 69:1338–1346.
29. Kerley JS, Olsen SL, Freemantle SJ, et al. (2001) Transcriptional activation of the nuclear receptor corepressor RIP140 by retinoic acid: a potential negative-feedback regulatory mechanism. Biochem Biophys Res Commun 285:969–975.
30. Kumar MB, Tarpey RW, Perdew GH (1999) Differential recruitment of coactivator RIP140 by Ah and estrogen receptors. Absence of a role for LXXLL motifs. J Biol Chem 274:22155–22164.
31. Wijayaratne AL, McDonnell DP (2001) The human estrogen receptor-alpha is a ubiquitinated protein whose stability is affected differentially by agonists, antagonists, and selective estrogen receptor modulators. J Biol Chem 276:35684–35692.
32. Alarid ET, Bakopoulos N, Solodin N (1999) Proteasome-mediated proteolysis of estrogen receptor: a novel component in autologous down-regulation. Mol Endocrinol 13:1522–1534.
33. El Khissiin A, Leclercq G (1999) Implication of proteasome in estrogen receptor degradation. FEBS Lett 448:160–16
34. Nawaz Z, Lonard DM, Dennis AP, et al. (1999) Proteasome-dependent degradation of the human estrogen receptor. Proc Natl Acad Sci USA 96:1858–1862.
35. Duong V, Boulle N, Daujat M, et al. (2007) Differential regulation of estrogen receptor turnover and transactivation by Mdm2 and stress-inducing agents. Cancer Res 67:5513–21.
36. Momand J, Jung D, Wilczynski S, et al. (1998) The *MDM2* gene amplification database. Nucleic Acids Res 26:3453–3459.

37. Honda R, Tanaka H, Yasuda H (1997) Oncoprotein MDM2 is a ubiquitin ligase E3 for tumor suppressor p53. FEBS Lett 420:25–27.
38. Haupt Y, Maya R, Kazaz A, et al. (1997) Mdm2 promotes the rapid degradation of p53. Nature 387:296–299.
39. Michael D, Oren M (2003) The p53-Mdm2 module and the ubiquitin system. Semin Cancer Biol 13:49–58.
40. Honda R, Yasuda H (2000) Activity of MDM2, a ubiquitin ligase, toward p53 or itself is dependent on the RING finger domain of the ligase. Oncogene 19:1473–1476.
41. Lonard DM, Nawaz Z, Smith CL, et al. (2000) The 26S proteasome is required for estrogen receptor-alpha and coactivator turnover and for efficient estrogen receptor-alpha transactivation. Mol Cell 5:939–948.

11
Dynamics of Estrogen Receptor-mediated Transcriptional Activation of Responsive Genes In Vivo: Apprehending Transcription in Four Dimensions

Raphaël Métivier, Guillaume Huet, Rozenn Gallais, Laurence Finot, Fabien Petit, Christophe Tiffoche, Yohann Mérot, Christine LePéron, George Reid, Graziella Penot, Florence Demay, Frank Gannon, Gilles Flouriot, and Gilles Salbert

Introduction

Estrogens, such as 17β-estradiol (E_2), are commonly recognized as pivotal hormones controlling female reproductive physiology (1), but they also exhibit pleiotropic actions in male reproductive development and physiology, bone and lipid metabolisms, and the maintenance of the cardiovascular and neuronal systems (2–4). These effects of E_2 are mainly transduced through specific receptors, the estrogen receptors (ERα and ERβ), although another protein, G protein-coupled receptor (GPR30) has been recently suggested to transduce some of the estrogenic responses (5). Further studies are awaited to identify the exact respective contribution of both pathways to E_2 signaling. ERs are dimeric, intranuclear, ligand-dependent transcription factors belonging to the superfamily of nuclear receptors (NRs) (6). ERs classically recognize defined palindromic target DNA sequences located within the promoter regions of estrogen responsive target genes (7, 8). Upon binding its ligand, ER undergoes drastic conformational changes (8) that generate surfaces that associate with transcriptional cofactors that in turn allow the recruitment to the target promoter and activation of the RNA polymerase II complex (Pol II) (9).

The inducible property of this system, offered by a ligand-mediated regulation of the transcriptome through specific receptors, provided recent studies with a major model for describing the kinetic properties of transcriptional regulation and its cell- and promoter-specificities. In summary, these studies demonstrate that accomplishing transcription is a complex process that requires the integration of several parameters: *cis*-acting factors (DNA sequence), chromatin structure, three-dimensional structures of proteins, *trans*-acting factors (transcriptional activators, associated complexes, and the basal Pol II complex), nucleus organization and time (10).

Cis-Elements: A Challenging Definition of Estrogen-Responsive Sequences

Emerging from early in vitro studies, binding of ERs to DNA in vivo was referred to principally occur on palindromic repetitions of their cognate response element (PuGGTCA) (11) called estrogen-responsive elements (ERE). Recruitment of ERs onto promoters of target genes is affected by the number of these EREs and their spacing. Furthermore, critical nucleotides affect the recognition of ERE by the zinc-finger motifs present within the DNA-binding domain of ERs (12). This paradigm was first challenged by the discovery that the estrogenic sensitivity of many cloned estrogen-responsive promoters does not depend upon EREs but rather upon indirect mobilization of ERs through AP-1 or Sp1 elements (13, 14). The use of bioinformatics, SAGE assays, and RNA hybridization on microarrays identified a number of E_2-responsive genes (15–17). Analysis of the proximal promoter regions of these genes confirms that the occurrence of canonical EREs or functional EREs concerns only a portion of E_2-responsive genes. Recently, novel insights into the mechanisms by which ERs are recruited onto DNA were reached using chromatin immunoprecipitation (ChIP)-on-Chips assays on whole chromosomes (18, 19): within distal promoter regions, the presence of pioneering factors such as FoxA1 facilitates the subsequent interaction of ER with EREs or other ER-recognition sites (19). On some promoters, c-*myc* is also proposed to stabilize the ERα complex onto DNA (20). Therefore, quite surprisingly, 20 years following the characterization of an ERE sequence, little is still known about the exact *cis*-elements involved in the estrogenic sensitivity of promoters. Completion of genome sequences, in combination with large-scale ChIPs arrays, offers now novel opportunities to get important insights into this major aspect of E_2-signaling.

Chromatin Organization and Spatio-Temporal Regulation of Transcription

Differential expression of the eukaryotic genome, as it occurs during cell differentiation or cell response to its microenvironment (i.e., hormones), requires precise modulations of its spatial organization (21, 22) as well as changes within the chromatin structure in itself (23). Indeed, chromatin grants stability and segregation of chromosomes during cell mitosis and meiosis, coordinated regulations of the expression of the genome (24, 25), but also provides a repressive environment for transcription through its basic organization: 146 bp of DNA wrapped around histone octamers including dimers of each of the core histones H2A, H2B, H3, and H4 (26) constitute nucleosomes that are the elemental "pearls" of chromatin.

Overcoming the structural restriction that chromatin has on gene expression is achieved through the modulation of particular marks that signal active/inactive domains. For instance, DNA methylation on cytosines within CpG dinucleotides is an epigenetic mark that impacts gene activity (27): hypomethylated DNA is correlated with active genes, while hypermethylated genes are silent (28). Specific posttranscriptional modifications of histones (acetylation, methylation, ubiquitylation, etc.) also affect nucleosomes condensation and concurrently signal given transcriptional states of chromatin. Most of these modifications are taking place on histone N-terminal tails, and are generated by enzymes such as histone acetyl transferases (HATs), histone deacetylases (HDACs), histone methyl transferases (HMTs), etc. (29). Changes brought by these modifications are affecting the electrostatic charge of nucleosomes and thus histone–histone and DNA–histones interactions. Furthermore, the association of given histone modifications within a chromatin region defines a histone code that signal particular states of chromatin: transcriptionally silent, poised, or active (30, 31). Finally, the plasticity of chromatin is also modulated by ATP-dependent proteins such as SWI/SNF or Mi2 that rearrange the nucleosomal organization along the chromatin fiber (32).

Initial Kinetic Aspects of NRs Transcriptional Regulation

Biochemical, in vitro, studies generated combinatorial models; the first bricks that built our understanding of the kinetics of transcriptional activation through the definition of an ordered sequence of recruitments required to complete and activate the Pol II complex (33, 34). The kinetic nature of NRs-mediated transcription has been initially substantiated by the discovery that these factors interact with many macromolecular complexes, some modulating histone acetylation, methylation, and chromatin remodeling: (1) HATs such as SRC1, CBP/p300; (2) HMTs such as CARM1; (3) nucleosome remodeling complexes such as SWI/SNF. Others such as complexes of the SMCC/TRAP/DRIP/ARC class facilitate the activation of Pol II complex (35). In the absence of ligand, or upon binding of antiestrogens such as 4-hydroxytamoxifen (OHT) in the case of ERα, NRs down-regulate transcription through the recruitment of corepressors exhibiting HDAC activity such as Sin3 and NuRD complexes (36, 37). Binding of agonist ligands induces, through structural rearrangement of the nuclear receptor (NR), the exchange of these corepressors for coactivators, indicating the dynamic nature of NRs-mediated modulation of transcription (38, 39). Ordered, successive, recruitments of these different cofactors by NRs were further suggested by the fact that many proteins of these macromolecular complexes with huge sterical hindrance interact with the same three-dimensional cleft of the surface of the receptor (40, 41). Furthermore, allosteric changes induced within interacting partners, such as ERα and TATA-binding protein (TBP), or between NRs and p300/cAMP-response element-binding protein (CBP), p/CAF and p160 define ordered series of interaction (42, 43). Resulting from the histone code notion, additional allosteric processes integrate chromatin remodeling with transcriptional regulation by NRs: specific histone modifications imposed by one factor sequentially impacts the mobilization of other proteins. This is exemplified by the recruitment of CBP, which is alleviated by methylation of H3 (44). Finally,

DNA sequence in itself allosterically influences the binding of cofactors, or NRs, as shown for ERs (45). Definitely, the dynamic intrinsic property of transcriptional processes was next illustrated by live cell microscopy methods that demonstrated that NRs and cofactors exhibit high mobility within the nucleus (46).

The establishment and maintenance of the pattern of CpGs methylation is ensured by DNA methyltransferases (Dnmt) that can act on hemimethylated CpGs generated during DNA replication, or de novo on unmethylated DNA (47). Localized DNA demethylation is occurring on specific genes upon their reactivation by NRs (48). However, these processes are still poorly understood. Indeed, although described in the literature, no enzymatic machineries able to demethylate methylated CpG are unequivocally characterized (49). Recent data have shown a cleavage of the DNA backbone resulting from a demethylation process initiated by a NR, allowing a DNA repair pathway to act (50). Clearly, additional data are needed to attain further understanding of the kinetics of these mechanisms and to assay their linkage with processes taking place within chromatin during transcriptional regulation.

Transcriptional Cycling of ERα onto Responsive Promoters

Transcriptional Cycles of the pS2/TFF1 Promoter. Number of studies using ChIP assays undertook the effort to study precise kinetics of the mechanisms involved in transcription initiation. The most detailed advances have been obtained using ERα transcriptional regulation of the pS2 (*TFF1*) gene in breast cancer derived cell-lines. Our main contribution to these analyses was to provide experimental conditions (a transcriptional blockade) in which cells are perfectly synchronized, allowing the description of the cyclical mobilization of unliganded receptor on gene promoters, such as pS2 (51). The main role of this unexpected continuous cycling of unliganded ERα would be to ensure a rapid incidence of transcription upon cell exposure to ligand. Indeed, these cycles generate a chromatin environment that is permissive for transcription (relocation of nucleosomes by ATP-dependent complex and histones modifications), though without attaining transcription (51, 52). This latter observation is important, since it confirms that the binding of E_2 is a critical event, leading to the recruitment of cofactors that are absolutely required for transcription initiation: transcription commitment, as it occurs in these instances, is not sufficient to provoke events leading to a gene transcriptional activity. In either presence or absence of E_2, a main feature of the cycles is the occurrence of combinatorial recruitments, as determined by sequential immuno-precipitations. Indeed, within the different complexes observed, many constituents occur such that a given enzymatic activity (HAT, HMT, etc.) is achieved by alternative redundant proteins.

In the presence of ligand, three types of cycle occur: an initial unproductive cycle, which is similar than those observed in the absence of E_2, followed by two different, alternating, transcriptionally productive cycles (51, 53). During these cycles, the recruitment of liganded ERα initiates the process of a sequential achievement of transcription of the *pS2* gene, which is attained through an ordered recruitment of ERα, then intermediate transcription factors such as SRC1, followed by the basal transcription machinery that in turn recruits and activates Pol II. This directed, sequential and ordered recruitment of factors led us to define the notion of a transcriptional clock, or rather of a transcriptional ratchet that acts to ensure and achieve expression of the *pS2* gene (53, 54). The difference between both types of transcriptionally productive cycles resides in the complete resetting of the chromatin organization of the pS2 promoter that occurs at the completion of every double cycle, correlating with the complete removal of the Pol II machinery (53). This is related to functional differences occurring during the clearance phase of these alternating transcriptionally productive cycles. Indeed, as now several other lines of evidence also question their exclusive role in achieving transcriptional repression, we demonstrated that the periodic limitation of the pS2 transcriptional cycles involves macromolecular complexes generally implicated in transcriptional repression (55). Activation of Pol II through phosphorylation on its C-terminal extension serves as a signal for the mobilization of complexes that direct termination of the cycle, such as SWI/SNF and HDACs. It is HDACs, in association with the SWI/SNF complex that restricts the engagement of ERα and cofactors at the end of the first productive cycle. In addition to these proteins, another ATP-dependent remodeling complex, NuRD, act at the end of every second productive cycle to completely reset the chromatin organization of the pS2 promoter. Therefore, an absence of recruitment of NuRD at the end of a given cycle is critical for the persistence of components of the Pol II machinery, such as TBP, over two cycles (53). We recently undertook experiments aiming to get more details into this specific recruitment of NuRD, and therefore insights into molecular mechanisms generating the transcriptional clock. These experiments indicate that DNA methylation on CpG generate the specific signals that target methylated CpG-binding proteins included within NuRD complex to chromatin (56) throughout the cycles (Métivier et al., in preparation). This process might be a general mechanism involved in transcriptional regulation, as indicated by results that we obtained through the development of dynamic DNA pull-downs (Reid et al., in preparation).

Ubiquitination of components of the assembled transcriptionally active complex and subsequent addressing to the proteasome mediated is also integral in the clearance phases of the pS2 promoter (51). Although ERα degradation and transcriptional activity can be dissociated in certain circumstances (57), there is adequate evidence to link proteasome activity and transcription processes (58, 59). As demonstrated with another NR, the RAR (60, 61), phosphorylation of ERα on critical residues such as its serine 118, is a major event that regulates its degradation, and its phosphorylation through cdk7, which is incorporated into Pol II machinery within $TF_{II}H$, is an integral feature of the transcriptional cycles (unpublished observations).

In conclusion, a raising number of experimental data now gave further details on the mechanisms underlying transcriptional cycling, all rising from specific modifications of DNA, chromatin structure, and posttranslational modifications of proteins. It, therefore, becomes apparent that these processes constitute the driving force between each steps of the transcriptional ratchet.

Generalization of the Occurrence of Transcriptional Cycles? ERα transcriptional cycles were evaluated on pS2, c-*myc*, cathepsin D, and cyclin D1 gene with remarkably consistent kinetics of association (periodicity of 40–60 min) on these promoters (37, 62–64). We extended these analyses on other promoters and found cyclicity to be an intrinsic feature of their regulation by ERα. Importantly, although cycling, each promoter exhibits some specificity in terms of cofactor engagement. For instance, TRAP/Mediator and p160 proteins associate at the same time on the CATD promoter (63), in contrast to the situation found with the pS2 promoter (53). Taking advantage of their dissimilar promoter structures, we also comparatively evaluated the features of the transcriptional cycle of the pS2 and PR_A promoters, and found specific recruitments to occur. Furthermore, emerging from our preliminary observations, these processes are directly correlated to the core promoter architectures. It is, therefore, likely that the association of the general transcription machinery and cofactors exhibits kinetics related to genetic and epigenetic information. By evaluating the cellular-specificity of these mechanisms (Huet et al, in preparation), with particular interest to the preferential use of ERα transactivation functions, we also demonstrate that intracellular pathways directly influence transcription mechanisms.

ChIP vs. Live Cell Imaging: An Outdated Controversy?

Real-time, single live cell imaging of nuclear receptors and cofactors tagged with fluorescent proteins also illustrated the dynamic nature of transcriptional activation: these proteins are highly mobile within the nucleus (65–67). Imaging of NR-directed transcription has been facilitated by the integration into the genome of arrays of promoters that locally concentrate responsive elements, visible as discrete loci when associated with fluorescent proteins. These analyses demonstrated that nuclear receptors and interacting cofactors exhibit rapid rates of exchange (68). This continuous sampling of responsive promoters led to the emergence of the "hit and run" model (69). These data apparently contrasted with the longer cycling times determined by kinetic ChIP assays. Taking into account the limitations of both approaches (antibody used and α-amanitin for the ChIP assays that do not assay the persistence of proteins onto DNA; uncertain synchrony of all elements present within the array in imaging experiments, leading to a potential under-estimation of the period of residence of fluorescent proteins onto DNA), we recently proposed a stochastic model that integrates both "hit and run" and transcriptional ratchet models (54). We postulate that transcriptionally productive complexes, with slower mobility, are only rarely formed on promoters: within its residence time, a bound factor can recruit another available and required cofactor. If this partner is not recruited, then the first factor dissociates. If it is recruited, then both of them would be stabilized on the promoter, "sensing" for the next recruitment to occur. Specific parameters, likely taking place at the level of the chromatin structure, or, as we now have evidence, on transcription factors themselves, act to determine direction of cycling. Hager and coworkers also proposed a "return to the template model," which is based on similar principles (70). Importantly, a major feature emerging from the mathematical modelization of the pS2 transcriptional cycling is that the more binding events occur, which would be detected by FRAP (fluorescence recovery after photo-bleaching), the more the model curves fit to the ChIP data (71). This definitely implies that both sets of data are consistent and complementary to describe transcription mechanistic and temporal details.

Concluding Remarks

Recent insights into the dynamic, cyclical nature of transcription present new perspectives and challenges. One major issue is the determination of how general is the phenomenon of promoters cycling? In addition to ERα cycling, kinetic ChIP analyses have determined the cyclical recruitment of transcriptionally competent complexes by other NRs onto cognate promoters (72–74). Integration of the other hierarchical regulatory elements of transcription (nucleus organization) into these models is now eagerly awaited.

References

1. Nilsson S, Makela S, Treuter E, et al. (2001) Mechanisms of estrogen action. Physiol Rev 81:1535–1565.
2. Lombardi G, Zarrilli S, Colao A, et al. (2001) Estrogens and health in males. Mol Cell Endocrinol 178:51–55.
3. Juul A (2001) The effects of oestrogens on linear bone growth. Hum Reprod Update 7:303–313.
4. Maggi A, Ciana P, Belcredito S, et al. (2004) Estrogens in the nervous system: mechanisms and nonreproductive functions. Annu Rev Physiol 66:291–313.
5. Revankar CM, Cimino DF, Sklar LA, et al. (2005) A transmembrane intracellular estrogen receptor mediates rapid cell signaling. Science 307:1625–1630.
6. Robinson-Rechavi M, Escriva Garcia H, Laudet V (2003) The nuclear receptor superfamily. J Cell Sci 116:585–586.
7. Klinge CM (2001) Estrogen receptor interaction with estrogen response elements. Nucleic Acids Res 29:2905–2919.
8. Pike AC (2006) Lessons learnt from structural studies of the oestrogen receptor. Best Pract Res Clin Endocrinol Metab 20:1–14.
9. Lemon BD, Freedman LP (1999) Nuclear receptor cofactors as chromatin remodelers. Curr Opin Genet Dev 9:499–504.
10. van Driel R, Fransz PF, Verschure PJ (2003) The eukaryotic genome: a system regulated at different hierarchical levels. J Cell Sci 116:4067–4075.
11. Krust A, Green S, Argos P, et al. (1986) The chicken oestrogen receptor sequence: homology with v-erbA and the human oestrogen and glucocorticoid receptors. Embo J 5:891–897.

12. Klinge CM, Jernigan SC, Smith SL, et al. (2001) Estrogen response element sequence impacts the conformation and transcriptional activity of estrogen receptor alpha. Mol Cell Endocrinol 174:151–166.
13. Kushner PJ, Agard DA, Greene GL, et al. (2000) Estrogen receptor pathways to AP-1. J Steroid Biochem Mol Biol 74:311–317.
14. Safe S (2001) Transcriptional activation of genes by 17 beta-estradiol through estrogen receptor-Sp1 interactions. Vitam Horm 62:231–252.
15. Charpentier AH, Bednarek AK, Daniel RL, et al. (2000) Effects of estrogen on global gene expression: identification of novel targets of estrogen action. Cancer Res 60:5977–5983.
16. Cicatiello L, Scafoglio C, Altucci L, et al. (2004) A genomic view of estrogen actions in human breast cancer cells by expression profiling of the hormone-responsive transcriptome. J Mol Endocrinol 32:719–775.
17. Kamalakaran S, Radhakrishnan SK, Beck WT (2005) Identification of estrogen-responsive genes using a genome-wide analysis of promoter elements for transcription factor binding sites. J Biol Chem 280:21491–21497.
18. Laganiere J, Deblois G, Lefebvre C, et al. (2005) From the cover: location analysis of estrogen receptor alpha target promoters reveals that FOXA1 defines a domain of the estrogen response. Proc Natl Acad Sci USA 102:11651–11656.
19. Carroll JS, Brown M (2006) Estrogen receptor target gene: an evolving concept. Mol Endocrinol 20:1707–1714.
20. Cheng AS, Jin VX, Fan M, et al. (2006) Combinatorial analysis of transcription factor partners reveals recruitment of c-MYC to estrogen receptor-alpha responsive promoters. Mol Cell 21:393–404.
21. Spector DL (2003) The dynamics of chromosome organization and gene regulation. Annu Rev Biochem 72:573–608.
22. Casolari JM, Brown CR, Drubin DA, et al. (2005) Developmentally induced changes in transcriptional program alter spatial organization across chromosomes. Genes Dev 19:1188–1198.
23. Cabal GG, Genovesio A, Rodriguez-Navarro S, et al. (2006) SAGA interacting factors confine sub-diffusion of transcribed genes to the nuclear envelope. Nature 441:770–773.
24. Perkins TJ, Hallett M, Glass L (2004) Inferring models of gene expression dynamics. J Theor Biol 230:289–299.
25. Snel B, van Noort V, Huynen MA (2004) Gene co-regulation is highly conserved in the evolution of eukaryotes and prokaryotes. Nucleic Acids Res 32:4725–4731.
26. Khorasanizadeh S (2004) The nucleosome: from genomic organization to genomic regulation. Cell 116:259–272.
27. Rountree MR, Bachman KE, Herman JG, et al. (2001) DNA methylation, chromatin inheritance, and cancer. Oncogene 20:3156–3165.
28. Razin A (1998) CpG methylation, chromatin structure and gene silencing-a three-way connection. Embo J 17:4905–4908.
29. Belandia B, Parker MG (2003) Nuclear receptors: a rendez-vous for chromatin remodeling factors. Cell 114:277–280.
30. Cosgrove MS, Boeke JD, Wolberger C (2004) Regulated nucleosome mobility and the histone code. Nat Struct Mol Biol 11:1037–1043.
31. Jenuwein T, Allis CD (2001) Translating the histone code. Science 293:1074–1080.
32. Becker PB, Horz W (2002) ATP-dependent nucleosome remodeling. Annu Rev Biochem 71:247–273.
33. McKenna NJ, O'Malley BW (2002) Combinatorial control of gene expression by nuclear receptors and coregulators. Cell 108:465–474.
34. Berk AJ (1999) Activation of RNA polymerase II transcription. Curr Opin Cell Biol 11:330–335.
35. Chen J, Kinyamu HK, Archer TK (2006) Changes in attitude, changes in latitude: nuclear receptors remodeling chromatin to regulate transcription. Mol Endocrinol 20:1–13.

36. Jepsen K, Rosenfeld MG (2002) Biological roles and mechanistic actions of co-repressor complexes. J Cell Sci 115:689–698.
37. Liu XF, Bagchi MK (2004) Recruitment of distinct chromatin-modifying complexes by tamoxifen-complexed estrogen receptor at natural target gene promoters in vivo. J Biol Chem 279:15050–15058.
38. Perissi V, Aggarwal A, Glass CK, et al. (2004) A corepressor/coactivator exchange complex required for transcriptional activation by nuclear receptors and other regulated transcription factors. Cell 116:511–526.
39. Lavinsky RM, Jepsen K, Heinzel T, et al. (1998) Diverse signaling pathways modulate nuclear receptor recruitment of N-CoR and SMRT complexes. Proc Natl Acad Sci USA 95:2920–2925.
40. Heery DM, Kalkhoven E, Hoare S, et al. (1997) A signature motif in transcriptional co-activators mediates binding to nuclear receptors. Nature 387:733–736.
41. Ren Y, Behre E, Ren Z, et al. (2000) Specific structural motifs determine TRAP220 interactions with nuclear hormone receptors. Mol Cell Biol 20:5433–5446.
42. Warnmark A, Wikstrom A, Wright AP, et al. (2001) The N-terminal regions of estrogen receptor alpha and beta are unstructured in vitro and show different TBP binding properties. J Biol Chem 276:45939–45944.
43. Demarest SJ, Martinez-Yamout M, Chung J, et al. (2002) Mutual synergistic folding in recruitment of CBP/p300 by p160 nuclear receptor coactivators. Nature 415:549–553.
44. Wang H, Huang ZQ, Xia L, et al. (2001) Methylation of histone H4 at arginine 3 facilitating transcriptional activation by nuclear hormone receptor. Science 293:853–857.
45. Klinge CM, Jernigan SC, Mattingly KA, et al. (2004) Estrogen response element-dependent regulation of transcriptional activation of estrogen receptors alpha and beta by coactivators and corepressors. J Mol Endocrinol 33:387–410.
46. Maruvada P, Baumann CT, Hager GL, et al. (2003) Dynamic shuttling and intranuclear mobility of nuclear hormone receptors. J Biol Chem 278:12425–12432.
47. Klose RJ, Bird AP (2006) Genomic DNA methylation: the mark and its mediators. Trends Biochem Sci 31:89–97.
48. Geiser M, Mattaj IW, Wilks AF, et al. (1983) Structure and sequence of the promoter area and of a 5′ upstream demethylation site of the estrogen-regulated chicken vitellogenin ii gene. J Biol Chem 258:9024–9030.
49. Kress C, Thomassin H, Grange T (2001) Local DNA demethylation in vertebrates: how could it be performed and targeted? FEBS Lett 494:135–140.
50. Kress C, Thomassin H, Grange T (2006) Active cytosine demethylation triggered by a nuclear receptor involves DNA strand breaks. Proc Natl Acad Sci USA 103:11112–11117.
51. Reid G, Hubner MR, Metivier R, et al. (2003) Cyclic, proteasome-mediated turnover of unliganded and liganded ERalpha on responsive promoters is an integral feature of estrogen signaling. Mol Cell 11:695–707.
52. Metivier R, Penot G, Carmouche RP, et al. (2004) Transcriptional complexes engaged by apo-estrogen receptor-alpha isoforms have divergent outcomes. Embo J 23:3653–3666.
53. Metivier R, Penot G, Hubner MR, et al. (2003) Estrogen receptor-alpha directs ordered, cyclical, and combinatorial recruitment of cofactors on a natural target promoter. Cell 115:751–763.
54. Metivier R, Reid G, Gannon F (2006) Transcription in four dimensions: nuclear receptor-directed initiation of gene expression. EMBO Rep 7:161–167.
55. Ma J (2005) Crossing the line between activation and repression. Trends Genet 21:54–59.
56. Zhang Y, Ng HH, Erdjument-Bromage H, et al. (1999) Analysis of the NuRD subunits reveals a histone deacetylase core complex and a connection with DNA methylation. Genes Dev 13:1924–1935.
57. Valley CC, Metivier R, Solodin NM, et al. (2005) Differential regulation of estrogen-inducible proteolysis and transcription by the estrogen receptor alpha N terminus. Mol Cell Biol 25:5417–5428.

58. Dhananjayan SC, Ismail A, Nawaz Z (2005) Ubiquitin and control of transcription. Essays Biochem 41:69–80.
59. Kinyamu HK, Chen J, Archer TK (2005) Linking the ubiquitin-proteasome pathway to chromatin remodeling/modification by nuclear receptors. J Mol Endocrinol 34:281–297.
60. Gaillard E, Bruck N, Brelivet Y, et al. (2006) Phosphorylation by PKA potentiates retinoic acid receptor alpha activity by means of increasing interaction with and phosphorylation by cyclin H/cdk7. Proc Natl Acad Sci USA 103:9548–9553.
61. Srinivas H, Juroske DM, Kalyankrishna S, et al. (2005) c-Jun N-terminal kinase contributes to aberrant retinoid signaling in lung cancer cells by phosphorylating and inducing proteasomal degradation of retinoic acid receptor alpha. Mol Cell Biol 25:1054–1069.
62. Shang Y, Hu X, DiRenzo J, et al. (2000) Cofactor dynamics and sufficiency in estrogen receptor-regulated transcription. Cell 103:843–852.
63. Burakov D, Crofts LA, Chang CP, et al. (2002) Reciprocal recruitment of DRIP/mediator and p160 coactivator complexes in vivo by estrogen receptor. J Biol Chem 277:14359–14362.
64. Mishra SK, Mazumdar A, Vadlamudi RK, et al. (2003) MICoA, a novel metastasis-associated protein 1 (MTA1) interacting protein coactivator, regulates estrogen receptor-alpha transactivation functions. J Biol Chem 278:19209–19219.
65. Stenoien DL, Patel K, Mancini MG, et al. (2001) FRAP reveals that mobility of oestrogen receptor-alpha is ligand- and proteasome-dependent. Nat Cell Biol 3:15–23.
66. Rayasam GV, Elbi C, Walker DA, et al. (2005) Ligand-specific dynamics of the progesterone receptor in living cells and during chromatin remodeling in vitro. Mol Cell Biol 25:2406–2418.
67. Feige JN, Gelman L, Tudor C, et al. (2005) Fluorescence imaging reveals the nuclear behavior of peroxisome proliferator-activated receptor/retinoid X receptor heterodimers in the absence and presence of ligand. J Biol Chem 280:17880–17890.
68. Stenoien DL, Nye AC, Mancini MG, et al. (2001) Ligand-mediated assembly and real-time cellular dynamics of estrogen receptor alpha-coactivator complexes in living cells. Mol Cell Biol 21:4404–4412.
69. McNally JG, Muller WG, Walker D, et al. (2000) The glucocorticoid receptor: rapid exchange with regulatory sites in living cells. Science 287:1262–1265.
70. Hager GL, Elbi C, Johnson TA, et al. (2006) Chromatin dynamics and the evolution of alternate promoter states. Chromosome Res 14:107–116.
71. Lemaire V, Lee CF, Lei J, et al. (2006) Sequential recruitment and combinatorial assembling of multiprotein complexes in transcriptional activation. Phys Rev Lett 96:198102.
72. Kang Z, Janne OA, Palvimo JJ (2004) Coregulator recruitment and histone modifications in transcriptional regulation by the androgen receptor. Mol Endocrinol 18:2633–2648.
73. Dunlop TW, Vaisanen S, Frank C, et al. (2005) The human peroxisome proliferator-activated receptor delta gene is a primary target of 1alpha,25-dihydroxyvitamin D3 and its nuclear receptor. J Mol Biol 349:248–260.
74. Vaisanen S, Dunlop TW, Sinkkonen L, et al. (2005) Spatio-temporal activation of chromatin on the human *CYP24* gene promoter in the presence of 1alpha,25-dihydroxyvitamin D3. J Mol Biol 350:65–77.

12 Expression of Estrogen Receptors α and β in Early Steps of Human Breast Carcinogenesis

Pascal Roger, Majida Esslimani-Sahla, Christophe Delfour, Gwendal Lazennec, Henri Rochefort, and Thierry Maudelonde

Introduction

Association of hormonal factors and preinvasive mammary lesions increase the risk of sporadic breast cancer (BC) incidence. Indeed, the mitogenic activity of estrogens and their role as a promoter of BC was confirmed by epidemiological, clinical, and experimental studies (1). Preinvasive mammary lesions frequently precede the development of invasive BC. They correspond to nonproliferative and proliferative benign breast disease without or with atypia and ductal carcinoma *in-situ* (DCIS) (2).

Because of the difficult access to benign mammary lesions with high risk, the biological pattern of the final steps in BC are more studied than the initial steps, both in vivo invasive BCs and in vitro cancer cell lines.

The multistep model of carcinogenesis established for colorectal cancer (3) might be useful in understanding mammary carcinogenesis starting from normal epithelial cells and ending as an invasive BC through high risk proliferative lesions and DCIS (4, 5).

To understand the early steps of human mammary carcinogenesis is crucial to establish BC prevention to stop the evolution to metastases. To this aim, one approach is to compare the expression of putative oncogenes and tumor suppressor genes in normal glands vs. high risk benign breast disease, and in situ carcinoma. Recent clinical trial showed that selective estrogen receptor modulators (SERM) are able to decrease BC incidence up to 50% (NSABP, CORE, STAR) (6–8). This is in favor of a major role of estrogen in early steps of breast carcinogenesis. Estrogen receptors *ERα* and *ERβ* genes are located, respectively, on chromosome 14 and 6 (9). ERα the marker of hormone dependency is responsible of the estrogen-induced mammary proliferation. ERβ has been proposed to inhibit ERα activity (10). A decrease of the ratio ERβ/ERα was described at the RNA level in estrogen-dependant carcinogenesis (11, 12). However, this evolution was not clear at the protein level and we did not know when this ratio is modified in the carcinogenesis. In invasive human BCs, the significance of ERβ is not clear since it is associated to Tamoxifen responsiveness (13) but it has a bad prognostic significance in the ERα negative basal-like cell cancers (14, 15). Furthermore, the significance of the ERβcx (ERβ2) variant of ERβ is unclear (16).

J.J. Li et al. (eds.), *Hormonal Carcinogenesis V.*

We have then studied expression of ERα, ERβ, and cathepsin D (CD), an estrogen- and growth factors-induced protease (17) overexpressed in aggressive BC (18), in mammary lesions having an increased risk of developing invasive BCs compared with adjacent "normal mammary glands", in order to specify at which steps these factors involved in the effects of estrogens are modified.

Populations

We successively studied these proteins on three different populations (Table 1) of tissue samples collected from the Montpellier University Hospital (CHU) and from the Montpellier Cancer Center. Samples were formol fixed and paraffin embedded.

Preinvasive mammary lesions are heterogeneous. They are usually constituted with mixed juxtaposed lesions with different associated risks. For each patient, we have focused only on lesions having the highest risk and on normal adjacent gland. ERs and CD were studied by immunohistochemistry. Their expression was quantified using image analyzer or by semiquantitative method in the different groups of preinvasive mammary lesions and in the adjacent normal glands. These groups have been constituted according to their risks of developing invasive BC.

Estrogen Receptor α in Preinvasive Mammary Lesions

In a first study, we showed that ERα expression increased with the risk of developing BC. Its expression was low in normal glands and in high grade DCIS, it increased in atypia, while it was in low and intermediate grade DCIS (Fig. 1a). Using tissue micro arrays, we recently confirmed on another population of preinvasive mammary lesions that ERα expression also increased in the stage of proliferative benign breast diseases without atypia (Fig. 2a). These data are consistent with other studies (19). Interestingly, ER2α level was also increased in a fraction of high grade DCIS (Roger et al., in preparation). When we compared ERα in a lesion with its corresponding normal glandular tissue, we found a significant lower expression in normal gland (Fig. 2a).

Table 1 Characteristics of studied populations

Patients	Lesions	Characteristics	Reference
118	130	Preinvasive mammary lesions, compared with adjacent normal glands	Roger 2000, 2001
39	39	In situ component of invasive carcinoma and adjacent normal glands	Esslimani-Sahla, 2005
112	119	Pure preinvasive mammary lesions and adjacent normal glands regrouped on tissue micro arrays	Roger, in preparation

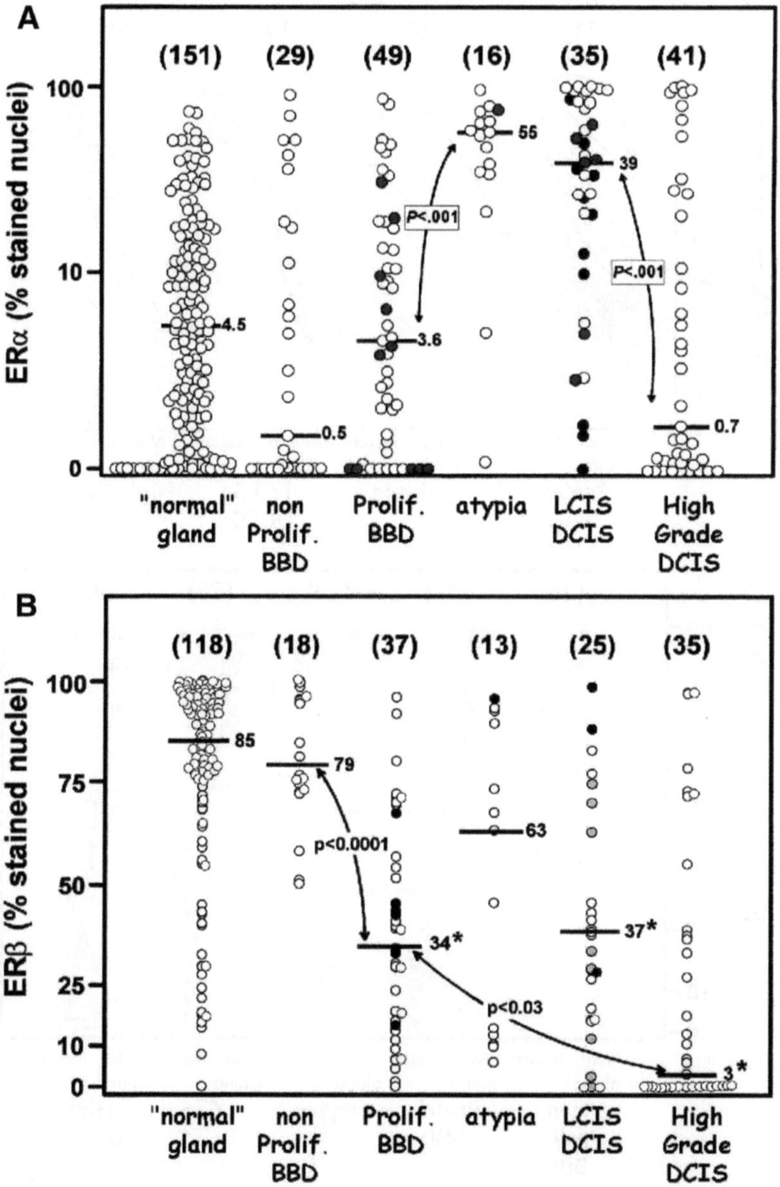

Fig. 1 Distribution of ERα$^+$ (**a**) and ERβ$^+$ epithelial cells (**b**). The percentage of stained nuclei is represented in normal glands (at the periphery of lesions), nonproliferative BBD, proliferative BBD without atypia, proliferative BBD with atypia, high grade DCIS, and in other CISs. The number of different samples is in parentheses. *Horizontal lines*, median value; *P* according to non parametric Kruskal-Wallis test. $^*P < 0.0001$ vs. normal gland group. The percentage of ERα-stained nuclei is represented on a log scale

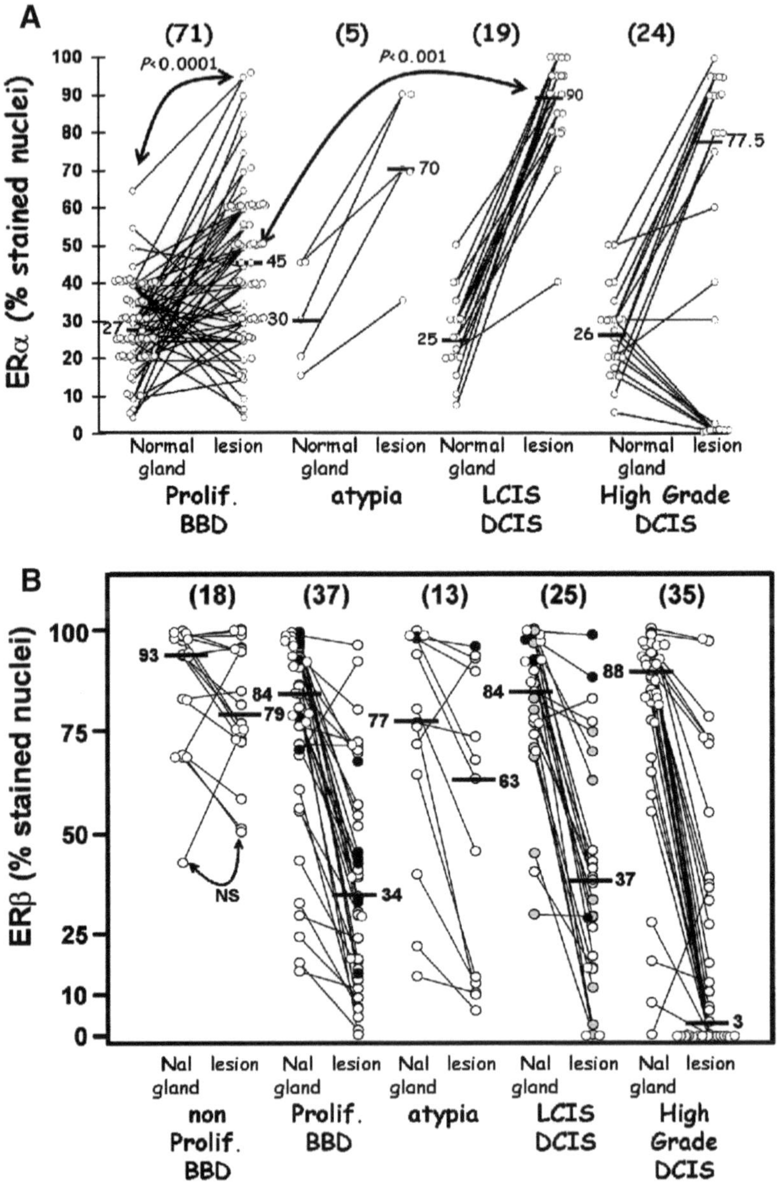

Fig. 2 (**a**) Comparison of ERα level (expressed by the % of stained nuclei) between normal adjacent glands and lesions using tissue microarrays (Roger, in preparation). (**b**) Comparison of ERβ level (expressed by % of stained nuclei) in normal adjacent glands vs. lesions on another population (reproduced from Roger, 2001). The number of different samples is in parentheses. *Horizontal lines*, median value. *P* between normal gland and lesions is significant according to the paired Wilcoxon test for all lesions except in nonproliferative BBD

Estrogen Receptors β in Preinvasive Mammary Lesions

Figure 3 shows a representative ERα and ERβ staining in normal glands and in low grade DCIS. In normal glands, ERβ was highly expressed both in luminal and basal cells, and in some stromal cells, whereas ERα was only expressed in some luminal cells. Conversely, low grade DCIS highly express ERα whereas ERβ was negative.

In preinvasive mammary lesions, in contrast to ERα, ERβ expression decreased significantly early in the stage of proliferative BBD without atypia (Fig. 1b). Compared with its adjacent uninvolved gland, ERβ was significantly lower in proliferative lesions and in DCIS than in the normal gland (Fig. 2b) (20).

Comparison Between ERβ and ERα Levels

When ERα and ERβ expression are compared in a ratio ERβ/ ERα, an early decreased from the stage of proliferative BBD without atypia is observed (20). The variation of these proteins in preinvasive mammary lesions compared with normal glands strongly suggests an increased sensitivity to estrogens that should facilitate the mitogenic activity of low concentrations of estrogens via ERα, particularly after

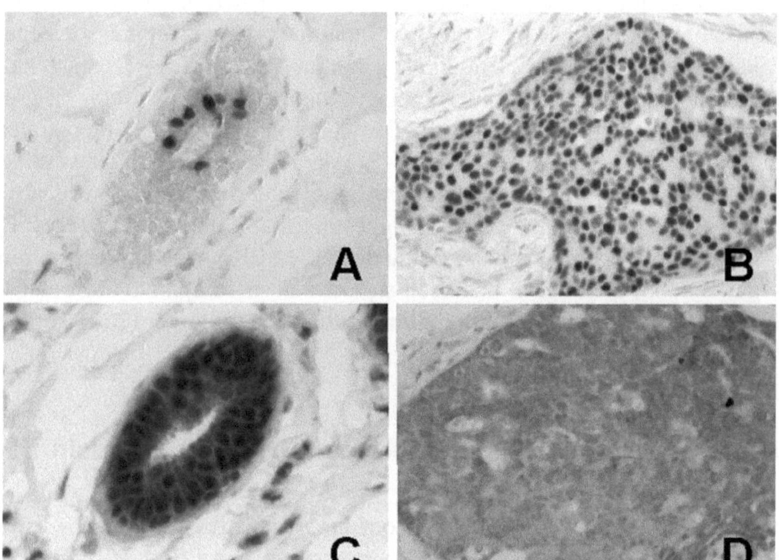

Fig. 3 Representative ERα **(a, b)** and ERβ **(c, d)** immunostaining in normal duct **(a, c)** and non high grade carcinoma in situ **(b, d)** showing the opposite variation of the ER during carcinogenesis

menopause. Furthermore, Shaaban et al. (21) have shown that a low ERβ/ERα ratio is associated with an increased risk in proliferative lesions.

ERβcx in Preinvasive Mammary Lesions

ERβcx (also named ERβ2), is a variant of ERβ wild type, by differential splicing on the exon eight. It does not bind estrogens, but inhibits transcriptional activity of ERα and ERβ1 by heterodimerization (10). By contrast to total ERβ, ERβcx expression level estimated by antibodies provided by Dr. J. Gustafsson was low in adjacent normal glands and increased in adjacent DCIS and invasive carcinoma (22).

In premalignant lesions, the expression of ERβcx also increased precociously in the earlier stage of proliferative BBD without atypia (not shown), contrasting with an early decreased in total ERβ expression (Fig. 1b). The mechanism and the consequence of the increase of this differential splicing in BC are still unknown.

ERβ: A Putative Tumor Suppressor

The fact that ERβ expression decreases during the early steps of mammary carcinogenesis whereas most other markers increase (as ERα) is consistent with the hypothesis that ERβ might be a putative mammary tumor suppressor gene (23) implicated in the early steps of mammary carcinogenesis. This decrease has also been observed in other hormone responsive organs such as ovary or prostate. Moreover, in vitro experiments performed in our laboratory have demonstrated that the reintroduction of ERβ in both $ER\alpha^-$ (24) and $ER\alpha^+$ BC cells leads to growth inhibition (Fig. 4, Lazennec et al., unpublished). Interestingly, ERβ was also able to decrease BC cell invasion potential, suggesting that it may be an interesting target for BC therapy. The mechanism of the decreased ERβ gene expression remains poorly understood, although it has been shown that in comparison to other tumor suppressor genes, there was a methylation on the ERβ promoter (25).

Estrogen- and Growth Factor-induced Protein Cathepsin D

CD is a protease, extensively studied in our laboratory. It is induced by estrogens in $ER\alpha^+$ cell lines and is constitutively over expressed in $ER\alpha^-$ cell lines (17). In vivo, its overexpression in invasive BC has poor prognosis significance (18, 26) but is not correlated with ERα or PR (27). In preinvasive mammary lesions, CD is mostly over expressed in high-grade DCIS, compared with proliferative BBD and

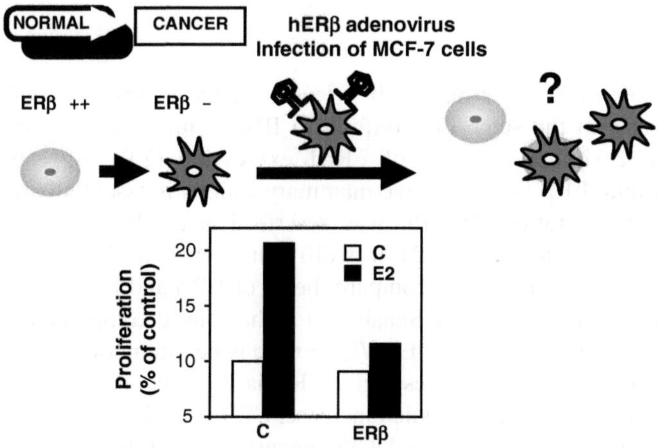

Fig. 4 To determine whether ERβ may modulate the proliferation of ERα$^+$ cells, MCF-7 cells were infected with Ad5 backbone adenovirus or Ad-hERβ viruses at MOI 50. MCF-7 cells were treated with ethanol vehicle or E_2 (10^{-9} M), 24 h after the beginning of the infection. Proliferation rate was determined by counting the cells at day 4. The number of control cells (infected with Ad5 and treated with control vehicle ethanol) was set to 100%

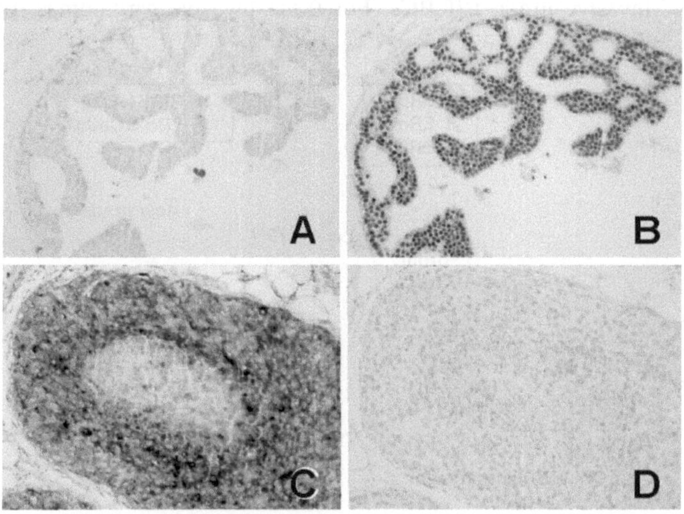

Fig. 5 Representative CD (**a, c**) and ERα (**b, d**) immunostaining in low grade DCIS (**a, b**) and high grade DCIS (**c, d**) illustrating the dissociation of evolution of the two markers

other DCIS (28). There is a dissociated expression of CD and ERα in DCIS. Low-grade DCIS highly expressed ERα but not CD, whereas high-grade DCIS over expressed CD and most often not ERα (Fig. 5).

Discussion

We have shown that ERα and ERβcx levels increase, whereas total ERβ level decreases early in the stage of proliferative BBD without atypia. Shaaban et al. have shown a similar decrease of ERβ1 expression (29), suggesting that the decrease of total ERβ in preinvasive mammary lesions is essentially due to ERβ1. A decrease in the ratio ERβ/ERα was described at the RNA level in estrogen-dependant carcinogenesis (11, 12). It is difficult to evaluate ERβ/ERα ratio at the protein level. However, we can compare the % of ERβ and ERα stained nuclei in lesions and normal glands (20). Shaaban et al. have used comparable method (29) and observed similar results: the ERβ/ERα ratio is declining during initial steps of mammary carcinogenesis. Interestingly, ERβ is the only marker that decreases, whereas most others we have studied increase. CD is mainly overexpressed in high-grade DCIS, and not in low-grade DCIS suggesting that tumor invasive characteristics are a later event.

The branching pathway of mammary carcinogenesis centered on overexpressed estrogen-associated markers (Fig. 6) is in agreement with other studies using different approaches (15, 30). Invasive breast carcinoma is believed to develop from epithelial cells of the terminal ductal lobular unit. Our data agree with two main pathways, one ERα$^+$ and the other ERα$^-$. The ERα$^+$ pathway leads toward the majority of invasive grade I/II BCs. For these invasive carcinomas, preinvasive

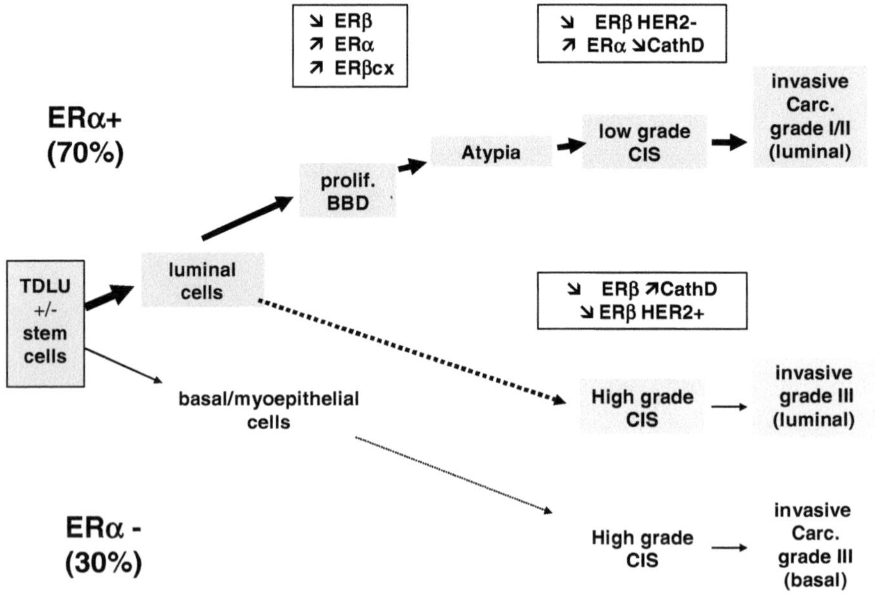

Fig. 6 Branching pathway of mammary carcinogenesis centered on overexpressed estrogen-associated markers. However, it is not totally excluded that some of the ERα$^-$ BC originate from ERα$^+$ BC as initially proposed to explain the evolution of some BC toward hormone resistance

mammary lesions are relatively well known (with proliferative BBD with or without atypia and low-grade DCIS). Along this way, ERα and ERβcx level increase in stage of proliferative BBD and ERβ decreases. For grade III invasive carcinomas, preinvasive stages are unknown except the high-grade DCIS stage. Finally low-grade DCIS are typically $ER\alpha^+$, and CD^-, whereas conversely high grade DCIS are $ER\alpha^-$, and CD overexpressed. This is in agreement with the different clusters defined from cDNA microarrays (15).

Conclusion

To conclude, we propose that total ERβ ought to be considered in mammary carcinogenesis as proposed by Gustafsson et al. (23). However, it is also clear that studies on cancer cell lines are obviously not sufficient to understand mammary carcinogenesis. We need target(s) other than HER2 neu in the $ER\alpha^-$ negative pathway of carcinogenesis since the estrogen induced CD is generally totally dissociated from ERα in patients.

References

1. The Endogenous Hormones and Breast Cancer Collaborative Group (2002) Endogenous sex hormones and breast cancer in postmenopausal women: reanalysis of nine prospective studies. J Natl Cancer Inst 94:606–16.
2. Tavassoli FA, Devilee P (2003) World Health Organization Classification of Tumours. Pathology and genetics of tumours of the breast and female genital organs. Lyon: IARC Press.
3. Vogelstein B, Kinzler KW (1993) The multistep nature of cancer. Trends Genet 9:138–41.
4. Dupont WD, Page DL (1985) Risk factors for breast cancer in women with proliferative breast disease. N Engl J Med 312:146–51.
5. Hartmann LC, Sellers TA, Frost MH, et al. (2005) Benign breast disease and the risk of breast cancer. N Engl J Med 353(3):229–37.
6. Fisher B, Costantino JP, Wickerham DL, et al. (1998) Tamoxifen for prevention of breast cancer: report of the National Surgical Adjuvant Breast and Bowel Project P-1 Study. J Natl Cancer Inst 16;90(18):1371–88.
7. Martino S, Cauley JA, Barrett-Connor E, et al. (2004) Continuing outcomes relevant to Evista: breast cancer incidence in postmenopausal osteoporotic women in a randomized trial of raloxifene. J Natl Cancer Inst 1;96(23):1751–61
8. Vogel VG, Costantino JP, Wickerham DL el al. (2006) Effects of tamoxifen vs raloxifene on the risk of developing invasive breast cancer and other disease outcomes: the NSABP Study of Tamoxifen and Raloxifene (STAR) P-2 trial. JAMA 21;295(23):2727–41.
9. Enmark E, Pelto-Huikko M, Grandien K, et al. (1997) Human estrogen receptor ß-gene structure, chromosomal localization, and expression pattern. J Clin Endocrinol Metab 82:4258–65.
10. Ogawa S, Inoue S, Watanabe T et al. (1998) Molecular cloning and characterization of human estrogen receptor betacx: a potential inhibitor of estrogen action in human. Nucleic Acids Res 26:3505–12.
11. Leyghe E, Dotzlaw H, Waton PH, et al. (1998) Altered estrogen receptor alpha and beta messenger RNA expression during human breast tumorigenesis. Cancer Res 1;58(15):3197–201.

12. Pujol P, Rey JM, Nirde P, et al. (1998) Differential expression of estrogen receptor alpha and beta as a potential marker of ovarian carcinogenesis. Cancer Res 58:5367–73.
13. Esslimani-Sahla M, Simony-Lafontaine J, Kramar A, et al. (2004) ERß level helps to predict Tam resistance in breast cancer but not its ERßcx variant. Clin Cancer Res 10:5769–76.
14. Skliris GP, Leygue E, Curtis–Snell L, et al. (2006) Expression of oestrogen receptor-beta in oestrogen receptor-alpha negative human breast tumours. Br J Cancer 95(5):616–26.
15. Sorlie T, Perou CM, Tibshirani R, et al. (2001) Gene expression patterns of breast carcinomas distinguish tumor subclasses with clinical implications. Proc Natl Acad Sci USA 98(19):10869–74.
16. Leygue E, Dotzlaw H, Watson PH, et al. (1999) Expression of estrogen receptor beta1, beta2, and beta5 messenger RNAs in human breast tissue. Cancer Res 15;59(6):1175–9.
17. Rochefort H, Cavailles V, Augereau P, et al. (1989) Overexpression and hormonal regulation of pro-cathepsin D in mammary and endometrial cancer. J Steroid Biochem 34(1–6):177–82.
18. Spyratos F, Maudelonde T, Brouillet JP, et al. (1989) An independent prognostic factor for metastasis of breast cancer. Lancet ii:1115–18.
19. Khan SA, Rogers MA, Khurana KK, et al. (1998) Estrogen receptor expression in benign breast epithelium and breast cancer risk. J Natl Cancer Inst 90:37–42.
20. Roger P, Esslimani Sahla M, Makela S, et al. (2001) Decreased expression of estrogen receptor beta protein in proliferative preinvasive mammary tumors. Cancer Res 61:2537–2541.
21. Shaaban AM, Jarvis C, Moore F, et al. (2005) Prognostic significance of estrogen receptor Beta in epithelial hyperplasia of usual type with known outcome. Am J Surg Pathol 29(12):1593–9.
22. Esslimani Sahla M, Kramar A, Simony-Lafontaine J, et al. (2005) Increased estrogen receptor beta cx expression during mammary carcinogenesis. Clin Cancer Res 11:3170–74.
23. Matthews J, Gustafsson JA (2003) Estrogen signaling: a subtle balance between ER alpha and ER beta. Mol Interv 3(5):281–92. Review.
24. Lazennec G, Bresson D, Lucas A, et al. (2001) ERß inhibits proliferation and invasion of breast cancer cells. Endocrinology 142:4120–30.
25. Zhao C, Lam EW-F, Sunters A, et al. (2003) Expression of estrogen receptor ß isoforms in normal breast epithelial cells and human breast cancer: regulation by methylation. Oncogene 22:7600–6.
26. Rochefort H, Garcia M, Glondu M, et al. (2000) Cathepsin D in breast cancer: Mechanisms and clinical applications. A 1999 overview. Clin Chim Acta 291(2):157–70.
27. Maudelonde T, Khalaf S, Garcia M, et al. (1988) Immunoenzymatic assay of Mr 52,OOO cathepsin D in 182 breast cancer cytosols. Low correlation with other prognostic parameters. Cancer Res 48:462–6.
28. Roger P, Daures JP, Maudelonde T, et al. (2000) Dissociated overexpression of cathepsin-D and estrogen receptor alpha in preinvasive mammary tumors. Hum Pathol 31(5):593–600.
29. Shaaban AM, O'Neill PA, Davies MP, et al. (2003) Declining estrogen receptor-beta expression defines malignant progression of human breast neoplasia. Am J Surg Pathol 27(12):1502–12.
30. Simpson PT, Reis-Filho JS, Gale T, et al. (2005) Molecular evolution of breast cancer. J Pathol 205(2):248–54. Review.
31. Brouillet JP, Dujardin MA, Chalbos D, et al. (2001) Analysis of the potential contribution of estrogen receptor (ER) α in ER cytosolic assay of breast cancer. Int J Cancer 95:205–8.
32. Clarke Rb, Howell A, Potten CS, et al. (1997) Dissociation between steroid receptor expression and cell proliferation in the human breast. Cancer Res 15;57(22):4987–91.
33. Rochefort H, Capony F, Garcia M, et al. (1987) Estrogen-induced lysosomal proteases secreted by breast cancer cells. A role in carcinogenesis? J Steroid Biochem 34:177–82.

Part 4
Risk Assessment and Relevant Early Biomarkers

13
Women's Health Initiative Studies of Postmenopausal Breast Cancer

Ross L. Prentice

Introduction

The Women's Health Initiative (WHI) is a major population research enterprise that aims to generate knowledge and recommendations pertinent to the health of postmenopausal women. The WHI is funded and administered through the USA National Institutes of Health, with the National Heart Lung and Blood Institute having a central role. The WHI is conducted through 40 Clinical Centers in the USA. Our group at the Fred Hutchinson Cancer Research Center in Seattle serves as Clinical Coordinating Center for the WHI, and as the site for one of the Clinical Centers.

The WHI is conducted among 161,808 postmenopausal women in the age range 50–79 at enrollment. Specialized recruitment efforts led to a racial/ethnic distribution among enrollees that is similar to that for postmenopausal women in the USA, and an age distribution was specified with a targeted 10, 20, 45, and 25% of enrollees having baseline age in 50–54, 55–59, 60–69, and 70–79 years, respectively. The centerpiece of the WHI is a multifaceted randomized controlled clinical trial (CT) among 68,132 women. The CT includes four randomized comparisons, each building on a substantial preceding history of basic research and observational epidemiologic investigation. Two CT components studied the health benefits and risks of postmenopausal hormone therapy (HT), either 0.625 mg day^{-1} continuous conjugated equine estrogen alone (E-alone) among women who were posthysterectomized or this same estrogen preparation plus 2.5 mg day^{-1} continuous medroxyprogesterone acetate (E + P) among women with a uterus. These two preparations were used by about 8 million and 6 million women, respectively, in the USA in 2002, with many additional millions of users worldwide. A large number of cohort and case–control studies provided consistent information to suggest that use of these agents would much reduce coronary heart disease (CHD) risk. In fact, beliefs concerning this benefit were so strongly held that reviewers of WHI concept proposals questioned whether a trial was necessary. Although CHD was designated as the primary outcome in the WHI HT trials, breast cancer (BC) was designated as the primary safety (adverse) outcome, as a substantial observational epidemiologic literature suggested an increase in BC risk with E-alone, and a larger increase with E + P. In the next section, we highlight the results of the HT trials, which have included important

J.J. Li et al. (eds.), *Hormonal Carcinogenesis V.*

surprises for both CHD and BC (1, 2). Some comments are also provided on ongoing studies to compare findings in the WHI CT with corresponding results from the WHI Observational Study (OS) conducted among 93,676 postmenopausal women from the same recruitment sources as the CT.

The CT also includes a dietary modification (DM) component that tests whether adopting a low-fat eating pattern in the 50–79 age range can reduce the risk of BC and colorectal cancer separately and CHD secondarily. This trial component was motivated by animal experiments dating back to the 1940s, along with subsequent international correlation and time trend studies. Analytic epidemiologic studies of the dietary fat and cancer hypotheses then followed and yielded mixed and inconsistent findings. As a result, when the DM trial was initiated, observational data sources were consistent with both an important public health benefit, or with no effect whatsoever, for a low-fat eating pattern. Here, an overview of the DM trial results on BC (3) will be described, as will ongoing efforts in WHI to understand and assimilate currently available data.

The final CT component tests whether dietary supplementation using a combination of calcium (1,000 mg day^{-1} elemental calcium) plus vitamin D (400 IU day^{-1}) will reduce the risk of fractures, and secondarily colorectal cancer. Principal results from this trial component have recently been presented (4, 5), but BC results have yet to be published. Hence, the calcium and vitamin D trial will not be discussed further here.

Postmenopausal Hormone Therapy and BC Risk

Table 1 shows principal results from both the E+P randomized controlled trial among 16,608 postmenopausal women, and the E-alone trial among 10,739 postmenopausal women. The E+P trial was stopped early in 2002 when it was judged that risks exceeded benefits over an average 5.6 years follow-up period. The

Table 1 Clinical outcomes in the WHI postmenopausal hormone therapy Trials (1, 2)

	E + P trial		E-alone trial	
Outcomes	Hazard ratio	95% CI	Hazard ratio	95% CI
Coronary heart disease	1.29	1.02–1.63	0.91	0.75–1.12
Stroke	1.41	1.07–1.85	1.39	1.10–1.77
Venous thromboembolism	2.11	1.58–2.82	1.33	0.99–1.79
Invasive breast cancer	1.26	1.00–1.59	0.77	0.59–1.01
Colorectal cancer	0.63	0.43–0.92	1.08	0.75–1.55
Endometrial cancer	0.83	0.47–1.47		
Hip fracture	0.66	0.45–0.98	0.61	0.41–0.91
Death due to other causes	0.92	0.74–1.14	1.08	0.88–1.32
Global index	1.15	1.03–1.28	1.01	0.91–1.12
Number of women	8506	8102	5310	5429
Follow-up time, mean (SD), mo	62.2 (16.1)	61.2 (15.0)	81.6 (19.3)	81.9 (19.7)

E-alone trial was stopped early in 2004, primarily based on an elevation in stroke incidence over an average 7.1 years of follow-up. The trigger for an early stoppage discussion in the E+P trial, under the CT monitoring plan, adopted by the WHI Data and Safety Monitoring Committee (DSMB), was an elevation in invasive BC, based on a protocol-specified weighted log rank test, with weights that increased linearly from zero at randomization to one at 10 years from randomization. The fact that the BC elevation was complemented by a "global index" that met statistical criteria in the unfavorable direction supported the early stopping recommendation by the DSMB. The global index for the E+P trial was defined as the time to the earliest of the events listed above it in Table 1 (except only the pulmonary embolism component of venous thromboembolism was included). Essentially, E+P use over a 5.6 year follow-up period led to an estimated 15% increase in the combined risk of CHD, stroke, pulmonary embolism, invasive BC, colorectal cancer, endometrial cancer, hip fracture, and death due to cancers other than those just listed. The biggest surprise with the E+P trial results was a borderline elevation in CHD risk, while a major reduction had been expected.

The benefit vs. risk profile from the E-alone trial is rather different from that for E+P, with the global index almost perfectly balanced, reflecting an elevation in stroke risk, and a reduction in hip fracture risk, that is of about the same magnitude as for E+P. At the time of early stoppage, the coronary heart disease hazard risk was trending in the favorable direction, but there was little potential for a significant reduction emerging by the planned end of the intervention period (3/31/05). A surprise in the E-alone trial was a nearly significant reduction in the invasive BC hazard ratio over the average 7.1 years of follow-up. This hazard ratio (95% confidence interval) of 0.77 (0.59, 1.01) contrasted with a wealth of observational data suggesting a moderate elevation in BC risk with postmenopausal estrogen therapy.

The initial publications resulting from the WHI hormone therapy trials were followed by papers providing detailed analysis of results for each of the major clinical outcomes, using fully-adjudicated clinical outcome data through the date of trial termination. Chlebowski et al. (6) make a number of important observations about the BC results in the E+P trial: The incidence of invasive BCs was higher in the E+P compared with the placebo group ($p=0.003$) based on 199 vs. 150 cases, as was true also for total (invasive plus noninvasive) BC ($p<0.001$) based on 245 vs. 185 cases. The invasive BCs diagnosed in the E+P group were similar in histology and grade to those in the placebo group, but they were larger (mean 1.7 vs. 1.5 cm, $p=0.04$) and were at a more advanced stage (regional/metastatic 25.4% vs. 16.0%, $p=0.04$). The percentage of women having abnormal mammograms was substantially higher in the E+P group compared with the placebo group, and the elevation in abnormal mammogram frequency was evident as early as 1 year from randomization (9.4% vs. 5.4%, $p<0.001$). Another interesting finding was that most of the elevation in BC risk occurred among women who had used postmenopausal hormone therapy (E-alone or E+P) prior to enrollment in the WHI. Specifically, the hazard ratio (95% CI) among women who had used postmenopausal hormones for 5 or more years prior to enrollment was 2.25 (1.00, 5.15), for less than 5 years was 1.70 (0.99, 2.91), compared with a HR (95% CI) of 1.09 (0.86, 1.39) among women who had not used postmenopausal hormones prior to WHI enrollment. However, there was evidence of a positive trend in hazard ratio with years from randomization in the E+P trial, both overall and within prior HT and no prior HT strata. Among women with prior HT, the estimated hazard ratio started at about one, and increased to two or more by trial termination, while among women without prior HT, the hazard ratio started at about 0.5 and increased to perhaps 1.5 by the end of the 5.6-year intervention period. These patterns led the WHI investigators to conclude that E+P may stimulate BC growth, largely based on increasing hazard ratios with increasing time from randomization. They also concluded that E+P may hinder diagnosis, based on the reduced rate of BC diagnoses early in the trial, in conjunction with the increased frequency of mammographic abnormalities and the larger tumor size at diagnosis in the E+P compared with the placebo group. A recent WHI report (7) looked in considerable detail at the E+P BC results in relation to prior HT, to determine the extent to which these patterns could be attributed to different characteristics between the prior HT and no prior HT groups, or to the total duration of E+P use including usage prior to randomization. This report concludes that durations of use only slightly larger than those in the E+P trial are associated with an increased risk, and called for studies involving longer-term follow-up and longer-term E+P exposure.

A detailed account of the BC data in the E-alone trial has also recently been presented (8): Following final adjudication, the hazard ratio (95% CI) for invasive BC was 0.80 (0.62, 1.04) based on 104 cases in the E-alone group and 133 in the placebo group over the 7.1 year average follow-up period. As with E+P the tumors diagnosed in the E-alone group were larger on average (1.80 cm) than in the placebo group (1.45 cm) ($p=0.03$). Also, the frequency of abnormal mammograms at 1 year from randomization was higher in the E-alone than in the placebo group (9.2% vs. 5.5%, $p<0.001$), a pattern that continued throughout the trial. However, the increase in abnormalities with E-alone mainly resulted in short interval follow-up, whereas, with E + P, there were also elevations in findings that were suspicious or suggestive of malignancy. Also, the hazard ratio (95% CI) among women without prior postmenopausal hormone use was 0.65 (0.46, 0.92) compared with 1.02 (0.70, 1.50) among women with prior hormone use. There was no clear trend in hazard ratio with years from E-alone randomization. The authors concluded that E-alone does not increase invasive BC risk over a 7.1-year period, but that E-alone increases mammographic abnormalities requiring short interval follow-up.

The interpretation of the two HT trials depends on the influence that E+P and E-alone have on BC diagnosis. E+P evidently increases BC risk in a duration-dependent fashion. The increasing hazard ratios as a function of years from randomization in conjunction with larger tumor sizes at diagnosis suggests that the use of E+P delays diagnosis, in which case HRs will be underestimated, at least in the short term. The similar increase in tumor size at diagnosis suggests a similar diagnostic delay hypothesis for E-alone. However, one might then expect a catch-up phase with hazard ratios above one following some years from E-alone initiation. Such a catch-up phase is not evident over the 7.1 years of average follow-up in the E-alone trial, but a pattern along these lines has been reported recently among long-term E-alone users in the Nurses Health Study (9).

It also seems plausible that diagnostic delay could be more influential among women who have never used postmenopausal hormone therapy and are at some years beyond the menopause. Mammography is quite likely to be very sensitive among such women, and such sensitivity may be reduced by even a short exposure to postmenopausal hormone therapy. This line of thinking could help to explain the lower hazard ratios for E-alone and E+P in the WHI trials compared with observational studies. For example, the Million Women Study Collaborators (10) report hazard ratios in the vicinity of 2.0 for E+P and 1.3 for E-alone, though their report is based on combined cross-sectional and longitudinal data.

In an attempt to more fully understand the E+P and E-alone effects on BC WHI investigators are currently analyzing data from each of the E+P and E-alone trials jointly with data from the pertinent subset of the WHI Observational Study. Women were recruited to the Clinical Trial and Observational Study from the same sources, and there were many common procedures, including personal interviews at baseline to ascertain prior hormone therapy histories, and common core aspects to clinical outcome ascertainment. These analyses allow a careful account of confounding factors, prior HT use history, and prior and ongoing mammographic screening history in comparing CT vs. OS hazard ratios, in examining hazard ratios for key subgroups, and in examining hazard ratios for longer-term E+P and E-alone use.

Low-Fat Dietary Pattern and Breast Cancer Risk

The hypothesis that a low-fat diet would reduce BC risk dates back to animal feeding experiments in the 1940s. These were followed by international correlation studies showing a positive association between per capita fat consumption and BC incidence, and eventually by a large number of analytic epidemiologic studies. For example, Howe et al. (11) reported a positive relationship between fat consumption and BC risk, based on data from 12 case–control studies that used a variety of dietary assessment methods in diverse populations. In contrast, Hunter et al. (12) found no significant association between the fat content of the diet and BC incidence in seven large cohort studies in Western populations, each using a food frequency questionnaire (FFQ) for dietary assessment. As these data were emerging, the National Cancer Institute initiated studies of a low-fat dietary pattern to demonstrate the feasibility of a full-scale primary prevention intervention trial (13) and to examine effects on intermediate markers, such as blood estradiol concentration (14). The lack of support from cohort studies caused the low-fat diet and BC hypothesis to become controversial. However, uncertainty about the effects of measurement error in dietary assessment on observational study findings implied that there was insufficient data to reject the low-fat diet and BC hypothesis and, additionally, the feasibility and intermediate outcome trials were encouraging. Hence, it was decided to proceed with a full-scale intervention trial as a part of the WHI. A 20% energy from fat diet was the major intervention target. Animal feeding studies indicating a particular role for polyunsaturated fat in stimulating mammary tumors argued for a focus on total fat reduction, even though allowing higher levels of polyunsaturated and monounsaturated fat may have advantages for reducing cardiovascular disease risk. Also, total energy reduction was not taken to be an intervention goal, even though a calorie reduction may have benefits for breast and other cancers, since a well developed and practical intervention program for long-term energy reduction was not available. The DM intervention program also included goals of increasing vegetables and fruits to five or more servings per day and increasing grain servings to six or more per day, in order to incorporate additional cancer prevention concepts from nutritional epidemiologic studies.

The WHI DM trial excluded women having percent energy from fat less than 32, as assessed by a FFQ, and excluded women with prior breast or colorectal cancer. The basic trial relative risk model projected that about half of the international variation in BC rates is attributable to the fat content of the diet, and assumed that the reduced BC hazard ratio following the adoption of a low-fat diet in the middle to later decades of life would be realized linearly over a 10-year period. These, along with assumptions about the % energy from fat difference between the intervention and control (usual diet) groups led to a projected 14% lower BC risk in the intervention compared with the control group over a planned 9-year average follow-up period. The trial design then called for a total sample size of 48,000 women, with 40% randomly assigned to the dietary intervention, and the projected power for detecting a BC incidence reduction in the intervention compared with the control group was 86%.

The trial proceeded by assigning intervention group women to groups of size 10–15 who met with project nutritionists to learn how to undertake and maintain a low-fat eating pattern. The group intervention integrated both nutritional and behavioral strategies. A total of 48,835 women enrolled in the trial.

Table 2 shows some key results on the dietary habits of intervention and control group women at 1 year from randomization, and on the difference in habits between the groups at 1, 3, and 6 years from randomization. A surprise was encountered in that the baseline percent energy from fat among trial enrollees was only about 35% in contrast to the 38–39% energy from fat that was anticipated from preceding feasibility studies and from the baseline FFQ % energy from fat exclusion. As a result the % energy from fat difference between the intervention and control groups was only about 70% of that projected and study power was reduced. Also, the average years from randomization to completion of the intervention phase of the trial turned out to be about 8.1 years rather than the targeted 9 years, these departures from design assumptions combined to give a projected BC rate in the intervention group that is only 8–9% less than that of the control group under the other design assumptions.

As shown in Table 3, the realized BC incidence over the 8.1 year trial follow-up period was 9% lower in the intervention vs. the control group, based on 1,727 incident cases, though the different just failed to be significant at conventional levels (log rank $p = 0.07$). BC mortality was 23% lower in the intervention vs. the comparison group, based on a small number (80) of BC deaths. The hazard ratios for total cancer incidence and mortality and total mortality were nonsignificantly below one for the intervention vs. control group, as was also the case for a global index defined as the time to the earliest of BC, colorectal cancer, CHD, or death from other causes.

Table 2 Mean (SD) of nutrient consumption and body weight by randomization group

	Year 1		Year 1	Year 3	Year 6
	Intervention	Control	Difference	Difference	Difference
Fat (% of calories)	24.3 (7.5)	35.1 (6.9)	−10.7[a] (7.0)	−9.5[a] (7.4)	−8.1[a] (7.8)
Total Fat (g)	40.8 (21.4)	63.0 (31.0)	−22.4[a] (31.1)	−20.1[a] (32.0)	−18.4[a] (33.5)
Fruits and Veg (sv/day)	5.1 (2.3)	3.9 (2.0)	1.2[a] (1.9)	1.3[a] (2.0)	1.1[a] (2.1)
Grains (sv/day)	5.1 (2.7)	4.2 (2.3)	0.9[a] (2.5)	0.7[a] (2.6)	0.4[a] (2.6)
Weight (kg)	74.4 (16.7)	76.3 (16.7)	−2.2[a] (8.4)	−1.3[a] (9.1)	−0.8[a] (9.4)

[a] Difference significant at $p < 0.001$ from a two sample t test

Table 3 Risk of breast cancer and other major clinical outcomes (3)

	Intervention Cases = 655 Annualized %	Comparison Cases = 1072 Annualized %	Hazard ratio (95% CI)	Unweighted p value
Breast cancer				
Incidence	0.42%	0.45%	0.91 (0.83–1.01)	0.07
Mortality	0.02%	0.02%	0.77 (0.48–1.22)	0.26
Total cancer				
Incidence	1.23%	1.28%	0.96 (0.91–1.02)	0.15
Mortality	0.28%	0.29%	0.95 (0.84–1.07)	0.41
Total mortality	0.60%	0.61%	0.98 (0.91–1.02)	70
Global index	1.30%	1.35%	0.96 (0.91–1.02)	0.16

The likelihood that the low-fat eating pattern intervention studied in WHI is on the right track for reducing BC risk is enhanced by the observation that blood estradiol concentrations were reduced in the intervention compared with the control group at 1 year from randomization and, importantly, by the fact that women in the upper half in baseline % energy from fat had larger fat reductions compared with corresponding control group women and experienced significantly larger (15–20%) BC risk reductions (3). Participating women are currently being followed for an additional five years to provide information on the longer term effects of 8.1 years of dietary intervention, and to provide more precise intervention vs. control group comparisons overall. An interesting observation from the BC data available thus far is that the hazard ratio differed significantly between progesterone receptor positive and progesterone receptor negative tumors ($p = 0.04$) with respective hazard ratios (95% confidence intervals [CIs]) of 0.96 (0.85, 1.09) for PR+ and 0.76 (0.63, 0.92) for PR− tumors.

Studies to Elucidate Hormone Therapy and Dietary Modification Trial Findings

For both the HT trials and the DM trial, WHI investigators have a series of studies underway toward more fully understanding the trial results, and their clinical and public health implications.

For the HT trials and BC, the ongoing combined analyses of CT and OS data have already been mentioned. Additionally, case–control studies are underway to examine the BC hazard ratio in relation to baseline plasma concentrations of steroid hormones, and to examine the extent to which the major increases in key concentrations following the initiation of E + P can provide an explanation for observed elevations in BC risk.

Also, a whole genome association study is underway to identify single nucleotide polymorphisms (SNPs) that relate to BC risk or that relate to E + P BC hazard ratios. This collaborative project with Perlegen Sciences uses Perlegen's 360,000 tag SNP set and includes about 2,200 BC cases and controls in a three-stage design (15), with a similar design also for CHD and stroke.

A proteomic study is also underway to identify novel proteins for which the concentration is changed as a result of E + P or E-alone, using an Intact Protein Analysis System (16). The intention is to follow this discovery effort, which involves a comparison of baseline and 1-year plasma samples among women assigned to E + P or E-alone in pools of size 10, by the development of specialized tests for candidate proteins for application to individual cases and controls in the E + P and E-alone trials, for BC as well as other major clinical outcomes.

For the DM trial, the continuing effort for BC focuses on clarifying whether, and the extent to which, a low-fat dietary pattern may reduce BC risk. For example, a case–control study within the control group in the DM trial has been recently reported (17). This study noted that a positive association between fat (adjusted for calories) and BC was evident if 4-day food records were used to assess diet, but not if an FFQ was used, as was also the case in an earlier study from the European Prospective Investigation of Cancer and Nutrition (18). This result has led to a call for a reevaluation of both methods and results in the observational study of diet and nutrition associations (19).

Understanding the measurement properties of dietary assessment methods used is likely to be fundamental to assessing and enhancing the validity of observational nutritional epidemiology studies, and to the validity of efforts to attribute intervention effects in the DM trial to specific dietary differences between the intervention and control groups. Hence, WHI investigators have undertaken a Nutrient Biomarker Study among 544 intervention and control women in the DM trial. So-called recovery biomarkers, such as the doubly-labeled water assessment of energy consumption and the urinary protein assessment of protein consumption that evidently adheres to a classical measurement model, will be used to develop measurement models for FFQ assessments. Calibrated assessments of nutrient intakes based on such measurement models will then be used as basic ingredients in dietary association studies in WHI and in explanatory analyses of DM intervention effects.

A corresponding biomarker study of both nutrient consumption and physical activity energy expenditure among 450 OS women is also at the early stages of implementation. This Nutrition and Physical Activity Assessment Study will include an attempt to separately estimate fat from carbohydrate consumption using biomarkers.

Summary

The WHI has provided randomized controlled trial data on interventions that are among the most important in relation to the health benefits and risks among postmenopausal women in the USA. Postmenopausal hormone therapy did not yield the anticipated major reduction in coronary heart disease, and was responsible for both stroke elevations and fracture reductions over average 5.6 (E+P) and 7.1 (E-alone) years intervention periods. BC hazard ratios for both E+P and E-alone were about as anticipated, based on preceding observational studies, among women who had used postmenopausal hormones prior to WHI enrollment, but were lower than expected among women without such prior exposures. Whether these lower hazard ratios reflect upward bias in the observational studies or detection lags among women who tended to be quite a few years from menopause when randomized in the WHI trials remains to be fully clarified. At any rate, E + P is associated with an elevated BC risk, while the effects of E-alone remain somewhat uncertain.

The WHI DM trial is taking place among women having baseline % energy from fat around 35, in contrast to the 38–39% projected. This has led to a smaller than anticipated intervention vs. control group difference in dietary habits, and to projected and realized BC incidence rates that were about 9% lower in the intervention vs. the control group. The likelihood that this lower rate reflects a meaningful reduction in BC risk is enhanced by a lower BC hazard ratio among the 50% of women who started relatively higher in percent energy from fat, and by cohort studies, including that among women in the DM control group, showing clear positive trends between food record percent energy from fat and BC incidence.

Acknowledgment The WHI is funded by the National Heart Lung and Blood Institute of the USA NIH. This report was partially supported by program grant CA53996 by the U.S. National Cancer Institute.

References

1. The Writing Group for the Women's Health Initiative Investigators (2002) Risks and benefits of estrogen plus progestin in healthy post-menopausal women: Principal results of the Women's Health Initiative randomized controlled trial. JAMA 288:321–333.
2. The Women's Health Initiative Steering Committee (2004) Effects of conjugated equine estrogen in postmenopausal women with hysterectomy. The Women's Health Initiative randomized controlled trial. JAMA 291:1701–1712.
3. Prentice RL, Caan B, Chlebowski RT, et al. (2006) Low-fat dietary pattern and risk of invasive breast cancer: the Women's Health Initiative Randomized Controlled Dietary Modification Trial. JAMA 295:629–642.
4. Jackson RD, LaCroix AZ, Gass M, et al. (2006) Calcium plus vitamin D supplementation and the risk for fractures. N Engl J Med 354:669–683.
5. Wactawski-Wende J, Kotchen JM, Anderson GL, et al. (2006) Calcium plus vitamin D supplementation and the risk of colorectal cancer. N Engl J Med 354:684–696.
6. Chlebowski RT, Hendrix SL, Langer RD, et al. (2003) Influence of estrogen plus progestin on breast cancer and mammography in healthy postmenopausal women: the Women's Health Initiative Randomized Trial. JAMA 289:3243–3253.

7. Anderson GL, Chlebowski RT, Rossouw JE, et al. (2006) Prior hormone therapy and breast cancer risk in the Women's Health Initiative randomized trial of estrogen plus progestin. Maturitas 55:103–115.
8. Stefanick ML, Anderson GL, Margolis KL, et al. (2006) Effects of conjugated equine estrogens on breast cancer and mammography screening in postmenopausal women with hysterectomy. JAMA 295:1647–1657.
9. Chen WY, Manson JE, Hankinson SE, et al. (2006) Unopposed estrogen therapy and the risk of invasive breast cancer. Arch Intern Med 166:1027–1032.
10. Beral V, Million Women Study Collaborators (2003) Breast cancer and hormone replacement therapy in the Million Women Study. Lancet 362:419–427.
11. Howe GR, Hirohata T, Hislop TG, et al. (1990) Dietary factors and risk of breast cancer: combined analysis of 12 case–control studies. J Natl Cancer Inst 82:561–569.
12. Hunter DJ, Spiegelman D, Adami HO, et al. (1996) Cohort studies of fat intake and the risk of breast cancer – a pooled analysis. N Engl J Med 334:356–361.
13. Insull W, Henderson MM, Prentice RL, et al. (1990) Results of a randomized feasibility study of a low-fat diet. Arch Intern Med 150:421–427.
14. Prentice RL, Thompson DJ, Clifford C, et al. (1990) Dietary fat reduction and plasma estradiol concentration among healthy postmenopausal women. J Natl Cancer Inst 82:129–134.
15. Prentice RL, Qi L (2006) Aspects of the design and analysis of high-dimensional SNP studies for disease risk estimation. Biostatistics 7:339–354.
16. Wang H, Clouthier SG, Galchev V, et al. (2005) Intact-protein-based high-resolution three-dimensional quantitative analysis system for proteome profiling of biological fluids. Mol Cell Proteomics 4:618–625.
17. Freedman LS, Potischman N, Kipnis V, et al. (2006) A comparison of two dietary instruments for evaluating the fat-breast cancer relationship. Int J Epid 35:1011–1021.
18. Bingham SA, Luben R, Welch A, et al. (2003) Are imprecise methods obscuring a relationship? Report from the EPIC Norfolk Prospective Cohort Study. Lancet 362:212–214.
19. Bingham SA, Day N (2006) Commentary: Fat and breast cancer: time to re-evaluate both methods and results? Int J Epid 35:1022–1024.

14
Circulating Levels of Sex Steroids and Prolactin in Premenopausal Women and Risk of Breast Cancer

Susan E. Hankinson

Introduction

The key role of the hormonal environment in breast cancer (BC) etiology is well documented. Reproductive factors such as parity and age at menopause influence BC risk and postmenopausal hormone use increases risk (1). Selective estrogen receptor modulators (SERMs), such as tamoxifen, decrease risk (2, 3) and aromatase inhibitors decrease BC recurrence. Further, the association of circulating estrogen and androgen levels with BC risk is now established among postmenopausal women (4–6). However, little is known of the relationships between endogenous hormones and BC risk in premenopausal women. Determining the association between circulating hormones and BC risk provides insight into etiology and may help identify high-risk women who would benefit from increased screening or chemoprevention.

In the current chapter, the possible role of estrogens, androgens, and prolactin in the etiology of BC in premenopausal women will be reviewed. Supporting data from laboratory experiments and the few previously published prospective studies of circulating levels will be briefly presented. Data from retrospective case–control studies will not be discussed as, in these studies, circulating hormone levels in the cases are evaluated after diagnosis and hence findings may be biased by the presence of the cancer or by cancer treatment. Finally, new findings on prolactin (7) and sex steroid concentrations, specifically 17β-estradiol (E_2), estrone, testosterone, and dehydroepiandrosterone sulfate DHEAS (8, 9) and BC risk from the large Nurses' Health Study (NHS) II cohort will be presented.

Background

Estrogen. Estrogen contributes to tumor growth by promoting the proliferation of cells with existing mutations or perhaps by increasing the opportunity for mutations (10). E_2, generally considered the most biologically active endogenous estrogen, circulates in the blood bound either to sex hormone-binding

J.J. Li et al. (eds.), *Hormonal Carcinogenesis V.*

globulin (SHBG) or albumin, or unbound ("free"). To date, prospective studies have produced mixed results, with nonsignificant positive associations reported in three (11–13) of five (11–15) small studies (< 80 cases each). The largest study to date, with 285 cases and 555 controls, was conducted within the European Prospective Investigation into Nutrition and Cancer (EPIC) cohort (16). Subjects were identified from seven European countries; a single blood sample was collected per woman and the day of collection within the menstrual cycle was recorded. No association was observed for either E_2 or estrone; however, because blood samples were collected across the menstrual cycle and the investigators had relatively limited ability to evaluate associations within specific parts of the cycle. The cyclic variation of premenopausal estrogen through the menstrual cycle complicates its investigation in epidemiologic studies; all previously published prospective studies included blood samples from women without restriction of phase or cycle day and, therefore, had limited power to examine phase-specific associations.

Androgens. Androgens have been hypothesized to increase BC risk either directly, by increasing the growth and proliferation of BC cells, or indirectly, by their conversion to estrogen (17). Both normal and malignant breast cells have androgen receptors and all the enzymes needed to convert androgens to estrogens. In animal and in vitro studies, androgens either increase or decrease cell proliferation, depending upon the model system. An effect of the adrenal androgen, DHEAS, has been hypothesized to depend on estrogen levels, such that before menopause, DHEAS exhibit antiestrogenic effects, but after menopause it is weakly estrogenic (18, 19).

Few prospective studies have evaluated the association of plasma testosterone or DHEAS levels and premenopausal BC risk. Results are quite consistent in showing an increased risk in women with higher testosterone levels (13, 16, 20). Although a significant positive association between DHEAS and risk has been consistently observed among postmenopausal women (4–6), results from prospective studies of premenopausal women are mixed, with positive (16), null (20, 21), and inverse (22, 23) associations observed; these differences may be related to the small size of several of the studies (20, 22, 23) as well as differences in the age of the populations assessed.

Prolactin. Prolactin is a polypeptide hormone that plays an important role in the proliferation and differentiation of normal mammary epithelium as well as in stimulating lactation (24, 25). Prolactin is produced primarily in the pituitary, although it is also locally produced in breast tissue (25) and prolactin receptors have been found on more than 50% of breast tumors (26). Prolactin administration also is well-documented to increase mammary tumor rates in mice (25). Cumulatively, substantial laboratory evidence suggests that prolactin could play a role in mammary carcinogenesis by promoting cell proliferation and survival, increasing cell motility, and supporting tumor vascularization (25). Prior prospective studies in premenopausal women have been small (n = 21–71 cases) and thus, not surprisingly, have not reported significant associations (11, 14, 27).

Nurses' Health Study II: Methods. The NHSII was established in 1989, when 116,678 female registered nurses completed and returned a questionnaire. The NHS cohort has been followed every 2 years since inception by questionnaire to update exposure variables and ascertain newly diagnosed disease. Between 1996 and 1999, 29,611 cohort members who were 32–54 years of age provided blood sample. Details of the collection have been previously published (7). In brief, women were sent a blood collection kit and asked to have blood samples drawn. Premenopausal women, who had not taken any type of hormones, been pregnant, or breastfed in the previous six months (n = 18,521), provided an initial 15 ml blood sample drawn on the third to fifth day of their menstrual cycle (follicular blood draw) and a second 30 ml blood sample drawn 7–9 days before the anticipated start of their next cycle (luteal blood draw; collectively called "timed samples"). Follicular samples were initially processed by the participants: the sample was placed in a refrigerator for 8–24 h and then the plasma was aliquoted into a labeled cryotube, which was frozen until the second blood collection. On the day of the luteal sample collection, the woman shipped both samples, via overnight courier and with an ice-pack, to our laboratory where the luteal blood draw was processed and stored in liquid nitrogen freezers. Women who were ineligible to provide timed blood samples (i.e., those with simple hysterectomy) provided a single 30 ml blood sample (an "untimed" sample). These samples were shipped and processed similar to the luteal timed samples.

Women providing timed blood samples completed a questionnaire that recorded the first day of the menstrual cycle during which the blood samples were drawn. They also completed and returned a postcard recording the first day of their next menstrual cycle, to determine the timing of the luteal phase blood draw. In addition, all women provided information on the date, time of day, and number of hours since last food intake for each blood draw, current weight, recent medication use, and current smoking status.

Women who provided a timed sample were considered to be premenopausal. Among women providing an untimed sample, a woman was considered to be premenopausal if she reported that her periods had not ceased or had a hysterectomy but had at least one ovary remaining and was ≤47 (for nonsmokers) or ≤45 (for smokers) years of age.

Cases were diagnosed with BC after blood collection but before June 1, 2003. In all, 235 premenopausal cases of BC were reported and confirmed. Time from blood draw to diagnosis ranged from 1 to 87 months (mean = 31 months). Cases were matched to two controls on age (years), menopausal status, month and year of blood draw, and ethnicity; additionally for each blood collection, we matched on time of day (± 2 h) and fasting status. Further, cases providing timed samples were matched on the luteal day of the blood collection (date of next period – date of luteal draw, ± 1 day). For each matching variable, > 90% of case–control pairs had exact matches.

Hormone assay methods have been described previously (7–9, 28). In brief, samples were assayed for E_2 and testosterone by radioimmunoassay (RIA) following extraction and celite column chromatography. Free E_2 and testosterone were calculated per Sodergard et al. (29). Sex hormone-binding globulin (SHBG) and progesterone were measured by chemiluminescent immunoassay with the Immulite auto-analyzer (Diagnostic Products, UK). Prolactin was measured using the AxSYM Immunoassay system (Abbott Diagnostics, Chicago, IL). The coefficients of variation (CVs) were 6–14% for all hormones except progesterone, which had a CV of 17%.

We examined follicular and luteal hormones separately. Quartile cut points were based on control distributions and determined separately by phase. Several samples had missing hormone values related to technical difficulties or low volume; the final sample size varied by phase and hormone. We used conditional logistic regression to estimate relative risks (RRs) and 95% confidence intervals (CIs). Multivariate models adjusted for body mass index (BMI) at age 18, family history of BC, ages at menarche and first birth, history of benign breast disease, and parity. Further adjustment for history of breastfeeding and past oral contraceptive use did not alter results. In stratified analyses, we used unconditional logistic regression, adjusting for matching factors. We evaluated estrogen ($ER\alpha^+$)/progesterone (PR^+) receptor cases ($n = 89$) separately, but could not evaluate other hormone receptor subtypes because there were so few cases. All p-values were two-sided tests and were considered statistically significant if ≤ 0.05.

Results

Both cases and controls were, on average, 42 years of age (SD = 3.8) and had a BMI of 25 kg m^{-2} (SD = 5). Follicular samples were collected on days 3.8 (1.0) and 3.9 (1.1) and luteal samples 7.7 (3.1) and 7.6 (2.9) days before the start of the next menstrual cycle for cases and controls, respectively.

Sex Steroids. Women with high follicular total and free E_2 levels had a significantly increased BC risk (4th vs. 1st quartile RR (95% CI) = 2.1 (1.1–4.1), p-trend = 0.08 and 2.4 (1.3–4.5), p-trend = 0.01, respectively) (Table 1). The associations were somewhat stronger for invasive cases. Other estrogens, in both phases of the menstrual cycle, were overall unassociated with disease. Similarly, progesterone was not related to BC risk.

Women with high testosterone levels in the follicular or luteal phase had a modest, nonsignificant increased BC risk (4th vs. 1st quartile RR (95% CI) = 1.3 (0.8–2.4), p-trend = 0.35 and 1.6 (0.9–2.8), p-trend = 0.10, respectively), although, particularly for follicular testosterone, the estimates did not increase linearly across quartiles (Table 1).

The associations were stronger and statistically significant when restricting to invasive tumors. Findings for free testosterone mirrored those for total testosterone (data not shown). We observed no association with DHEAS and risk either overall or among invasive cases only. Associations were essentially unchanged when we excluded the subset of women with anovulatory cycles (progesterone

Table 1 Multivariable relative risks (95% confidence intervals) of breast cancer by quartile of prediagnostic plasma sex steroid and prolactin levels, Nurses' Health Study II

Plasma hormone	Q1	Q2	Q3	Q4	*p*-trend
Estradiol (pg mL^{-1})					
Follicular levels	≤29	>29–44	>44–66	>66	
All cases	1.0	2.0 (1.1–3.6)	1.7 (1.0–3.2)	2.1 (1.1–4.1)	0.08
Invasive cases	1.0	2.9 (1.5–5.7)	2.0 (1.0–4.0)	2.7 (1.3–5.4)	0.07
Luteal levels	≤90	>90–120	>120–159	>159	
All cases	1.0	1.2 (0.7–2.3)	1.8 (1.0–3.3)	1.0 (0.5–1.9)	>0.99
Invasive cases	1.0	1.3 (0.7–2.6)	2.0 (1.1–3.8)	0.9 (0.4–1.9)	0.91
Free Estradiol (pg mL^{-1})					
Follicular levels	≤0.40	>0.40–0.56	>0.56–0.80	>0.80	
All cases	1.0	1.6 (0.9–2.9)	2.0 (1.1–3.6)	2.4 (1.3–4.5)	0.01
Invasive cases	1.0	1.8 (0.9–3.6)	2.1 (1.0–4.1)	2.7 (1.4–5.3)	0.01
Luteal levels	≤1.18	>1.18–1.59	>1.59–2.07	>2.07	
All cases	1.0	1.5 (0.8–2.8)	1.4 (0.8–2.6)	1.5 (0.8–2.8)	0.30
Invasive cases	1.0	1.4 (0.7–2.8)	1.3 (0.7–2.6)	1.3 (0.7–2.7)	0.56
Estrone (pg mL^{-1})					
Follicular levels	≤31	>31–39	>39–49	>49	
All cases	1.0	1.1 (0.6–1.8)	1.3 (0.8–2.1)	1.2 (0.7–1.9)	0.48
Invasive cases	1.0	1.3 (0.7–2.3)	1.2 (0.7–2.2)	1.4 (0.8–2.6)	0.25
Luteal levels	≤61	>61–79	>79–99	>99	
All cases	1.0	0.5 (0.3–0.9)	0.5 (0.3–0.9)	0.6 (0.3–1.0)	0.13
Invasive cases	1.0	0.6 (0.3–1.1)	0.5 (0.3–1.0)	0.7 (0.4–1.3)	0.30
Testosterone (ng dL^{-1})					
Follicular levels	≤15	>15–20	>20–26	>26	
All cases	1.0	1.3 (0.8–2.2)	1.4 (0.8–2.3)	1.3 (0.8–2.4)	0.35
Invasive cases	1.0	1.9 (1.0–3.4)	1.6 (0.8–3.0)	1.8 (0.9–3.4)	0.17
Luteal levels	≤20	>20–26	>26–32	>32	
All cases	1.0	1.3 (0.8–2.3)	1.4 (0.8–2.3)	1.6 (0.9–2.8)	0.10
Invasive cases	1.0	1.6 (0.9–3.1)	1.3 (0.7–2.4)	2.0 (1.1–3.6)	0.05
DHEAS (μg dL^{-1})					
Levels	≤52	53–74	75–103	≥104	
All cases	1.0	0.7 (0.4–1.2)	1.1 (0.6–1.8)	1.2 (0.7–2.1)	0.10
Invasive cases	1.0	0.9 (0.6–1.4)	1.2 (0.7–1.9)	1.4 (0.9–2.2)	0.07
Prolactin (ng mL^{-1})					
Levels[a]					
All cases	1.0	1.2 (0.7–1.9)	0.9 (0.6–1.5)	1.5 (1.0–2.5)	0.06
Invasive cases	1.0	1.2 (0.7–2.0)	0.9 (0.5–1.6)	1.6 (0.9–2.7)	0.09

[a]Batch specific cutpoints used: batch 1 ≤11.5, >11.5–15.3, >15.3–19.4, >19.4; batch 2 ≤11.0, >11.0–15.4, >15.4–21.2, >21.2

<400 ng ml^{-1}). None of the associations varied substantially by time between blood collection and diagnosis. With one exception, stratification by time since blood collection, age, BMI, or past oral contraceptive use did not substantially change any of the results. The association of DHEAS with BC significantly differed by age (*p*-interaction = 0.03). The RR comparing the top vs. bottom quartile was 0.6 (95% CI: 0.3, 1.3, *p*-trend = 0.48) for women <45-years old and 2.4 (95% CI: 1.1, 5.2, *p*-trend = 0.005) for women ≥45-years old.

When assessed using ERα$^+$/PR$^+$ cases only, in general, the associations appeared somewhat stronger. For example, relative risks, comparing the top to bottom quartile of levels, for follicular total and free E_2, and luteal testosterone were 2.7 (1.2–6.0), 2.8 (1.3–6.2), and 2.9 (1.4–6.0), respectively. DHEAS also was positively associated with ERα$^+$/PR$^+$ BC (comparable RR: 1.8, 95% CI: 1.0, 3.2, *p*-trend = 0.02).

Prolactin. We observed a modest positive association of borderline significance between prolactin and BC risk. Because the associations with BC were similar in each phase and prolactin levels did not vary substantially from the follicular to luteal phases, we used the average of the two phases for women both samples. The relative risk, comparing top to bottom quartiles was 1.5 (95% CI = 1.0–2.5). Findings were similar among invasive cases only, and appeared slightly stronger for ER$^+$/PR$^+$ cases (comparable RR = 1.9; 95% CI = 1.0–3.7). The prolactin/BC associations tended to be stronger in the first four years following blood collection, although the difference was not statistically significant (i.e., top to bottom quartile association by time from blood draw to diagnosis: <2 yr RR = 1.6; 3–3.9 RR = 1.8; >3.9 years RR = 1.1).

Discussion

We observed positive associations between BC risk and circulating levels of follicular total and free E_2, testosterone, and prolactin. Associations tended to be stronger among invasive cases and the small subset of ER$^+$/PR$^+$ tumors. For DHEAS, the association appeared to vary by age; no such interaction was noted for other hormones. No associations were observed for estrone or luteal phase estradiol. In contrast to our findings, no association was observed overall or by menstrual cycle phase in the recent, large prospective EPIC study (16).

However, although cases and controls were matched on menstrual cycle day, because blood samples were collected on any day of the menstrual cycle, these investigators had relatively limited power to assess associations within the early follicular phase.

There are several possibilities why we may have found a positive association with follicular E_2 but little association with luteal E_2 or estrone. First, early follicular E_2 levels may better reflect breast tissue estrogen exposure, as a greater proportion of follicular (vs. luteal) E_2 would derive from adipose vs. the ovary (29). Compared with the FSH-driven ovarian aromatase expression, aromatase in both adipose and normal breast tissue is regulated by a different promoter that is activated by factors such as IL-6 and TNFα. The low correlation we observed between follicular and luteal estradiol levels (*r*, total E_2 = 0.13; free E_2 = –0.02) provides some indirect support for this hypothesis. Alternatively, the follicular phase may be a more relevant period of exposure, given that women in later premenopause, as are included in this study, have longer follicular and slightly shorter luteal phases (30, 31). Finally, estrogen activity may differ between the low- and high-progesterone environments of the follicular and luteal phases. For example, E_2 increases whereas progesterone decreases antiapoptotic proteins (32, 33), such that apoptosis in lobuloalveolar cells is higher in the luteal vs. the follicular phase (34). Thus, the proliferative effects of high estrogen levels in the luteal phase may be offset by apoptosis. Given the range of potential hypotheses and that this is the first study to observe this association, the relationship between premenopausal estrogen levels and BC risk needs to be examined further.

Higher testosterone have been associated with increased invasive BC risk among premenopausal women in several prior prospective studies, with a nonsignificant association in one small study (13) and significant associations in two others (370 and 65 cases) (16, 20). Although our overall results with testosterone are somewhat weak, when we restricted our analyses to invasive cases, we observed associations similar to these previous reports. We provided the first assessment by tumor characteristics; the positive associations appeared stronger for invasive and $ER\alpha^+/PR^+$ tumors, although case numbers were limited.

The androgen DHEAS has been hypothesized to possibly act as an estrogen antagonist in a high estrogen environment (such as in premenopausal women) via competitive binding of the ERα, but act as a weak estrogen agonist in a low estrogen environment (17, 35). Alternatively, women who develop BC have been hypothesized to not have the normal age-related decline in DHEAS levels, but instead to have relatively constant levels over time, such that in early to mid-premenopausal years women with BC have lower DHEAS levels than normal controls but in late-premenopausal and postmenopausal years cases have higher levels (35).

Results from prospective studies of premenopausal BC and adrenal androgens generally have been contradictory. Three smaller studies (15–116 cases) reported an inverse (22, 23) or nonsignificantly positive association (20). In two larger studies, no association was reported between DHEAS and BC in older premenopausal women (21), while in EPIC a significant positive association was reported for DHEAS (6). Our study reported a positive association, primarily in $ER\alpha^+/PR^+$ tumors and among older women; we saw a suggestion of an inverse association in younger women. Although the reasons for the differences in study results are unclear, the larger studies generally suggested a positive association between DHEA and DHEAS levels in premenopausal women, despite the fact that this population has very high estrogen levels; hence epidemiologic data do not support the laboratory data indicating a possible antiestrogenic influence of DHEAS in premenopausal women.

Three previous small prospective studies in premenopausal women (n = 21–71 cases) did not find any significant associations between circulating prolactin concentrations and BC (11, 14, 27). The number of cases in these studies was very small, precluding the ability to detect even a moderate to strong association between premenopausal levels of prolactin and BC risk. We observed a positive association in the Nurses' Health Study II that was of similar magnitude to that observed among postmenopausal women in the Nurses' Health Study (NHS) (36).

Cumulatively, recent data from the NHS II cohort provide support for the important role that endogenous hormone in premenopausal play in influencing BC risk. However, given the relatively small number of cases in this study, and the small body of data on this topic worldwide, further large prospective investigations in this arena are needed.

References

1. Hankinson SE, Colditz GA, Willett WC (2004) Towards an integrated model for breast cancer etiology: the lifelong interplay of genes, lifestyle, and hormones. Breast Cancer Res 6:213–8.
2. Cummings SR, Eckert S, Krueger KA, et al. (1999) The effect of raloxifene on risk of breast cancer in postmenopausal women: results from the MORE randomized trial. Multiple Outcomes of Raloxifene Evaluation. JAMA 281:2189–97.
3. Cuzick J, Powles T, Veronesi U, et al. (2003) Overview of the main outcomes in breast-cancer prevention trials. Lancet 361:296–300.
4. Key T, Appleby P, Barnes I, et al. (2002) Endogenous sex hormones and breast cancer in postmenopausal women: reanalysis of nine prospective studies. J Natl Cancer Inst 94:606–16.
5. Missmer SA, Eliassen AH, Barbieri RL, et al. (2004) Endogenous estrogen, androgen, and progesterone concentrations and breast cancer risk among postmenopausal women. J Natl Cancer Inst 96:1856–65.
6. Kaaks R, Rinaldi S, Key TJ, et al. (2005) Postmenopausal serum androgens, oestrogens and breast cancer risk: the European prospective investigation into cancer and nutrition. Endocr Relat Cancer 12:1071–82.
7. Tworoger SS, Sluss P, Hankinson SE (2006) Association between plasma prolactin concentrations and risk of breast cancer among predominately premenopausal women. Cancer Res 66:2476–82.
8. Eliassen AH, Missmer SA, Tworoger SS, et al. (2006) Endogenous steroid hormone concentrations and risk of breast cancer among premenopausal women. JNCI 98:1406–15.
9. Tworoger SS, Missmer SA, Eliassen AH, et al. (2006) The association of plasma DHEA and DHEA sulfate with breast cancer risk in predominantly premenopausal women. Cancer Epidemiol Biomarkers Prev 15:967–71.
10. Henderson BE, Feigelson HS (2000) Hormonal carcinogenesis. Carcinogenesis 21:427–33.
11. Kabuto M, Akiba S, Stevens RG, et al. (2000) A prospective study of estradiol and breast cancer in Japanese women. Cancer Epidemiol Biomarkers Prev 9:575–9.
12. Rosenberg CR, Pasternack BS, Shore RE, et al. (1994) Premenopausal estradiol levels and the risk of breast cancer: a new method of controlling for day of the menstrual cycle. Am J Epidemiol 140:518–25.
13. Thomas HV, Key TJ, Allen DS, et al. (1997) A prospective study of endogenous serum hormone concentrations and breast cancer risk in premenopausal women on the island of Guernsey. Br J Cancer 75:1075–9.
14. Helzlsouer KJ, Alberg AJ, Bush TL, et al. (1994) A prospective study of endogenous hormones and breast cancer. Cancer Detect Prev 18:79–85.
15. Wysowski DK, Comstock GW, Helsing KJ, et al. (1987) Sex hormone levels in serum in relation to the development of breast cancer. Am J Epidemiol 125:791–9.
16. Kaaks R, Berrino F, Key T, et al. (2005) Serum sex steroids in premenopausal women and breast cancer risk within the European Prospective Investigation into Cancer and Nutrition (EPIC). J Natl Cancer Inst 97:755–65.

17. Liao DJ, Dickson RB (2002) Roles of androgens in the development, growth, and carcinogenesis of the mammary gland. J Steroid Biochem Mol Biol 80:175–89.
18. Ebeling P, Koivisto VA (1994) Physiological importance of dehydroepiandrosterone. Lancet 343:1479–81.
19. Seymour-Munn K, Adams J (1983) Estrogenic effects of 5-androstene-3 beta, 17 beta-diol at physiological concentrations and its possible implication in the etiology of breast cancer. Endocrinology 112:486–91.
20. Micheli A, Muti P, Secreto G, et al. (2004) Endogenous sex hormones and subsequent breast cancer in premenopausal women. Int J Cancer 112:312–8.
21. Page JH, Colditz GA, Rifai N, et al. (2004) Plasma adrenal androgens and risk of breast cancer in premenopausal women. Cancer Epidemiol Biomarkers Prev 13:1032–6.
22. Wang DY, Allen DS, De Stavola BL, et al. (2000) Urinary androgens and breast cancer risk: results from a long-term prospective study based in Guernsey. Br J Cancer 82:1577–84.
23. Helzlsouer KJ, Gordon GB, Alberg AJ, et al. (1992) Relationship of prediagnostic serum levels of dehydroepiandrosterone and dehydroepiandrosterone sulfate to the risk of developing premenopausal breast cancer. Cancer Res 52:1–4.
24. Freeman ME, Kanyicska B, Lerant A, et al. (2000) Prolactin: structure, function, and regulation of secretion. Physiol Rev 80:1523–631.
25. Clevenger CV, Furth PA, Hankinson SE, et al. (2003) The role of prolactin in mammary carcinoma. Endocr Rev 24:1–27.
26. Partridge RK, Hahnel R (1979) Prolactin receptors in human breast carcinoma. Cancer 43:643–6.
27. Wang DY, De Stavola BL, Bulbrook RD, et al. (1992) Relationship of blood prolactin levels and the risk of subsequent breast cancer. Int J Epidemiol 21:214–21.
28. Hankinson SE, Willett WC, Manson JE, et al. (1998) Plasma sex steroid hormone levels and risk of breast cancer in postmenopausal women. J Natl Cancer Inst 90:1292–9.
29. Sodergard R, Backstrom T, Shanbhag V, et al. (1982) Calculation of free and bound fractions of testosterone and estradiol-17 beta to human plasma proteins at body temperature. J Steroid Biochem 16:801–10.
30. Santoro N, Lasley B, McConnell D, et al. (2004) Body size and ethnicity are associated with menstrual cycle alterations in women in the early menopausal transition: The Study of Women's Health across the Nation (SWAN) Daily Hormone Study. J Clin Endocrinol Metab 89:2622–31.
31. Landgren BM, Collins A, Csemiczky G, et al. (2004) Menopause transition: Annual changes in serum hormonal patterns over the menstrual cycle in women during a nine-year period prior to menopause. J Clin Endocrinol Metab 89:2763–9.
32. Desreux J, Kebers F, Noel A, et al. (2003) Effects of a progestogen on normal human breast epithelial cell apoptosis in vitro and in vivo. Breast 12:142–9.
33. Gompel A, Somai S, Chaouat M, et al. (2000) Hormonal regulation of apoptosis in breast cells and tissues. Steroids 65:593–8.
34. Stute P, Wood CE, Kaplan JR, et al. (2004) Cyclic changes in the mammary gland of cynomolgus macaques. Fertil Steril 82 (Suppl 3):1160–70.
35. Adams JB (1998) Adrenal androgens and human breast cancer: a new appraisal. Breast Cancer Res Treat 51:183–8.
36. Tworoger SS, Eliassen AH, Rosner B, et al. (2004) Plasma prolactin concentrations and risk of postmenopausal breast cancer. Cancer Res 64:6814–9.

15
Ovulation and Ovarian Cancer

Stephen G. Hillier, Michael T. Rae, and Oliver Gubbay

Introduction

Ovarian cancer (OC) most frequently arises from the ovarian surface epithelium (OSE), which comprises a single layer of mesothelial, squamous-to-cuboidal cells covering the entire surface of the ovary (1). This dynamic cellular layer and underlying basement membrane is breached and repaired each time a follicle ovulates, which can happen up to around 400 times in an average woman's lifetime. It is, therefore, perhaps not surprising that there is a positive association between ovulation and OC and that a majority of OCs arise from the OSE. OC has genetic and environmental aetiologies, and there is growing evidence for inflammatory involvement as well. Ovulation is a natural inflammatory process, the suppression of which by pregnancy, breast-feeding, or oral contraception reduces OC risk. On the other hand, environmental factors and medical conditions associated with ovarian inflammation such as use of talc, endometriosis, ovarian cysts, and hyperthyroidism increase OC risk (2). If inflammation promotes cancer (3, 4), we argue that antiinflammation is quite likely to be protective. In this chapter, we rehearse evidence that inflammation is integral to ovulation and consider how associated antiinflammatory mechanisms might impact OC initiation and progression.

Ovulation: Inflammation

Ovulation bears cardinal features of an acute inflammatory response: vasodilatation, increased vascular permeability, and cellular infiltration driven by locally produced proinflammatory cytokines. The mid-cycle luteinising hormone (LH) surge initiates a cascade of biochemical changes in follicular cells and macrophages, leading to dissolution of the follicle wall and shedding of the oocyte. These changes include increased proinflammatory cytokine production, progesterone production, formation of prostanoids essential for ovulation, histamine release, and increased proteolysis, culminating in focal proteolysis of apical follicle wall, overlying connective tissue, and OSE to allow oocyte release (5, 6).

J.J. Li et al. (eds.), *Hormonal Carcinogenesis V.*

In the absence of pregnancy or lactation, ovulation is normally regular and frequent so mechanisms must exist to localise and limit tissue trauma and minimise emergent ovarian disease. During ovulation, OSE cells are exposed to inflammatory mediators capable of inducing genetic changes that predispose to neoplasia (7, 8) and the elimination of OSE cells overlying the ovulatory follicle via apoptosis is a likely oncoprotective mechanism brought into play before postovulatory reepithelialisation (9, 10). Inflammation, essential for wound healing but inherently harmful, must therefore be contained and rapidly resolved. Its deficient resolution could help explain the high incidence of ovarian tumors that develop from the OSE.

Postovulatory Healing

A healthy wound response entails an orchestrated sequence of events involving inflammation, reepithelisation, and connective tissue remodeling and maturation (11), central to which is controlled deposition and removal of collagen (12). Studies of the cytokine-regulated matrix hydrolyzing enzymes such as urokinase type plasminogen activator (uPA) and matrixmetalloproteinase-9 (MMP9) in OSE cells have led to the suggestion that loss of the basement membrane could provide a potential mechanistic link between ovulation and OC risk (13). Further evidence for pro-/antiinflammatory modulation of extracellular matrix (ECM) remodeling by OSE cells comes from studies of lysyl oxidase (LOX) mRNA expression in vitro, revealing stimulation by IL-1 (Interleukin-1) and suppression by cortisol (14). Besides its primary enzymic function to initiate cross-linkage of collagen and fibrils necessary for deposition of insoluble ECM (15), LOX is also directly implicated in tumorigenesis, being essential for hypoxia-induced metastasis of human breast, head, and neck tumors (16).

Inflammation and Ovarian Cancer

Cancers frequently arise at sites of chronic inflammation and inflammation may affect several levels of the cancer process relevant to ovary including initiation, progression, and metastasis (NCI Think Tanks in Cancer Biology: Inflammation and Cancer http://dcb.nci.nih.gov/thinktank/Executive_Summary_of_Inflammation_and_Cancer_Think_Tank.cfm). The inflammatory response at ovulation induces extensive tissue remodeling and accompanying death of OSE cells (7, 17). Mutations responsible for generating cancerous cells are believed to arise from DNA replication/repair errors during subsequent rounds of OSE cell proliferation. The rate of mutation is further believed to increase in the presence of toxic oxidants released during the inflammatory response (18).

Inflammatory signals produced during ovulation include the inflammatory cytokines TNFα and IL-1. Interleukin-1 and TNFα signal via membrane-associated receptors on target cells to increase the expression of genes with roles in inflammation (19). The type I IL-1 receptor (ILR) is a member the toll-like receptor (TLR) super-family involved in signal transduction during inflammation and host defence. Ligand binding of ILR activates postreceptor signaling networks that in turn activate stress-related transcription factors such as activated protein-1 (AP-1) and nuclear factor-κB (NFκB). NFκB and AP-1 are particularly important in immune and inflammatory responses because they localise the expression of genes encoding chemoattractants, cytokines, cytokine receptors, cell adhesion molecules, and matrix-degrading MMPs. Gene products stimulated by AP-1 and NFκB include the inducible cyclooxygenase-2 isozyme (COX2) that catalyzes the formation of prostanoids essential for ovulation and MMP9, which executes tissue breakdown necessary for oocyte release. Research indicates that each of these gene products is involved in positive regulation of tumor growth, progression and metastasis, and both COX2- and MMP9-deficient mice are tumour resistant (20, 21). AP-1 and NFκB are also molecular targets of antiinflammatory steroid action and long-term use of nonsteroidal antiinflammatory agents reduces the risk of many cancers, including ovary. All of which supports the concept that antiinflammatory agents negatively regulate key events in carcinogenesis (22, 23).

Ovulation: Anti-inflammation

OSE cell responses to inflammatory mediators include increased capacity to generate cortisol, thereby acquiring the potential to quell inflammation (8). Although ovarian cells are incapable of de novo glucocorticoid biosynthesis, they express genes (*11BHSD1* and *11BHSD2*) encoding 11β-hydroxysteroid dehydrogenase (11βHSD) enzyme proteins that interconvert cortisone and cortisol secreted by the adrenal gland. Of the two known 11βHSD isoforms, 11βHSD1 principally converts cortisone to cortisol while 11βHSD2 inactivates cortisol to cortisone. Thereby differential expression of 11βHSD1 and 11βHSD2 can determine local cortisol availability to engage glucocorticoid receptors (GR) and activate antiinflammatory GR (24). Within ovarian follicles granulosa cells express mainly 11βHSD2 mRNA but little or no 11βHSD1 mRNA until shortly before ovulation. However, once ovulation is triggered the 11βHSD2 mRNA level is suppressed and 11βHSD1 mRNA enhanced (25). This shift in potential for glucocorticoid metabolism from inactivation to activation is reflected in increased ability of granulosa cells to undertake predominantly reductive (cortisone→cortisol) metabolism and generate substantially raised concentrations of cortisol in follicular fluid (26). Treatment of granulosa cells and OSE cells with ovulation-associated cytokines such as IL-1 or TNFα causes similar up-regulated expression of 11βHSD1 mRNA and enzymic activity in vitro (26–28). Thus, cytokine-enhanced expression of 11βHSD1 may positively regulate the local availability of cortisol for antiinflammatory signaling at the site of follicular rupture in vivo. Importantly, cortisol not only suppresses the formation of potentially deleterious gene products such as COX2 and MMP9 in inflamed OSE cells but it simultaneously up-regulates expression of 11βHSD1 and GR (29). This provides a mechanism for feed-forward amplification of antiinflammatory signaling that might operate in vivo (Fig. 1).

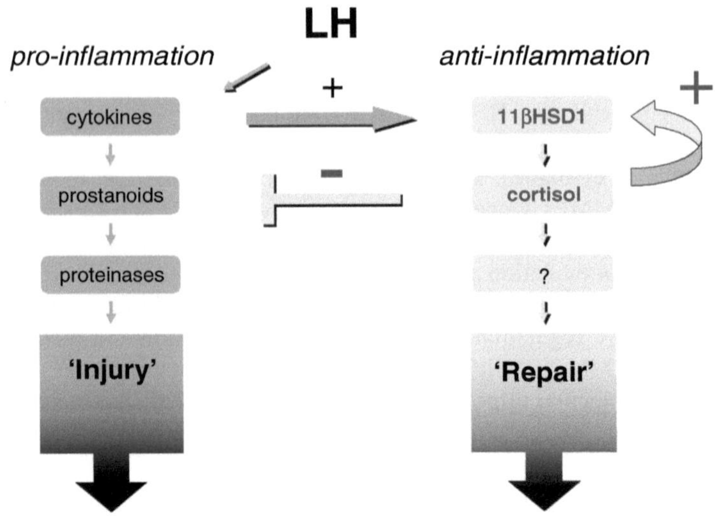

Fig. 1 Proposed role for locally activated cortisol in resolving inflammation of the OSE associated with ovulation. The LH surge induces ovulation via the initiation of an acute inflammatory reaction in the apical wall of the preovulatory follicle and overlying OSE, mediated by locally produced cytokines and prostaglandins. Proinflammatory mediators such as IL-1 and TNFα induce *HSD11B1* gene expression in OSE cells, increasing the local generation of cortisol from bioinert cortisone by the encoded 11βHSD1 enzyme. Locally activated cortisol binds to GR and initiates downstream antiinflammatory signaling in the OSE, including suppression of prostaglandin synthesis (i.e., inhibition of *COX2* expression). Cortisol further up-regulates cytokine-induced expression of *HSD11B1* and *GR* in the OSE, thereby amplifying antiinflammatory action at the site of natural injury (follicle rupture) and promoting subsequent tissue repair. Redrawn with permission from reference 8

Anti-inflammation and Ovarian Cancer

The association between chronic inflammation and OC implies an inflammatory basis for the progression of benign into malignant reproductive conditions (3, 30) and strengthens the concept that deficient antiinflammatory signaling at ovulation might promote ovarian cancer. Building on our hypothesis that cortisol activation by 11βHSD1 antagonises ovulation-associated inflammation (8), we asked whether 11βHSD1 expression is deficient in cell lines derived from patients with ovarian adenocarcinoma (31). It was also of interest to assess 11βHSD2, since this 11βHSD isotype is present at increased levels in tumors originating from breast, colon, adipose, adrenal, and pituitary tissue (32). We found that cultured human OSE cells possessed substantially higher levels of 11βHSD1 mRNA relative to the SKOV-3, BG-1, PEO-4, and PEO-14 OC cell lines studied (Fig. 2). Conversely, the cancer cell lines possessed substantially higher levels of 11βHSD2 mRNA than OSE cells. Moreover, 11βHSD1 mRNA expression in OSE cells was highly responsive to IL-1 but much less so in the cancer cells (Fig. 3) (33). The IL-1-responsive 11βHSD2

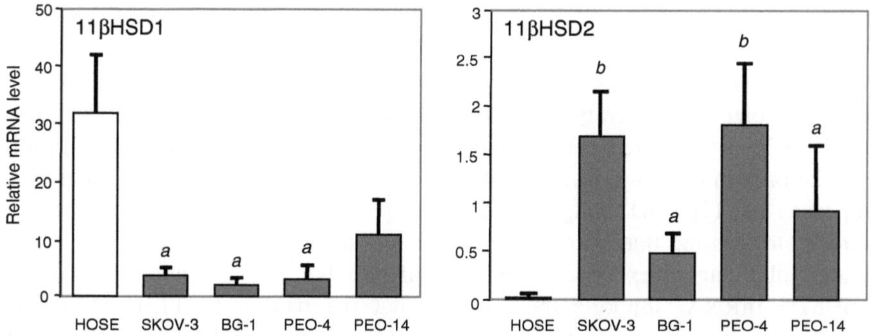

Fig. 2 Dysregulated 11βHSD expression in human ovarian cancer cells. 11βHSD1 and 11βHSD2 mRNA levels were compared among normal human OSE (HOSE) cells and a series of human OC lines, popularly used as in vitro models to gain mechanistic insight on human ovarian cancer: SKOV-3, BG-1, PEO-4, and PEO-14 cells. Levels of mRNA in total RNA from cultured cell monolayers were determined by quantitative (TAQMAN) RT-PCR standardized to 18S ribosomal RNA an internal control and normalised to a reference sample. Data ($n = 4$) are presented as fold-increase (mean ± SEM). Letters denote significance relative to HOSE cells: *a*: $P < 0.01$ and *b*: $P < 0.001$. Reproduced with permission from reference 33

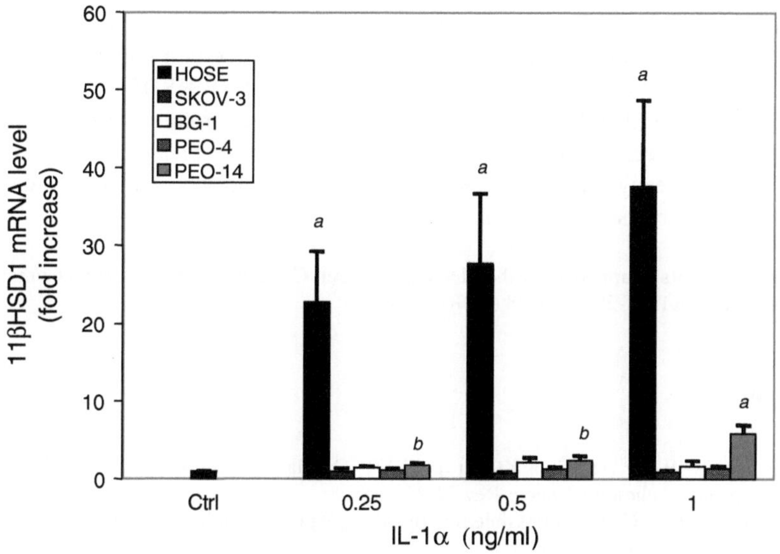

Fig. 3 OC cells show deficient *11βHSD* gene response to inflammation. Normal human OSE (HOSE) cells and human OC (HOC) lines were cultured for 48 h under identical conditions in presence of increasing concentrations of IL-1α, as indicated. Levels of mRNA were determined by quantitative (TAQMAN) RT-PCR standardized to 18S ribosomal RNA using an internal control and normalized to untreated cells. Data ($n = 6$) are presented as mean ± SEM. Letters denote significance above control (ctrl): *a*: $P < 0.05$ and *b*: $P < 0.01$ Reproduced with permission from reference 33

mRNA readout was less informative, with only very mild elevation in OSE cells and inconsistent results for cancer cells (33).

On the basis of their observation of divergent effects of 11βHSD1 and 11βHSD2 expression on cell proliferation in vitro, Rabbitt et al. (32) suggested that the ability of 11βHSD1 to generate cortisol might act as an autocrine antiproliferative, prodifferentiation stimulus in normal adult tissues. In contrast, the cortisol-inactivating properties of 11βHSD2 might lead to proproliferative effects, particularly in tumors. Our finding that normal OSE cells, unlike cell lines derived from ovarian tumors, abundantly express 11βHSD1 mRNA but are essentially devoid of 11βHSD2 mRNA, supports the concept that the pattern of 11βHSD isoform expression is a defining feature of neoplastic cellular transformation that could have particular relevance to the ovary (33). Clearly, it is necessary to determine whether these observations translate to primary ovarian tumor cells before their potential clinical relevance can be assessed.

Concluding Remarks

Ovulation is a natural injury-repair process in which localised and limited inflammation promotes rapid, high-fidelity, postovulatory surface healing. Factors related to OSE inflammation are associated with increased OC risk (2, 3) and antiinflammatory agents inhibit tumor invasion and protease production by ovarian carcinoma cell lines (34). Therefore, it is of interest that unlike normal OSE cells certain human ovarian carcinoma cell lines do not show up-regulated expression of 11βHSD1 when challenged with IL-1 in vitro (33). It remains to be determined whether similar loss of cytokine responsiveness occurs in primary OC cells. In which case, as prereceptor regulators of endogenous and exogenous antiinflammatory corticosteroid action, 11βHSDs could be potential OC markers or mediators.

Acknowledgments Supported by the Medical Research Council and the European Commission EU Contract EUK1-CT-2002-00128 (EURISKED).

References

1. Auersperg N, Wong AS, Choi KC, et al. (2001) Ovarian surface epithelium: biology, endocrinology, and pathology. Endocr Rev 22:255–88.
2. Ness RB, Cottreau C (1999) Possible role of ovarian epithelial inflammation in ovarian cancer. J Natl Cancer Inst 91:1459–67.
3. Balkwill F, Charles KA, Mantovani A (2005) Smoldering and polarized inflammation in the initiation and promotion of malignant disease. Cancer Cell 7:211–17.
4. Fleming JS, Beaugie CR, Haviv I, et al. (2006) Incessant ovulation, inflammation and epithelial ovarian carcinogenesis: revisiting old hypotheses. Mol Cell Endocrinol 247:4–21.
5. Espey LL (1994) Current status of the hypothesis that mammalian ovulation is comparable to an inflammatory reaction. Biol Reprod 50:233–8.

6. Richards JS, Russell DL, Ochsner S, et al. (2002) Ovulation: new dimensions and new regulators of the inflammatory-like response. Annu Rev Physiol 64:69–92.
7. Murdoch WJ, Wilken C, Young DA (1999) Sequence of apoptosis and inflammatory necrosis within the formative ovulatory site of sheep follicles. J Reprod Fertil 117:325–9.
8. Rae MT, Hillier SG (2005) Steroid signalling in the ovarian surface epithelium. Trends Endocrinol Metab 16:327–33.
9. Murdoch WJ (1995) Programmed cell death in preovulatory ovine follicles. Biol Reprod 53:8–12.
10. Ghahremani M, Foghi A, Dorrington JH (1999) Etiology of ovarian cancer: a proposed mechanism. Med Hypotheses 52:23–6.
11. Kirsner RS, Eaglstein WH (1993) The wound healing process. Dermatol Clin 11:629–40.
12. Verrecchia F, Mauviel A (2004) TGF-beta and TNF-alpha: antagonistic cytokines controlling type I collagen gene expression. Cell Signal 16:873–80.
13. Yang WL, Godwin AK, Xu XX (2004) Tumor necrosis factor-alpha-induced matrix proteolytic enzyme production and basement membrane remodeling by human ovarian surface epithelial cells: molecular basis linking ovulation and cancer risk. Cancer Res 64:1534–40.
14. Rae MT, Niven D, Ross A et al. (2004) Steroid signalling in human ovarian surface epithelial cells: the response to interleukin-1alpha determined by microarray analysis. J Endocrinol 183:19–28.
15. Kagan HM, Li W (2003) Lysyl oxidase: properties, specificity, and biological roles inside and outside of the cell. J Cell Biochem 88:660–72.
16. Erler JT, Bennewith KL, Nicolau M, et al. (2006) Lysyl oxidase is essential for hypoxia-induced metastasis. Nature 440:1222–6.
17. Bukulmez O, Arici A (2000) Leukocytes in ovarian function. Hum Reprod Update 6:1–15.
18. Ness RB, Cottreau C (1999) Possible role of ovarian epithelial inflammation in ovarian cancer. J Natl Cancer Inst 91:1459–673.
19. Karin M (2006) Nuclear factor-kappaB in cancer development and progression. Nature 441:431–6.
20. Coussens LM, Tinkle CL, Hanahan D, et al. (2000) MMP-9 supplied by bone marrow-derived cells contributes to skin carcinogenesis. Cell 1:481–90.
21. Tiano HF, Loftin CD, Akunda J, et al. (2002) Deficiency of either cyclooxygenase (COX)-1 or COX-2 alters epidermal differentiation and reduces mouse skin tumorigenesis. Cancer Res 62:3395–401.
22. Aggarwal BB, Shishodia S, Sandur SK, et al. (2006) Inflammation and cancer: How hot is the link? Biochem Pharmacol 72:1605–21.
23. Grosch S, Maier TJ, Schiffmann S, Geisslinger G, et al. (2006) Cyclooxygenase-2 (COX-2)-independent anticarcinogenic effects of selective COX-2 inhibitors. J Natl Cancer Inst 98:736–47.
24. Tomlinson JW, Walker EA, Bujalska IJ, et al. (2004) 11beta-hydroxysteroid dehydrogenase type 1: a tissue-specific regulator of glucocorticoid response. Endocr Rev 25:831–66.
25. Tetsuka M, Thomas FJ, Thomas MJ, et al. (1997) Differential expression of messenger ribonucleic acids encoding 11beta-hydroxysteroid dehydrogenase types 1 and 2 in human granulosa cells. J Clin Endocrinol Metab 82:2006–9.
26. Yong PY, Thong KJ, Andrew R, et al. (2000) Development-related increase in cortisol biosynthesis by human granulosa cells. J Clin Endocrinol Metab 85:728–33.
27. Tetsuka M, Haines LC, Milne M, et al. (1999) Regulation of 11beta-hydroxysteroid dehydrogenase type 1 gene expression by LH and interleukin 1beta in cultured rat granulosa cells. J Endocrinol 163:417–23.
28. Yong PY, Harlow C, Thong KJ, et al. (2002) Regulation of 11beta-hydroxysteroid dehydrogenase type 1 gene expression in human ovarian surface epithelial cells by interleukin-1. Hum Reprod 1:2300–6. Erratum in: Hum Reprod 17:3009.
29. Rae MT, Niven D, Critchley HO, et al. (2004) Antiinflammatory steroid action in human ovarian surface epithelial cells. J Clin Endocrinol Metab 89:4538–44.

30. Fleming JS, Beaugie CR, Haviv I, et al. (2006) Incessant ovulation, inflammation and epithelial ovarian carcinogenesis: revisiting old hypotheses. Mol Cell Endocrinol 247:4–21.
31. Langdon SP, Lawrie SS, Hay FG, et al. (1988) Characterization and properties of nine human ovarian adenocarcinoma cell lines. Cancer Res 48:6166–72.
32. Rabbitt EH, Gittoes NJ, Stewart PM, et al. (2003) 11beta-hydroxysteroid dehydrogenases, cell proliferation and malignancy. J Steroid Biochem Mol Biol 85:415–21.
33. Gubbay O, Guo W, Rae MT, et al. (2005) Inflammation-associated gene expression is altered between normal human ovarian surface epithelial cells and cell lines derived from ovarian adenocarcinomas. Br J Cancer 92:1927–33.
34. Altinoz MA, Korkmaz R (2004) NF-kappaB, macrophage migration inhibitory factor and cyclooxygenase-inhibitions as likely mechanisms behind the acetaminophen- and NSAID-prevention of the ovarian cancer. Neoplasma 51:239–47.

16
Prostate Cancer Susceptibility Loci: Finding the Genes

Elanie A. Ostrander and Bo Johannesson

Introduction

In 2006 alone, nearly 234,460 men in the USA will be diagnosed with prostate cancer (PC). In addition, about 27,000 will die *from*, rather than *with*, their disease (1). Although there is considerable variability in disease incidence by age, race, and family history, about 70% of all cases are ≥65 years (yrs) at diagnosis, and the median age at diagnosis in the USA is 68 yrs (2). The disease is more frequent among African American (AA) than Caucasians; Asian men have the lowest reported incidence (3, 4).

Segregation analyses were the first to suggest a strong hereditary component for PC, particularly among younger men. Two similar studies based on ascertainment of family history through probands treated with radical prostatectomy showed evidence for the dominant transmission of a rare high-risk allele/s (population prevalence of 0.3–0.6%), with carriers having an 88–89% risk of getting PC by age 85, compared with 3–5% in noncarriers (5, 6). Carter et al. have suggested that the cumulative proportion of PC cases within the U.S. population that is attributable to high-risk susceptibility alleles is 43% for men diagnosed ≤55 yrs, 34% for men ≤70 yrs, and 9% for men ≤85 yrs (5). By comparison, population-based studies from Sweden (6) and Australia (7, 8) estimate a higher population prevalence of carriers (1.1–1.67%) and a lower lifetime incidence (63–79%). The latter studies also suggest that 23% of all PC cases diagnosed <65 yrs may be due to inherited mutations in susceptibility genes (6).

In addition to evidence for autosomal dominant Mendelian inheritance, several studies also support an X-linked or recessive model with higher relative risks for PC in men with affected siblings compared with affected fathers. A segregation analysis from Australia found that the best fitting models included a dominantly inherited increased risk that was greater at younger ages (penetrance of 70% by age 80) and a recessively inherited or X-linked increased risk that was greater at older ages of diagnosis (8). The latter study also found that all two-locus models gave better fit than single-locus models, suggesting that multiple loci are responsible for the disease.

Several types of epidemiological studies also present compelling evidence for the existence of PC susceptibility loci. Both case–control and cohort studies show

J.J. Li et al. (eds.), *Hormonal Carcinogenesis V.*

that having a first-degree relative with PC increases a man's risk of being diagnosed with the disease by two- to threefold relative to those without a family history (9). If the relative is diagnosed before age 65 (RR = 5.9) or if there are ≥3 affected first-degree relatives (RR = 10.9), the risk is increases significantly (9–11). Twin studies report higher concordance rates for monozygotic (19–27%) than dizygotic (4–7%) twins (7, 12, 13), with the largest study reporting a relative risk (RR) for PC of 12.3 (95%, CI 8.4–18.1) in monozygotic twins (13).

Efforts to identify susceptibility loci for hereditary PC (HPC) have been ongoing for several years (14–30). At the heart of the problem is the extreme locus and disease heterogeneity that are hallmarks of the disease (31–35). With over a dozen genome scans completed to date, suggestive evidence for loci has been described on nearly every chromosome, and efforts to replicate results using seemingly similar data sets has been challenging (31–35). Yet, in recent years, it appears that progress is finally being made. Strategies for overcoming the problem of locus heterogeneity and ultimately identifying PC loci are the focus of this discussion.

Genome Wide Scans for Prostate Cancer Susceptibility

In 1993, Carter et al. (36) provided a definition for HPC based on families meeting at least one of the following criteria: (1) PC in ≥3 first-degree relatives, (2) PC in three successive generations of the maternal or paternal lineages, or (3) two first-degree relatives affected at age ≤55 yrs. We, like other researchers, have used these criteria to ascertain families for genome-wide scans aimed at finding PC susceptibility loci. In aggregate, the community has published the results from over a dozen genome wide scans (37–39). This has allowed the community to make several predictions regarding the genetic nature of PC in high-risk families. These are as follows: (1) PC susceptibility is caused by mutations in multiple genes; (2) Different genes are likely associated with distinct population frequencies in various ethnic groups; and (3) Distinct models of Mendelian inheritance and various levels of penetrance are associated with different loci.

Among data sets of hereditary families, several genes have been suggested as causative. However, there has been extreme difficulty in both confirming linkage results, as well as results derived from candidate genes analysis of selected populations. Among the most frequently debated genes are *HPC2/ELAC2* (40), *HPC1/RNASEL* (41, 42), and *BRCA2* (43). For each, both confirmatory and lack of replication studies are reported. In some cases, such as that of *HPC1/RNASEL*, replication studies have focused exclusively on isolated populations such as Ashkenazi Jewish patients or Finnish men, in an effort to reduce problems associated with heterogeneity (44–46).

In late 2003, eight genome-wide scans for PC susceptibility, including our own of 255 families, were published (21). The aggregate results are summarized in a review by Easton et al. (33, 38) and shown in Fig. 1. The eight scans include 1,293 families with multiple cases of PC. Across all studies, 11 peaks with LOD scores >2.0 were

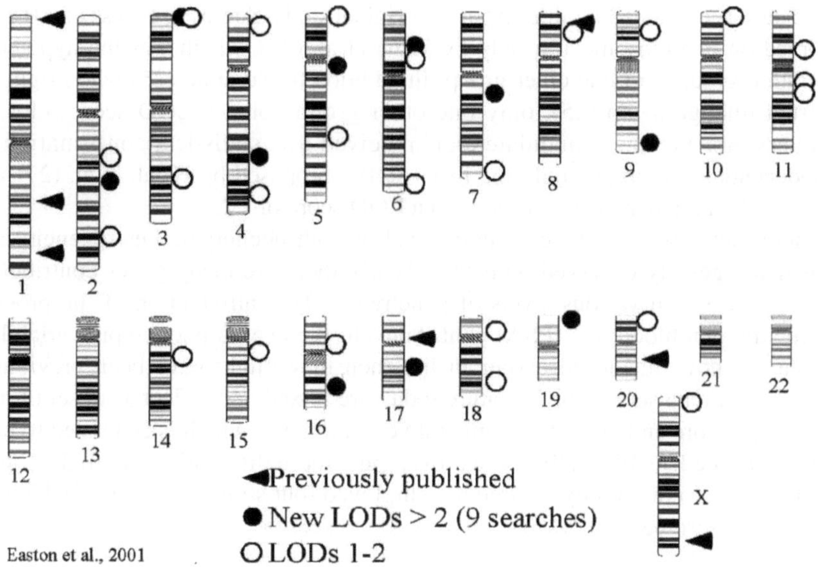

Fig. 1 Prostate cancer linkage peaks

observed, identifying regions on chromosomes 2, 3, 4, 5, 6, 7, 9, 16, 17, 19, and 20 as containing putative susceptibility loci. However, other than a replication by the group describing the initial results on chromosome 20, none of these peaks coincided precisely with any of the previously reported linkage peaks. The most promising were the results on chromosome 19 by Wiklund et al. (24), which is close to the peak on chromosome 19 originally reported by reported by Hsieh et al. (27).

Loci Mapped to Date

Of even greater surprise was the fact that no chromosomal region was reported with a significance level of 2.0 or greater by more than one study. In addition, only one study demonstrated a LOD score ≥ 3.0 that of Cunningham et al. (19), which reported a LOD score of 4.77 on chromosome 20q13. The same group had originally mapped HPC20 and the data in the replication study included a subset of the original mapping families. The highest reported LOD score by any other group on chromosome 20 was LOD = 1.02.

The results from other candidate loci were similarly discouraging. A total of 24 additional LOD scores > 1.0 were reported, identifying every chromosome as carrying a putative locus except 12, 13, 21, and 22. No LOD scores > 1.0 were observed at the HPC2 locus on 17p or the HPCX locus on the X chromosome. Indeed, LOD scores in excess of 1.0 were noted only by Xu et al. (25) and Cunningham et al. (19). Again, both data sets contained families from the original study. Scores in excess of 1.0 were observed on chromosome 16q only by Witte et al. (47), which was the hypothesis generating group, and one other group. In addition to Xu et al. (25) who originally reported linkage to 8p (25), only one other group noted a LOD score >1.0. In summary, none of the "candidate loci" received true statistical confirmation by an independent group and only the locus at 19p proposed by Hsieh et al. (27) was replicated by an independent group with a LOD score of ≥2.0.

These facts speak to the enormous level of both phenotypic heterogeneity and locus heterogeneity observed with PC. Clearly there are many genes contributing to the disease with various levels of penetrance. The introduction of the prostate specific antigen blood test (PSA) contributes to the overall phenotypic variability, as man are now diagnosed earlier in life then they might have been previously. Indeed, most data sets in the literature today are mixed and reflect a subset of men diagnosed before and after PSA came into common use, which is estimated to have occurred in the late 1980s. To try and overcome these difficulties and find PC loci, the PC research community at large has employed four strategies, each of which are discussed in turn below.

Metaanalysis and the International Consortium of Prostate Cancer Genetics

The PC mapping community has formed an international working group termed the International Consortium of PC Genetics (ICPCG). The goal of the ICPCG is to work together, often sharing data prior to publication, to generate large meta data sets with increased power for tackling problems related to PC susceptibility. The advantage of this strategy is that is allows extensive subclassification and stratification, while still retaining sufficient numbers in each strata such that statistically meaningful analysis can be done. Appropriate corrections for multiple corrections can be made, and accurate results still achieved.

Thus far, the group has focused on replication studies on chromosome 1 and 20 (48, 49). In the case of HPC1, the community has shown that only very large families with no evidence for linkage to the X chromosome likely can attribute their disease to mutations at HPC1. On chromosome 20, the consortium finds little evidence for replication in a dataset of over 1,200 families, calling into question the original finding (49). More recently, the ICPCG has also undertaken studies focused on families with an excess of aggressive disease (50).

Clinical Features of Disease

Several studies have focused exclusively on aggressive disease (51). While many men die *with* PC, it is those that die *of* PC that are the most clinically interesting and the ones that the research community most wishes to study. Initial studies focused on using Gleason score as a measure of aggressiveness. Gleason score is an assessment of tumor grade (52). To obtain a Gleason score, several regions of the tumor are independently scored by a pathologist and assigned a number of 1–5, representing well to poorly differentiated patterns. The higher the score the worst the prognosis (53). The two predominant scores are then added together to give a summary score between 2 and 10, with most tumors falling in the range of Gleason 5–7.

Some studies have treated Gleason score as a quantitative trait for outcome of disease aggressiveness, since it is reported to be a good predictor of survival (53). Others have treated the Gleason score as a covariate, using it to explain locus heterogeneity. For example, Witte et al. analyzed grade as a quantitative trait on 513 men from concordant sibships. They reported evidence for linkage on chromosomes 5q31–33, 7q32, and 19q12–13.11 (54). Using the same data set as Witte (54), Goddard et al. (18) used a variety of factors as covariates including sum of the sib-pair Gleason scores, mean family age at diagnosis, existence of male-to-male transmission (which argues against X linkage), and the number of affected first-degree relatives. In doing so, they detected linkage at three previously reported loci (1q24–25, 1q42.2–43, and 4q) and found linkage at Xq12–13 (LOD score 3.06, $p = 0.00053$), adjacent to the androgen receptor. In addition, they identified five other loci with LOD scores ≥ 2.5.

Interestingly, the loci at 1q24, Xq12, and chromosome 5 were evident only when Gleason score was considered as one of the covariates. In the absence of the covariates, results were weak to nonexistence. Others have found this approach to be similarly informative (55–57). Some of the strongest data using Gleason score as a measure of tumor aggressiveness have come from Slager et al. in a genome scan of 161 sib-pairs (58). They not only strongly confirm the linkage results for chromosome 19q ($p < 0.00001$), but report evidence for linkage on chromosome 4 ($p = 0.00012$). In a subsequent and independent study, involving 175 brother pairs from 103 families, the same group found evidence for linkage at 6q23 ($p = 0.0009$), 1p13-q21, and 5p13-q11 (59).

As more studies have been undertaken, it has become clear that a precise definition of aggressiveness is needed. Most recent studies have used a definition of aggressive disease (50, 60, 61) that includes families in which at least two genotyped men had at least one of the following disease features: regional- or distant-stage disease (based on pathology if a radical prostatectomy has been done, including T3, T4, N1, or M1, otherwise data from clinical staging are accepted); a Gleason score at diagnosis of ≥ 7 (poorly differentiated grade if no Gleason score is available); a pretreatment PSA score of 20 ng ml^{-1} or higher; and if deceased, death from metastatic PC at <65 yrs.

Using the above criteria, we reported suggestive evidence for linkage on chromosome 22 (dominant HLOD = 2.18) (61). We utilized clinical data from 784 affected men from 248 HPC families for whom a genomic screen had been previously performed (21). Disease characteristics described earlier were used to classify affected men into categories of clinically insignificant, moderate, or aggressive PC. Only men with aggressive disease were coded as affected in the linkage analysis. Suggestive linkage was observed at chromosome 22q11.1 (dominant HLOD = 2.75) and at 22q12.3-q13.1 using a recessive model of inheritance (HLOD = 1.90). Other studies have reported at least nominal evidence for linkage on chromosome 22. For instance, Lange et al. reported a LOD score of 1.87 at 45 cM in a set of 16 African American families and a LOD = 1.87 at 51 cM in 79 families with four or more affected men (22). A subset of younger-age-at-onset families from Utah (HLOD = 2.42) also gave evidence for linkage in this region (29).

Other Cancers

Although the earlier criteria have proven useful for dealing with the locus heterogeneity problem, they do not fully solve the problem. One additional strategy is that of examining high-risk PC families for an excess of other diseases and then using the resulting group of families in an isolated analysis.

Johannesson et al. have done that with high risk prostate–kidney cancer families (62). An association between these prostate and kidney cancer has been suggested by at least two studies. In a Swedish study, Grönberg et al. examined 1,364 relatives of 62 HPC families for the incidence of other cancers and found a significant association for kidney cancer (Standardized Incidence Ratio, SIR = 2.51; 95%, CI 1.15–4.77) (63). A second Swedish study utilizing the nation-wide Family Cancer Database reported a familial relative risk of 1.3 (95% CI 1.0–1.76) for kidney cancer in offspring of PC cases (64).

We selected a set of 15 prostate–kidney cancer families from among the 154 families on which we had completed a genome wide scan using 441 microsatellite markers (21) The 15 families all reported both primary PC and primary kidney cancer. All kidney cancer cases were confirmed by either death certificate or medical records. To be eligible for the study, the kidney case had to be either a PC case himself or a first-degree relative of a PC case. There are 191 (99 genotyped) individuals and all families were of Caucasian ancestry. Ten of the kidney cancer cases also had PC, and the other five kidney cancer cases were first-degree relatives of PC cases.

Although we found no statistically significant evidence for linkage in the initial analysis, we found two regions of suggestive linkage at 11q12 and 4q21, with HLOD scores of 2.59 and 2.10, respectively. The primary result on chromosome 11 was strengthened after excluding two families with members with transitional cell carcinoma (TCC). This was a valid strategy, as TCC is a very different and much rarer disease then in renal cell carcinoma, which was reported by the majority of families.

The nonparametric analysis revealed a Kong and Cox p-value of 0.004 for marker D11S1290 at 11p11.2. The 8 cM region between 11p11.2 and 11q12.2 was refined by the addition of 16 additional markers. The subset of HPC families with a median age of diagnosis >65 yrs demonstrated the strongest evidence for linkage (HLOD = 2.50). The p-values from nonparametric analysis ranged from 0.004 to 0.05 across five contiguous markers. There are some provocative candidate genes in the area, including the gene for *prostate-specific membrane antigen (PSMA)*. However, mutation screening of that gene is made difficult by the presence of another gene, termed *PSMA-Like,* that is located at 11q14.3 and shares 98% homology with the coding sequence of the *PSMA* gene itself. Studies pursing this interesting question continue in our laboratory and others.

Studies of Isolated Populations

Studies of isolated populations have frequently been undertaken in cancer genetics as a way to both deal with locus heterogeneity as well as identify founder mutations for specific tumor suppressor genes. In the case of PC, studies to date have focused on populations from Iceland (65–67), Finland (68), and Ashkenazi Jewish men (69–73).

In our collaborative group, Friedrichsen et al. have examined a genome wide scan in a set of 36 families of Ashkenazi Jewish origin (74). The 36 Jewish families represent a combined dataset of 17 Jewish families from the Fred Hutchinson Cancer Research Center (FHCRC)-based *PROGRESS* dataset, and 19 Ashkenazi Jewish families collected at Johns Hopkins University (JHU). All available family members, including 94 affected men, were genotyped using a set of microsatellite markers distributed across the genome at an average of about 8–10 cM density. To combine the two datasets, only markers present in the UCSC genome browser April 2003 assembly (http://genome.ucsc.edu/) were used (*PROGRESS* 421 markers, JHU 398 markers) and map order and distance between markers were taken from the UCSC map.

Since no segregation analysis has been done exclusively on Ashkenazi Jewish men, and models of inheritance were thus hard to predict, the data were analyzed primarily using nonparametric multipoint methods. The strongest signal in this data set was a significant linkage peak at 7q11–21, associated with a nonparametric linkage (NPL) score of 3.01 ($p = 0.0013$). Simulation analysis indicate that this corresponds to a genome-wide empirical $p = 0.006$. Empirical p values were calculated using the computer program Merlin (75), which was used to generate and analyze 1,000 replicates of the entire genome from the original dataset of 36 Jewish families.

After genotyping additional markers within the 7q11–21 peak, the NPL score increased to 3.35 ($p = 0.0004$) at marker D7S634 with an allele-sharing LOD of 3.12 ($p = 0.00007$). Detailed SNP analysis is underway in an attempt to find a shared haplotype that is over represented among affected vs. unaffected men within the Jewish families. Within that haplotype should lay the susceptibility gene and variant of interest.

We had noted a minor signal on chromosome 7q in our initial genome scan (21). We were thus curious as to the degree to which Jewish families accounted for that result. In the 254 *PROGRESS* families, we previously reported an HLOD of 2.25 (LOD = 1.55) at marker D7S2212 on 7q21 using a recessive parametric model. In a nonparametric analysis, we reported an NPL score of 1.79 ($p = 0.038$) in the same region. Analysis of the 237 non-Jewish families from the *PROGRESS* dataset yielded an NPL score of 1.11 ($p = 0.134$), revealing no evidence for linkage. This suggested clearly that the majority of result in the original genome-wide scan for the *PROGRESS* families was due to the presence of a modest number of Jewish families.

Similar conclusions were reached by investigators at JHU. In the genome-wide study of 188 JHU families by Xu et al., the strongest result on 7q was an allele-sharing LOD of 1.63 with marker D7S486, which is adjacent to the region of interest (25). When 17 of the 19 JHU Ashkenazi Jewish families were analyzed using D7S486, the allele-sharing LOD was only 0.04, suggesting that the Johns Hopkins-collected Ashkenazi Jewish families do not contribute significantly to the results previously reported for 7q22.

Our analysis of 36 Jewish families also highlighted regions on chromosomes 1q31–32, 2p11, 3q27–28, 14q12, and 20q11 with p values of 0.02–0.06. The strongest of these was at 14q12 ($p = 0.02$). Other minor peaks with an NPL p value ≤ 0.05 included 3q27–28 ($p = 0.03$) and 20q11 ($p = 0.04$). Although these may represent other loci that contribute to PC in Jewish, and perhaps non Jewish families, further investigation is clearly needed to draw definitive conclusions.

Summary

Studies to date suggest that PC is a genetically very heterogeneous disease. High-risk families, in which multiple men are affected likely, reflect the contributions of a number of genes, some that are rare and highly penetrant, while others are more common and weakly penetrant. In this review, we have discussed only the first type of loci, and found that the identification of such genomic regions is a formidable problem. Replication between seemingly similar data sets is weak, likely reflecting the older age of onset associated with the disease, the inability to collect affected individuals from more than two generations in a family, and the variation seen in disease presentation, in addition to the underlying locus heterogeneity. Indeed, the definition of PC is ever changing, as diagnostic criteria and tools for pinpointing early lesions improve.

Are we making progress? Clearly the answer is yes. The ability to divide large data sets into homogenous subset of families likely to share common genetic underpinnings has improved power to identify loci and reproducibility between loci is now more common. Indeed, several groups report linkage to loci on chromosomes 1, 17, 19, and 22. Key to our continued success is our ever increasing ability to understand the disease. Identifying the subset of men who are likely to get clinically significant disease is the goal of genetic studies like these, and identifying the underlying loci is the key for developing diagnostics. The willingness of the community to work together has been an important factor in the successes the community has enjoyed to date, and will likely be as important as we move forward to untangle the genetics of this complex and common disorder.

Acknowledgments We and our many colleagues are grateful to the many men who continue to participate in PC genetic studies across the world. The authors thank the National Human Genome Research Institute of the National Institutes of Health for their continued support.

References

1. Jemal A, Siegel R, Ward E, et al. (2006) Cancer Statistics, 2006. CA Cancer J Clin, *56:* 106–130.
2. Ries L, Harkins D, Krapcho M, et al. (2005) SEER Cancer Statistics Review, 1975–2003. Bethesda MD: National Cancer Institute.
3. Wingo PA, Bolden S, Tong T, et al. (1996) Cancer statistics for African Americans, 1996. CA Cancer J Clin, *46:* 113–125.
4. Walker B, Figgs LH, Zahm S (1995) Differences in cancer incidence, mortality, and survival between African Americans and whites. Environ Health Perspect, *103:* 275–182.
5. Carter BS, Beaty TH, Steinberg GD, et al. (1992) Mendelian inheritance of familial prostate cancer. Proc Natl Acad Sci USA, *89:* 3367–3371.
6. Schaid DJ, McDonnell SK, Blute ML, et al. (1998) Evidence for autosomal dominant inheritance of prostate cancer. Am J Hum Genet, *62:* 1425–1438.
7. Grönberg H, Damber L, Damber, JE, et al. (1997) Segregation analysis of prostate cancer in Sweden: Support for dominant inheritance. Am J Epidemiol, *146:* 552–557.
8. Cui J, Staples MP, Hopper JL, et al. (2001) Segregation analyses of 1,476 population-based australian families affected by prostate cancer. Am J Hum Genet, *68:* 1207–1218.
9. Stanford JL, Ostrander EA (2001) Familial prostate cancer. Epidemiol Rev, *23:* 19–23.
10. Steinberg GD, Carter BS, Beaty TH, et al. (1990) Family history and the risk of prostate cancer. Prostate, *17:* 337–347.
11. Cannon L, Bishop DT, Skolnick M, et al. (1982) Genetic epidemiology of prostate cancer in the Utah Mormon genealogy. Cancer Surv, *1:* 47–69.
12. Page WF, Braun MM, Partin AW, et al. (1997) Heredity and prostate cancer: a study of World War II veteran twins. Prostate, *33:* 240–245.
13. Lichtenstein P, Holm NV, Verkasalo PK, et al. (2000) Environmental and heritable factors in the causation of cancer–analyses of cohorts of twins from Sweden, Denmark, and Finland [see comments]. N Engl J Med, *343:* 78–85.
14. Smith JR, Freije D, Carpten JD, et al. (1996) Major susceptibility locus for prostate cancer on chromosome 1 suggested by a genome-wide search. Science, *274:* 1371–1374.
15. Suarez BK, Lin J, Burmester JK, et al. (2000) A genome screen of multiplex sibships with prostate cancer. Am J Hum Genet, *66:* 933–944.
16. Gibbs M, Stanford JL, Jarvik GP, et al. (2000) A genomic scan of families with prostate cancer identifies multiple regions of interest. Am J Hum Genet, *67:* 100–109.
17. Suarez BK, Lin J, Witte JS, et al. (2000) Replication linkage study for prostate cancer susceptibility genes. Prostate, *45:* 106–114.
18. Goddard KA, Witte JS, Suarez BK, et al. (2001) M. Model-free linkage analysis with covariates confirms linkage of prostate cancer to chromosomes 1 and 4. Am J Hum Genet, *68:* 1197–1206.
19. Cunningham JM, McDonnell SK, Marks A, et al. (2003) Genome linkage screen for prostate cancer susceptiblity loci: Results from the Mayo Clinic Familial Prostate Cancer Study. Prostate, *57:* 335–346.
20. Edwards S, Meitz J, Eles R, et al. (2003) Results of a genome-wide linkage analysis in prostate cancer families ascertained through the ACTANE consortium. Prostate, *57:* 270–279.

21. Janer M, Friedrichsen DM, Stanford JL, et al. (2003) Genomic scan of 254 hereditary prostate cancer families. Prostate, *57:* 309–319.
22. Lange EM, Gillanders EM, Davis CC, et al. (2003) Genome-wide scan for prostate cancer susceptibility genes using families from the University of Michigan prostate cancer genetics project finds evidence for linkage on chromosome 17 near BRCA1. Prostate, *57:* 326–334.
23. Schleutker J, Baffoe-Bonnie AB, Gillanders E, et al. (2003) Genome-wide scan for linkage in finnish hereditary prostate cancer (HPC) families identifies novel susceptibility loci at 11q14 and 3p25–26. Prostate, *57:* 280–289.
24. Wiklund F, Jonsson BA, Goransson I, et al. (2003) Linkage analysis of prostate cancer susceptibility: confirmation of linkage at 8p22–23. Hum Genet, *112:* 414–418.
25. Xu J, Gillanders EM, Wiley KE, et al. (2003) Genome-wide scan for prostate cancer susceptilbity genes in the Johns Hopkins Hereditary prostate cancer families. Prostate, *57:* 320–325.
26. Matsui H, Suzuki K, Ohtake N, et al. (2004) Genomewide linkage analysis of familial prostate cancer in the Japanese population. J Hum Genet, *49:* 9–15.
27. Hsieh CL, Oakley-Girvan I, Balise RR, et al. (2001) A genome screen of families with multiple cases of prostate cancer: evidence of genetic heterogeneity. Am J Hum Genet, *69:* 148–158.
28. Gillanders EM, Xu J, Chang BL, et al. (2004) Combined genome-wide scan for prostate cancer susceptibility genes. J Natl Cancer Inst, *96:* 1240–1247.
29. Camp NJ, Farnham JM, Cannon Albright LA (2005) Genomic search for prostate cancer predisposition loci in Utah pedigrees. Prostate, *65:* 365–374.
30. Xu J, Dimitrov L, Chang BL, et al. (2005) A combined genomewide linkage scan of 1,233 families for prostate cancer-susceptibility genes conducted by the international consortium for prostate cancer genetics. Am J Hum Genet, *77:* 219–229.
31. Ostrander EA, Stanford JL (2000) Genetics of prostate cancer: too many loci, too few genes. Am J Hum Genet, *67:* 1367–1375.
32. Ostrander EA, Markianos K, Stanford JL (2004) Finding prostate cancer susceptibility genes. Annu Rev Genomics Hum Genet, *5:* 151–175.
33. Easton DF, Schaid DJ, Whittemore AS, et al. (2003) Where are the prostate cancer genes? A summary of eight genome wide searches. Prostate, *57:* 261–269.
34. Verhage BA, Kiemeney LA (2003) Inherited predisposition to prostate cancer. Eur J Epidemiol, *18:* 1027–1036.
35. Schaid DJ (2004) The complex genetic epidemiology of prostate cancer. Hum Mol Genet, *13:* R103-R121.
36. Carter BS, Bova GS, Beaty TH, et al. (1993) Hereditary prostate cancer: epidemiologic and clinical features. J Urol, *150:* 797–802.
37. Singh R, Eeles RA, Durocher F, et al. (2000) High risk genes predisposing to prostate cancer development-do they exist? Prostate Cancer Prostatic Dis, *3:* 241–247.
38. Easton D, Peto J (1990) The contribution of inherited predisposition to cancer incidence. Cancer Surv, *9:* 395–416.
39. Nupponen NN, Carpten JD (2001) Prostate cancer susceptibility genes: many studies, many results, no answers. Cancer Metastasis Rev, *20:* 155–164.
40. Tavtigian SV, Simard J, Teng DH, et al. (2001) A candidate prostate cancer susceptibility gene at chromosome 17p. Nat Genet, *27:* 172–180.
41. Carpten J, Nupponen N, Isaacs S, et al. (2002) Germline mutations in the ribonuclease L gene in families showing linkage with HPC1. Nat Genet, *30:* 181–184.
42. Xu J, Zheng SL, Carpten JD, et al. (2001) Evaluation of linkage and association of HPC2/ELAC2 in patients with familial or sporadic prostate cancer. Am J Hum Genet, *68:* 901–911.
43. Edwards SM, Kote-Jarai Z, Meitz J, et al. (2003) Two percent of men with early-onset prostate cancer harbor germline mutations in the *BRCA2* gene. Am J Hum Genet, *72:* 1–12.
44. Rokman A, Baffoe-Bonnie AB, Gillanders E, et al. (2005) Hereditary prostate cancer in Finland: fine-mapping validates 3p26 as a major predisposition locus. Hum Genet, *116:* 43–50.

45. Rennert H, Bercovich D, Hubert A, et al. (2002) A novel founder mutation in the *RNASEL* gene, 471delAAAG, is associated with prostate cancer in Ashkenazi Jews. Am J Hum Genet, *71:* 981–984.
46. Rokman A, Ikonen T, Seppala EH, et al. (2002) Germline alterations of the *RNASEL* gene, a candidate *HPC1* gene at 1q25, in patients and families with prostate cancer. Am J Hum Genet, *70:* 1299–1304.
47. Witte JS, Suarez BK, Thiel B, et al. (2003) Genome-wide scan of brothers: replication and fine mapping of prostate cancer susceptibility and aggressiveness loci. Prostate, *57:* 298–308.
48. Xu J (2000) *ICPCG* Combined analysis of hereditary prostate cancer linkage to 1q24–25: results from 772 hereditary prostate cancer families from the International Consortium for Prostate Cancer Genetics. Am J Hum Genet, *66:* 945–957.
49. Schaid DJ, Chang BL (2005) Description of the International Consortium For Prostate Cancer Genetics, and failure to replicate linkage of hereditary prostate cancer to 20q13. Prostate, *63:* 276–290.
50. Schaid DJ, McDonnell SK, Zarfas KE, et al. (2006) Pooled genome linkage scan of aggressive prostate cancer: results from the International Consortium for Prostate Cancer Genetics. Hum Genet 120:471–85.
51. Ostrander EA, Kwon EM, Stanford JL (2006) Genetic susceptibility to aggressive prostate cancer. Cancer Epidemiol Biomarkers Prev, *15:* 1761–1764.
52. Gleason D (1997) Histologic grading and clinical staging of prostatic carcinoma. Philadelphia: Lea and Febiger.
53. Gleason DF (1992) Histologic grading of prostate cancer: a perspective. Hum Pathol, *23:* 273–279.
54. Witte JS, Goddard KA, Conti DV, et al. (2000) Genomewide scan for prostate cancer-aggressiveness loci. Am J Hum Genet, *67:* 92–99.
55. Neville PJ, Conti DV, Paris PL, et al. (2002) Prostate cancer aggressiveness locus on chromosome 7q32–q33 identified by linkage and allelic imbalance studies. Neoplasia, *4:* 424–431.
56. Neville PJ, Conti DV, Krumroy LM, et al. (2003) Prostate cancer aggressiveness locus on chromosome segment 19q12-q13.1 identified by linkage and allelic imbalance studies. Genes Chromosomes Cancer, *36:* 332–339.
57. Paiss T, Worner S, Kurtz F, et al. (2003) Linkage of aggressive prostate cancer to chromosome 7q31–33 in German prostate cancer families. Eur J Hum Genet, *11:* 17–22.
58. Slager SL, Schaid DJ, Cunningham JM, et al. (2003) Confirmation of linkage of prostate cancer aggressiveness with chromosome 19q. Am J Hum Genet, *72:* 759–762.
59. Slager SL, Zarfas KE, Brown WM, et al. (2006) Genome-wide linkage scan for prostate cancer aggressiveness loci using families from the University of Michigan Prostate Cancer Genetics Project. Prostate, *66:* 173–179.
60. Chang B, Isaacs S, Wiley K, et al. (2005) Genome-wide screen for prostate cancer susceptibility genes in men with clinically significant disease. Prostate, *64:* 356–361.
61. Stanford JL, McDonnell SK, Friedrichsen DM, et al. (2006) Prostate cancer and genetic susceptibility: a genome scan incorporating disease aggressiveness. Prostate, *66:* 317–325.
62. Johanneson B, Deutsch K, McIntosh L, et al. Suggestive genetic linkage to chromosome 11q12.1 in hereditary prostate cancer families with primary kidney cancer. Prostate (in press).
63. Gronberg H, Bergh A, Damber JE, et al. (2000) Cancer risk in families with hereditary prostate carcinoma. Cancer, *89:* 1315–1321.
64. Vaittinen P, Hemminki K (1999) Familial cancer risks in offspring from discordant parental cancers. Int J Cancer, *81:* 12–19.
65. Amundadottir LT, Sulem P, Gudmundsson J, et al. (2006) A common variant associated with prostate cancer in European and African populations. Nat Genet, *38:* 652–658.
66. Eldon BJ, Jonsson E, Tomasson J, et al. (2003) Familial risk of prostate cancer in Iceland. BJU Int, *92:* 915–919.

67. Jonsson E, Sigbjarnarson HP, Tomasson J, et al. (2006) Adenocarcinoma of the prostate in Iceland: a population-based study of stage, Gleason grade, treatment and long-term survival in males diagnosed between 1983 and 1987. Scand J Urol Nephrol, *40:* 265–271.
68. Rokman A, Baffoe-Bonnie AB, Gillanders E, et al. (2005) Hereditary prostate cancer in Finland: fine-mapping validates 3p26 as a major predisposition locus. Hum Genet, *116:* 43–50.
69. Hamel N, Kotar K, Foulkes WD (2003) Founder mutations in BRCA1/2 are not frequent in Canadian Ashkenazi Jewish men with prostate cancer. BMC Med Genet, *4:* 7.
70. Orr-Urtreger A, Bar-Shira A, Bercovich D, et al. (2006) RNASEL mutation screening and association study in Ashkenazi and non-Ashkenazi prostate cancer patients. Cancer Epidemiol Biomarkers Prev, *15:* 474–479.
71. Poynter JN, Cooney KA, Bonner JD, et al. (2006) APC I1307K and the risk of prostate cancer. Cancer Epidemiol Biomarkers Prev, *15:* 468–473.
72. Bar-Shira A, Matarasso N, Rosner S, et al. (2006) Mutation screening and association study of the candidate prostate cancer susceptibility genes *MSR1*, *PTEN*, and *KLF6*. Prostate, *66:* 1052–1060.
73. Dagan E, Laitman Y, Levanon N, et al. (2006) The 471delAAAG Mutation and C353T Polymorphism in the *RNASEL* Gene in Sporadic and Inherited Cancer in Israel. Fam Cancer, *31:* 31.
74. Friedrichsen DM, Stanford JL, Isaacs SD, et al. (2004) Identification of a prostate cancer susceptibility locus on chromosome 7q11–21 in Jewish families. Proc Natl Acad Sci USA, *101:* 1939–1944.
75. Cook Jr, EH (2002) Merlin: faster linkage analysis with improved genotyping error detection. Pharmacogenomics J, *2:* 139–140.

Part 5
Novel Strategies for Preventing and Treatment of Endocrine-related Cancers

17 Strategy for the Application of Therapy in Prostate Cancer

Christopher J. Logothetis

Introduction

The rational application of integrated multidisciplinary therapy of prostate cancer (PC) requires an understanding of the determinants of cancer progression and response to therapy to successfully treat common adult cancers. In the successful examples of therapy application agents (chemotherapy, biologics) and other modalities (radiation, surgery) are applied on the basis of clinical biology during specific intervals in progression. Thus integrating the knowledge of clinical biology with an appreciation for the determinants is central to therapy development.

Examples of multidisciplinary therapies selectively applied, based on a rational strategy, and developed, based on an understanding of the underlying biology, are germ cell cancers and leukemia. In both diseases, the agents (chemotherapy, biologics) and other treatment modalities (radiation, surgery) were chosen on the basis of clinical biology. If this paradigm can be generalized to prostate cancer and the development of successful therapy for human PC is to follow this paradigm, a categorization based on the understanding of prostate carcinogenesis will be required.

Treatments must take advantage of disease-specific opportunities for intervention as they clinically present. The widespread clinical application of prostate-specific antigen (PSA) levels has increased the opportunities for intervention by creating new "disease states" (1). The development of that can "therapeutically modulate" the target.

The elements required to efficiently develop effective therapy strategies for cancer are: (1) a biological framework for the application of therapy, (2) surrogate markers of relevant biologic events, (3) methodologies to validate laboratory-based observations and accelerate the application of new therapies, and (4) new therapy targets and therapeutic agents. A useful categorization of cancers into relevant groups has a central facilitating role in the development and application of new therapy (2).

The existing anatomic-based classification of PC has served radiotherapists and surgeons well. The classification has promoted a view of PC that has facilitated the development of therapies whose effectiveness is determined by tumor volume and

J.J. Li et al. (eds.), *Hormonal Carcinogenesis V.*

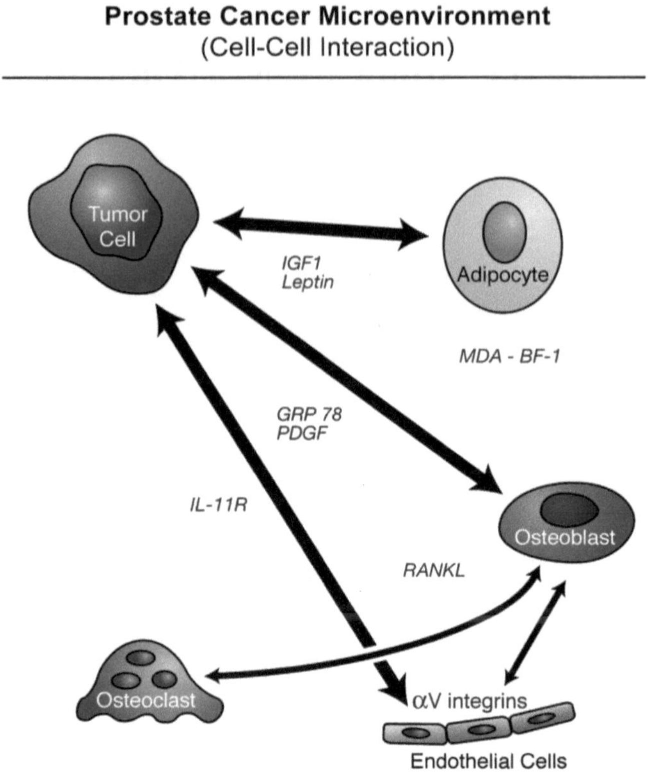

Fig. 1 Cell–cell interaction at the center of prostate cancer progression

degree of dissemination. However, the anatomic-based classification is not useful for the development of therapies aimed at controlling other aspects of PC.

The paracrine regulation of PC cell partners in the progression of disease is central to the biology of the cancer. The remarkably predictable phenotype of clinical PC with the initial effective control by androgen ablation followed by the dominance of the bone–epithelial interaction has led us to study the role of host and tumor cell partners at the center of this (Fig. 1) (2).

Predisposition to Prostate Cancer Progression (Amphotropic Phase)

The development of a broader therapeutic framework that includes events implicated in predisposition to prostate carcinogenesis is necessary because clinicians will need such a framework to apply prevention strategies and exploit the newly identified therapy targets.

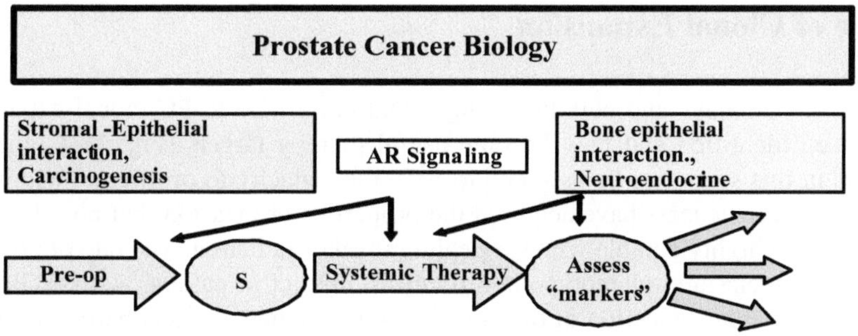

Fig. 2 Prostate cancer biology

Factors that increase the risk of developing PC are being identified (5–6). Strom et al. have noted that cancer progression to the aggressive phenotype is increased within obese patients or those with a high fat diet (7). Available clinicopathological correlative data support the view that prostatic intraepithelial neoplasia III (PIN III) is a precursor lesion of human PC (8). Investigators have also described inflammatory lesions that appear to correlate with an increased risk for PC and are mechanistically implicated in carcinogenesis (9). It has been postulated that epigenetic events when superimposed on a genetic predisposition and interaction with the environment contribute to prostate carcinogenesis. We have focused on the understanding of the host–tumor cell interactions that are implicated in PC progression. These interactions are at the center of the PC malignant phenotype (Fig. 2).

Therapy for Amphotropic Phase

The assumption that governs the selection of therapy for the amphotropic phase is that all of the processes contributing to an increased predisposition are either reversible or can be meaningfully modulated. The national clinical Selinium, Vitamin E, Chemoprevention Trial (SELECT) currently underway will provide substantial information regarding the biology of early PC and support or refute ongoing assumptions. Dietary intervention has also been proposed as a means of prevention.

Dr. Kim et al. have initiated a strategy to apply a fixed pre-op dose of anti-androgen therapy in hope of distinguishing tumors with the potential to become aggressive from the remainder (10). The prostate is an organ that can be readily visualized. An approach similar to that which has been successfully used in daily practice to image the prostate for the delivery of brachytherapy is being employed to deliver intraprostatically novel agents to inhibit the progression of PC. It is likely that the population targeted for chemoprevention is as heterogeneous as that with advanced PC, and the design of future studies may need to take this into account.

Phase of Clonal Expansion

Molecular pathways and cellular changes that contribute to PC clonal expansion have been identified and may become useful therapy targets (Fig. 1). Clinically, cancers in this stage can be assumed to have the capacity to proliferate within the prostate. These cancers have acquired the properties of invasion, but not of metastasis, and are highly curable with surgical removal or radiation. The rate of progression to the metastatic phenotype is difficult to predict in patients with a Gleason score of 7, making selection of therapy difficult. The pathological hallmark of this stage of PC progression is low-grade (Gleason score ≤6), low-volume, and organ-confined cancer (11–14). Defining this population with a more reproducible molecular characterization is an area of intense research interest.

Treatment for Patient in the Clonal Expansion Phase (Localized Disease with Low Risk for Metastases)

The management of patients in the clonal expansion phase is principally influenced by the recognition that the process is irreversible but confined to the prostate. Because there are no therapies that can arrest or reverse this process (such as differentiation therapy or angiogenesis inhibition), anatomically targeted cytotoxic therapy is the treatment of choice. Existing therapies in this category are radiation therapy, surgery, and cryotherapy. Single-modality therapy is effective because the cancer is anatomically confined to the prostate. Some patients whose cancer is in this category may be considered for "watchful waiting" if the cancer survival is predicted to be longer than the patient's survival because of comorbid conditions. Critical to the development of therapy for this category of patient are the refinement of existing predictors and the development of new ones. The clinical challenges in this category of patient are centered on two issues: morbidity of therapy and proper selection of patients for ablative therapy. Patients in this category will benefit most from the efforts to develop alternatives and reduce the morbidity of radiation or surgery.

Heterotopic Phase (Potential for Metastases)

In the most advanced phase of PC progression, the cancer has acquired the capacity to develop metastasis. The distinguishing biologic feature of this category of prostate cancer is the existence of a host–epithelium interaction supportive of cancer growth in heterotopic sites. Thus, these cancers have escaped the constraints of the prostate microenvironment and acquired the capacity to adapt to a heterotopic environment. In this setting, targeting the tumor microenvironment becomes a critical aspect and an overarching therapy strategy. The laboratory hallmarks of this clinical stage include the expression of markers of stromal–epithelial communication. Candidate markers in the category that are currently being investigated include serum interleukin-6, serum vascular endothelial growth factor (VEGF), urinary basic fibroblast growth factor (bFGF), and platelet-derived growth factor receptor (PDGFr). Markers that have been traditionally used for clinical staging include lactate dehydrogenase (LDH) and alkaline phosphatase.

The current treatment for patients in this category includes cytotoxic agents and androgen ablation. Androgen ablation is the main therapy and appears more effective the earlier in the course of the illness that it is applied. However, it is often not used early because of the side effects of continuous and sustained androgen ablation. Chemotherapy is being applied increasingly as clinical evidence has shown that some of the widely employed agents can achieve significant palliation (Fig. 3) when used in combination for patients with prostate cancer (15, 16).

To improve effectiveness, strategies targeting the bone–epithelial interaction have been proposed and tested. The clinical studies have been supported by laboratory-based investigation that implicates specific pathways in PC progression (17–21). Further clinical benefit may be achieved by integrating existing treatments and bone-targeting therapy (22).

The hypothesis is that combining systemic cytotoxic therapy with agents targeting the factors implicated in host–epithelial interaction of neoplastic growth in heterotopic sites will increase the effectiveness of therapy. Host response factors may include angiogenesis, metalloproteases, and paracrine loops [epidermal growth factor receptor/tumor-derived growth factor beta (EGF-TGFβ, PDGFr)]. An integrated research strategy incorporating experimental systems and clinical investigation is the approach we have adopted to develop integrated therapy for bone metastases.

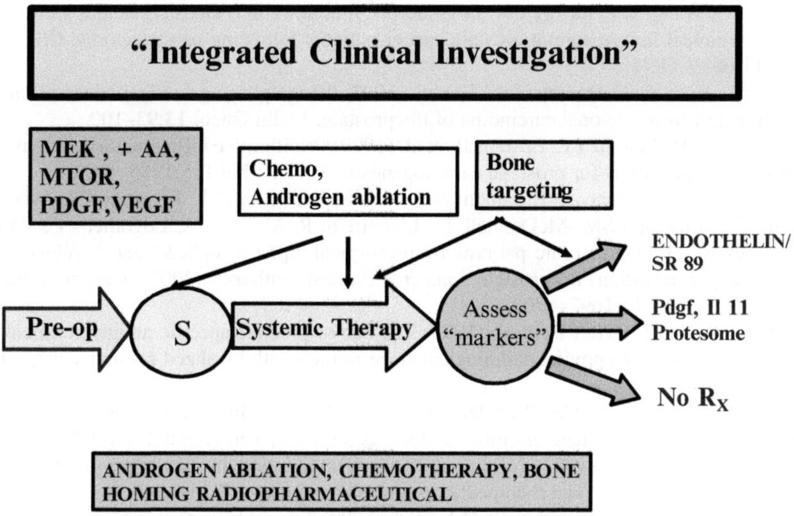

Fig. 3 Treatment strategies for heterotopic phase (metastatic phase)

Conclusion

The development of an integrated treatment strategy using the combination of cytotoxic and molecular-targeted therapy requires an understanding of the underlying biology. The development of clinical methodologies that will allow us to efficiently assess the efficacy of individuals "targeted" therapeutics and develop rational combinations is an essential component of a therapy development strategy. The strategy for linked PC progression, to clinical state of progression, and finally to therapy options may result in a strategy both for the applied study of PC (Fig. 2) and rational application for therapy (Fig. 3). It is assumed that a biologic-based application of therapy will increase the prospects for the control of the cancer.

References

1. Scher HI, Mazumdar M, Kelly WK (1996) Clinical trials in relapsed prostatcancer: Defining the target. J Natl Cancer Inst 88:1623–1634.
2. Logothetis CJ (2003) The case for a biologically based classification of prostate cancer. Semin Oncol 30:562–566.
3. Lee WH, Isaacs WB, Bova GS, et al. (1997) CG island methylation changes near the *GSTP1* gene in prostatic carcinoma cells detected using the polymerase chain reaction: A new prostate cancer biomarker. Cancer Epidemiol Biomarkers Prev 6:443–450.
4. Nelson WG, DeWeese TL, De Marzo AM, et al. (2002) Prostate cancer prevention, in Kantoff PW, Carroll PR, D'Amico AW (eds): Prostate Cancer: Principles and Practice. Philadelphia, PA, Lippincott Williams & Wilkins, pp. 67–72.
5. Mohan S, Baylink DJ (1991) Bone growth factors. Clin Orthop 263:30–48.
6. Whitmore Willet F (1990) Natural history of low-stage prostatic cancer and the impact of early detection. Urol Clin North Am 17:689–697.
7. Strom SS, Wang X, Pettaway CA, Logothetis CJ, et al. (2005) Obesity, weight gain and risk of biochemical failure among prostate cancer patients following prostatectomy. Clin Cancer Res 11:6889–6894.
8. Harlan L, Brawley O, Pommerenke F, et al. (1995) Geographic, age and racial variation in the treatment of local/regional carcinoma of the prostate. J Clin Oncol 13:93–100.
9. De Marzo AM, Marchi VL, Epstein JI, et al. (1999) Proliferative inflammatory atrophy of the prostate: Implications for prostatic carcinogenesis. Am J Pathol 155:1985–1992.
10. Kim J*, Sun P*, Troncoso P, Sabichi AL, Babaian RJ, Pisters LL, Pettaway CA, Wood CG, Lam YW, Lippman SM, McDonnell TJ, Lieberman R, Mobley JA, Logothetis CJ, Ho SM. Changes in serum proteomic patterns by presurgical alpha-tocopherol and L-selenomethionine supplementation in prostate cancer (*co-first authors) (2005) Cancer Epidemiol Biomarkers Prev 14:1697–1702.
11. Partin AW, Yoo J, Carter B, et al. (1993) The use of prostate specific antigen, clinical stage and Gleason score to predict pathological stage in men with localized prostate cancer. J Urol I50:110–114.
12. Kattan MW, Stapleton AM, Wheeler TM, et al. (1997) Evaluation of a momogram used to predict the pathologic stage of clinically localized prostate carcinoma. Cancer 79:528–537.
13. Di Sant' Agnese PA (1992) Neuroendocrine differentiation in carcinoma of the prostate: Diagnostic, prognostic, and therapeutic implications. Cancer 70:254–268.
14. Pound CR, Partin AW, Eisenberger MA, et al. (1999) Natural history of progression after PSA elevation following radical prostatectomy. JAMA 281:1591–1597.

15. Ellerhorst JA, Tu SM, Amato RJ, et al. (1997) Phase II trial of alternating weekly chemohormonal therapy for patients with androgen-independent prostate cancer. Clin Cancer Res 3:2371–2376.
16. Millikan RE, Baez L, Banerjee T, et al. (2001) Randomized phase 2 trial of ketoconazole and ketoconazole/doxorubicin in androgen-independent prostate cancer. Urol Concol 6:111–115.
17. Glotzman D (1997) Mechanisms of the development of osteoblastic metastasis. Cancer 80:1581–1587 (suppl).
18. Navone NM, Labate ME, Troncoso P, et al. (1999) p53 mutations in prostate cancer bone metastases suggest that selected p53 mutants in the primary site define foci with metastatic potential. J Urol 161:304–308.
19. Antwerp DJV, Martin SJ, Kafri T, et al. (1996) Suppression of the TNFα-induced apoptosis by NFκB. Nature 274:787–789.
20. McDonnell TJ, Troncoso P, Brishay SM, et al. (1992) Expression of the proto-oncogene bcl-2 in the prostate and its association with the emergence of androgen-independent prostate cancer. Cancer Res 52:6940–6944.
21. Macoska JA, Trybus TM, Sakr WA, et al. (1994) Fluorescence in situ hybridization analysis of pallelic loss and chromosomal instability in human prostate cancer. Cancer Res 54:3824–3830.
22. Tu SM, Millikan RE, Mengistu B, et al. (2001) Bone-targeted therapy for advanced androgen-independent carcinoma of the prostate: A randomized phase II trial. Lancet 357:336–341.

18
Response and Resistance to the Endocrine Prevention of Breast Cancer

Anthony Howell, Nigel J. Bundred, Jack Cuzick, D. Craig Allred, and Robert Clarke

Introduction

Current endocrine approaches to breast cancer (BC) prevention are targeted mainly at the estrogen receptor alpha (ERα) (1). Endocrine prevention depends upon either reducing the concentration of estrogen reaching $ER\alpha^+$ breast epithelial cells [estrogen deprivation (ED)] by, for example, ovarian suppression in premenopausal women, or aromatase inhibition in postmenopausal women, or by blocking the interaction of estrogen with ERα by selective estrogen receptor modulators (SERMs) such as tamoxifen (Tam) and raloxifene (Ral). Observational studies (2–11) and randomised controlled trials (12–25) indicate that treatment by ED and SERMS reduces the risk of subsequent BC by approximately 50% and that this effect is prolonged (16, 18, 20, 26). In the human breast, ERα is detectable in epithelial cells of lobules and ducts and not in the stroma (27). However, it is not clear whether normal, premalignant or malignant epithelial cells are the targets for endocrine prevention. In this chapter, we summarise the results of endocrine prevention trials to date and what is known about the response of the three types of potential target structures to estrogen stimulation and inhibition by ED and SERMS.

Clinical Endocrine Prevention Studies

Data concerning the effectiveness of BC endocrine prevention comes from several sources. One is observational studies comparing women who have had, for example, an oophorectomy or Tam with those who have not (2–11), another is estimating the change in incidence of contralateral BC in women treated with adjuvant endocrine therapy for ipsilateral BC (5, 6, 9, 12–14) and a third is randomised trials comparing an endocrine intervention with placebo (or another endocrine intervention) in women at increased risk or at population risk of BC (15–25). Clearly randomised controlled trials are the most important studies and several have and are being conducted since the first study, the Royal Marsden Trial of Tam vs. placebo, was initiated in 1986 (15, 16). The trials, their recruitment periods, and numbers of women randomised are shown in Fig. 1.

J.J. Li et al. (eds.), *Hormonal Carcinogenesis V.*

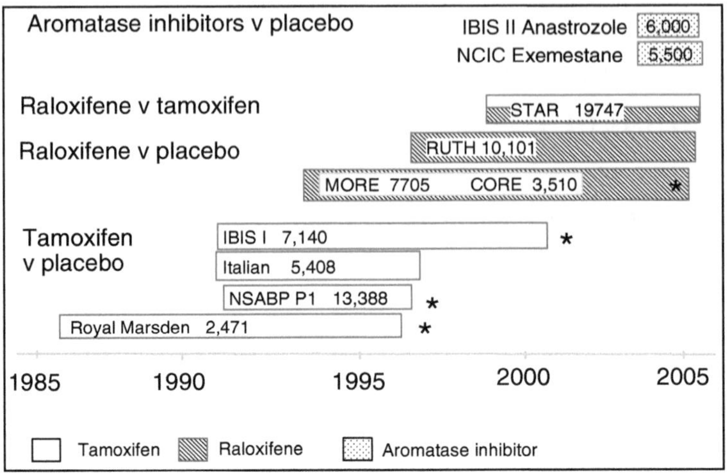

Fig. 1 Recruitment periods of the randomised controlled trials of endocrine prevention vs. placebo or other endocrine intervention. With the exception of the Royal Marsden Trial where women were treated for up to 8 yrs the treatment period with SERMs and AIs is for 5 yrs. Abbreviations: *NSABP* National Surgical Adjuvant Breast Project, *IBIS* International Breast Intervention Study, *MORE* Multiple Outcomes of Raloxifene Evaluation, *CORE* Continued Outcomes Relevant to Evista, *RUTH* Raloxifene Use and the Heart, *STAR* Study of Tam and Raloxifene, *NCIC* National Cancer Institute of Canada

Long-term follow up studies of women who underwent bilateral oophorectomy before the menopause for benign indications show that their BC risk is reduced by more than half if the operation was performed before the age of 40 (2–4). More recent studies where BC incidence after prophylactic oophorectomy or Tam treatment in women with mutations in the *BRCA1* and *BRCA2* genes compared with no operation also indicates about a 50% reduction in risk (Fig. 2) (5–11). In postmenopausal women, there are no results of prevention studies using aromatase inhibitors (AIs) given to reduce breast and circulating oestrogen concentrations but two studies are in progress at present (IBIS-II studying anastrozole vs. placebo in the UK and centers outside North America and MAP3 studying exemestane vs. placebo in North America, Fig. 1) (14). However, data from all adjuvant trials, which compared an AI (anastrozole, letrozole or exemestane) with Tam, indicated that AI-induced ED reduces (or prevents) contralateral BC by 42% (95% CI 12–62%; $p = 0.01$) compared with Tam (14). It is of interest that the first indication that Tam might prevent BC was also that contralateral BC was reduced by half in adjuvant trials which compared Tam with placebo (12). Subsequently four prevention trials comparing Tam with placebo were initiated (15, 17, 19, 25). An overview analysis of the results of these trials showed an overall risk reduction of 38% (95% CI 28–46%; $p < 0.0001$) at about 5 years of follow up (Fig. 2) (25). Longer follow up of two of these trials at a median of 3–5 yrs after cessation of Tam indicates that the preventive effect of Tam continues (or "carries over") after the cessation of therapy (16, 18). The SERM, Ral, has also been compared with placebo in three trials: one in women with osteoporosis (MORE/CORE) (21, 22), one in women with or at risk of cardiac

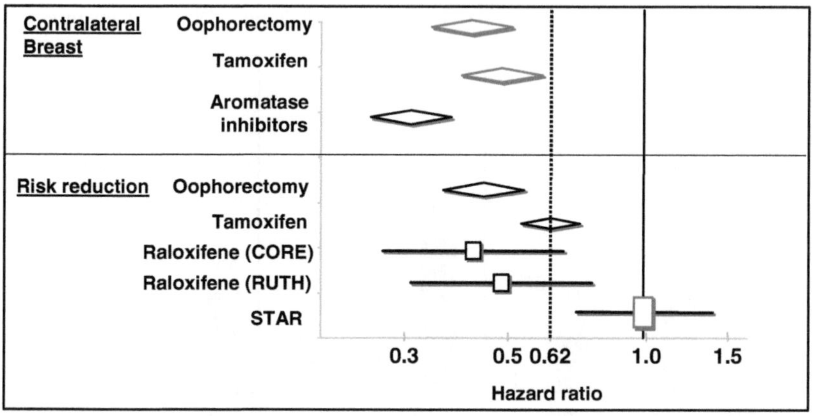

Fig. 2 Results of prevention studies using ED and SERMS. *Contralateral breast:* studies where treatment was given after surgery for BC and its effect on BC incidence in the contralateral breast was reported. *Risk reduction:* randomised studies comparing an intervention with nil or another intervention (as in the STAR trial). The STAR trial hazard ratio is 1 since Tam and Ral were equally active. However, it is estimated that the reduction in risk was 50% compared with estimated controls. *Open Diamonds* represent overview analyses of more than one trial and *open squares* represent individual studies. *Superscript numbers* indicate the main references to the studies summarised in this diagram

disease (RUTH) (23), and another where Ral was compared with Tam in women at high BC risk (STAR) (24). The MORE/CORE (in the MORE trial Ral was given for 4 yrs: the CORE trial was an extension of MORE treating with four more years of Ral) showed a risk reduction of 66% (HR 0.34; 95% CI 0.22–0.50; $p < 0.001$). The RUTH trial of 5 yrs of Ral vs. placebo showed a risk reduction by Ral of 44% (HR 0.56; 95% CI 0.38–0.83; $p = 0.003$). The STAR trial showed that Ral was as effective as Tam with an estimated reduction compared with a comparable group of untreated women of similar risk of 50% (Fig. 2) (24).

Thus, in summary, both ED and SERMS reduce BC risk. Because AIs appear superior to Tam in preventing contralateral BC, they may prove to be the agents of choice in postmenopausal women. Tam appears as effective as ovarian suppression in premenopausal women in all four Tam prevention trials and may be the agent of choice for this group of women, since there are no data concerning the use of Ral before the menopause. Ral has the advantage of maintaining bone density in postmenopausal women. It is possible that AIs are superior in efficacy to Ral but this hypothesis will be formally tested in the SELECT trial in the USA and Canada where the AI, letrozole, is being compared with Ral in women at increased BC risk by the NSABP.

Tumor Type Prevented by SERMS

The studies outlined earlier indicate that endocrine therapies prevent invasive BCs. Three studies (NSABP1 (19), MORE/CORE (22), and RUTH (23) have reported the ERα status of the invasive tumors that arose during treatment or placebo (Fig. 3).

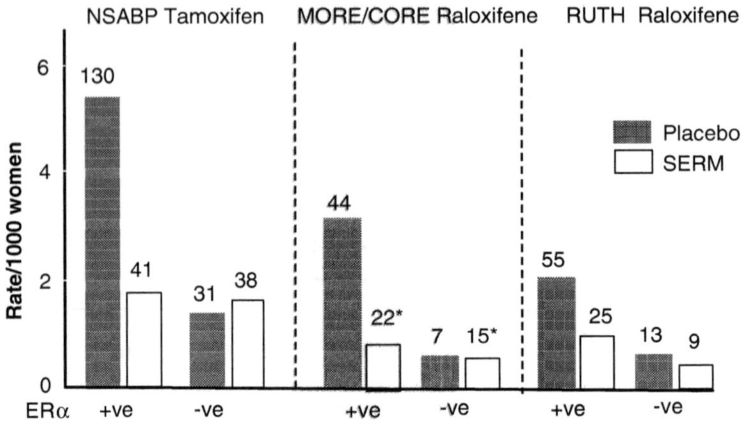

Fig. 3 ERα status in tumors that arose during the NSABP P1, MORE/CORE and RUTH prevention trials. The reduction in risk was seen only in ERα⁺ tumors. *The MORE/CORE trial was a three arm study with two active arms (Ral 60 and 120 mg) and thus there are relatively more tumors in the treatment arms

Compared with placebo there was no reduction in the incidence of ERα⁺ tumors but reductions in the incidence of ERα⁺ tumors were 69% in the NSABP Tam trial (HR 0.31; 95% CI 0.22–0.45; Fig. 3) (19), 76% in the MORE/CORE trial (HR 0.24; 95% CI 0.15–0.40; $p < 0.001$) (22), and 55% in the RUTH trial (HR 0.45; 95% CI 0.28–0.72. $p < 0.001$) (23). The corresponding figures for the incidence of ERα⁻ tumors was HR 1.22 (95% CI 0.74–2.03) for the NSABP trial, HR 1.06 (95% CI 0.43–2.59) for the MORE/CORE trial, and HR 1.44 (95% CI 0.61–3.36) for the RUTH trial. A surprising result is that Tam appears to reduce the risk of breast carriers in women with mutations in the *BRCA1* and *BRCA2* genes (5, 6). Since most tumors that arise in women with BRCA1 mutations are ERα⁻, it is remarkable that Tam is effective in this clinical situation but not in women destined to develop ERα⁻ tumors in noncarriers.

The prevention studies have also reported on the effect of treatment on the incidence of ductal carcinoma in situ (DCIS) in addition to invasive carcinoma. Tam is associated with a 50% reduction in DCIS in the NSABP P1 study (19) but, surprisingly, Ral is associated with a lesser reduction than Tam in the STAR trial (HR 1.40; 95% CI 0.98–2.00) (24) and no significant reductions at all in the MORE/CORE and RUTH trials (22, 23).

Targets for Prevention

Since, in humans, the stroma of the breast either does not contain ERα⁺ cells or has low levels of ERα expression, the major target for endocrine therapy is the ERα⁺ cell within the epithelium of the breast. In the normal breast lobule approximately

20% of cells are ER^+ but this proportion increases in most $ER\alpha^+$ premalignant lesions and tumors (27–32, 36). Careful histological studies have shown that it is likely that most BC arise in cells at the junction between terminal ducts and the lobule of the breast (34). However, Wellings et al. (35) reported that by careful subgross examination of whole breasts that virtually all normal lobules of the breast did not have histological abnormalities consistent with the cell changes found in precursors of cancers. Instead they demonstrated that abnormal changes in the epithelium appeared to arise in greatly enlarged lobules many times the size of the normal terminal duct lobular unit (TDLU) but which arise originally from normal TDLUs. Wellings called these enlarged TDLUs atypical lobules and demonstrated that all degrees of atypia from near normal looking epithelial cells through hyperplasia, atypical hyperplasia to the appearances carcinoma in situ (CIS) could be found within atypical lobules (35). These enlarged lobules have been referred to by many names over the years, including blunt duct adenosis and columnar cell lesions most recently. Allred et al. refer to them descriptively as hyperplastic enlarged lobular units (HELU) in order to distinguish them from TDLU (32). This group showed that HELU contained greater proportions of $ER\alpha^+$ cells and were more proliferative than TDLU as judged by the greater proportion of Ki-67$^+$ cells and reduced numbers of apoptotic cells (32). Other studies have also shown increased proportions of $ER\alpha^+$ cells in precursor lesions (28, 29, 31, 36, 37). In the TDLU, $ER\alpha^+$ cells rarely divide but appear to induce proliferation in adjacent $ER\alpha^-$ cells whereas in tumors many $ER\alpha^+$ cells can be shown to divide and this phenomenon was also demonstrated by Lee et al. in HELU and other groups in invasive cancers (27). These observations suggest that some TDLU enlarge (Wellings reported about 30–40 in the premenopuasal normal breast) and develop atypical epithelium which, in some, leads to CIS and invasive cancers (34, 35). Thus, the targets for prevention may be $ER\alpha^+$ cells in TDLU, in atypical lobules (which may have a wide spectrum of premalignant lesions) or invasive cancers or any combination of these three types of structure. It is possible that different approaches to endocrine prevention and also prevention initiated at different times of life may target different structures. For example, the fact that Ral does not appear to reduce DCIS appreciably suggests that it may target small invasive lesions whereas ovarian suppression at the age of 30, which reduced lifetime risk of breast cancer by about 70% (26), may reduce proliferation in TDLU and prevent the proposed formation of atypical lobules.

Experimental Studies Concerning Potential Targets for Endocrine Prevention: The Normal Breast, Premalignant Lesions (Atypical Lobules), and Invasive Tumors

Experimental studies to evaluate the responsiveness of estrogen, ED, and SERMS on TDLU, atypical lobules (HELU/CIS), and invasive cancers have been performed by a variety of methods. The major endpoint of most of the studies has been changes in epithelial cell proliferation, usually using Ki-67 immunostaining, but also labeled thymidine incorporation and only these methods will be reported here. Breast tissue containing TDLU may be biopsied at various times during the menstrual cycle and after the menopause and proliferation assessed at that time in relation to serum hormones or transplanted into immuno-suppressed mice treated with hormones and antihormones. Comparative studies of TDLU and HELU before and after age 50 have been performed by Lee et al. (32). CIS can also be studied in the immuno-suppressed mouse model but invasive cancer grows poorly in this situation. Instead most studies on invasive cancer have been performed in women with primary invasive tumors of the breast, where the effect of an intervention is assessed in the interval between diagnostic biopsy and later surgical tumor excision. In all the studies outlined earlier, histological confirmation of the type of breast lesion is mandatory. Using needle aspirates this is not possible and this type of study will not be reported here.

The effects of treatment with estrogen withdrawal and SERMs on the normal breast, premalignant lesions including CIS and invasive cancers are outlined in Table 1. These studies indicate that with some exceptions, mainly caused by gaps in our knowledge, all three types of structure responds to all three treatments.

Responsiveness of the Normal Breast. Premenopausal TDLU increase proliferation during the luteal estrogenic phase of the menstrual cycle and this response is enhanced by the additional oestrogenic effect of the oral contraceptive pill (38–42). Histologically normal TDLU, whether derived from women with or without a family history of BC or carrying BRCA1 or BRCA2 mutations, also have proliferative responses when transplanted into estrogen-treated immuno-suppressed mice (30, 39). Proliferation of TDLU is further increased during the first trimester of pregnancy but declines to low levels at lactation (when the cells become entirely $ER\alpha^{-}$) and the breast develop relative resistance to oestrogenic stimuli (40, 41). Proliferation declines further after the menopause and the TDLU appears to become resistant to estrogen, since estrogen replacement therapy does not result in a significant increase in Ki-67 labeling even after five or more years of use and combined HRT only causes proliferation after prolonged periods (43–45). The decline in proliferation of TDLU after the menopause is well documented (31, 32), and there is little evidence that further reduction in proliferation can be induced by AIs or SERMs (46, 47). However, Tam, Ral, and fulvestrant (Ful) all reduce proliferation in premenopausal TDLU (30, 48–50).

Table 1 Data available concerning the responsiveness of TDLU

	Treatment			Menop.	HRT	AI	Tam	Ral	Ful
TDLU	+ (30, 38, 39)	+(42)	=(43–45)	– (31, 32)	ND	= (46, 47)	–(30,48,49)	– (50)	–(30)
AL/CIS	+ (51–52)	ND	ND	– (32)	– (53)	ND	= (54)	ND	= (52)
Invasive	ND	ND	ND	=		– (59)	– (59)	– (60)	– (61)

SERD – selective ER down regulator, *HRT*, hormone replacement therapy

AL/CIS and invasive cancers to estrogen, estrogen depletion and SERMS/SERDS. + Stimulatory effect, - Inhibitory effect and = No Effect. ND – No available Data.

Responsiveness of Premalignant Lesions. A marked response to estrogen administration can be demonstrated when ERα$^+$ CIS is transplanted into the flanks of immune-deprived mice (51–52). There are no data on whether premalignant lesions respond to the oral contraceptive pill in premenopausal women or ERT or HRT (estrogen or HRT) in postmenopausal women. However, there is a modest fall in the number of Ki-67$^+$ epithelial cells in HELU after the menopause (from 7.32% ± 0.74% to 5.11% ± 0.73%; a 30% reduction) suggesting responsiveness to oestrogen the ED of the menopause (32). However, it is likely that ERα$^+$ CIS is stimulated by estrogen since cessation of HRT in postmenopausal women is associated with a reduction in proliferation of 49% in ERα$^+$ DCIS (53). Response to SERMs of premalignant lesions has not been well studied but Mohsin et al. (54) saw no reduction in proliferation after one year of Tam treatment and Ful seemed inactive against CIS in the immune-deprived mouse model (52).

Responsiveness of Invasive Malignant Tumors. Because of their lack of "take" in immune-deprived mice and the ethical problems of giving estrogen, OCP, or HRT, there are, as far as we are aware, no data on the responsiveness of ERα$^+$ invasive breast tumors to estrogen, although we would predict that they would respond since, although there does not appear to be a reduction of proliferation across the menopause (33), withdrawal of HRT is associated with 34% drop in proliferation (55). The lack of reduction of proliferation of ERα$^+$ invasive tumors after the menopause is consistent with data from several groups (56) that tissue estrogens remain at or near premenopausal concentrations in postmenopausal breast tissue. Administration of modern AIs reduces tumor proliferation by up to 60% (57–59). Multiple prospective studies also show that the SERMs, Tam and Ral, and the selective estrogen receptor downregulator (SERD) Ful all reduce proliferation in most tumors over the short periods of the experiments of 2 weeks to 3 months (58–60). Although responsiveness can be demonstrated in most tumors over a 2-week period, there is evidence of return (or escape) of the antiproliferative response to baseline values at 3 months indicating rapid onset of resistance in some (58, 59).

Summary

The data from observational studies and clinical trials indicates that it is possible to prevent BC for prolonged periods using various endocrine manipulations. Ovarian suppression is thought to give lifelong protection (26) and recent data indicate that the effectiveness of Tam continues after cessation of treatment at 5–8 years (16, 18). It is clear from three randomised trials that SERMs prevent ERα$^+$ tumors only (19, 22, 23) in women at increased risk (19) and at population risk (22, 23) of BC entered into these trials. The data from the Ral trials also suggests that this agent appears less effective than Tam in preventing DCIS (22–24). This is surprising since a large proportion of DCIS is ERα$^+$. Equally surprising is the effectiveness of oophorectomy and Tam in mutation carriers, particularly BRCA1, which is associated with ERα+ tumors (5–11). The fact that ERT can be given without apparently abrogating the effect of oophorectomy and also to naturally postmenopausal women without increasing BC risk (62, 63) suggests that cyclical estrogen or estrogen + progestin are important for BC initiation and/or progression.

The question arises whether the information we have concerning the responsiveness of ERα+ cells in TDLU, premalignant lesions, and invasive cancers give an indication of the targets for endocrine prevention. Data summarised in Table 1 indicate that TDLU are responsive to estrogen, ED (31, 32, 46, 47), and SERMs/SERDs (30, 48–50) in premenopausal women and there may be the targets for the preventative effect of early oophorectomy particularly in BRCA1 carriers where we have demonstrated endocrine responsiveness of TDLU, which at this heterozygote stage are ERα+ (2–11, 30).

The decline in numbers of atypical lobules in breasts without invasive cancer suggests that these are targets for the 'preventive' effect of the menopause, as suggested by Wellings (1, 35). The data also suggest that ERα+ DCIS is responsive to estrogen and ED supporting premalignant lesions is a target as does the data from the NSABP P1 trial indicating a marked preventive effect of Tam in women previously diagnosed with atypical ductal hyperplasia and a preventative effect on CIS (19).

References

1. Howell A, Clarke RB, Evans G, et al. Estrogen deprivation for breast cancer prevention. Recent Results Cancer Res (in press).
2. Feinleib M (1968) Breast cancer and artificial menopause: a cohort study. J Natl Cancer Inst 41: 315–329.
3. Hirayama T, Wynder EL (1962) A study of the epidemiology of cancer of the breast II. The influence of hysterectomy, Cancer 5: 28–38.
4. Trichopoulos D, MacMahon B, Cole P (1972) Menopause and breast cancer risk. J Natl Cancer Inst 48: 605–613.
5. Narod SA, Brunet JS, Ghadirian P, et al. (2000) Tamoxifen and risk of contralateral breast cancer in BRCA1 and BRCA2 mutation carriers: a case control study. Hereditary Breast Cancer Clinical Study Group, Lancet 356: 1876–1881.
6. Gronwald J, Tung N, Foulkes WD, et al. (2006) Tamoxifen and contralateral breast cancer in BRCA1 and BRCA2 carriers: an update. Int J Cancer 118: 2281–2284.
7. Rebbeck TR, Lynch HT, Neuhausen SL (2002) Prophylactic oophorectomy in carriers of BRCA1 or BRCA2 mutations. N Engl J Med 346: 1616–1622.
8. Kauff ND, Satagopan JM, Robson ME, et al. (2002) Risk reducing salpino-oophorectomy in women with a BRCA1 and BRCA2 mutation. N Engl J Med 346: 1609–1615.
9. Metcalfe K, Lynch HT, Ghadirian P, et al. (2004) Contralateral breast cancer in BRCA1 and BRCA2 mutation carriers. J Clin Oncol 22: 2328–2335.
10. Olson JE, Sellers TA, Iturria SJ, et al. (2004) Bilateral oophorectomy and breast cancer risk reduction among women with a family history. Cancer Detect Prev 28: 357–360.
11. Eisen A, Lubinski J, Klijn J, et al. (2005) Breast cancer risk following bilateral oophorectomy in BRCA1 and BRCA2 mutation carriers: an international case control study. J Clin Oncol 23: 7491–7496.
12. Cuzick J, Baum M (1985) Tamoxifen and contralateral breast cancer. Lancet 2: 282.

13. Early Breast Cancer Trialists' Collaborative Group (2005) Effects of chemotherapy and hormonal therapy for early breast cancer on recurrence and 15 year survival: an overview of the randomised trials. Lancet 365: 1687–1717.
14. Cuzick J (2005) Aromatase inhibitors for breast cancer prevention. J Clin Oncol 23: 1636–1643.
15. Powles TJ, Eeles R, Ashley S, et al. (1998) Interim analysis of the incidence of breast cancer in the Royal Marsden Hospital tamoxifen randomised chemoprevention trial. Lancet 352: 98–101.
16. Powles TJ, Ashley S, Tidy VA, et al. 20 year follow up of the Royal Marsden tamoxifen breast cancer prevention trial. JNCI (in press).
17. Cuzick J, Forbes J, Edwards R, et al. (2002) First results from the International Breast Cancer Intervention Study (IBIS-I): a randomised prevention trial. Lancet 360: 817–824.
18. Cuzick J, Sestak I, Forbes J, et al. Long-term results of tamoxifen prophylaxis for breast cancer – 96 month follow up of the IBIS-I study, JNCI (in press).
19. Fisher B, Costantino JP, Wickerham DL, et al. (1998) Tamoxifen for prevention of breast cancer: report of the National Surgical Adjuvant Breast and Bowel Project P-1 study. JNCI 90: 1371–1388.
20. Fisher B, Costantino JP, Wickerham DL, et al. (2005) Tamoxifen for the prevention of breast cancer: current status of the National Surgical Adjuvant Breast and Bowel Project P-1 study. JNCI 97: 1652–1662.
21. Cummings SR, Eckert S, Krueger KA (1999) The effect of Raloxifene on risk of breast cancer in postmenopausal women. Results from the MORE randomised trial. JAMA 281: 2189–2197.
22. Martino S, Cauley JA, Barrett-Connor E, et al. (2004) Continuing outcomes relevant to evista: breast cancer incidence in postmenopausal osteoporotic women in a randomised trial of raloxifene. JNCI 96: 1751–1761.
23. Barrett-Connor E, Mosca L, Collins P, et al. (2006) Effects of raloxifene on cardiovascular events and breast cancer in postmenopausal women. N Engl J Med 355: 125–137.
24. Vogel VG, Costantino JP, Wickerham DL, et al. (2006) Effects of tamoxifen v raloxifene on the risk of developing invasive breast cancer and other disease outcomes. The NSABP study of tamoxifen and raloxifene (STAR) P-2 trial, JAMA 295: 2727–2741.
25. Cuzick J, Powles T, Veronesi U, et al. (2003) Overview of the main outcomes in breast cancer prevention trials. Lancet 361: 296–300.
26. Pike MC, Spicer DV (1993) The chemoprevention of breast cancer by reducing sex steroid exposure: perspectives from epidemiology. J Cell Biochem Suppl 17G: 26–36.
27. Clarke RB, Howell A, Potten CS, et al. (1997) Dissociation between steroid receptor expression and cell proliferation in the human breast. Cancer Res 57: 4987–4991.
28. Shaaban AM, Sloane JP, West CR, et al. (2002) Breast cancer risk in usual ductal hyperplasia is defined by estrogen receptor-α and Ki-67 expression. Am J Pathol 160: 597–604.
29. Dessauvagie BF, Zhao W, Heel-Miller KA, et al. (2006) Characterisation of columnar cell lesions of the breast: immunophenotypic analysis of columnar alteration of lobules with prominent apical snouts and secretions. Hum Pathol 38: 284–292.
30. Bramley M, Clarke RB, Howell A, et al. (2006) Effects of oestrogens and anti-oestrogens on normal breast tissue from women bearing BRCA1 and BRCA2 mutations. Br J Cancer 94: 1021–1028.
31. Shoker BS, Jarvis C, Clarke RB, et al. (1999) Estrogen receptor-positive proliferating cells in the normal and precancerous breast. Am J Pathol 155: 1811–1815.
32. Lee S, Mohsin SK, Mao S, et al. (2005) Hormones, receptors and growth in hyperplastic enlarged lobular units: early potential precursors of breast cancer. Breast Cancer Res 8: 1–9.
33. Zavagno G, Meggiolaro F, Pluchinotta A, et al. (2000) Influence of age and menopausal status on pathologic and biologic features of breast cancer. Breast 9: 320–328.
34. Wellings SR (1980) A hypothesis of the origin of human breast cancer from the terminal ductal lobular unit. Pathol Res Pract 515–535.
35. Wellings SR, Jensen HM, Marcum RG (1975) An atlas of subgross pathology of the human breast with special reference to possible precancerous lesions. JNCI 55: 231–273.

36. Kim H-J, Jung W-H, Kim D-Y, et al. (2000) Expression of cyclins in ductal hyperplasia, atypical ductal hyperplasia and ductal carcinoma in-situ of the breast. Yonsei Med J 41: 345–353.
37. Allred DC, Mohsin SK, Fuqua SAW (2001) Histological and biological evolution of human premalignant breast disease. Endocr Relat Cancer 8: 47–61.
38. Potten CS, Watson RJ, Williams GT, et al. (1988) The effect of age and menstrual cycle upon proliferative activity of the normal human breast. Br J Cancer 58: 163–170.
39. Laidlaw IJ, Clarke RB, Howell A, et al. (1995) The proliferation of normal human breast tissue implanted into athymic nude mice is stimulated by estrogen but not progesterone. Endocrinology 136: 164–171.
40. Battersby S, Anderson T (1988) Proliferative and secretory activity in the pregnant and lactating human breast. Virchows Archiv A Pathol Anat 413: 189–196.
41. Feuerhake F, Sigg W, Hofter EA, et al. (2003) Cell proliferation, apoptosis and expression of bcl-2 and Bax in non-lactating human breast epithelium in relation to the menstrual cycle and reproductive history. Breast Cancer Res Treat 77: 37–48.
42. Williams G, Anderson E, Howell A, et al. (1991) Oral contraceptive (OCP) use increases proliferation and decreases oestrogen receptor content of epithelial cells in the normal human breast. Int J Cancer 48: 206–210.
43. Hargreaves DF, Knox F, Swindell R, et al. (1998) Epithelial proliferation and hormone receptor status in the normal postmenopausal breast and the effects of hormone replacement therapy. Br J Cancer 78: 945–949.
44. Dobson RRH, Chan CK, Knox F, et al. (2000) The effects of prolonged HRT treatment in normal postmenopausal breast epithelium. Breast Cancer Res Treat 64: 106.
45. Hofseth LJ, Raafat AM, Osuch JR, et al. (1999) Hormone replacement therapy with estrogen or estrogen plus medroxyprogesterone acetate is associated with increased epithelial proliferation in the normal postmenopausal breast. J Clin Endo Metab 84: 4559–4565.
46. Harper-Wynne C, Ross G, Sacks N, et al. (2002) Effects of the aromatase inhibitor letrozole on normal breast epithelial cell proliferation and metabolic indices in postmenopausal women: a pilot study for breast cancer prevention. Cancer Epidemiol Biomarkers Prev 11: 614–621.
47. Walker KJ, Price-Thomas JM, Candlish W, et al. (1991) Influence of the antioestrogen tamoxifen on normal breast tissue. Br J Cancer 64: 764–768.
48. de Lima GR, Facina G, Shida JY, et al. (2003) Effects of low dose tamoxifen on normal breast tissue from premenopausal women. Eur J Cancer 39: 891–898.
49. Bernardes JR Jr, Nonogaki S, Seixas MT, et al. (1999) Effect of a half dose of tamoxifen on proliferative activity in normal breast tissue. Int J Gynaecol Obstet 67: 33–38.
50. Borges da Silva B, Lopes IMS, Gebrim LH (2006) Effects of raloxifene on normal breast tissue from premenopausal women. Breast Cancer Res Treat 95: 99–103.
51. Holland PA, Knox WF, Potten CS, et al. (1997) Assessment of hormone dependence of comedo ductal carcinoma in-situ of the breast. JNCI 89: 1059–1064.
52. Gandhi A, Holland PA, Knox WF, et al. (2000) Effects of a pure antiestrogen on apoptosis and proliferation within human breast ductal carcinoma in-situ. Cancer Res 60: 4284–4288.
53. Boland GP, McKeown A, Chan KC, et al. (2003) Biological response to hormonal manipulation in oestrogen receptor positive ductal carcinoma in-situ of the breast. Br J Cancer 89: 277–283.
54. Mohsin SK, Allred DC, Osborne CK, et al. (2005) Morphological and immunophenotypic markers as surrogate endpoints of tamoxifen effect for prevention of breast cancer. Breast Cancer Res Treat 94: 205–211.
55. Prasad R, Boland GP, Cramer A, et al. (2003) Short term biologic response to withdrawal of hormone replacement therapy in patients with invasive breast carcinoma. Cancer 98: 2539–2546.
56. Geisler J (2003) Breast cancer tissue estrogens and their manipulation with aromatase inhibitors and inactivators. J Steroid Biochem Mol Biol 86: 245–253.
57. Geisler J, Detre S, Berntsen H, et al. (2001) Influence of neoadjuvant anastrozole (arimidex) on intratumoral estrogen levels and proliferation markers in patients with locally advanced breast cancer. Clin Cancer Res 7: 1230–1236.
58. Miller WR, White S, Dixon JM, et al. (2006) Proliferation, steroid receptor and clinical/pathological response in breast cancer treated with letrozole. Br J Cancer 94: 1051–1056.

59. Dowsett M, Ebbs SR, Dixon JM, et al. (2005) Biomarker changes during neoadjuvant anastrozole, tamoxifen or the combination: influence of hormonal status and HER-2 in breast cancer – a study from the IMPACT trialists. J Clin Oncol 23: 2477–2492.
60. Dowsett M, Bundred NJ, Decensi A, et al. (2001) Effect of raloxifene on breast cancer cell Ki-67 and apoptosis: a double-blind, placebo-controlled, randomised clinical trial in postmenopausal patients. Cancer Epidemiol Biomarkers Prev 10: 961–966.
61. Robertson JF, Nicholson RI, Bundred NJ, et al. (2001) Comparison of the short term biological effects of 7α-[9-(4,4,5,5,5-pentafluoropentylsulfinyl)-nonyl]estra-1,3,5, (10)-triene-3,17β-diol (faslodex) versus tamoxifen in postmenopausal women with primary breast cancer. Cancer Res 61: 6739–6746.
62. Stefanick ML, Anderson GL, Margolis KL, et al. (2006) Effects of conjugated equine estrogens on breast cancer and mammography screening in postmenopausal women with hysterectomy. JAMA 295: 1647–1657.
63. Kotsopoulos J, Lubinski J, Neuhausen SL, et al. (2006) Hormone replacement therapy and the risk of ovarian cancer in BRCA1 and BRCA2 mutation carriers. Gynecol Oncol 100: 83–88.

19
Estrogen-Induced Breast Oncogenesis: Modulation by an Aurora Kinase Inhibitor

Sara Antonia Li, Luke K.T. Lam, Nayaz Ahmed, Adrianne E. Hontz, and Jonathan J. Li

Introduction

Breast cancer (BC) alone accounts for about 32% of all cancers occurring in women in industrialized countries, and thus, is clearly an immense world wide public health concern. More than 90% of all human BC cases are sporadic or nonfamilial with an equally high percentage of these cases being ductal breast carcinomas, the rest are lobular. This latter distinction is particularly important since ductal BCs are highly aneuploid, while lobular BCs are mainly diploid. While the presence of estrogen receptor (ERα) is nearly a ubiquitous feature of sporadic BCs; about 55–73%, aneuploidy, not the presence of $ER\alpha^+$, is its most defining characteristic (65–90%) (1–3). Moreover, the detection of high aneuploid frequencies in a preinvasive stage, ductal carcinoma in situ (DCIS), strongly implicates that this molecular alteration has a primary role in the ontogeny and progression of early sporadic ductal BCs.

Currently, it is widely accepted that ovarian hormones, particularly estrogen, is the major etiologic agent affecting BC risk. This has led to a compilation of well-established BC risk factors (4, 5) all of which are related to elevated serum estrogen levels in normal cycling women. These risk factors are summarized in Fig. 1. The majority of risk factors are related to premenopausal women, whereas only obesity and combined hormone replacement therapy (CHRT) but not estrogen-replacement therapy (ERT) are significant risk factors in postmenopausal women (Fig. 1). Despite these compelling data, the precise mechanism(s) whereby estrogen elicits these oncogenic effects in breast tissue remain elusive. Nevertheless, we believe significant progress has been made in unravelling this long standing conundrum (6).

Estrogen-Elicited ACI Rat Ductal Breast Carcinomas. The histopathological alterations and molecular changes seen in the ACI rat mammary gland after 17β-estradiol (E_2) treatment strikingly resemble the early development and progression seen in human primary ductal breast neoplasms (6–9). In female ACI rats, these changes are elicited at E_2 serum concentrations of only 50–120 pg ml^{-1}, a range approaching the upper E_2 limit (10–45 pg ml^{-1}) detected in the serum of normal cycling rats (10), and coincides with the total and free E_2 levels found in the follicular phase in normal women (66–100 pg ml^{-1}), found to significantly increase BC risk in women (11).

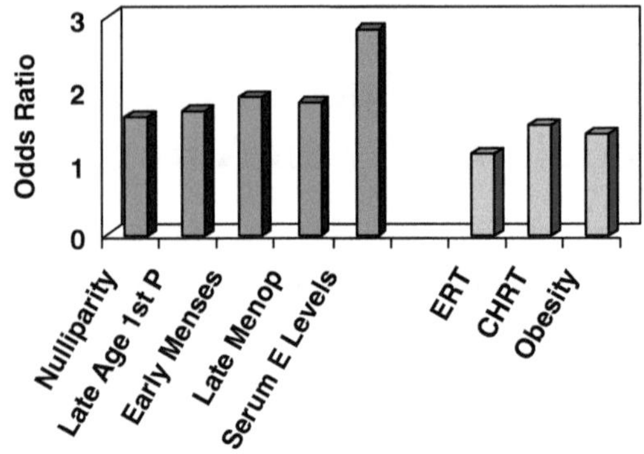

Fig. 1 Established risk factors for sporadic breast cancer

The E_2 serum levels in ACI rats, although administered in a sustained manner, is comparable to E_2 serum concentrations to those observed in normal cycling women (11–13).

The normal cycling ACI rat mammary gland is largely composed of fat cells interspersed with ductal epithelial cells (Fig. 2A). Hyperplasia of the breast epithelial cells is an expected occurrence when the mammary gland is exposed to periodic or sustained estrogen and progesterone, both ovarian hormones being mitogenic in this tissue. The premalignant stages developing after E_2 treatment in female ACI rats exhibit various stages of focal dysplasias with stage 3 resembling atypical ductal hyperplasia (ADH). The earliest dysplastic stage is shown in Fig. 2B. At about 4.3 months of E_2 treatment, ductal carcinoma in situ (DCIS) begins to appear. The most common histopathologic type of DCIS found in E_2 treated ACI rat mammary glands is the cribiform type (Fig. 2C), which is coincidentally the most common DCIS seen in preinvasive stages leading to sporadic ductal BC in women. The comedo type of DCIS was also detected but was eight to ten times less frequent (Fig. 2D). Other DCIS subtypes including solid and papillary (Fig. 2E) were seen in E_2-treated ACI rats and have similar frequencies to those found in human BC development. After 5.0 months, invasive ductal breast tumors were commonly detected (Fig. 2F).

We have proposed a novel paradigm (6–9) based on our studies in the estrogen-elicited female ACI rat breast tumor model (Fig. 3). Possibly, the most common over-expressed and amplified protein/gene in human sporadic ductal BC is MYC/c-*myc* mediated by estrogen. A similar over-expression and amplification of this protein/gene has also been detected in early preinvasive stages in E_2-treated mammary glands in female ACI rats via ERα. This in turn elicits a deregulation of certain cell cycle entities, including cyclins D1 and E1, their respective binding kinases, cdk4 and cdk2, and MDM2. Sustained overexpression of Aurora (Aur) A and B results by either direct action of E_2 via ERα or as a consequence of overexpression of MYC and cyclin E cdk2.

Recent evidence indicates that members of a mitotic kinase family, AurA and AurB, are regulated by estrogen (14) and its sustained overexpression in E_2-induced breast tumors precipitates centrosome amplification, which results in the missegregation of

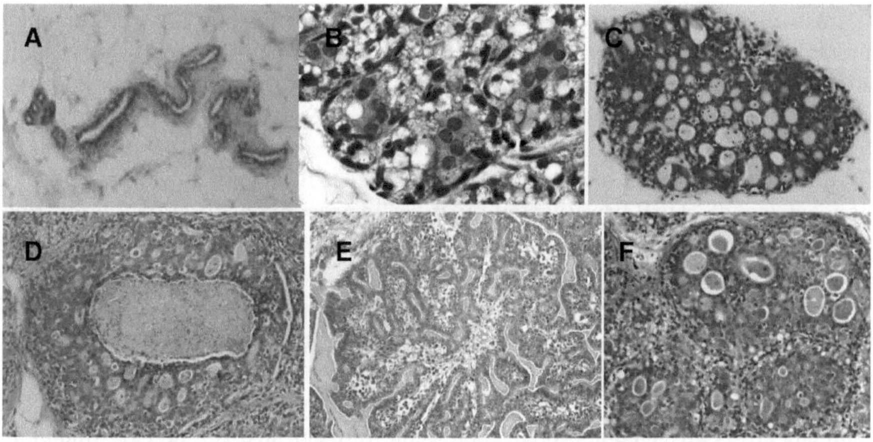

Fig. 2 Female ACI rats. A Control untreated mammary gland. B Incipient lesion, 3.0 months E_2. C Cribiform DCIS, 4.5 months E_2. D Comedo DCIS, 5.0 months E_2. E Papillary DCIS, 5.0 months E_2. f Invasive ductal BC, 6.0 months E_2. Magnification ×40

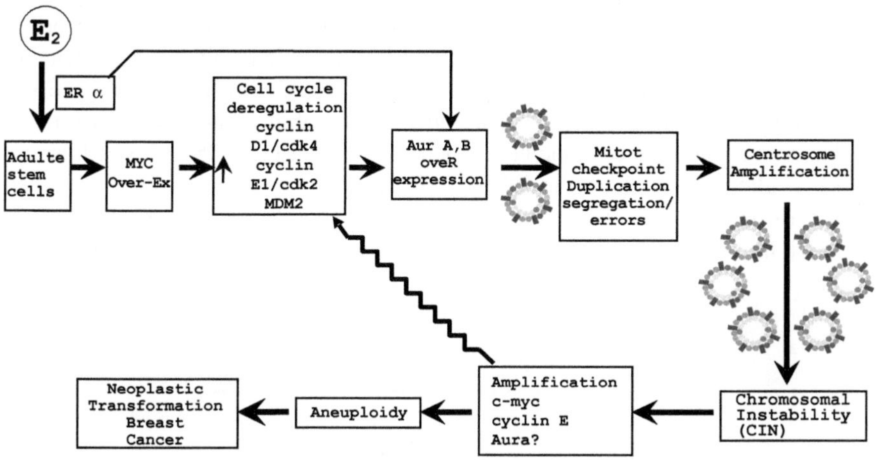

Fig. 3 A novel paradigm for estrogen-driven breast oncogenesis

chromosomes and aneuploidy. Since this process is estrogen-driven, clones are selected from the developing aneuploidy resulting in gene amplification, particularly c-*myc*. This cascade of molecular changes eventually progresses to a malignant breast phenotype. These molecular alterations seen in ACI rat E_2-mediated mammary oncogenesis mirror similar events in the development of estrogen-dependent human sporadic ductal BCs (15–19).

Aurora Kinases. During cell division, the Aurora (Aur) family of serine/threonine mitotic kinases, particularly A and B, has a crucial role in proper centrosome duplication, maturation, and separation, spindle assembly and stability, chromosome condensation and segregation, and cytokinesis (19–23). AurA overexpression elicits neoplastic transformation in mammalian cells, both in vitro and in vivo (19, 22), designating it as an oncoprotein. Although AurB has not as yet proven to be oncogenic per se, AurB has essential functions in proper condensation, segregation, and cytokinesis by regulating microtubule kinetochore associations. While AurA is localized to the pericentriolar material of the centrosomes in prophase, AurA is localized between sister centrosomes on the inner centromere from late G2 through metaphase. AurB is concentrated in the spindle midzone and in the cell cortex at the site of cleavage-fusion ingression.

We reported that during E_2-induced mammary tumorigenesis, at 4.0 months of E_2 treatment, AurA mRNA and protein expression were elevated 1.4- and 1.5-fold in the ACI rat breast compared with cholesterol-treated, age-matched controls. Additionally, a 7.2-fold rise in AurA protein was also detected in ACI rat breast tumors. We have now found that AurA activity, at this time period when premalignant lesions (DCIS) are most abundant in the mammary gland, rises 2.45-fold and a 1.64-fold increase in Aur-A activity in primary breast neoplasms (Fig. 4).

Centrosome Amplification. The centrosome is the microtubule organizing center (MTOC) for the nucleation of microtubule arrays and therefore necessary for proper cell division. During mitosis, the centrosome is responsible for the equal segregation of chromosomes in establishing the bipolar spindle (24). Centrosome amplification is quite likely the primary course of chromosomal instability and aneuploidy in the development of most cancers, including those of the breast (23, 25). This cellular and molecular alteration is characterized by an increase in centrosome size and volume because of accumulated γ-tubulin, centrin, and pericentrin; all components of the pericentriolar matrix. Importantly, there is an increase in centrosome number, ele-

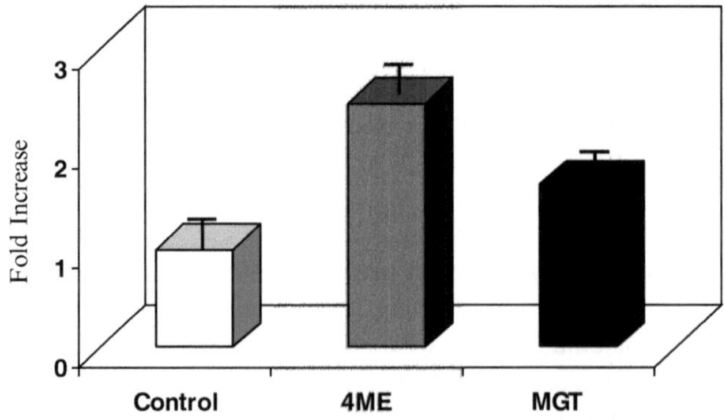

Fig. 4 Aurora A activity in control cholesterol (Control) and 4.0-month E_2-treated (4ME) mammary glands, and E_2-induced mammary gland tumors (MGT)

vated microtubule nucleation capacity, and a rise in the level of phosphorylated centrosomal and other mitotic-associated proteins (18).

In human ductal BC, centrosome amplification is a common occurrence (>80%), with three to eight centrosomes per tumor cell (18, 26). The level of centrosome amplification showed a positive correlation with chromosomal instability and aneuploidy. Amplified centrosomes were found in high frequency in premalignant stages and primary ductal breast tumors (Fig. 5) in female ACI rats elicited primarily by estrogens (6). The finding that centrosome amplification occurs in the earliest stages of E_2-induced ACI rat mammary tumorigenesis strongly suggests its involvement in the oncogenic process as predicted by Boveri (27).

Small Molecule Intervention in Estrogen-Induced Breast Oncogenesis. MK-0457 (Vertex) is a potent and selective-inhibitor of Aurora kinases with an apparent inhi-

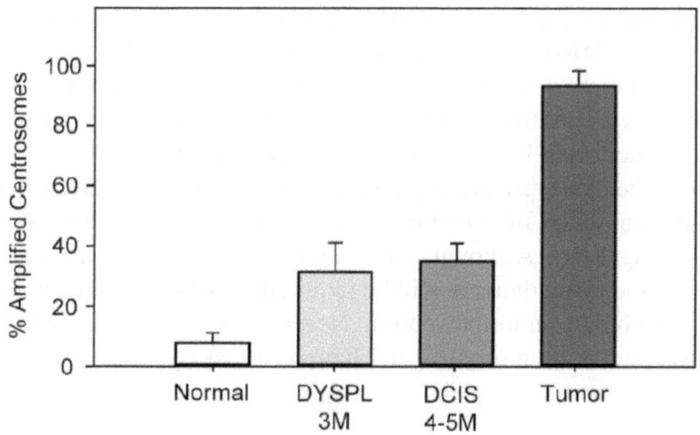

Fig. 5 % of amplified centrosomes relative to untreated mammary gland controls in ductal cells after 3.0-mo E_2 treatment showing dysplasia (DYSPL) and in DCIS after 4.0- and 5.0-mo of E_2 treatment, respectively. Centrosome amplification significantly increased 30 to 38% in DYSPL and DCIS, and up to 90% in tumor samples compared to control untreated samples.

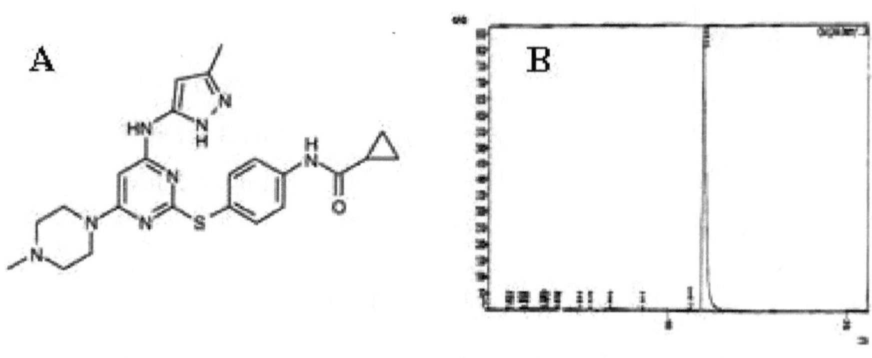

Fig. 6 **A** Structure of MK-0457. **B** HPLC profile of MK-0457. Using a C18 reverse phase column, a single peak was identified. Its identity was further confirmed by LC/MS

bition [Ki(app)] values of 0.6 < 18 < 4.6 nM for AurA, AurB, and AurC, respectively (Fig. 6a) (28). Using human breast tumor cell lines, MK-0457 inhibitory efficacy was ZR-75-1 (ERα$^+$) > MCF-7 (ERα$^+$) > MDA-MB-231 (ERα$^-$). MK-0457 was synthesized and shown to have a purity of 99.3% (Fig. 6b).

Additionally, we found that using recombinant AurA, AurB, and AurC, our MK-0457 preparation exhibited a Ki(app) of 0.7, 1.1, and 1.5 nM for these Aurora kinases, respectively (Fig. 7). Since AurA exhibited sustained overexpression and activity in premalignant stages and in primary breast tumors mediated by estrogen in female ACI rats (6), this finding suggested to us that estrogen-induced mammary tumorigenesis in female ACI rats could be inhibited.

In these initial studies with MK-0457, the small molecule AurA inhibitor was administered at 2.0 mg per 100 g BW every other day for 12 days with an interval period of 6 days of no treatment. This regimen was repeated for 5.5 months. No differences in body weights (~160–190 g) was detected in animal groups receiving E_2 alone compared with E_2 + MK-0457 during the initial 4.7 month-treatment. However, the E_2 + MK-0457-treatment group exhibited modest weight loss in the last month of treatment compared with the E_2-treated alone group (Fig. 8). Although tumor incidence was unaffected, the E_2 alone and E_2 + MK-0457 groups exhibited a significant reduction in both tumor multiplicity and size (~60%).

In summary, the data presented represent an initial first step in the chemoprevention of estrogen-driven ductal BC targeting specific entities, Aur-A and Aur-B, without directly involving ERα. Although additional refinements in our small molecule intervention protocol and delivery will be required, it is anticipated that this will lead to more effective breast tumor prevention. Nevertheless, this approach holds promise for a non-ERα intervention in estrogen-elicited BC oncogenesis.

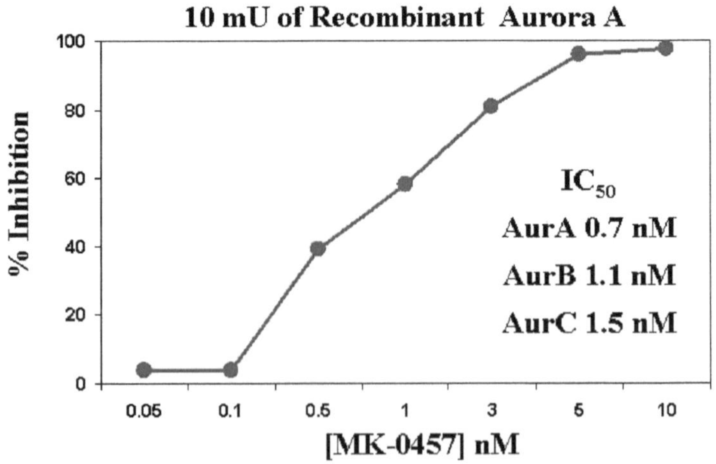

Fig. 7 In vitro kinase assay. The inhibitory effects of MK-0457 was determined using 10 mU of purified AurA, AurB, and AurC. Only the AurA is shown in this figure

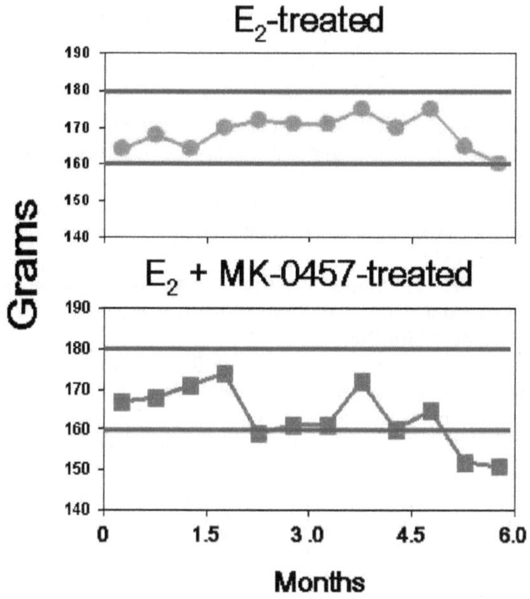

Fig. 8 Body weight fluctuations during E_2 and E_2 + MK-0457 treatments

Acknowledgments This study was supported by the Susan G. Komen for the Cure Grant No BCTR42906 to J.J. Li. We also wish to acknowledge Ms. Stephanie Schock for her excellent editorial assistance.

References

1. Arnerlov C, Emdin SO, Cajander S, et al. (2001) Intratumoral variations in DNA ploidy and s-phase fraction in human breast cancer. Anal Cell Pathol, 23, 21–8.
2. Makris A, Allred DC, Powles TJ, et al. (1997) Cytological evaluation of biological prognostic markers from primary breast carcinomas. Breast Cancer Res Treat, 44, 65–74.
3. Leal CB, Schmitt FC, Bento MJ, et al. (1995) Ductal carcinoma in situ of the breast. Histologic categorization and its relationship to ploidy and immunohistochemical expression of hormone receptors, p53, and c-erbB-2 protein. Cancer, 75, 2123–31.
4. Feigelson HS, Henderson BE (1996) Estrogens and breast cancer. Carcinogenesis, 17, 2279–84.
5. Adami HO, Persson I, Ekbom A, et al. (1995) The aetiology and pathogenesis of human breast cancer. Mutat Res, 333, 29–35.
6. Li JJ, Weroha, SJ, Lingle WL, et al. (2004) Estrogen mediates Aurora-A overexpression, centrosome amplification, chromosomal instability, and breast cancer in female ACI rats. Proc Natl Acad Sci USA, 101, 18123–8.
7. Li SA, Weroha SJ, Tawfik O et al. (2002) Prevention of solely estrogen-induced mammary tumors in female aci rats by tamoxifen: evidence for estrogen receptor mediation. J Endocrinol, 175, 297–305.

8. Li JJ, Li SA (2007) Deciphering the conundrum of estrogen-driven breast cancer: aurora kinase deregulation. Springer Verlag, New York.
9. Weroha SJ, Li SA, Tawfik O, et al. (2006) Overexpression of cyclins D1 and D3 during estrogen-induced breast oncogenesis in female ACI rats. Carcinogenesis, 27, 491–8.
10. Butcher RL, Collins WE, Fugo NW (1974) Plasma concentration of LH, FSH, prolactin, progesterone and estradiol-17beta throughout the 4-day estrous cycle of the rat. Endocrinology, 94, 1704–8.
11. Eliassen AH, Missmer SA, Tworoger SS et al. (2006) Endogenous steroid hormone concentrations and risk of breast cancer: does the association vary by a woman's predicted breast cancer risk? J Clin Oncol, 24, 1823–30.
12. Chetrite GS, Cortes-Prieto J, Philippe JC, et al. (2000) Comparison of estrogen concentrations, estrone sulfatase and aromatase activities in normal, and in cancerous, human breast tissues. J Steroid Biochem Mol Biol, 72, 23–7.
13. Vermeulen A, Deslypere JP, Paridaens R, et al. (1986) Aromatase, 17 beta-hydroxysteroid dehydrogenase and intratissular sex hormone concentrations in cancerous and normal glandular breast tissue in postmenopausal women. Eur J Cancer Clin Oncol, 22, 515–25.
14. Hontz AE, Li SA, Lingle WL, et al. (2007) Aurora a and B overexpression and centrosome amplification in early estrogen-induced tumor foci in the Syrian hamster kidney: implications for chromosomal instability, aneuploidy, and neoplasia. Cancer Res, 67, 2957–63.
15. Dubik D, Dembinski TC, Shiu RP (1987) Stimulation of c-myc oncogene expression associated with estrogen-induced proliferation of human breast cancer cells. Cancer Res, 47, 6517–21.
16. Sutherland RL, Musgrove EA (2004) Cyclins and breast cancer. J Mammary Gland Biol Neoplasia, 9, 95–104.
17. Tanaka T, Kimura M, Matsunaga K, et al. (1999) Centrosomal kinase AIK1 is overexpressed in invasive ductal carcinoma of the breast. Cancer Res, 59, 2041–4.
18. Lingle WL, Lutz WH, Ingle JN, et al. (1998) Centrosome hypertrophy in human breast tumors: implications for genomic stability and cell polarity. Proc Natl Acad Sci USA, 95, 2950–5.
19. Zhou H, Kuang J, Zhong L, et al. (1998) Tumour amplified kinase STK15/BTAK induces centrosome amplification, aneuploidy and transformation. Nat Genet, 20, 189–93.
20. Marumoto T, Zhang D, Saya H (2005) Aurora-A – a guardian of poles. Nat Rev Cancer, 5, 42–50.
21. Meraldi P, Honda R, Nigg EA (2002) Aurora-A overexpression reveals tetraploidization as a major route to centrosome amplification in p53–/– cells. Embo J, 21, 483–92.
22. Brinkley BR, Goepfert TM (1998) Supernumerary centrosomes and cancer: Boveri's hypothesis resurrected. Cell Motil Cytoskeleton, 41, 281–8.
23. Li JJ, Li SA (2006) Mitotic kinases: the key to duplication, segregation, and cytokinesis errors, chromosomal instability, and oncogenesis. Pharmacol Ther, 111, 974–84.
24. Salisbury JL (2001) The contribution of epigenetic changes to abnormal centrosomes and genomic instability in breast cancer. J Mammary Gland Biol Neoplasia, 6, 203–12.
25. Brinkley BR (2001) Managing the centrosome numbers game: from chaos to stability in cancer cell division. Trends Cell Biol, 11, 18–21.
26. Lingle WL, Barrett SL, Negron VC, et al. (2002) Centrosome amplification drives chromosomal instability in breast tumor development. Proc Natl Acad Sci USA, 99, 1978–83.
27. Bovari T (1914) The origin of malignant tumours. Williams and Wikins, Baltimore.
28. Harrington EA, Bebbington D, Moore J, et al. (2004) VX-680, a potent and selective small-molecule inhibitor of the Aurora kinases, suppresses tumor growth in vivo. Nat Med, 10, 262–7.

Part 6
Hormone Dependency Versus Hormone Independency

20
Castration-Recurrent Prostate Cancer Is Not Androgen-Independent

James L. Mohler

Significance

In the USA in 2006, an estimated 234,460 new cases of prostate cancer (PC) will be diagnosed and 27,350 men will die from PC (1). Despite the increased use of digital rectal examination and serum prostate-specific antigen (PSA) measurement for early detection, ~30% of men treated with curative intent suffer PC recurrence. These men and those who present with locally advanced or metastatic PC can be palliated by androgen deprivation therapy (ADT), a treatment that remains unimproved since its discovery more than 60 years ago (2). Over 80% of men with disseminated PC demonstrate clinical or biochemical response that is associated with a mean life expectancy of ~3.5 years in contrast to nonresponders or untreated patients who live an average of 9 months. Regardless of the androgen responsiveness of incurable PC, almost all patients succumb to castration-recurrent PC because it responds poorly to all known therapies.

AR Protein Expression in Castration-Recurrent CaP

Androgen receptor (AR) immunostaining of 19 specimens of castration-recurrent PC [% positive nuclei, 83.7 ± 11.6, and mean optical density (MOD), 0.284 ± 0.115] was similar to 16 specimens of benign prostate (% positive nuclei, 77.3 ± 13.0, and MOD, 0.315 ± 0.044) ($p = 0.25$ for % positive nuclei and 0.48 for MOD) (Fig. 1) (3). These findings, measured using an AR monoclonal antibody and automated image analysis, were similar to earlier reports that used qualitative methods (4, 5). High levels of AR expression in castration-recurrent PC in the absence of testicular androgens provides the potential for enhanced AR sensitivity to available androgens or alternate mechanisms of activation that would allow AR to remain central to growth regulation of castration-recurrent PC (6–9). The central question becomes *"How is AR activated after medical or surgical castration?"*

J.J. Li et al. (eds.), *Hormonal Carcinogenesis V.*

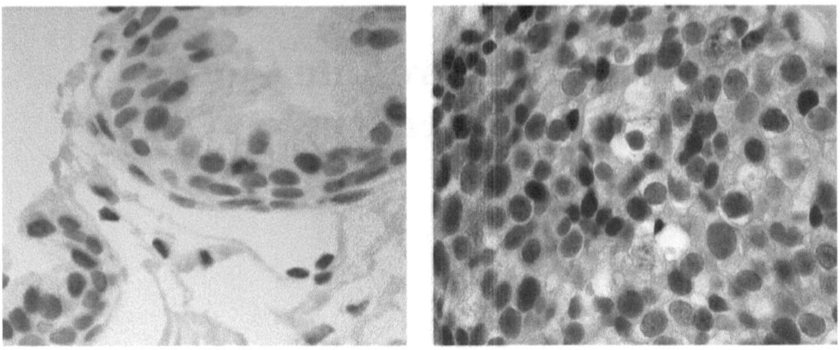

Fig. 1 AR protein expression is similar in androgen-stimulated benign prostate (*left*) and castration-recurrent PC (*right*)

AR Mutations in Castration-Recurrent Prostate Cancer

At the molecular level, AR mutations have been reported with frequencies ranging from 0 (10) to 44% (11) in androgen-stimulated PC, and 0 (12) to 50% (13) in castration-recurrent PC. Most investigators use a single method to search for mutations and since evaluation of exon A is technically difficult, investigators evaluate only exons B-H, although exon A may harbor many mutations (11). Specimens of castration-recurrent PC from 25 men were used to test whether the AR mutation frequency variation in castration-recurrent PC resulted from methodological differences. Mutation analysis used denaturing gradient gel electrophoresis (DGGE) of all exons except the first fragment of exon A [4 PCR products span exon A (A1–4)], single-strand conformational polymorphism (SSCP), and direct sequencing of all exons. The three mutational analysis methods were similar in sensitivity; i.e., the frequency of LNCaP mutation was 10% by DGGE and direct sequencing, and 20% by SSCP (14). A silent mutation was found in exon F (800C→T). A second patient had two changes: the LNCaP mutation (877T→A) in exon H and a CAG repeat deletion from 25 in genomic DNA to 10. The mutation was confirmed by cloning. Neither mutation was present in peripheral blood mononuclear cells nor in the original androgen-stimulated PC. A third patient had CAG repeat expansion from 21 to 26 and GGN repeat deletion from 23 to 10 in castration-recurrent PC compared with his original PC. SSCP suggested mutations (1) in fragment A3 and 4 in exon E) not confirmed by direct sequencing. A consensus has developed that AR mutations are infrequent (6, 7), although in bone metastases, they may occur with a frequency as high as 30% when ADT includes antiandrogens (15), mutagenesis rates are high in general. When characterized functionally, most of the mutant ARs retain transcriptional activity in response to androgens and some have altered steroid-binding specificity that changes the spectrum of ligands capable of activating AR (13, 16–19).

AR Amplification in Castration-Recurrent Prostate Cancer

Chen et al. used in vitro and in vivo models to suggest that castration-recurrent PC results from increased expression of AR protein through AR gene amplification that allows expression of androgen-regulated genes despite castrate levels of serum androgens (20). This hypothesis was tested clinically using a tissue microarray constructed from 24 samples of castration-recurrent PC and six samples of benign prostate (21). Eight (33%) castration-recurrent PCs exhibited AR gene amplification, a frequency similar to that reported by others. AR was immunostained more intensely in PC with amplified AR (MOD 0.36 ± 0.07) than nonamplified AR (MOD 0.24 ± 0.09) ($p < 0.01$) but AR immunostaining intensity was unrelated to the degree of AR amplification. A single laboratory has reported on AR gene amplification and survival; a survival advantage was found for amplified patients in their first two reports but not in their most recent publication (22). We found that AR gene amplification was unrelated to duration of survival after ADT.

Ligand-Independent AR Activation

AR is activated by IL-6 (23). Phosphorylation of the coactivator SRC-1 is regulated by IL-6 causing protein interaction between the N-terminal domain of AR and SRC-1 (24). Growth factor kinase signaling pathways may activate AR directly or sensitize AR by regulation of coactivator interaction with AR (25–29). Evidence is strong for HER-2, but HER-2 receptor levels are low in castration-recurrent PC at the protein and mRNA levels; however, HER-2 amplification was not detected in any of 39 androgen-stimulated or castration-recurrent PC specimens tested (44). "Ligand-independent" AR activation may occur when AR is sensitized to low androgen levels by growth factors, change in AR coregulator profiles, or microenvironmental hypoxia. For example, the neuropeptide growth factor bombesin synergizes with 10 pmol dihydrotestosterone (DHT) to activate AR in PC-3 cells that overexpress transfected AR (30).

AR Activating Levels of Androgens in Castration-Recurrent Prostate Cancer

A group of 21 patients aged 57–86 years demonstrated clinical evidence of castration-recurrent PC (3) (Fig. 2 and Table 1). All underwent transurethral prostatectomy for urinary retention from local recurrence that occurred from 7–92 months after medical (10 men) or surgical (11 men) ADT. Histologic examination revealed poorly differentiated carcinoma (Gleason sum 8–10) that represented an

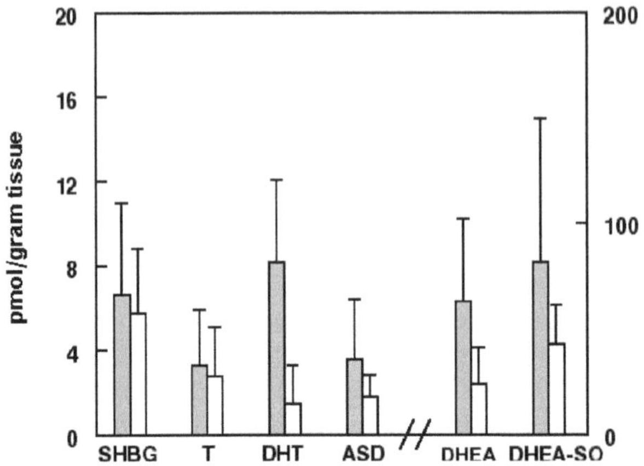

Fig. 2 Tissue androgen levels in castration-recurrent PC (*open bars*) vs. androgen-stimulated benign prostate

Table 1 Androgen levels in castration-recurrent PC

LC/MS/MS			RIA		
Titus 2005			**Page 2006**		
	T	**DHT**		**T**	**DHT**
AS-BP (n = 18)	2.75	13.7	AS-BP (n = 4)	1.84	9.26
RCaP (n = 18, 37 m)	3.75	1.25	LHRH+T (n = 4, 1 m)	1.38	6.8
Nishiyama 2004			LHRH (n=4, 1 m)	0.56	1.94
	T	**DHT**	**Mohler 2004**		
AS-BP (n = 30)	-	18.7		**T**	**DHT**
CaP (n = 30, 3–6 m)	-	4.65	AS-BP (n = 30)	3.26	8.13
Mizokami 2004			RCaP (n = 15, 37 m)	2.78	1.45
	T	**DHT**	**Labrie 1989**		
AS-BP (n = 15)	-	8.53		**T**	**DHT**
CaP (n = 15, 3–6 m)	-	2.13	Human CaP (n = ?)	-	18.6
			Orch (n = 5, 2–12 m)	-	9.29
			Orch + flu (n = 4, 2 m)	-	ND
			Rat prostate	-	14.6
			Orch, orch ± flu	-	ND
			Guinea pig prostate	-	32.4
			Orch, orch ± flu	-	ND
			Geller 1979		
				T	**DHT**
			AS-BP (n = 17)	-	17.6
			CaP orch ± DES (n=9)	-	4.47
			CaP DES 1 mg (n = 6)	-	12.4

average of 92% (range 72–99%) of the cross-sectional area of the tissue sections. To compare these tissues to androgen-stimulated prostate tissue, frozen specimens of benign prostate tissue were obtained from radical prostatectomy specimens. The frozen tissues were assayed for total levels of T, DHT, androstenedione (ASD), dehydroepiandrosterone (DHEA), DHEA-sulfate (DHEA-SO_4), estradiol, sex hormone-binding globulin (SHBG), and PSA. Although tissue levels of DHT, DHEA, and ASD were lower in castration-recurrent PC from men undergoing ADT than in benign prostate from untreated men ($p<0.01$), DHT tissue levels averaged 1.45 nM in castration-recurrent PC and 8.14 nM in benign prostate. Tissue levels of testosterone (T) were similar in castration-recurrent PC (2.78 nM) and benign prostate (3.27 nM) ($p=0.21$). Tissue levels of PSA in castration-recurrent PC were approximately 1/10 the level measured in benign prostate ($p<0.000001$). Castration-recurrent PC tissue levels of androgens, estradiol, SHBG, and PSA did not differ between three patients who received flutamide and 12 patients who did not. In particular, tissue levels of DHT were similar ($p=0.29$) in both groups (flutamide, 3.75 ± 3.58 pmol/g tissue, range 0.40–7.53 pmol/g tissue; no flutamide, 0.87 ± 0.53 pmol /g tissue, range 0.37–2.17 pmol /g tissue).

These results obtained using radioimmunoassay (RIA) were surprising. Therefore, the data were confirmed using mass spectrometry (MS) (31). A prostate tissue homogenization and androgen extraction protocol and a liquid chromatography (LC)/electrospray ionization (ESI)/MS analytic method were developed in collaboration with Dr. K. Tomer, Director, NIEHS MS Facility. T levels were similar in castration-recurrent PC (3.75 pmol/g tissue) and benign prostate (2.75 pmol/g tissue, $p=0.30$). DHT levels in castration-recurrent PC (1.25 pmol/g tissue) were less than in benign prostate (13.7 pmol /g, tissue $p<0.0001$), although, in most specimens of castration-recurrent PC, DHT levels were sufficient for AR activation. DHT levels in castration-recurrent PC compared with benign prostate decreased 91% by MS and 82% by RIA.

Are Tissue Androgen Levels Really Elevated During ADT?

Dr. Labrie et al. gained widespread recognition for their work on tissue androgen levels during ADT measured by RIA. PC tissue DHT levels decreased from 5.24 ng / g tissue in noncastrated men 55–68 years of age to 2.7 ng /g tissue in five men who castrated 2–12 months before radical prostatectomy (32). Among four castrated men receiving flutamide, 250 mg three times daily for 2 months prior to prostatectomy, tissue DHT was undetectable. It was postulated that flutamide, competing for high affinity DHT binding to AR, decreased prostate DHT levels by increasing its degradation. These data led to the use of "total androgen blockade" where tissue DHT was eliminated using antiandrogens (33). However, a metaanalysis of clinical trials comparing LH-RH agonists and anti-androgens vs. LH-RH agonists alone (34) and a study comparing orchiectomy and anti-androgens vs. orchiectomy alone (35) demonstrated no survival benefit to combination therapy. Careful review of older literature and recent findings cast further doubt on Labrie's hypothesis.

In 1979, Geller et al. published an analysis of tissue androgen levels in prostate and non-androgen target tissues with a specific emphasis upon tissue steroid levels as markers of tumor differentiation and adequacy of anti-androgen therapy (36).

Tissues procured by transurethral resection of the prostate were assayed by RIA. They reported that 1 mg of DES did not adequately suppress tissue DHT levels since the levels remained intermediate between androgen-stimulated benign prostate and prostate tissue procured from castrated men. They concluded that their findings "support the long suspected theoretical role of andrenal cortical androgens as biologically important sources of DHT in relapse of PC." The clinical importance of their findings were obscured by Labrie's assertion that coadministration of anti-androgens cured PC by reducing tissue DHT levels to 0, a finding that was based upon experimental data in a total of four men. Interest and attention on Geller's original hypothesis has been rekindled by our RIA and MS data from men with castration-recurrent PC. MS findings are supported by two recent reports. Mizokami et al. (37) showed that average tissue DHT levels measured by LC/MS/MS decreased 75% in prostatectomy specimens obtained after 3–6 months of ADT, and Nishiyama et al. (38) found tissue DHT concentration decreased 75% in prostate tissue from 30 men receiving ADT for 6 months. Recently, Page et al. (39) reported that tissue levels of RIA testicular androgens in benign prostate may be sufficient for AR activation as early as 1 month after castration! Twelve men underwent prostate biopsies on day 28, four men received placebos, a long acting LHRH antagonist, acyline, or acycline and T. In four men who received acycline, T, and DHT, prostate tissue levels decreased by 70 and 80%, respectively. Despite this decrease in prostate tissue levels of androgens, IHC revealed no detectable differences among the three groups in cellular proliferation, apoptosis, and PSA or AR expression. This report is especially important for two reasons. First, it suggests that benign prostate recovers the ability to produce testicular androgens as soon as 1 month after institution of ADT. Second, normal tissue homeostasis was recovered 1 month after ADT. Yet, in benign prostate, prostate volume remains reduced forever; ADT cures benign prostate enlargement. In contrast, PC cells must develop the ability to use these tissue androgens to recur as castration recurrent PC to kill the patient.

Clinical Relevance of AR Activating Levels of Tissue Androgens

Are levels of ~3 nM T and DHT measured in castration-recurrent PC tissues sufficient to activate AR? Simard et al. (40) were the first to suggest that residual DHT in prostatic tissue after castration was androgenic. On the basis of traditional transient transfection experiments in PC cell lines, 1 nM T efficiently activates most androgen-regulated reporter genes. We (41) and others (42) have shown that the "supersensitive" AR is activated in castration-recurrent PC cell lines by pM DHT. The presence of PSA in these specimens of castration-recurrent PC and in serum of patients is consistent with the presence of an activated AR, although PSA levels in castration-recurrent CaP tissue were only 7.6% of levels in benign tissue. Stege et al. (43) reported a PSA level of 4,973 μg/g tissue (assuming 1 mg DNA per gram tissue) in aspirated benign prostate, which was similar to the level we found in benign prostate (3,198 μg per gram tissue). They reported a tissue PSA level of 458 μg per gram tissue in PC from noncastrated patients that were similar to the level we measured for castration-recurrent PC (297 μg per gram tissue). In transurethral resection specimens, Yang et al. (44) reported tissue PSA levels of 1952.27 μg/g protein in benign prostate, and 583.75 μg/g protein in PC from noncastrated patients. Since we and others obtained similar PSA levels in androgen-stimulated benign prostate, the similar PSA levels measured by us in castration-recurrent PC and those of others in androgen-stimulated PC and benign prostate suggest that the AR is activated in all tissues despite the castrate serum levels of androgens.

Adrenal Androgens may be the Source of Prostate Tissue DHT

Belanger et al. (32) suggested that persistent levels of prostatic DHT after castration alone resulted from metabolism of adrenal-derived DHEA, DHEA-SO_4, and ASD in prostate tissue. Serum DHEA-SO_4 levels can be 300–500 times the concentration of DHEA, and the sulfatase present in human prostate converts DHEA-SO_4 to DHEA (45). In the only report of tissue levels of DHEA, nonhyperplastic tissue specimens obtained by open prostatectomy contained 90 pmol mg^{-1} DNA (equivalent to 90 nM DHEA assuming 1 mg DNA per gram tissue) (45). These levels of DHEA cause detectable activation of AR in cotransfection assays (46). Moreover, small amounts of DHT have been reported to be formed from DHEA and DHEA-SO_4 in benign prostate (47).

Some preliminary data support the possibility that conversion of DHEA and DHEA-SO_4 to DHT in castration-recurrent PC contributes to AR activation. Thirty-six tissue homogenates were made from 12 samples each of frozen operative specimens of benign prostate, androgen-stimulated PC, and castration-recurrent PC. [^{3}H]-ASD appeared as [^{3}H]-DHT in all three tissue types (thin layer chromotography, data not shown) suggesting that the androgen metabolic enzymes present in androgen-stimulated benign prostate and PC remain in castration-recurrent PC. The androgen metabolism pathway from adrenal androgens to DHT appears present in the CWR-R1 cell line that was generated from a castration-recurrent CWR22 human xenograft tumor. An average of 5% of [^{14}C]-DHEA appeared as [^{14}C]-DHT in three experiments (Fig. 3).

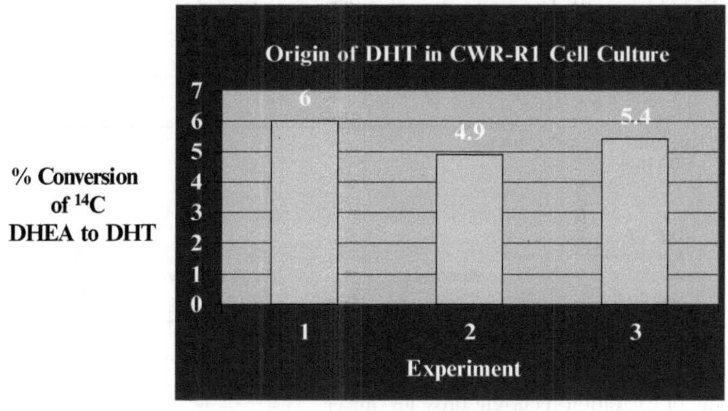

Fig. 3 % Conversion of [^{14}C]-DHEA to DHT in three individual experiments

These observations are consistent with recent reports of upregulation of androgen metabolism enzymes (45, 47) during ADT that allow T formation from adrenal androgens or androgen metabolites.

Model of Castration-Recurrent Prostate Cancer

The benign and malignant prostate is stimulated by circulating and tissue levels of androgens from puberty through adulthood. Both benign and malignant prostate epithelium grows slowly; rates of apoptosis and cellular proliferation are similar. Circulating T is reduced to DHT, the preferred AR ligand. Castration reduces circulating T and DHT to castrate levels that remain so indefinitely. Both benign and malignant prostate respond to this insult with a massive wave of apoptotic cell death, which peaks on day two (29) after castration (48). Both castration-recurrent PC (3, 31) and benign prostate (39) demonstrate tissue androgen levels sufficient for AR activation soon after castration. When PC recurs clinically after castration, PSA begins to rise in patients and xenograft models (48) and, for unknown reasons, castration-recurrent PC begins to grow again whereas benign prostate hyperplasia remains permanently dormant (Fig. 4).

Conclusion

AR remains active in growth signaling despite castrate levels of circulating androgens (49). AR protein and AR-regulated proteins are expressed in PC that recurs during ADT in both primary (3, 4, 50, 51) and bone metastases (52, 53). The substrates and metabolic pathways (54) responsible for maintenance of functional tissue levels of T and DHT in castration-recurrent PC remain to be clarified. New therapies that target AR directly (8) and prevent the formation of androgens within PC tissue (55) may offer novel approaches to prolong remission or induce reremission of castration-recurrent PC.

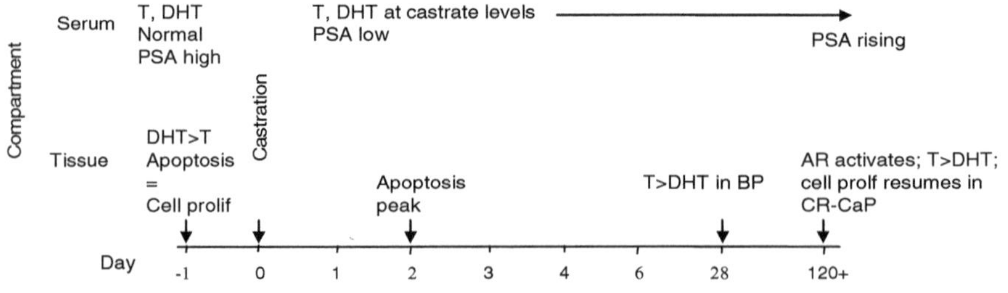

Fig. 4 Model of castration-recurrent prostate cancer

Summary

An American man is diagnosed with prostate cancer (PC) every 3 min and dies from the disease every 17 min. Although androgen receptor (AR) expression is diminished following androgen deprivation therapy (ADT) that induces clinical remission in most patients, castration-recurrent PC expresses levels of AR protein similar to those found in androgen-stimulated PC and benign prostate. This observation suggests that the AR may be as important for growth regulation in castration-recurrent PC, as it is in androgen-stimulated PC and benign hyperplasia. Neither ligand-independence, point mutations, glutamine and/or glycine repeat expansion nor amplification have explained AR activation in most cases of castration-recurrent PC. Castration-recurrent PC tissue has levels of testosterone (T) similar to androgen-stimulated benign prostate and levels of dihydrotestosterone (DHT), the most active androgen for AR activation that are approximately 10% of androgen-stimulated benign prostate. These levels of tissue androgens appear capable of activating the AR since prostate-specific antigen (PSA), the classic androgen-regulated gene, is expressed at similar tissue levels in castration-recurrent and androgen-stimulated PC. These startling findings suggest a paradigm shift; *PC that recurs during ADT is not androgen-independent.*

References

1. Jemal A, Siegel R, Ward E, et al. (2006) Cancer statistics, 2006. CA Cancer J Clin 56: 106–130.
2. Huggins C, Hodges CV (2002) Studies on prostatic cancer: I. The effect of castration, of estrogen and of androgen injection on serum phosphatases in metastatic carcinoma of the prostate. 1941. J Urol 168: 9–12.
3. Mohler JL, Gregory CW, Ford OH III, et al. (2004) The androgen axis in recurrent prostate cancer. Clin Cancer Res 10: 440–448.
4. van der Kwast TH, Schalken J, Ruizeveld de Winter JA, et al. (1991) Androgen receptors in endocrine-therapy-resistant human prostate cancer. Int J Cancer 48: 189–193.
5. Visakorpi T, Hyytinen E, Koivisto P, et al. (1995) In vivo amplification of the androgen receptor gene and progression of human prostate cancer. Nat Genet 9: 401–406.
6. Feldman BJ, Feldman D (2001) The development of androgen-independent prostate cancer. Nat Rev Cancer 1: 34–45.
7. Gelmann EP (2002) Molecular biology of the androgen receptor. J Clin Oncol 20: 3001–3015.
8. Grossmann ME, Huang H, Tindall DJ (2001) Androgen receptor signaling in androgen-refractory prostate cancer. J Natl Cancer Inst 93: 1687–1697.
9. Sadar MD, Hussain M, Bruchovsky N (1999) Prostate cancer: molecular biology of early progression to androgen independence. Endocr Relat Cancer 6: 487–502.
10. Takahashi H, Furusato M, Allsbrook WC, Jr, et al. (1995) Prevalence of androgen receptor gene mutations in latent prostatic carcinomas from Japanese men. Cancer Res 55: 1621–1624.
11. Tilley WD, Buchanan G, Hickey TE, et al. (1996) Mutations in the androgen receptor gene are associated with progression of human prostate cancer to androgen independence. Clin Cancer Res 2: 277–285.

12. Ruizeveld de Winter JA, Janssen PJ, Sleddens HM, et al. (1994) Androgen receptor status in localized and locally progressive hormone refractory human prostate cancer. Am J Pathol 144: 735–746.
13. Taplin ME, Bubley GJ, Shuster TD, et al. (1995) Mutation of the androgen-receptor gene in metastatic androgen-independent prostate cancer. N Engl J Med 332: 1393–1398.
14. Greene SL, Stockton P, Kozyreva OG, et al. (2003) Mutational analysis of the androgen receptor using laser capture microdissection and direct sequencing. In: Terrain D (ed.) Cancer Cell Signalling: Methods and Protocols. Humana Press, Totowa, NJ, pp. 287–302.
15. Taplin ME, Bubley GJ, Ko YJ, et al. (1999) Selection for androgen receptor mutations in prostate cancers treated with androgen antagonist. Cancer Res 59: 2511–2515.
16. Shi XB, Ma AH, Xia L, et al. (2002) Functional analysis of 44 mutant androgen receptors from human prostate cancer. Cancer Res 62: 1496–1502.
17. Culig Z, Hobisch A, Cronauer MV, et al. (1993) Mutant androgen receptor detected in an advanced-stage prostatic carcinoma is activated by adrenal androgens and progesterone. Mol Endocrinol 7: 1541–1550.
18. Peterziel H, Culig Z, Stober J, et al. (1995) Mutant androgen receptors in prostatic tumors distinguish between amino-acid-sequence requirements for transactivation and ligand binding. Int J Cancer 63: 544–550.
19. Tan J, Sharief Y, Hamil KG, et al. (1997) Dehydroepiandrosterone activates mutant androgen receptors expressed in the androgen-dependent human prostate cancer xenograft CWR22 and LNCaP cells. Mol Endocrinol 11: 450–459.
20. Chen CD, Welsbie DS, Tran C, et al. (2004) Molecular determinants of resistance to antiandrogen therapy. Nat Med 10: 33–39.
21. Ford OH, 3rd, Gregory CW, Kim D, et al. (2003) Androgen receptor gene amplification and protein expression in recurrent prostate cancer. J Urol 170: 1817–1821.
22. Palmberg C, Koivisto P, Kakkola L, et al. (2000) Androgen receptor gene amplification at primary progression predicts response to combined androgen blockade as second line therapy for advanced prostate cancer. J Urol 164: 1992–1995.
23. Ueda T, Bruchovsky N, Sadar MD (2002) Activation of the androgen receptor N-terminal domain by interleukin-6 via MAPK and STAT3 signal transduction pathways. J Biol Chem 277: 7076–7085.
24. Ueda T, Mawji NR, Bruchovsky N, et al. (2002) Ligand-independent activation of the androgen receptor by interleukin-6 and the role of steroid receptor coactivator-1 in prostate cancer cells. J Biol Chem 277: 38087–38094.
25. Yeh S, Lin HK, Kang HY, et al. (1999) From HER2/Neu signal cascade to androgen receptor and its coactivators: a novel pathway by induction of androgen target genes through MAP kinase in prostate cancer cells. Proc Natl Acad Sci USA 96: 5458–5463.
26. Liu Y, Majumder S, McCall W, et al. (2005) Inhibition of HER-2/neu kinase impairs androgen receptor recruitment to the androgen responsive enhancer. Cancer Res 65: 3404–3409.
27. Mahajan NP, Whang YE, Mohler JL, et al. (2005) Activated tyrosine kinase Ack1 promotes prostate tumorigenesis: role of Ack1 in polyubiquitination of tumor suppressor Wwox. Cancer Res 65: 10514–10523.
28. Park SY, Kim YJ, Gao AC, et al. (2006) Hypoxia increases androgen receptor activity in prostate cancer cells. Cancer Res 66: 5121–5129.
29. Smitherman AB, Gregory CW, Mohler JL (2003) Apoptosis levels increase after castration in the CWR22 human prostate cancer xenograft. Prostate 57: 24–31.
30. Dai J, Shen R, Sumitomo M, et al. (2002) Synergistic activation of the androgen receptor by bombesin and low-dose androgen. Clin Cancer Res 8: 2399–2405.
31. Titus MA, Schell MJ, Lih FB, et al. (2005) Testosterone and dihydrotestosterone tissue levels in recurrent prostate cancer. Clin Cancer Res 11: 4653–4657.
32. Belanger B, Belanger A, Labrie F, et al. (1989) Comparison of residual C-19 steroids in plasma and prostatic tissue of human, rat and guinea pig after castration: unique importance of extratesticular androgens in men. J Steroid Biochem 32: 695–698.

33. Labrie F, Dupont A, Belanger A, et al. (1982) New hormonal therapy in prostatic carcinoma: combined treatment with an LHRH agonist and an antiandrogen. Clin Invest Med 5: 267–275.
34. Prostate Cancer Trialists' Collaborative Group (1995) Maximum androgen blockade in advanced prostate cancer: an overview of 22 randomized trials with 3283 deaths in 5710 patients. Lancet 265–269.
35. Eisenberger MA, Blumenstein BA, Crawford ED, et al. (1998) Bilateral orchiectomy with or without flutamide for metastatic prostate cancer. N Engl J Med 339: 1036–1042.
36. Geller J, Albert J, Loza D (1979) Steroid levels in cancer of the prostate –markers of tumour differentiation and adequacy of anti-androgen therapy. J Steroid Biochem 11: 631–636.
37. Mizokami A, Koh E, Fujita H, et al. (2004) The adrenal androgen androstenediol is present in prostate cancer tissue after androgen deprivation therapy and activates mutated androgen receptor. Cancer Res 64: 765–771.
38. Nishiyama T, Hashimoto Y, Takahashi K (2004) The influence of androgen deprivation therapy on dihydrotestosterone levels in the prostatic tissue of patients with prostate cancer. Clin Cancer Res 10: 7121–7126.
39. Page ST, Lin DW, Mostaghel EA, et al. (2006) Persistent Intraprostatic Androgen Concentrations after Medical Castration in Healthy Men. J Clin Endocrinol Metab 91: 3850–3856.
40. Simard J, Luthy I, Guay J, et al. (1986) Characteristics of interaction of the antiandrogen flutamide with the androgen receptor in various target tissues. Mol Cell Endocrinol 44: 261–270.
41. Gregory CW, Johnson RT, Jr, Mohler JL, et al. (2001) Androgen receptor stabilization in recurrent prostate cancer is associated with hypersensitivity to low androgen. Cancer Res 61: 2892–2898.
42. Culig Z, Hoffmann J, Erdel M, et al. (1999) Switch from antagonist to agonist of the androgen receptor bicalutamide is associated with prostate tumour progression in a new model system. Br J Cancer 81: 242–251.
43. Stege R, Tribukait B, Lundh B, et al. (1992) Quantitative estimation of tissue prostate specific antigen, deoxyribonucleic acid ploidy and cytological grade in fine needle aspiration biopsies for prognosis of hormonally treated prostatic carcinoma. J Urol 148: 833–837.
44. Yang Y, Chisholm GD, Habib FK (1992) The distribution of PSA, cathepsin-D, and pS2 in BPH and cancer of the prostate. Prostate 21: 201–208.
45. Bartsch W, Klein H, Schiemann U, et al. (1990) Enzymes of androgen formation and degradation in the human prostate. Ann N Y Acad Sci 595: 53–66.
46. Gregory CW, He B, Johnson RT, et al. (2001) A mechanism for androgen receptor-mediated prostate cancer recurrence after androgen deprivation therapy. Cancer Res 61: 4315–4319.
47. Harper ME, Pike A, Peeling WB, et al. (1974) Steroids of adrenal origin metabolized by human prostatic tissue both in vivo and in vitro. J Endocrinol 60: 117–125.
48. Kim D, Gregory CW, French FS, et al. (2002) Androgen receptor expression and cellular proliferation during transition from androgen-dependent to recurrent growth after castration in the CWR22 prostate cancer xenograft. Am J Pathol 160: 219–226.
49. Taplin ME, Balk SP (2004) Androgen receptor: a key molecule in the progression of prostate cancer to hormone independence. J Cell Biochem 91: 483–490.
50. Edwards J, Krishna NS, Grigor KM, et al. (2003) Androgen receptor gene amplification and protein expression in hormone refractory prostate cancer. Br J Cancer 89: 552–556.
51. Linja MJ, Savinainen KJ, Saramaki OR, et al. (2001) Amplification and overexpression of androgen receptor gene in hormone-refractory prostate cancer. Cancer Res 61: 3550–3555.
52. Brown RS, Edwards J, Dogan A, et al. (2002) Amplification of the androgen receptor gene in bone metastases from hormone-refractory prostate cancer. J Pathol 198: 237–244.
53. Hobisch A, Culig Z, Radmayr C, et al. (1995) Distant metastases from prostatic carcinoma express androgen receptor protein. Cancer Res 55: 3068–3072.

54. Hsing AW, Reichardt JK, Stanczyk FZ (2002) Hormones and prostate cancer: current perspectives and future directions. Prostate 52: 213–235.
55. Stanbrough M, Bubley GJ, Ross K, et al. (2006) Increased expression of genes converting adrenal androgens to testosterone in androgen-independent prostate cancer. Cancer Res 66: 2815–2825.

21 Estrogen-Receptor-Related Receptors and Hormone-Dependent Cancers

Olivia Lanvin, Stéphanie Bianco, and Jean-Marc Vanacker

The NR Superfamily and Estrogen-Receptor-Related Receptors

Nuclear receptors (NRs) comprise a superfamily of ligand-dependent transcription factors that all share similar modular structures (1). A highly conserved DNA-binding domain (DBD), located toward the center of the NR, mediates recognition of specific response elements (REs) within the promoters of target genes. The C-terminus contains a less conserved ligand-binding domain (LBD), which is responsible for ligand-recognition, homo- or hetero-dimerization, interaction with co-modulators, and activation of target gene transcription. A number of NRs, termed "orphan" receptors, display features similar to those found in the "classical" (i.e., "liganded") receptors, including a putative LBD, but for which no natural ligand has been characterized (2). Whether such ligands indeed exist, or whether orphan receptors can activate transcription in a constitutive manner is still subject to debate.

The 48 known human NRs have been grouped into seven phylogenetic families (3). Specifically, members of NR family three (NR3), which include estrogen receptors (ER) α and β, and all other steroid receptors [androgen (AR), progesterone (PR), mineralocorticoid (MR), and glucocorticoid receptors (GR)] are particularly relevant in hormone-dependent cancers. The dependence of prostate cancer on AR activity, and breast, ovarian, and uterine cancers on ERα and β and the PR is well-known. The NR3 family also includes three orphan receptors. Estrogen-receptor-related receptor (ERR) α and β were isolated in the late 1980s on the basis of cDNA hybridization studies (4). ERRγ was isolated 10 years later using yeast two hybrid screening technology (5). Since then, ERRs have been identified in various vertebrates, and a unique ERR homolog has been cloned in diverse invertebrates, supporting the hypothesis that a common predecessor existed for all three ERR genes (6) (Fig. 1).

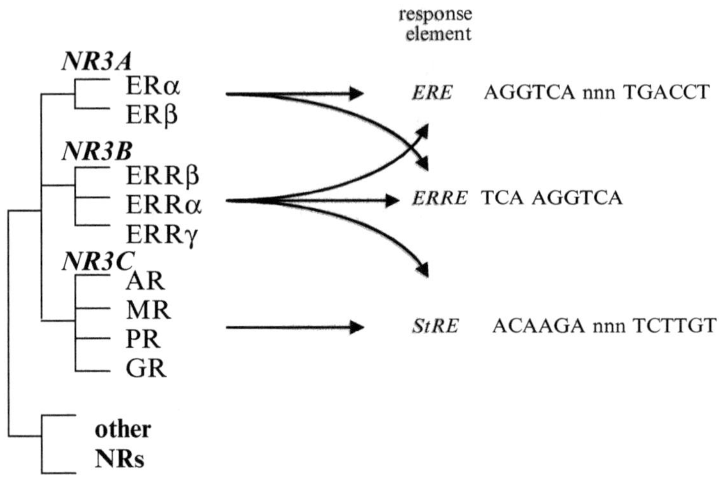

Fig. 1 The NR3 family of nuclear receptors. ERR receptors are as much related to the ER as to the SR. The specific response elements are indicated. Arrows indicate a functional relationship

Expression and Known Function of ERRs

ERR expression patterns have been determined in the mouse and a few physiological functions have been attributed to them. For instance, ERRβ is expressed early during embryonic development, and studies using ERRβ knock-out (KO) mice have revealed that ERRβ is required for placentation and primordial germ cell migration (7). ERRα is broadly expressed both in the embryo and in the adult (4, 8). ERRαKO mice are viable and fertile, and do not display gross abnormalities at the anatomical level (9). However, in comparison to their wild-type littermates, ERRαKO mice have a reduced body and fat mass and are resistant to high fat diet-induced obesity. This evidence suggests that ERRα functions to control energy balance (10–11). Indeed, ERRαKO mice display deficient lipid synthesis and absorption (12), as well as enhanced fat catabolism (9). In addition, through interaction with the coactivator PGC1-α, ERRα also enhances mitochondrial biogenesis (13), a typical feature of slow-twitch (i.e., lipid consuming) muscle fibers. A role for ERRα in bone differentiation and maintenance has also been strongly suggested, though this remains to be formally demonstrated in vivo (14). Finally, work in the zebrafish has demonstrated that ERRα, or at least its fish homolog, is involved in the regulation of cellular movements during embryonic development (15). Whether this is also true in mammals remains to be demonstrated. ERRγ is also broadly expressed in the mouse embryo and in adult mice, although to a lesser extent than ERRα (16). As yet, in-vivo functions have not been attributed to this receptor and ERRγKO mice have not been described.

ERR Ligands

Although significant, the sequence identity between ERs and ERRs is relatively low in the LBD (~30%) (17). Accordingly, none of the ERRs can bind or be regulated by natural estrogens. Crystallographic studies predict that the ERRγ LBD spontaneously adopts an "active" conformation (18) and thus regulates transcription in a constitutive manner, an observation that has been extended to all three ERRs. This is consistent with the very small hydrophobic pocket of ERRα, as determined by 3D structure (19). Despite these potential structural hindrances to natural ligand binding, researchers are actively searching for synthetic compounds that could either potentiate or counteract the transcriptional activities of ERR receptors. Molecules that act with a certain degree of specificity as antagonists, or rather "inverse agonists," of ERRs have been identified. For example, in vitro, 4-OH tamoxifen binds to ERRβ and ERRγ, but not to ERRα (20), and deactivates ERRγ in cell-based assays. Diethylstilbestrol acts as a pan-ERR inverse agonist (21). On the other hand, various phytoestrogens, i.e., daidzein and genistein, further enhance the transcriptional activities of ERRα and ERRβ, but not of ERRγ (22). However, using these molecules as tools to study the functions of ERRs is difficult, as they are well-known modulators of ER activity, and their IC_{50}s are considerably higher for the ERRs than for the ERs. Considerable progress has been made more recently with the characterization of XCT790, a synthetic thiadiazoleacrylamide, which appears to specifically inhibit ERRα (23), while not affecting the activities of several other NRs tested, including the other ERR subfamily members. A number of ERRα target genes have been identified, as a result of having their expression down-regulated by XCT790 in an ERRα-dependent manner. These include ERRα itself (24), monoamine oxidase (MAO)-A and -B (23, 25), and medium chain acyl-dehydrogenase (MCAD) (9). However such observations are mitigated by the fact that to achieve such inhibition, XCT790 must be used at relatively high doses (binding to ERRα at IC_{50} – 0.37 μM) and potential unrelated effects of this compound can not yet be discounted.

Modulation of Steroid-Response by ERRs

The particularly striking ~70% sequence identity between ERs and ERRs within their DBDs leads to the prediction that they recognize and bind to common promoter sequences. Indeed, while ERRs bind to their own specific response elements (ERRE) (26–28), it has been shown that they also can recognize classical estrogen-response elements (EREs), at least in vitro (29–30). Published reports indicate that ERRs can act on EREs to either activate or repress expression (31), for reasons that are not yet well understood, but may depend on cell type or ER status. ERα, but not ERβ, has been shown to bind to and activate transcription through the ERRE (29). For example, both ERα and ERRα act through common ERREs present within the osteopontin gene promoter. In the case of the lactoferrin gene, both ERRα and ERα act through different promoter binding sites and cooperate to insure a maximal response (28). In-vitro evidence suggests that this response requires physical interaction between the receptors. Both ERRα and ERα have also been shown to target the breast cancer biomarker pS2 (32).

Through its effects on gene expression, ERRα can affect estrogen signaling in other ways. By up-regulating SULT2A1 (DHEA-sulfonating enzyme) expression (33), ERRα may indirectly stabilize plasma DHEA levels, thus increasing the bioavailability of this androgen precursor. In turn, androgens can be converted to estrogens by the aromatase enzyme (cyp19), whose expression is also up-regulated by ERRα (34). Whether increasing the local or circulating level of estrogens is a major or minor role for ERRα remains to be determined, since ERRαKO mice did not display noticeable estrogen deficiencies. Conversely, the expression of ERRα has been shown to be positively regulated by the synthetic estrogen DES in the mouse uterus (35). Taken together, these data suggest that ERRα could act as an amplifier of estrogen signaling.

Phylogenetical studies have indicated that ERR receptors are as closely related to the ERs as to the other steroid receptors forming NR family 3 including the AR (3). On this basis, our laboratory has investigated the potential for cross-talk between ERRs and androgen signaling. Our unpublished data demonstrate that the expression of AR responsive genes in LNCaP (human androgen-responsive prostate cancer) cells can be down-regulated by XCT790. These AR targets include PSA (prostate-specific antigen, a marker of prostate cancer), Nkx3.1, and Kallikrein 2 (36). XCT790 does not modulate the expression of AR at the RNA or protein level, nor does it block androgen responsiveness (i.e., dihydrotestosterone can still enhance gene expression in the presence of XCT790), suggesting that XCT790 does not act on the AR itself. Steroid-responsive promoters, such as PSA, probasin, MMTV, or multimerized androgen-response elements, cloned in front of a minimal promoter respond to all ERRs, but not to ERs, upon transient cotransfection in various cell lines. Interestingly this effect also occurs in the presence of steroid antagonists, such as bicalutamide or RU486. This indicates that optimal silencing of steroid target genes not only requires inhibition of steroid receptor activity but also that of ERRs. We thus conclude that ERRs may activate the same genes as do estrogens or steroids including androgens. In other words, ERRs might exert molecular effects similar to those of its related receptors (ER and AR), although in a ligand-independent manner. Given the implications of ER and AR in breast and prostate cancers, respectively, the effect of ERRs in these tumors may thus be relevant.

ERRs and Cancer

The expression of ERR genes, in particular ERRα, has been examined in several human cancer types (37). ERRα is more highly expressed in colon cancer than in normal colon tissue (38). Furthermore the degree of ERRα expression correlates with tumor grade. ERRγ was equally detected in normal and transformed tissue, whereas ERRβ was not expressed in either. ERRα is also expressed in the prostate with a slight signal decrease in neoplastic vs. hyperplastic tissues (39).

In ovarian cancer, ERRα was also found to be overexpressed compared to normal tissue (40). ERRα expression increased with advanced tumor stage and correlated with low survival. This receptor may constitute a marker of poor prognosis. The prognostic value of ERR expression is not yet clear in breast cancer, considering the results of two independent studies. Examining RNA expression in a sample of 38 breast tumors, Ariazi et al. (41) saw an inverse correlation between the expression of ERRα and that of ERα/PR. On the other hand, expression of ERRγ correlated positively with that of ERα/PR. In tumors, the expression of ERRα correlated positively with that of HER2 and HER3. Given their known prognostic value in breast cancer when $ER\alpha^+/PR^+$ and $HER2^-/HER3^-$, these results suggested that elevated ERRγ expression was a favorable marker, whereas elevated ERRα expression was not. Suzuki et al. (42) studied 102 breast cancers at the protein level and saw an inverse correlation between ERRα expression and disease-free survival. No correlation was found between ERRα and ERα/PR. However, their observed association between ERα and pS2 expression (significant when considering all tumors) was lost when only the set of $ERR\alpha^+$ tumors was analyzed. This may suggest that, with high ERRα expression, ERα may become dispensable for high pS2 expression, reinforcing the hypothesis of ERRα/ERα cross-talk.

Since ERRα expression appears to be correlated with cancer aggressiveness, this suggests that it may be involved in the regulation of proliferation and metastasis. On this basis, our laboratory has started to analyze the proliferative effect of ERRα. Our yet unpublished data show that treatment of MCF7 cells with XCT790 reduces estrogen-induced proliferation, as efficiently as tamoxifen (Fig. 2).

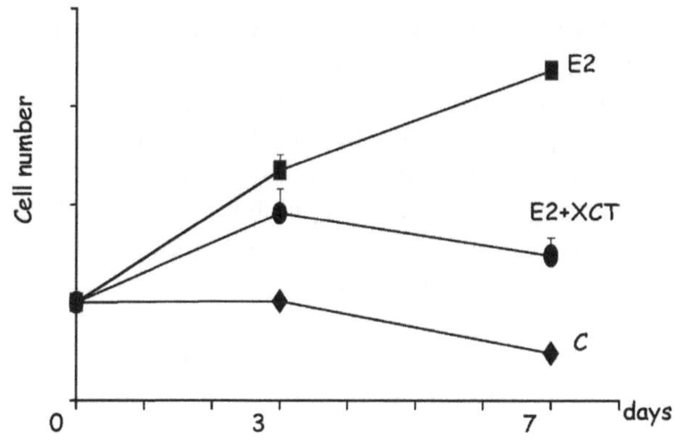

Fig. 2 Effect of XCT790 (XCT) on estrogen (E_2)-induced proliferation of MCF7 cells. C, unstimulated control

The precise mechanism by which XCT790 functions is presently unknown. However, it is likely to affect the cell cycle and not apoptosis. Indeed, estrogens enhance the number of cells in the S phase of the cycle, whereas co-treatment with XCT790 restores the G1 phase block observed in the absence of estrogens. This is consistent with the previous observation that ERR modulates the expression of p21WAF/Cip1 promoter (43). This again suggests that ERRα could exert similar effects as ERα. However, the XCT790 antiproliferative effect is not limited to estrogen-induced cell division. Indeed, XCT790 blocks the proliferation of MDA-MB231 cells, which grow in an estrogen-independent manner. The androgen-induced proliferation of LNCaP cells is also abolished by XCT790 treatment, and EGF-induced proliferation of LNCaP cells is greatly reduced. EGF has been shown to activate the transcriptional capacities of ERRα through a phosphorylation-dependent mechanism (44). XCT790 also reduced the proliferation of PC3, an androgen-independent prostate cancer cell line. Given these results, down-modulating ERRα with XCT790 thus appears to be a potent strategy to reduce cell proliferation, although the exact mechanisms by which XCT790 works need further study. Interestingly, the antiproliferative effects of XCT790 may not only be limited to hormone-dependent cells but may also extend to cells/tumors that have escaped hormonal control, for which established treatments are scarce.

Perspectives and Pending Questions

The progress made in understanding the biologic functions of ERRs, specifically ERRα, leaves many important questions that remain unanswered. For instance, as pointed out above, the mechanisms through which ERRα and its antagonist XCT790 act to regulate proliferation are presently unknown and demand characterization. Does ERRα regulate the same set of genes as ER or AR, as predicted by the cross-talk hypothesis, or does it act through completely different mechanisms? In this respect, it will also be interesting to determine whether ERRα contributes to the acquisition of hormone-resistance, as its tendency to be over-expressed in tumors with a poor prognosis can lead to such hypothesis. The role of the ERRα in the normal mammary gland and prostate is also unclear at this time, especially since an altered phenotype was not observed in these organs in ERRαKO mice (9). Redundancy with the other ERR receptors could be invoked, but this also raises the question of the function of ERRβ and ERRγ in these organs, with respect to proliferation. Finally, our previous work has shown that ERRα was involved in the regulation of cellular movements during zebrafish gastrulation (15). If this function is conserved in mammals, then ERRα might contribute to invasion of the adjacent tissue by cancerous cells and also to metastasis, a hypothesis to be tested.

Acknowledgments Work in the authors' laboratory was made possible by grants from the Ligue Contre le Cancer (Comités Drôme, Languedoc-Roussillon and Loire), the Association pour la Recherche sur le Cancer, and the Association pour la Recherche sur les Tumeurs de la Prostate.

References

1. Chambon P (2005) The nuclear receptor superfamily: a personal retrospect on the first two decades. Mol Endocrinol 19:1418–28.
2. Giguère V (1999) Orphan nuclear receptors: from gene to function. Endocrine Rev 20: 689–725.
3. Laudet V (1997) Evolution of the nuclear receptor superfamily: early diversification from an ancestral orphan receptor. J Mol Endocrinol 19:207–26.
4. Giguère V, Yang N, Segui P, et al. (1988) Identification for a new class of steroid hormone receptors. Nature 331:91–4.
5. Hong H, Yang L, Stallcup MR (1999) Hormone-independent transcriptional activation and coactivator binding by novel orphan nuclear receptor ERR3. J Biol Chem 274:22618–26.
6. Bardet PL, Laudet V, Vanacker JM (2006) Studying non-mammalian models? Not a fool's ERRand! Trends Endocrinol Metab 17:166–71.
7. Luo J, Sladek R, Bader JA, et al. (1997) Placental abnormalities in mouse embryos lacking the orphan nuclear receptor ERRα. Nature 388:778–82.
8. Bonnelye E, Vanacker JM, Spruyt N, et al. (1997) Expression of the estrogen-related receptor 1 (ERR-1) orphan receptor during mouse development. Mech Dev 65:71–85.
9. Luo J, Sladek R, Carrier J, et al. (2003) Reduced fat mass in mice lacking orphan nuclear receptor estrogen-related receptor α. Mol Cell Biol 23:7947–56
10. Huss JM, Torra IP, Staels B, et al. (2004) Estrogen-related receptor alpha directs peroxisome proliferator-activated receptor alpha signaling in the transcriptional control of energy metabolism in cardiac and skeletal muscle. Mol Cell Biol 24:9079–91.
11. Schreiber SN, Knutti D, Brogli K, et al. (2003) The transcriptional coactivator PGC-1 regulates the expression and activity of the orphan nuclear receptor estrogen-related receptor alpha (ERRalpha). J Biol Chem 278:9013–8.
12. Carrier JL, Deblois G, Champigny C, et al. (2004) Estrogen-related receptor α (ERRα) is a transcriptional regulator of apolipoprotein A-IV and controls lipid handling in the intestine. J Biol Chem 279:52052–8.
13. Schreiber SN, Emter R, Hock MB, et al. (2004) The estrogen-related receptor α (ERRα) functions in PPARgamma coactivator 1-α (PGC1-α)-induced mitochondrial biogenesis. Proc Natl Acad Sci USA 101:6472–7.
14. Bonnelye E, Aubin JE (2005) Estrogen receptor-related receptor alpha: a mediator of estrogen response in bone. J Clin Endocrinol Metab 90:3115–21.
15. Bardet PL, Horard B, Laudet V, et al. (2005) The ERRalpha orphan nuclear receptor controls morphogenetic movements during zebrafish gastrulation. Dev Biol 281:102–11.
16. Susens U, Hermans-Borgmeyer I, Borgmeyer U (2000) Alternative splicing and expression of the mouse estrogen receptor-related receptor gamma. Biochem Biophys Res Commun 2000 267:532–5.
17. Horard B, Vanacker JM (2003) Estrogen receptor-related receptors: orphan receptors desperately seeking a ligand. J Mol Endocrinol 31:349–57.
18. Greschik H, Wurtz JM, Sanglier S, et al. (2002) Structural and functional evidence for ligand-independent transcriptional activation by the estrogen-related receptor 3. Mol Cell 9:303–313.
19. Kallen J, Schlaeppi JM, Bitsch F (2004) Evidence for ligand-independent transcriptional activation of the human estrogen-related receptor α (ERRα): crystal structure of ERRα ligand binding domain in complex with peroxisome proliferator-activated receptor coactivator-1α. J Biol Chem 279:49330–7.
20. Coward P, Lee D, Hull MV, et al. (2001) 4-Hydroxytamoxifen binds to and deactivates the estrogen-related receptor gamma. Proc Natl Acad Sci USA 98:8880–4.
21. Tremblay GB, Kunath T, Bergeron D, et al. (2001) Diethylstilbestrol regulates trophoblast stem cell differentiation as a ligand of orphan nuclear receptor ERRβ. Genes Dev 15:833–8.
22. Suetsugi M, Su L, Karlsberg K, et al. (2003) Flavone and isoflavone phytoestrogens are agonists of estrogen-related receptors. Mol Cancer Res 1:981–91.

23. Willy PJ, Murray IR, Qian J, et al. (2004) Regulation of PPARγ coactivator 1α (PGC-1α) signaling by an estrogen-related receptor α (ERRα) ligand. Proc Natl Acad Sci USA 101:8912–17.
24. Laganière J, Tremblay GB, Dufour CR, et al. (2004) A polymorphic autoregulatory hormone response element in the human estrogen-related receptor alpha (ERRα) promoter dictates peroxisome proliferator-activated receptor γ coactivator-1α control of ERRα expression. J Biol Chem 279:18504–10.
25. Zhang Z, Chen K, Shih JC, et al. (2006) Estrogen-related receptors-stimulated monoamine oxidase B promoter activity is down-regulated by estrogen receptors. Mol Endocrinol 20:1547–61.
26. Yang N, Shigeta H, Shi H, et al. (1996) Estrogen-related receptor, hERR1, modulates estrogen receptor mediated response of human lactoferrin gene promoter. J Biol Chem 271:5795–804.
27. Johnston SD, Liu X, Zuo F, et al. (1997) Estrogen-related receptor α 1 functionally binds as a monomer to extended half-site sequences including ones contained within estrogen-response elemens. Mol Endocrinol 11:342–52.
28. Bonnelye E, Vanacker JM, Dittmar T, et al. (1997) The ERR-1 orphan receptor is a transcriptional activator expressed during bone development. Mol Endocrinol 11:905–16.
29. Vanacker JM, Pettersson K, Gustafsson JA, et al. (1999) Transcriptional targets shared by estrogen-receptor-related receptors (ERRs) and estrogen receptor (ER)α, but not ERβ. EMBO J 18:4270–9.
30. Pettersson K, Svensson K, Mattsson R, et al. (1996) Expression of a novel member of estrogen response element-binding nuclear receptors is restricted to the early stages of chorion formation during mouse embryogenesis. Mech Dev 54:211–23.
31. Kraus KJ, Ariazi EA, Farrell ML, et al. (2002) Estrogen-related receptor alpha-1 actively antagonizes estrogen-receptor regulated transcription in MCF-7 mammary cells. J Biol Chem 277:24826–34.
32. Lu D, Kiriyama Y, Lee KY, et al. (2001) Transcriptional regulation of the estrogen-inducible pS2 breast cancer marker gene by the ERR family of orphan nuclear receptors. Cancer Res 61:6755–61.
33. Seely J, Amigh KS, Suzuki T, et al. (2005) Transcriptional regulation of dehydroepiandrosterone sulfotransferase (SULT2A1) by estrogen-related receptor alpha. Endocrinology 146:3605–13.
34. Yang C, Zhou D, Chen S (1998) Modulation of aromatase expression in the breast tissue by ERRα-1 orphan receptor. Cancer Res 58:5696–700.
35. Shigeta H, Zuo W, Yang N, et al. (1997) The mouse estrogen-receptor related orphan receptor α 1: molecular cloning and estrogen responsiveness. J Mol Endocrinol 19:299–309.
36. Nelson PS, Clegg N, Arnold H, et al. (2002) The program of androgen-responsive genes in neoplastic prostate epithelium. Proc Natl Acad Sci USA 99:11890–5.
37. Ariazi EA, Jordan VC (2006) Estrogen-related receptors as emerging targets in cancer and metabolic disorders. Curr Topics Med Chem 6:181–93.
38. Cavallini A, Notarnicola M, Giannini R, et al. (2005) Oestrogen receptor-related receptor alpha (ERRα) and oestrogen receptors (ERα and ERβ) exhibit different gene expression in human colorectal tumor progression. Eur J Cancer 41:1487–1494.
39. Cheung CP, Yu S, Wong KB, et al. (2005) Expression and functional study of estrogen receptor-related receptors in human prostatic cells and tissues. J Clin Endocrinol Metab 90:1830–44.
40. Sun P, Sehouli J, Denkert C, et al. (2005) Expression of estrogen receptor-related receptors, a subfamily of orphan nuclear receptors, as new tumor biomarkers in ovarian cancer cells. J Mol Med 83: 457–67.
41. Ariazi EA, Clark GM, Mertz JE (2002) Estrogen-related receptor α and estrogen-related receptor γ associate with unfavorable and favorable biomarkers, respectively, in human breast cancer. Cancer Res 62:6510–6518.

42. Suzuki T, Miki Y, Moriya T, et al. (2004) Estrogen-related receptor α in human breast carcinoma as a potent prognostic factor. Cancer Res 64: 4670–6.
43. Castet A, Herledan A, Bonnet S, et al. (2006) Receptor-interacting protein 140 differentially regulates estrogen receptor-related receptor transactivation depending on target genes. Mol Endocrinol 20: 1035–47.
44. Barry JB, Giguère V (2005) Epidermal growth factor-induced signaling in breast cancer cells results in selective target gene activation by orphan nuclear receptor estrogen-related receptor alpha. Cancer Res 65:6120–9.

22
Androgen Receptor Coactivators and Prostate Cancer

Irina U. Agoulnik and Nancy L. Weigel

Introduction

Among USA men, prostate cancer (PC) is the second most common cause of death from cancer. Thus, the etiology, prevention, and treatment of the disease are a major health concern. Development and differentiation of the prostate is androgen dependent and PC, too, is androgen dependent (1). Consequently, some form of androgen deprivation is the primary treatment for metastatic PC. Although effective initially in reducing tumor burden, the tumors become resistant to androgen deprivation and recur within a relatively short period of time. The actions of androgens are mediated by the androgen receptor (AR) a hormone-activated transcription factor, which belongs to the large nuclear receptor superfamily of ligand-activated transcription factors (2, 3). AR differs from many of the other receptors in that it has two natural endogenous ligands. Testosterone (T) (Fig. 1) is the major circulating androgen and is the major hormone in most tissues. T is produced in the testis and is converted to 5α-dihydrotestosterone (DHT) (Fig. 1) by the enzyme 5α-reductase in the prostate as well as in selected other tissues including skin. DHT is a higher affinity ligand and is functionally the most important androgen in the prostate. In addition, there are a number of androgen metabolites including DHEA and androstenediol, which have much lower affinities for AR. Although these androgens are not thought to play a major role in AR action in androgen-repleted males, they may activate the AR when levels of T and DHT are reduced as a result of androgen ablation therapy.

Despite the frequency of failure of androgen ablation therapy indicating that the prostate tumors have developed means to grow without normal circulating levels of androgens, there is good evidence that many of these tumors remain dependent upon AR. Indeed, treatment failure is often detected due to increasing serum levels of PSA (prostate-specific antigen), an androgen-regulated protease expressed in normal prostate and in PC. This has led to interest in understanding how AR functions and in identifying the changes that permit AR to function despite reduced levels of androgens.

J.J. Li et al. (eds.), *Hormonal Carcinogenesis V.*

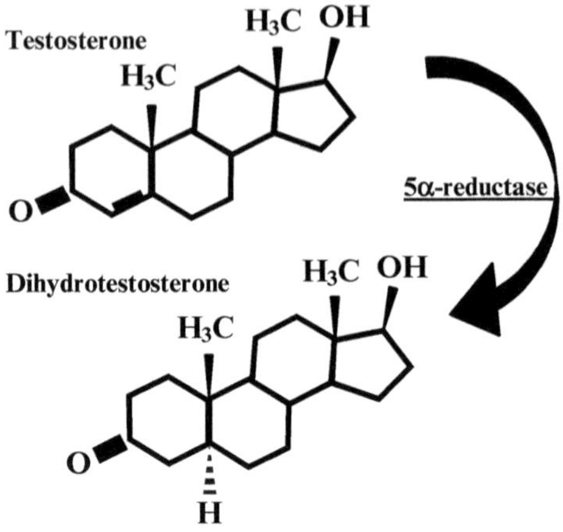

Fig. 1 The structures of testosterone and 5α-dihydrotestosterone

AR Structure and Function

Similar to other steroid receptor family members, AR contains a central DNA binding domain (DBD) consisting of two Zn finger motifs, a carboxyl terminal hormone binding domain (HBD) linked to the DBD by a hinge region that contains a nuclear localization signal and an amino-terminal region that is important for full transcriptional activation (3) (Fig. 2).

Unique to the AR is the variable length polyglutamine repeat in the amino terminus. A shorter polyglutamine repeat appears to make the AR more transcriptionally active. Thus, there has been much interest in determining whether a short polyglutamine repeat increases risk for PC, but most studies have failed to find a correlation. Very long polyglutamine repeats (>40) cause spinal bulbar muscular atrophy (SBMA) also known as Kennedy's disease (4). The variation in the polyglutamine repeat leads to variation in the reported length of AR. It is most commonly referred to as containing 919 amino acids and the description of the location of mutations is based on the assumption of a total length of 919. In addition to hormone-binding and DNA-binding activity, receptors must recruit coactivators that modify chromatin structure and facilitate transcription through regions termed activation functions. All steroid receptors contain a hormone-dependent activation function, AF2, in the HBD and at least one additional domain in the amino-terminus termed AF1. The relative importance of these regions is receptor dependent and may also be cell type dependent as well as target gene dependent. Evidence to date indicates that AR is more highly dependent upon its AF1 function than most steroid receptors (5). Deletion of the HBD yields a constitutively active AR (6). In the absence of hormone, AR is located in the cytoplasm and each AR molecule is associated with a heat shock protein (hsp) complex that maintains the capacity of AR to bind hormone and protects the receptor from proteolysis (Fig. 3). Upon hormone binding, AR dissociates from the heat shock protein complex,

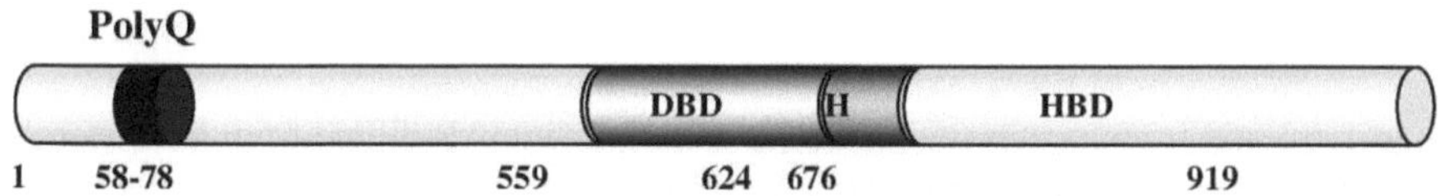

Fig. 2 Structure of the androgen receptor. The AR contains a carboxyl-terminal hormone binding domain (HBD) linked to its DNA binding domain (DBD) by a hinge region (H). The amino terminal region contains a variable length polyglutamine (PolyQ) repeat. Numbers indicate boundaries of domains based on an assumed length of 919 amino acids

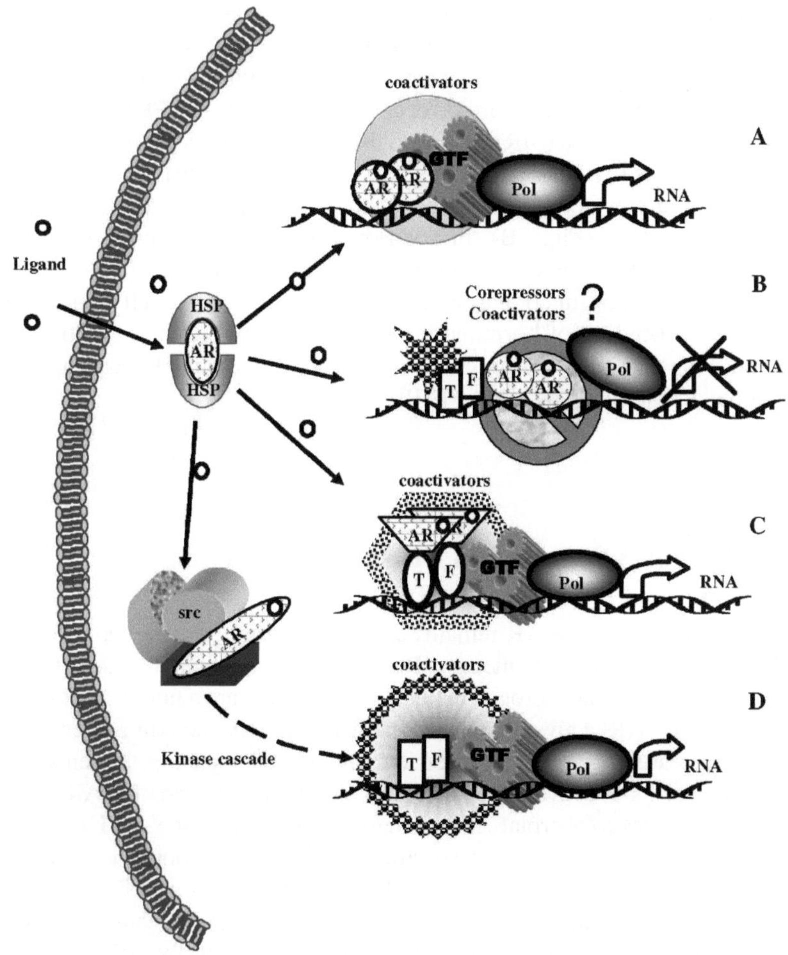

Fig. 3 AR action. In the absence of ligand, AR is localized to the cytoplasm and is associated with heat shock proteins. Hormone binding results in dissociation of AR from heat shock proteins. A: ARs dimerize, bind to DNA response elements recruiting coactivators and stimulating transcription. B: AR bind to DNA at negative regulatory elements preventing transcription through as yet undefined mechanisms, which may include recruitment of corepressors or interfering with binding of other transcription factors. C: ARs also potentiate transcription through interactions with other transcription factors independent of direct DNA binding. D: Ligand bind AR can interact with kinases including src activating a kinase cascade that results in activation of p42/p44 MAPK and phosphorylation of transcription factors and coactivators regulating additional genes

dimerizes, and translocates to the nuclear compartment. AR differs from other steroid receptors in that the major dimerization interface between the two monomers is not an interaction between two HBDs measured using a transiently transfected reporter. Rather, it appears that regions in the amino terminus of AR interact with the HBD through a region that overlaps with a region corresponding to the major AF2 coactivator interface in other steroid receptors (7). Consistent with this, the AR HBD crystallizes as a monomer (8), whereas other steroid receptor HBDs crystallize as homodimers.

In the best characterized mechanism of regulating transcription, the AR dimer binds to an androgen response element (ARE) in the DNA and recruits a series of coactivators necessary for chromatin remodeling and induction of transcription (3) (Fig. 3A). AR and the related steroid receptors, progesterone (PR), glucocorticoid (GR), and mineralocorticoid (MR) all bind to the same consensus palindrome separated by three nucleotides (9). There are many variations of this consensus sequence. AR binding results in elevated transcription of many target genes, but also represses transcription of others (Fig. 3B). In addition AR regulates transcription through protein–protein interactions with other transcription factors (Fig. 3C). Finally, there is evidence that AR can function as an extranuclear activator of cell signaling pathways (Fig. 3D). Binding of hormone induces association of some fraction of AR with src and, in some cases, estrogen receptor (ER) activating a kinase cascade that results in the activation of p42/p44 MAPK (10).

Potentiation of AR Action in Prostate Cancer

There is good evidence that AR remains a factor in many tumors that have become resistant to androgen ablation therapy. Resistance is often initially detected by an increase in serum PSA, an androgen regulated gene. There are a number of factors that can contribute to AR activation under these conditions including increased AR expression, increased coactivator expression, altered cell signaling that enhances the activity of AR or its coactivators or increases in alternate ligands for AR (11). One factor that contributes to aberrant AR activation is mutation of the AR HBD resulting in a broadening of the ligand-binding specificity of AR (12). Although the overall % of AR mutations in PC (<10% – >25% depending on the study) is insufficient to account for the majority of resistant tumors, tumors from patients who fail treatment with an AR antagonist, flutamide, frequently contain a mutation that permits the AR to utilize flutamide as an agonist. The most common mutation is T877A, which has been detected in a number of patients by several independent investigators; it is also found in LNCaP PC cells. This mutation permits AR to utilize a variety of steroids as agonists.

In an effort to understand the differences between androgen-dependent and androgen ablation-resistant tumors, a number of investigators have examined mRNA expression patterns in the two tumor types both in authentic human tumors as well as in xenograft models. The most consistent difference is the detection of elevated levels of AR mRNA (13, 14). There is evidence that elevated AR protein levels correlate with increased risk of recurrence of primary tumors as well as evidence for enhanced AR protein expression in androgen independent tumors (15). In cell culture studies, inactivation of AR using AR antibodies, or elimination of AR using ribozymes or siRNA shows that androgen "independent" cell lines remain AR dependent (16, 17) despite their ability to grow in castrated mice or in medium depleted of androgens by charcoal stripping.

There are a number of studies implicating altered cell signaling in activating AR in the absence of added androgens. Early studies showed that growth factors such as KGF and EGF or activators of cAMP signaling such as forskolin could induce AR-dependent transcription of reporter genes or of PSA (18, 19). Subsequent studies showed that the cytokine, IL-6, also activates AR in LNCaP PC cells (20). Whether these studies truly demonstrate hormone-independent activation of AR or simply permit AR to use the extremely low levels of residual hormone in charcoal stripped serum is unknown. In vivo, androgen ablation treatments typically target testicular androgens without depleting adrenal androgens. Earlier studies suggested and more recent studies have confirmed that despite the dramatic decrease in serum T upon treatment, tumor levels of T and DHT remain significant although substantially reduced (21). Thus, there has been significant interest in identifying factors that allow AR to function despite the reduced levels of androgens.

Among the factors that have the potential to facilitate AR activation are AR coactivators. A large number of candidate coactivators have been identified (22). Artificial overexpression of each of these enhances AR-dependent transcriptional activation of an AR-dependent reporter and each candidate either interacts directly with AR or is found in a protein complex that directly interacts with AR. Determining the contribution of the endogenous levels of these proteins to the activity of endogenous AR is an active area of research, but most have not been thoroughly characterized. One reason for the intense interest in this area is the possibility that increased levels of coactivators might potentiate the activity of AR at low residual levels of hormone or allow AR to use alternate ligands. As shown in Fig. 4,

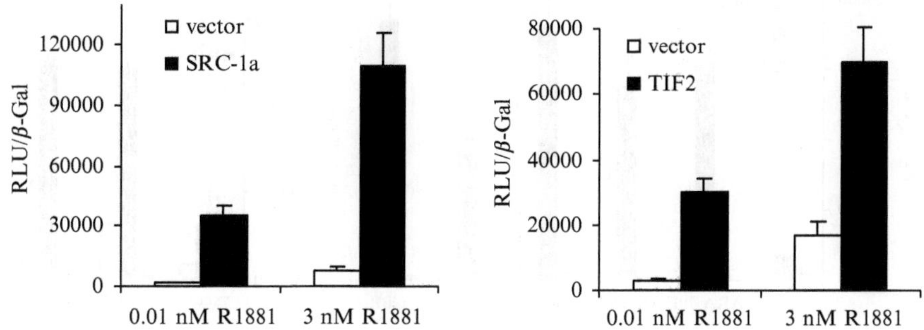

Fig. 4 A: HeLa cells were transfected with 400 ng of the AR responsive reporter, GRE_2-E1b-Luc, 30 ng of pCR3.1-β-Galactosidase, 5 ng pCR3.1-AR, and either 400 ng of pCR3.1 vector or 400 ng pCR3.1-SRC-1a per well in a 6-well plate. After transfection cells were treated as indicated overnight, lysed, and assayed for luciferase and β-galactosidase activity. B: HeLa cells were transfected as in A except 400 ng of pCR3.1 vector or 400 ng of pCR3.1-TIF2 were used

cells transfected with AR and coactivator expression plasmids and an AR-responsive reporter, artificial overexpression of either SRC-1 or TIF2 not only potentiates the very low AR activity at low levels of R1881, but under appropriate conditions the fold increase in activity induced by overexpression of the coactivator is even greater than at high levels of hormone. That overexpression of coactivators might broaden the range of ligands for AR is also a concern. In an early study, DU145 cells transiently transfected with an AR expression vector, an MMTV reporter, and various coactivators, ARA70 overexpression greatly potentiated the ability of estradiol to activate AR, but little effect was observed with SRC-1 or ACTR (SRC-3) (23). However, the ability of ARA70 to broaden ligand specificity is not unique to this coactivator. In HeLa cells transiently transfected with an AR expression vector, an AR responsive luciferase reporter, and various coactivators, many coactivators enhanced estradiol-dependent activity (Fig. 5). Although a fragment of ARA70 (ARA70N) was extremely effective in inducing estradiol-dependent activity, full length ARA70 (ARA70F) was much less effective. Gregory et al. also found that overexpression of TIF2 strongly enhances the ability of estradiol and androgen metabolites to potentiate AR action in transfected CV1 cells (24).

Two major approaches have been utilized to assess the role of coactivators in AR function and in PC. The first is to express a dominant negative form of the coactivator or to use siRNA to reduce expression of the coactivator and to determine the effects on androgen action, cell growth, and in some cases, xenograft tumor growth. The second is to measure expression levels in normal prostate and PC and seek to correlate the expression levels either with cancer or aggressiveness of tumors. A major complication in interpreting the cell growth and tumor expression data in the context of AR action is that many of the coactivators are used by many transcription factors, while others have functions independent of their actions as coactivators. Thus, the protein may be required for cell growth independent of AR.

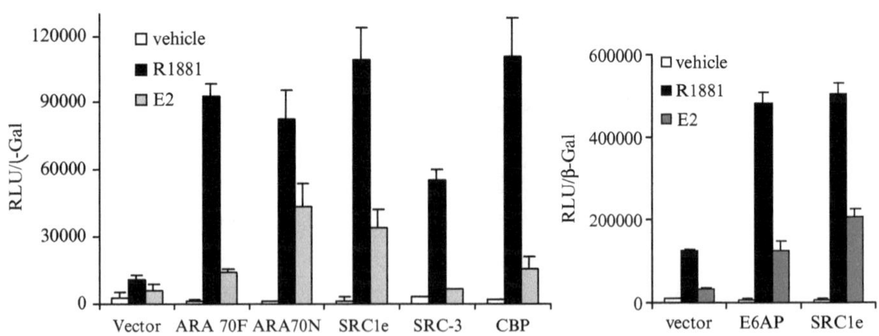

Fig. 5 DU145 cells were transfected with 400 ng GRE_2-E1b-Luc, 30 ng of pCR3.1-β-Galactosidase, 5 ng pCR3.1-AR, and either 400 ng of vector or 400 ng of indicated coactivator expression plasmid. Cells were treated overnight with either vehicle, 3 nM R1881, or 10 nM estradiol. Luciferase activity was normalized for β-galactosidase activity. CBP, CREB binding protein: E6AP, E6-associated protein

The p160 coactivator family members (SRC-1, SRC-2/TIF2/GRIP1, and SRC-3/ AIB1/ACTR) were the first steroid receptor coactivators identified (25). These proteins all have intrinsic histone acetyltransferase (HAT) activity and/or recruit other HATs to promoters facilitating histone acetylation as well as acetylation of other proteins. Mice null for SRC-1 exhibit decreased androgen-dependent prostate growth suggesting that SRC-1 plays a role in the growth of normal prostate (26). In a study of samples from 8 androgen-dependent PCs and 8 recurrent PCs after androgen deprivation, both coactivators were overexpressed in recurrent cancers relative to BPH (benign prostatic hyperplasia) and SRC-1 was also increased in androgen-dependent PC (24). In a more comprehensive analysis of SRC-1 expression in a tissue microarray containing more than 500 clinically localized primary tumors, elevated SRC-1 expression correlated with characteristics of tumor aggressiveness, but the mean staining in normal versus tumor samples was not different (16). The extent of expression in both normal and tumor was quite variable. Interestingly, patients with high levels of expression in their tumors were much more likely to also have high expression in their normal tissues, suggesting that high levels may predispose patients to aggressive cancers. That SRC-1 is important for proliferation of AR-positive PC cells was shown by depleting SRC-1 mRNA and protein using antisense oligonucleotides or siRNA and measuring [^{3}H] thymidine incorporation. Proliferation was reduced in both androgen-dependent LNCaP cells and in androgen independent, but AR-dependent C4-2 cells. In contrast, reducing SRC-1 expression in AR-negative PC-3 or DU145 cells had no effect on proliferation. Reducing SRC-1 expression reduced PSA expression and enhanced the expression of maspin, a gene whose expression is repressed by AR. Thus, SRC-1 is required both for induction and for repression of target genes by AR.

Expression of TIF2 has also been analyzed in a large PC tissue microarray. High TIF2 levels correlated with a shorter time to PSA recurrence, increased tumor proliferation (Ki67) and reduced apoptosis (TUNEL) (27). When TIF2 and AR expression were both examined, the level of TIF2 had no correlation with time to recurrence in patients with relatively low levels of AR. However, in patients with high levels of AR expression, high levels of TIF2 correlated with a shorter time to recurrence. A comparison of TIF2 expression in 9 transurethral sections from patients who had failed androgen ablation therapy uniformly showed extremely high levels of staining (27) consistent with the western blot analyses in the earlier study (24). In studying TIF2 expression in PC cell lines, Agoulnik, et al. (27) found that TIF2 expression is repressed by androgens and that this is a primary response. Thus, the extremely high levels of TIF2 in androgen ablated patients may be a direct result of the androgen ablation. Analyses of the effects of depleting TIF2 expression in PC cells have yielded mixed results. Agoulnik et al. (27) reported that reducing TIF2 expression in cells grown in serum containing medium reduced [^{3}H]-thymidine incorporation in LNCaP and PC-3 cells suggesting both AR-dependent and AR-independent actions. Reducing TIF2 expression also reduced induction of positively regulated genes including PSA, but had no effect on AR-dependent repression of maspin. Wang et al. (28) failed to detect a reduction in PSA in LNCaP cells depleted of TIF2, while Gregory et al. (29) found that depleting TIF2 in the CWR-R1 PC cell line reduced AR-dependent induction of a PSA luciferase reporter. An adenovirus expressing TIF2 RNAi had no effect on the growth of LNCaP C4-2B or CWR22Rv1 cells, but the degree of TIF2 reduction during the course of the experiments (7–8 days) was not shown (30). Under the same conditions, SRC-3/ACTR RNAi greatly reduced cell and tumor growth.

SRC-3 (p/CIP, AIB1, ACTR, RAC3, and TRAM-1) plays a role in both AR-positive and AR-negative PC. Initial studies showed that overexpression of SRC-3 in LNCaP and PC-3 cells increased AKT expression and activity and cell size (31). Subsequent studies using siRNA revealed that SRC-3 was required for optimal growth of LNCaP, PC-3, and DU145 cells and for xenograft tumor growth (32). Consistent with this, SRC-3 is overexpressed in PC and this expression correlates with higher levels of Ki67 staining and lower levels of TUNEL staining (apoptosis).

Studies of other coactivators are more limited. CARM1 (coactivator-associated arginine methyltransferase) is overexpressed in androgen ablation resistant, but not androgen-dependent cancer. Reducing CARM1 expression reduces PC cell growth, induces apoptosis, and decreases expression of PSA (33). BAF57, a component of the SWI/SNF chromatin remodeling complex, is required for AR-dependent activity and AR-dependent PC cell growth (34). Dominant negative forms of ARA54 (35) and ARA55 (36) coactivators decrease expression of PSA and decrease cell growth in LNCaP cells. ARA70 is overexpressed in PC and decreasing ARA70 expression with siRNA reduces the ability of antagonists to function as agonists in LNCaP cells (37). Numerous other candidate coactivators have been identified and partially characterized, but their contributions to AR action have not been fully evaluated.

Although much attention has been directed towards the role of coactivators in AR action and PC, there is evidence that corepressors regulate agonist-dependent activity as well as facilitating the actions of antagonists. Corepressors reduce AR activity through a variety of mechanisms including prevention of appropriate nuclear localization and DNA binding, recruiting HDACs that deacetylate chromatin and other proteins required for transcriptional activation, competing with coactivators for binding to AR, and reducing amino terminal/carboxyl terminal interactions required for optimal transcriptional activation of some target genes (38). Both SMRT and NCoR inhibit AR actions under some conditions (39). DAX-1, an orphan nuclear receptor, is also a repressor of AR action (39). Interestingly cyclin D1a is an AR corepressor (40, 41), although it functions as an estrogen receptor coactivator. Consistent with its role as a corepressor, AR activity is lowest at the G1/S boundary where cyclin D1a is expressed (42). Other corepressors include HDACs, and less well characterized proteins including Hey1 and FoxH1 (43–45).

Conclusion

The actions of AR are regulated by a variety of factors including the cellular milieu of corepressors and coactivators. The relative abundance of these proteins and their activation states influence the sensitivity of AR to androgens and the ability of AR to utilize ligands other than T and DHT. This flexibility of AR highlights the need to develop alternate methods for blocking AR activity either through developing a high affinity antagonist that prevents dissociation of AR from heat shock protein complexes or through identifying means to inhibit the potent AF1 activity of AR.

References

1. Arnold JT, Isaacs JT (2002) Mechanisms involved in the progression of androgen-independent prostate cancers: it is not only the cancer cell's fault. Endocr Relat Cancer 9:61–73.
2. Evans RM (1988) The steroid and thyroid hormone receptor superfamily. Science 240:889–895.
3. McEwan IJ (2004) Molecular mechanisms of androgen receptor-mediated gene regulation: structure–function analysis of the AF-1 domain. Endocr Relat Cancer 11:281–293.
4. Walcott JL, Merry DE (2002) Trinucleotide repeat disease. The androgen receptor in spinal and bulbar muscular atrophy. Vitam Horm 65:127–147.
5. Bevan CL, Hoare S, Claessens F, et al. (1999) The AF1 and AF2 domains of the androgen receptor interact with distinct regions of SRC1. Mol Cell Biol 19:8383–8392.
6. Simental JA, Sar M, Lane MV, et al. (1991) Transcriptional activation and nuclear targeting signals of the human androgen receptor. J Biol Chem 266:510–518.
7. He B, Kemppainen JA, Wilson EM (2000) FXXLF and WXXLF sequences mediate the NH2-terminal interaction with the ligand binding domain of the androgen receptor. J Biol Chem 275:22986–22994.
8. Sack JS, Kish KF, Wang C, et al. (2006) Crystallographic structures of the ligand-binding domains of the androgen receptor and its T877A mutant complexed with the natural agonist dihydrotestosterone. Proc Natl Acad Sci USA 98:4904–4909.
9. Beato M, Chalepakis G, Schauer M, et al. (1989) DNA regulatory elements for steroid hormones. J Steroid Biochem 32:737–747.
10. Migliaccio A, Castoria G, Di Domenico M, et al. (2000) Steroid-induced androgen receptor–oestradiol receptor β-Src complex triggers prostate cancer cell proliferation. EMBO J 19:5406–5417.
11. Agoulnik IU, Weigel NL (2006) Androgen receptor action in hormone-dependent and recurrent prostate cancer. J Cell Biochem Apr 17 [Epub ahead of print]:
12. Lamb DJ, Weigel NL, Marcelli M (2001) Androgen receptors and their biology. Vitam Horm 62:199–230.
13. Holzbeierlein J, Lal P, LaTulippe E, et al. (2004) Gene expression analysis of human prostate carcinoma during hormonal therapy identifies androgen-responsive genes and mechanisms of therapy resistance. Am J Pathol 164:217–227.
14. Balk SP (2002) Androgen receptor as a target in androgen-independent prostate cancer. Urology 60:132–139.
15. Li R, Wheeler T, Dai H, et al. (2004) High level of androgen receptor is associated with aggressive clinicopathologic features and decreased biochemical recurrence-free survival in prostate cancer patients treated with radical prostatectomy. Am J Surg Pathol 28:928–934.
16. Agoulnik IU, Vaid A, Bingman IWE, et al. (2005) Role of SRC-1 in the promotion of prostate cancer cell growth and tumor progression. Cancer Res 65:7959–7967.
17. Zegarra-Moro OL, Schmidt LJ, Huang H, et al. (2002) Disruption of androgen receptor function inhibits proliferation of androgen-refractory prostate cancer cells. Cancer Res 62:1008–1013.
18. Culig Z, Hobisch A, Cronauer MV, et al. (1994) Androgen receptor activation in prostatic tumor cell lines by insulin-like growth factor-I, keratinocyte growth factor, and epidermal growth factor. Cancer Res 54:5474–5478.

19. Nazareth LV, Weigel NL (1996) Activation of the human androgen receptor through a protein kinase A signaling pathway. J Biol Chem 271:19900–19907.
20. Hobisch A, Eder IE, Putz T, et al. (1998) Interleukin-6 regulates prostate-specific protein expression in prostate carcinoma cells by activation of the androgen receptor. Cancer Res 58:4640–4645.
21. Titus MA, Schell MJ, Lih FB, et al. (2005) Testosterone and dihydrotestosterone tissue levels in recurrent prostate cancer. Clin Cancer Res 11:4653–4657.
22. Heinlein CA, Chang C (2002) Androgen receptor (AR) coregulators: an overview. Endocr Rev 23:175–200.
23. Yeh S-H, Miyamoto H, Shima H, et al. (1998) From estrogen to androgen receptor: a new pathway for sex hormones in prostate. Proc Natl Acad Sci USA 95:5527–5532.
24. Gregory CW, He B, Johnson RT, et al. (2001) A mechanism for androgen receptor-mediated prostate cancer recurrence after androgen deprivation therapy. Cancer Res 61:4315–4319.
25. McKenna NJ, Xu J, Nawaz Z, et al. (1999) Nuclear receptor coactivators: multiple enzymes, multiple complexes, multiple functions. J Steroid Biochem Mol Biol 69:3–12.
26. Xu J, Qiu Y, DeMayo FJ, et al. (1998) Partial hormone resistance in mice with disruption of the steroid receptor coactivator-1 (SRC-1) gene. Science 279:1922–1925.
27. Agoulnik IU, Vaid A, Nakka M, et al. (2006) Androgens modulate expression of TIF2, an androgen receptor coactivator whose expression level correlates with early biochemical recurrence in prostate cancer. Cancer Res 66:10594–602.
28. Wang Q, Carroll JS, Brown M (2005) Spatial and temporal recruitment of androgen receptor and its coactivators involves chromosomal looping and polymerase tracking. Mol Cell 19:631–642.
29. Gregory CW, Fei X, Ponguta LA, et al. (2004) Epidermal growth factor increases coactivation of the androgen receptor in recurrent prostate cancer. J Biol Chem 279:7119–7130.
30. Zou JX, Zhong Z, Shi XB, et al. (2006) ACTR/AIB1/SRC-3 and androgen receptor control prostate cancer cell proliferation and tumor growth through direct control of cell cycle genes. Prostate [Epub ahead of print]:
31. Zhou G, Hashimoto Y, Kwak I, et al. (2003) Role of the steroid receptor coactivator SRC-3 in cell growth. Mol Cell Biol 23:7742–7755.
32. Zhou HJ, Yan J, Luo W, et al. (2005) SRC-3 is required for prostate cancer cell proliferation and survival. Cancer Res 65:7976–7983.
33. Majumder S, Liu Y, Ford OH, et al. (2006) Involvement of arginine methyltransferase CARM1 in androgen receptor function and prostate cancer cell viability. Prostate 66:1292–1301.
34. Link KA, Burd CJ, Williams E, et al. (2005) BAF57 governs androgen receptor action and androgen-dependent proliferation through SWI/SNF. Mol Cell Biol 25:2200–2215.
35. Miyamoto H, Rahman M, Takatera H, et al. (2002) A dominant-negative mutant of androgen receptor coregulator ARA54 inhibits androgen receptor-mediated prostate cancer growth. J Biol Chem 277:4609–4617.
36. Rahman MM, Miyamoto H, Lardy H, et al. (2003) Inactivation of androgen receptor coregulator ARA55 inhibits androgen receptor activity and agonist effect of antiandrogens in prostate cancer cells. Proc Natl Acad Sci USA 100:5124–5129.
37. Rahman MM, Miyamoto H, Takatera H, et al. (2003) Reducing the agonist activity of antiandrogens by a dominant-negative androgen receptor coregulator ARA70 in prostate cancer cells. J Biol Chem 278:19619–19626.
38. Wang L, Hsu CL, Chang C (2005) Androgen receptor corepressors: an overview. Prostate 63:117–130.
39. Agoulnik IU, Krause WC, Bingman WEI, et al. (2003) Repressors of androgen and progesterone receptor action. J Biol Chem 278:31136–31148.
40. Burd CJ, Petre CE, Morey LM, et al. (2006) Cyclin D1b variant influences prostate cancer growth through aberrant androgen receptor regulation. Proc Natl Acad Sci USA 103:2190–2195.
41. Burd CJ, Petre CE, Moghadam H, et al. (2005) Cyclin D1 binding to the androgen receptor (AR) NH2-terminal domain inhibits activation function 2 association and reveals dual roles for AR corepression. Mol Endocrinol 19:607–620.

42. Martinez ED, Danielsen M (2002) Loss of androgen receptor transcriptional activity at the G(1)/S transition. J Biol Chem 277:29719–29729.
43. Karvonen U, Janne OA, Palvimo JJ (2006) Androgen receptor regulates nuclear trafficking and nuclear domain residency of corepressor HDAC7 in a ligand-dependent fashion. Exp Cell Res [Epub ahead of print]:
44. Chen G, Nomura M, Morinago H, et al. (2005) Modulation of androgen receptor transactivation by FoxH1. A newly identified androgen receptor corepressor. J Biol Chem 280:36355–36363.
45. Belandia B, Powell SM, Garcia-Pedrero JM, et al. (2005) Hey1, a mediator of notch signaling, is an androgen receptor corepressor. Mol Cell Biol 25:1425–1436.

Part 7
Emerging Fields: Hormones and Colorectal and Lung Cancers

23 Why Hormones Protect Against Large Bowel Cancer: Old Ideas, New Evidence

Polly A. Newcomb, Gaia Pocobelli, and Victoria Chia

Introduction

Colorectal cancer (CRC) is second only to lung cancer as the leading cause of cancer mortality in men and women, and remains a significant cause of morbidity. Accumulating evidence from observational, clinical, and laboratory studies indicates that hormones reduce large bowel cancer risk. Recently, questions have arisen regarding which hormones are germane, and in what milieu they are relevant. In this chapter, we consider the genesis of this hypothesis, its studies in humans, and biological pathways that might explain recent and past findings, including new genetic and epigenetic mechanisms. Other perturbations in metabolizing enzymes responsible for estrogen metabolisms and the conversion of progesterone to estrogens may be relevant, although specific polymorphisms contributing to variations in hormone concentrations have not been directly related to CRC risk (1). We also provide some new evidence on the long-term effects of hormones on specific subtypes of large bowel tumors, and on CRC mortality, further supporting putative roles for hormone effects.

Colorectal Cancer Occurs at Different Rates in Men and Women

Historically CRC occurred with approximately equal frequency in men and women, however, in the USA the incidence rate of CRC in men now exceeds that of women by about one quarter (2). CRC incidence rate declines were most marked for women through 1985, but thereafter decreased equally in both sexes. As with most cancers, CRC risk increases with increasing age; the slope of the rise in CRC incidence is particularly steep – age to the power of 4 or 5. Thus, nearly 95% of CRCs in women occur after menopause. There are also differences in the distribution of colon cancer site by gender; incidence rates of proximal colon cancer are greater in women than men (3), and currently among women, proximal colon cancer occurs more frequently than distal colon cancer (4).

J.J. Li et al. (eds.), *Hormonal Carcinogenesis V.*

Additional observations suggest that hormones are important in CRC. Fraumeni observed that nuns had elevated mortality from this disease, as well as other hormonally related cancers such as breast and ovarian (5). Indeed, breast and CRC mortality rates are positively correlated both within populations and internationally between populations, which is consistent with these cancers having some common etiologies (3).

Epidemiologic Risk Factors Vary by Sex

Major risk factors for CRC, obesity, physical activity, and smoking, appear to differ by gender; indeed, there is evidence that these factors have a hormonal basis (6). In general, obesity is associated with an increased risk of CRC. In men, this relationship is consistently positive, yet in women, the association is generally weaker or null (7). This suggests that in the fat the peripheral aromatization of adrenal androgens to circulating estrogens (8) may confer a benefit beyond the other deleterious effects of obesity (9). These changes may be reflected in increases in insulin and insulin-like growth factors (10). However, the magnitude of this effect has not been completely consistent, perhaps depending upon the physical activity of the study population, which may also result in altered concentrations of circulating hormones (11). Smoking, too, results in decreased circulating estrogens (12), and women who smoke may derive a greater benefit from estrogens than women who do not (13). Thus, the effects of many factors appear to be modified by the hormonal background.

Studies of Exogenous Hormones and Large Bowel Cancer

A major perturbation in the hormonal milieu of the postmenopausal women is the common use of estrogen (E-only) or combined estrogen and progestin (EP). Observational evidence has been remarkably consistent regarding an inverse relationship between exogenous hormones and CRC in women (Fig. 1). Recent pooled meta-analyses have found a reduced relative risk for CRC according to postmenopausal hormone (PMH) use (from 20 to 40%) (14, 15). The greatest risk reduction for CRC was observed for current PMH users. Similar inverse associations were observed for studies of PMH and colorectal adenomas, established precursors of CRC (16), although fewer studies were available (17–19). No differences were apparent for type of preparation (E-only or EP), suggesting that estrogen is the active agent. However, in these other studies the use of EP was low and statistical power to identify differences was limited (14). Some have questioned whether PMH users simply represent a population that was healthier than women not using PMH (20, 21), and that the inverse association was, instead, due to other attributes of the users rather than the PMH itself (20, 22).

Any doubts regarding this relationship between PMH and CRC were resolved by the report of the Women's Health Initiative (WHI) (23). In this randomized controlled

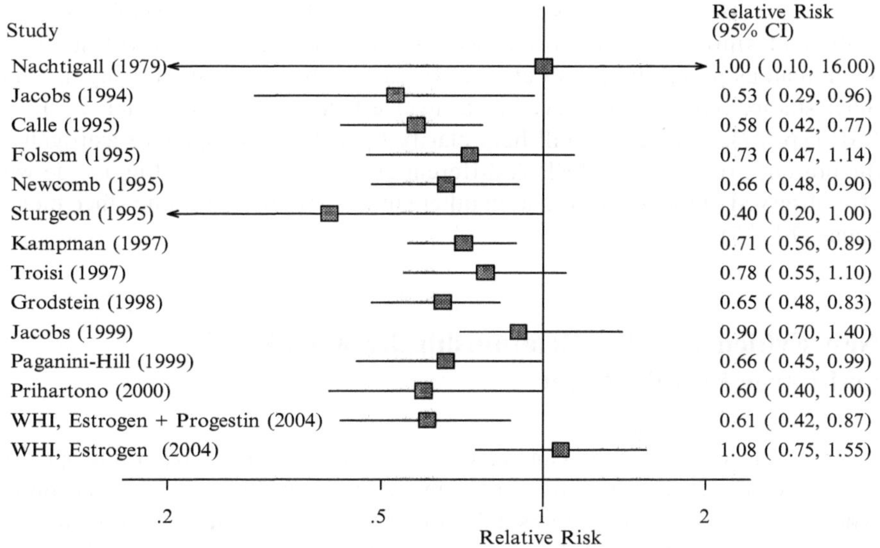

Fig. 1 Relative risks of large bowel cancer associated with current versus never use of PMHs (age- or multivariable-adjusted). (Calle (1995) and Sturgeon (1995) are relative risk of death from large bowel cancer)

trial, 8,506 women were assigned to 0.625 mg/d conjugated equine estrogen plus 2.5 mg/d medroxyprogesterone acetate in one pill; placebo was received by 8,102 women. After an average 5.2 years of follow-up, there were 45 cases of CRC in the treatment group and 67 in the placebo arm [hazard ratio (HR) 0.63, nominal 95% confidence interval (CI) 0.43–0.92] (4). Soon after, the WHI terminated the E-only arm of the hormone study based upon a risk profile similar to that observed in the combined arm (24). However, the HR for CRC was 1.08, based on 62 cases in the E-only one arm and this result varied by age ($p = 0.048$). There were important differences in the women in the two study arms that may have affected the results. Specifically, the E-only study enrolled slightly older women who did not have a uterus. Further study of the WHI population, and other larger observational studies that include current and former users, will be important in understanding whether the EP difference persists, and in what subgroup it is most pronounced.

Are Some Women Using PMH at Lower (or Higher) Risk?

More information on the effects of PMH in subgroups of users is needed to identify women who may be at different risk of CRC because of current or past PMH use, for example, whether the effect of PMH is stronger in women with lower endogenous levels of estrogen, such as lean women. This effect modification was reported by Newcomb and Storer (25) for invasive CRC; the relative risk for the lowest body mass index (BMI) quartiles was 0.28 (95% CI 0.13–0.59) and similar modification was observed in the case-control study of adenomatous polyps by Potter et al. (18). Although not statistically significant, other studies have also observed a stronger effect of PMH in women with low BMI (26, 27). Again, this will help clarify the role of endogenous estrogen in the presence of E-only and EP. A different effect of PMH according to age was also observed in the WHI (4) and in other studies according to time since menopause (14, 25).

New Evidence on the Relationship Between PMH and Large Bowel Cancer

A large observational study can answer questions that remain regarding specific patterns of use and subgroups at risk, including user characteristics and tumor types, such as microsatellite-unstable (MSI) lesions, that may be more strongly associated with different PMH patterns (28). MSI is the phenotypic hallmark of loss of DNA mismatch repair functions, either due to genetic alterations or to epigenetic silencing. To evaluate these outstanding issues, we conducted a population-based case-control study among women ages 50–74 years with incident CRC (n = 1,030) and controls (n = 1,074) randomly selected from population lists (29).

For all types of PMH use, we found a statistically significant reduction in CRC risk only among current users of PMH (OR = 0.8, 95% CI, 0.6–0.9), particularly among those who had use of 5 or more years. When stratified by type of preparation, the analysis showed no clear associations between E-only use (31% of controls) and CRC risk, regardless of recency or duration. For women who used combination EP preparations only (13% of controls), there was no association between former use and CRC; however current users had a 40% reduction in risk (95% CI, 0.5–0.9), although this association appeared to be only present among women who used them for 5 or more years. There was a statistically significant difference in the risk estimates for women who took E-only preparations and women who took EP (p = 0.01).

Among cases, MSI-H (microsatellite instability in at least 30% of loci tested) status was observed in 23.7% of tumors. There were no associations between MSI-H tumors and PMH use, but for MSI-L (microsatellite instability in less than 30% of loci tested) or microsatellite-stable (MSS) tumors; there was a 20% reduction in risk associated with any PMH use, and a 40% (95% CI, 0.4–0.9) reduction in risk associated with EP use (Table 1). Use of EP was also associated with a 40% (95% CI, 0.4–1.0) reduction in tumors with a localized stage at diagnosis. For tumors diagnosed at a distant stage, E-only and EP were associated with a statistically significant 40–50% reduction in risk. EP, but not E-only, use was consistently associated with a 30–40% reduction in risk, regardless of tumor site.

Table 1 Association between ever use of postmenopausal hormones and CRC risk by tumor characteristics

	All PMH		E-only		EP-only	
	Cases	OR (95% CI)	Cases	OR (95% CI)	Cases	OR (95% CI)
MSI						
MSI-H	83	1.0 (0.7–1.5)	48	0.8 (0.5–1.3)	15	0.7 (0.4–1.4)
MSI-L/MSS	228	0.8 (0.6–1.0)	118	1.2 (0.8–1.6)	56	0.6 (0.4–0.9)
Stage at diagnosis						
Localized	237	0.9 (0.7–1.2)	128	1.1 (0.8–1.5)	183	0.6 (0.4–1.0)
Regional	261	0.9 (0.7–1.2)	144	1.1 (0.8–1.4)	64	0.7 (0.5–1.1)
Distant	48	0.5 (0.3–0.7)	19	0.4 (0.2–0.7)	16	0.5 (0.3–1.0)
Site of tumor						
Rectum	104	0.8 (0.6–1.2)	58	0.9 (0.6–1.4)	25	0.6 (0.4–1.1)
Colon	392	0.9 (0.7–1.1)	217	1.1 (0.8–1.4)	93	0.7 (0.5–1.0)
Proximal	264	1.0 (0.8–1.3)	148	1.2 (0.9–1.6)	59	0.7 (0.5–1.1)
Distal	128	0.7 (0.5–1.0)	69	0.8 (0.6–1.2)	34	0.6 (0.4–1.0)

OR adjusted for age at diagnosis, BMI, adult onset diabetes, smoking status, regular NSAID use, colorectal screening sigmoidoscopy history within the past 10 years, physical activity, and family history of CRC

Is PMH Associated with Mortality from Large Bowel Cancer?

The population of CRC survivors is increasing, as more patients are diagnosed at earlier stages and successful treatment is initiated (30); over one half million women in the USA are currently living with CRC (31), and nearly 40% of these women will have ever used PMH (32). There is some evidence to suggest that PMH use, consistently associated with reduced CRC incidence, may be associated with reduced mortality as well. Two recent studies show that women who used PMH at the time of diagnosis had a 40% lower risk of mortality compared to women who had not (33, 34). Mandelson et al. found this reduction in mortality was limited to women with distal disease (Newcomb PA, et al, in preparation). Data from our population-based Wisconsin Women's Health Study, although finding no differences in risk of all cause or CRC mortality between PMH users and nonusers (34), suggest that large bowel cancer that arises among PMH users may be the result of a different biological pathway than cancer in nonusers.

It is also plausible that other factors associated with elevated circulating estrogens in postmenopausal women, such as obesity, might be the relevant exposure associated with improved survival. Studies in cohorts of healthy women have found conflicting results, from minimal (35) to threefold (36) increases in risk associated with the highest BMI compared to the lowest. Using our own Wisconsin Women's Health Study, we found that women who were obese (BMI $\geq 30\,kg/m^2$) had a twofold increase in risk of colon cancer death compared to women of normal weight (BMI 20–24.9 kg/m^2) (37).

Within the colon, survival has been shown to be better in individuals who have cancer in the proximal colon (from the cecum to the splenic flexure) compared to those with cancer in the distal colon (from the splenic flexure to the rectosigmoid junction) (38). A possible explanation for a survival advantage for proximal disease may be due to genomic differences. Microsatellite instability (MSI) is more common among proximal colon cancers, and although tumor characteristics are generally worse (i.e., mucinous histology and poor differentiation), stage at diagnosis is more favorably distributed (39). This is particularly evident in among older patients (40, 41). From a recent meta-analysis, pooled results estimated a 35% reduction in risk of death after CRC if the patients had a tumor with high MSI (42). Thus, the results of several small studies are inconsistent regarding a role for PMH in reducing proximal mortality, and likely MSI-H lesions. We posit however, that morality from lesions that arise in the background of hormonal replacement will not be reduced.

Hormones Affect the Biology of Colorectal Carcinogenesis

Although some aspects of colon carcinogenesis are related to DNA damaging agents and processes, emerging evidence suggests that hormones, specifically estrogens and progesterone, are relevant. CRC can arise via several molecular pathways (6) which may well reflect different environmental exposures. The biological actions of hormones are myriad and are regulated by, among other things, their circulating concentrations, the conversion to more active or less active derivatives, and relevant receptor concentrations in the target tissue (43). Estrogen, in the presence of estrogen receptor (ER) is generally thought to act as a proliferative signal. Recent studies have also shown there can be estrogen dependent growth inhibition in the colon. Given the range of estrogen action and the nature of colorectal carcinogenesis (44, 45), several hormone-related mechanisms, both indirect and direct, could be germane to the development of large bowel cancer (6), although the underlying pathways by which estrogen and progesterone affect colorectal carcinogenesis have not been established. Mechanisms such as reduced secondary bile acids, decreased growth enhancing effects of insulin like growth factors (IGFs), and maintenance of transcription and expression of the estrogen and progesterone receptor genes, may have all have a tumor suppressive effect. The latter may be due to genetic or epigenetic effects.

As an indirect mechanism, first posited by McMichael and Potter (46), bile acid secretion may be favorably influenced by hormones (both endogenous and exogenous), leading to reduced risk of CRC. There is both laboratory evidence, where bile acids appear to cause proliferation and promote colon cancer in rats, (47), and human data (48) to support this theory. It is not clear whether exogenous hormones necessarily result in decreased bile acid secretion, however, this does suggest that exogenous hormone use may result in decreased synthesis of bile acids and lipids.

Alterations in IGF physiology may also play a role in the relationship between hormonal factors and CRC risk. IGF-1 is a potent mitogen and an antiapoptotic agent, the action of which is partly regulated by IGF binding protein-3 (IGFBP-3). In postmenopausal women, both our own epidemiologic (49) and in vitro studies have shown that endogenous and exogenous estrogens reduce serum IGF-1 concentrations (50). Recent epidemiologic studies, evaluating the relationship between circulating IGF-1 and IGFBP-3 concentrations and CRC risk, support a role of IGFs in tumor genesis although the magnitude and direction of these relations have been inconsistent (51). We also found that polymorphisms in the IGF genes may be responsible for the interindividual variability in IGF levels, and there may also be interactions with hormonal factors, such as PMH use and BMI. For example, we observed that the association of current PMH use with reduced CRC risk was observed only among women with the *IGFBP-3 2133GG* genotype (52). These results also suggest that heterogeneity in genotypes may be partially responsible for the consistently noted sex differences in CRC and its association with risk factors such as obesity.

Hormones, including exogenous preparations of postmenopausal estrogen and progestin, may have a direct effect on the colorectal epithelium. Steroid receptors are a prerequisite for hormone responsiveness and there is substantial data that demonstrates ER and progesterone receptor (PR) presence in the colon (53). There are two known isoforms of the ER, ERα, and ERβ (54), and both are present in the colorectum, and activate the same genes in response to estradiol (54). Data support a protective role for both ERβ (55), where the decline in ERβ with age was associated with loss of differentiation in malignant CRC cells, and ERα (56). Several polymorphisms in the ERs have been associated with functional responses to PMH, but no studies have evaluated the relation between these polymorphisms and CRC.

The biological mechanisms underlying an effect of progesterone in the colon are less clear, although they may be synergistically amplifying estrogen's effects. For example, the isozyme of 17β-hydroxy steroid dehydrogenase induced by progesterone predominantly catalyzes the conversion of the less potent estrone to the more potent estradiol (57). Such an effect has been hypothesized to explain the increased risk of breast cancer associated with E + P beyond that of E-alone (58). PR expression may be increased by estrogen binding (59), and in return, progesterone can also competitively bind to ERs in the colon (60).

Transcription and expression may also be modified by epigenetic events that do not involve changes in DNA nucleotide sequences. Substantial evidence indicates estrogen may be an important factor in the pathway resulting in the occurrence of widespread hypermethylation, the CpG island methylator phenotype (CIMP) (61). ER methylation increases with age and is a central feature of CRC (62). Potter suggested that the inverse relationship between PMH and cancer may be a consequence of replacing the declining exogenous estrogen, and thus result in (or reverse) the likelihood that ER will be silenced by methylation (63). Estrogen may also specifically play a role in the epigenetic silencing of the *hMLH1* DNA mismatch repair gene, via hypermethylation of its promoter region. In series of CRC cases, an estimated 28–58% were CIMP+ (64, 65), however, these results are based upon small clinical series. In vitro and animal studies have suggested, too, that estrogen intervention reduces DNA methylation of specific genes and restores the protective methylation patterns in colon and various other tissues (64).

Future Possibilities

PMH use has dramatically declined following the publication of the HERS and WHI reports (32). Current recommendations limit PMH to short term use for the treatment of acute menopausal symptoms. However, since a significant number of women, perhaps 50% of women aged 50–69 years have ever used PMH, and a significant proportion of women will continue to use them, research should continue on the biological basis for the effects, subgroups likely to benefit from its use, and the impact of these preparations on mortality after a diagnosis of CRC.

Acknowledgement Expert technical assistance was generously provided by Deanne Young.

References

1. Thompson PA, Ambrosone C (2000) Molecular epidemiology of genetic polymorphisms in estrogen metabolizing enzymes in human breast cancer. J Natl Cancer Inst Monogr 125–34.
2. Surveillance, Epidemiology, and End Results (SEER) Program (www.seer.cancer.gov) SEER*Stat Database: Incidence – SEER 17 Regs Public-Use, Nov 2005 Sub (1973–2003 varying), National Cancer Institute, DCCPS, Surveillance Research Program, Cancer Statistics Branch, released April 2006, based on the November 2005 submission.
3. McMichael AJ, Potter JD (1983) Do intrinsic sex differences in lower alimentary tract physiology influence the sex-specific risks of bowel cancer and other biliary and intestinal diseases? Am J Epidemiol 118: 620–7.
4. Chlebowski R, Wactawski-Wende J, Ritenbaugh C, et al (2004) Estrogen plus progestin and colorectal cancer in postmenopausal women. N Engl J Med 350: 991–1004.
5. Fraumeni JF Jr., Lloyd JW, Smith EM, et al (1969) Cancer mortality among nuns: role of marital status in etiology of neoplastic disease in women. J Natl Cancer Inst 42: 455–68.
6. Potter JD (1999) Colorectal cancer: molecules and populations. J Natl Cancer Inst 91: 916–32.
7. Gunter MJ, Leitzmann MF (2006) Obesity and colorectal cancer: epidemiology, mechanisms and candidate genes. J Nutr Biochem 17: 145–56.
8. Siiteri PK (1987) Adipose tissue as a source of hormones. Am J Clin Nutr 45: 277–82.
9. Slattery ML, Ballard-Barbash R, Edwards S, et al (2003) Body mass index and colon cancer: an evaluation of the modifying effects of estrogen (United States). Cancer Causes Control 14: 75–84.
10. Morimoto LM, Newcomb PA, White E, et al (2005) Insulin-like growth factor polymorphisms and colorectal cancer risk. Cancer Epidemiol Biomarkers Prev 14: 1204–11.
11. Slattery ML, Edwards SL, Ma KN, et al (1997) Physical activity and colon cancer: a public health perspective. Ann Epidemiol 7: 137–45.
12. MacMahon B, Trichopoulos D, Cole P, et al (1982) Cigarette smoking and urinary estrogens. N Engl J Med 307: 1062–5.

13. Newcomb PA, Storer BE, Marcus PM (1995) Cigarette smoking in relation to risk of large bowel cancer in women. Cancer Res 55: 4906–9.
14. Grodstein F, Newcomb PA, Stampfer MJ (1999) Postmenopausal hormone therapy and the risk of colorectal cancer: a review and meta-analysis. Am J Med 106: 574–82.
15. La Vecchia C, Gallus S, Fernandez E (2005) Hormone replacement therapy and colorectal cancer: an update. J Br Menopause Soc 11: 166–72.
16. Leslie A, Carey FA, Pratt NR, et al (2002) The colorectal adenoma-carcinoma sequence. Br J Surg 89: 845–60.
17. Terry MB, Neugut AI, Bostick RM, et al (2002) Risk factors for advanced colorectal adenomas: a pooled analysis. Cancer Epidemiol Biomarkers Prev 11: 622–9.
18. Potter JD, Bostick RM, Grandits GA, et al (1996) Hormone replacement therapy is associated with lower risk of adenomatous polyps of the large bowel: the Minnesota Cancer Prevention Research Unit Case-Control Study. Cancer Epidemiol Biomarkers Prev 5: 779–84.
19. Morimoto LM, Newcomb PA, Ulrich CM, et al (2002) Risk factors for hyperplastic and adenomatous polyps: evidence for malignant potential? Cancer Epidemiol Biomarkers Prev 11: 1012–8.
20. Matthews KA, Kuller LH, Wing RR, et al (1996) Prior to use of estrogen replacement therapy, are users healthier than nonusers? Am J Epidemiol 143: 971–8.
21. Barrett-Connor E (1991) Postmenopausal estrogen and prevention bias. Ann Intern Med 115: 455–6.
22. Sturgeon SR, Schairer C, Brinton LA, et al (1995) Evidence of a healthy estrogen user survivor effect. Epidemiology 6: 227–31.
23. Rossouw JE, Anderson GL, Prentice RL, et al (2002) Risks and benefits of estrogen plus progestin in healthy postmenopausal women: principal results From the Women's Health Initiative randomized controlled trial. JAMA 288: 321–33.
24. Anderson GL, Limacher M, Assaf AR, et al (2004) Effects of conjugated equine estrogen in postmenopausal women with hysterectomy: the Women's Health Initiative randomized controlled trial. JAMA 291: 1701–12.
25. Newcomb PA, Storer BE (1995) Postmenopausal hormone use and risk of large-bowel cancer. J Natl Cancer Inst 87: 1067–71.
26. Kampman E, Potter JD, Slattery ML, et al (1997) Hormone replacement therapy, reproductive history, and colon cancer: a multicenter, case-control study in the United States. Cancer Causes Control 8: 146–58.
27. Pischon T, Lahmann PH, Boeing H, et al (2006) Body size and risk of colon and rectal cancer in the European Prospective Investigation Into Cancer and Nutrition (EPIC). J Natl Cancer Inst 98: 920–31.
28. Slattery ML, Potter JD, Curtin K, et al (2001) Estrogens reduce and withdrawal of estrogens increase risk of microsatellite instability-positive colon cancer. Cancer Res 61: 126–30.
29. Newcomb PA, Zheng Y, Chia VM, et al (2007) Estrogen plus progestin use and the risk of colorectal cancer in postmenopausal women cancer res 67:7534–9.
30. Chu KC, Tarone RE, Chow WH, et al (1995) Colorectal cancer trends by race and anatomic subsites, 1975–1991. Arch Fam Med 4: 849–56.
31. Surveillance, Epidemiology, and End Results (SEER) Program (www.seer.cancer.gov). Prevalence database: "US Estimated Complete Prevalence Counts on 1/1/2003". National Cancer Institute, DCCPS, Surveillance Research Program, Statistical Research and Applications Branch, released April 2006, based on the November 2005 SEER data submission.
32. Hersh AL, Stefanick ML, Stafford RS (2004) National use of postmenopausal hormone therapy: annual trends and response to recent evidence. JAMA 291: 47–53.
33. Slattery ML, Anderson K, Samowitz W, et al (1999) Hormone replacement therapy and improved survival among postmenopausal women diagnosed with colon cancer. Cancer Causes Control 10: 467–73.

34. Mandelson MT, Miglioretti D, Newcomb PA, et al (2003) Hormone replacement therapy in relation to survival in women diagnosed with colon cancer. Cancer Causes Control 14: 979–84.
35. Murphy TK, Calle EE, Rodriguez C, et al (2000) Body mass index and colon cancer mortality in a large prospective study. Am J Epidemiol 152: 847–54.
36. Tamakoshi K, Wakai K, Kojima M, et al (2004) A prospective study of body size and colon cancer mortality in Japan: The JACC Study. Int J Obes Relat Metab Disord 28: 551–8.
37. Doria-Rose VP, Newcomb PA, Morimoto LM, et al (2006) Body mass index and the risk of death following the diagnosis of colorectal cancer in postmenopausal women (United States). Cancer Causes Control 17: 63–70.
38. Gervaz P, Bouzourene H, Cerottini JP, et al (2001) Dukes B colorectal cancer: distinct genetic categories and clinical outcome based on proximal or distal tumor location. Dis Colon Rectum 44: 364–72.
39. Kim H, Jen J, Vogelstein B, et al (1994) Clinical and pathological characteristics of sporadic colorectal carcinomas with DNA replication errors in microsatellite sequences. Am J Pathol 145: 148–56.
40. Gryfe R, Kim H, Hsieh ET (2000) Tumor microsatellite instability and clinical outcome in young patients with colorectal cancer. N Engl J Med 342: 69–77.
41. Samowitz WS, Curtin K, Ma KN (2001) Microsatellite instability in sporadic colon cancer is associated with an improved prognosis at the population level. Cancer Epidemiol Biomarkers Prev 10: 917–23.
42. Popat S, Hubner R, Houlston RS (2005) Systematic review of microsatellite instability and colorectal cancer prognosis. J Clin Oncol 23: 609–18.
43. Foley EF, Jazaeri AA, Shupnik MA, et al (2000) Selective loss of estrogen receptor beta in malignant human colon. Cancer Res 60: 245–8.
44. Fearon ER, Vogelstein B (1990) A genetic model of colorectal tumorigenesis. Cell 61: 759–67.
45. Morson BC (1974) Evolution of cancer of the colon and rectum. Cancer 34: 845.
46. McMichael AJ, Potter JD (1980) Reproduction, endogenous and exogenous sex hormones, and colon cancer: a review and hypothesis. J Natl Cancer Inst 65: 1201–7.
47. Reddy BS, Watanabe K, Weisburger JH, et al (1977) Promoting effect of bile acids in colon carcinogenesis in germ-free and conventional F344 rats. Cancer Res 37: 3238–42.
48. McKeown-Eyssen G (1994) Epidemiology of colorectal cancer revisited: are serum triglycerides and/or plasma glucose associated with risk? Cancer Epidemiol Biomarkers Prev 3: 687–95.
49. Morimoto LM, Newcomb PA, White E, et al (2005) Variation in plasma insulin-like growth factor-1 and insulin-like growth factor binding protein-3: personal and lifestyle factors (United States). Cancer Causes Control 16: 917–27.
50. Heald A, Selby PL, White A, et al (2000) Progestins abrogate estrogen-induced changes in the insulin-like growth factor axis. Am J Obstet Gynecol 183: 593–600.
51. Renehan AG, Wille-Jorgensen P (2004) Systematic review, Cochrane and Colorectal Disease. Colorectal Dis 6: 73–4.
52. Morimoto LM, Newcomb PA, White E, et al (2005) Variation in plasma insulin-like growth factor-1 and insulin-like growth factor binding protein-3: genetic factors. Cancer Epidemiol Biomarkers Prev 14: 1394–401.
53. Bardin A, Boulle N, Lazennec G, et al (2004) Loss of ERbeta expression as a common step in estrogen-dependent tumor progression. Endocr Relat Cancer 11: 537–51.
54. McDonnell DP, Norris JD (2002) Connections and regulation of the human estrogen receptor. Science 296: 1642–4.
55. Konstantinopoulos PA, Kominea A, Vandoros G, et al (2003) Oestrogen receptor beta (ERbeta) is abundantly expressed in normal colonic mucosa, but declines in colon adenocarcinoma paralleling the tumour's dedifferentiation. Eur J Cancer 39: 1251–8.
56. Potter JD, Cerhan JR, Sellers TA, et al (1995) Progesterone and estrogen receptors and mammary neoplasia in the Iowa Women's Health Study: how many kinds of breast cancer are there? Cancer Epidemiol Biomarkers Prev 4: 319–26.

57. Poutanen M, Isomaa V, Peltoketo H, et al (1995) Role of 17 beta-hydroxysteroid dehydrogenase type 1 in endocrine and intracrine estradiol biosynthesis. J Steroid Biochem Mol Biol 55: 525–32.
58. Schairer C, Lubin J, Troisi R, et al (2000) Menopausal estrogen and estrogen–progestin replacement therapy and breast cancer risk. JAMA 283: 485–91.
59. Peterson CM (2000) Estrogen and progesterone receptors: an overview from the year 2000. J Soc Gynecol Investig 7: S3–7.
60. Di Leo A, Messa C, Cavallini A, et al (2001) Estrogens and colorectal cancer. Curr Drug Targets Immune Endocr Metabol Disord 1: 1–12.
61. Baylin SB and Herman JG (2000) DNA hypermethylation in tumorigenesis: epigenetics joints genetics. Trends Genet 16: 168–74.
62. Issa JP, Ottaviano YL, Celano P (1994) Methylation of the oestrogen receptor CpG island link ageing and neoplasia in human colon. Nature Genet 7: 536–40.
63. Potter JD (1995) Hormones and colon cancer. J Natl Cancer Inst 87: 1039–40.
64. Hawkins N, Norrie M, Cheong K, et al (2002) CpG island methylation in sporadic colorectal cancers and its relationship to microsatellite instability. Gastroenterology 122: 1376–87.
65. Whitehall VL, Wynter CV, Walsh MD, et al (2002) Morphological and molecular heterogeneity within nonmicrosatellite instability-high colorectal cancer. Cancer Res 62: 6011–4.

24
Vitamin D_3 and Colorectal Cancer

M.J. Larriba, N. Valle, S. Álvarez, and A. Muñoz

Introduction

Vitamin D_3 intake and sunlight exposure have a protective effect against colorectal cancer (CRC) and other neoplasias. In line with epidemiological data, the most active metabolite of vitamin D_3,1α, 25-dihydroxyvitamin D_3, [1α,25$(OH)_2D_3$] has antiproliferative, prodifferentiation and proapoptotic effects on cultured human colon cancer cells and antitumoral action in colon cancer mouse models. Intense research is ongoing to identify 1α,25$(OH)_2D_3$ target genes and the mechanisms responsible for its antitumoral actions, as well as to define treatments based on 1α,25$(OH)_2D_3$ analogs for clinical use. However, colon cancer cells can develop resistance to 1α,25$(OH)_2D_3$ action due to aberrant expression of the enzymes responsible for its synthesis and catabolism or to the down-regulation of vitamin D receptor, a ligand-regulated transcription factor that mediates most 1α,25$(OH)_2D_3$ actions.

Colorectal Cancer: An Overview

CRC is the second cause of cancer death in the world and more than 900,000 new cases are diagnosed each year (1). Surgical resection is the therapy of choice for localized tumors, but there is no satisfactory treatment when surgery is not curative or for advanced colon cancer. Only a low percentage of CRC (between 5 and 10%) is attributable to hereditary factors, as is the case of two syndromes: familial adenomatous polyposis (FAP) and hereditary nonpolyposis colorectal cancer (HNPCC).

In 1990, Fearon and Vogelstein suggested that CRC is the result of the accumulation of alterations in certain oncogenes and tumor suppressor genes involved in proliferation and cell death control (2). The mutation of a gene involved in the canonical Wnt/β-catenin pathway (*adenomatous poliposis coli/APC* or, less commonly, *CTNNB1*/β-catenin or *AXIN*) seems to be sufficient for adenoma initiation, while mutation of *K-RAS* or *B-RAF* and subsequent alterations in the transforming growth factor (TGF-β) pathway confer additional malignant features to the adenoma. The progression towards carcinoma involves the inactivation of *TP53* gene in 50%

J.J. Li et al. (eds.), *Hormonal Carcinogenesis V.*

of colon tumors. This is known as the suppressor pathway (1). In addition, a second via for CRC, the mutator pathway, is driven by the epigenetic silencing or the mutation of genes involved in DNA mismatch repair (MMR) (1).

The canonical Wnt/β-catenin pathway controls cell fate during embryonic development, regulates homeostasis in adult self-renewing tissues and has been recently implicated in the initiation and progression of several human cancers (3). In the absence of extracellular secreted proteins of the Wnt family, β-catenin binds to a destruction complex that includes the scaffolding proteins APC and Axin and two kinases, casein kinase I (CKI) and glycogen synthase kinase (GSK)-3β. These kinases phosphorylate β-catenin, which is subsequently degraded by the proteasome (1, 3). Wnt proteins bind to Frizzled/LRP (low density lipoprotein receptor-related protein) receptors resulting in the inhibition of β-catenin phosphorylation. As a consequence, β-catenin is stabilized, forms nuclear transcription complexes with members of the TCF/LEF (T cell factor/lymphoid-enhancing factor) family and induces the expression of genes involved in proliferation and invasiveness (1, 3).

The mutations of *APC*, *CTNNB1*/β-catenin or *AXIN* found in colon tumors prevent β-catenin phosphorylation. Therefore, β-catenin accumulates in the cytoplasm independently of Wnt signaling, translocates into the nucleus and activates the expression of its target genes promoting CRC progression (1, 3).

Vitamin D_3: Metabolism and Biological Actions

Vit D_3 can be both obtained from the diet (10%) and synthesized in the skin from 7-dehydrocholesterol by UV-B sunlight action (90%). Vit D_3 metabolic activation begins in the liver, where it is hydroxylated by the vit D_3-25-hydroxylase (CYP27A1) to render 25-hydroxyvit D_3 (25(OH)D_3), the most abundant form of vit D_3 in the body (4). In the kidney, the 25(OH)D_3-1α-hydroxylase (CYP27B1) transforms 25(OH)D_3 into 1α,25-dihydroxyvit D_3 (1α,25(OH)$_2$$D_3$), the most potent metabolite of vit D_3 and that accounts for most of its biological actions. In addition, other tissues express CYP27B1 and can autocrinally synthesize 1α,25(OH)$_2$$D_3$ from serum 25(OH)D_3 (4). The catabolism of vit D_3 and its metabolites is initiated by the 25(OH)D_3-24-hydroxylase (CYP24A1). Subsequent oxidations generate metabolites less active but still produce calcitroic acid, which is biologically inert and is excreted (4).

1α,25(OH)$_2$$D_3$ is a pleiotropic hormone that in addition to its classical actions on bone biology and calcium and phosphate homeostasis has antiproliferative, proapoptotic and prodifferentiation effects on different cell types (5). These 1α,25(OH)$_2$$D_3$ actions suggest its use for the treatment of hyperproliferative disorders such as psoriasis and cancer. However, the toxicity, due to the hypercalcemic effect of 1α,25(OH)$_2$$D_3$ at therapeutic doses, prevents the above-mentioned treatment. More than 400 analogs with less hypercalcemic effect are at different preclinical and clinical development stages. In cancer, few phase I and II clinical trials have been reported and with effectiveness on a low percentage of patients (5).

Vitamin D Receptor Mediates 1α,25(OH)$_2$D$_3$ Action

1α,25(OH)$_2$D$_3$ exerts most of its biological effects through the interaction with a member of the superfamily of nuclear receptors, the vit D receptor (VDR). VDR forms heterodimers with a member of the same family, the retinoid X receptor (RXR), and primarily acts regulating the expression of genes that contain specific DNA sequences known as vit D response elements (VDRE) in their promoters. One or more VDRE have been identified in the promoters of many 1α,25(OH)$_2$D$_3$ target genes such as *CYP24A1*, osteopontin, osteocalcin, insulin-like growth factor-binding proteins (IGFBP)-1, -3, -5, p21$^{WAF1/CIP1}$ and cyclin C (4).

In the absence of 1α,25(OH)$_2$D$_3$, the VDR/RXR heterodimer is bound to the DNA at VDRE and recruits corepressor proteins and histone deacetylases that promote chromatin compaction and target genes are silenced. The binding of 1α,25(OH)$_2$D$_3$ to VDR leads to the switch of corepressors for coactivators, histone acetyltransferases and chromatin-remodeling complexes that together mediate chromatin decompaction. Subsequently, coactivators are substituted by the VDR-interacting proteins complex (DRIP/TRAP) that links VDR to the basal transcriptional machinery and promotes gene activation (4). Additionally, VDR activated by 1α,25(OH)$_2$D$_3$ can repress gene expression through a less characterized mechanism.

1α,25(OH)$_2$D$_3$ can also induce rapid responses (sec/min/hr) independent of gene expression. These nongenomic effects include opening of voltage-gated Ca^{2+} and Cl^- channels and changes in the activity of certain enzymes (5). Their relation to 1α,25(OH)$_2$D$_3$ antitumoral action is unclear.

Epidemiological Data and Studies in Animal Models Show a Beneficial Effect of 1α,25(OH)$_2$D$_3$ Against Colorectal Cancer

Several epidemiological findings show an inverse correlation between serum concentrations of 25(OH)D$_3$ and the risk of CRC in humans. Moreover, high levels of 25(OH)D$_3$ are associated with a lower proliferation of colonic epithelial cells (6). Accordingly, animals fed a Western-style diet (high fat and phosphate and low vit D$_3$ and calcium content) show hyperproliferation of colonic epithelial cells and colon-crypt hyperplasia. These effects are markedly suppressed when this diet is supplemented with calcium and vit D$_3$ (5). The effects of a high fat diet have also been tested in mutant *Apc*^{min} mice, a model for human FAP. The high number of polyps developed by these mice increases when the mice are fed a Western-style diet and mouse survival diminishes (5). Remarkably, the total tumor load of *Apc*^{min} mice decreases 36% when mice are treated with the 1α,25(OH)$_2$D$_3$ analog Ro26-9114 (7).

In addition, 1α,25(OH)$_2$D$_3$ analogs inhibit tumor incidence and the rate of spontaneous metastases induced by chemical carcinogens in mice (5). Our group observed that the 1α,25(OH)$_2$D$_3$ analog EB1089 inhibits the growth of tumors generated in immunosupressed mice by a subpopulation of SW480 cells that express VDR (SW480-ADH cells), but not of those generated by a more tumorigenic subpopulation that do not express VDR (SW480-R cells) (eight and unpublished data). Other studies have also reported the anticancer activity of 1α,25$(OH)_2D_3$ and its analogs against xenografted tumors generated by several different human colon cancer cells (5). Additionally, *Vdr*-deficient mice display an increase of proliferation and oxidative stress in the colon. However, $Vdr^{-/-}$ mice do not present high rates of spontaneous colon cancer (5).

In summary, the results obtained with 1α,25$(OH)_2D_3$ and its analogs in different CRC animal models demonstrate their ability to prevent tumor formation, cause tumor regression and avoid metastasis development.

Antitumoral Effects of 1α,25$(OH)_2D_3$ and its Analogs on Cultured Human Colorectal Cancer Cells

The antitumoral activity of 1α,25$(OH)_2D_3$ and its analogs against CRC is mostly based on the inhibition of proliferation and the induction of differentiation or apoptosis (Fig. 1).

The induction of cell cycle arrest by 1α,25$(OH)_2D_3$ usually results from the upregulation of the cyclin dependent kinase inhibitors $p21^{WAF1/CIP1}$ and $p27^{KIP1}$. $p21^{WAF1/CIP1}$ gene has several VDRE and is induced by 1α,25$(OH)_2D_3$ in many cell types (5, 9). By contrast, the induction of $p27^{KIP1}$ gene, which lacks VDRE, is mediated by the transcription factors NF-Y and Sp1 and by protein stabilization mechanisms (10, 11). Recently, it has been demonstrated that the repression of the transcriptional regulator Id2 is necessary for cycle arrest, $p21^{WAF1/CIP1}$ induction and cyclin A inhibition by 1α,25$(OH)_2D_3$ in SW480-ADH cells (12).

Growth arrest of colon cancer cells by 1α,25$(OH)_2D_3$ involves also the interference of several signaling pathways (Fig. 1). 1α,25$(OH)_2D_3$ counteracts the growth stimulatory effect of epidermal growth factor (EGF) through the blockade of EGF receptor (EGFR) expression and the promotion of EGFR ligand-induced internalization (13). Additionally, EGF reduces VDR expression and the activation of EGF pathway could allow colon carcinoma cells to avoid 1α,25$(OH)_2D_3$ actions (5, 14). Furthermore, 1α,25$(OH)_2D_3$ inhibits the secretion of the potent mitogen insulin-like growth factor (IGF)-II in colon cancer cells and induces the expression of its negative regulator IGFBP-6 (15).

The proliferation of healthy colon cells is inhibited by TGF-β. Progression to carcinoma is accompanied by inactivation of this pathway and creates resistance to TGF-β antiproliferative action in most human colon cancer cells. 1α,25$(OH)_2D_3$ is able to restore the sensitivity of these cells to TGF-β by inducing type I TGF-β receptor expression (16). Additionally, the TGF-β signaling downstream protein SMAD3 binds to SRC1 and acts as a coactivator of VDR, thus cooperating in 1α,25$(OH)_2D_3$ actions (17).

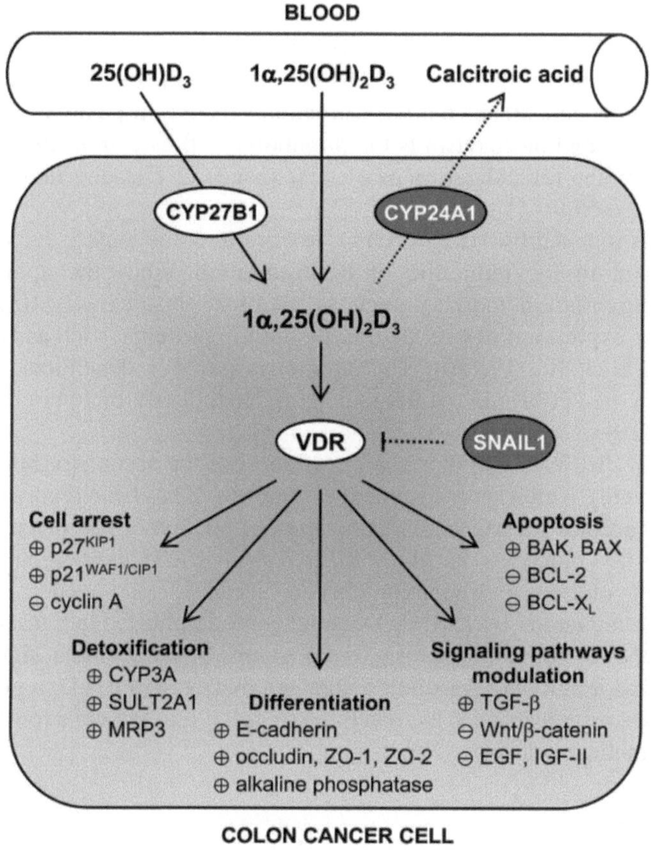

Fig. 1 Schematic representation of $1\alpha,25(OH)_2D_3$ action in human colon cancer cells. Colon cells receive $1\alpha,25(OH)_2D_3$ from the bloodstream and also synthesize it from serum $25(OH)D_3$ by CYP27B1 action. Through its binding to VDR, $1\alpha,25(OH)_2D_3$ regulates the expression of many genes involved in different cellular functions related to its antitumoral action. Alteration of the expression and activity of the enzymes regulating $1\alpha,25(OH)_2D_3$ synthesis (CYP27B1) and catabolism (CYP24A1) have been found in colon cancer and cause the reduction of $1\alpha,25(OH)_2D_3$ cell availability. Similarly, an increase in SNAIL1 transcription factor expression promotes VDR down regulation and blocks $1\alpha,25(OH)_2D_3$ action in colon

$1\alpha,25(OH)_2D_3$ promotes strong enterocytic differentiation of colon cancer cells (5). Our group demonstrated that $1\alpha,25(OH)_2D_3$ and several analogs induce epithelial differentiation in the VDR-positive SW480-ADH human colon cancer cells, while they do not affect the VDR-negative SW480-R cells. This effect is linked to the induction of E-cadherin, the main component of *adherens junctions*, and other adhesion proteins (18). Furthermore, $1\alpha,25(OH)_2D_3$ and its analogs inhibit the Wnt/β-catenin pathway and silence its target genes, which contributes to maintain the differentiated phenotype (Fig. 1). The first mechanism for 1α,25$(OH)_2D_3$ antagonism on Wnt/β-catenin pathway is the direct interaction between ligand activated VDR and β-catenin in the nucleus, which prevents the formation of TCF/β-catenin complexes. This mechanism is E-cadherin-independent, since it is observed in E-cadherin-positive and -negative cells (18). A second mechanism is the promotion of β-catenin nuclear export and plasma membrane relocalization as a consequence of E-cadherin accumulation at *adherens junctions* (18).

Apoptosis induced by 1α,25$(OH)_2D_3$ in several colon cancer cells appears to be subsequent to the induction of differentiation, since the apoptotic cells express differentiation markers such as alkaline phosphatase. 1α,25$(OH)_2D_3$ regulates the expression of pro- and antiapoptotic proteins, such as BAK, BAX, BCL-2, and BCL-X_L (19, 20). This apoptosis is p53-independent, allowing a therapy with 1α,25$(OH)_2D_3$ or its analogs independently of tumor *TP53* mutation status (19).

A high-fat diet leads to colonic accumulation of the secondary bile acid LCA (lithocholic acid), which acts as a potent carcinogen. It has been reported that VDR function as receptor for LCA (21). Activation of VDR by LCA or 1α,25$(OH)_2D_3$ transcriptionally induces *CYP3A*, *SULT2A1*, and *MRP3* expression, which are implicated in colon LCA elimination (14).

Moreover, numerous 1α,25$(OH)_2D_3$ target genes implicated in transcription, cell adhesion, DNA synthesis, apoptosis, redox status, or intracellular signaling have been identified in microarrays studies, showing that 1α,25$(OH)_2D_3$ regulates many genetic events responsible for the reversion of colon tumoral cells toward a differentiated epithelial phenotype (20).

VDR Expression and 1α,25$(OH)_2D_3$ Metabolism in Normal and Tumoral Colon

Tissue responsiveness to 1α,25$(OH)_2D_3$ depends mainly on VDR expression and the activity of the enzymes regulating 1α,25$(OH)_2D_3$ synthesis and degradation. In addition, it has been shown that aberrant expression of certain VDR co-repressors causes 1α,25$(OH)_2D_3$ insensitivity in prostate and breast cancer cells (22), although similar findings have not been reported in colon cancer.

Normal colon expresses CYP27B1 and VDR and, thus, can synthesize 1α,25$(OH)_2D_3$ from serum 25$(OH)D_3$ and response to it. The expression of CYP27B1 and VDR are induced at early stages of colon cancer progression, suggesting that autocrine synthesis of 1α,25$(OH)_2D_3$ can be a tumor strategy to control its own progression (23) (Fig. 1). However, CYP27B1 and VDR expression decrease in advanced colon tumors, while CYP24A1 expression and activity are induced (23). These data suggest that advanced colon tumors are likely resistant to the antitumoral effects of endogenous 1α,25$(OH)_2D_3$ and also to the treatment with 1α,25$(OH)_2D_3$ or its analogs.

Transcriptional Regulation of the Human *VDR* Gene: SNAIL1 and Other Transcription Factors

Human *VDR* gene structure is complex. Differential promoter usage and alternative splicing of the untranslated exons 1a–1f give rise to 14 different human *VDR* transcripts. The most abundant and ubiquitous VDR transcripts are those that begin in exon 1a. Therefore, the upstream region of that exon is considered the main human *VDR* gene promoter (24).

VDR down-regulation during colon cancer progression was first described more than a decade ago, but its molecular basis has remained elusive. Recently, our group revealed that the transcription factor SNAIL1 represses VDR expression in human colon cancer cells (8) (Fig. 1). We observed an inverse correlation between SNAIL1 and VDR expression in a panel of colon cancer cell lines (unpublished data). This led us to analyze the promoter region of exon 1a of the human *VDR* gene searching for putative SNAIL1 binding sites. Three putative sites were identified within the first 600 nucleotides of *VDR* promoter. The binding of SNAIL1 to them was verified by electrophoretic mobility shift assays and chromatin immunoprecipitation. In addition, we observed that the three binding sites mediate the repressive effect of SNAIL1 on *VDR* gene expression (8). Accordingly, overexpression of SNAIL1 in human colon cancer cells decreases VDR expression and blocks $1\alpha,25(OH)_2D_3$ target genes regulation and its prodifferentiation action (8). Moreover, SNAIL1 inhibited the antitumoral effect of the EB1089 analog in xenografted mice (8).

SNAIL1 overexpression in epithelial cells induces the acquisition of a mesenchymal fibroblastic phenotype (epithelial-to-mesenchymal transition) (25). Aberrant SNAIL1 overexpression has been found in different types of carcinomas and it is frequently associated with E-cadherin downregulation, dedifferentiation, invasiveness or metastasis (25). The analysis of *SNAIL1* and *VDR* RNA expression in normal and tumoral biopsies from colon cancer patients showed that SNAIL1 is upregulated in colon tumors and that its upregulation correlates with a reduction of *VDR* expression (8, 26). These data indicate that SNAIL1 induction is probably responsible for VDR downregulation during colon cancer progression and suggest that tumors with high SNAIL1 expression would be resistant to the treatment with $1\alpha,25(OH)_2D_3$ analogs. Since SNAIL1 is upregulated in advanced tumors (associated with acquisition of migratory and invasive properties), our data support the preferential use of $1\alpha,25(OH)_2D_3$ analogs as chemopreventive for high risk patients or as chemotherapeutic for patients with low grade tumors (27).

Several transcription factors have been reported to induce *VDR* gene transcription and sensitize cells to $1\alpha,25(OH)_2D_3$: Sp1, the intestine-specific transcription factor Cdx-2, Wilms' tumor suppressor (WT1), ZEB1, ZEB2, C-MYB, and the AP-1 complex (24). In addition, it has been recently reported that the transcription factor p53 and the p53 family members p63γ, p73α, and p73β activate human *VDR* gene transcription (28, 29). These transcription factors promote cell arrest and apoptosis in response to different genotoxic stresses such as certain chemotherapeutic drugs. It is therefore expected that the increase in the levels of p53 family members by chemotherapy could also lead to an induction of VDR expression that, subsequently, may enhance cell sensitivity to the antitumoral effects of 1α,25$(OH)_2D_3$ or its analogs (28, 29).

Modulation of Colonic VDR Expression and 1α,25$(OH)_2D_3$ Level by Estrogens, Butyrate, and CYP24A1 Inhibitors

Several studies have identified compounds that increase VDR expression and/or 1α,25$(OH)_2D_3$ level in colon cancer and, therefore, could overcome the 1α,25$(OH)_2D_3$ resistance of advanced colon tumors. 1α,25$(OH)_2D_3$ synthesis and VDR expression can be regulated by estrogens. 17β-estradiol and certain phytoestrogens induce VDR and CYP27B1 expression while reduce CYP24A1 levels in human colon cancer cells in vitro and in the mouse colon. In addition, the antitumoral action of estrogens in a colon cancer mice model has been associated with an increase in VDR expression (30). This may explain the low colon cancer incidence in people who consume a phytoestrogens-rich diet or with a high estrogenic background (30). Since it has been reported that estrogens reduce SNAIL1 expression in breast cancer (25), SNAIL1 repression may mediate the upregulation of VDR by estrogens.

Butyrate is a short chain fatty acid normally produced by intestinal bacteria. Butyrate increases VDR expression in human colon cancer cells and synergistically enhance 1α,25$(OH)_2D_3$ antiproliferative and prodifferentiation action (31). Therefore, the treatment of advanced colon tumors with butyrate or estrogens would increase VDR levels and avoid the resistance to 1α,25$(OH)_2D_3$ analogs.

The inhibition of 1α,25$(OH)_2D_3$ catabolism is another useful strategy to increase 1α,25$(OH)_2D_3$ levels in colon cancer. Several CYP24A1 inhibitors that prolong 1α,25$(OH)_2D_3$ biological lifetime have been developed. These compounds have the potential to be used in combination with low doses of 1α,25$(OH)_2D_3$ or its analogs allowing an effective chemotherapy while minimizing or even avoiding hypercalcemia (32).

Concluding Remarks

Epidemiological data and studies in animal models and cultured human colon cancer cells support the preventive and therapeutic effect of 1α,25$(OH)_2D_3$ against CRC. These results prompted the research on safe ways for human 1α,25$(OH)_2D_3$ administration and also the design of nonhypercalcemic analogs. However, initial clinical trials show that these treatments have effectiveness only in a low percentage of patients, and lead to suppose that this resistance could be due to tumoral alteration of proteins involved in 1α,25$(OH)_2D_3$ signaling. In line with this, the transcription factor SNAIL1 represses VDR expression in colon cancer and it has also been reported that CYP27B1 expression decreases and CYP24A1 increases in this neoplasia. Further studies are needed to better characterize the mechanisms responsible for 1α,25$(OH)_2D_3$ resistance to design improved therapies and to select the more suitable patients for treatment with 1α,25$(OH)_2D_3$ or its analogs.

Acknowledgments The work of the authors is supported by Grants from Fundación de Investigación Médica Mutua Madrileña and Ministerio de Educación y Ciencia (SAF04-1015) of Spain.

References

1. Sancho E, Batlle E, Clevers H (2004) Signaling pathways in intestinal development and cancer. Annu Rev Cell Dev Biol 20:695–723.
2. Fearon ER, Vogelstein B (1990) A genetic model for colorectal tumorigenesis. Cell 61:759–767.
3. Gregorieff A, Clevers H (2005) Wnt signaling in the intestinal epithelium: from endoderm to cancer. Genes Dev 19:877–890.
4. Sutton ALM, MacDonald PN (2003) Vitamin D: more than a "bone-a-fide" hormone. Mol Endocrinol 17:777–791.
5. Ordóñez-Morán P, Larriba MJ, Pendás-Franco N, et al. (2005) Vitamin D and cancer: an update of in vitro and in vivo data. Front Biosci 10:2723–2749.
6. Garland CF, Garland FC, Gorham ED, et al (2006) The role of vitamin D in cancer prevention. Am J Public Health 96:252–261.
7. Huerta S, Irwin RW, Heber D, et al. (2002) 1α,25-$(OH)_2D_3$ and its synthetic analogue decrease tumor load in the Apc^{min} mouse. Cancer Res 62:741–746.
8. Pálmer HG, Larriba MJ, García JM, et al. (2004) The transcription factor SNAIL represses vitamin D receptor expression and responsiveness in human colon cancer. Nat Med 10:917–919.
9. Saramaki A, Banwell CM, Campbell MJ, et al. (2006) Regulation of the human *p21*$^{(waf1/cip1)}$ gene promoter via multiple binding sites for p53 and the vitamin D_3 receptor. Nucleic Acids Res 34:543–554.
10. Lin R, Wang TT, Miller WH, et al (2003) Inhibition of F-Box protein $p45^{SKP2}$ expression and stabilization of cyclin-dependent kinase inhibitor $p27^{KIP1}$ in vitamin D analog-treated cancer cells. Endocrinology 144:749–753.
11. Cheng HT, Chen JY, Huang YC, et al. (2006) Functional role of VDR in the activation of $p27^{Kip1}$ by the VDR/Sp1 complex. J Cell Biochem 98(6):1450–1456.
12. Fernández-García NI, Pálmer HG, García M, et al (2005) 1α,25-Dihydroxyvitamin D_3 regulates the expression of *Id1* and *Id2* genes and the angiogenic phenotype of human colon carcinoma cells. Oncogene 24:6533–6544.
13. Tong W-M, Hofer H, Ellinger A, et al. (1999) Mechanism of antimitogenic action of vitamin D in human colon carcinoma cells: relevance for suppression of epidermal growth factor-stimulated cell growth. Oncol Res 11:77–84.
14. González-Sancho JM, Larriba MJ, Ordónez-Morán P, et al (2006) Effects of 1α,25-dihydroxyvitamin D_3 in human colon cancer cells. Anticancer Res 26:2669–2682.
15. Oh YS, Kim EJ, Schaffer BS, et al. (2001) Synthetic low-calcaemic vitamin D_3 analogues inhibit secretion of insulin-like growth factor II and stimulate production of insulin-like growth factor-binding protein-6 in conjunction with growth suppression of HT-29 colon cancer cells. Mol Cell Endocrinol 183:141–149.
16. Chen A, Davis BH, Sitrin MD, et al. (2002) Transforming growth factor-β 1 signaling contributes to Caco-2 cell growth inhibition induced by 1,25$(OH)_2D_3$. Am J Physiol Gastrointest Liver Physiol 283:G864–G874.

17. Yanagisawa J, Yanagi Y, Masuhiro Y, et al. (1999) Convergence of transforming growth factor-β and vitamin D signaling pathways on SMAD transcriptional coactivators. Science 283:1317–1321.
18. Pálmer HG, González-Sancho JM, Espada J, et al. (2001) Vitamin D_3 promotes the differentiation of colon carcinoma cells by the induction of E-cadherin and the inhibition of β-catenin signaling. J Cell Biol 154:369–387.
19. Díaz GD, Paraskeva C, Thomas MG, et al. (2000) Apoptosis is induced by the active metabolite of vitamin D_3 and its analogue EB1089 in colorectal adenoma and carcinoma cells: possible implications for prevention and therapy. Cancer Res 60:2304–2312.
20. Pálmer HG, Sánchez-Carbayo M, Ordóñez-Morán P, et al. (2003) Genetic signatures of differentiation induced by 1α,25-dihydroxyvitamin D_3 in human colon cancer cells. Cancer Res 63:7799–7806.
21. Makishima M, Lu TT, Xie W, et al. (2002) Vitamin D receptor as an intestinal bile acid sensor. Science 296:1313–1316.
22. Banwell CM, MacCartney DP, Guy M, et al. (2006) Altered nuclear receptor corepressor expression attenuates vitamin D receptor signaling in breast cancer cells. Clin Cancer Res 12:2004–2013.
23. Cross HS, Kállay E, Khorchide M, et al. (2003) Regulation of extrarenal synthesis of 1,25-dihydroxyvitamin D_3-relevance for colonic cancer prevention and therapy. Mol Aspects Med 24:459–465.
24. Esteban LM, Eisman JA, Gardiner EM (2005) Vitamin D receptor promoter and regulation of receptor expression. In: Feldman D, Pike JW, Glorieux FH (eds.). Vitamin D. Burlington: Elsevier Academic Press, pp. 1571–1597.
25. Barrallo-Gimeno A, Nieto MA (2005) The Snail genes as inducers of cell movement and survival: implications in development and cancer. Development 132:3151–3161.
26. Peña C, García JM, Silva J, et al. (2005) E-cadherin and vitamin D receptor regulation by SNAIL and ZEB1 in colon cancer: clinicopathological correlations. Hum Mol Genet 14:3361–3370.
27. Larriba MJ, Muñoz A (2005) SNAIL *vs* vitamin D receptor expression in colon cancer: therapeutics implications. Br J Cancer 92:985–989.
28. Kommagani R, Caserta TM, Kadakia MP (2006) Identification of vitamin D receptor as a target of p63. Oncogene 25:3745–3751.
29. Maruyama R, Aoki F, Toyota M, et al. (2006) Comparative genome analysis identifies the vitamin D receptor gene as a direct target of p53-mediated transcriptional activation. Cancer Res 66:4574–4583.
30. Lechner D, Kállay E, Cross HS (2005) Phytoestrogens and colorectal cancer prevention. Vitam Horm 70:169–198.
31. Gaschott T, Stein J (2003) Short-chain fatty acids and colon cancer cells: the vitamin D receptor-butyrate connection. Recent Results Cancer Res 164:247–257.
32. Schuster I, Egger H, Reddy GS, et al. (2003) Combination of vitamin D metabolites with selective inhibitors of vitamin D metabolism. Recent Results Cancer Res 164:169–188.

25
Estrogen-Signaling Pathways in Lung Cancer

D.C. Márquez-Garbán and R.J. Pietras

Introduction

Lung cancer (LC) is the most common cause of mortality from malignancy throughout the world, causing an estimated 1.2 million deaths annually (1). In the USA, it constitutes the leading cause of cancer death among men and women, exceeding the combined number of deaths from breast, prostate, and colon cancers. Despite advances in chemotherapy treatments, the five-year survival rate for LC has not significantly increased in the past 25 years, remaining at 15%, as compared with 61% for colon cancer, 86% for breast cancer, and 96% for prostate cancer. Cigarette smoking remains the primary risk factor for LC with 85–90% of all LC patients having smoked cigarettes at some time in their lives. Remaining LC cases occur in nonsmokers, mainly women. Of note, there has been a dramatic increase in the number of LC cases in women mainly as a consequence of increased prevalence of smoking. This phenomenon has been defined as a "full blown epidemic" by the US Surgeon General (2), with a 600% increase in the death rate from 1930 to 1997. Moreover, it is projected that mortality from LC will continue to increase as the at-risk populations increases.

Gender differences in the behavior of the disease suggest a role of hormones in the etiology and progression of LC. In addition to smoking, other established risk factors for development of LC in men have also been confirmed in women. These include a family history of LC and a history of prior lung disease. There is still controversy as to whether women are more or less susceptible to the carcinogenic effects of cigarette smoke than men (3–6). In addition to risk factors shared with men, some LC risk factors may be unique to women. In a study published by Taioli and Wynder (7), the authors postulated that estrogens may be a relevant factor in the promotion phase of lung carcinogenesis, in particular for development of adenocarcinoma, among women. There are also differences in the relative distribution of LC histological features between men and women that are not explained by differences in smoking patterns. Gender disparities in adenocarcinoma, which accounts for 75% of LC in females, implicate hormones in LC. Estrogen circulating levels are naturally higher in women than men, and this may increase their susceptibility to LC.

J.J. Li et al. (eds.), *Hormonal Carcinogenesis V.*

Estrogens in Lung Cancer Biology

Recent studies provide compelling evidence that estrogen-signaling pathways play an important role in normal lung biology and in controlling the growth of LC (8–12). Furthermore, estrogen status appears to be a significant factor in LC in women, with evidence that exogenous and endogenous estrogens may play a role in development of LC, especially adenocarcinoma (2, 7, 13, 14). In limited studies, estrogen levels are found to be elevated in female LC patients as compared with those in women without LC (15). In addition, estrogen biosynthesis due to activity of aromatase has been reported in lung (9, 10, 16), suggesting that estrogens are produced locally in women and men and could affect tumor development (9, 10).

The histologic distribution of LC among women is distinctly different from that among men (17, 18). Women are much more likely than men to be diagnosed with lung adenocarcinoma rather than squamous cell carcinoma which is more common in men. Never smokers that develop LC almost uniformly have adenocarcinoma and are also 2.5 times more likely to be female than male (19). Adenocarcinoma is the most common form of LC in young persons, women of all ages and never smokers (2). These differences in the relative distribution of histologic types of LC between men and women are not explained only by differences in smoking patterns, thus allowing for the possibility that estrogens may play a causative role in LC.

Estrogens may play a role in LC tumorigenesis through a receptor-mediated mechanism. Estrogen receptors were noted to be present in lung in early reports (see Table 1). Pasqualini et al. demonstrated specific estrogen binding in the lung of fetal and newborn guinea pigs (20), and Khosla et al. described the effect of estradiol-17β in stimulating production of lung surfactant in the fetal rabbit (21). Nonetheless, in the ensuing literature, there have been conflicting reports on the

Table 1 Role of estrogens in lung biology

Year	Selected observations
1978	Estrogen binding sites in lung from rats and from fetal and newborn guinea pig (20, 66)
1979	Estradiol-17β stimulates production of lung surfactant in fetal rabbit (21)
1985	Specific ER protein is present in human NSCLC cells
1996	Cloning of novel ERβ gene
1999	Estradiol-17β-Estradiol stimulates NO production by normal lung bronchial cells (29)
2001	ERβ expressed in normal and premalignant lung, with role in lung carcinogenesis (25)
2002	Several reports of ERα and ERβ in NSCLC cell lines and tumor specimens (8, 22, 26, 27) and biologic responses to estrogen stimulation (8)
2003	ERβ knockout mouse model suggests role of ERβ in normal lung development (31)
2005	Aromatase present and active in NSCLC cells and tumors; and aromatase inhibitors reduce tumor growth in vitro and in vivo (9, 10)
	Combination therapy with EGFR tyrosine kinase inhibitor and antiestrogen elicits improved antitumor efficacy in NSCLC in vitro and in vivo (28)
	NSCLC cells respond to estrogen and to antiestrogens with distinct alterations in endogenous lung gene expression (12)
2006	High-affinity, low-capacity estrogen-binding receptor protein confirmed to be present in human NSCLC cells (11)

presence of estrogen receptors (ER), ERα and β, in human lung tissues and cell lines (22–27). More recent work has more convincingly demonstrated that both ERα and β mRNA and proteins are present and functional in normal and malignant lung cells as wells as in nonsmall cell LC (NSCLC) tumor specimens from the clinic (8, 10, 12, 28, 29). Moreover, these receptors appear to play important biologic roles in lung and respond to antiestrogens.

The effects of estrogens in lung were previously thought to be indirect. The discovery of ERβ (30) (see Table 1) and development of its knockout model demonstrated that estradiol-17β (E_2) has direct actions on the lung. A comparison of the lungs of wild type versus ERβ null (βERKO) mice revealed decreased numbers of alveoli in adult female ERβ$^{-/-}$ mice and decreased surfactant, platelet-derived growth factor A and granulocyte-macrophage colony stimulating factor (31). More recently, ERβ was reported to be necessary for the maintenance of the extracellular matrix composition in lung, with loss of ERβ leading to abnormal lung structure and systemic hypoxia. This evidence suggests a role for ERβ in lung homeostasis (32).

Nuclear and extranuclear ER forms were both postulated in early concepts of steroid action in target cells, such as breast and ovary (33), with the transcriptional activity of estrogen mediated by high affinity ER in cell nuclei (34). On estrogen binding in target cells, ER is phosphorylated and undergoes a conformational

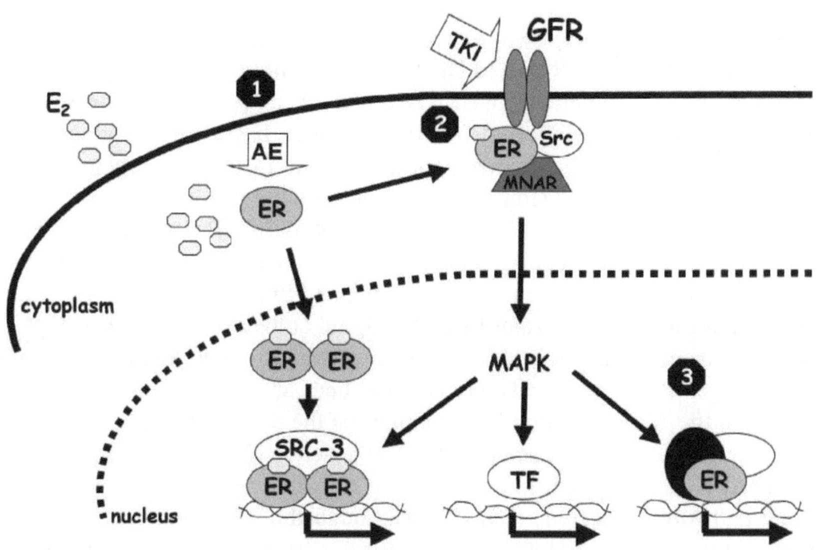

Fig. 1 Steroid and growth factor receptor signaling pathways in lung cancer target cells ER$^+$ cells may undergo *ligand-dependent receptor activation* as shown in pathway 1 (classical ER signaling); downstream signaling by estrogen due to interaction with *nonnuclear ER's* as in pathway 2 (extranuclear ER signaling); or *ligand-independent receptor activation* as in pathway 3, via growth factor receptor (GFR)-induced ER phosphorylation and activation (EGFR/HER-induced signaling). Antiestrogens (AE) act by interfering with estrogen (E2) binding to ER, inhibiting transcription of target genes. Tyrosine kinase inhibitors (TKI) such as gefitinib (Iressa) or erlotinib (Tarceva) inhibit EGFR activation and signal transduction

change that allows receptor dimerization and association of estrogen-ER complexes with specific estrogen response elements (ERE) in DNA, leading to transcription. Nuclear actions of estrogen are dependent, in part, on the subtype of receptor (ERα, ERβ), the gene promoter, and the coactivator/corepressor proteins that modulate transcription (34, 35). In addition, ER also regulates gene expression without direct binding to DNA. This occurs by protein–protein interaction with other transcription factors, such as AP-1 (36), and with extranuclear signaling complexes (Fig. 1) that, in turn, modulate downstream gene transcription. Extranuclear signaling, such as MAPK activation, has a rapid onset and is mediated by ER that is either in or tethered to cell membrane (37–39).

In lung, as in breast, current evidence indicates that extranuclear ER derives from the same transcript as nuclear ER but may undergo posttranslational modification to promote membrane association (10, 37, 40–42). Further, both nuclear and extranuclear ER forms appear to act in concert with growth factor signaling pathways to promote lung tumor growth and survival (43–45) (see below).

Interactions Between Estrogen Receptors and Growth Factor Receptors

The EGFR/HER family of growth factor receptors, including the epidermal growth factor receptor (EGFR, HER-1) and HER-2, is also implicated in LC pathogenesis. Thus, in addition to direct effects of ERα as a nuclear transcription factor (see Fig. 1, pathway 1), there is growing evidence that estrogen has a proliferative effect in lung via interactions with various peptide growth factors and their receptors, including EGFR and HER-2 (Fig. 1), as well as possibly insulin-like growth factor, TGF-α and TGF-β receptors. In particular, EGFR and HER-2 receptors appear to regulate ligand-independent and ligand-dependent activation of ERα and ERβ (Fig. 1). Molecular details of interaction between ER and EGFR/HER receptors are now emerging, and ER is an important locus for signal convergence (44–46) (Fig. 1).

Of note, in gene knockout mice lacking ERα, both estrogen- and EGF-stimulated growth in target tissues are blocked (43). It has been shown that phosphorylation of serine-118 in the AF-1 domain of ERα enhances the transcriptional activity of this domain. This serine-118 is phosphorylated by the action of mitogen-activated protein kinase (MAPK) (47), which, in turn, may be activated by growth factor-mediated signaling pathways triggered by EGFR. Although the AF-1 domains of ERα and ERβ are significantly divergent, the serine residue at position-118 in ERα is structurally and functionally conserved in the ERβ protein (48), suggesting its biologic importance for cell functions.

ER can also directly or indirectly, through growth factors, activate phosphatidylinositol-3-kinase and MAPK pathways to regulate cell proliferation. ER likely mediates gene transcripts by integrating signals from growth factor-activated pathways as well as from steroid binding. Emerging evidence points to the importance of extranuclear ER in initiating signals for cell growth and survival. It remains to be determined how ER couples with membrane signaling complexes such as EGFR. Recent reports suggest that posttranslational ER modification may promote membrane targeting (10, 49–51) but association of ER with other adaptor or signaling proteins may also occur (52–54).

EGFR gene mutations are reported to associate with differential responses to gefitinib (55, 56). Such mutations associate with ethnicity, smoking status, tumor histology, and they may be more prevalent in females than males. Although men and women may be differentially exposed to certain environmental factors, another important difference is exposure to estrogen. The notion of a link between EGFR and steroid receptors was suggested before. Activation of EGFR was first associated with tumorigenesis with the discovery that EGFR gene is a retroviral oncogene of the erythroblastosis virus (v-erb B) (57). Further, the v-erb A gene, derived from the thyroid hormone nuclear receptor gene, is contained in the same virus. Coselection of these genes, erb A and erb B, by the transforming retrovirus raises the possibility that this combination is important for cell replication and transformation and suggests functional links between members of the nuclear receptor family (erb A) and the transforming activity of the erb B family (EGFR, HER-2). In view of its known oncogenic role in human breast cancer, ER, a member of the erb A gene family, is a prime suspect for a nuclear receptor cooperating with EGFR activation in NSCLC (58). If biologic interactions between ER and EGFR/HER promote LC growth, this signaling axis could offer a critical new target to treat NSCLC (10, 14, 28, 59, 60).

Aromatase in Human Nonsmall Cell Lung Cancer

It is well known that estrogen has an important role in breast cancer development, and several studies have shown that the concentration of estrogen in breast tissue is higher than in plasma in postmenopausal women. This is due to aromatase-mediated biosynthesis of estrogen that takes place in breast (61). Aromatase is a cytochrome P-450 enzyme complex found in many tissues (62), and it is responsible for the final, rate-limiting step in estrogen biosynthesis, catalyzing conversion of androstenedione and testosterone into estrone and estradiol, respectively (Fig. 2). Similar to breast, emerging evidence suggests that aromatase is also expressed in lung (9, 10), thus suggesting that estrogens may be produced locally in women and men and could affect lung tumor development. Furthermore, aromatase inhibitors may be useful to block estrogen biosynthesis in LCs.

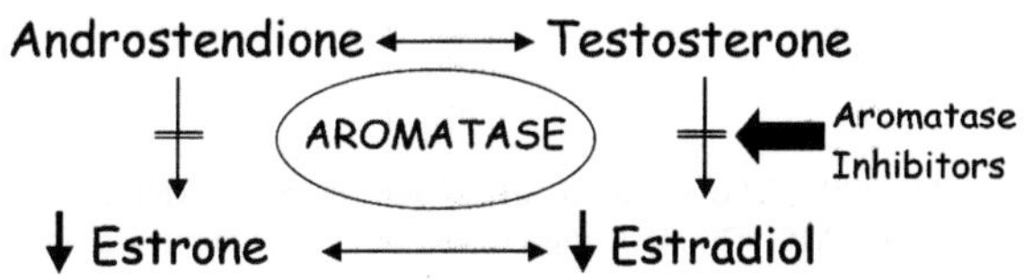

Fig. 2 Diagram of the primary biosynthetic pathways for local estrogen production in lung tumors

Current aromatase inhibitors are divided in two main classes, steroidal agents that bind the substrate-binding site of the enzyme (exemestane) and nonsteroidal compounds (anastrazole, letrozole) that interact competitively with the heme-group of cytochrome P-450 components of aromatase (62). Nonsteroidal aromatase inhibitors, such as letrozole and anastrozole, improve outcome in breast cancer patients and are FDA-approved as therapy for hormone receptor-positive, metastatic breast cancer in postmenopausal women (9, 10).

Conclusion

There are notable differences in the biology, natural history and response to therapy between men and women with NSCLC. At this time, there is an urgent need to determine how the estrogenic hormonal milieu may influence lung homeostasis, as well as lung tumor development and growth. Based on current evidence, it seems likely that estrogens are involved in some aspects of lung carcinogenesis, either by acting as ligands to activate ER and promote cell proliferation or by other indirect metabolic pathways (63). These estrogenic pathways may be responsible, in part, for reported sex differences in lung carcinogenesis.

The success of endocrine therapy for human cancer, such as breast malignancies, depends on close regulation of cell growth by steroids such as estrogens. Antiestrogen therapy has been highly effective in improving outcomes for patients with $ER\alpha^+$ breast cancer. At present, LC patients have few therapeutic options, but the potential antitumor efficacy of endocrine therapy in LC could offer a previously unsuspected option for treatment of this disease in the future. If ER is indeed involved in lung tumorigenesis, inhibiting lung tumor growth may be possible with antihormone therapy in combination with other available therapies. This hypothesis deserves to be tested in the clinic (64, 65).

In conclusion, further studies to assess the notion that activation of ER and associated growth factor receptors promotes NSCLC growth are required. Results of this work will contribute further to our understanding of the distribution and interactions of ER forms associated with nuclear and extranuclear sites in NSCLC cells. We hope that these studies will offer new options to stop lung tumor growth, such as multitargeted therapy with selected combinations of antiestrogens, aromatase inhibitors and/or EGFR/HER receptor antagonists.

References

1. Mackay J, Eriksen M (2002) The Tobacco Atlas. Geneva, Switzerland.
2. Patel JD, Bach PB, Kris MG (2004) Lung cancer in US women: a contemporary epidemic. JAMA 291(14): 1763–8.
3. Henschke CI, Miettinen OS (2004) Women's susceptibility to tobacco carcinogens. Lung Cancer 43(1): 1–5.

4. Zang EA, Wynder EL (1996) Differences in lung cancer risk between men and women: examination of the evidence. J Natl Cancer Inst 88(3–4): 183–92.
5. Bach PB et al (2003) Variations in lung cancer risk among smokers. J Natl Cancer Inst 95(6): 470–8.
6. Bain C, et al. (2004) Lung cancer rates in men and women with comparable histories of smoking. J Natl Cancer Inst 96(11): 826–34.
7. Taioli E, Wynder EL (1994) Re: Endocrine factors and adenocarcinoma of the lung in women. J Natl Cancer Inst 86(11): 869–70.
8. Stabile LP, et al. (2002) Human non-small cell lung tumors and cells derived from normal lung express both estrogen receptor alpha and beta and show biological responses to estrogen. Cancer Res 62(7): 2141–50.
9. Weinberg OK et al. (2005) Aromatase inhibitors in human lung cancer therapy. Cancer Res 65(24): 11287–91.
10. Pietras RJ et al. (2005) Estrogen and growth factor receptor interactions in human breast and non-small cell lung cancer cells. Steroids 70(5–7): 372–81.
11. Dougherty SM et al. (2006) Gender difference in the activity but not expression of estrogen receptors alpha and beta in human lung adenocarcinoma cells. Endocr Relat Cancer 13(1): 113–34.
12. Hershberger PA et al. (2005) Regulation of endogenous gene expression in human non-small cell lung cancer cells by estrogen receptor ligands. Cancer Res 65(4): 1598–605.
13. Siegfried JM (2001) Women and lung cancer: does oestrogen play a role? Lancet Oncol 2(8): 506–13.
14. Stabile LP, Siegfried JM (2004) Estrogen receptor pathways in lung cancer. Curr Oncol Rep 6(4): 259–67.
15. Tiuriunova AM et al. (1986) Hormonal balance in women with lung cancer and its changes after combined treatment. Vopr Onkol, 1986. 32: 2–30.
16. Pezzi V et al (2003) Profiling transcript levels for steroidogenic enzymes in fetal tissues. J Steroid Biochem Mol Biol. 87(2–3): 181–9.
17. Radzikowska E, Glaz P, Roszkowski K (2002) Lung cancer in women: age, smoking, histology, performance status, stage, initial treatment and survival. Population-based study of 20 561 cases. Ann Oncol. 13(7): 1087–93.
18. Travis WD et al. (1996) United States lung carcinoma incidence trends: declining for most histologic types among males, increasing among females. Cancer 77(12): 2464–70.
19. Patel JD (2005) Lung cancer in women. J Clin Oncol 23(14): p 3212–8.
20. Pasqualini JR et al. (1978) Specific binding of estrogens in different fetal tissues of guinea pig during fetal development. Cancer Res 38(11 Pt 2): 4246–50.
21. Khosla SS, Rooney SA (1979) Stimulation of fetal lung surfactant production by administration of 17beta-estradiol to the maternal rabbit. Am J Obstet Gynecol 133(2): 213–6.
22. Radzikowska E, Langfort R, Giedronowicz D (2002) Estrogen and progesterone receptors in non small cell lung cancer patients. Ann Thorac Cardiovasc Surg 8(2): 69–73.
23. Croxtall JD et al. (1994) Tamoxifen inhibits growth of oestrogen receptor-negative A549 cells. Biochem Pharmacol 47(2): 197–202.
24. Caltagirone S et al. (1997) Interaction with type II estrogen binding sites and antiproliferative activity of tamoxifen and quercetin in human non-small-cell lung cancer. Am J Respir Cell Mol Biol 17(1): 51–9.
25. Omoto Y et al. (2001) Expression, function, and clinical implications of the estrogen receptor beta in human lung cancers. Biochem Biophys Res Commun 285(2): 340–7.
26. Mollerup S et al. (2002) Expression of estrogen receptors alpha and beta in human lung tissue and cell lines. Lung Cancer 37(2): 153–9.
27. Fasco MJ, Hurteau GJ, Spivack SD (2002) Gender-dependent expression of alpha and beta estrogen receptors in human nontumor and tumor lung tissue. Mol Cell Endocrinol 188(1–2): 125–40.
28. Stabile LP et al (2005) Combined targeting of the estrogen receptor and the epidermal growth factor receptor in non-small cell lung cancer shows enhanced antiproliferative effects. Cancer Res 65(4): 1459–70.

29. Kirsch EA et al. (1999) Estrogen acutely stimulates endothelial nitric oxide synthase in H441 human airway epithelial cells. Am J Respir Cell Mol Biol 20(4): 658–66.
30. Kuiper GG et al. (1996) Cloning of a novel receptor expressed in rat prostate and ovary. Proc Natl Acad Sci USA 93(12): 5925–30.
31. Patrone C et al. (2003) Regulation of postnatal lung development and homeostasis by estrogen receptor beta. Mol Cell Biol 23(23): 8542–52.
32. Morani A et al. (2006) Lung dysfunction causes systemic hypoxia in estrogen receptor beta knockout (ERbeta-/-) mice. Proc Natl Acad Sci USA 103(18): 7165–9.
33. Jensen EV, DeSombre ER (1973) Estrogen-receptor interaction. Science 182(108): 126–34.
34. Evans RM (1988) The steroid and thyroid hormone receptor superfamily. Science 240(4854): 889–95.
35. Katzenellenbogen BS, Frasor J (2004) Therapeutic targeting in the estrogen receptor hormonal pathway. Semin Oncol 31(1 Suppl 3): 28–38.
36. Srivastava S et al. (1999) Estrogen decreases TNF gene expression by blocking JNK activity and the resulting production of c-Jun and JunD. J Clin Invest 104(4): 503–13.
37. Marquez DC et al. (2006) Estrogen receptors in membrane lipid rafts and signal transduction in breast cancer. Mol Cell Endocrinol 246(1–2): 91–100.
38. Marquez DC, Pietras RJ (2001) Membrane-associated binding sites for estrogen contribute to growth regulation of human breast cancer cells. Oncogene 20(39): 5420–30.
39. Levin ER (2005) Integration of the extranuclear and nuclear actions of estrogen. Mol Endocrinol 19(8): 1951–9.
40. Marquez DC, Pietras RJ (2002) Membrane-associated estrogen receptors and breast cancer, in The Identities of Membrane Steroid Receptors, C.S. Watson, Editor, Kluwer Academic Publisher, Boston. p. 1–10.
41. Norfleet AM et al. (2000) Antibodies to the estrogen receptor-alpha modulate rapid prolactin release from rat pituitary tumor cells through plasma membrane estrogen receptors. Faseb J 14(1): 157–65.
42. Razandi M et al. (1999) Cell membrane and nuclear estrogen receptors (ERs) originate from a single transcript: studies of ERalpha and ERbeta expressed in Chinese hamster ovary cells. Mol Endocrinol 13(2): 307–19.
43. Curtis SW et al. (1996) Physiological coupling of growth factor and steroid receptor signaling pathways: estrogen receptor knockout mice lack estrogen-like response to epidermal growth factor. Proc Natl Acad Sci USA 93(22): 12626–30.
44. Marquez DC et al. (2001) Epidermal growth factor receptor and tyrosine phosphorylation of estrogen receptor. Endocrine 16(2): 73–81.
45. Pietras RJ (2003) Interactions between estrogen and growth factor receptors in human breast cancers and the tumor-associated vasculature. Breast J 9(5): 361–73.
46. Gruber CJ et al. (2002) Production and actions of estrogens. N Engl J Med 346(5): 340–52.
47. Kato S et al. (1995) Activation of the estrogen receptor through phosphorylation by mitogen-activated protein kinase. Science 270(5241): 1491–4.
48. Dechering K, Boersma C, Mosselman S (2000) Estrogen receptors alpha and beta: two receptors of a kind? Curr Med Chem 7(5): 561–76.
49. Li L, Haynes MP, Bender JR (2003) Plasma membrane localization and function of the estrogen receptor alpha variant (ER46) in human endothelial cells. Proc Natl Acad Sci USA 100(8): 4807–12.
50. Acconcia F et al. (2004) S-palmitoylation modulates human estrogen receptor-alpha functions. Biochem Biophys Res Commun 316(3): 878–83.
51. Acconcia F et al. (2003) Does palmitoylation target estrogen receptors to plasma membrane caveolae? IUBMB Life 55(1): 33–5.
52. Wong CW et al. (2002) Estrogen receptor-interacting protein that modulates its nongenomic activity-crosstalk with Src/Erk phosphorylation cascade. Proc Natl Acad Sci USA 99(23): 14783–8.
53. Song RX et al. (2004) The role of Shc and insulin-like growth factor 1 receptor in mediating the translocation of estrogen receptor alpha to the plasma membrane. Proc Natl Acad Sci USA 101(7): 2076–81.

54. Pedram A et al. (2002) Integration of the non-genomic and genomic actions of estrogen. Membrane-initiated signaling by steroid to transcription and cell biology. J Biol Chem 277(52): 50768–75.
55. Lynch TJ et al. (2004) Activating mutations in the epidermal growth factor receptor underlying responsiveness of non-small-cell lung cancer to gefitinib. N Engl J Med 350(21): 2129–39.
56. Paez JG et al. (2004) EGFR mutations in lung cancer: correlation with clinical response to gefitinib therapy. Science 304(5676): 1497–500.
57. Beug H, Graf T (1989) Co-operation between viral oncogenes in avian erythroid and myeloid leukaemia. Eur J Clin Invest 19(6): 491–502.
58. Sellers WR, Meyerson M (2005) EGFR gene mutations: a call for global × global views of cancer. J Natl Cancer Inst 97(5): 326–8.
59. Pietras R, Marquez D, Chen H-W, Ayala R, Ramos L, Slamon D (2003) Improved antitumor receptor therapy with Herceptin and Faslodez for dual targeting of HER-2 and estrogen receptor signaling pathways in human breast cancers with overexpression of HER-2/neu gene. Breast Cancer Res 82(Suppl 1): 12–13.
60. Pietras RJ et al. (1995) HER-2 tyrosine kinase pathway targets estrogen receptor and promotes hormone-independent growth in human breast cancer cells. Oncogene 10(12): 2435–46.
61. Brodie A, Long B, Lu Q (1998) Aromatase expression in the human breast. Breast Cancer Res Treat 49(Suppl 1): S85–91; discussion S109–19.
62. Brodie A (2003) Aromatase inhibitor development and hormone therapy: a perspective. Semin Oncol 30(4 Suppl 14): 12–22.
63. Yager JD, Liehr JG (1996) Molecular mechanisms of estrogen carcinogenesis. Annu Rev Pharmacol Toxicol 36: 203–32.
64. Dubey S, Siegfried JM, Traynor AM (2006) Non-small-cell lung cancer and breast carcinoma: chemotherapy and beyond. Lancet Oncol 7(5): 416–24.
65. Traynor A, Schiller J, Stabile L, Kolesar J, Belani C, Hoang T, Dubey S, Eickhoff J, Marcotte S, Siegfried J (2005) Combination therapy with gefitinib and fulvestrant (G/F) for women with non-small cell lung cancer (NSCLC) in ASCO.
66. Morishige WK, Uetake CA (1978) Receptors for androgen and estrogen in the rat lung. Endocrinology 102(6): 1827–37.

COMMUNICATIONS

Session I
Human Derived Studies

Breast Cancer

Noninvasive Measurement of the Electrical Properties of Breast Epithelium During the Menstrual Cycle: A Potential Biomarker for Breast Cancer Risk

Richard J. Davies, Mary K. Brumfield, and Maribeth Pierce

Summary Breast epithelia maintain an ionic and electrical gradient across the ductal lumen, which can be measured using ductal epithelial impedance spectroscopy (DEIS). Furthermore, the breast remodels itself through a secretory, proliferative, and involutional cycle each month in premenopausal women, which changes the electrical properties of the epithelium. DEIS was used to examine the electrical properties of breast epithelia in 14 normal women at multiple points during 1–2 menstrual cycles with IRB/patient consent. The epithelial resistance (Re) measured 159 ± 26 kΩ (mean ± SEM) in week 1 (n = 55) of menstrual cycle, when the epithelium is single layered and atrophic, increased to 217 ± 20 kΩ ($p < 0.05$) by weeks 2–3 (n = 129), as the epithelium becomes more proliferative and multilayered, and reached 263 ± 30 kΩ ($p = 0.002$) by week 4 (n = 75). The capacitance of the epithelium increased between week 2 and 3 of the cycle from 0.32 ± 0.02 μF (microFarads) to 0.51 ± 0.05 μF ($p = 0.003$), when the epithelial cell mass is greatest. The characteristic frequency (fC) of the epithelium, a property related to morphology and transport function, decreases progressively through weeks 1–4 of menstrual cycle from 17 ± 3 to 11 ± 2, 10 ± 1, and finally to 7 ± 1 Hz respectively ($p = 0.003$). This study demonstrates that the electrical properties of breast epithelium are dependent on the time during the menstrual cycle, and likely represents morphological changes in breast epithelia associated with the menstrual cycle. Patients at risk for breast cancer may have an aberrant epithelial response to estrogen and progesterone. Electrical measurements of the epithelium using DEIS, during the menstrual cycle, may represent a novel biomarker for assessing BC risk.

Introduction

Breast cancer (BC) develops in the epithelia lining the terminal-ductal-lobular-units (TDLUs). During menstrual cycle the breast is under hormonal control with the epithelium undergoing changes in protein expression, cell kinetics, and morphology (1). During week 1, the epithelium is single layered and atrophic, with a sharp luminal border, and no distinction between epithelial and myoepithelial cells. By week 2 of cycle, the epithelium becomes double-layered. In week 3, there is an increase in the number of TDLUs with secretions and alveolar budding; the basal layer becomes vacuolated with columnar appearing epithelial cells and basophilic cytoplasm. In week 4, extensive vacuolation occurs within the TDLUs with frequent mitoses and apoptoses observed. There is also marked stromal edema and an increase in the inflammatory cell population.

An inverse relationship has been observed between serum estradiol (E_2) and estrogen receptor (ERα) of breast epithelium in women without BC, which is dependent on position in menstrual cycle (2, 3). This relationship has not been observed in breast epithelium derived from women with BC. In addition, progesterone receptor (PR) was positively correlated with serum E_2 levels in BC cases, but not in controls. Epithelial proliferation was inversely correlated to ERα in controls, but was positively related in BC cases. These observations have lead to the suggestion that that the normal epithelium of women with BC may display an aberrant response to E_2 with ERα up-regulating in the luteal phase of menstrual cycle, whereas it down-regulates in breast epithelium from women without BC (4, 5). The effect of this aberrant response on breast epithelial morphology is unknown.

Breast epithelium can only be directly accessed by invasive procedures such as biopsy, ductoscopy, and ductal lavage; and imaging studies do not have the resolution to evaluate the morphology of breast epithelium. Electrical approaches may be used to characterize breast tissue without the limitations of the above-referenced approaches. Epithelia lining the ducts can be modeled as an electrical circuit with resistors and capacitors in series and parallel. The skin has a high impedance, which obscures the dielectric and resistance properties of the underlying epithelial and stromal elements (6). Furthermore previous attempts to use electrical or impedance techniques to characterize breast tissue do not probe the ductal epithelium, where BC originates (7–11). Therefore, we have developed a new technique, ductal epithelial impedance spectroscopy (DEIS), which avoids the high impedance of the overlying skin and noninvasively probes the ductal epithelium to characterize the electrical signature of breast epithelia during menstrual cycle.

Methodology

With IRB and patient consent, electrical contact with ductal epithelium was established noninvasively, using a specially designed nipple sensor in 14 healthy women, with a median age of 25 (range 21–40), at multiple points during 1–2 menstrual cycles. Skin impedance was reduced using alcohol wipes and keratin plugs removed from the nipple using a dekeratinizing agent to open up the duct ostia. Measurements were made between the nipple sensor and skin surface electrodes. Transepithelial potential (TEP) and DEIS measurements were obtained over a frequency range of 0.1–60 KHz and processed using a frequency response analyzer and sine-wave correlation technique. Impedance data were collected using ZPlot and analyzed using ZView (Scribner Associates, Inc.). Figure 1 illustrates the measurement set-up. Data were analyzed using a *t* test, Mann-Whitney test, or ANOVA as appropriate. Figure 1

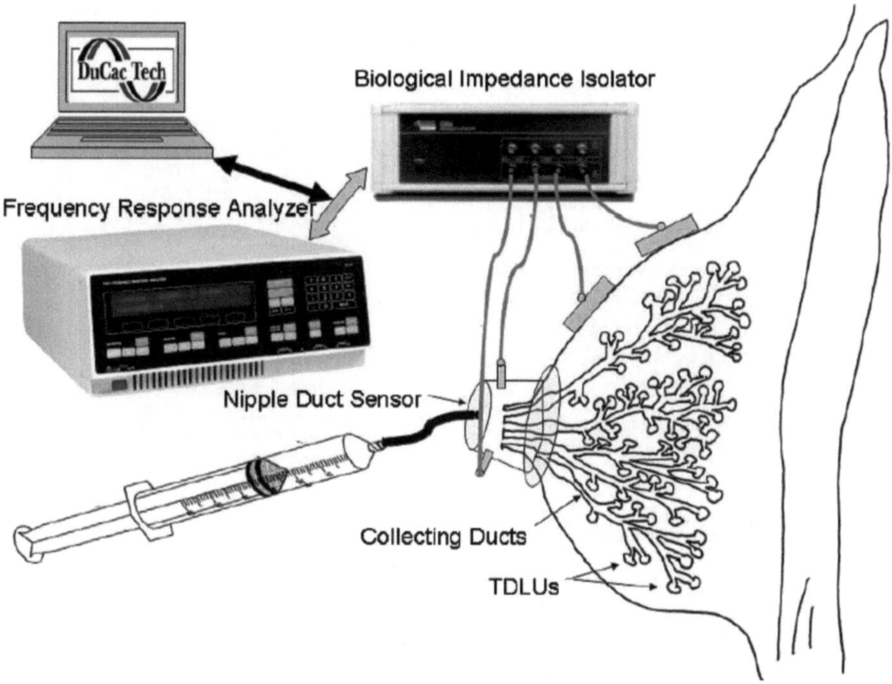

Fig. 1 Measurement set-up for ductal–epithelial impedance spectroscopy (DEIS)

depicts a nipple sensor, which is applied to the nipple after removal of keratin plugs with NuPrep®. A saline-filled syringe is used to remove air and apply 100 mm Hg negative suction, which opens up the nipple duct ostia. The skin surface is cleaned with alcohol wipes before applying low offset Ag-AgCl electrodes. A frequency response analyzer (Solartron 1250) is used to pass a series of sine waves logarithmically spaced between 60 KHz and 0.1 Hz, between one of the nipple electrodes and the electrode on the outer aspect of the breast. The voltage drop and phase are measured between the other nipple sensor electrode and the inner skin surface electrode, which are also used to measure the transepithelial potential after each frequency sweep. Each of four quadrants was measured in every subject. All electrical connections to the patient were made through a biological impedance isolator (Solarton 1294) in compliance with ISO-601.

Results

Figure 2 illustrates typical Nyquist impedance curves obtained using the device illustrated in Fig. 1. Nyquist impedance curves plot the real component of impedance, resistance (Z'), against the imaginary component, capacitive reactance (Z'').

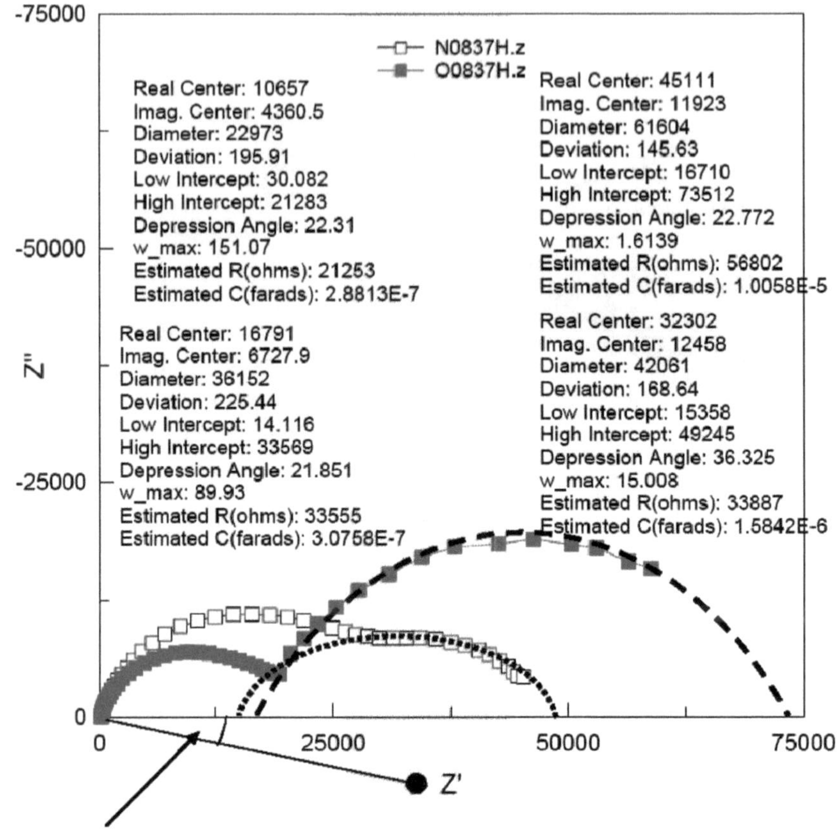

Fig. 2 Typical Nyquist plots of breast epithelium obtained using DEIS. Z′ x-axis resistance in ohms. Z″ y-axis reactance in ohms. Impedance parameters derived from curve fitting described in text

Figure 2 illustrates typical Nyquist plots of breast impedance. The impedance is frequency dependent with the impedance lowest at the lower left corner of the plot with impedance increasing as the frequency of the applied sinusoidal wave form decreases and the curve moves to the right. Two types of curve were observed; those with a clear "double hump" indicating two separate time constants for the 2 RC (resistor–capacitor) components of the circuit, and a single suppressed semicircle where the separation of the time constants is less evident. With the second type curve there still appears two distinct curves, which are partly fused because the RC time constants are similar.

The high and low frequency curves separate at ≤6 Hertz (Hz). Best-fit impedance parameters were estimated from the fit of the high and low frequency data points to a semicircle, which is usually suppressed so that the locus falls below the *x*-axis.

The dotted line indicates the best fit of the low frequency data points of N0837H (open squares) to a semicircle with the derived impedance parameters in the lower right corner of the graph. The resistance (*Re*) is derived from the difference between the high and low intercept of the semicircle with the *x*-axis. The capacitance (*C*) is derived from the capacitive reactance (Z'') where $Z'' = 1/2\pi fC$, and f = frequency. Depression angle is the depression of the locus of the semicircular arc when a line is drawn between the lower intercept of the semicircular arc on the *x*-axis and the locus of the semicircular arc, and the angle it makes with the *x*-axis. *W*_max is the characteristic frequency in radians per second, or Hertz and is defined as the frequency at the highest point on the semicircular arc.

The dashed line indicates the best fit of the low frequency data points of O0837H (closed squares) to a semicircle with the derived impedance parameters in the upper right corner of the graph; the data at the upper and lower left hand corners of the graph are the derived impedance parameters from the high frequency components of the Nyquist plots (i.e., curves on the left side of the plots) for O0837H and N0837H, respectively.

Transepithelial Potential (TEP). TEP was −76.4 ± 1.4 mV (mean ± SEM) at week 1 ($n = 55$) of menstrual cycle and appeared to depolarize to −72.5 ± 1.5 mV by week 2 ($n = 69$), −75.94 ± 1.2 mV by week 3 ($n = 60$), and −73.1 ± 1.2 mV by week 4 ($n = 75$). However, this did not reach statistical significance ($p = 0.089$).

Transepithelial Resistance (*Re*). The total *Re* increased from 159 ± 26 kΩ (mean ± SEM) in week 1 ($n = 55$) of menstrual cycle to 217 ± 20 kΩ ($p < 0.05$) by weeks 2–3 ($n = 129$), and reached 263 ± 30 kΩ ($p = 0.002$) by week 4 ($n = 75$).The transepithelial resistance at high frequency (60 KHz–6 Hz) was significantly higher in week 4 of cycle compared with week 1, with an *Re* of 189 ± 21 KΩ compared with 116 ± 18 KΩ ($p = 0.012$). Greater differences in *Re* were observed in the low frequency part of the curve (<6 Hz) comparing week 4 and 1 of the cycle, with values of 274 ± 34 KΩ compared with 177 ± 31 KΩ ($p = 0.027$).

Transepithelial Capacitance (*C*). Although a trend of increasing *C* was observed in the high frequency curve of the Nyquist plot in weeks 1 through 4 of cycle, with values of 0.39 ± 0.04 µF (microfarads) in week 1, to 0.40 ± 0.04 µF in week 2, to 0.44 ± 0.05 µF in week 3, and to 0.51 ± 0.05 µF in week 4, this did not reach statistical significance ($p = 0.308$). The *C* of the epithelium in the low frequency component of the Nyquist plots (<6 Hz) was 0.31 ± 0.11 µF in week 1, increased between week 2 and 3 of the cycle from 0.32 ± 0.02 µF to 0.51 ± 0.05 µF ($p = 0.003$), and fell to 0.41 ± 0.08 µF in week 4.

Normalized Resistance and Capacitance. The *Re* and *C* were not normally distributed and would vary between different cycles. Figure 3 demonstrates *Re* (total) and *C* (low-frequency) normalized to the *Re* and *C* of the preceding menstrual cycle. When measurement were made starting early in cycle measurements they were normalized to the last set of measurements made in the first cycle just before menstruation. Figure 3 demonstrates that normalized *Re* was lowest during week 1 of the cycle and progressively increased until week 4; 0.54 ± 0.12 (week 1); 0.95 ± 0.20 (week 2); 1.17 ± 0.16 (week 3) ($p < 0.004$); 1.60 ± 0.40 (week 4)

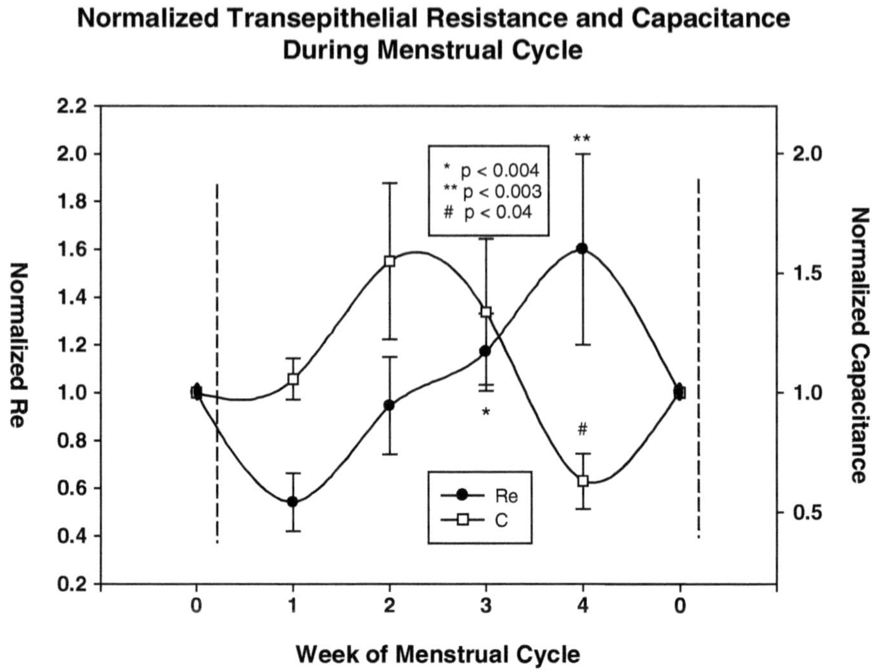

Fig. 3 Normalized impedance parameters Re and C (low frequency) obtained during weeks 1–4 through menstrual cycle. Results expressed as mean ± SEM

($p < 0.003$), and began to fall late in week 4 before menstruation. Normalized *C* reaches a peak between week 2 and 3, and falls during the week 4 with values; 1.06 ± 0.09 (week 1); 1.55 ± 0.33 (week 2); 1.34 ± 0.31 (week 3); 0.63 ± 0.12 (week 4) ($p < 0.04$).

Characteristic Frequency (f_C). In the high frequency region (60 KHz–6 Hz) of the Nyquist plots, the f_C of the epithelium, a property related to morphology and transport function, decreased progressively through weeks 1–4 of the menstrual cycle from 17.3 ± 3.0, 10.9 ± 1.6, 9.7 ± 1.5, 7.0 ± 1.1 Hz, respectively, ($p = 0.003$). In the low frequency region (<6 Hz) of the Nyquist plots, the f_C of the epithelium did not change significantly through weeks 1–4 of the menstrual cycle with values of 3.9 ± 0.3, 3.7 ± 0.3, 3.8 ± 0.3, 3.1 ± 0.3 Hz, respectively ($p = 0.059$).

Depression Angle (σ). The σ of the high frequency arc did not change significantly through weeks 1–4 of menstrual cycle with values of 15.9 ± 0.8, 16.7 ± 0.7, 15.0 ± 0.8, 15.3 ± 0.7°, respectively ($p = 0.482$). The σ of the low frequency arc decreased through weeks 1–4 of menstrual cycle with values of 23.7 ± 1.2, 21.0 ± 1.2, 21.2 ± 1.8, 18.4 ± 1.3, respectively ($p = 0.034$).

Discussion

We have previously demonstrated that BC can be diagnosed noninvasively with DEIS (12), but have not examined whether epithelial impedance varies during menstrual cycle until the present study. Other investigators have noted changes in impedance during menstrual cycle (13, 14). However, these studies by virtue of their surface measurements did not examine the breast epithelia. They measure changes in current density or impedance through the whole breast, or skin impedance, which significantly influence purely surface measurements (6).

The *Re* and *C* of the high and low frequency components were not normally distributed and showed significant variation between subjects. However, within the same subject *Re* tended to increase during cycle and *C* peaked between week 2 and 3 of the cycle. Figure 3 demonstrates the effect of normalizing the *Re* and *C* so that variability between subjects does not influence the observed cyclical change in impedance measured through multiple menstrual cycles in different individuals.

This study demonstrates that the electrical properties of the breast epithelium are cyclical during a 28-day period, and may represent morphological changes in breast epithelia associated with menstrual cycle. The increase in *Re* during cycle likely represents an increase in the thickness of the epithelium, which is under cyclical hormonal control, with the decrease representing a thinning of the ductal lining and a break-down of the tight-junctions before menstruation. *C*, which is related to the epithelial cell mass, increased close to the proliferative peak, and also fell at the time of menstruation. The f_C and σ of the impedance curves were also observed to change during menstrual cycle. These parameters may provide independent information about the morphology of breast epithelia during menstrual cycle (15). For example it has been suggested that suppression of the locus of the semicircular arc in the Nyquist plots may be related to the morphology of the extracellular space and cytoskeleton (15). The vacuolation, inflammation, apoptosis, and breakdown of tight junctions observed in breast epithelium before menstruation (1) may explain the decrease in the low frequency σ observed in this study.

Patients at risk for BC may have an aberrant epithelial response to estrogen and progesterone with persistent proliferative morphology throughout menstrual cycle (4). The observation that physiological changes in breast epithelial morphology may be monitored using an electrophysiological approach suggests that abnormal proliferative changes (i.e., hyperplasia or ADH), may also be identified using a similar technique. Noninvasive electrical measurements of the epithelium, using DEIS, may represent a novel biomarker for assessing BC risk by evaluating the electrical correlates of abnormal epithelial morphology and how it varies during menstrual cycle.

Acknowledgments The authors acknowledge and thank the women who volunteered their time for participation in this study, and the preparation of the figures by CA Davies.

References

1. Ramakrishnan R, Khan SA, Badve S (2002) Morphological changes in breast tissue with menstrual cycle. Mod Pathol 15(12):1348–56.
2. Khan SA, Rogers MA, Obando JA, et al. (1994) Estrogen receptor expression of benign breast epithelium and its association with breast cancer. Cancer Res 54(4):993–7.
3. Khan SA, Yee KA, Kaplan C, et al. (2002) Estrogen receptor alpha expression in normal human breast epithelium is consistent over time. Int J Cancer 102(4):334–7.
4. Khan SA, Sachdeva A, Naim S, et al. (1999) The normal breast epithelium of women with breast cancer displays an aberrant response to estradiol. Cancer Epidemiol Biomarkers Prev 8(10):867–72.
5. Khan SA, Rogers MA, Khurana KK, et al. (2000) Oestrogen receptor expression in normal breast epithelium. Eur J Cancer 36:S27–S28.
6. Ultchin Y, Nachaliel U, Ori A (2002) Indirect calculation of breast tissue impedance values. Physiol Meas 23(1):177–82.
7. Cuzick J, Holland R, Barth V, et al. (1998) Electropotential measurements as a new diagnostic modality for breast cancer. Lancet 352(9125):359–63.
8. Hope TA, Iles SE (2004) Technology review: the use of electrical impedance scanning in the detection of breast cancer. Breast Cancer Res 6(2):69–74.
9. Malich A, Bohm T, Facius M, et al. (2001) Additional value of electrical impedance scanning: experience of 240 histologically-proven breast lesions. Eur J Cancer 37(18):2324–30.
10. Stojadinovic A, Nissan A, Gallimidi Z, et al. (2005) Electrical impedance scanning for the early detection of breast cancer in young women: preliminary results of a multicenter prospective clinical trial. J Clin Oncol 23(12):2703–15.
11. Zou Y, Guo Z (2003) A review of electrical impedance techniques for breast cancer detection. Med Eng Phys 25(2):79–90.
12. Davies RJ, Davisson TH (2005) Epithelial impedance spectroscopy: a new approach to breast cancer diagnosis. Breast Cancer Res Treat 94(S87):2012.
13. Hartov A, Soni NK, Paulsen KD (2004) Variation in breast EIT measurements due to menstrual cycle. Physiol Meas 25(1):295–9.
14. Perlet C, Kessler M, Lenington S, et al. (2000). Electrical impedance measurement of the breast: effect of hormonal changes associated with the menstrual cycle. Eur Radiol 10(10): 1550–4.
15. Ivorra A, Genesca M, Sola A, et al. (2005) Bioimpedance dispersion width as a parameter to monitor living tissues. Physiol Meas 26(2):S165–S173.

Is the Mannose-6-Phosphate/Insulin-Like Growth Factor 2 Receptor Coded by a Breast Cancer Suppressor Gene?

Guy Joseph Lemamy, Majida Esslimani Sahla, Marie Laurence Berthe, and Pascal Roger

Summary The multifunctional growth factor mannose-6-phosphate/insulin-like growth factor 2 receptor (M6P/IGF2-R) binds proteins sharing M6P signals, including cathepsins and IGF2. It is involved in targeting newly synthesized mannose-6-phosphorylated lysosomal enzymes, activating transforming growth factor β (TGFβ), and neutralising the mitogen IGF2 by transporting it to lysosomes. The M6P/IGF2-R was proposed as being coded by a tumor suppressor gene. We measured gene expression at the protein level by quantitative immunohistochemistry, using chicken high affinity IgY antibodies directed against human M6P/IGF2-R. Chicken immunization was performed with human purified M6P/IGF2-R, and IgY antibodies were extracted from egg yolk by polyethylene glycol precipitation method. The biosensor analysis showed that IgY antibodies bind M6P/IGF2-R with high affinity (K_d = 7.5 nM). Quantitative immunohistochemical studies in sections from invasive breast carcinoma and ductal carcinoma in situ (DCIS) indicated various levels (from 5 to 400 units) of the M6P/IGF2-R protein, which did not correlate with tumor size, histological grade, estrogen and progesterone receptors. Moreover, the M6P/IGF2-R level was increased in DCIS relative to adjacent normal tissue ($p < 0.005$) and then decreased in invasive carcinoma compared with DCIS ($p < 0.02$). The hypothesis of tumor suppressor gene is not supported by these studies. However, it is not excluded for a small proportion of the tumors. Its assay might help to complement the cathepsin D assay to predict breast cancer prognosis and physiopathology.

Introduction

The M6P/IGF2-R is a multifunctional receptor that binds proteins sharing mannose-6-phosphate signals and IGF2 (1). It is involved in the routing of newly synthesized mannose-6-phosphorylated lysosomal enzymes, (including several cathepsins), in degradation of the mitogen IGF2 by transporting it to lysosomes, in activation of transforming growth factor beta (TGFβ), a potent inhibitor for most cell types (2). In mammary cancer cell lines, procathepsin D secretion is increased (3, 4), suggesting an alteration of *M6P/IGF2-R* gene, mapped to chromosome $6q_{25-27}$ (5).

J.J. Li et al. (eds.), *Hormonal Carcinogenesis V.*

Clinical studies of *M6P/IGF2-R*, performed at the gene level, have shown frequent loss of heterozygosity (LOH) and mutations in the remaining allele in human cancers such as hepatocarcinomas (6), breast (7, 8), and ovarian (9) cancers. Therefore, this gene has been proposed as tumor suppressor (8), and consequently it could be expected to have low protein levels. Since the *M6P/IGF2-R* is highly conserved throughout several species including humans (60–90% homology), we choose chicken as host to produce antibodies against human liver and placenta receptors on the basis of their ability to mount a vigorous humoral response against highly conserved proteins in mammals (10). To assess the potential value of *M6P/IGF2-R* gene expression as a breast cancer (BC) prognostic biomarker according to the tumor suppressor gene hypothesis, we measured the *M6P/IGF2-R* gene expression at the protein level in two studies. The first, in 40 invasive BC (11), and the second, in 42 benign breast diseases (BBD), 61 ductal carcinoma in situ (DCIS), and 133 invasive BC (12) by quantitative immunohistochemistry (IHC) using high affinity chicken antibodies directed against human purified M6P/IGF2-R and used in clinical cancer studies (11).

Methodology

Human Samples (University Hospitals of Montpellier and Nîmes). The first study included 40 invasive BC collected by mastectomy. The second one included 95 patients with BBD or CIS (95) from a multicenter prospective study (University Hospitals of Montpellier and Nîmes). The patients were 30- to 79-year old, with a median of 51 years. The tissues were collected during breast surgery for diagnosis/therapeutic purposes and embedded in paraffin. The invasive BC samples consisted of 133 tissue sections collected in liquid N. Frozen serial sections were IHC assayed. Among them, 41 contained normal mammary duct structures. Only breast tumors without clinical distant metastasis were included. The patients were 29–88 years with a median of 56 years. Written consent was obtained for all patients. The studies were approved by the local ethical committee.

Immunohistochemistry. M6P/IG2-R quantitative IHC was performed using purified chicken polyclonal IgY-415 directed against human M6P/IGF2-R previously produced from egg's yolk and characterized (11). Antibody specificity and the validity of the IHC were determined by Western blot analysis and incubation with purified receptor, respectively (11). M6P/IGF2-R IHC assays in frozen and paraffin sections were performed as previously described (11, 12). Staining specificity was tested using nonspecific IgY from nonimmunized hen.

IHC Quantification. A computerized image analyser SAMBA 2005 (TITN Alcatel, Grenoble, France) was used to quantify M6P/IGF2-R, as previously described (13). The results were expressed by a quantitative IHC (QIC) score = [(% of surface stained in epithelial cells) × (mean staining intensity) × 10] expressed in arbitrary units (AU). M6P/IGF2-R staining was quantified in the lesions and adjacent normal ducts and lobules. Reproducibility of M6P/IGF2-R staining among experiments was evaluated on MDA-MB231 cell pellets and found to be slightly variable within a QIC score range of 100–150 AU for the frozen section and 80–100 AU for the paraffin-embedded sections.

Statistical Analysis. The nonparametric Wilcoxon test was used for the paired samples. Correlations between the former variables were estimated using the Spearman test. The p value significance was set at 0.05.

Results

In the presence of the negative control antibody, no brown staining was observed (Fig. 1a).

In the first study (11), in 10 out of 21 breast cancer tissue containing normal mammary duct structures, breast cancer cells had a lower M6P/IGF2-R concentration than normal peritumoral ducts. The staining was found mostly in the perinuclear region and corresponded to the Golgi network (11). In the second study (12), the same Golgi network staining was also observed in frozen sections. The same pattern was observed using paraffin sections of mammary lesions, with cellular cytoplasmic staining being observed in both epithelial and stromal cells (Fig. 1b–e). In normal lobules (Fig. 1d) and ducts (Fig. 1e), staining was found in cytoplasm and mostly apical in epithelial polarised cells and also present in some of the stromal cells (Fig. 1b–e). In normal lobules (Fig. 1d) and ducts (Fig. 1e), staining was found in cytoplasm, mainly apical in epithelial polarised cells, and also in some of the stromal cells. The extent and staining intensity in epithelial cells were quantified using a QIC score and compared with QIC scores in adjacent normal tissues.

In the first study, 10/21 BC tissue contained normal mammary ducts. Since there was a large variation in the M6P/IGF2-R levels according to individual patient samples, in the same study and in the same patient, we also compared the receptor level in the lesion and the normal adjacent glands in the attempt to detect variations due to the neoplastic process. Figure 2 depicts the M6P/IGF2-R levels in benign lesions and adjacent normal glands. These values varied according to individual patient samples and were highly correlated using the Spearman test ($r = 0.675$; $P > 0.0001$).

There was no influence of the type of lesion on the value in adjacent normal tissues. In a limited number of patients, however, as shown in Fig. 2, there was a strong variation (~2.0-fold) in expression in the lesion compared with the normal glands. The intensity of the staining in the lesions was substantially increased in a few patients with both benign and malignant lesions, and mainly in high-grade DCIS. By contrast, a large decreased level of staining was observed in the cancerous tissues, in 10% of cases with CIS and 24% of invasive BC. This decrease suggested a late alteration of M6P/IGF2-R in cancer cells and was compatible with the tumor suppressor gene hypothesis.

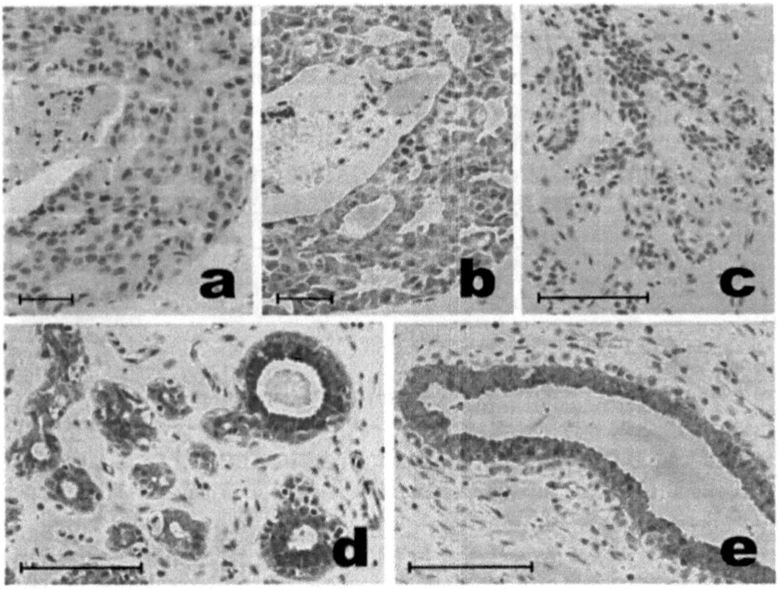

Fig. 1 IHC of M6P/IGF2-R in paraffin embedded sections. Specificity of IHC staining of the M6P/IGF2-R. Adjacent sections from a high-grade DCIS were treated with M6P/IGF2-R 415 IgY (b) or with a nonspecific IgY (**a**). Strong M6P/IGF2-R staining was observed in the cytoplasm of a high-grade DCIS. (**b**) Notice the weak staining in adjacent normal glands of the same section. (**c**). Normal lobules (**d**) and ducts (**e**) at the periphery of lesions from individual patients showing strong staining of the apical pole of the epithelial cells and staining in stromal cells. Bar = 50 μm. Reproduced by permission from (12)

Discussion

The results obtained by quantitative IHC performed in sections from benign breast diseases, in situ carcinomas, and invasive carcinomas showed various levels of M6P/IGF2-R (from 5 to 400 units), which did correlate with others prognosis factors such as tumor size, histological grade, ER and PR (data not shown). The M6P/IGF2-R concentration was markedly increased or decreased relative to normal tissue in some lesions with a significant proportion of invasive BCs showing a decrease of expression.

We found that the M6P/IGF2-R level was significantly increased in DCIS relative to adjacent normal tissue ($p < 0.005$) and decreased in invasive BC compared with DCIS ($p < 0.02$). May be because of the limited number of samples, our data do not support the hypothesis of a frequent and earlier inactivation of *M6P/IGF2-R* gene in breast (7) or liver (14) cancers. A similar increase of *M6P/IGF2-R* has been reported in human primary thyroid neoplasms (15). Clinical BC studies performed at the gene level did not show any gene amplification (16), although RNA levels were increased

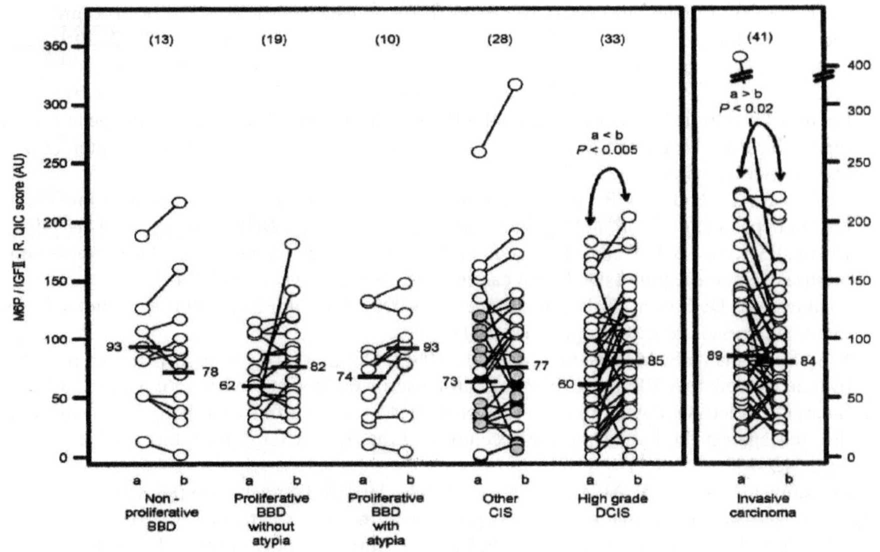

Fig. 2 Comparison of M6P/IGF2-R level between adjacent normal breast tissue (**a**) and lesions (**b**). *First panel*: Comparison of M6P/IGF2-R level between normal adjacent glands and noninvasive lesions in paraffin sections. Benign breast disease (BBD), high-grade DCIS: Other carcinomas in situ include lobular CIS, low-grade, and intermediate DCIS. *Second panel*: Comparison of M6P/IGF2-R level between normal adjacent glands and invasive BCs in the frozen sections. The number of samples is in brackets. Bar: median value. *P* value between a and b according to the paired nonparametric Wilcoxon test is indicated when significantly increased (high-grade DCIS) or decreased (invasive BCs)

(17). The various levels of M6P/IGF2-R expression may be due to several mechanisms, such as different regulation of gene or protein expression stimulated by E2 or IGFs (18); down-regulation by E2 (19); gene alteration leading to loss of heterozygosity (6, 7); somatic gene deletion on one allele, since point mutations of *M6P/IGF2-R* gene are rare events in BCs (20). These pathways may be worth investigating by basic and clinical studies on a large number of patients. To determine the prognostic significance of the M6P/IGF2-R in mammary carcinogenesis, it may be useful to complement this assay with the expression of cathepsin D, a prognostic marker for BC metastasis (21), since the RNA levels were found to correlate in breast tumors (17).

References

1. Morgan DO, Edman JC, Standring DN, et al. (1987). Insulin-like growth factor 2 receptor as a multifunctional binding protein. Nature 329: 301–307.
2. Dennis PA, Rifkin DB (1991). Cellular activation of latent transforming growth inhibitor β requires binding to the cation-independent mannose-6-phosphate/insulin-like growth factor 2 receptor. Proc. Natl. Acad. Sci. USA 88: 580–584.

3. Rochefort H, Capony F, Garcia M, et al. (1987). Estrogen-induced lysosomal proteases secreted by breast cancer cells. A role in carcinogenesis. J. Cell Biochem. 35: 17–29.
4. Capony F, Rougeot C, Montcourrier P, et al. (1989) Increased secretion altered processing and glycosylation of pro-cathepsin D in human mammary cancer cells. Cancer Res. 49: 3904–3909.
5. Laureys G, Barton GE, Ullrich A, et al. (1988). Chromosomal mapping of the gene for the type 2 Insulin-like growth factor/cation independent mannose-6-phosphate receptor in man and mouse. Genomics 3: 224–229.
6. De Souza AT, Hankins GR, Washington MK, et al. (1995). M6P/IGF2-R gene is mutated in human hepatocellular carcinomas with loss of heterozygosity. Nature Genet. 11: 447–449.
7. Chappel SA, Walsh T, Walker RA, et al. (1997). Loss of heterozygosity at chromosome 6q in preinvasive and early invasive breast carcinomas. Brit. J. Cancer 75: 1324–1329.
8. Hankins GR, De Souza AT, Bentley RC, et al. (1996). M6P/IGF2 receptor: a candidate breast tumor suppressor gene. Oncogene 12: 2003–2009.
9. Saito S, Sirahama S, Matsushima M, et al. (1996). Definition of commonly deleted region in ovarian cancers to a 300-kb segment of chromosome 6q27. Cancer Res. 56: 5586–5589.
10. Stuart CA, Pietrzyk RA, Furlanetto RW, et al. (1988). High affinity antibody from hen's eggs directed against the human insulin receptor and the human IGF1 receptor. Anal. Biochem. 173: 142–150.
11. Lemamy GJ, Roger P, Mani JC, et al. (1999). High-affinity antibodies from hen's-egg yolks against human mannose-6-phosphate/Insulin-like growth factor 2 receptor (M6P/IGF2-R): characterization and potential use in clinical cancer studies. Int. J. Cancer 80: 896–902.
12. Berthe ML, Esslimani Sahla M, Roger P, et al. (2003) Mannose-6-phosphate/Insulin-like growth factor 2 receptor expression levels during the progression from normal human mammary tissue to invasive breast carcinomas. Eur. J. Cancer 39: 635–642.
13. Maudelonde T, Brouillet JP, Roger P, et al. (1992). Immunostaining of cathepsin D in breast cancer: quantification by computerised image analysis and correlation with cytolsolic assay. Eur. J. Cancer 28 A: 1686–1691.
14. Yamada T, De Souza AT, Finkelstein S, et al. (1997). Loss of the gene encoding mannose-6-phosphate/inslin-like growth factor 2 receptor is an early event in liver carcinogenesis. Proc. Natl. Acad. Sci. (Wash) 94: 10351–10355.
15. Yashiro T, Tsushima T, Murakami H, et al. (1991). Insulin-like growth factor 2 (IGF2)/mannose-6-phosphate receptors are increased in primary human thyroid neoplasms. Eur. J. Cancer 27: 699–703.
16. Hébert E, Herbelin C, Bougnoux P (1994). Analysis of the IGF2 receptor gene copy number in breast carcinoma. Br. J. Cancer 69: 120–124.
17. Zhao Y, Escot C, Maudelonde T, et al. (1993). Correlation between mannose-6-phosphate/IGF2 receptor and cathepsin D RNA levels by in situ hybridisation in benign and malignant mammary tumors. Cancer Res. 53: 2901–2905.
18. Confort C, Rochefort H, Vignon F (1995). Insulin-like growth factors (IGFs) stimulate the release of alpha 1- antichymotrypsin and soluble IGF-II/mannose 6-phosphate receptor from MCF7 breast cancer cells. Endocrinology 136: 3759–3766.
19. Mathieu M, Vignon F, Capony F, et al. (1991). Estradiol down-regulates the mannose-6-phosphate/insulin-like growth factor 2 receptor gene and induces cathepsin D in breast cancer cells: a receptor saturation mechanism to increase the secretion of lysosomal pro-enzymes. Mol. Endocrinol. 5: 815–822.
20. Rey JM, Theillet C, Brouillet JP, et al. (2000). Stable amino-acid sequence of the mannose-6-phosphate/Insulin-like growth factor 2 receptor in ovarian carcinomas with loss of heterozygosity in breast cancer cell lines. Int. J. Cancer 85: 466–473.
21. Spyratos F, Brouillet JP, Defrenne A, et al. (1989). Cathepsin D: an independent prognostic factor for metastasis of breast cancer. The Lancet 11: 1115–1118.

Her-2/*neu*-induced "Cytokine Signature" in Breast Cancer

Alejandro Vazquez-Martin, Ramon Colomer, and Javier A. Menendez

Summary Identification of genes/proteins that are differentially expressed in HER2-overexpressing breast carcinomas (BC) is essential in elucidating the mechanistic basis of their increased metastastic potential. With the goal of identifying a unique HER2-induced "cytokine signature" in BC, Human Cytokine Array III (RayBiotech, Inc.) simultaneously detecting 42 cytokines and growth factors on one membrane was used to determine the profile of cytokines in conditioned media obtained from MCF-7/Her2–18 cells, a MCF-7-derived clone engineered to stably express the full-length human HER2 cDNA, and from the MCF-7/neo control sub-line. We identified two inflammatory and proangiogenic CXC chemokines with at least a 10-fold increased expression in MCF-7/Her2–18 transfectants when compared with matched control MCF-7/neo cells: CXCL8 (IL-8; interleukin-8) and CXCL1 and (GRO; growth-related oncogene). HER2 up-regulation of IL-8 and GRO was validated by ELISA and further confirmed by switching off the HER2 signaling. Treatment with the tyrosine kinase inhibitor gefitinib (Iressa™) returned the expression levels of IL-8 and GRO back to the baseline observed in HER2-negative MCF-7 BC cells. Moreover, IL-8 and GRO circulating levels were significantly higher in sera from HER2-positive BC patients. These findings reveal for the first time that (a) Enhanced synthesis and secretion of members of the IL-8/GRO chemokine family, which have recently been linked to estrogen receptor (ER) inaction, increased cell invasion, and angiogenesis, may represent a new pathway involved in the metastatic progression and endocrine resistance of HER2-overexpressing breast carcinomas; and (b) Circulating levels of IL-8 and GRO cytokines may represent novel biomarkers monitoring BC responses to endocrine treatments and/or HER2-targeted therapies.

Introduction

The *HER2* oncogene codes for the transmembrane tyrosine kinase receptor p185$^{\text{Her-2/}neu}$ and, at present, represents one of the most important oncogenes in breast cancer (BC) (1–13). Aberrant expression of *HER2* triggers the activation of multiple downstream signal transduction pathways, including the PI3′-K)/AKT/PTEN

J.J. Li et al. (eds.), *Hormonal Carcinogenesis V.*

pathway and the Ras/Raf/MAPK pathway, which are essential in inducing increased cell proliferation and differentiation, decreasing apoptosis, and enhancing tumor cell motility and angiogenesis. Although these signaling pathways emanating from HER2 have been extendedly characterized, much less is known about the specific genes/proteins regulated by HER2 that contribute to its tumorigenic effects (14–20).

Alteration of cytokine levels is associated with cancer progression, response to chemotherapy, and metastatic status, and are emerging as potential factors that may contribute to key autocrine or paracrine loops in BC etiology and metastatic phenotype (21–23). With the goal of identifying those cytokines playing key roles in HER2-driven human BC progression, we herein took advantage of the recently developed RayBio™ Human Cytokine Array III that simultaneously detects 42 cytokines and growth factors on one membrane. Using conditioned media from MCF-7 BC cells, which endogenously express low levels of HER2, before and after reexpression of HER2 (i.e., MCF-7/*neo* and MCF-7/Her2–18 cells, respectively) as well as sera obtained retrospectively from metastatic BC patients, we present data to suggest that enhanced protein synthesis and secretion of CXCL8 (IL-8; interleukin-8) and CXCL1 (GRO; growth-related Oncogene), two members of the CXC chemokine family, which have recently been linked to estrogen receptor (ER) inaction, increased cell invasion and angiogenesis (24–31), may represent a new pathway involved in the metastatic progression of HER2-positive BC disease.

Results

In-Vitro Identification of IL-8 and GRO as HER2-Related BC Cytokines Using Chemokine Antibody Array Technology. We first applied cytokine antibody array system to identify the “key cytokine(s)” associated with HER2-driven BC progression (Fig. 1). By using a RayBio™ Human Cytokine Array III (Fig. 2), we simultaneously screened the expression of 42 cytokines when BC cells naturally expressing physiological levels of HER2 (i.e., MCF-7 BC cells) were engineered to overexpress *HER2* gene (i.e., MCF-7/Her2–18 transfectants, which are known to express 45 times the level of HER2 than parental MCF-7 cells or the MCF-7/*neo* control subline).

We found that solely four different cytokines were significantly up-regulated in MCF-7 cells engineered to overexpress HER2. Thus, HER2 overexpression induced a dramatic >10.0-fold increase in the expression of IL-8 and GRO, while inducing a highly significant 5.0–10.0-fold increase in the expression of GRO-α, one of the 3 GRO isotypes. A noteworthy up-regulation of VEGF, a well-characterized proangiogenic effect driven by HER2 overexpression (8–10), was confirmed in our experimental system.

Pharmacological Inhibition of HER2-Driven Cellular Signaling Knock-Down IL-8 and GRO Expression. There was a clear correlation between the

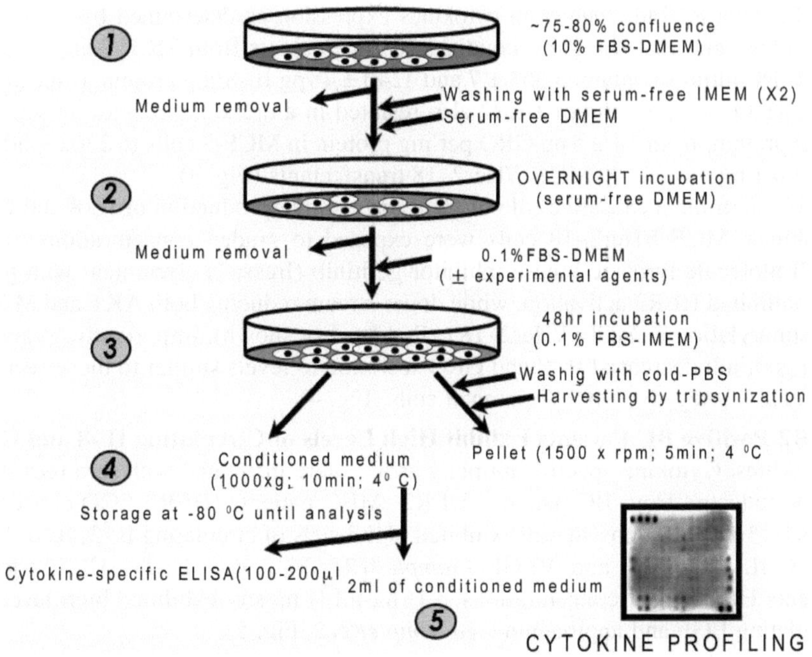

Fig. 1 Preparation of conditioned medium (CM) from human BC cells. MCF-7 BC cells stably overexpressed HER2 oncogene (MCF-7/Her2–18 clone) and the matched control MCF-7/*neo* cells were kindly provided by Prof. Mien-Chie Hung (The University of Texas, M. D. Anderson Cancer Center, Houston, TX, USA)

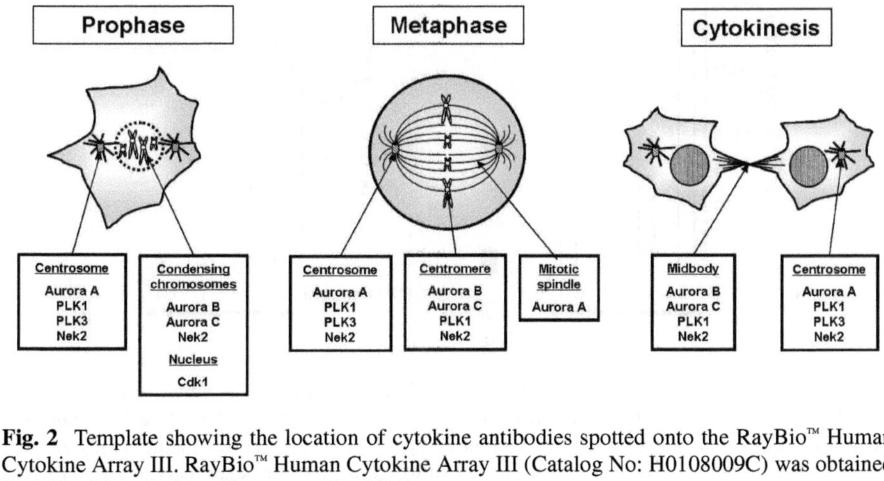

Fig. 2 Template showing the location of cytokine antibodies spotted onto the RayBio™ Human Cytokine Array III. RayBio™ Human Cytokine Array III (Catalog No: H0108009C) was obtained from RayBiotech, Inc. (Norcross, GA, USA)

HER2-induced fold-changes in cytokines expression as determined by both array technology and ELISA. Thus, conditioned supernatants from MCF-7 and MCF-7/Her2–18 cultures contained 353 ± 7 and 4240 ± 48 pg IL-8/mg protein, respectively (Fig. 4). Overexpression of *HER2* also resulted in a dramatic increase in secreted GRO protein, from 14 ± 5 pg GRO per mg protein in MCF-7 cells to 2302 ± 300 pg GRO per mg protein in MCF-7/Her2–18 transfectants (Fig. 3).

To determine a causative role of HER2 in the overproduction of IL-8 and GRO cytokines, MCF-7/Her2–18 cells were exposed to graded concentrations of the small molecule tyrosine kinase inhibitor gefitinib (Iressa™). Treatment with gefitinib inhibited HER2 activation, while down-stream reducing both AKT and MAPK phosphorylation in MCF-7/Her2–18 cells (data not shown). Importantly, treatment with gefitinib decreased IL-8 and GRO secretion to levels similar to those secreted from HER2-negative MCF-7 parental cells (Fig. 4).

HER2-Positive BC Patients Exhibit High Levels of Circulating IL-8 and GRO Cytokines. Cytokine-specific antibody arrays were incubated with sera retrospectively collected from BC patients. HER2^{+} MBC patients (HER2 ECD concentrations ≥ 15 ng ml^{-1}) consistently exhibited high levels of circulating IL-3, IL-6, IL-8, IL-13, IL-15, GRO, and VEGF (*Sample#121*, Fig. 5), whereas HER2^{-} MBC patients HER2 ECD concentrations < 15 ng ml^{-1}) mostly exhibited high levels of circulating EGF and angiogenin (*see Sample#212*, Fig. 5).

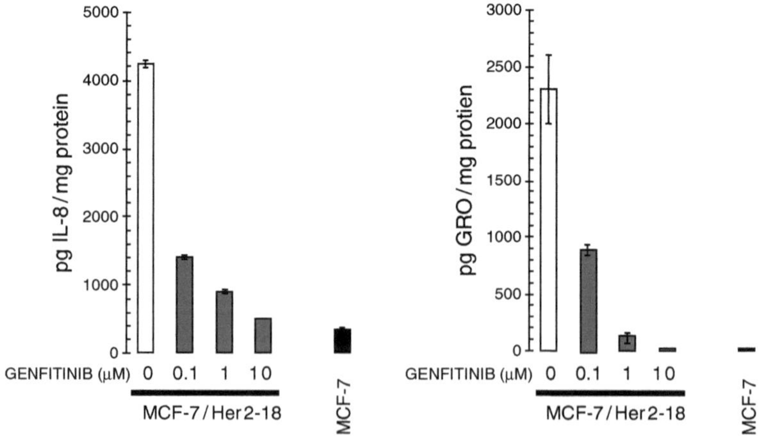

Fig. 3 Detection and modulation of cytokines secreted from MCF-7 cells by overexpression of HER2. *Top panels*: 48 h CM prepared from an equal number of MCF-8/Her2–18 cells (*right*) and MCF-7/*neo* cells (*left*) was assayed for cytokine content by the protein array methodology. *Bottom panels*: Densitometric data were arbitrarily expressed as red for extremely high (>10-fold increase) when compared with those found in MCF-7/*neo* matched control cells

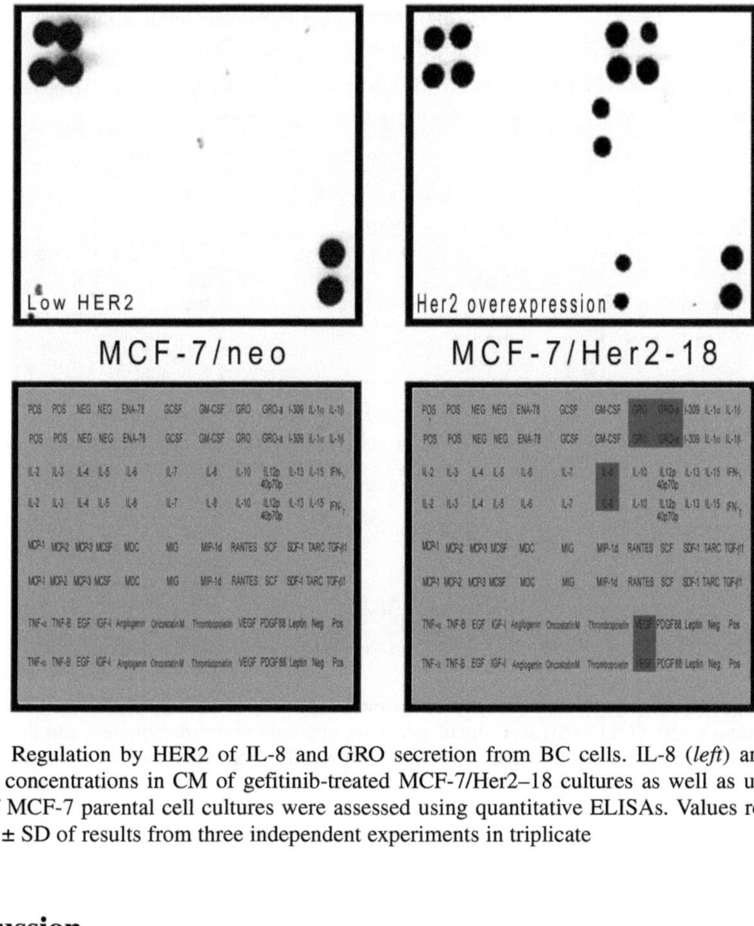

Fig. 4 Regulation by HER2 of IL-8 and GRO secretion from BC cells. IL-8 (*left*) and GRO (*right*) concentrations in CM of gefitinib-treated MCF-7/Her2–18 cultures as well as untreated CM of MCF-7 parental cell cultures were assessed using quantitative ELISAs. Values represent means ± SD of results from three independent experiments in triplicate

Discussion

Cytokines are emerging as potential factors that could contribute to the progression of BC (21–23). Because HER2-overexpressing cancer cells display distinct phenotypes to those observed in HER2-negative cancer cells, we here hypothesized that such phenotypes may be a result of differential expression of inflammatory and proangiogenic chemokines (21–32). Using a recently developed proteomic technique capable to simultaneously detect expression levels of multiple cytokines, we here describe that HER2 reexpression in BC cells endogenously expressing low levels of HER2 (i.e., MCF-7 → MCF-7/Her2–18 transition) leads to the specific up-regulation of solely IL-8 and CXCL1 GRO, two members of the CXC family of chemokines. Given that gefitinib (Iressa™)-induced blockade of HER2 signaling returned IL-8 and GRO expression levels back to the baseline observed in parental HER2-negative MCF-7 cells, it is reasonable to suggest that hypersecretion of IL-8 and GRO is a previously unrecognized molecular feature that specifically accompanies HER2-enhanced metastatic phenotype in BC disease.

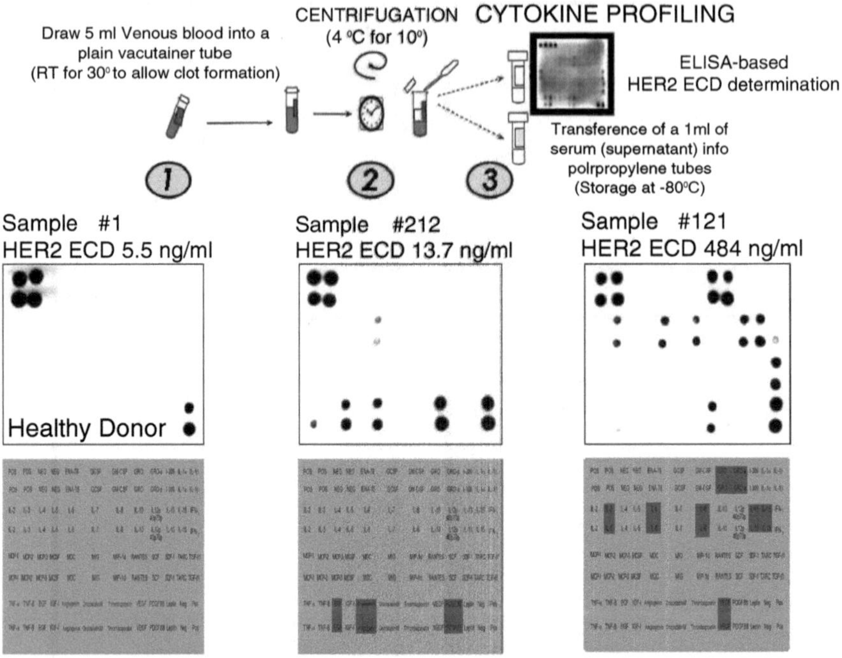

Fig. 5 Detection of cytokine expression from metastatic BC (MBC) patient's sera. Undiluted MBC patient's sera (1 ml/each) were incubated with cytokine array membranes, and the signals detected as described. Densitometric data were expressed as darkened areas = extremely high (>10-fold increase) when compared with those found in healthy donor's sera

Young et al. were pioneers observing that BC cells, especially ER$^+$ MCF-7 cells, can respond to IL-8 and GRO chemolines and suggested a potential role for these molecules in the process of tumor cell migration, invasion, and metastasis (32). Lin et al. recently identified IL-8 as a key factor involved in BC invasion and angiogenesis, but not in BC proliferation (28). Interestingly, IL-8 overexpression in invasive BC cells inversely correlated with ER status. Thus, ER$^+$ BC cells expressed low levels of IL-8, IL-8 overexpression naturally occurred in ER$^-$ BC cells, and exogenous expression of ER in ER$^-$ cells decreased IL-8 levels (27–30). Li and Sidell determined that highly invasive and ER$^-$ MDA-MB-231 BC cells secrete a number of cytokines known to regulate cellular growth and motility (31). One such cytokine, GRO, which was previously not described in BC, was found to mediate the invasive potential of MDA-MB-231 cells. Similarly to IL-8, GRO was not involved in the proliferation rate of metastatic BC cells. These findings, altogether, strongly suggest that HER2-induced "cytokine signature" in ER$^+$ MCF-7/Her2–18 cells appears to be molecularly equivalent to that observed in ER$^-$ BC cells. Signal transduction mediated by HER2 can partially overcome the estrogen dependence of "ER$^+$" BC cells for growth and that HER2 overexpression confers a selective advantage to such cell in the absence of estrogen (33, 34). Indeed, increased ER-HER2 cross-talk has been recognized as a main molecular mechanism underlying tamoxifen (TAM) resistance in patients receiving adjuvant TAM whose tumors express high levels of both HER2 and ER, while treatments such as gefitinib blocking receptor cross-talk efficiently restore TAM antitumor effects (35, 36). Future research should elucidate whether or not IL-8 and GRO chemokines actively contribute to the process of HER2-promoted estrogen-independence and antiestrogen-resistance in BC disease.

Despite the small sample size of patients' sera profiled in this study, it was obvious that IL-8, GRO, and GROα circulating levels were significantly higher in HER2-positive MBC patients when compared with those found in HER2-negative patients. If enhanced synthesis and secretion of members of the IL-8/GRO chemokines family, which have recently been linked to ER inaction, increased cell invasion and angiogenesis (24–31), in fact represents a new pathway involved in the metastatic progression and endocrine resistance of HER2-overexpressing BCs, circulating levels of IL-8/GRO cytokines may represent novel biomarkers monitoring BC responses to endocrine treatments and/or HER2-targeted therapies. Moreover, considering that neutralization of IL-8 or GRO by functional antibodies has been found to specifically block cell migration and invasion of metastatic BC cell lines in vitro (28, 30, 31), development of therapeutic molecules targeting IL-8/GRO chemokines and/or human IL-8/GRO receptors may represent a novel avenue in the management of $HER2^+$ BCs. Our study further supports the notion that comparative cytokine mapping of conditioned media from tumor-derived BC cell lines and sera from BC patients should represent a valuable discovery tool to identify potential targets involved in BC progression.

References

1. Slamon DJ, Clark GM, Wong SG, et al. (1987) Human breast cancer: correlation of relapse and survival with amplification of the HER-2/neu oncogene. Science 235:177–182.
2. Slamon DJ, Godolphin W, Jones LA, et al. (1989) Studies of the HER-2/neu proto-oncogene in human breast and ovarian cancer. Science 244:707–712.
3. Di Fiore PP, Pierce JH, Kraus MH, et al. (1987) erbB-2 is a potent oncogene when overexpressed in NIH-3T3 cells. Science 237:78–182.
4. Pierce JH, Arnstein P, Di Marco E, et al. (1991) Oncogenic potential of erbB-2 in human mammary epithelial cells. Oncogene 6:1189–1194.
5. DiGiovanna MP, Chu P, Davison TL, et al. (2002) Active signaling by HER-2/neu in a subpopulation of HER-2/neu-overexpressing ductal carcinoma in situ: clinicopathological correlates. Cancer Res 62:6667–6673.
6. Xu R, Perle MA, Inghirami G, et al. (2002) Amplification of Her-2/neu in Her-2/neu-overexpressing and nonexpressing breast carcinomas and their synchronous benign, premalignant, and metastatic lesions detected by FISH in archival material. Mod Pathol 15:116–124.
7. Alroy I, Yarden Y (2000) Biochemistry of HER2 oncogenesis in breast cancer. Breast Dis 11:31–48.
8. Yarden Y, Sliwkowski MX (2001) Untangling the ErbB signalling network. Nat Rev Mol Cell Biol 2:127–137.

9. Harari D, Yarden Y (2000) Molecular mechanisms underlying ErbB2/HER2 action in breast cancer. Oncogene 19:6102–6114.
10. Yarden Y (2001) Biology of HER2 and its importance in breast cancer. Oncology 61 (Suppl 2): 1–13.
11. Yu D, Hung M-C (2000) Role of erbB2 in breast cancer chemosensitivity. BioEssays 22:673–680.
12. Pietras RJ, Arboleda J, Reese DM, et al. (1995) HER-2 tyrosine kinase pathway targets estrogen receptor and promotes hormone-independent growth in human breast cancer cells. Oncogene 10:2435–2446.
13. Alaoui-Jamali MA, Paterson J, Al Moustafa A.E, Yen L (1997) The role of ErbB2 tyrosine kinase receptor in cellular intrinsic chemoresistance: mechanisms and implications. Biochem Cell Biol 75:315–325.
14. Wilson KS, Roberts H, Leek R, et al. (2002) Differential gene expression patterns in HER2/*neu*-positive and -negative breast cancer cell lines and tissues. Am J Pathol 161:1171–1185.
15. Mackay A, Jones C, Dexter T, et al. (2003) cDNA microarray analysis of genes associated with ErbB2 (HER2/*neu*) overexpression in human mammary luminal epithelial cells. Oncogene 22:2680–2688.
16. White SL, Gharbi S, Bertani MF, et al. (2004) Cellular responses to ErbB-2 overexpression in human mammary luminal epithelial cells: comparison of mRNA and protein expression. Br J Cancer 90:173–181.
17. Kumar-Sinha C, Ignatoski KW, Lippman ME, et al. (2003) Transcriptome analysis of HER2 reveals a molecular connection to fatty acid synthesis. Cancer Res 63:132–139.
18. Gharbi S, Gaffney P, Yang A, et al. (2002) Evaluation of two-dimensional differential gel electrophoresis for proteomic expression analysis of a model breast cancer cell system. Mol Cell Proteomics 1:91–98.
19. Zhang D-H, Tai LK, Wong LL, et al. (2005) Proteomics of breast cancer: enhanced expression of cytokeratin19 in HER-2/*neu*-positive breast tumors. Proteomics 5:1797–1805.
20. Zhang D, Tai LK, Wong LL, et al. (2005) Proteomic study reveals that proteins involved in metabolic and detoxification pathways are highly expressed in HER-2/neu-positive breast cancer. Mol Cell Proteomics 4:1686–1696.
21. Walser TC, Fulton AM (2004) The role of chemokines in the biology and therapy of breast cancer. Breast Dis 20:137–143.
22. Moore MA (2001) The role of chemoattraction in cancer metastases. BioEssays 23:674–676.
23. Murphy PM (2001) Chemokines and the molecular basis of cancer metastasis. N Engl J Med 345:833–835.
24. Benoy IH, Salgado R, Van Dam P, et al. (2004) Increased serum interleukin-8 in patients with early and metastatic breast cancer correlates with early dissemination and survival. Clin Cancer Res 10:7157–7162.
25. De Larco JE, Wuertz BR, Rosner KA, et al. (2001) A potential role for interleukin-8 in the metastatic phenotype of breast carcinoma cells. Am J Pathol 158:639–646.
26. Prest SJ, Rees RC, Murdoch C, et al. (1999) Chemokines induce the cellular migration of MCF-7 human breast carcinoma cells: subpopulations of tumour cells display positive and negative chemotaxis and differential in vivo growth potentials. Clin Exp Metastasis 17:389–396.
27. Freund A, Chauveau C, Brouillet JP, et al. (2003) IL-8 expression and its possible relationship with estrogen-receptor-negative status of breast cancer cells. Oncogene 22:256–265.
28. Lin Y, Huang R, Chen L, et al. (2004) Identification of interleukin-8 as estrogen receptor-regulated factor involved in breast cancer invasion and angiogenesis by protein arrays. Int J Cancer 109:507–515.
29. Freund A, Jolivel V, Durand S, et al. (2004) Mechanisms underlying differential expression of interleukin-8 in breast cancer cells. Oncogene 23:6105–6114.
30. Lin Y, Wang SM, Lu WM, Huang RP (2005) Effect of interleukin-8 in cell invasion and proliferation of human breast cancer. Zhonghua Wai Ke Za Zhi 43:1541–1544.

31. Li J, Sidell N (2005) Growth-related oncogene produced in human breast cancer cells and regulated by Syk protein-tyrosine kinase. Int J Cancer 117:14–20.
32. Youngs SJ, Ali SA, Taub DD, Rees RC (1997) Chemokines induce migrational responses in human breast carcinoma cell lines. Int J Cancer 71:257–266.
33. Benz CC, Scott GK, Sarup JC, et al. (1992) Estrogen-dependent, tamoxifen-resistant tumorigenic growth of MCF-7 cells transfected with HER2/neu. Breast Cancer Res Treat 24:85–95.
34. Liu Y, el-Ashry D, Chen D, et al. (1995) MCF-7 breast cancer cells overexpressing transfected c-erbB-2 have an in vitro growth advantage in estrogen-depleted conditions and reduced estrogen-dependence and tamoxifen-sensitivity in vivo. Breast Cancer Res Treat 34:97–117.
35. Shou J, Massarweh S, Osborne CK, et al. (2004) Mechanisms of tamoxifen resistance: increased estrogen receptor-HER2/neu cross-talk in ER/HER2-positive breast cancer. J Natl Cancer Inst 96:926–935.
36. Kurokawa H, Arteaga CL (2001) Inhibition of erbB (HER) tyrosine kinases as a strategy to abrogate antiestrogen resistance in human breast cancer. Clin Cancer Res 7(12 Suppl): 4436–4442.

Involvement of a JAK/STAT Pathway Inhibitor: Cytokine Inducible SH2 Containing Protein in Breast Cancer

Sahra Borgés, Elara Moudilou, Cécile Vouyovitch, Jean Chiesa, Peter Lobie, Hichem Mertani, and Mireille Raccurt

Summary Cytokines and growth factors are responsible for inducing the expression of suppressor of cytokine signaling (SOCS) and cytokine-inducible SH2 containing (CIS) proteins. SOCS and CIS proteins are negative regulators of the JAK/STAT pathway, and exert their physiological effects by suppressing the tyrosine kinase activity of cytokine receptors and inhibiting STAT activation. Growth hormone (GH) is considered as a true cytokine and its local production directly contributes to tumor progression. In an initial study, we have found that CIS expression is increased in human breast cancer in proliferative areas corresponding to high level of GH synthesis. The results of the study presented here confirm the presence of a negative feed back loop in MCF7 cells stably transfected with the hGH gene (MCF-hGH). Real-time PCR analysis showed that gene expression levels of CIS were increased by 80% in MCF-hGH cells as compared to control cell line. Similarly, we have found that the level of CIS gene expression is increased by 50% in primary cultures of human breast cancer, reinforcing the pathophysiological impact of CIS. We previously demonstrated that increasing levels of transfected CIS resulted in strong activation of the mitogen-activated protein (MAP) kinase pathway. Thus, CIS protein has been hypothesized as acting like an activator of the MAPK pathway and an inhibitor of the differentiated cells functions mediated through the JAK/STAT pathway. In the present study, we demonstrate the role of CIS protein in tumor progression in particular its positive effects on cell proliferation and colony formation.

Introduction

Hormones, growth factors, and cytokines although essential to the physiological processes of cellular differentiation and proliferation can however support tumor progression when their synthesis or signaling pathways are deregulated (1). Recent highlights in the understanding of the molecular mechanism controlling these signaling pathways defined three major groups of proteins which play an essential role in transmitting signal from the cell membrane to target gene in the nucleus. Under normal physiological conditions, the Janus kinases (JAK) activation results in the recruitment and phosphorylation of the signal transducer and activator of

J.J. Li et al. (eds.), *Hormonal Carcinogenesis V.*

transcription (STAT) molecules and subsequently induces the synthesis of their endogenous inhibitors, the suppressor of cytokine signaling (SOCS) (2). There is increasing evidence that defective cross talking between these proteins is involved in human disorders including breast cancer (BC).

In a previous work, we demonstrated the expression of human growth hormone (hGH) in the normal human mammary gland epithelium and a largely increased expression of this gene in metastatic BC cells (3). In addition, autocrine production of hGH in human BC cells increased proliferation, decreased apoptosis and altered the cellular morphology (4). Moreover forced expression of hGH in immortalized mammary epithelial cells resulted in their oncogenic transformation (5). The precise mechanism controlling growth hormone (GH) signaling cascade suppression remains unknown even if recent studies have shown that expression of cytokine-inducible SH2 (CIS) gene is strongly and preferentially induced upon GH stimulation (6). CIS is the first identified SOCS family member and a well-known STAT5 target gene. Given the crucial roles of the JAK/STAT signaling pathway in carcinogenesis, SOCS proteins have been suggested to function as a tumor suppressor. Are SOCS/CIS proteins real tumor suppressor? To date, several conflicting results have emerged. Indeed, SOCS1 has been shown to inhibit cellular transformation mediated by several oncogens such as *v-Abl*, *BCR-Abl*, and *HPV-E7* (7). Restoration of SOCS3 expression in nonsmall-cell cancer cells induces apoptosis and suppresses cell growth (8). Conversely, enhanced expression of SOCS1 appears to correlate with tumor invasion, tumor thickness, and tumor stage in human melanoma (9). Similarly, constitutive expression of SOCS1 correlates with shorter survival time and a poor cytogenic response to IFN-α in patients with chronic myeloid leukemia (10). Finally, we demonstrated a consistent overexpression of CIS transcripts in hyper-proliferative areas of human BC and an enhanced mitogen-activated protein kinase (MAPK) activity after transient transfection of CIS expression plasmid in CHO cell line (11) Thus, a coordinated increase in MAPK activation and STAT5 could allow CIS protein to function as a switch promoting proliferation while blocking breast cell differentiation.

To verify this hypothesis, we have first investigated the expression of the different SOCS and CIS proteins in MCF7 cell line stably transfected with an expression plasmid encoding the hGH gene. Then we have demonstrated the capacity of CIS overexpression to promote human breast carcinoma cell progression.

Results

CIS is Overexpressed in Response to "Autocrine" GH. We first examined expression and localization of the different SOCS/CIS proteins by confocal laser scanning microscopy in MCF-hGH transfected cells. These cells synthesized hGH and secreted hGH in the medium (100 pM hGH into 2 ml of media over a 24 h period). In control cell line, the ATG start site in the expression plasmid containing the hGH gene was mutated to TTG, these cells are designated MCF-Control (12).

GH receptor is present in MCF-hGH as well as in MCF-Control cells and it is essential for autocrine GH response. We have used rabbit or goat antibodies raised against human SOCS1, SOCS2, SOCS3, and CIS proteins. No immunofluorescence was observed if the specific antibody was omitted. It is evident from Fig. 1 that the subcellular localization is specific for each SOCS/CIS protein. While SOCS2 immunoreactivity was mainly concentrated in the nucleus, SOCS1, SOCS3, and CIS were detected in a nucleocytoplasmic distribution. By confocal laser scanning microscopy and quantitative analysis of the integrated fluorescence (Fig. 1), we show that no significant difference could be observed in SOCS1 immunoreactivity obtained for the two cell types even if we considered the nuclear and cytoplasmic distribution. For SOCS2, maximal fluorescence intensity was observed in the nucleus but there was no significant difference between MCF-hGH and MCF-Control cells. On the contrary, a significant increase of SOCS3 and CIS fluorescence intensity was observed in MCF-hGH compared to MCF-Control. This increase was limited to 50% for SOCS3 but could achieve more than 250% for CIS fluorescence intensity in the nucleus.

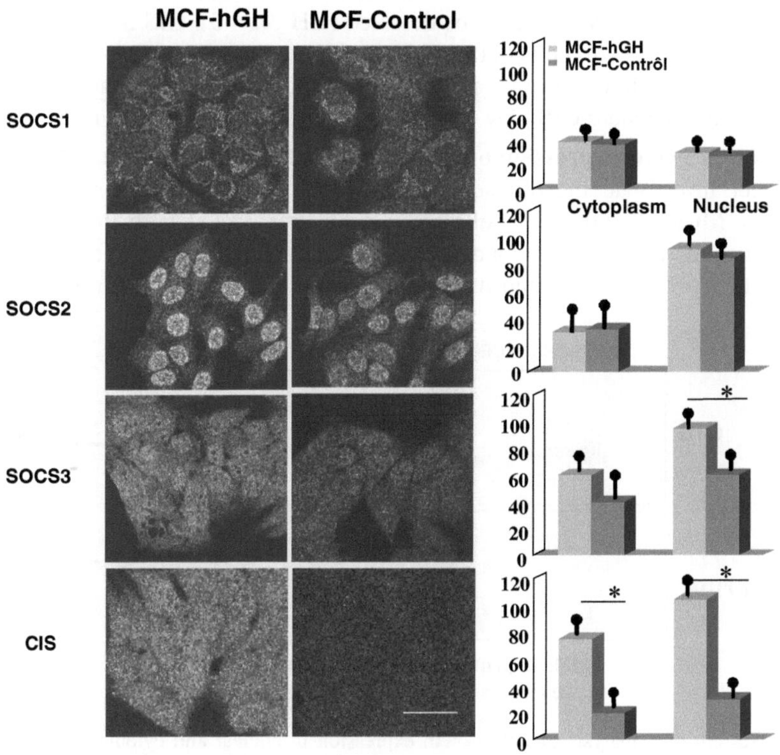

Fig. 1 Confocal laser scanning microscopic detection and quantitative analysis of SOCS1, SOCS2, SOCS3, and CIS immunofluorescence intensity obtained in MCF-hGH and MCF-Control cells

In agreement with these results, Western blot analysis of cytoplasmic and nuclear proteins extracted from MCF-hGH and MCF-Control showed also that CIS protein with an apparent molecular mass of 32 kDa was more abundant in the nuclear extracts of MCF-hGH cells (Fig. 2). In order to determine if expression of the CIS protein was dependant on serum factors, the BC cells were deprived of serum for 12 h before immunofluorescence and western blot analysis.

We then investigated the regulation of CIS transcripts in MCF-hGH and MCF-Control cells by using real-time PCR employing CIS specific primers. Total RNA was isolated from serum deprived confluent cells. As expected, CIS gene expression level was increased by 80% in MCF-hGH cells as compared to control cell line (Mann–Whithney test; $p \leq 0.001$) (Fig. 3).

To check the physiopathological impact of CIS protein induction on breast carcinoma development, we have investigated primary cultures obtained from human breast carcinoma. We have already demonstrated GH expression in 50% of a large series of BC resections. This GH synthesis remains steady in primary culture as shown on Fig. 4 after GH (red fluorescence) and its receptor GHR (green fluorescence) double immunolabeling.

Real-time PCR analysis of CIS transcripts was performed on primary culture derived from BCs recognized to synthesize GH (GH^+, $n = 5$) in comparison to those in which no GH expression could be detected (GH^-, $n = 5$). CIS mRNA induction was increased of 50% in cells obtained from GH^+ tumors (Fig. 5). These data confirm the involvement of CIS overexpression in cancer development and highlight a specific pathway where CIS play a crucial role in GH signaling down regulation.

Effects of CIS Overexpression on Phenotype and Carcinomatous Cell Growth. To further investigate the effects of CIS overexpression on carcinomatous cell functions, MCF7 "wild type" cells were stably transfected with the human CIS gene and called MCF-hCIS. Control cells were transfected with the empty expression vector (pcDNA). Immunofluorescence using anti-CIS antibody revealed the

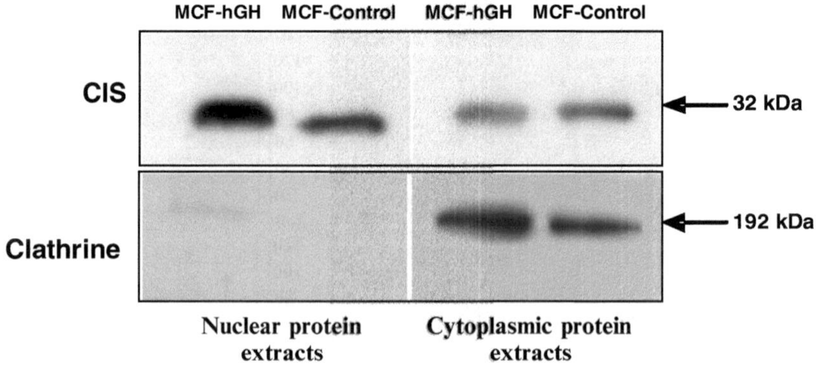

Fig. 2 Western blot analysis of CIS protein expression in nuclear and cytoplasmic extracts of MCF-hGH and MCF-Control cells. A goat polyclonal antibody directed against the terminal sequence of human CIS was used for detection. Clathrine detection in the cytoplasmic fraction only confirmed our protein extracts quality

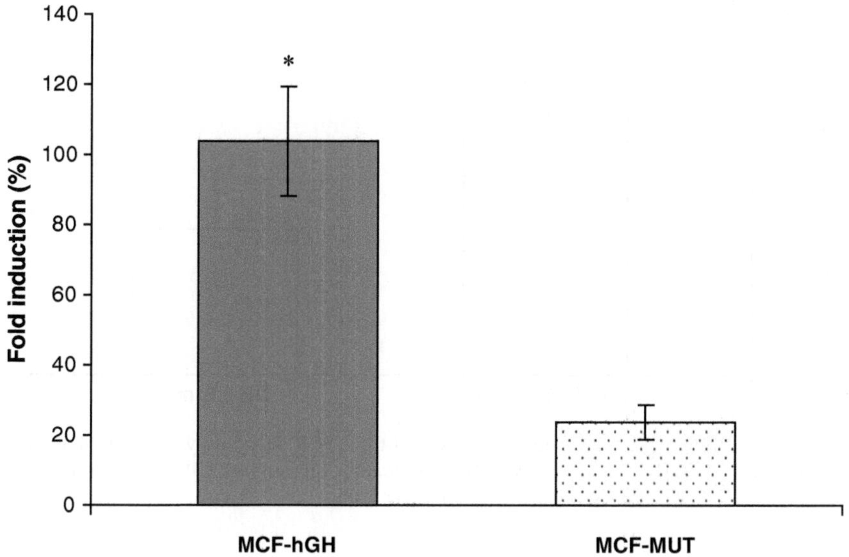

Fig. 3 Effects of GH autocrine stimulation on CIS mRNA expression in MCF-hGH and MCF-Control cells. Amplification curves were plotted as fluorescence signal against cycle number, the first turning point (crossing point) was obtained for each samples by using the Second Derivative Maximum Methods (Roche Molecular Biochemicals). Induction of CIS transcripts are shown as mean ± SE of three different experiments. CIS mRNA was significantly increased (80%) in MCF-hGH cells compared to MCF-Control cells. (*) significant difference $p < 0.05$

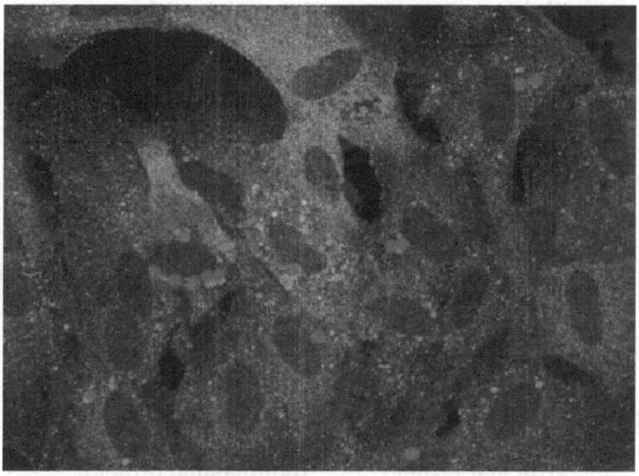

Fig. 4 GH and GHR double immunofluorescence on primary cultures of human BC. Rabbit polyclonal GH and monoclonal GHR antibodies were, respectively, detected with Cy3 and Alexa 488 fluorochroms

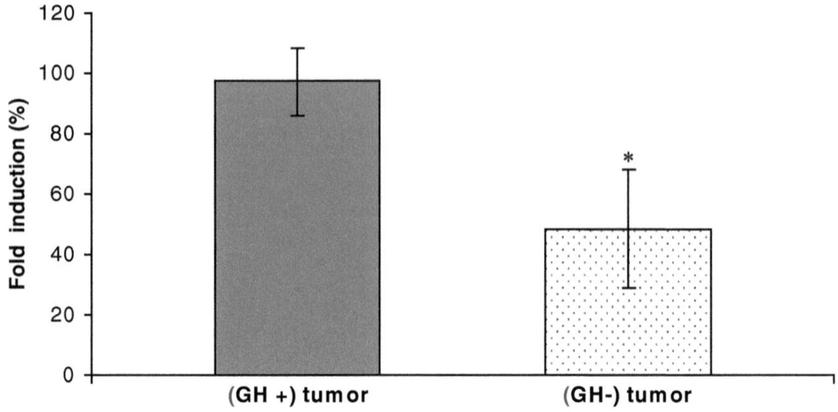

Fig. 5 CIS mRNA expression determined by real-time PCR in primary cultures of BCs with (GH^+) or without (GH^-) synthesis. A 50% significantly increase of CIS expression could be observed in (GH^+) BCs. (*) significant difference $p < 0.05$

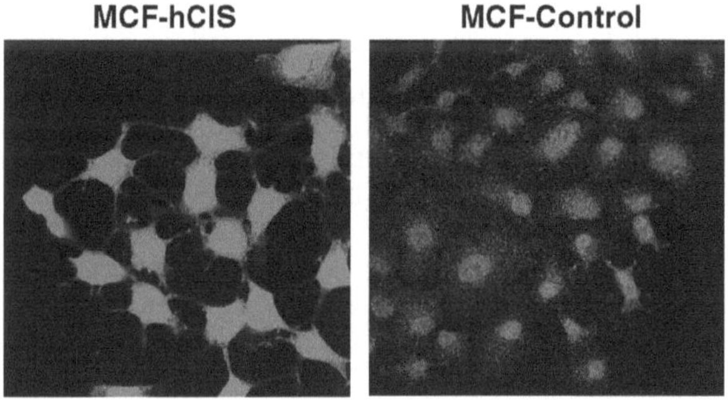

Fig. 6 CIS protein immunofluorescence detection in stably transfected MCF-hCIS and MCF-Control cells

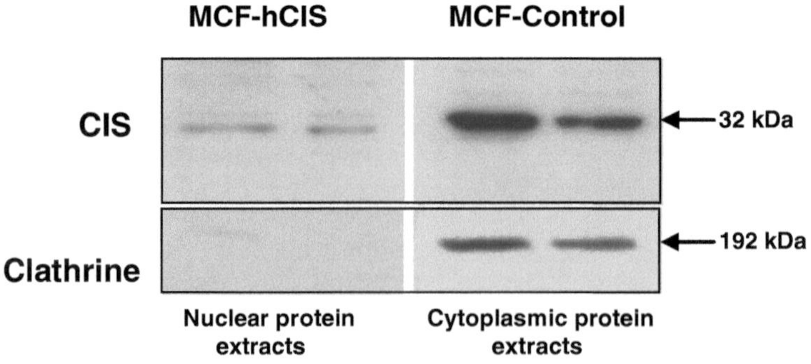

Fig. 7 Western blot analysis of CIS protein expression in nuclear and cytoplasmic extracts of MCF-hCIS and MCF-Control cells

constitutive expression of CIS in MCF7-Control cells and its overexpression in MCF7-hCIS cells. Phenotypic transformations were evidenced with rapid cell extensions development (Fig. 6).

Western blot analysis revealed a major band of 32 kDa corresponding to CIS protein in the cytoplasmic proteins extracted from MCF-hCIS cells (Fig. 7). This result confirmed our previous fluorescence microscopic observation. We have first characterized the consequences of CIS overexpression by phenotypical studies of MCF-hCIS and MCF-Control cells. CIS overexpression increased cell proliferation with a maximal and significant increase (30%) after 48 h (Fig. 8). After 2 weeks, the colony number of CIS transfected cells was significantly increased (30%) compared with that of control cells (Fig. 9) showing that CIS protein overexpression enhanced anchorage independent growth. The morphological aspect of

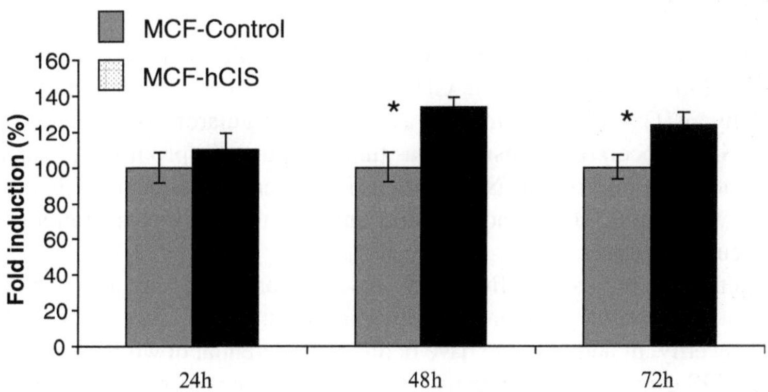

Fig. 8 Cell proliferation of MCF-hCIS and MCF-Control cells using a proliferation detection kit (Cell titer 96 Promega)

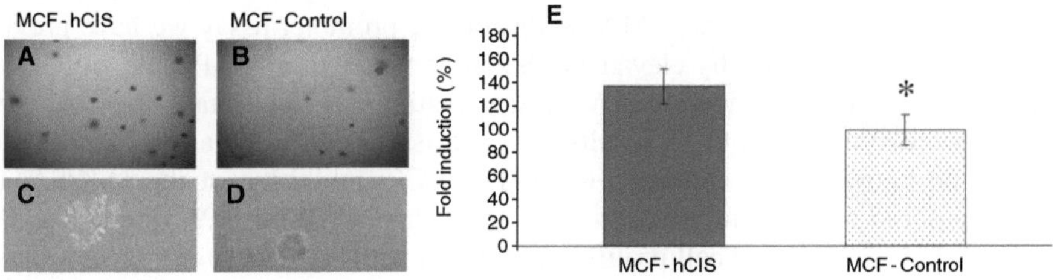

Fig. 9 Soft agar colony formation assay was used to estimate the number of colonies formed by MCF-hCIS (**a**) and MCF-Control (**b**) cells. At low magnification (×10), colonies were counted in a minimum of three wells (**a, b**) and the diameter of the colonies was measured at higher magnification (×20) (**c, d**) using appropriate software. The quantitative analysis showed an increase in the number of colonies from approximately 30% in cells MCF7-CIS (**e**)

the colonies was different in MCF-hCIS compared with MCF-Control cultures, their diameter seemed to be more important but this increase was not significant because of the evident heterogeneity of their respective size. Nevertheless, the loose aspect of CIS transfected cells colonies could suggest modifications of cell adhesion proteins which represent true indicators of tumoral state (Fig. 9).

Discussion

In this study we have identified CIS proteins as the major negative regulators induced by autocrine hGH production in human mammary carcinoma cell line. The ability of CIS to bind to tyrosine-phosphorylated GHR in vitro has been demonstrated using recombinant fragment of the cytoplasmic domain of the GHR (13). The specific sites on the GHR required for the interaction with CIS have been mapped to the distal cytoplasmic part of the receptor which is also the major binding site for STAT5b (13). Thus, CIS inhibits STAT5 phosphorylation by competitively binding to STAT5-docking sites on the GHR and blocks further activation of the JAK/STAT pathway. Given their function as negative regulators of JAK/STAT pathway, SOCS have been first considered as tumor suppressor promoting resistance to cytokine-induced cell growth. Nevertheless increased SOCS expression correlates positively with tumor stage and invasion and shorter survival in human chronic myeloid leukemia (14).

In agreement with this later finding we have demonstrated here an overexpression of CIS protein in primary cultures obtained from human BC resections of different grades of severity. In addition, we have defined a differential down regulation mechanism since CIS overexpression directly correlates with the presence of GH synthesis in carcinomatous cells. Since 50% of BCs expressed proliferative hGH, it is essential to investigate GH down regulation pathway in tumors with a particular attention on CIS protein.

The forced CIS expression in stably transfected MCF7 cells enhanced their proliferation rate and their capacity to form colonies in soft agar, suggesting that aberrant negative regulation of the JAK/STAT pathway plays an important role in malignant progression. Using a MAPK responsive promoter assay we have previously shown that chronically elevated CIS is able to enhance MAPK activity (11). Other studies have also shown that MAPK activation is an important consequence of CIS overexpression in CD4 T cells (15) and is thought to be a result of direct association between CIS and PKC theta. Thus, CIS inhibition of the JAK/STAT signaling could redirect molecular activation of the MAPK and PI-3 kinase, both implicated in survival and cell proliferation. What is the relevance of that cytokine “signal switch” in the pathogenesis of human BC? Studies in progress will identify the molecular mechanisms underlying the relationship between tumor growth and CIS protein overexpression.

References

1. Ali S, Coombes RC (2002) Endocrine-responsive breast cancer and strategies for combating resistance. Nat Rev Cancer 2:101–12.
2. Imada K, Leonard WJ (2000) The JAK/STATpathway. Mol Immunol. 37:1–11.
3. Raccurt M, Lobie PE, Moudilou E, et al. (2002) High stromal and epithelial hGH gene expression is associated with proliferative disorders of the mammary gland. J. Endocrinol. 175: 307–18.
4. Mukhina S, Mertani HC, Guo K, et al. (2004) Phenotypic conversion of human mammary carcinoma cells by autocrine human growth hormone. Proc Natl Acad Sci USA 101:14992–3.
5. Zhu T, Starling-Emerald B, Zhang X, et al. (2005) p44/42 MAP kinase-dependent regulation of catalase by autocrine human growth hormone protects human mammary carcinoma cells from oxidative stress-induced apoptosis. Oncogene 24:3774–85.
6. Du L, G. Frick P, Tai LR, et al. (2003) Interaction of the growth hormone receptor with cytokine-induced Src homology domain 2 protein in rat adipocytes. Endocrinology 144:868–76.
7. Kamio M, Yoshida T, Ogata H, et al. (2004) SOCS corrected inhibits HPV-E7-mediated transformation by inducing degradation of E7 protein. Oncogene 23:3107–15.
8. He B, You L, Xu Z, et al. (2004) Activity of the suppressor of cytokine signaling-3 promoter in human non-small-cell lung cancer. Clin Lung Cancer 5:366–70.
9. Li Z, Metze D, Nashan D, et al. (2004) Expression of SOCS-1, suppressor of cytokine signalling-1, in human melanoma. J Invest Dermatol 123:737–45.
10. Roman-Gomez J, Jimenez-Velasco A, Castillejo JA, et al. (2004) The suppressor of cytokine signaling-1 is constitutively expressed in chronic myeloid leukemia and correlates with poor cytogenetic response to interferon-alpha. Haematologica 89:42–8.
11. Raccurt M, Ping Tam, Lau P, et al. (2003) Suppressor of cytokine signalling gene expression is elevated in breast carcinoma. Brit J Cancer 89:524–32.
12. Liu N, Mertani HC, Norstedt G, et al. (1997) Mode of the autocrine/paracrine mechanism of growth hormone action. Exp Cell Res 237:196–206.
13. Ram PA, Waxman DJ (2000) Role of the cytokine-inducible SH2 protein CIS in desensitization of STAT5b signalling by continuous growth hormone. J Biol Chem 275:39487–96.
14. Cooper JC, Boustead JN, Yu CL (2006) Characterization of STAT5B phosphorylation correlating with expression of cytokine-inducible SH2-containing protein (CIS). Cell Signal 18:851–60.
15. Campbell IL (2005) Cytokine-mediated inflammation, tumorigenesis, and disease-associated JAK/STAT/SOCS signaling circuits in the CNS. Brain Res Brain Res Rev 48:166–77.

In Silico and In Vitro Analysis of Small Breast Epithelial Mucin as a Marker for Bone Marrow Micrometastasis in Breast Cancer

Manuel Valladares Ayerbes, Silvia Díaz-Prado, Daniel Ayude, Rosario G. Campelo, Pilar Iglesias, Mar Haz, Vanessa Medina, Isabel Gallegos, Maria Quindós, and Luis Antón Aparicio

Summary Molecular signatures associated with malignant phenotype would be useful for detection of micrometastatic carcinoma cells. The small breast epithelial mucin (SBEM) gene is predicted to code for a low molecular weight glycoprotein. To evaluate its potential role as a marker for bone marrow (BM) micrometastasis in breast cancer (BC) patients, we have studied in silico and in vitro expression profiles of SBEM gene. Digital SBEM expression in libraries obtained from normal and neoplastic tissues and cell -lines (CL) were displayed and counted on the SAGE Anatomic Viewer. Profiles for cytokeratin-19 and mammaglobin (hMAM), commonly targets used for detection of disseminated BC cells were obtained and compared with SBEM data. Human breast and haematopoietic cancer CL and normal BM were examined by RT-PCR for SBEM and hMAM. Bioinformatics tools were used to gain further insights about the biological role of SBEM in normal breast and BC. Genes with expression patterns in breast libraries correlating with SBEM were identified using two-dimensional display. SBEM tag was detected in 40 libraries (21 BC; 8 non-cancerous breast tissues). Intermediate to high expression was found on 15/21 BC libraries and 7/8 non-tumor breast tissue. SBEM tag count was correlated with ERBB2 (0.662), hMAM (0.409), and RRM2 (−0.379). A model system based on RT-PCR for SBEM mRNA was highly sensitive and specific in order to detect isolated tumor cells. Our results demonstrate that SBEM mRNA may be an imp ortant marker for targeting BC micrometastasis.

Introduction

Although relative survival from breast cancer (BC) in women improved steadily in all European countries in the latest years, age-standardized 5-year relative survival remains in 60–82.6% (1). Metastatic hematogenous spreading is one of the most important factors affecting the prognosis of carcinoma patients, including BC. Circulating tumor cells and occult metastasis (micrometastasis) are considered early events in the progression of BC. Detection of carcinoma cells in the blood or minimal deposits in distant organs as bone marrow could be important to identify patients at high risk of relapse or disease progression (2). PCR amplification of

J.J. Li et al. (eds.), *Hormonal Carcinogenesis V.*

tissue or tumor selective mRNA is the most powerful tool for detection of this circulating or micrometastatic cells. Cytokeratins and mammaglobin are among the most frequents mRNA markers used in different reverse-transcriptase polymerase-chain reaction (RT-PCR) assays in BC patients. However down-regulation of mRNA marker in tumor cells (3) or low-level transcription of selected target in the hematopoietic compartment (4) could compromise both sensitivity and specificity of molecular methods. Selection of novel breast-specific transcripts and development of multimarker RT-PCR assays are clearly outstanding research questions. In this context we have evaluated the potential role for small breast epithelial mucin (SBEM) as a marker for bone marrow (BM) micrometastasis in BC. The SBEM gene [Genbank (#AF414087)] was identified by Miksicek, et al. (5) using the cDNA xProfiler tool. SBEM is similar to proteins B511s (6) and BS106 (7). SBEM gene is predicted to code for a low molecular weight glycoprotein with a specific patterns of expression, limited to breast and salivary glands.

Materials and Methods

In Silico Expression Profiles: Serial Analysis of Gene Expression (SAGE). We used an in silico analysis approach to examine SBEM gene expression in normal and cancerous tissues and cell lines. All available published SAGE data were used for analysis of SBEM gene expression. We obtained a mapping of UniGene cluster (Hs.348419) to NlaIII tags from the SAGE tag to Gene Mapping (SAGEmap) search tool (8) available at the NCBI Web site http://www.ncbi.nlm.nih.gov/SAGE/. Expression levels are displayed as blots with different densities and corrected as tag/million (tpm) to facilitate evaluation. (9) Digital SBEM gene expression profiles were analyzed using SAGE Genie tools (http://cgap.nci.nih.gov/). SAGE Genie automatically identifies SAGE tags and provides a link between gene names and SAGE transcript levels (counts). SBEM transcript expression in different libraries obtained from normal and tumor tissues and cell lines were displayed and counted on the SAGE Anatomic Viewer. Libraries were constructed by using *Nla*III as the anchoring enzyme and *Bsm*FI as the tagging enzyme. In addition, in silico expression profiles for cytokeratin-19 and mammaglobin 1 (hMAM, secretoglobin, family 2A, member 2) two commonly targets mRNA used for detection of disseminated BC cells were obtained. These results were compared with SBEM transcript expression. In order to gain further insights about the biological role of SBEM in normal breast tissue and BC, bioinformatics tools were used.

Virtual Northern. Monochromatic SAGE/cDNA virtual northern for SBEM, mammaglobin and CK-19 were accessed from the gene info tool of the CGAP site. It provides an output indicating the relative abundance of each expressed sequence tag (EST) and SAGE sequence. Spot images represent expression level of the gene. For each combination of tissue and histology (normal vs. cancer), expression is computed by dividing the number of ESTs or SAGE tags representing the gene divided by the total number of ESTs or SAGE tags in all libraries with the given tissue/histology. This ratio is then multiplied by 200,000, giving the number of ESTs or SAGE tags per 200,000. To measure the significance of differences in transcript expression, the method had been described (10).

Cell Lines. BC-derived cell lines (BCCL) MCF-7, MDA-MB468, T47D, BT-549, and PM1 and hematopoietic cell lines (HCL) Jurkat, KG1 and K562, were grown in RPMI-1640 supplemented with 10% (vol/vol) fetal calf serum, L-glutamine, penicillin, streptomycin, and amphotericin at 37°C in 5% CO_2. Cells from adherent cultures were recovered with trypsin-EDTA or nonenzymatic cell dissociating reagent (SIGMA). From each cell line at 50–70% confluence 10^6 cells were obtained for RNA isolation.

RNA Extraction and RT-PCR. Purification of RNA from cell cultures was performed with High Pure RNA Isolation Kit (Roche) as suggested by the manufacturer. Total RNA was treated with DNase I and it was quantified at a wavelength of 260 nm using a spectrophotometer. The reverse transcription was performed using the SuperScript II First-Strand Synthesis System for RT-PCR (Invitrogen) using random hexamers as manufacturer's instructions. PCR amplifications were carried out with AmpliTaq Gold PCR Master Mix (Applied Biosystems). First round PCR amplification of SBEM mRNA was performed using specific primers (SBEM-U-O 5′CTT TGA AGC ATT TTT GTC TGT G3′ and SBEM-L-O 5′AAG GTA AGT AGT TGG ATG AAA T3′) and described by Miksicek (5). In the nested PCR new primers pair (SBEM-S-I 5′TGA TCT TCA GGT CAC CAC CA3′ and SBEM-A-I 5′TGG ATA CGT GTC AGC TGG AG3′) designed using software available on Internet was used (11). First round PCR was performed in 50 μL of reaction mixture containing 2 μL of template cDNA, deionized water, outer SBEM primers U and L, 1 μL 20 μM of each, and PCR Master Mix (2X) 25 μL. In SBEM nested reaction 1 μL of first round PCR template and 0.5 μL at 20 μM of each inner SBEM primers pair was used. For first round SBEM PCR amplification, an initial activation at 95°C for 5 min was used, followed by 35 cycles of 95°C 30 s, 54°C 1 min, and 72°C 1 min. Finally, last extension was at 72°C for 7 min. For SBEM nested reaction, an initial activation at 95°C for 2 min was used, followed by 20 cycles of 95°C 40 s, 62°C 15 s and 72°C 20 s. Last extension was at 72°C for 7 min. First round and nested-PCR amplification of hMAM mRNA was performed using specific primers described previously (12). hMAM first round and nested PCR were performed as described for SBEM with the exception that for nested PCR 1 μL of 1/100 dilution of first round PCR template was used. For first round hMAM PCR amplification, an initial activation at 95°C for 5 min was used, followed by 35 cycles of 95°C 30 s, 57°C 1 min, and 72°C 1 min. Finally, last extension was at 72°C for 7 min. For hMAM nested reaction, an initial activation at 95°C for 5 min was used, followed by 15 cycles of 95°C 30 s, 60°C 1 min, and 72°C 1 min. Last extension was at 72°C for 7 min. PCR products were electrophoresed through agarose gel and stained with 5% ethidium bromide. β-2 microglobulin serves as a positive control target. Negative controls were included in each experiment.

Analysis of mRNA Markers in Normal Human Bone Marrow. Total RNA acids isolated from human normal BMs (BM, $n=23$) were purchased from BD

Biosciences-Clontech and were examined by RT-PCR for the expression of SBEM and hMAM transcripts. cDNA synthesis was carried out as previously described using different amounts of RNA (up to 2 μg). PCR amplifications were carried out with AmpliTaq Gold PCR Master Mix (Applied Biosystems) as indicated.

Assay Detection Sensitivity and Specificity. Serial RNA dilution analysis was used to assess detection performance of the assay. Total RNA was isolated and purified from the cancer cell line MDA-MB468 and then serially diluted in molecular biology grade water. To additionally assess the detection sensitivity and specificity of the assay, an in vitro model was set up by serially diluting RNA from MDA-MB468 tumor cells in pooled normal human BM derived RNA. One-round and nested RT-PCR were performed for each marker on serially diluted RNA.

DNA Sequencing. PCR products were used as template DNA. Products were purified by enzymatic method (ExoSAP-It, Amersham USB). DNA sequencing was performed in a reference facility on ABI 3700 (Applied Biosystems) using Big Dye Terminators. Sense SBEM and hMAM-specific primers were used.

Results

In Silico Expression Profiles: SAGE. We obtained a mapping of UniGene cluster (Hs.348419) to NlaIII tags from the SAGEmap. Tag CTTCCTGTGA (Ref seq nm-058173.1) was selected. The output file shown expression levels are displayed as blots with different densities and corrected as tpm to facilitate evaluation. Tag CTTCCTGTGA was found in 21 mRNA-source sequences. Of these 20 clustered in UniGene Hs.348419 (LOC118430 small breast epithelial mucin). SBEM sequence tag was detected in 40 libraries, including 21 BC libraries, eight libraries obtained from noncancerous breast tissues, and 11 libraries from other sources. Intermediate or high value (13) for gene expression (cut-off value ≥37 tpm) was found on 15/21 BC libraries and 7/8 nontumor breast tissue. Moderate SBEM expression (≥37 tpm) was only present in four libraries from nonmammary tissues. Interestingly it is included a library developed from white blood cells obtained from a breast carcinoma patient. Digital SBEM gene expression profiles were analyzed using SAGE Genie tools. SBEM transcript expression in different libraries obtained from normal and tumor tissues and cell lines were displayed and counted on the SAGE Anatomic Viewer. SBEM was found in 48 libraries; 31 of them were obtained from mammary tissues. SBEM was expressed in 89 and 67% of libraries derived from normal breast and BC, respectively. Moderate to high expression was found in 41% of the libraries from BC. In addition, in silico expression profiles for cytokeratin-19 and mammaglobin 1 (hMAM, secretoglobin, family 2A, member 2) two commonly targets mRNA used for detection of disseminated BC cells were obtained. These results (Table 1) were compared with SBEM transcript expression.

Table 1 In silico expression profiles results for cytokeratin-19, mammaglobin 1 and SBEM

Gene	SAGE tag	Libraries	Positivity (%)	Low expression (%)	Moderate-to-High expression (%)	Mean Density-TAGS per 200,000
SBEM	CTTCCTGTGA	Normal breast	8/9(89)	2/9(22)	6/9(67)	210,33
		Breast Cancer	18/27(67)	7/27(26)	11/27(41)	211,3
KRT19	GACATCAAGT	Normal breast	3/9(78)	0/9	7/9(78)	183,609
		Breast Cancer	24/27(89)	2/27(7.5)	22/23(81.5)	136,48
SCGB2A2	TTTATTTTAA	Normal breast	7/9(78)	2/9(22)	5/9(56)	10,259
		Breast Cancer	16/27(59)	6/27(22)	10/27(37)	54,15

Tissue		EST Data		SAGE Data		EST Data			SAGE Data		
		Normal	Cancer	Normal	Cancer	Normal	Cancer	P	Normal	Cancer	P
SBEM	ALL TISSUES					13/ 2014309	12/ 1970906	0.45	455/4270902	1707/ 8895803	0.00
	bone marrow				–	0/14453	0/20958	–	0/204563	–	–
	mammary gland					4/39573	11/79977	0.32	450/538122	1693/ 1418238	0.00
	white blood cells	–	–		–	–	–	–	0/79858	–	–
hMAM	ALL TISSUES					1/ 2271660	20/ 2063711	0.00	111/7080391	424/ 11023868	0.00
	bone marrow					0/14831	0/21185	–	0/204563	0/16808	–
	mammary gland					0/44822	18/81550	0.00	22/445680	324/1679255	0.00
	white blood cells	–	–		–	–	–	–	1/645820	–	–
CK19	ALL TISSUES					452/ 2271660	1405/ 2063711	0.00	1154/7080391	3171/ 11023868	0.00
	bone marrow					0/14831	0/21185	–	0/204563	0/16808	–
	mammary gland					62/44822	138/ 81550	0.09	146/445680	1177/ 1679255	0.00
	white blood cells	–	–		–	–	–	–	11/645820	–	–

Fig. 1 Monochromatic SAGE/cDNA Virtual Northern for SBEM, hMAM, and CK19 in different tissues, including mammary gland and hematopoietic tissues (BM and WBC)

Virtual Northern. Monochromatic SAGE/cDNA Virtual Northern for SBEM, hMAM, and CK19 were accessed from the Gene Info tool of the CGAP site. We analyzed the relative abundance of each EST and SAGE sequence in different tissues, including mammary gland and hematopoietic tissues (BM and WBC). We observed high expression of SBEM, CK19, and hMAM in mammary gland. These reflect the lack of SBEM sequence in BM and WBC compartments (Fig. 1).

Biological Role for Small Breast Epithelial Mucin. In order to gain further insights about the biological role of SBEM in normal breast tissue and BC, bioinformatics tools were used. First, the expression of SBEM in a series of human breast carcinomas SAGE libraries (n = 27) was quantified and correlated with the tags numbers of different molecular markers associated with BC progression. Expression of SBEM was correlated (Spearman's rho) to ERBB2 (0.662: $p=0.000$), hMAM (0.409; $p=0.034$) and RRM2 (–0.379; $p = 0.051$). No significant correlations with SBEM expression were found for estrogen receptor α (ESR1), CK19, STAT1, EGFR, FLT1, HIF1A, FGF 18, GSTM3, TP53, PTTG1, and EpCAM.

In addition, tags from normal breast tissue and breast cancer SAGE data (available from CGAP SAGE Genie) correlating SBEM sequence tag expression were identified and displayed in array format. The color spots are based on normalized values. Correlation coefficients, means, and standard deviations that we display were based on the set of unnormalized values. Results are shown in Fig. 2.

Analysis of mRNA Markers in Cell Lines and Bone Marrow. BC-derived cell lines (MCF-7, MDA-MB468, T47D, BT-549, and PM1) and hematopoietic cell lines

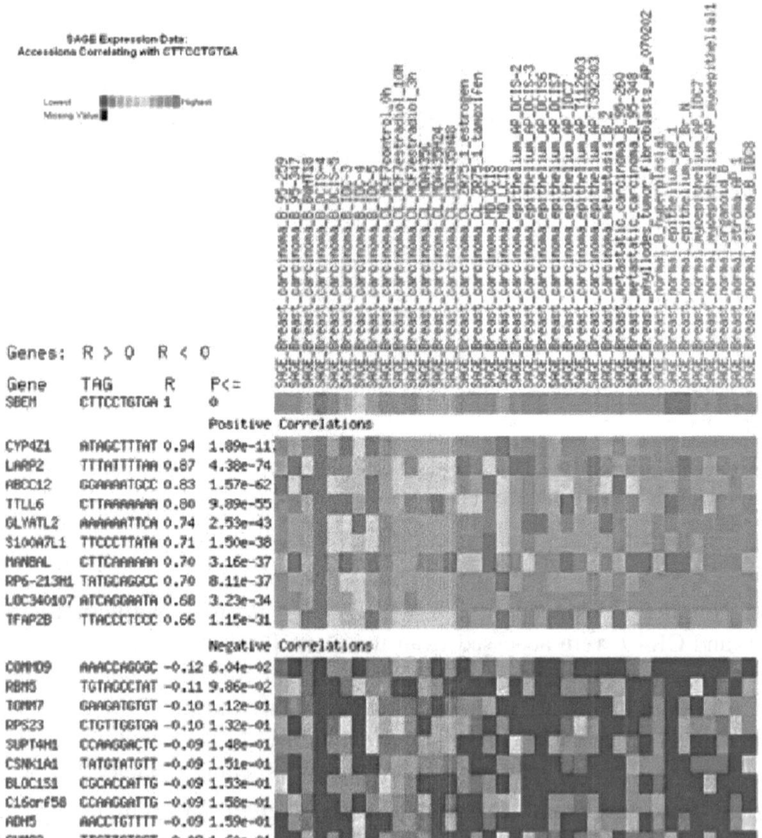

Fig. 2 Tags from normal breast tissue and breast cancer SAGE data correlating SBEM sequence tag expression were identified and displayed in array format

(Jurkat, KG1, and K562) were assessed for hMAM and SBEM mRNA expression by first round and nested RT-PCR. Specific transcript for hMAM was detected in 3/5 BCCL both in first and nested RT-PCR. SBEM mRNA was detected in 2/5 BCCL in first round PCR (Fig. 3) and in 5/5 using nested approach. In both cases, hMAM and SBEM mRNA was not detected in all hematopoietic cell lines assessed. We analyzed SBEM and hMAM mRNA expression in three different pools of normal human bone marrow ($n = 23$). We did not obtain neither SBEM nor hMAM mRNA expression by RT-PCR approach in all pools tested. Low expression was detected on nested RT-PCR in one BM pool ($n = 7$) for both markers.

Sensitivity and Specificity of Nested RT-PCR for SBEM and hMAM. RNA from the cancer cell line MDA-MB468 was serially diluted in molecular biology grade water. One-round and nested RT-PCR were performed for each marker on serially diluted RNA. Using SBEM nested approach we could detect the presence of two cancer cells (Fig. 4) whereas in hMAM nested approach we detected up to one cancer cell.

To additionally asses the detection sensitivity and specificity of the assay, we performed an in vitro model by serially diluting RNA from MDA-MB468 tumor

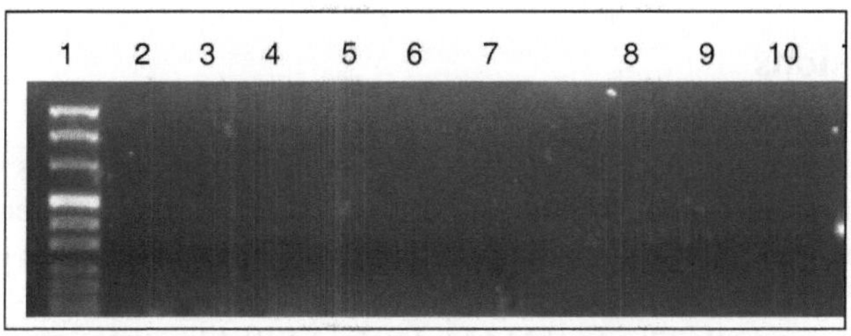

Fig. 3 PCR results for SBEM mRNA expression in different BC-derived cell lines. 1 MW marker, 2–6 (–) controls of BT-549, T47D, PM1, MDA-MB468, and MCF cancer cell lines, respectively, without SuperScript, 7(–) control without RNA, 8–10, BT-549, T47D, and PMI cancer cell lines, respectively

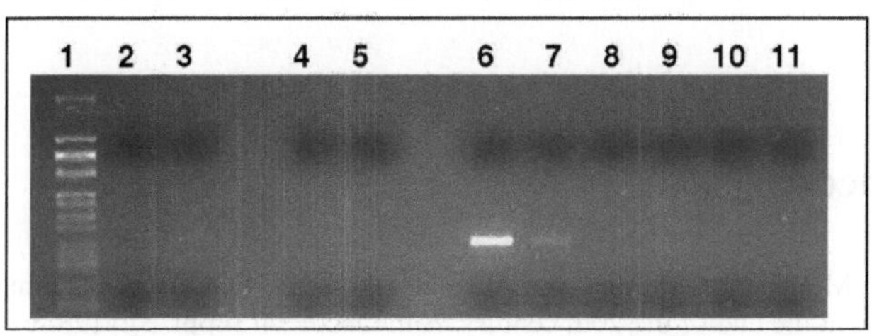

Fig. 4 RT-PCR for SBEM mRNA detection. 1 MW marker, 2 (–) control of normal human BM without SuperScript, 3 normal human BM, 4 MDA-MB468 (–) control without SuperScript, 5 (–) control without RNA, 6 MDA-MB468 not diluted, 7, 8, 9, 10, and 11 MDA-MB468 dilutions at 1:100, 1:1000, 1:10.000, 1:100.000, and 1:1.000.000, respectively

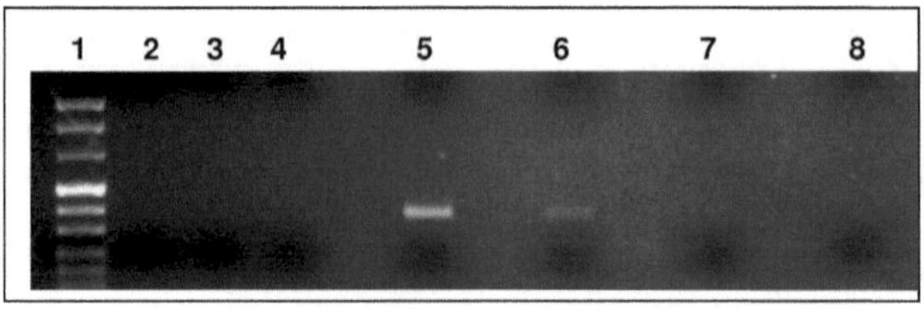

Fig. 5 In vitro model by serially diluting RNA from MDA-MB468 tumor cells in pooled normal human BM derived RNA 1 MW marker, 2, and 3 (–) controls of normal human BM, and MDA-MB468 cancer cell line, respectively, without SuperScript, 4 (–) control without RNA, 5, 6, 7, and 8 pool normal human BM + 1:10, 1:100, 1:1.000, 1:10.000 MDA-MB468 dilutions, respectively. 9 normal human BM

cells in pooled normal human BM derived RNA. One-round and nested RT-PCR were performed for each marker on serially diluted RNA. For SBEM nested approach we could detect up to one cancer cell among 1 μg/μL of normal BM RNA (Fig. 5). In case of hMAM we could also detect up to one cancer cell among 1 μg/μL of normal BM RNA.

Conclusions

Bioinformatics approach based on SAGE and EST data confirms the selective and high expression of SBEM both in normal and BC tissues. Moreover, SBEM was over expressed in BC comparing normal mammary gland based on SAGE counts ($p<0.005$). In addition moderate to high expression was found in 41% of BC libraries comparing with 37% for hMAM. Lack of SBEM expression in hematopoietic-derived libraries was confirmed by our in silico analysis. Nested RT-PCR for SBEM mRNA was highly sensitive and specific in order to detect isolated tumor cells in a model system. Our results probe that SBEM mRNA could serve as a marker for targeting BC micrometastasis.

Acknowledgments This work was supported by a Grant of the Universidade da Coruña 2000-5435256011.

References

1. Coleman MP, Gatta G, Verdecchia A, et al. (2003) EUROCARE-3 summary: cancer survival in Europe at the end of the 20th century. Ann Oncol 14(Suppl. 5):v128–v149.
2. Braun S, Naume B (2005) Circulating and disseminated tumor cells. review article. J Clin Oncol 23:1623–1626.
3. Woelfle U, Cloos J, Sauter G, et al. (2003) Molecular signature associated with bone marrow micrometastasis in human breast cancer. Cancer Res 63:5679–5684.

4. Kruger WH, Jung R, Detlefsen B, et al. (2001) Interference of cytokeratin-20 and mammaglobin-reverse-transcriptase polymerase chain assays designed for the detection of disseminated cancer cells. Med Oncol 18:33–38.
5. Miksicek RJ, Myal Y, Watson PH, et al. (2002) Identification of a novel breast- and salivary gland-specific, mucin-like gene strongly expressed in normal and tumor human mammary epithelium. Cancer Res 62:2736–2740.
6. Houghton RL, Dillon DC, Molesh DA, et al. (2001) Transcriptional complementarity in breast cancer: application to detection of circulating tumor cells. Mol Diagn 6:79–91.
7. Roberts L, Russell J, Friedman P, et al. (2002) Identification and immunohistochemical characterization of a mucin-like glycoprotein expressed in early stage breast carcinoma. Tumour Biol. 23:263–278.
8. Lash AE, Tolstoshev CM, Wagner L, et al. (2000). SAGEmap: a public gene expression resource. Genome Res 10:1051–1060.
9. Boon K, Elison C, Greenhut SF, et al. (2002) An anatomy of normal and malignant gene expression. Proc Natl Acad Sci 99:11287–11292.
10. Lal A, Lash AE, Altschul SF, et al. (1999) A public database for gene expression in human cancers. Cancer Res 59: 5403–5407.
11. Rozen S, Skaletsky HJ (2000) Primer3 on the WWW for general users and for biologist programmers. In: Krawetz S, Misener S, editors. Bioinformatics methods and protocols: methods in molecular biology. Humana Press, Totowa, NJ, pp. 365–386.
12. Grunewald K, Haun M, Urbanek M, et al. (2000) Mammaglobin gene expression: a superior marker of breast cancer cells in peripheral blood in comparison to epidermal-growth-factor receptor and cytokeratin-19. Lab Invest 80:1071–1077.
13. Lercher MJ, Urrutia AO, Hurst LD (2002) Clustering of housekeeping genes provides a unified model of gene order in the human genome. Nat Genet 31:180–183.

Estrogen-regulated Cut-off Values of pS2 and Cathepsin D Expression in Breast Carcinomas

Milan Markićević, Aleksandra Petrović, Ksenija Kanjer, Zora Nešković-Konstantinović, and Dragica Nikolić-Vukosavljević

Summary The purpose of the study was to asses the expression of estrogen-induced pS2 and cathepsin D (CD) that might facilitate biological subgrouping of patients with breast carcinomas (BC) and its potential applicability in clinical oncology. The study included 226 patients with histologically verified BC. Clinico-pathological findings were classified according to age, menopausal status, tumor size, histologic grade, and regional lymph node status. Estrogen and progesterone receptors (ER and PR), as well as CD and pS2 protein concentrations were assayed on the same cytosolic extract in accordance with the recommendation of EORTC. Statistically significant direct correlations were observed between CD expression and axillary node status and between pS2 expression and histologic grade, while the expression of both proteins was related to both ER and PR status. Baseline levels of CD expression were found in patients with SR-negative status and node-negative or tumors less than 2 cm. Unfavorable carcinoma subgroups, in relation to pS2 expression, were defined as pre- and postmenopausal carcinomas with histologic grade III. The highest CD level observed in SR-negative unfavorable subgroups (38.7 pmol/mg) and the highest pS2 level observed in ER^- unfavorable subgroups (14.7 ng/mg) were considered as the cut-off values. These values defined estrogen-regulated expression of CD and pS2 protein that might enable the identification of patients at high risk of disease progression, for whom more aggressive adjuvant approach would be warranted, as well as the identification of patients whose prognosis is so good that adjuvant therapy would not be cost-beneficial.

Introduction

Breast cancer (BC) is heterogeneous disease and its determinants are largely unknown. The identification of prognostic/predictive factors associated with different clinical features of primary BC, after surgery, remains one of the major goals in oncology. In addition to classical prognostic/predictive factors such as age, tumor size, axillary node status, histologic tumor grade, and type, steroid receptors (SR) status is accepted to provide potentially relevant information regarding natural or clinical course of disease (1). The fact that some BC patients with SR-negative

J.J. Li et al. (eds.), *Hormonal Carcinogenesis V.*

status respond to endocrine treatment and that certain number of tumors will recur after such treatment in spite of SR-positivity emphasize the need for identifying markers complementary to SR status. Hence, attention has been directed to estrogen-regulated proteins, including pS2 and cathepsin D (CD). It was assumed that estrogen-regulated proteins may be indicators of a functional signal transduction pathway through which tumor cells respond to estrogen (or antiestrogen) stimulation.

pS2 is a 6.4 kDa polypeptide of 60 amino acids secreted by MCF-7 BC cells and many human BC (2). It has been shown that the pS2 gene contains a complex promoter/enhancer region and that pS2 protein may be constitutive product as well as estrogen-regulated product in breast carcinoma (3). The physiological role of this cysteine-rich protein in breast tissue remains unclear to date. pS2 appears to be positively correlated with ER (4), associated with a good prognosis (5) and a predictor of response to endocrine treatment of primary and metastatic BC (6).

CD is a lysosomal aspartic endoprotease that is ubiquitously distributed in all cells at low concentration (7). CD is synthesized as an inactive 52 kDa proenzyme that, after proteolytical processing, yields the mature active form composed of heavy (34 kDa) and light (14 kDa) chains (8). Its expression is under control of a complex promoter region of CD gene that has mixed structure of housekeeping genes and those of hormone-regulated tissue-specific genes that include a TATA sequence (9). Therefore, CD protein may be both constitutively expressed from TATA-independent start sites and overexpressed, as a consequence of transcription stimulation by estrogens (10). Its overexpression was observed both in BC cell lines and in human breast tumors, at the mRNA and protein level. It was believed that the main role of CD was to degrade protein (11), but many other biological functions of CD were recognized (12). CD level in primary BC has been demonstrated as an independent marker of poor prognosis associated with increased risk for metastasis and shorter survival times (13).

In the current study, our aim was to assess estrogen-regulated pS2 and CD expression cut-off values that might facilitate biological subgrouping of patients with BC and its potential applicability in clinical oncology.

Results

The range of expression of CD was between 8.3 and 237.8 pmol/mg, while the range of pS2 expression was between 1.5 and 135.8 ng/mg. The corresponding median values were 37.7 pmol/mg and 9 ng/mg, respectively. The distribution of CD as well as the distribution of pS2 showed a cluster of breast carcinomas with small quantitative values and many carcinomas with larger, but variable quantitative values.

The expression of CD was related neither to age nor to menopausal status. No significant association was found between CD expression and tumor size, histologic grade or histologic type. Patients with positive lymph node status (pN+) had significantly higher expression of CD protein in relation to patients with negative lymph node status (pN0, $p<0.001$). The relationships between expression of CD and steroid hormone receptor status revealed that positive status of estrogen or progesterone receptors (ER or PR) was related to significantly higher expression of CD ($p<0.001$ and $p=0.014$, respectively).

Comparison of CD expression within matching axillary status (pN0 or pN+) revealed no association with tumor size [tumors <2 cm (pT1) vs. tumors ≥2 cm (pT2–3)] while comparison of CD expression within matching tumor size (pT1 or pT2–3) revealed significantly higher CD content in node-positive compared to node-negative tumors but only for those equal to or larger than 2 cm ($p<0.001$). Figure 1 displays CD levels in node-negative and node-positive carcinomas of different steroid receptor status.

Lymph node status did not influence CD expression within neither ER^+PR^+ nor ER^+PR^- carcinomas while among carcinomas with ER^-PR^- status, those with node-positive status had significantly higher CD levels compared to carcinomas with node-negative status ($p<0.001$). Among patients with the same nodal status, those with ER^-PR^- status had significantly lower expression of CD in relation to patients with ER^+PR^+ ($p<0.001$ for pN0 and $p=0.048$ for pN+) or ER^+PR^- carcinomas

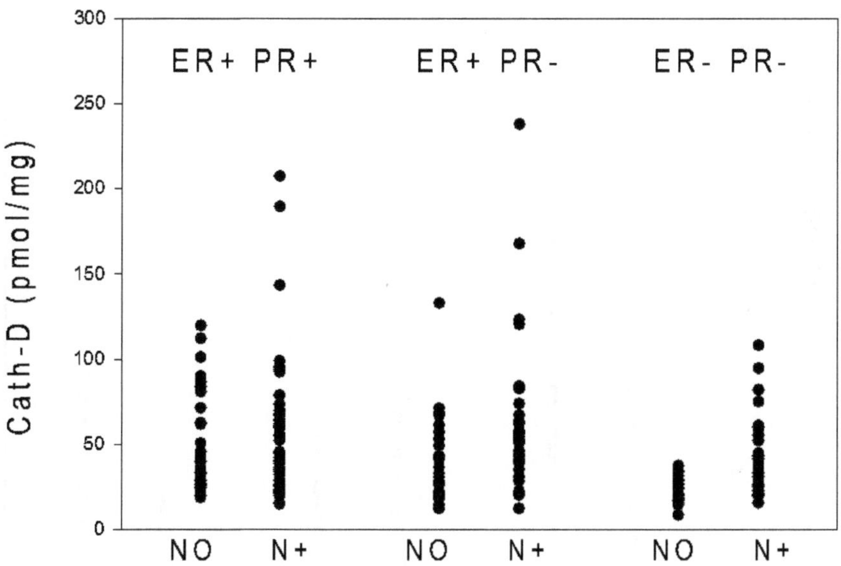

n	37	44	25	31	27	31
Range (pmol/mg)	18,6-119,6	14,7-207,2	12,1-132,7	12,2-237,8	8,3-17,3	15,4-107,9
Median (pmol/mg)	39,5	44,8	36,4	50,7	19,4	31,7

Fig. 1 CD content in node-negative and node-positive tumors of different SR status

(p <0.001 for pN0 and p=0.026 for pN+). Similar to nodal status, tumor size was related to CD expression only in ER^-PR^- tumors (p=0.047), Fig. 2.

Furthermore, the analysis of CD contents in tumors of the same size revealed that for pT1 tumors, it was significantly lower in ER^-PR^- tumors compared to ER^+PR^+ (p<0.001) or ER^+PR^- tumors (p<0.001) while it was significantly lower in ER^-PR^- in relation to ER^+PR^+ pT2–3 tumors (p<0.001). It is important to point out that the highest CD level observed in steroid receptors-negative and nodal status or tumor size favorable subgroups of tumors was less than 39 pmol/mg.

Similar to CD expression, the pS2 expression was related neither to age nor to menopausal status. Furthermore, it was not related to lymph node status, tumor size and histologic type, while it was related to histologic grade in such manner that it was significantly lower in tumors with histologic grade III when compared to tumors with histologic grade I and II. As for the expression of CD, positive status of ER or PR was associated to significantly higher expression of pS2 (p<0.001 for both SRs).

Analysis of pS2 levels showed that there were significantly higher pS2 quantitative values in tumors with histologic grade I or II compared to tumors with histologic grade III, within both pre- and postmenopausal patients (p=0.043 and p= 0.004, respectively), Fig. 3.

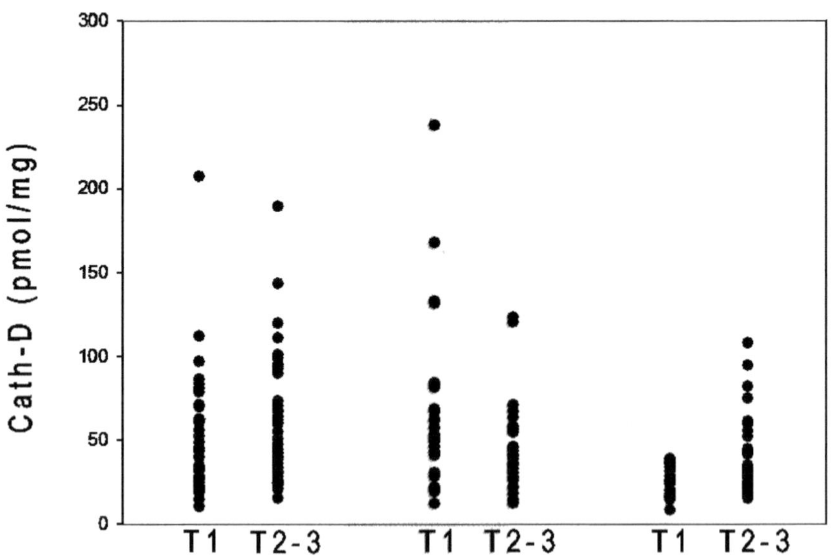

n	42	49	33	27	24	38
Range (pmol/mg)	10,4-207,2	15,3-189,3	12,1-237,8	12,2-123,1	8,3-38,7	15,2-107,9
Median (pmol/mg)	37,2	44,1	51,8	36,4	24,5	28,5

Fig. 2 CD content in tumors less then 2 cm and tumors equal to or greater then 2 cm

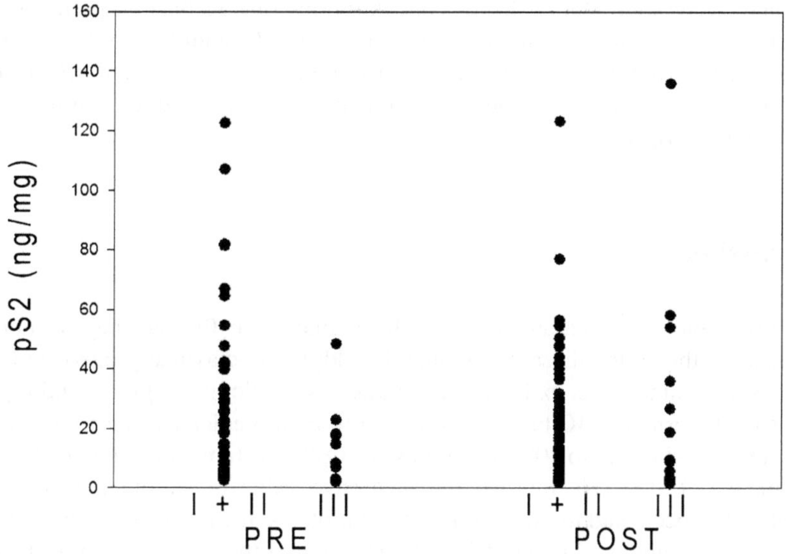

n	63	14	112	28
Range (pmol/mg	2,4-122,5	2,0-48,3	1,6-123,1	1,5-135,8
Median (pmol/mg)	13,6	7,6	10,7	2,8

Fig. 3 Histologic grade-related pS2 quantitative values within pre- and postmenopausal patients

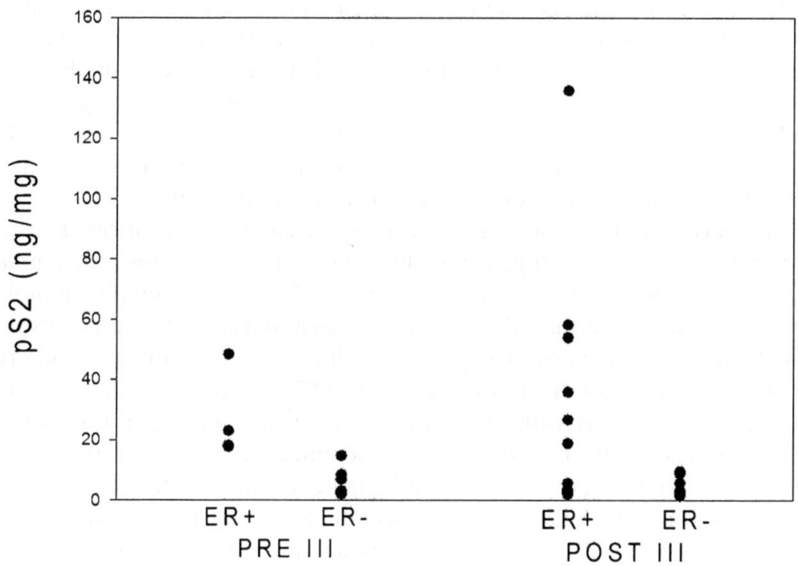

n	5	9	16	12
Range (pmol/mg)	17,7-48,3	2,0-14,7	1,9-135,8	1,5-9,6
Median (pmol/mg)	18,1	3,0	3,4	2,6

Fig. 4 ER status-related pS2 quantitative values within pre- and postmenopausal patients bearing tumors with histologic grade III

There were significantly lower concentrations of pS2 only in premenopausal patients bearing tumors with negative ER status. In addition, pS2 quantitative values in ER^- tumors did not exceed 15 ng/mg. ER-status related pS2 levels in pre- and postmenopausal patients bearing tumors with histologic grade III are presented in Fig. 4.

Discussion

Numerous biological parameters have been studied in BC but they do not fully account for the varied disease outcome. In addition to classical prognostic factors, such as age, tumor size, axillary node status, histologic tumor grade, and type, SR status of the primary BC have been proven to be a predictor of response to endocrine therapy since up to 80% of patients with ER^+PR^+ tumors respond to endocrine treatment (14, 15). To improve the predictive value of steroid receptor status, attention has been paid to estrogen-regulated proteins. It was supposed that estrogen-regulated proteins, including pS2 and CD among others, may be indicators of a functional signal transduction pathway through which tumor cells respond to estrogen stimulation. From a tumor biology point of view, considering possible clinical application, it is important to determine cut-off value for defining estrogen- vs nonestrogen-regulated expression of a protein in BC. In this study, we aimed to assess the estrogen-regulated cut-off values for pS2 and CD expression.

Our results showed that CD expression is not associated to age, menopausal status, tumor size, histologic type, or grade, but is positively associated to axillary node status and both ER and PR. This finding is in accordance with the results of Gion et al. (16). Positive correlations were observed between $pS2^+$ and ER and PR levels. In addition, tumors with grade I and II expressed higher levels of pS2 then tumors with grade III, as reported previously (6, 17, 18). Our further analyses were directed by these and findings from our previous studies (19–21).

Data revealed that adjustment for menopausal status confirmed histologic grade-related expression of pS2: histologic grade I and II vs. histologic grade III, but due to the extensive overlap in individual ps2 values no significant biological differences could be postulated. Therefore, we analyzed ER status-related expression of pS2 within pre- and postmenopausal carcinomas with histologic grade III. We found statistically significant differences in pS2 quantitative values of ER^+ and ER^- breast carcinomas with pS2 levels lower than 15 ng/mg in ER^- subgroups. This value indicates the cut-off value for discrimination of estrogen-regulated from nonestrogen-regulated expression of pS2 and confirms our previous finding.

When we analyzed CD data, the association between CD and axillary node status in tumors equal to or greater then 2 cm was found to be a statistically significant, but not biologically meaningful due to the wide range of individual values among subgroups. Additionally, we noted that SR status determines axillary node- and tumor size-related expression of CD. Negative SR status was generally associated with lower levels of CD in breast carcinoma subgroups adjusted for both axillary node status and tumor size. It is important to point out that expression of CD in SR status-negative BCs with negative nodal status or smaller then 2 cm did not exceed value of 39 pmol/mg indicating that CD could be expressed at the baseline level which is probably not under estrogen control. Hence, this value may be assumed as the cut-off value for estrogen-related expression of CD.

It remains open question whether proposed cut-off values may help in identifying patients whose prognosis is so good that adjuvant therapy after local surgery would not be cost-beneficial or patients at high risk of disease progression for whom more aggressive adjuvant treatment may be suggested. It is clear that controversial results, obtained by various methods and based on different cut-off values, complicate direct comparison between studies and may, partially, explain the reluctance to use pS2 and CD expression levels as prognostic/predictive markers in clinical settings.

Acknowledgments This study was supported by a grant from the Ministry of Science and Environment Protection, Republic of Serbia, contract No. 145018: "Molecular biomarkers of growth, invasiveness and metastasis of breast carcinoma: biological and clinical aspects."

References

1. National Institutes of Health Consensus Development Conference Statement (2001) Adjuvant therapy for BC, November 1–3, 2000. J Natl Cancer Inst 93:979–989.
2. Stack G, Kumar V, Green S, et al. (1988) Structure and function of the pS2 gene and estrogen receptor in human BC cells. In: Lippmann ME, Dickson RB, editors. BC. Cellular and molecular biology. Boston: Kluwer, pp. 185–206.
3. Giamarchi C, Solanas M, Chailleux C, et al. (1999) Chromatin structure of the regulatory regions of pS2 and cathepsin D genes in hormone-dependent and -independent BC cell lines. Oncogene 18:533–541.
4. Gillesby BE, Zacharewski TR (1999) pS2 (TTF1) levels in human BC tumor samples: correlation with clinical and histological prognostic markers. Breast Cancer Res Treat 56:253–265.
5. Thompson AM, Hawkins RA, Elton RA, et al. (1993) pS2 is an independent factor of good prognosis in primary BC. Br J Cancer 68:93–96.
6. Foekens JA, van Putten WLJ, Portengen H, et al. (1993) Prognostic value of pS2 and cathepsin D in 710 human primary breast tumors: multivariate analysis. J Clin Oncol 11:899–908.
7. Barett AJ (1970) Cathepsin D: purification of isoenzymes from human and chicken liver. Biochem J 117:601–607.
8. Richo G, Conner GE (1991) Proteolytic activation of human procathepsin D. Adv Exp Med Biol 306:289–296.
9. Cavailles V, Augereau P, Rochefort H (1993) Cathepsin D gene is controlled by a mixed promoter, and estrogens stimulate only TATA-dependent transcription in BC cells. Proc Natl Acad Sci 90:203–207.
10. Westley BR, May FE (1987) Oestrogen regulates cathepsin D mRNA levels in oestrogen responsive human BC cells. Nucl Acids Res 15:3773–3786.
11. Briozzo P, Morisset M, Capony F, et al. (1988) In vitro degradation of extracellular matrix with Mr 52,000 cathepsin D secreted by BC cells. Cancer Res 48:3688–3692.
12. Liaudet-Coopman E, Beaujouin M, Derocq D (2006) Cathepsin D: newly discovered functions of a long-standing aspartic protease in cancer and apoptosis. Cancer Lett 237:167–179.

13. Foekens JA, Look MP, Bolt-de-Vries J, et al. (1999) Cathepsin-D in primary BC: prognostic evaluation involving 2810 patients. Br J Cancer 79:300–307
14. Fitzgibbons PL, Page DL, Weaver D, et al. (2000) Prognostic factors in BC. College of American Pathologist Consensus Statement 1999. Arch Pathol Lab Med 124:966–978.
15. Wittliff JL, Pasic R, Bland KI (1991) Steroid and peptide hormone receptors identified in breast tissue. In: Bland KI, Copelano EM, editors. The breast: comprehensive management of benign and malignant disease. Philadelphia: WB Saunders, pp. 900–936.
16. Gion M, Mione R, Dittadi R, et al. (1995) Relationship between cathepsin D and other pathological and biological parameters in 1752 patients with primary BC. Eur J Cancer 31A:671–677.
17. Bolufer P, Torregrosa MD, Gomez L, et al. (1996) Specific oncological contribution of cathepsin D and pS2 in human BC: their relationship with TNM status, estradiol receptors, epidermal growth factor receptor and neu amplification. Clin Chim Acta 274:89–103.
18. Gillesby BE, Zacharewski TR (1999) pS2 (TTF1) levels in human BC tumor samples: correlation with clinical and histological prognostic markers. Breast Cancer Res Treat 56:253–265.
19. Nikolić-Vukosavljević D, Grujić-Adanja G, Janković R, et al. (1998) pS2 in breast carcinomas: association with steroid hormone receptor status. Tumori 84:691–694.
20. Nikolić-Vukosavljević D, Grujić-Adanja G, Janković R (2001) pS2 protein in breast carcinomas: cut-off value of estrogen-regulated expression. Neoplasma 48:1–6.
21. Nikolić-Vukosavljević D, Grujić-Adanja G, Nastić-Mirić D (1998) Cathepsin D: association between TN-stage and steroid receptor status of breast carcinoma. Tumor Biol 19:329–334.

Molecular Signatures of Estrogen Receptor-Associated Genes in Breast Cancer Predict Clinical Outcome

James L. Wittliff, Traci L. Kruer, Sarah A. Andres, and Irina Smolenkova

Summary Our goal is to identify new molecular targets for drug design and improve understanding of the molecular basis of clinical behavior and therapeutic response of breast cancer (BC). Pure populations of BC cells were procured by laser capture microdissection (LCM) from deidentified tissue specimens. RNA from either LCM-procured cells or whole tissue sections was extracted, purified, and quantified by RT-qPCR using β-actin for relative quantification. RNA was amplified, Cy5-labeled, and hybridized for microarray. Spectrophotometric and BioAnalyzer™ analyses evaluated aRNA yield, purity, and transcript length for gene microarray. Unsupervised and supervised methods selected 7 000 genes with significant variation. Expression profiles of BC cells were dominated by genes associated with estrogen receptor-α (ERα) status; over 3 000 genes were identified as differentially expressed between $ER\alpha^+$ and $ER\alpha^-$ BC cells. Other prominent gene expression patterns divided $ER\alpha^+$ BCs into subgroups, which were associated with significantly different clinical outcomes ($p < 0.01$). While exploiting larger gene sets derived from LCM-cells and reports using whole tissues, a preliminary 14 gene subset was selected by UniGene Cluster analysis. Additionally, ERE-binding proteins (ERE-BP) were detected by EMSA, which were not recognized by ERα antibodies. Kaplan-Meier analysis indicated that patients with ERE-BP positive BCs had lower over-all survival than those with ERE-BP negative cancers. Collectively, these results will establish molecular signatures for assessing clinical features of BC and aid in the selection of molecular targets for drug development.

Introduction

Progress toward understanding human cancer at the molecular level has been extraordinary since the National Cancer Act in 1971. The Human Genome Project, increased research of genomic and proteomic techniques, and data mining approaches have revealed many fundamentals of human cell biology and alterations occurring during malignant transformation. The magnitude of the problem is illustrated by the fact that almost 1.4 million individuals will be diagnosed with

J.J. Li et al. (eds.), *Hormonal Carcinogenesis V.*

some form of cancer in year 2006 and more than 560 000 Americans will die from this dreaded disease, according to the National Cancer Institute and the American Cancer Society.

Human tissue collection, handling, and analyses present specific problems for clinically reliable genomic and proteomic testing (1) unlike conventional studies with tissues or homogeneous cell lines grown in culture (2, 3). As noted previously (1–3), collection and processing of human tissue biopsies have focused on their clinical purpose (i.e., diagnosis, staging, prognosis, therapy selection) with little emphasis on sampling and cryopreservation for sophisticated genomic (i.e., microarrays) and proteomic analyses (i.e., protein chips). Cellular heterogeneity of a tissue section may result in misleading molecular findings complicating these analyses (4–7). Therefore, a reproducible method for obtaining homogeneous cell populations from normal tissue or from cancer biopsies was required (6) in order to obtain cell-specific information from molecular analyses.

Results

Laser Capture Microdissection and RNA Isolation. Figure 1 illustrates the protocol used for gene expression profiling of deidentified frozen tissue from an IRB-approved biorepository. Using laser capture microdissection (LCM), pure populations of breast cancer (BC) cells were procured for RNA extraction and isolation as described previously (1, 7). Briefly, 5–7 µm frozen tissue sections were placed on uncharged slides, followed by fixation in 70% ethanol, hydration in nuclease-free water, staining with hematoxylin and eosin, dehydration in 100% ethanol, and a final incubation in xylenes before processing by LCM (Fig. 2). Using

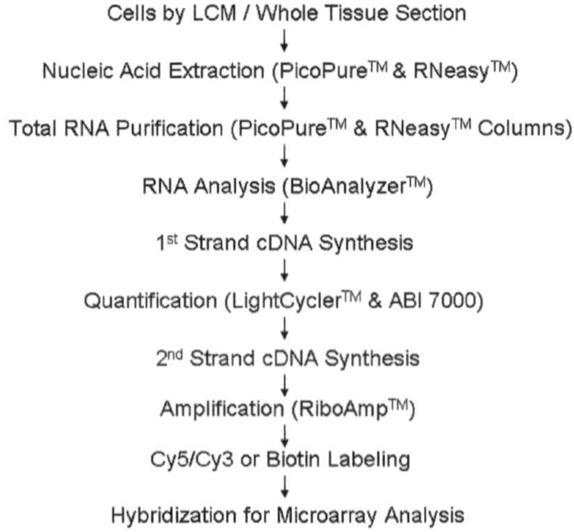

Fig. 1 Protocol for gene expression analyses of frozen tissue

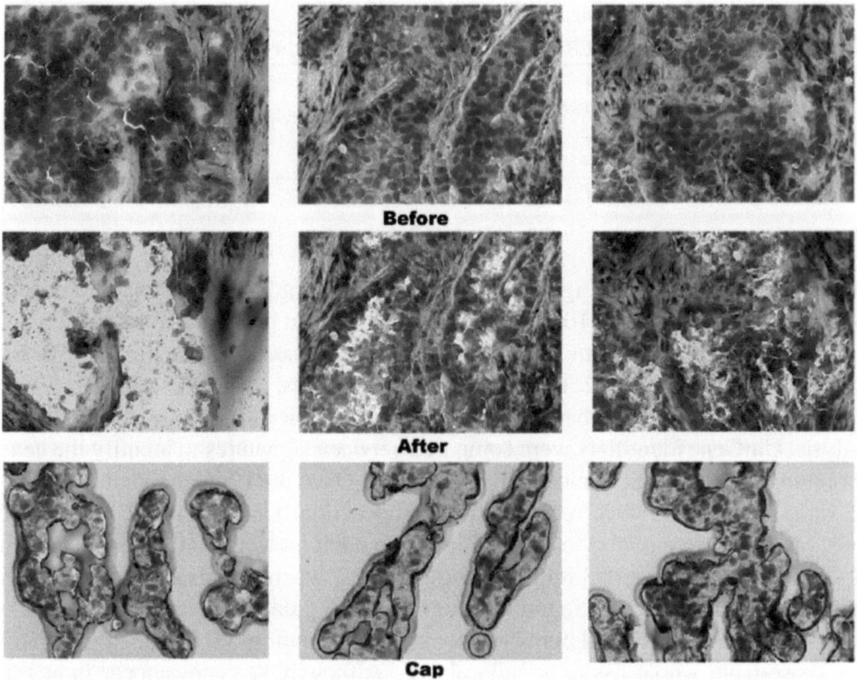

Fig. 2 LCM procurement of carcinoma cells from three separate regions in a single biopsy of human infiltrating ductal carcinoma of the breast

PicoPure™ RNA extraction reagents, 2–10 ng total RNA were retrieved from 500–1 000 BC cells captured by LCM from frozen sections.

Microarray Analyses. To confirm reproducibility of LCM processing, triplicate caps were obtained from a single biopsy (Fig. 2), followed by RNA isolation, amplification, and hybridization for microarray analyses. High correlations ($r > 0.95$) from microarray results were observed between all three independent captures. Gene expression profiles from laser captured cells were also compared with those from whole tissue sections of the same frozen biopsies. As expected, LCM procured cells revealed unique molecular signatures compared with those of matched intact tissue sections, illustrating the significance of cellular contamination in whole tissue (8, 9). By relating microarray results with tumor marker data from cancer specimens, gene expression profiles for both estrogen receptor-alpha (ERα) and HER-2/neu oncoprotein positive biopsies were identified. Molecular signatures of four distinct BC subtypes were discerned, two correlating with $ER\alpha^+$ and two with $ER\alpha^-$ cancers. Using Kaplan-Meier survival analyses, it was observed that patients with $ER\alpha^+$ cancers with subtype B had much lower disease-free survival (DFS) than those with $ER\alpha^+$-subtype A group (Table 1). The higher recurrence in the B subtype was similar to that of BCs with the $ER\alpha^-$-subtype C. Additionally, survival differences were observed between the two $ER\alpha^-$-subtypes (Table 1).

Table 1 ERα status and DFS probability

Subtype	ERα status	% DFS probability at 60 months
A	+	72
B	+	63
C	–	56
D	–	46

Gene List Comparisons. Figure 3 depicts the overall goal of developing a concise molecular signature from a 10–15 gene subset that could be used as a routine clinical test. Examining results from 11 published gene expression signatures from breast cancer (8–18), reporting 2 604 total gene sequences, UniGene identifers were obtained from the CBI database and compiled into a Microsoft Access® database for analysis. UniGene identifiers were compared between signatures to identify the genes correlating with relevant clinical behavior of BC. Once the UniGene identifiers from the various gene lists were recorded and composed into a Microsoft Excel spreadsheet, and they were sorted to create a list from each molecular signature. The gene list data were then imported into Microsoft Access, where the lists were matched to identify genes that appeared in multiple signatures. A greater emphasis was placed on those signatures that utilized pure carcinoma cell populations (8, 9, 12, 16) compared with those from whole tissue or cell cultures. Although 32 genes appear in at least three molecular signatures analyzed (8–18), only 14 of those were found in our studies utilizing pure carcinoma cells (8, 9). Validation of expression of these 14 genes was accomplished by RT-qPCR for correlations with clinical characteristics.

Expression Profiling of the Candidate Gene Subset. Total RNA extraction of cells from frozen whole tissue sections (1, 7) was accomplished with the RNeasy® Mini Kit. Integrity of RNA was analyzed using the Bioanalyzer™ 2100. Intact total RNA was reverse transcribed in a solution composed of 50 mM Tris-HCl buffer containing 37.5 mM KCl, 1.5 mM $MgCl_2$, 10 mM DTT, 0.5 mM dNTPs, 40 units RNasin®, and 200 units Superscript RT III®. RNA quantification and analyses were performed using triplicate cDNA preparations with real-time quantitative PCR (1, 7) in duplicate wells using the ABI Prism® 7900HT. Universal human reference RNA was reverse transcribed and amplified simultaneously with test samples as both a positive control and as a standard for quantification of RNA using β-actin as a reference gene. Relative gene expression was calculated using the ΔΔCt method.

In this preliminary study, 11/14 genes appear to predict disease-free or overall survival in these BC patients when analyzed independently. Kaplan-Meier regression shown in Fig. 4 depicts overall survival differences observed in patients with expression levels either above or below the median for one of the genes in the candidate subset. As shown, patients with higher expression (above the median) of this gene appear to have a higher probability of overall survival than those with lower expression (below the median). In an effort to understand the biological activities of genes from the four ERα-associated molecular signatures composed of a total of 200 genes (Table 1), we are using gene ontology (GO) mapping (Fig. 5). We surmise

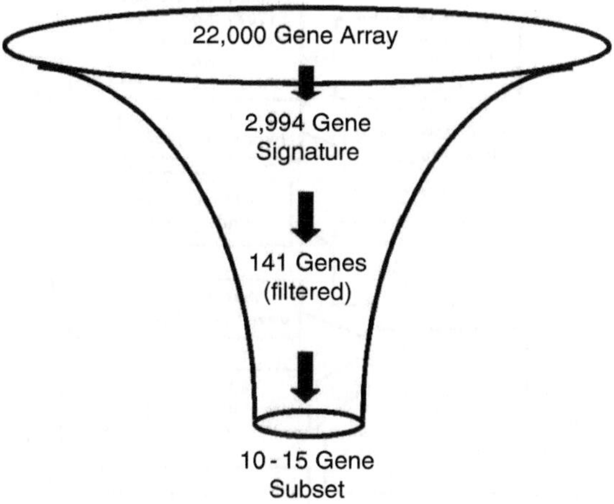

Fig. 3 Filtering of global gene expression profiles to obtain a manageable gene subset for clinical use in breast cancer management

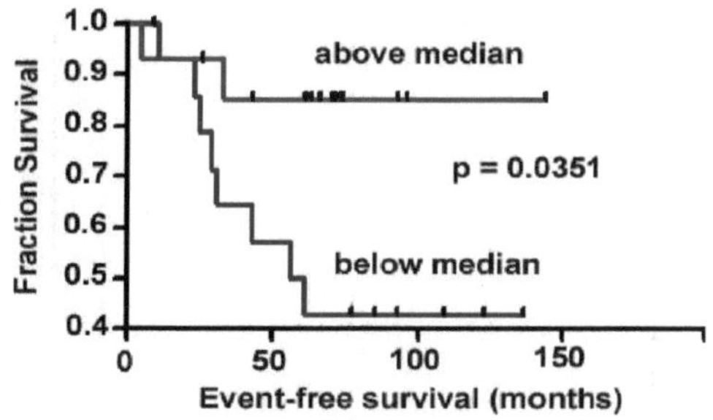

Fig. 4 Kaplan-Meier regression of event-free survival probability for patients with gene expression above or below the median for a selected gene in 14 gene subset

that gene selection will be greatly improved since the molecular signatures were compiled from results with pure populations of LCM-procured BC cells.

Distribution of ERE-Binding Proteins in Breast Cancer Extracts. While investigating the properties of ERα protein isoforms in cancer extracts to determine whether the molecular signatures reflecting different clinical outcomes were associated with ERα mutants, we discovered non-ERα proteins that recognized ERE sequences. To determine distribution and expression levels of ERE-binding proteins (ERE-BP), cytosols were prepared in 10 mM Tris-HCl buffer, pH 7.4 as previously described (2, 19) from various BC biopsies and analyzed by electrophoretic mobility shift assay, EMSA (19, 20). As shown (Fig. 6), many cancer extracts exhibited varied abundance of these ERE-binding proteins that migrated more rapidly than intact recombinant human ERα (19, 20). Furthermore, certain cytosols contained multiple ERE-binding protein species with different migration properties (i.e., lanes 1 and 13, Fig. 6). One sample (lane 11) also displayed a slower migrating band, which appears to represent hERα. Levels of expression of

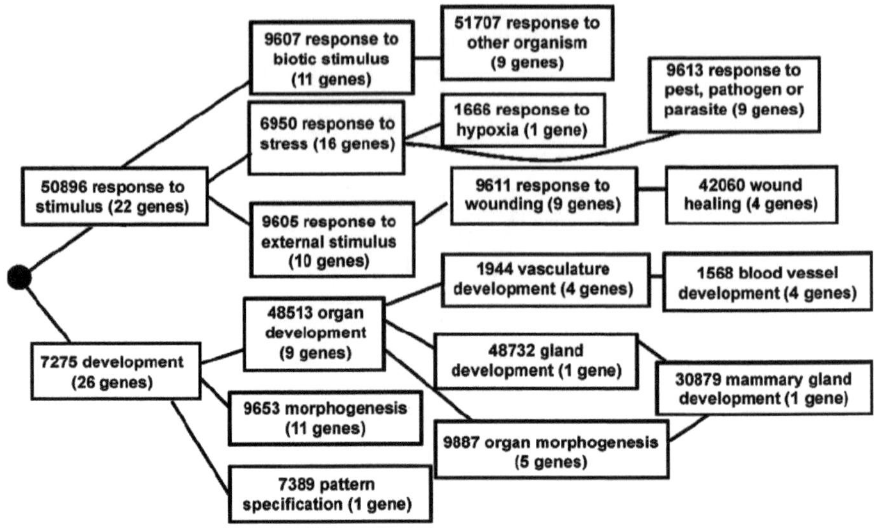

Fig. 5 Gene ontology (GO) map of genes associated with expression of ERα in breast cancers

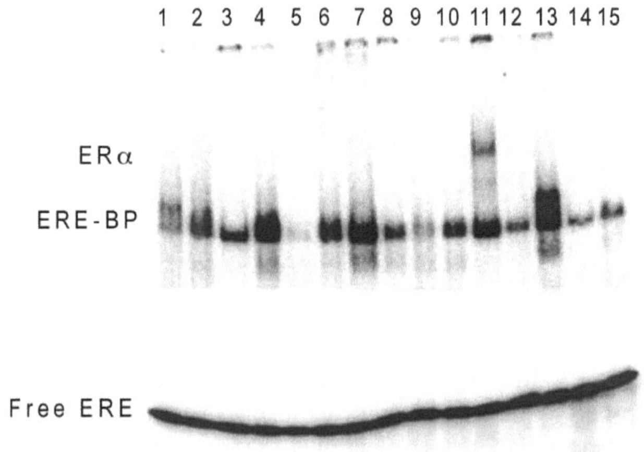

Fig. 6 Distribution of VitA2 ERE-binding proteins in breast cancer extracts indicated by EMSA. Location of intact recombinant ERα is indicated as a reference

ERE-binding protein were determined by scanning each lane of the EMSA gels and using Optiquant™ imaging software for quantification (19, 20). There was no apparent correlation between the level of ERE-BP and the amount of ERα in a cancer biopsy, suggesting the expression of the proteins recognizing ERE sequences is unrelated to that of ERα proteins.

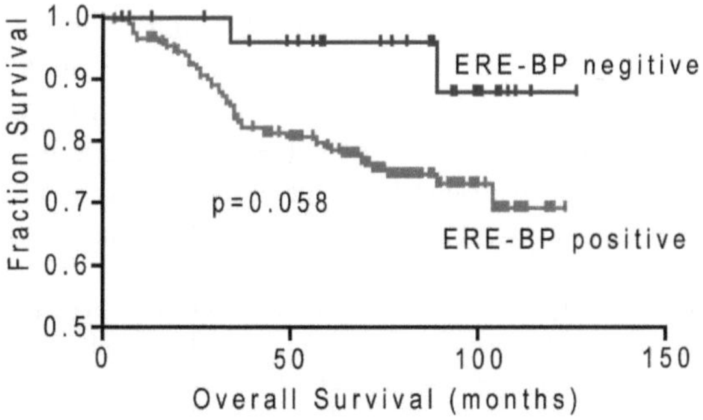

Fig. 7 Kaplan–Meier regression of overall survival probability in breast cancer patients that were either ERE$^+$ or ERE$^-$

As further evidence that the ERE-binding proteins detected in breast cancer were not fragments of the ERα protein, super-shift assays were performed with six antibodies prepared against different domains of the ERα protein. As predicted, rhERα exhibited super-shift with each antibody verifying its identity; however, none of these antibodies brought about a super-shift of the novel ERE-binding protein bands, when incubated with cancer cytosols (manuscript in preparation). These data suggest that the ERE-binding proteins do not contain epitopes recognized by antibodies prepared against various regions of the human ERα protein.

Influence of ERE-BP Expression on Overall Survival of Breast Cancer Patients. Kaplan-Meier regression analyses were performed on 229 BC patients defined as either ERE-BP$^+$ or ERE-BP$^-$ (Fig. 7). As shown, decreased survival probability was observed in the ERE-BP$^+$ patients, suggesting that these novel proteins may be prognostic markers for BC.

Discussion

Use of tumor markers for clinical diagnoses has significant limitations since patients are grouped into generic categories (i.e., ERα$^+$, ERα$^-$) which do not consistently reflect the prognosis of the individual patient (Table 1). These original genomic signatures, when correlated with tumor-related characteristics and clinical outcome, clearly indicate the importance of examining pure populations of cells collected by LCM for genomic and proteomic studies. Disease-free survival of BC patients exhibiting the molecular signatures of the four subtypes (Table 1) clearly illustrates the potential of gene expression profiling in the future for individualized patient care.

Preliminary correlations of a candidate 14 gene subset with clinical outcome reveal promising results, to be confirmed in an expanded population. Correlations of various gene combinations are being analyzed to discern a smaller number of genes necessary to predict clinical outcome for establishment in the routine clinical laboratory.

Finally the analyses of ERE-binding proteins expressed in certain BCs suggest these previously undiscovered proteins may have prognostic significance. Identification and biological significance of these ERE-binding proteins are being investigated to determine whether they are related to the molecular signatures recognizing estrogen receptor-associated genes. The goal is to improve the ability to predict clinical behavior of BC while identifying new targets for drug design.

Acknowledgments This research was supported in part by grants from Arcturus Bioscience, Inc. and the Phi Beta Psi Charity Trust. The authors wish to thank D. Alan Kerr II for his consultation, as well as Neil Mucci and Dr. Jon Wetzel, Asterand, PLC, for their assistance with tissue specimens.

References

1. Wittliff JL, Erlander MG (2002) Laser capture microdissection and its applications in genomics and proteomics. Methods Enzymol 356:12–25.
2. Wittliff JL, Pasic R, Bland KI (1998) Steroid and Peptide Hormone Receptors: Methods, Quality Control and Clinical Use. In: Bland KI, Copeland III EM, editors. Breast: Comprehensive Management of Benign and Malignant Diseases. Philadelphia, PA: W.B. Saunders, Co., p. 458–498.
3. Cole KA, Krizman DB, Emmert-Buck MR (1999) The genetics of cancer – a 3D model. Nat Genet 21:38–41.
4. Bonner RF, Emmert-Buck M, Cole K, et al. (1997) Laser capture microdissection: molecular analysis of tissue. Science 278(5342):1481–1483.
5. Emmert-Buck MR, Bonner RF, Smith PD, et al. (1996) Laser capture microdissection. Science 274 (5289):998–1001.
6. Simone NL, Bonner RF, Gillespie JW, et al. (1998) Laser-capture microdissection: opening the microscopic frontier to molecular analysis. Trends Genet 14(7):272–276.
7. Wittliff JL, Kunitake ST, Chu SS, Travis JC (2000) Applications of laser capture microdissection in genomics and proteomics. J Clin Ligand Assay 23:66–73.
8. Ma XJ, Wang W, Salunga R, et al. (2003) Gene expression associated with clinical outcome in breast cancer via laser capture microdissection. Breast Cancer Res Treat 82(S15).
9. Wittliff JL, Ma XJ, Wang W, et al. (2003) Expression of estrogen receptor-associated genes in breast cancer cells procured by laser capture microdissection correlate with clinical outcome. Jensen Symposium, 81.
10. van't Veer JL, Dai H, Van de Vijver MJ, et al. (2002) Gene expression profiling predicts clinical outcome of breast cancer. Nature 415(6871):530–536.
11. Kang Y, Siegel PM, Shu W, et al. (2003) A multigenic program mediating breast cancer metastasis to bone. Cancer Cell 3(6):537–549.
12. Ma XJ, Salunga R, Tuggle JT, et al. (2003) Gene expression profiles of human breast cancer progression. Proc Natl Acad Sci USA 100(10):5974–5979.
13. Ramaswamy S, Ross KN, Lander ES, Golub TR (2003) A molecular signature of metastasis in primary solid tumors. Nat Genet 33(1):49–54.
14. Sorlie T, Tibshirani R, Parker J, et al. Repeated Observation of breast tumor subtypes in independent gene expression data sets. Proc Natl Acad Sci USA 100:8418–8423.
15. Sotiriou C, Neo S-Y, McShane LM, et al. (2003) Breast cancer classification and prognosis based on gene expression profiles from a population-based study. Proc Natl Acad Sci USA 100:10393–10398.
16. Ma XJ, Wang Z, Ryan PD, et al. (2004) A two-gene expression ratio predicts clinical outcome in breast cancer patients treated with tamoxifen. Cancer Cell 5(6):607–616.

17. Jansen MPHM, Foekens JA, van Staveren IL, et al. (2005) Molecular classification of tamoxifen-resistant breast carcinomas by gene expression profiling. J Clin Oncol 23:732–740.
18. Wang Y, Klijn JG, Zhang Y, et al. (2005) Gene-expression profiles to predict distant metastasis of lymph-node-negative primary breast cancer. Lancet 365(9460):671–679.
19. Graumann K, Wittliff JL, Raffelsberger W, et al. (1996) Structural and functional analysis of N-terminal point mutants of the human estrogen receptor. J Steroid Biochem Mol Biol 57(5–6):293–300.
20. Wittliff JL, Wenz LL, Dong J, et al. (1990) Expression and characterization of an active human estrogen receptor as a ubiquitin fusion protein from *Escherichia coli*. J Biol Chem 265(35):22016–22022.

Comparative Analysis of SNP in Estrogen-metabolizing Enzymes for Ovarian, Endometrial, and Breast Cancers in Novosibirsk, Russia

Lyudmila F. Gulyaeva, Olga N. Mikhailova, Vladimir O. Pustylnyak, Inessa V. Kim IV, Alexei V. Gerasimov, Sergey E. Krasilnikov, Maxim L. Filipenko, and Eugene V. Pechkovsky

Summary We estimated the frequency of CYP1A1, CYP1A2, CYP1B1, CYP19, and SULT1A1 allelic variants in a female population of the Novosibirsk district and their association with the elevated risk of breast (BC), ovarian (OC), and endometrial (EC) cancers. Significant differences (OR = 2.34, $p = 0.0002$) in the allele distributions for CYP1A1 M1 polymorphism between patients with BC ($n = 118$) and controls ($n = 180$) were found. No significant difference in both genotype and allele distributions for CYP1A1 polymorphisms in patients with OC ($n = 96$) and EC ($n = 154$) was observed. Remarkable differences in the allele and genotype distributions for CYP1A2*1F polymorphism in patients with BC or OC were found (OR = 0.26, $p = 0.0000005$ and OR = 0.34, $p = 0.00000002$). There were no differences for this polymorphism in women with EC. In patients with BC no significant differences were found in genotype and allele distributions for R264C polymorphism in the *CYP19* gene. The frequency of a mutant CYP19 heterozygote genotype C/T was higher in patients with OC and EC compared with healthy women (OR = 3.87, $p = 0.001$ and OR = 3.73, $p = 0.0004$, respectively). Comparison of allele frequencies revealed a deficiency of an allele A of SULT1A1*2 in patients with OC (OR = 0.64, p = 0.019) compared with controls. No differences were found in the genotype and allele distributions for SULT1A1 polymorphism between patients with BC and EC and controls. In addition, there were no difference in allele and genotype distributions for CYP1B1 119G→T polymorphism between BC and control. In conclusion, these results support the hypothesis that susceptibility gene alleles of estrogen-metabolizing enzymes may differentially influence risk for woman hormone-dependent cancers.

Introduction

Hormone-dependent cancer such as breast, ovarian, and endometrial cancers (BC, EC, and OC) shows a growing incidence rate all over the world including Russia. In the city of Novosibirsk, the incidence of BC, EC, and OC is approximately 80, 25, and 12/100 000 women. Currently, a common strategy to study mechanisms of hormone-dependent cancer is impairment of estrogen metabolism in general and

17β-estradiol (E_2) in particular, resulting in enhanced circulating hormone levels in blood or at local target tissues. At the present time, the ability of estrogens to stimulate the growth of a number of endocrine cancers is well established (1, 2). In most of human BC, OC, and ECs, estrogens, especially E_2, have been shown to contribute greatly to the growth and development of these tumors and some of these cancers require estrogen for their continued growth (3, 4). Excessive exposure to endogenous and exogenous estrogens increases cell division that increases cancer risk (5). Estrogens are produced by conversion of androgens after a series of complex biochemical reactions. The key role in these reactions belongs to the aromatase enzyme CYP19 (6). Estrogens are metabolized by several enzymes, including cytochrome P450s. The role of such P450 isozymes such as CYP1A1, CYP1A2, and CYP1B1 is to oxidize estrogens resulting in the formation of substrates for phase II metabolism of xenobiotics (7). Further utilization of hormone metabolites is carried out via sulfotransferase (8). Any breach in any of these systems may lead to significant changes in estrogen levels that may result in the development of malignant tumors.

A crucial mechanism for the development of all above mentioned cancers is excessive production of E_2. However, the development of BC, EC, or OC may occur by different mechanisms. One of which may be impairment of a a particular metabolite that may promote the preferential appearance of a hormone-dependent cancer. To prove this hypothesis, we studied genetic polymorphism present in several estrogen-metabolizing enzymes: CYP1A1, CYP1A2, CYP1B1, CYP19, and SULT1A in a woman population of the Novosibirsk region (Russia) aged 45 ± 17 years with BC, OC, and EC ($n = 358$). As a control group, women of the same age group without gynecological diseases ($n = 180$) were also studied. Clinical diagnosis was provided by board-certified gynecologists–oncologists from the Regional Clinical Oncological Hospital of Novosibirsk. Prior to enrollment for the study, its aims were fully explained and informed consent was obtained from each patient. The study protocol was reviewed and approved by the appropriate Institutional Review Boards.

Results

For detection of germ line mutations in estrogen-metabolizing enzymes, the PCR-RFLP method was used. The following functional polymorphisms were studied: the amino acid substitution Ile→Val increasing the enzyme activity of CYP1A1 by a few times (9); the substitution C→A in the 734 position for *CYP1A2* gene resulting in significant decrease in the protein activity (10); the functional polymorphism G→T in the codon 119 for the *CYP1B1* gene associated with BC risk (11) and a nucleotide substitution G638→A resulting in the Arg213His substitution for the SULT1A1 sulfotransferase leading to significant decrease (as high as 85%) of enzyme activity (12); the substitution C→A in the 264 codon for *CYP19* gene (Arg264Cys polymorphism) changing the enzyme stability (13). All polymorphisms examined in the study were explored with regard to whether the genotype frequencies observed were in agreement with those predicted by the Hardy–Weinberg equilibrium. χ^2 analysis was used to test deviation of genotype distributions from Hardy–Weinberg equilibrium.

The genotype and allele distributions for CYP1A1, CYP1A2, CYP19, CYP1B1, and SULT1A1 polymorphisms for both patients and controls are presented in Tables 1 and 2. Significant differences in the allele distributions for CYP1A1 M1 polymorphism between patients with BC and controls were found (OR = 2.34, $p = 0.0002$). A frequency of mutant genotypes T/C + C/C was also higher in BC patients (OR = 2.2, $p = 0.003$). There were no significant differences in both genotype and allele distributions for CYP1A1 polymorphisms in patients with OC and EC.

Remarkable differences in allele and genotype distribution for CYP1A2*1F polymorphism in patients with BC were found. There was a deficiency of a mutant allele A of CYP1A2*1F in BC patients (OR = 0.26, $p = 0.0000005$) as well as genotypes C/A + A/A (OR = 0.03, $p = 0.00000003$). Similar results were obtained for OC (OR = 0.34, $p = 0.00000002$ for alleles frequencies and OR = 0.10, $p = 0.001$ for C/A + A/A genotypes).

On the other hand, there were no differences for this polymorphism in women with EC. In patients with BC, there were no significant differences in both genotype and allele distributions for CYP19 polymorphism compared with OC and EC. The frequency of mutant CYP19 heterozygote genotype C/T was higher in patients with OC and EC compared with healthy women (OR = 3.87, $p = 0.001$ and OR = 3.73, $p = 0.0004$, respectively). Comparison of the allele frequencies revealed a deficiency of the allele A of SULT1A1*2 in patients with OC (OR = 0.64, $p = 0.019$) compared with controls. No differences were found in the genotype and allele distributions for SULT1A1 polymorphism between patients with BC and EC and controls. There is also no difference in allele and genotype distributions for CYP1B1 119G→T polymorphism between BC and control.

Discussion

At the present time, the hormonal origin of different forms of BC as well as female reproductive organs has been proven. It is also beyond any doubt that excessive production of estrogens, in particular E_2, is needed for the development and growth of tumors. There are many reasons for overproduction of hormones. One of the key reasons is errors in estrogen metabolism that may be due to both endogenous and environmental factors. Herein, we studied genetic factors, in particular genes of steroid-metabolizing enzymes that may play an important role in the susceptibility to hormonal carcinogenesis. We found that the frequency distribution for mutant alleles and genotypes of some of these enzymes is different for the studied cancers.

It is worth noting that a high frequency of the mutant allele CYP1A1 was observed only in BC patients. It was demonstrated that Msp1 genetic polymorphism (M1) in this enzyme might be a susceptibility factor for BC in Taiwan (14).

Table 1 Alleles and genotype frequencies of CYP1A1, CYP1A2, and polymorphisms

	Breast Cancer			Ovarian Cancer			Endometrial Cancer			
CYP1A1	n	OR (95% CI)	P-*value*[a]	n	OR (95% CI)	P-*value*[a]	n	OR (95% CI)	P-*value*[a]	Controls
Alleles freq.										
T	182			165			276			316
C	54	2.34 (1.49–3.66)	**0.0002**	27	1.72 (0.99–2.99)	0.314	32	0.91 (0.56–1.49)	0.802	40
Gentotype freq.										
T/T	74			69			122			148
T/C	34	1.7 (0.99–2.90)	0.069	27	1.44 (0.82–2.54)	0.237	32	0.97 (0.57–1.63)	0.999	40
C/C	10		**0.00002**	0		1.000	0		1.000	0
T/C + *C/C*	44	2.2 (1.31–3.66)	**0.003**	27	1.44 (0.82–2.54)	0.237	32	0.97 (0.57–1.63)	0.999	40
Total	**118**			**96**			**154**			**178**
CYP1A2										
Alleles freq.										
C	81			72			72			62
A	105	0.26 (0.18–0.40)	**0.0000005**	120	0.34 (0.23–0.51)	**0.0000000**	260	0.75 (0.51–1.09)	0.149	298
Gentotype freq.										
C/C	22			9			4			2
C/A	38	0.05 (0.01–0.26)	**0.00004**	54	0.20 (0.04–1.00)	0.054	64	0.55 (0.09–3.12)	0.999	58
A/A	33	0.02 (0.005–0.11)	**0.0000005**	33	0.06 (0.01–0.30)	**0.00008**	98	0.40 (0.07–2.27)	1.000	120
C/A + *A/A*	71	0.03 (0.008–0.15)	**.00000003**	87	0.10 (0.02–0.51)	**0.001**	162	0.45 (0.08–2.51)	0.999	178
Total	**93**			**96**			**166**			**180**

n - number of individuals in each category; bold values are statistically significant; [a]Fisher exact two-tailed P-values.

Table 2 Alleles and genotype frequencies of CYP19 and SULT1A1 polymorphisms

	Breast Cancer			Ovarian Cancer			Endometrial Cancer			
CYP1A19	n	OR (95% CI)	P-*value*[a]	n	OR (95% CI)	P-*value*[a]	n	OR (95% CI)	P-*value*[a]	Controls
Alleles freq.										
C	208			174			280			346
T	8	0.73 (0.31–1.73)	0.540	18	1.98 (1.00–3.91)	**0.047**	28	1.92 (1.04–3.54)	0.108	18
Gentotype freq.										
C/C	100			78			126			168
C/T	8	1.34 (0.51–3.51)	0.618	18	3.87 (1.71–8.78)	**0.001**	28	3.73 (1.74–7.96)	**0.0004**	10
T/T	0		0.300	0		0.312	0		0.140	4
C/T + *T/T*	8	0.96 (0.39–2.36)	0.999	18	2.76 (1.31–5.85)	**0.009**	28	2.66 (1.34–5.27)	**0.0004**	14
Total	**108**			**96**			**154**			**182**
SULT1A1										
Alleles freq.										
G	86			120			156			187
A	78	0.98 (0.67–1.41)	0.925	72	0.64 (0.45–0.92)	**0.019**	168	1.16 (0.60–1.57)	0.358	173
Gentotype freq.										
G/G	23			45			48			63
G/A	40	1.79 (0.96–3.34)	0.087	30	0.68 (0.38–1.23)	0.240	60	1.29 (0.76–2.16)	0.358	61
A/A	19	0.92 (0.45–1.88)	0.859	21	0.52 (0.27–0.98)	0.061	54	1.26 (0.74–2.15)	0.419	56
G/A + *A/A*	59	1.38 (0.78–2.44)	0.320	51	0.61 (0.36–1.01)	0.069	114	1.27 (0.81–2.01)	0.300	117
Total	**82**			**96**			**162**			**180**

Li et al. (15) suggested that CYP1A1 M1-containing and M3-containing genotypes increased BC risk associated with a long duration (>20 years) of cigarette smoking, but the effects of the CYP1A1 genotype appear to be quite weak. In our case, mutant genotype T/C, especially T/C + C/C, is an appropriate candidate for studying the contribution of genetic factors to BC susceptibility. The most remarkable differences were revealed for CYP1A2*1F polymorphism in both BC and OC patients. Recently, Long et al. showed that CYP1A2*1F polymorphism may not be related to BC risk (16). In our study, this polymorphism relates to risk of BC and OC, but not EC development. It can imply that reducing the formation of 2-OH metabolite of E_2 may be important for these cancers.

Aromatase or CYP19 is a major enzyme in estrogens synthesis (2) and genetic polymorphism in this enzyme contributes to the development of many hormonal cancers (17). In our study R264C polymorphism in *CYP19* gene was not associated with BC susceptibility but associated with OC and EC susceptibility (Table 2). Thus, this polymorphism may be important for these cancers.

Conjugation of estrogens and their metabolites is an important step for steroid degradation. We did not show any remarkable results in studied Arg213His polymorphism for SULT1A1. Only in the case of OC, we showed the higher frequency of G allele of SULT1A1*2 compared with healthy women. These results confirmed our previous data about this polymorphism for hormone-dependent cancers (18). It is also possible that studied SULT1A1 polymorphism does not play significant role for risk of BC and EC where effects of hormonal SULT1E1 are probably higher.

While studying BC, we have not revealed any clear distinctions between the frequency of alleles and genotypes for CYP1B1 (Table 3). At the moment there is evidence supporting the postulated role of CYP1B1 and 4-OH-E_2 in tumor initiation and growth in estrogen-induced carcinogenesis (19). Moreover, the 4-OH-lation activities of the variant enzymes, especially polymorphisms on codons 119 and 432, were two to fourfold higher than the wild-type enzyme (20). Apparently, this metabolism is not decisive for BC in the examined female group. Other studies also

Table 3 Alleles and genotype frequencies of CYP1B1 polymorphisms[a]

		Breast Cancer		
CYP1B1	*n*	OR (95% CI)	*P* value[b]	Control
Alleles freq.				
G	196			186
T	38	0.66 (0.42–1.05)	0.103	54
Genotype freq.				
G/G	84			78
G/T	28	0.86 (0.47–1.57)	0.649	30
T/T	5	0.38 (0.13–1.14)	0.124	12
G/T + *T/T*	33	0.72 (0.42–1.26)	0.267	42
Total	**117**			**120**

[a] Ovarian and endometrial cancers have not been determined; *n* – number of individuals in each category; bold values are statistically significant

[b] Fisher exact two-tailed *P* values

did not reveal an association of this polymorphism with risk of BC (21). How this polymorphism affects development of OC and EC is still under our study. In 2003, Japanese authors reported for the first time that rare polymorphisms at codons 119 and 432 of the *CYP1B1* gene have higher risk for EC (22) though for OC there was no association (23).

Our results suggest that despite common hormonal origin of tumors under study their development may be affected by various genetically determined features like *CYP1A1*, *CYP1A2*, *CYP19*, and *SULT1A1* genes. It is quite probable that studied estrogen-metabolizing enzymes contribute to formation of different tumors not in the same way and this could be important for identification of prognostic markers for these cancers.

References

1. White R, Parker MJ (1998) Molecular mechanisms of steroid hormone action. Endocr Relat Cancer 5:1–14.
2. Simpson ER (2003) Sources of estrogen and their importance. J Steroid Biochem Mol Biol 86: 225–230.
3. Sasano H, Harada N (1998) Intratumoral aromatase in human breast, endometrial, and ovarian malignancies. Endocrine Rew 19:593–607.
4. Anderson E (2001) Ovarian steroids and control of proliferation in the normal human breast. Breast 4:273–278.
5. Doisneau-Sixou SF, Sergio CM, Carroll JS, et al. (2003) Estrogen and antiestrogen regulation of cell cycle progression in breast cancer cells. Endocr Relat Cancer 10:179–186.
6. Bulun SE, Sebastian S, Takayama K, et al. (2003) The human *CYP19* (aromatase P450) gene: update on physiologic roles and genomic organization of promoters. J Steroid Biochem Mol Biol 86:219–224.
7. Tsuchiya Y, Nakajima M, Yokoi T (2005) Cytochrome P450-mediated metabolism of estrogens and its regulation in human. Cancer Lett 227:115–124.
8. Strott CA (2002) Sulfonation and molecular action. Endocr Rev 23:703–732.
9. Kawajiri K, Nakachi K (1990) Identification of genetically high risk individuals to lung cancer by DNA polymorphism of the cytochrome P450 1A1 gene. FEBS Lett 263:131–133.
10. Goodman MT, Sachse C, Bhambra U, et al. (2003) Polymorphisms in the cytochrome P450 CYP1A2 gene (*CYP1A2*) in colorectal cancer patients and controls: allele frequencies, linkage disequilibrium and influence on caffeine metabolism. Br J Clin Pharmacol 55:68–76.
11. Bailey LR, Roodi N, Dupont WD, Parl FF (1998) Association of cytochrome P450 1B1 (CYP1B1) polymorphism with steroid receptor status in breast cancer. Cancer Res 58:5038–5041.
12. Raftogianis R, Creveling C, Weinshilboum R, Weisz J (2000) Estrogen metabolism by conjugation. J Natl Cancer Inst Monogr 27:113–124.
13. Kagawa N, Hori H, Waterman MR, Yoshioka S (2004) Characterization of stable human aromatase expressed in *E. coli*. Steroids 69:235–243.
14. Huang CS, Shen CY, Chang KJ, et al. (1999) Cytochrome P4501A1 polymorphism as a susceptibility factor for breast cancer in postmenopausal Chinese women in Taiwan. Br J Cancer 80:1838–1843.
15. Li Yu, Millikan RC, Bell DA, et al. (2004) Cigarette smoking, cytochrome P4501A1 polymorphisms, and breast cancer among African-American and white women. Breast Cancer Res 6:R460–R473.
16. Long JR, Egan KM, Dunning L, et al (2006) Population-based case-control study of AhR (aryl hydrocarbon receptor) and CYP1A2 polymorphisms and breast cancer risk. Pharmacogenet Genomics 16:237–243.

17. Ma CX, Adjei AA, Salavaggione OE, Coronel J (2005) Human aromatase: gene resequencing and functional genomics. Cancer Res 65:11071–11082.
18. Mikhailova ON, Gulyaeva LF, Prudnikov AV, et al. (2006) Estrogen-metabolizing gene polymorphisms in the assessment of female hormone-dependent cancer risk. Pharmacogenomics J 6:189–193.
19. Hanna IH, Dawling S, Roodi N, et al. (2000) Cytochrome P450 1B1 (CYP1B1) pharmacogenetics: association of polymorphisms with functional differences in estrogen hydroxylation activity. Cancer Res 60:3440–3444.
20. Shimada T, Watanabe J, Kawajiri K, et al. (1999) Catalytic properties of polymorphic human cytochrome P450 1B1 variants. Carcinogenesis 20:1607–1614.
21. Rylander-Rudqvist T, Wedren S, Granath F, et al. (2003) *Cytochrome P450 1B1* gene polymorphisms and postmenopausal breast cancer risk. Carcinogenesis 24:1533–1539.
22. Sasaki M, Tanaka Y, Kaneuchi M, et al. (2003) *CYP1B1* gene polymorphisms have higher risk for endometrial cancer, and positive correlations with estrogen receptor α and estrogen receptor β expressions. Cancer Res 63:3913–3918.
23. Sellers TA, Schildkraut JM, Pankratz VS, et al. (2005) Estrogen bioactivation, genetic polymorphisms, and ovarian cancer. Cancer Epidemiol Biomarkers Prev 14:2536–2543.

IRF-1 Promotes Apoptosis in p53-damaged Basal-type Human Mammary Epithelial Cells: A Model for Early Basal-type Mammary Carcinogenesis

Michelle L. Bowie, Catherine Ibarra, and Victoria L. Seewalt

Summary Mammary gland homeostasis is regulated by both endogenous and exogenous signals, creating a balance between proliferation and apoptosis. It is thought that breast cancer develops from the acquisition of multiple genetic changes. The function of tumor suppressor p53 is fequently lost in cancers; however, not all cells that lose p53 progress to become invasive cancer. We have developed a model of early mammary carcinogenesis to investigate some of the internal and external signaling pathways that target the elimination ot normal basal-type human mammary epithelial cells (HMECs) that acutely acquire p53-damage. Here, we show that both tamoxifen (Tam) and three-dimensional prepared extracellular matrix culture (3-D rECM) induce apoptosis in HMEC cells with acute loss of p53 [*p53($^-$) HMECs] through induction of interferon regulatory factor-1 (IRF-1). Tam and rECM signaling in *p53($^-$) HMECs (1) promotes the recruitment of a STAT1/CBP complex to the *IRF-1 promoter*, (2) upregulates IRF-1, (3) activates caspase-1 and -3, and (4) induces apoptosis. Suppression of IRF-1 with siRNA oligos inhibited both Tam- and rECM-induced apoptosis. These observations demonstrate that IRF-1 plays a critical role in eliminating p53-damaged cells, and may play a more global role in mammary gland homeostasis.

Introduction

Breast cancer (BC) is thought to be a multistep process resulting in the accumulation of genetic damage. Not all cells, however, that acquire damage go on to become invasive cancer. Instead they are thought to be eliminated by apoptosis. The tumor suppressor p53 is frequently lost in cancers and plays a pivotal role in apoptotic signaling, yet not all cells with loss of p53 progress to become cancer. Inactivation of p53 is frequently observed in basal-type epithelial BCs (1, 2). Young African American women and BRCA1 mutation carriers are predominantly affected by this cancer type (2, 3). Clinically, basal-type BCs are characterized as ER/PR($^-$/$^-$), Her2/neu($^-$), p53($^-$), and cytokeratin 5/6($^+$/$^+$) (3). Given the poor prognosis associated with basal-type BC, improved understanding of the early biology of this cancer could lead to development of effective prevention strategies. Here, we identify a role for interferon regulatory factor-1 (IRF-1) in targeting the elimination of p53-damaged basal-type mammary epithelial cells.

IRF-1 was originally identified as a transcription factor that binds to the "enhancer-like" elements of the human interferon beta (*IFN-β*) promoter, and is known to promote apoptosis in response to viral infections (4). IRF-1 has been shown to be cell cycle regulated and to be induced by a number of different cytokines, including IFN-α, -β, and -γ, TNF-α, IL-6, and prolactin (5). Upon cytokine stimulation, the family of protein tyrosine kinases, Jak/STAT, are activated and transcriptionally regulate the *IRF-1* gene. In turn, IRF-1 induces the expression of additional interferon-stimulated genes (ISGs) through complex formation between interferon consensus sequence 2/gamma-interferon-activated sequence (ICS2/GAS) element and STAT1 (6).

In the absence of functional p53, IRF-1 promotes apoptosis in response to DNA damage (7, 8). Recently, IRF-1 was shown to mediate growth arrest and apoptosis in BC cell lines (9, 10). Tamura et al. showed that p53-independent DNA damage induced apoptosis in mouse T lymphocytes dependent on IRF-1. In addition, ectopic expression of IRF-1 induces cell death in human BC cell lines (11). More recently, IRF-1 has been implicated in p53-independent apoptotic signaling in BC cell lines treated with type II interferon (IFN-γ) and the antiestrogen ICI 182,780, respectively (7, 12). Simultaneous abrogation of IRF-1 and p53 in mice resulted in significant increase in spontaneous tumor formation compared with mice lacking p53 alone (13). These observations suggest that IRF-1 plays a role in p53-independent apoptotic signaling in mouse and human mammary epithelial cells.

There is a growing body of evidence that suggests the importance of IRF-1 in mammary gland homeostasis and hormone responsiveness. In addition to its growth inhibitory effects, IRF-1 has been shown to play a critical role in mammary gland involution in the rat (10, 13). Reduced IRF-1 expression or gene mutations are frequently observed in multiple cancers, including BC (14, 15). Although IRF-1 expression is observed in normal mammary epithelial cells, expression was lost in high-grade ductal carcinoma in-situ and invasive BCs (14). We have shown that therapeutic levels of tamoxifen (Tam) (1.0 μM) promote apoptosis in p53-damaged human mammary epithelial cells (HMECs) through rapid, "nonclassic" signaling that is initiated at the cell membrane surface and through IRF-1 induction (16–18). In addition, treatment with ICI 182,780 (a pure estrogen antagonist) also results in induction of IRF-1 (12, 19). These observations suggest a role for IRF-1 in mammary gland homeostasis and hormone-signaling.

The focus of this study is to elucidate the functional significance of IRF-1 in the elimination of p53-damaged basal-type mammary epithelial cells, as mediated by Tam and three-dimensional extracellular matrix culture (3-D rECM). The results presented here indicate that loss of IRF-1 signaling would increase the survival of a p53-damaged mammary epithelial cell and provide evidence that loss of IRF-1 is a short-term marker of early basal-type BC risk.

Results

Building a Model of p53 Loss in Basal-Type HMECs. The primary HMEC strain used for these studies was AG11132/172R/AA7, a cytokeratin 5/6($^{+/+}$), Her2/neu($^{-}$), basal cytokeratin ($^{+}$) HMEC strain with low ER/PR expression that was derived from the breast tissue of a young African American woman (20). Results were confirmed in two additional HMEC strains obtained from young women that were also cytokeratin 5/6($^{+/+}$) and exhibited low ER/PR expression. In these basal-type HMECs, we modeled the acute loss of p53 function using expression of HPV-16 E6 (*p53($^{-}$) HMEC-E6) to study the role of Tam and rECM signaling in mediating apoptosis in basal cytokeratin ($^{+}$) HMEC strains. In previous studies, we confirmed that p53-specific antisense ODNs and HPV-16 E6 resulted in the same apoptosis-sensitive phenotype (21).

3-hD culture is an adaptation of the large-scale culture system developed by the laboratory of Mina Bissell (22). This contrasts our prior studies employing small scale 3-D rECM culture where HMECs are plated as single cell suspension in semi-solid growth factor-depleted rECM (21, 23). For the experiments described here, large-scale 3-hD culture was chosen over small-scale 3-D culture to precisely synchronize our cells for promoter recruitment and differential gene expression studies.

Tam and rECM Culture Induce Expression of IRF-1. As seen in Fig. 1, IRF-1 protein was induced in *p53($^{-}$) HMEC-E6 cells by 1.0 μM Tam at 30 min and maximally (2.3-fold) by 3 h. Similar results were observed for 3-hD rECM by 1 h (2.5-fold) (data not shown). These data indicate that Tam and 3-hD rECM converge to upregulate IRF-1 expression in basal-type HMECs.

Tam and rECM Promote Tecruitment of CBP to the GAS Element of the *IRF-1* Promoter. STAT1 and CBP are both known transcriptional regulators of type II interferon signaling (24). Chromatin immunoprecipitation studies (ChIP) were preformed to test for STAT1 and CBP binding to the *IRF-1* GAS element. STAT1 was bound to the *IRF-1* GAS element at baseline in basal-type *p53($^{-}$) HMEC-E6 cells but not in p53($^{+}$) HMEC-LX controls (Fig. 2, data not shown). Tam-treatment of *p53($^{-}$) HMECs or growth in 3-hD rECM promoted CBP recruitment to the *IRF-1* GAS element (Fig. 2A, B). In contrast, neither STAT1 nor CBP were bound or

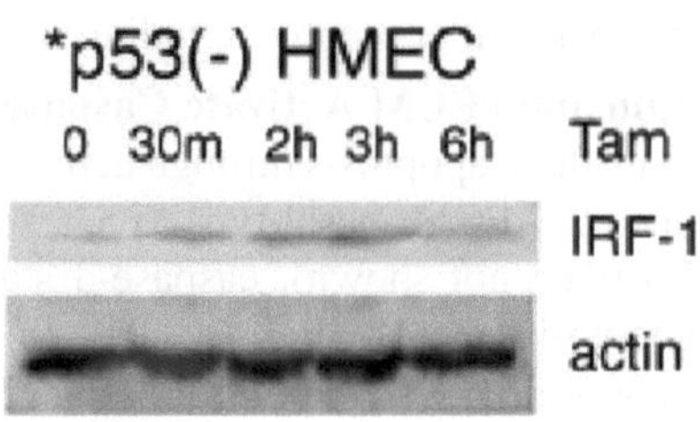

Fig. 1 Western analysis demonstrates that Tam promotes induction of IRF-1 protein in *p53($^{-}$) HMEC-E6 cells

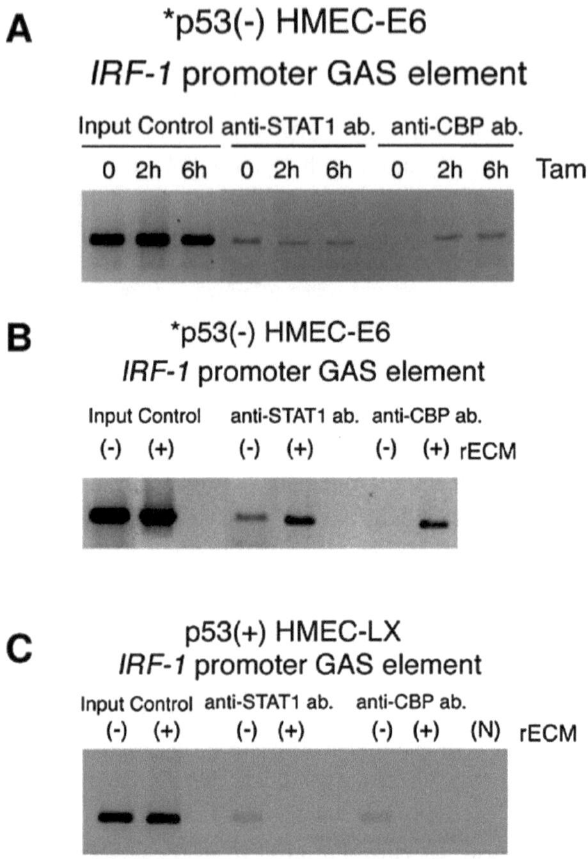

Fig. 2 ChIP studies were performed to test for STAT1 and CBP recruitment to the *IRF-1* GAS element as a function of Tam treatment and growth in 3-hD rECM culture. Basal-type *p53(−) HMEC-E6 cells were treated with 1.0 μM Tam (**A**) for 0–6 h or grown in 3-hD rECM culture (**B**) for 0 (–) and 2 h (+). For comparison, p53(+) HMEC-LX cells grown in 3h-rECM were also tested (**C**)

recruited to the *IRF-1* GAS element in p53(+) HMEC-LX vector control cells (Fig. 2C, data not shown). These observations show that binding of STAT1 and CBP to the *IRF-1* GAS element is observed during Tam and 3-hD rECM p53-independent apoptotic signaling in basal-type HMECs.

Tam and rECM Activate Caspase-1/-3 and Induce Apoptosis. IRF-1 is thought to mediate apoptosis through activation of caspase-1/-3 (25). In the presence of Tam or 3-hD rECM, basal-type *p53(−) HMEC-E6 cells activated caspase-1 starting at 3 h (data not shown), caspase-3 starting at 12 h (Fig. 3A, B), and effector-phase apoptosis at 24 h (Fig. 3C, D). In contrast, Tam did not induce caspase-1/-3 in p53(+) HMEC-LX controls (Fig. 3A, C, data not shown). Pretreatment with a

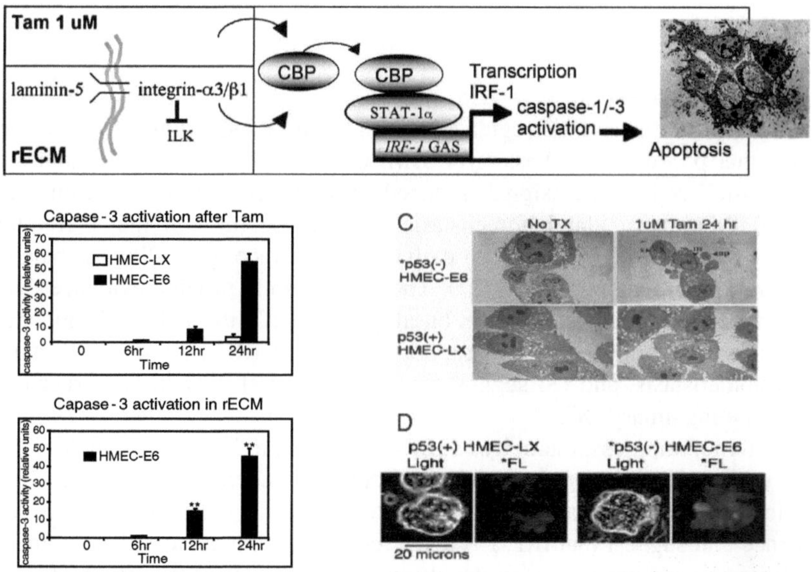

Fig. 3 Tam- and rECM-mediated caspase-3 activation and apoptosis. (**A**) Caspase-3 is activated in Tam-treated basal-type *p53(⁻) HMEC-E6 cells but not in p53(⁺) HMEC-LX controls. (**B**) Caspase-3 is induced in basal-type *p53(⁻) HMEC-E6 cells grown in 3-hD rECM culture starting at 12 h and maximally at 24 h. *, statistically significant compared with NS IgG and control, $p < 0.01$. (**C**) Apoptosis in Tam-treated basal-type *p53(⁻) HMEC-E6 cells is evidenced by cell shrinkage *(s)*, formation of apoptotic bodies containing cellular organelles (***ap***), and margination of chromatin ***(m)***. No apoptosis was observed in p53(⁺) HMEC-LX controls grown in 3hD-rECM. (**D**) Qualitative caspase-3 activation visualized on a fluorescence microscope. Light = total cells and FL= fluorescent positive cells. p53(⁻) HMEC-LXSN control cells showed no caspase-3 activation

caspase-1 inhibitor block Tam-mediated caspase-1 activation and apoptosis (data not shown). Pretreatment with alpha-3 and beta-1 integrin-blocking antibodies inhibited caspase-1 activation in *p53(⁻) HMEC-E6 cells cultured in 3-hD rECM (data not shown). In contrast, p53(⁺) HMEC-LX control cells did not activate the effector caspase-3 or undergo apoptosis (Fig. 3D). These observations further support that Tam and 3-hD rECM induce p53-independent apoptosis and caspase-activation in basal-type *p53(⁻) HMEC-E6.

Suppression of IRF-1 Blocks Tam- and rECM-Mediated Caspase-1 and -3 Activation. We tested whether suppression of IRF-1 would block Tam- and rECM-induced apoptosis in *p53(⁻) HMEC-E6 cells. Suppression of IRF-1 was confirmed at 12 h. Pretreatment of basal-type *p53(⁻) HMEC-E6 cells with two different IRF-1 siRNA oligos blocked activation of caspase-1/-3 by both Tam treatment and 3-hD rECM culture. These data show that IRF-1 expression is required for both Tam- and rECM-mediated activation of caspase-1/-3 and suggest the following model:

Discussion

Our studies demonstrate that Tam- and rECM-mediated apoptosis in basal-type *p53(⁻) HMEC-E6 cells is regulated through IRF-1 induction. We have recently shown that p53-damaged basal-type HMECs undergo apoptosis through a rapid, "nonclassic" cell surface signal, induced by therapeutic levels of Tam (1.0 μM) (16–18). Mammary gland homeostasis is regulated through the interactions of extracellular matrix and the epithelial cell surface providing a balance of proliferation and apoptosis (26, 27). These studies (1) provide evidence that both Tam and ECM induce apoptosis in basal-type HMECs through a p53-independent pathway mediated by IRF-1, (2) support the role of IRF-1 signaling in mammary gland homeostasis, and (3) suggest that loss of IRF-1 may be a marker of early basal-type mammary BC.

The tumor suppressor activity of IRF-1 is related to its ability to induce apoptosis in damaged cells, both in a p53-dependent and independent manner (7). IRF-1 was initially identified as an IFN-responsive gene (11). There are many lines of evidence that suggest that IRF-1 is a key regulator of mammary gland homeostasis and hormone responsiveness in the absence of IFN treatment or production. Here, we demonstrate that both Tam and rECM promote apoptosis in basal-type p53-damaged HMECs through IRF-1 induction and caspase-1/-3 activation. Our studies show that the proapoptotic effects of rECM and Tam involve signaling pathways that converge on IRF-1. Our studies are consistent with other reports that highlight the importance of IRF-1 signaling in mammary gland homeostasis (14–16, 28).

We employed AG11132/172R/AA7 HMEC as a model of basal-type epithelial BC derived from a breast tissue of young African American woman with low ER/PR expression, cytokeratin 5/6 ($^{+/+}$), Her2/neu($^{-}$), and basal cytokeratin ($^{+}$) (20). To model loss of p53 function, we retrovirally transfected our basal cytokeratin($^{+}$) HMEC strain with HPV-16 E6 to study the role of rECM signaling in mediating apoptosis. Previously, we confirmed that p53-specific antisense ODNs and HPV-16 E6 shared similar apoptosis-sensitive phenotype (21). By modeling loss of p53 function in our ER/PR-"poor", Her2/neu($^{-}$), basal cytokeratin($^{+}$) HMECs derived from a young African American woman, we are providing a potentially relevant model of "early" basal-type breast carcinogenesis. Importantly, information gained from this model may provide insights into the early biology of basal-type BCs.

Our studies also provide evidence that IRF-1 promotes p53-independent apoptosis in acutely damaged basal-type HMECs. IRF-1 and p53 are thought to participate in parallel damage response pathways. Therefore, if basal-type mammary epithelial cells are acutely damaged through loss of p53 function, IRF-1 signaling represents an alternative pathway to eliminate p53-damaged cells. A subsequent loss of IRF-1 function in basal cytokeratin($^{+}$) mammary epithelial cells may provide a cellular environment in which the risk of invasive basal-type BC increases. These observations have important clinical implications, especially with regards to prognosis and treatment strategies. Therefore, understanding the precise role of IRF-1 in mediating p53–independent apoptosis will further aid in early detection and the development of biomarkers to predict short-term risk of basal-type BC.

References

1. Borresen-Dale AL (2003) TP53 and breast cancer. Hum Mutat *21:* 292–300.
2. Perou CM, Sorlie T, Eisen MB, et al. (2000) Molecular portraits of human breast tumours. Nature *406*: 747–752.
3. Foulkes WD, Stefansson IM, Chappuis PO, et al. (2003) Germline BRCA1 mutations and a basal epithelial phenotype in breast cancer. J Natl Cancer Inst *95*: 1482–1485.
4. Miyamoto M, Fujita T, Kimura Y, et al. (1988) Regulated expression of a gene encoding a nuclear factor, IRF-1, specifically binds to *IFN-beta* gene regulatory elements. Cell *54*: 903–913.
5. Taniguchi T, Ogasawara K, Takaoka A, et al. (2001) IRF family of transcription factors as regulators of host defense. Annu Rev Immunol *19*: 623–655.
6. Chatterjee-Kishore M, Kishore R, Hicklin DJ, et al. (1998) Different requirements for signal transducer and activator of transcription 1alpha and interferon regulatory factor 1 in the regulation of low molecular mass polypeptide 2 and transporter associated with antigen processing 1 gene expression. J. Biol. Chem. *273*: 16177–16183.
7. Porta C, Hadj-Slimane R, Nejmeddine M, et al. (2005) Interferons alpha and gamma induce p53-dependent and p53-independent apoptosis, respectively. Oncogene *24*: 605–615.
8. Tamura T, Ishihara M, Lamphier MS, et al. (1995) An Irf-1-dependent pathway of DNA damage-induced apoptosis in mitogen-activated T-lymphocytes. Nature *376*: 596–599.
9. Hoshiya Y, Gupta V, Kawakubo H, et al. (2003) Mullerian inhibiting substance promotes interferon gamma-induced gene expression and apoptosis in breast cancer cells. J Biol Chem *278:* 51703–51712.
10. Kim PK, Armstrong M, Liu Y, et al. (2004) IRF-1 expression induces apoptosis and inhibits tumor growth in mouse mammary cancer cells in vitro and in vivo. Oncogene *23*: 1125–1135.
11. Pizzoferrato E, Liu Y, Gambotto A, et al. (2004) Ectopic expression of interferon regulatory factor-1 promotes human breast cancer cell death and results in reduced expression of survivin. Cancer Res *64:* 8381–8388.
12. Bouker KB, Skaar TC, Fernandez DR, et al. (2004) Interferon regulatory factor-1 mediates the proapoptotic but not cell cycle arrest effects of the steroidal antiestrogen ICI 182,780 (faslodex, fulvestrant). Cancer Res *64:* 4030–4039.
13. Nozawa H, Oda E, Nakao K, et al. (1999) Loss of transcription factor IRF-1 affects tumor susceptibility in mice carrying the Ha-ras transgene or nullizygosity for p53. Genes Dev *13:* 1240–1245.
14. Doherty GM, Boucher L, Sorenson K, et al. (2001) Interferon regulatory factor expression in human breast cancer. Ann Surg *233:* 623–629.
15. Connett JM, Badri L, Giordano TJ, et al. (2005) Interferon regulatory factor 1 (IRF-1) and IRF-2 expression in breast cancer tissue microarrays. J Interferon Cytokine Res *25:* 587–594.
16. Bowie ML, Dietze EC, Delrow J, et al. (2004) Interferon-regulatory factor-1 is critical for tamoxifen-mediated apoptosis in human mammary epithelial cells. Oncogene *23*: 8743–8755.
17. Dietze EC, Caldwell LE, Grupin SL, et al. (2001) Tamoxifen but not 4-hydroxytamoxifen initiates apoptosis in p53(–) normal human mammary epithelial cells by inducing mitochondrial depolarization. J Biol Chem *276:* 5384–5394.
18. Dietze EC, Troch MM, Bean GR, et al. (2004) Tamoxifen and tamoxifen ethyl bromide induce apoptosis in acutely damaged mammary epithelial cells through modulation of AKT activity. Oncogene *23:* 3851–3862.

19. Gu Z, Lee RY, Skaar TC, et al. (2002) Association of interferon regulatory factor-1, nucleophosmin, nuclear factor-kappaB, and cyclic AMP response element binding with acquired resistance to Faslodex (ICI 182,780). Cancer Res *62*: 3428–3437.
20. Seewaldt VL, Mrozek K, Dietze EC, et al. (2001) Human papillomavirus type 16 E6 inactivation of p53 in normal human mammary epithelial cells promotes tamoxifen-mediated apoptosis. Cancer Res *61:* 616–624.
21. Seewaldt VL, Mrozek K, Sigle R, Dietze EC, Heine K, Hockenbery DM, Hobbs KB, Caldwell LE (2001) Suppression of p53 function in normal human mammary epithelial cells increases sensitivity to extracellular matrix-induced apoptosis. J Cell Biol *155:* 471–486.
22. Roskelley CD, Desprez PY, Bissell MJ (1994) Extracellular matrix-dependent tissue-specific gene expression in mammary epithelial cells requires both physical and biochemical signal transduction. Proc Natl Acad Sci USA *91:* 12378–12382.
23. Dietze EC, Bowie ML, Mrozek K, et al. (2005) CREB-binding protein regulates apoptosis and growth of HMECs grown in reconstituted ECM via laminin-5. J Cell Sci *118*: 5005–5022.
24. Horvai AE, Xu L, Korzus E, et al. (1997) Nuclear integration of JAK/STAT and Ras/AP-1 signaling by CBP and p300. Proc Natl Acad Sci USA *94:* 1074–1079.
25. Karlsen AE, Pavlovic D, Nielsen K, et al. (2000) Interferon-gamma induces interleukin-1 converting enzyme expression in pancreatic islets by an interferon regulatory factor-1-dependent mechanism. J Clin Endocrinol Metab *85*: 830–836.
26. Farrelly N, Lee YJ, Oliver J, et al. (1999) Extracellular matrix regulates apoptosis in mammary epithelium through a control on insulin signaling. J Cell Biol *144*: 1337–348.
27. Stupack DG, Cheresh DA (2002) Get a ligand, get a life: integrins, signaling and cell survival. J Cell Sci *115:* 3729–3738.
28. Bouker KB, Skaar TC, Riggins RB, et al. (2005) Interferon regulatory factor-1 (IRF-1) exhibits tumor suppressor activities in breast cancer associated with caspase activation and induction of apoptosis. Carcinogenesis *26*: 1527–1535.

Lung Cancer

Steroid Receptor and Growth Factor Receptor Expression in Human Nonsmall Cell Lung Cancers Using Cells Procured by Laser-capture Microdissection

Alan Kerr, II, James F. Eliason, and James L. Wittliff

Summary Few biomarkers exist for management of nonsmall cell lung cancers (NSCLC), although estrogen receptor (ERα and ERβ) and EGF receptor (EGFR) expression has been related to clinical outcome (1–6). To circumvent problems of cellular heterogeneity in whole tissue, relative gene expression of ERα, ERβ, EGFR, and HER-2 (c-erb-B2) was examined in pure lung carcinoma (LC) cells and normal epithelia by LCM. Cell-specific RNA was isolated and purified for RT-qPCR and microarray. Comparison of NSCLC cells to normal epithelia indicated increased levels of mRNA expression of ERβ, ERα, EGFR, and HER-2 by 31%, 38%, 54%, and 62%, respectively, in LCs. The majority of NSCLC exhibiting low ERα and high HER-2 expression were from smokers. Although there was no correlation between ERβ or EGFR expression and smoking history, there appeared to be an inverse relationship between levels of ERβ and EGFR mRNAs in normal and neoplastic lung. Additionally, microarray analyses of LCM cells revealed >2,000 genes significantly altered in LC compared with normal epithelia. Herein, differences in NSCLC gene expression and normal lung cells were noted between specimens from gender and smoking groups. Microarray data revealed ERα expression was associated with alterations in <20 genes while ERβ expression revealed >500 associated genes, suggesting a more prominent role for ERβ in lung. HER-2 mRNA levels appeared associated with >1,000 genes, while EGFR mRNA levels were associated with far fewer genes. Collectively, results suggest quantitative genomic analyses of pure cell populations allow more accurate interpretation of LC status, which is being correlated with clinical outcome.

Introduction

Although the human lung is not a classical endocrine responsive organ, expression and activities of estrogen receptors (ERα and ERβ) both in cell lines and human lung carcinoma (LC) specimens have been reported by several investigators, particularly in nonsmall cell lung cancers, (NSCLC) (1, 2, 5, 6). Also, evidence of biological response to estradiol treatment in NSCLC cells has been reported recently (3). Anti-estrogen treatment induced antiproliferative effects in NSCLC

J.J. Li et al. (eds.), *Hormonal Carcinogenesis V.*

cells, and combined treatment with growth factor receptor inhibitors (i.e., Gefitinib™ and Erlotinib™) further reduced growth (3). The appearance of ERβ protein in the nuclei of both male and female LCs has been suggested as a prognostic indicator (5, 6). Although nuclear ERα has rarely been detected, there are reports of cytoplasmic staining of ERα as an indicator of poor prognosis (4, 5). Kawai et al. (4) reported combined over-expression of EGFR and ERα, determined by immunohistochemistry, correlated with poor clinical outcome, with greater predictive value than over-expression of each individual analyte.

For decades, there have been both basic science and clinical studies observing pathway interaction between growth factor receptor and ER signaling in breast cancer (7–12). Because EGFR and ER protein isoforms appear to exhibit significant activities in LC, similar mechanisms may govern the interaction between these two signaling pathways, thereby providing new approaches for combined targeted therapy.

The goal of this investigation was to evaluate *ER* gene expression in human NSCLC specimens and compare those to EGFR and HER-2 expression levels in order to examine their correlation with patient parameters. RT-qPCR was used to determine relative levels of ER and growth factor receptor mRNA species in pure carcinoma cells procured by LCM. Global gene expression profiles were also developed using microarray analyses (Affymetrix) with RNA from pure carcinoma cells to determine pathways that may be influenced by expression of these putative tumor markers.

Results

Laser Capture Microdissection and RNA Isolation. To perform LCM on deidentified tissue specimens, 5–7 μm frozen sections mounted on uncoated glass slides were handled according to procedures established (13–15). Samples used for this preliminary study (13 NSCLC and 6 normal) were provided by Asterand, PLC. LCM was performed on 5 μm frozen tissue sections supplied on uncharged glass slides. Prior to LCM, structural integrity of tissue sections was assessed after staining with hematoxylin and eosin (Fig. 1). Briefly, frozen sections were fixed in 70% ethanol, followed by hydration in water. Then ~100 μl hematoxylin (Richard-Allan Scientific) was added to each section by filtered syringe and incubated for 5–10 s before a wash in fresh water. Slides were then transferred to 70% ethanol, followed by Eosin Y (ThermoShandon) incubation. After final dehydration in 95–100% ethanol, slides were incubated in xylene and allowed to air dry before performing LCM. Studies conducted in our laboratory (14–15) utilize the PixCell IIe™ LCM System (Arcturus Engineering, Inc.). Cells of interest were microdissected using CapSure™ LCM Caps with the intact cells collected on the transfer film. Total RNA from 3–4,000 cells was isolated from LCM extracts using PicoPure™ reagents (Arcturus Bioscience). Briefly, cells adhering to the LCM cap were lysed by incubation for 30 min at 42 °C in Extraction Buffer™ (XB), followed by binding of the

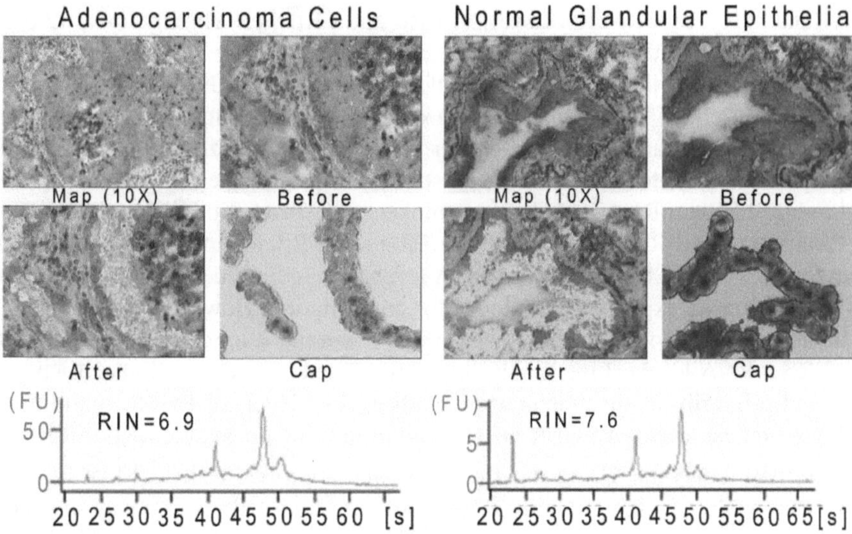

Fig. 1 Representative LCM-procurement of carcinoma (*top left*) and normal epithelia (*top right*) cells. Shown below are the Bioanalyzer profiles of the purified RNA with coinciding RNA integrity number (RIN) indicating intact RNA after LCM

RNA onto a spin column. Once bound, total RNA samples were washed and eluted, employing an on-column DNAse digestion (Qiagen) for 15 min at room temperature. RNA concentration and quality were evaluated using Bioanalyzer™ reagents (Agilent) for determination of 18S and 28S rRNA species. As shown in Fig. 1, intact RNA of sufficient quality was obtained from 3–4,000 microdissected cells.

Biomarker Gene Expression. Reverse transcription was performed using SuperScript™ II reverse transcriptase (Invitrogen) with oligo (dT) primers at 5 ng per reaction. Primer annealing was performed at 70 °C for 5 min, followed by addition of enzyme mix and incubation at 42 °C for 90 min. Real-time PCR was performed in an ABI Prism™ using SYBR™ Green detection (Applied Biosystems). cDNA was diluted in 2 ng μl^{-1} polyinositol (Sigma) to stabilize the template and added to each 20 μl reaction with forward and reverse primers (300 nM final concentration) selected by Primer Express™ (Applied Biosystems). Reactions were performed in duplicate wells for each gene (i.e., ERα, ERβ, EGFR, and HER-2) using β actin as a housekeeping gene. Relative gene expression levels were determined using the ΔΔCt method with human universal reference RNA (Stratagene) used as a calibrator. Experiments were performed in triplicate to calculate a mean ± SD. Both no template and RNA (no RT) controls were used to determine possible DNA contamination.

Both EGFR and HER-2 mRNA levels were amplified in a majority of the LCs (54% with a range of two to sevenfold and 62% with a range of two to tenfold, respectively), compared with normal epithelia. In addition, increased gene expression of both ERα and ERβ was observed in certain tumors compared with normal

(38% and 31%, respectively). Levels of ERα and ERβ were slightly lower (2.0- to 4.5-fold) relative to those of EGFR and HER-2 measured by RT-qPCR. However, in these specimens, the elevated levels of ERα and ERβ mRNAs were significant compared with the normal lung tissue ($p < 0.02$). Figure 2 shows the relative gene expression of ERα and HER-2 in cancer specimens normalized to that of normal lung epithelia. Preliminary results suggest a higher prevalence of increased HER-2 and reduced ERα mRNA in LCs from smokers compared with nonsmokers.

Comparing EGFR expression to both ERα and ERβ, an apparent positive relationship between EGFR and ERα mRNA levels was indicated, although this observation requires an expanded sample set for confirmation. However, there appeared to be an inverse association in either cancer or normal lung cells between EGFR and ERβ mRNA levels (Fig. 3). As shown, increased gene expression of ERβ was observed primarily in specimens with relatively low levels of EGFR, while those with the highest levels of EGFR mRNA had little or no detectable ERβ mRNA.

Microarray Analyses. Using our microgenomic approach with LCM of 3–4,000 cells, each yield of >10 ng quality RNA (indicated by a RIN > 4.0) was subjected to two rounds of amplification using GeneChip™ expression reagents (Affymetrix). T7-oligo (dT) primers were used in the first round reverse transcription, followed by a second strand synthesis with DNA polymerase and an in vitro transcription (IVT) using T7 RNA polymerase. The second round of amplification included biotin incorporation into the cRNA generated from the second round IVT. Second round cRNA yield was measured by absorbance at 260 nm, and sufficient transcript

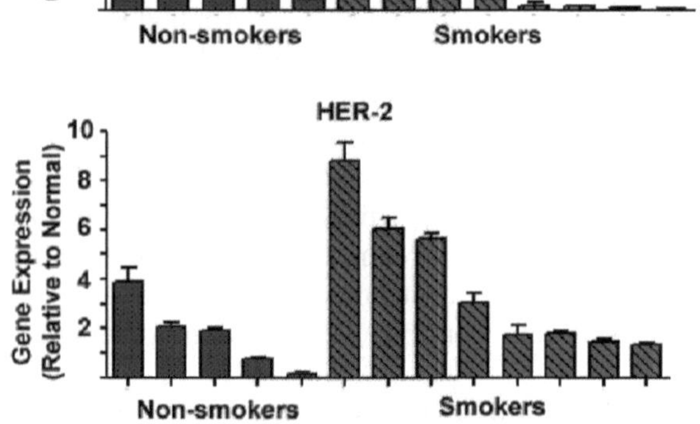

Fig. 2 RT-qPCR analyses using LCM-procured cells to illustrate relative gene expression of ERα and HER-2 in lung cancers compared to normal lung. Results shown represent mean ± SD from triplicate real-time PCR experiments

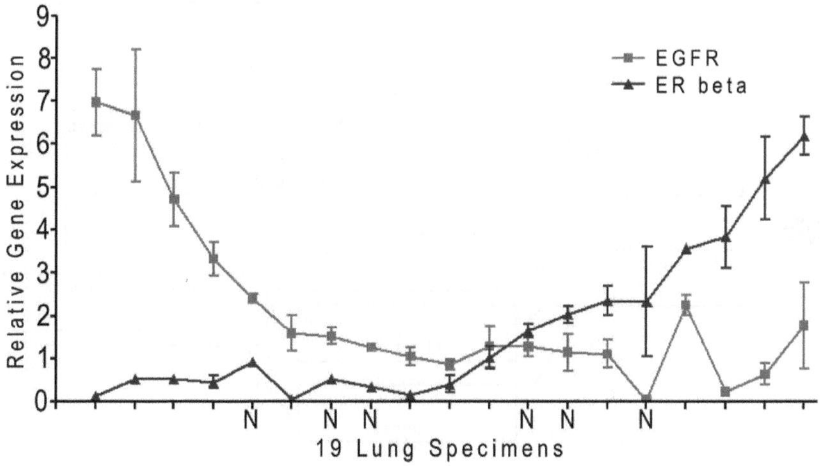

Fig. 3 Relative gene expression of EGFR and ERβ in both cancer and normal (N) lung epithelia procured by LCM. qPCR results represented as mean ± SD from triplicate assays

length (>500 bases) was confirmed by Bioanalyzer™ analysis. The cRNA (20 μg) from each selected sample was fragmented at 90 °C for 30 min and 15 μg fragmented cRNA was added to each hybridization mix (Affymetrix). Following overnight hybridization at 42 °C on HG U133 plus 2.0 microarray chips (Affymetrix), chips were washed and stained with streptavidin-capturing solution for detection of labeled transcripts.

Using 22 probes for each transcript, 11 perfect match and 11 mismatch, expression levels for each gene were calculated by averaging the normalized intensity of each probe set and subtracting the mismatch intensity from the perfect match. Since the two pathways have significant effects on overall gene expression patterns in vivo, global gene expression profiles of pure lung cell populations procured by LCM allow identification of unique molecular signatures associated with expression of the four analytes. The Affymetrix GeneChip U133 platform was used in conjunction with statistical software to determine genes significantly altered in each biomarker subtype.

Principal component analysis (PCA) of the global gene expression profiles obtained from the microarray analyses was performed. Gene expression profiles of cancer and normal epithelia clustered independently, indicating clear differences distinguishing the two groups. Two-way ANOVA analysis of the cancer compared with normal lung specimens revealed nearly 2,000 genes significantly altered (>2.0-fold) between the two groups ($p < 0.01$). Applying a false-discovery rate (FDR) correction reduced the number of significant genes to 171 ($p < 7.81893$ e-005), with 137 of these genes being altered >2.0-fold. Figure 4 shows the gene ontology (GO) map of some of the biological processes altered in the lung cancer specimens compared with normal lung epithelia.

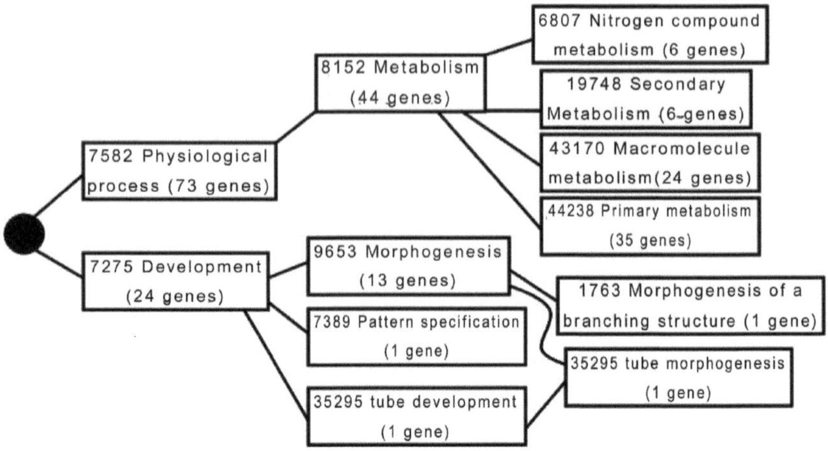

Fig. 4 Gene ontology map of genes (grouped by biological process) altered in LCs compared with normal lung

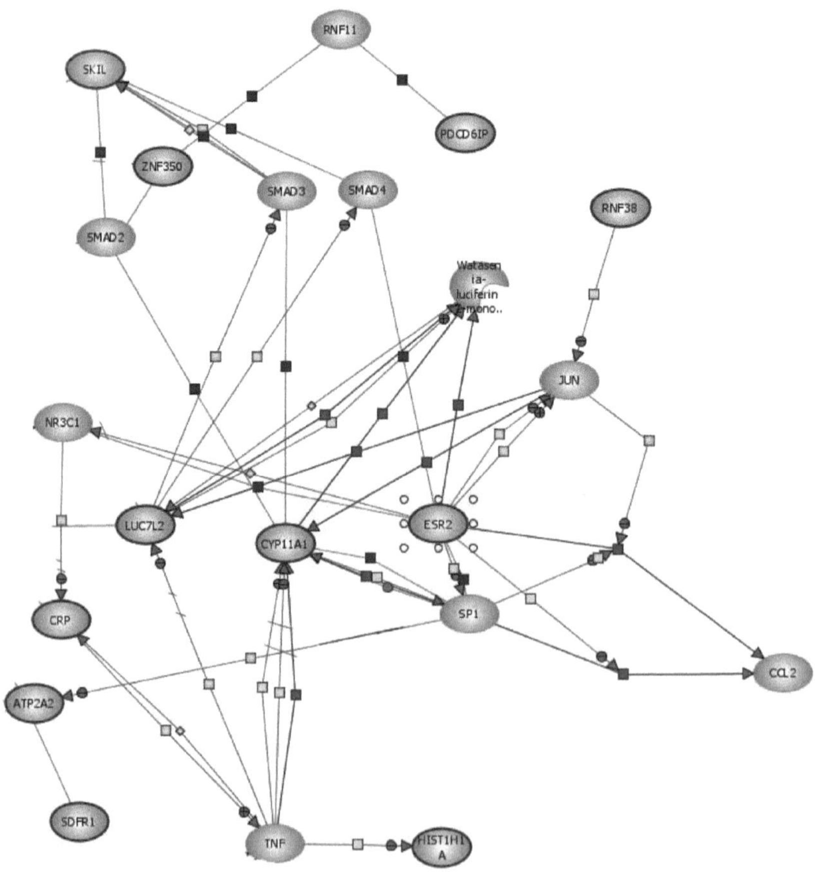

Fig. 5 Pathway interaction network of selected genes associated with ERβ expression

To determine which genes in the global expression profiles are associated with signaling pathways associated with expression of each of the four biomarkers (ERα, ERβ, EGFR, and HER-2) analyzed by RT-qPCR, Pearson correlations were utilized. Interestingly, less than 20 genes exhibited altered expression related to mRNA levels of ERα; in comparison, ERβ mRNA levels were associated with alterations in nearly 500 genes. Figure 5 illustrates some of the interactions among genes associated with ERβ expression as identified by PathwayArchitect™ (Stratagene). Although several genes were influenced directly by ERβ activity, a number of genes in this network appear to be regulated by targets of ERβ signaling, suggesting another level of hormonal control previously unrevealed.

Discussion

In the present study, we optimized the technique of LCM for gene expression profiling in lung tissue using both RT-qPCR for specific gene transcripts, as well as RNA amplification and microarray for global analysis of gene expression. We observed increased levels of both ERα and ERβ mRNA transcripts in NSCLC cells compared with normal lung in this preliminary examination. The inverse relationship between ERβ and EGFR mRNA levels appears similar to observations reported for breast cancer between ERα and EGFR protein expression (7, 8). If ERβ is indeed the more prominent isoform in NSCLC, then it may have similar effects on cell signaling and proliferation as ERα has in breast carcinoma cells.

The power of LCM is most significant when performing microarray analyses. Since global gene expression profiling is sensitive to thousands of mRNA species, those from contaminating cell types in these lung cancer biopsies may significantly alter a gene expression profile when using whole tissue. Since many other cell types express genes also expressed in cancer cells, such as those for growth factors, angiogenesis factors and proteases, it is essential to filter these gene patterns from contaminating cells by procuring pure cell populations for microarray analyses in conjunction with bioinformatic algorithms.

Current efforts include expanding the sample set to confirm the qPCR and microarray results and selecting additional genes of interest from the microarray data for validation by qPCR. These preliminary results support our earlier contention that LCM-procured cells of a specific type, RNA extraction, and measurements of gene expression by qPCR and microarray provide a more accurate assessment of a cell's molecular signature. Furthermore, qPCR and microarray results of pure lung cell types suggested altered levels of gene expression of both ER isoforms in NSCLC compared with normal cells as well as genes associated with several other signaling pathways. Using this approach, our goal is to develop distinct molecular signatures of NSCLC and identify new biomarkers that aid in the assessment of prognosis for lung cancer patients.

Acknowledgments These studies were supported in part by grants from the Kentucky Lung Cancer Research Program and the Phi Beta Psi Charity Trust. We are appreciative of the cooperation of Mr. Neil Mucci and Dr. Jon Wetzel of Asterand, PLC for providing tissue sections.

References

1. Su JM, Hsu HK, Chang H, et al. (1996) Expression of estrogen and progesterone receptors in non-small-cell lung cancer: immunohistochemical study. Anticancer Res 16(6B):3803–6.
2. Fasco MJ, Hurteau GJ, Spivack SD (2002) Gender-dependent expression of alpha and beta estrogen receptors in human nontumor and tumor lung tissue. Mol Cell Endocrinol 188(1–2):125–40.
3. Stabile LP, Davis AL, Gubish CT, et al. (2002) Human non-small cell lung tumors and cells derived from normal lung express both estrogen receptor alpha and beta and show biological responses to estrogen. Cancer Res 62(7):2141–50.
4. Kawai H, Ishii A, Washiya K, et al. (2005) Combined overexpression of EGFR and estrogen receptor alpha correlates with a poor outcome in lung cancer. Anticancer Res 25(6C):4693–8.
5. Kawai H, Ishii A, Washiya K, et al. (2005) Estrogen receptor alpha and beta are prognostic factors in non-small cell lung cancer. Clin Cancer Res 11(14):5084–9.
6. Schwartz AG, Prysak GM, Murphy V, et al. (2005) Nuclear estrogen receptor beta in lung cancer: expression and survival differences by sex. Clin Cancer Res 11(20):7280–7.
7. Fitzpatrick SL, Brightwell J, Wittliff JL, et al. (1984) Epidermal growth factor binding by breast tumor biopsies and relationship to estrogen receptor and progestin receptor levels. Cancer Res 44(8):3448–53.
8. Fekete M, Wittliff JL, Schally AV (1989) Characteristics and distribution of receptors for [D-TRP6]-luteinizing hormone-releasing hormone, somatostatin, epidermal growth factor, and sex steroids in 500 biopsy samples of human breast cancer. J Clin Lab Anal 3(3):137–47.
9. Gee JM, Harper ME, Hutcheson IR, et al. (2003) The antiepidermal growth factor receptor agent gefitinib (ZD1839/Iressa) improves antihormone response and prevents development of resistance in breast cancer in vitro. Endocrinology 144(11):5105–17.
10. Hutcheson IR, Knowlden JM, Madden TA, et al. (2003) Oestrogen receptor-mediated modulation of the EGFR/MAPK pathway in tamoxifen-resistant MCF-7 cells. Breast Cancer Res Treat 81(1):81–93.
11. Levin ER (2003) Bidirectional signaling between the estrogen receptor and the epidermal growth factor receptor. Mol Endocrinol 17(3):309–17.
12. Nicholson RI, Hutcheson IR, Britton D, et al. (2005) Growth factor signalling networks in breast cancer and resistance to endocrine agents: new therapeutic strategies. J Steroid Biochem Mol Biol 93(2–5):257–62.
13. Wittliff JL, Durant JR, Fisher B (1981) Methods of steroid receptor analyses and their quality control in the clinical laboratory. Prog Clin Biol Res 74:397–411.
14. Wittliff JL, Kunitake ST, Chu SS, Travis JC (2000) Applications of laser capture microdissection in genomics and proteomics. J Clin Ligand Assay 23:66–73.
15. Wittliff JL, Erlander MG (2002) Laser capture microdissection and its applications in genomics and proteomics. Methods Enzymol 356:12–25.

Reproductive Endometrial Cancer

Blood Levels of Organochlorine Pesticide Residues and Risk of Reproductive Tract Cancer Among Women from Jaipur, India

Vibha Mathur, Placheril J. John, Inderpal Soni, and Pradeep Bhatnagar

Summary Residues of organochlorine pesticides are integral part of our environment. Because of their strong lipophilic and non-biodegradable nature, organisms at higher trophic levels in the food chain tend to accumulate them. The aim of the present study was to assess the influence of organochlorine pesticides upon the occurrence of reproductive tract cancers in women from Jaipur, India. Blood samples were collected from 150 females. In that group, 100 females suffered from reproductive tract cancers like cervical, uterine, vaginal and ovarian cancers, while the rest did not suffer from cancers or any other major disease and were treated as control group. The collected blood samples were subjected to pesticide extraction and analyzed with the help of gas chromatography. The pesticides detected were benzene hexa chlororide and its isomers, dieldrin, heptachlor, dichloro diphenyl trichloro ethane and its metabolites. The data obtained indicate that the organochlorine pesticide residue levels were significantly higher in all the cancer patients as compared with the control group.

Introduction

The effect of pesticides in human health and safety of the environment have been of global concern. The widespread use and misuse of the early pesticides created an awareness of the potential hazard to the health of agriculturists. The large scale usage of these pesticides has led to their nearly ubiquitous presence in the environment including food, water, soil, and even air. However, most of the chemicals that are used as pesticides are not highly specific but are toxic to many nontarget organisms including humans. The major threat of contamination is by the organochlorines as they are lipid soluble and nonbiodegradable. Organisms at the top of the food chain are most adversely affected as these pollutants accumulate in maximum quantity in them through the process of biomagnifications. These chemicals are reported to be carcinogenic, mutagenic, and teratogenic and are also possessing estrogenic activity (1).

The organochlorine pesticides have been suggested to have potential as endocrine-disruptive xenobiotics (2, 3). The hypothesis that endocrine disruption can cause

J.J. Li et al. (eds.), *Hormonal Carcinogenesis V.*

cancer in humans is based on the causal association between diethyl stilbestrol (DES) exposure of pregnant women and clear cell adenocarcinoma of the vagina and cervix in their female offspring, hormone-related risk factors for breast and uterine cancer. Young women who developed cancer of the vagina were most likely to have had mothers who used DES during pregnancy to avoid miscarriage than mothers who did not use the drug (4). The study conducted by Tiemann et al. (5) showed that the pesticides DDT (dichloro diphenyl trichloro ethane), MXC, and g-BHC (benzene hexa chlororide) at concentrations between 41 and 200 μM inhibited DNA synthesis of cultured bovine oviductal endosalpingeal and uterine cells in the order DDT>MXC>BHC, in comparison to nonexposed controls. Arnold et al. (6) reported that mixtures of weakly estrogenic or nonestroenic pesticides (dieldrin, endosulphan, toxaphene, and chlordane) and hydroxylated polychlorinated biphenyls resulted in synergistic increases in estrogen receptor binding and reporter gene expression in transfection facilitated yeast and endometrial carcinoma derived cell cultures.

Although their chemical structures do not closely resemble those of steroid hormones such as estrogen, progesterone, or testosterone, a number of chlorinated pesticides strongly mimic estrogen in the body. The present study was, therefore, planned to evaluate the levels of organochlorine pesticide residues and the influence of these pesticides upon the occurrence of reproductive tract cancers in the women of Jaipur, Rajasthan, India. The blood samples of these patients were analyzed for pesticide residues. Parameters like age, parity, and diet were also taken into consideration. The organochlorine pesticides analyzed were DDT and its metabolites DDD and DDE, heptachlor, dieldrin, and isomers of BHC (α, β, γ).

Material and Methods

Sample Collection. The study included 100 females suffering from various reproductive tract cancers and 50 controls not suffering from any cancer from the Birla Cancer Institute, Sawai Man Singh (SMS) Hospital, Jaipur, Rajasthan, India. Blood samples from these females were collected to evaluate their body burden. They were asked to fill in questionnaire, which was divided into two sections. The first section included general information like age, health status, family history, economic status, dietary habits, and obstetrical and menstrual history. The second section included information regarding use and accidental exposure to the pesticides either at home or at work place.

Extraction and Purification of Pesticides. Pesticides were extracted and separated from the samples by the method of Bush et al. (7) with some modifications. To 2 ml blood, methanol (5 ml) and 1:1 diethyl ether/hexane (8 ml) were added and the contents were hand shaken for 3 min in a stoppered glass sample tube. The contents were then centrifuged for 10 min at 3000 rpm. The upper layer of the solvent was recovered using micropipette with disposable tips. The blood samples were subjected to reextraction by adding 8 ml of 1:1 diethyl ether/hexane. Purification was done using a glass column packed with prewashed glass wool followed by 5 g of activated silica gel, overlaid with 5 g of anhydrous sodium sulphate. The packed column was first rinsed with hexane, extracted, and finally rinsed with 1% methanol in hexane.

Extract Storage and Analysis. The elutes were completely evaporated using a rotatory vacuum evaporator. The resultant extracts were then dissolved in 2 ml of hexane and were kept in glass vials at 4 °C in the refrigerator until analysis. The qualitative and quantitative estimations of organochlorine pesticides were carried out by gas chromatography. A Hewlett Packard (HP) 5890 series II gas chromatograph equipped with 63Ni-foil electron capture detector (ECD) coupled with an integrator, HP 3396-A was used for the analysis of the samples. Environmental Protection Agency (EPA) method 608 with some modifications was followed for the experimental conditions of gas chromatography, which were as follows: Column: Capillary column, HP-1 (methyl silicon gum) 10 m × 0.54 mm × 2.65 μm. Injection parameters: Temperature: 250 °C; Pneumatics: Split less mode; Purge delay: 1 min; Carrier gas: Nitrogen; Column head pressure: 10 psi, 2.4 ml min^{-1} (constant flow); Oven parameters: Initial temperature: 80 °C (1 min); Ramps: First ramp: 30 °C min^{-1} to 190 °C; Second ramp: 6 °C min^{-1} to 300 °C; Final temperature: 300 °C (2 min); Detector Temperature: 330 °C.

Aliquot, 1 μl of the final extract, were injected into the column with the help of a 10 μl Hamilton syringe. By comparing their retention times with those of the standard, the different pesticide peaks were identified; quantitative estimations of the pesticides were based on the peak areas that were obtained from the integrator.

Results

All the blood samples that were analyzed for pesticide residues were found laced with α, β, γ isomers of BHC, dieldrin, heptachlor, *p,p′*-DDT and their metabolites *p,p′*-DDD and *p,p′*-DDE. In patients with reproductive tract cancers, the total organochlorine pesticides were 8.71 ± 0.920 mg l^{-1}. The total pesticide burden in cervical cancer patients was 10.891 ± 1.533 mg l^{-1}, in uterine cancer patients was 7.506 ± 2.331 mg l^{-1}, 6.303 ± 1.579 mg l^{-1} observed in vaginal cancer patients followed by ovarian cancer patients with a total pesticide burden of 7.577 ± 1.858 mg l^{-1}. All the pesticides analyzed were found significantly higher in patients with reproductive tract cancers (Table 1).

The total organochlorine pesticide residue in the reproductive tract cancer patients by age was as follows: From 21–30 years, 11.159 ± 2.686 mg l^{-1}. From 31–40 years, 8.519 ± 1.802 mg l^{-1}. From 41–50 years, 7.221 ± 1.282 mg l^{-1}. Pesticides analyzed by different cancer category, with reference to particular age groups, and the levels of individual pesticides are given in Tables 2, 3, and 4.

Table 1 Pesticide residues in normal and in women with reproductive tract cancers

Pesticides mg l^{-1}	Normal females (n = 50)	Females with reproductive tract cancers (n = 100) Cervical (n = 43)	Uterine (n = 18)	Vaginal (n = 22)	Ovarian (n = 17)	Total (R. T.) pesticides
α-BHC	0.156 ± 0.071	0.505 ± 0.094	0.498 ± 0.156	0.439 ± 0.104	0.383 ± 0.105	0.469 ± 0.056
γ-BHC	0 .088 ± 0.047	0.465 ± 0.077	0.713 ± 0.195	0.340 ± 0.064	0 .464 ± 0.154	0.483 ± 0.057
β-BHC	0 .080 ± 0.030	0.832 ± 0.199	0.746 ± 0.281	0.429 ± 0.115	0.502 ± 0.123	0.672 ± 0.105
HEPTACHLOR	0.084 ± 0.050	1.429 ± 0.394	1.748 ± 1.495	0.537 ± 0.132	0.775 ± 0.134	1.179 ± 0.314
ALDRIN	0.115 ± 0.043	2.685 ± 0.594	1.151 ± 0.359	1.788 ± 0.902	2.578 ± 1.310	2.194 ± 0.396
DDE	0.047 ± 0.018	1.293 ± 0.246	1.194 ± 0.309	0.830 ± 0.157	1.020 ± 0.265	1.127 ± 0.132
DDD	0.249 ± 0.059	1.577 ± 0.602	1.316 ± 0.925	0.608 ± 0.197	0.672 ± 0.180	1.163 ± 0.311
DDT	1.034 ± 0.221	2.145 ± 0.372	1.376 ± 0.429	1.369 ± 0.480	1.186 ± 0.373	1.673 ± 0.217
Σ-BHC	0.325 ± 0.134	1.782 ± 0.259	1.959 ± 0.493	1.169 ± 0.153	1.349 ± 0.193	1.605 ± 0.151
Σ-DDT	1.332 ± 0.239	5.017 ± 0.831	2.652 ± 0.593	2.808 ± 0.627	2.872 ± 0.659	3.741 ± 0.424
TOTAL PESTICIDES	1.857 ± 0.311	10.891 ± 1.533	7.506 ± 2.331	6.303 ± 1.579	7.577 ± 1.858	8.710 ± 0.920

Table 2 Pesticide residues in normal and in women with reproductive tract cancers (age group 21–30)

Pesticides mg l^{-1}	Normal females N = 20	Females with reproductive tract cancers (n = 17) Cervical (n = 5)	Uterine (n = 3)	Vaginal (n = 2)	Ovarian (n = 7)	Total pesticides
α-BHC	0.087 ± 0.042	0.676 ± 0.193	0.824 ± 0.799	0.043 ± 0.011	0.212 ± 0.084	0.436 ± 0.152
γ-BHC	0.042 ± 0.017	0.652 ± 0.180	0.092 ± 0.060	0.317 ± 0.287	0.451 ± 0.251	0.431 ± 0.122
β-BHC	0.070 ± 0.047	0.713 ± 0.273	0.829 ± 0.815	0.320 ± 0.054	0.644 ± 0.219	0.659 ± 0.170
HEPTACHLOR	0.164 ± 0.123	2.404 ± 1.701	0.148 ± 0.090	0.345 ± 0.066	1.020 ± 0.286	1.194 ± 0.520
ALDRIN	0.095 ± 0.057	4.779 ± 2.361	0.692 ± 0.340	1.280 ± 0.020	4.332 ± 3.125	3.462 ± 1.446

(continued)

Table 2 (continued)

Pesticides mg l^{-1}	Normal females $N = 20$	Females with reproductive tract cancers ($n = 17$)				
		Cervical ($n = 5$)	Uterine ($n = 3$)	Vaginal ($n = 2$)	Ovarian ($n = 7$)	Total pesticides
DDE	0.058 ± 0.042	1.340 ± 0.443	1.460 ± 0.862	1.419 ± 0.583	1.437 ± 0.578	1.410 ± 0.291
DDD	0.309 ± 0.117	4.845 ± 4.688	0.239 ± 0.107	1.517 ± 0.065	0.716 ± 0.249	1.940 ± 1.361
DDT	0.739 ± 0.201	1.635 ± 0.873	0.567 ± 0.558	0.985 ± 0.255	1.831 ± 0.797	1.451 ± 0.419
Σ-BHC	0.200 ± 0.077	2.042 ± 0.322	1.745 ± 1.672	0.681 ± 0.361	1.308 ± 0.298	1.527 ± 0.309
Σ-DDT	1.108 ± 0.229	7.822 ± 4.684	2.267 ± 1.515	3.921 ± 0.773	3.967 ± 1.301	4.795 ± 1.480
TOTAL PESTICIDES	1.568 ± 0.257	16.847 ± 6.735	4.853 ± 2.976	6.227 ± 1.180	10.629 ± 4.274	11.159 ± 2.686

Table 3 Pesticide residues in normal and in women with reproductive tract cancers (age group 31–40)

Pesticides mg l^{-1}	Normal females ($N = 14$)	Females with reproductive tract cancers ($n = 26$)				
		Cervical ($n = 12$)	Uterine ($n = 6$)	Vaginal ($n = 5$)	Ovarian ($n = 3$)	Total pesticides
α- BHC	0.327 ± 0.241	0.572 ± 0.216	0.576 ± 0.239	0.275 ± 0.109	0.618 ± 0.408	0.521 ± 0.121
γ- BHC	0.173 ± 0.159	0.388 ± 0.113	1.005 ± 0.490	0.503 ± 0.222	0.277 ± 0.140	0.539 ± 0.134
β- BHC	0.141 ± 0.079	0.640 ± 0.171	1.381 ± 0.708	0.169 ± 0.077	0.451 ± 0.252	0.699 ± 0.191
HEPTACHLOR	0.049 ± 0.028	1.145 ± 0.632	4.714 ± 4.487	0.506 ± 0.112	0.792 ± 0.057	1.807 ± 1.055
ALDRIN	0.195 ± 0.124	1.555 ± 0.448	1.581 ± 0.820	0.694 ± 0.188	1.176 ± 0.275	1.352 ± 0.281
DDE	0.044 ± 0.024	1.638 ± 0.708	1.851 ± 0.633	1.015 ± 0.288	0.493 ± 0.239	1.428 ± 0.362
DDD	0.228 ± 0.089	1.221 ± 0.655	2.902 ± 2.798	0.329 ± 0.176	0.514 ± 0.151	1.356 ± 0.695
DDT	1.232 ± 0.582	2.374 ± 0.896	1.991 ± 0.817	0.444 ± 0.158	0.192 ± 0.080	1.662 ± 0.475
Σ- BHC	0.639 ± 0.455	1.595 ± 0.465	2.963 ± 1.214	0.948 ± 0.323	1.346 ± 0.464	1.761 ± 0.334
Σ- DDT	1.505 ± 0.626	1.354 ± 0.166	3.036 ± 0.784	1.788 ± 0.348	1.199 ± 0.113	3.598 ± 0.825
TOTAL PESTICIDES	2.389 ± 0.907	4.416 ± 0.899	12.296 ± 6.368	3.937 ± 0.566	4.536 ± 0.774	8.519 ± 1.802

Table 4 Pesticide residues in normal females and those with cancers (age group 41–50)

Pesticides mg l^{-1}	Normal females $n = 12$	Females with reproductive tract cancers ($n = 30$) Cervical ($N = 15$)	Uterine ($N = 5$)	Vaginal ($N = 8$)	Ovarian ($N = 2$)	Total pesticides
α-BHC	0.125 ± 0.066	0.268 ± 0.070	0.444 ± 0.204	0.797 ± 0.223	0.167 ± 0.042	0.432 ± 0.085
γ-BHC	0.081 ± 0.066	0.306 ± 0.095	0.610 ± 0.242	0.313 ± 0.051	1.377 ± 0.899	0.430 ± 0.090
β-BHC	0.054 ± 0.039	0.647 ± 0.123	0.398 ± 0.179	0.396 ± 0.169	0.926 ± 0.550	0.557 ± 0.088
HEPTACHLOR	0.016 ± 0.010	0.622 ± 0.132	0.364 ± 0.155	0.441 ± 0.187	0.257 ± 0.220	0.505 ± 0.087
ALDRIN	0.093 ± 0.055	2.105 ± 0.480	1.585 ± 0.798	3.297 ± 2.470	1.360 ± 1.151	2.286 ± 0.693
DDE	0.036 ± 0.013	0.780 ± 0.210	1.085 ± 0.521	0.633 ± 0.277	0.114 ± 0.093	0.747 ± 0.153
DDD	0.250 ± 0 .112	0.808 ± 0.332	0.584 ± 0.346	0.752 ± 0.446	1.092 ± 0.665	0.774 ± 0.208
DDT	1.565 ± 0.511	1.702 ± 0.466	1.675 ± 1.157	2.578 ± 1.222	0.613 ± 0.147	1.858 ± 0.434
Σ-BHC	0.261 ± 0.115	1.222 ± 0.155	1.453 ± 0.404	1.393 ± 0.189	2.471 ± 0.307	1.389 ± 0.123
Σ-DDT	1.855 ± 0.543	3.298 ± 0.697	3.344 ± 1.776	3.918 ± 1.559	1.820 ± 0.719	3.384 ± 0.595
TOTAL PESTICIDES	2.226 ± 0.580	6.581 ± 1.018	6.747 ± 2.789	9.090 ± 4.257	5.908 ± 1.957	7.221 ± 1.282

Discussion

A study, conducted by the consumer education and research society, Ahmedabab, revealed the presence of pesticides DDT, lindane, aldrin, and dieldrin, all highly toxic and suspected carcinogens, in ten branded and three loose samples of wheat flour collected from all over the country. Another study, conducted by the Central Food Technological Research Institute (CFTRI), Mysore, revealed 30% of the 20 random flour samples contained BHC and DDT residues above tolerance limits prescribed jointly by FAO/WHO (8).

The results of the present study revealed that the older women, age group 41–50 years, were more prone to cancers of reproductive tract. But the pesticide residues were highest in women with reproductive tract cancer ages 21–30 years. Nair and Pillai (9) also mentioned that BHC burden was lower in this age group.

Over the years, a better understanding of estrogen and its metabolites, coupled with findings on the estrogen-like nature of chlorinated pesticides has led researchers to suggest how chlorinated pesticides might cause cancer. Both biochemical and epidemiological findings suggest that chlorinated pesticides can trigger cancer by adversely affecting the metabolism of estrogen (10). Although their chemical structures do not closely resemble estrogen, chlorinated pesticides strongly mimic estrogen in the body. Like estrogen, DDT, methoxychlor, or chlordane promote the implantation of embryos in rats and maintained their pregnancies. Kepone, heptachlor, and chlordane also promoted proliferation of breast tumor cells, as estrogens do (11). These exogenous estrogens do not directly involve estrogen receptors, but can affect estrogen-mediated responses, including blocking the ability of endogenous estrogen to interact with the estrogen receptor, reacting directly or indirectly with estrogen carrier proteins, or with free or bound endogenous estrogen to change circulating levels. This alters the amount of endogenous estrogen produced and also the number of available estrogen receptors (12). It is suggested that estrogens operate as promoters or cocarcinogens rather than initiators. Studies conducted by Harris et al. (13) have shown that estrogens have mitotic effects on breast epithelial cells.

The laboratory data do not provide compelling evidence for an organochlorine-dependent estrogen mediated effect on reproductive cancers, although they do not exclude the possibility of such an effect. Since the organochlorine pesticides are weak estrogens, one has to accommodate much stronger endogenous or exogenous estrogenic stimulus, as well as the possible competition for estrogen receptors.

The results of this study indicate that the presence of organochlorine pesticides in the blood is a matter of great concern as low dose exposure of these chemicals in human beings may lead to health effects such as immunosuppression, hormone disruption, reproductive abnormalities, and cancer.

The occurrence of DDT, BHC, dieldrin, and heptachlor in blood of general population confirms the exposure of these people to the pesticides either directly or indirectly. The studied cases were not exposed to these chemicals directly. The exposure was mainly through three possible ways: (1) measures to control disease spreading insects, (2) residues in the environment, and (3) through residues in food.

In the light of our findings, it may be stated that environmental chemical contamination with organochlorine residues may be an important etiologic factor in reproductive tract cancers. Given the widespread dissemination of organochlorine insecticides in the environment and the food chain, the implications are far-reaching for public health intervention worldwide. Therefore, stricter regulations may be deployed and such measures have to be weighed against the benefits of pesticide use.

Acknowledgments Authors gratefully acknowledge financial assistance from the Department of Science and Technology, Rajasthan (7/3/R&D/1999/20287/dt.22.12.1999) and Dr. R. G. Sharma, Birla Cancer Institute, Sawai Man Singh (SMS) Hospital, Jaipur, Rajasthan, India.

References

1. Falck F, Ricci A, Wolff MS, et al. (1992) Pesticides and polychlorinated biphenyl residues in human breast lipids and their relation to breast cancer. Arch Environ Health 47:143–146.
2. Colborn T, Vom Saal FS, Sole AM (1993) Developmental Effects of endocrine-disrupting chemicals in wildlife and humans. Environ Health Perspect 101:378–384.

3. Turner KJ and Sharpe RM (1997) Environmental oestrogens – Present understanding. Rev Reprod 2:69–73.
4. Herbst AL, Ulfelder H, Poskanzer DC (1971) Adenocarcinoma of the vagina. Association of maternal diethylstilbestrol therapy with tumor appearance in young women. N Engl J Med 284:878–881.
5. Tiemann U, Schneider F, Tuchscherer A (1996). Effects of ornganochlorine pesticides on DNA synthesis of cultured oviductal and uterine cells and on estrogen receptor of uterine tissue from heifers. Arch Toxicol 70:490–496.
6. Arnold SF, Klotz DM, Collins BM, et al. (1996) Synergistic activation of estrogen receptor with combinations of environmental chemicals. Science 272:1489–1492.
7. Bush B, Snow J, Koblintz R (1984) Polychlorobiphenyl (PCB) congeners *p,p′*-DDE and hexachlorobenzene in maternal and fetal cord blood from mothrs inupstate. New York. Arch Environ Contam Toxicol 13:517–527.
8. Agarwal, A (2000): Contaminated wheat. Down to Earth, Aug. 31. 22.
9. Nair A, Pillai MKK (1992) Trends in ambient levels of DDT and HCH residues in human and the environment of Delhi, India. The Sci Total Environ 121:145–157.
10. Patlak M (1996) Estrogens may link pesticides. Breast Cancer Environ Sci Tech News 30(5):210A–211A.
11. Davis DL, Bradlow HL, Wolff M, et al. Medical hypothesis: Xenoestrogens as preventable causes of breast cancer. Environ Health Perspect 101(5):372.
12. Golden RJ, Noller KL, Ernstoff LT, et al. (1998) Environmental endocrine modualators and human health: an Assessment of the biological evidence. Crit Rev Toxicol 28(2):109–227.
13. Harris JR, Lippman ME, Veronesi V (1992). Breast cancer, Part 1 of 3, New Eng J Med 319–327.

Prostate Cancer

AKT Regulates Androgen Receptor-dependent Growth and PSA Expression in Prostate Cancer

Margarita Mikhailova, Yu Wang, Roble Bedolla, Xiao-Hua Lu, Jeffrey I. Kreisberg, and Paramita M. Ghosh

Summary Recurrent prostate cancer (PC) is usually treated with androgen deprivation therapy, which, despite initial success, eventually fails due to the development of androgen-independent PC. Androgen deprivation stimulates a significant increase in the phosphorylation (activation) of Akt, a serine/threonine kinase, which regulates cell growth and survival. Hence, we asked whether the increase in Akt phosphorylation contributes to the development of androgen independence. Akt regulates transcriptional activity of the androgen receptor (AR), and our data show that Akt-stimulated AR transcriptional activity is dependent on androgen-binding to the AR. PC proliferation has both androgen-sensitive and insensitive components. The androgen sensitive component is Akt-dependent, while the androgen-insensitive is not. However, Akt-induced cell survival is largely AR independent, suggesting that the cell stimulates Akt phosphorylation when subjected to androgen deprivation as an alternate pathway to maintain survival.

Introduction

Since prostate tumors are initially dependent on androgens for growth and survival, recurrent prostate cancer (PC) is usually treated by androgen deprivation therapy (1). Patients usually respond to such treatment in the first stages of the disease, but frequently relapse indicative of the development of androgen-independent PC (AIPC) (2). The phosphatidylinositol 3-kinase (PI3K)/Akt pathway is a major cell growth and survival regulating pathway in human PC development and progression. Activation of PI3K results in the phosphorylation/activation of several downstream targets, including Akt. Akt activation was previously shown to be markedly increased in androgen-independent cells compared with the parental androgen-dependent cells in vitro (3). In this report, we investigate the role the increased Akt phosphorylation plays in the progression of PC from an androgen-dependent to an androgen-independent state.

We and others showed that Akt activation was necessary for the proliferation of PC cells (4, 5), in addition to its well-known role as a mediator of cell survival (6). Expression of constitutively active Akt in androgen-dependent LNCaP cells resulted

J.J. Li et al. (eds.), *Hormonal Carcinogenesis V.*

in a sixfold increase in xenograft tumor growth (7). Androgens regulate growth and survival through the activation of the androgen receptor (AR), which is present in androgen-dependent and most androgen-independent tumors (8, 9). Akt has been shown to regulate AR expression and transcriptional activity at the posttranscriptional and posttranslational levels (10–13). A recent study demonstrated direct synergy between Akt and AR signaling sufficient to initiate and progress native adult murine prostatic epithelium to frank carcinoma and override the effect of androgen ablation (14). However, the role of Akt in the progression to androgen independence is not currently known. Akt may promote androgen-independent growth and survival by inducing ligand-independent AR transcriptional activity, as suggested by some studies (12), or by bypassing the AR completely, as suggested in an earlier review (15). The purpose of the studies described here was to identify the mechanism by which Akt promotes androgen-independent PC growth and survival.

Time Course of the Increase in Akt Phosphorylation in LNCaP Cells by Androgen Withdrawal

The tumor-suppressor PTEN is an important negative regulator of the PI3K/Akt pathway (12). Since LNCaP cells are PTEN (⁻), they express a basal level of Akt phosphorylation in the presence of complete medium (RPMI 1640 + 10% FBS without added androgens). When these cells are cultured in androgen-reduced medium (phenol-red free RPMI 1640 with 10% charcoal stripped FBS), increased Akt phosphorylation is observed as early as 5 days, and increases over time (Fig. 1A, upper band) despite a lack of any significant change in total Akt levels (Fig. 1A, lower band). Consistent with this observation, androgen-independent sublines of LNCaP cells, C4-2 and LNCaP-AI, had increased Akt phosphorylation compared with LNCaP cells (Fig. 1B). LNCaP-AI cells were developed in our laboratory by

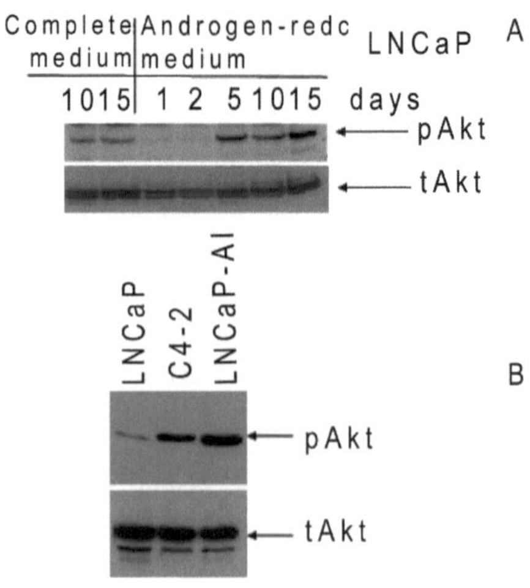

Fig. 1 (**A**) LNCaP cells cultured in complete or androgen-reduced medium for 0–5 days. Cells in androgen-reduced medium displayed increased Akt phosphorylation (pAkt, *upper panel*) but not total Akt (tAkt, *lower panel*) compared with those cultured in complete medium. (**B**) Increased Akt activation in AI LNCaP sublines. LNCaP cells and its AI sublines C4-2 and LNCaP-AI were examined for phosphorylated Akt (pAkt) (*upper panel*) and total Akt (*lower panels*)

prolonged culture of LNCaP (ATCC) in androgen-reduced medium, while C4-2 cells (Urocor, Oklahoma City, OK) were originally developed by others from LNCaP xenografts in castrated mice (13).

Characterization of Androgen-independent Sublines of LNCaP Cells

Compared with the parental cell line, both C4-2 and LNCaP-AI cell lines had higher proliferation rates (Fig. 2A), and increased AR transcriptional activity as measured by luciferase assay on a human PSA promoter tagged to a luciferase construct (hPSA-luc) (4) (Fig. 2B). LNCaP and C4-2 cells, but not LNCaP-AI

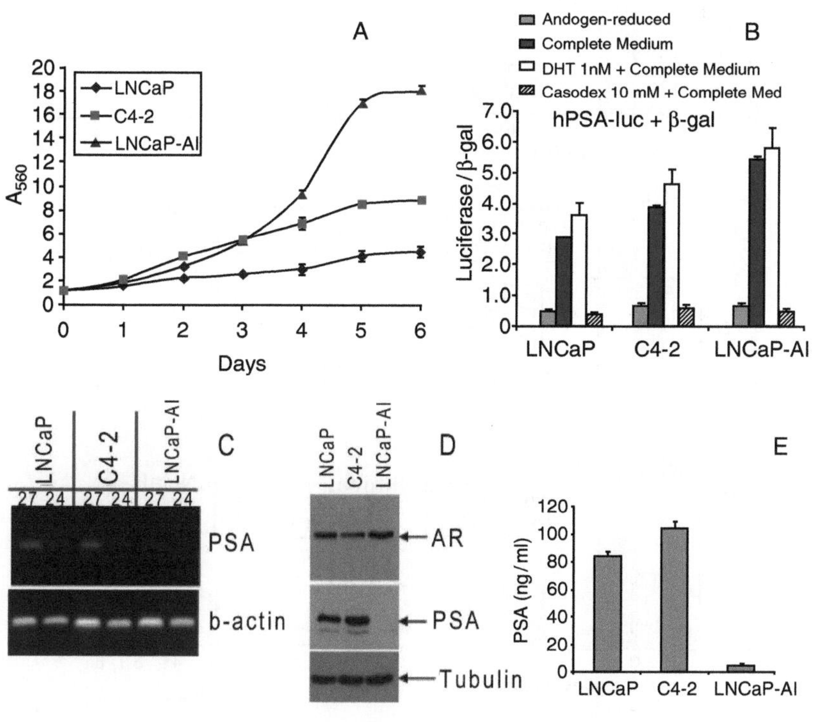

Fig. 2 (**A**) Comparison of growth rates over 6 days by MTT assay (4). Data were normalized to the cell number at origin and represent mean ± SD, $n = 3$. (**B**) AR transcriptional activity in the three cell lines measured by luciferase assay on hPSA-luc. Data shown indicate mean ± S.D. ($n = 3$) normalized to β-gal. (**C**) RT-PCR showing 27 and 24 cycles of amplification using primers for PSA (*upper band*) or β-actin (*lower band*). (**D**) Western blots comparing AR (*upper band*) and PSA levels (*middle band*). Tubulin (*lower band*) was used as loading control. (**E**) Measurement of secreted PSA levels in the same cell lines by ELISA in the extracellular media

expressed PSA (Figs. 2C–E), despite a functional AR in LNCaP-AI cells, stimulated by androgens and inhibited by the anti-androgen Casodex, a competitive inhibitor of androgen binding to the AR (Fig. 2B).

Despite being androgen-independent, both LNCaP sublines were androgen-sensitive as determined by increased proliferation in response to DHT (Fig. 3A). However, growth of LNCaP cells was inhibited at 10 μM Casodex (Fig. 3, upper panel), whereas C4-2 cells (Fig. 3, middle panel) and LNCaP-AI (Fig. 3, lower panel) were not affected. Interestingly, in the AI sublines, Casodex inhibited androgen-stimulated, but not basal proliferation.

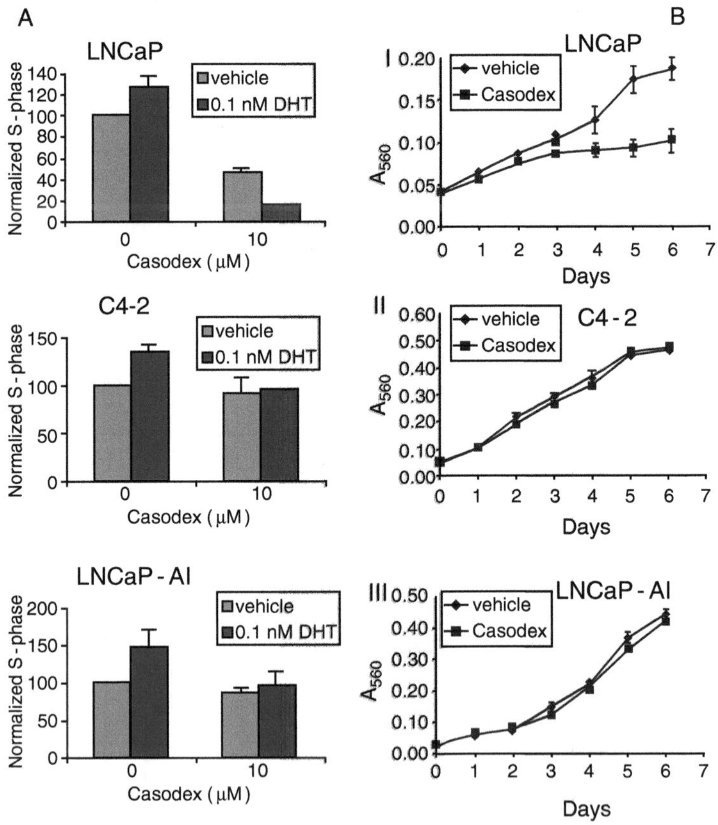

Fig. 3 (**A**) Flow cytometric analysis of the effect of Casodex on proliferation of LNCaP (*upper*), C4-2 (*middle*), and LNCaP-AI (*lower*) cells treated with DHT for 48 h. Proliferation rates were determined by the fraction of cells in S-phase. All data shown are representative of at least three individual experiments. (**B**) MTT assay to determine proliferation rates of LNCaP, C4-2, and LNCaP-AI cells in the presence/absence of Casodex. Cells were plated in 24-well plates and treated ± 10 μM Casodex for 0–6 days. Data indicate mean ± S.D. (n = 3)

Akt Regulates Androgen-dependent AR Transcriptional Activity

Previous studies had shown that in androgen reduced medium, transfection of a constitutively active form of Akt caused an increase in the basal levels of AR activity even in the absence of androgens (12). Since androgen reduced media still contain a considerable level of androgens, we investigated whether Akt sensitized the cells to lower levels of androgens, or induced AR transcriptional activity in the absence of androgen binding to the AR, such as with Casodex treatment. To measure AR transcriptional activity, we used luciferase constructs tagged to MMTV and PSA promoters, known to be regulated by the AR. Expression of a plasmid expressing a dominant negative Akt suppressed transcriptional activity on pMMTV-luc (Fig. 4A), whereas expression of a constitutively active Akt (myr-Akt) enhanced

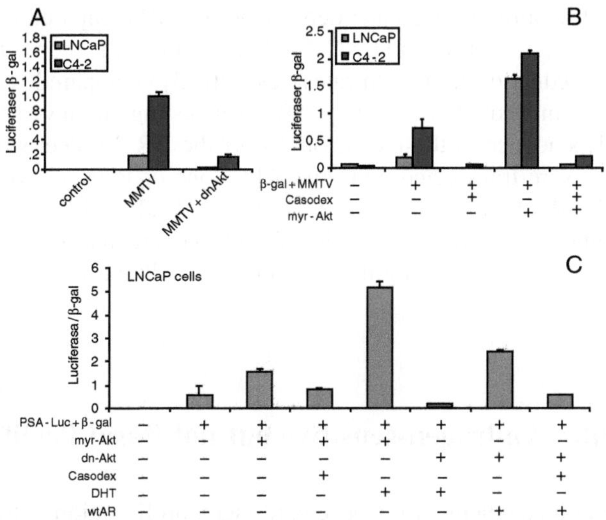

Fig. 4 (**A**) Dominant negative Akt inhibits AR transcriptional activity. LNCaP and C4-2 cells were transiently transfected with pMMTV-luc and β-gal alone or together with a plasmid expressing a dominant negative Akt (dn-Akt). Expression of dn-Akt inhibited AR transcriptional activity on pMMTV-luc. (**B**) Expression of constitutively active (myristoylated) Akt (myr-Akt) enhanced AR transcriptional activity in an androgen-dependent manner. LNCaP and C4-2 cells were transiently transfected with pMMTV-luc + β-gal alone or together with myr-Akt and treated with vehicle (DMSO) or 10 μM Casodex. (**C**) LNCaP cells were transfected with hPSA-luc + β-gal alone or together with myr-Akt or dn-Akt and treated with 10 μM Casodex or 1 nM DHT for 48 h. Data indicate mean ± S.D. of three individual readings

it (Fig. 4B). We had previously shown the successful expression of the mutant Akt plasmids in these cells (4). The effects of myr-Akt on both pMMTV and hPSA-luc were inhibited by 10 μM Casodex (Fig. 4B, C). DHT significantly enhanced the effects of myr-Akt but failed to rescue the cells from the effects of dn-Akt on AR transcriptional activity. Cotransfection of wild type AR (wtAR) rescued AR transcriptional activity from the suppression by dn-Akt but treatment with Casodex negated this effect. Hence, our results show that Akt induced AR transcriptional activity requires androgen binding to the AR.

Akt Regulates Cell Survival in an AR-independent Manner

Since Akt regulates AR transcriptional activity, and AR is needed for cell growth and survival in both androgen-dependent and independent cells (13–14), we investigated whether Akt-induced growth and survival is mediated by the AR. Akt expression was downregulated by RNA inhibition using an Akt1-specific siRNA duplex described elsewhere (15). Both the Akt1 siRNA and a pool of four different AR siRNA duplexes decreased endogenous PSA expression (Fig. 5A). LNCaP and LNCaP-AI proliferation had a greater dependence on AR compared with C4-2 cells (Fig. 5B). In contrast, LNCaP and LNCaP-AI had a smaller response to Akt siRNA (33% and 36% reduction in growth rates, respectively) compared with C4-2 cells (54% reduction in growth rate) (Fig. 5B), suggesting an inverse relationship between the dependence of these cells on Akt or the AR. In other studies, LNCaP and C4-2 cells were transfected with myr-Akt alone or together with AR siRNA (Fig. 5D). AR siRNA induced apoptosis in both LNCaP and C4-2 cells, whereas myr-Akt prevented this effect. AR siRNA failed to overcome the protective effect of myr-Akt (Fig. 5D), suggesting that Akt-induced cell survival was independent of the AR.

Akt Regulates Androgen-sensitive but not Basal Proliferation

Finally, we examined whether the Casodex resistant proliferation of the AI sublines was caused by increased Akt phosphorylation. DHT-stimulated proliferation was inhibited by transfection with dominant negative Akt (Fig. 6A) (earlier), but expression of Akt1 siRNA did not affect the response of cells to Casodex (Fig. 6B). Thus, androgen-sensitive proliferation was regulated by Akt but basal proliferation was not.

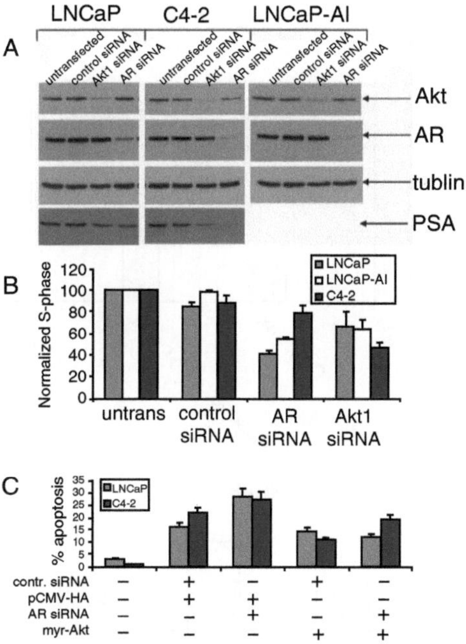

Fig. 5 (**A**) Downregulation of Akt and AR using 50pM/dish of control siRNA, Akt1 siRNA, or AR siRNA. Expression of β-tubulin (T) was used for equal loading. (**B**) Flow cytometric analysis to compare the effects of Akt siRNA and AR siRNA. The percentage of cells in S-phase was determined by MODFIT (Verity software) and all data normalized to untransfected controls for each cell line. The mean percentage of cells in S-phase for the untransfected cells in each cell line were as follows: LNCaP: 19.89 ± 4.87; C4-2: 26.42 ± 5.86; LNCaP-AI: 32.37 ± 3.37. Data demonstrate changes from the untransfected control and represent mean ± SE ($n = 4$). (**C**) LNCaP and C4-2 cells were transfected with vector or myr-Akt, together with control siRNA or AR siRNA. Apoptosis was measured by flow cytometry in propidium iodide stained LNCaP and C4-2 cells

Conclusions

Our data show that Akt cannot stimulate AR transcriptional activity in the absence of androgen-binding to the AR. However, we noticed that Akt-dependent cells had a lower dependence on the AR and vice versa. Further, Akt-induced cell survival is largely AR independent. Taken together; these results suggest that PC cells stimulate Akt phosphorylation as an alternate pathway to maintain survival when subjected to androgen deprivation. Proliferation of AR positive AI sublines of LNCaP cells has an androgen-sensitive and an androgen-insensitive component. Our data show that Akt regulates AR-dependent proliferation (stimulated by androgens and inhibited by Casodex) but not basal proliferation, which is not inhibited by Casodex in these

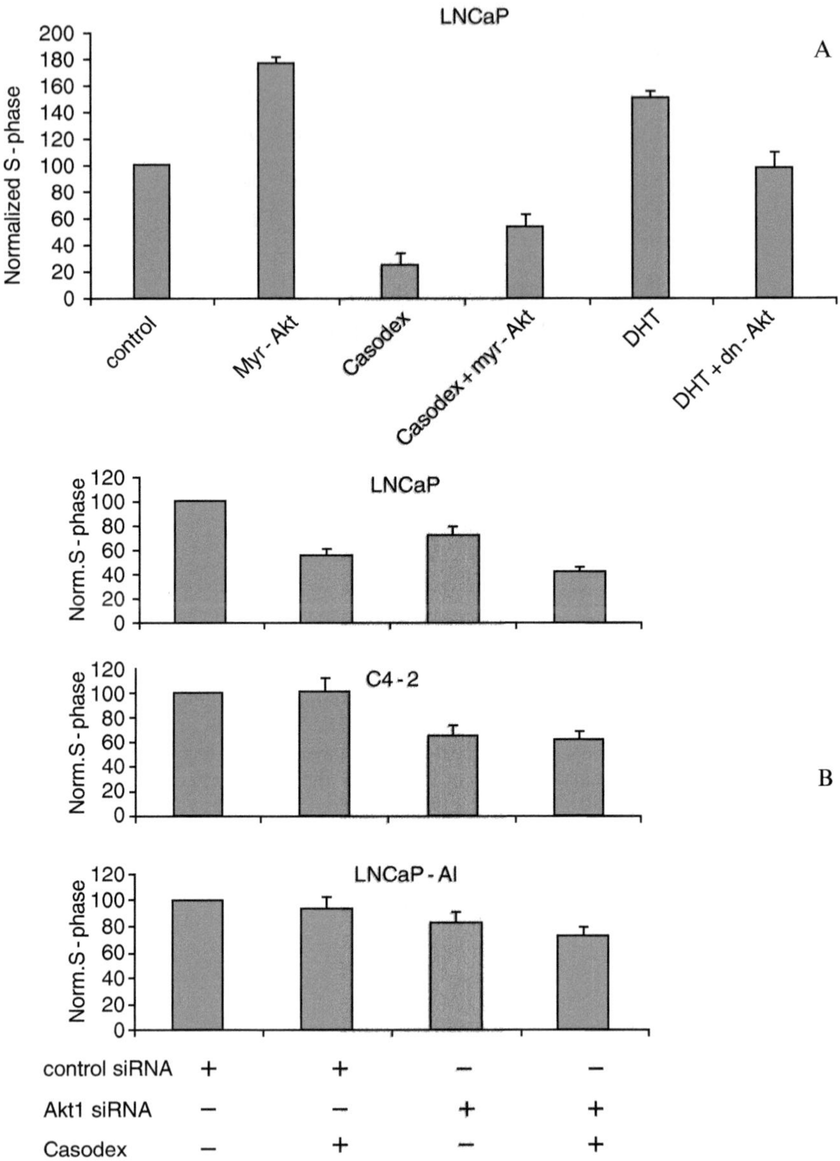

Fig. 6 (**A**) LNCaP cells were transfected with an empty vector (control, pCMV6-HA), myr-Akt or dn-Akt. Cells were treated with vehicle, 10 μM Casodex or 1 nM DHT for 48 h. Results were normalized to the percentage of cells in S-phase in control LNCaP cells. Bars represent mean ± SE ($n = 3$). (**B**) Cells were transfected with 50 pmol of a control siRNA (nonspecific) or an Akt1 specific siRNA duplex, and treated with 10 μM Casodex or vehicle (DMSO) for 48 h. Proliferation rates were determined by flow cytometry as the fraction of cells in S-phase and normalized to that of the untransfected cell (considered 100)

cells. Thus, the major contribution of Akt to androgen-independent PC appears to be the promotion of AR-receptor independent survival in addition to the stimulation of AR-dependent proliferation and AR transcriptional activity.

Acknowledgments We thank Dr. Barry Furr, AstraZeneca, Cheshire, UK, for the gift of bicalutamide (Casodex), and Dr. Thomas Franke, Columbia University, New York, NY, for pCMV-6-myr-Akt-HA and pCMV-6-Akt-K179M plasmids. This work was supported by a Merit Award from the Department of Veterans Affairs and Award CA109057 from the National Cancer Institute.

References

1. Edwards J, Bartlett JM (2005) The androgen receptor and signal-transduction pathways in hormone-refractory prostate cancer. Part 1: Modifications to the androgen receptor. BJU Int 95, 1320–1326.
2. Petrylak D (2005) Therapeutic options in androgen-independent prostate cancer: building on docetaxel. BJU Int 96 (Suppl 2), 41–46.
3. Murillo H, Huang H, Schmidt LJ, et al. (2001) Role of PI3K signaling in survival and progression of LNCaP prostate cancer cells to the androgen refractory state. Endocrinology 142, 4795–4805.
4. Ghosh PM, Malik SN, Bedolla RG, et al. (2005) Signal transduction pathways in androgen-dependent and -independent prostate cancer cell proliferation. Endocr Relat Cancer 12, 119–134.
5. Majumder PK, Sellers WR (2005) Akt-regulated pathways in prostate cancer. Oncogene 24, 7465–7474.
6. Graff JR, Konicek BW, McNulty AM, et al. (2000) Increased AKT activity contributes to prostate cancer progression by dramatically accelerating prostate tumor growth and diminishing p27Kip1 expression. J Biol Chem 275, 24500–24505.
7. Culig Z, Hobisch A, Bartsch G, et al. (2000) Androgen receptor – an update of mechanisms of action in prostate cancer. Urol Res 28, 211–219.
8. Lin HK, Yeh S, Kang HY, et al. (2001) Akt suppresses androgen-induced apoptosis by phosphorylating and inhibiting androgen receptor. Proc Natl Acad Sci USA 98, 7200–7205.
9. Wen Y, Hu MC, Makino K, et al. (2000) HER-2/neu promotes androgen-independent survival and growth of prostate cancer cells through the Akt pathway. Cancer Res 60, 6841–6845.
10. Xin L, Teitell MA, Lawson DA, et al. (2006) Progression of prostate cancer by synergy of AKT with genotropic and nongenotropic actions of the androgen receptor. Proc Natl Acad Sci USA 103, 7789–7794.
11. Feldman BJ, Feldman D (2001) The development of androgen-independent prostate cancer. Nat Rev Cancer 1, 34–45.
12. Wu X, Senechal K, Neshat MS, et al. (1998) The PTEN/MMAC1 tumor suppressor phosphatase functions as a negative regulator of the phosphoinositide 3-kinase/Akt pathway. Proc Natl Acad Sci USA 95, 15587–15591.
13. Thalmann GN, Anezinis PE, Chang SM, et al. (1994) Androgen-independent cancer progression and bone metastasis in the LNCaP model of human prostate cancer. Cancer Res 54, 2577–2581.
14. Haag P, Bektic J, Bartsch, et al. (2005) Androgen receptor down regulation by small interference RNA induces cell growth inhibition in androgen sensitive as well as in androgen independent prostate cancer cells. J Steroid Biochem Mol Biol 96, 251–258.
15. Czauderna F, Fechtner M, Aygun H, et al. (2003) Functional studies of the PI(3)-kinase signalling pathway employing synthetic and expressed siRNA. Nucleic Acids Res 31, 670–682.

Session II
Cell/Molecular Biology and Hormone Metabolism

Cell/Molecular Biology

Animal Models

Expression of Selected Aurora A Kinase Substrates in Solely Estrogen-induced Ectopic Uterine Stem Cell Tumors in the Syrian Hamster Kidney

Adrianne E. Hontz, Sara A. Li, Jeffrey L. Salisbury, Wilma L. Lingle, and Jonathan J. Li

Summary Sustained over-expression of Aurora A (AurA), centrosome amplification, chromosomal instability, and aneuploidy are salient features that occur in high frequency in human breast premalignant stages and in primary ductal breast cancer (BC), as well as in 17β-estradiol (E_2)-induced oncogenesis in animal models. We have reported that AurA/B protein expression increases 8.7- and 4.6-fold, respectively, in primary E_2-induced male Syrian hamster uterine stem cell-like tumors of the kidney (EUTK) when compared with cholesterol-treated control kidneys. Upon a 10-day E_2-withdrawal or coadministration of tamoxifen citrate, a 78–79% and 81–64% reduction in AurA/B protein expression, respectively, were observed in primary tumors when compared with tumors from animals continuously exposed to E_2. These data indicate that AurA/B expression is regulated by estrogens via estrogen receptor α. To determine whether this E_2-induced over-expression of the Aur kinases may contribute to the alterations observed during oncogenesis via their phosphorylation of specific substrates, we analyzed the protein expression of histone H3 and targeting protein for Xklp2 (TPX2). Histone H3 and TPX2 were significantly over-expressed 3.7- and 1.6-fold, respectively, in E_2-induced tumors when compared with cholesterol-treated control kidney samples. Immunohistochemistry revealed that TPX2 protein expression was essentially confined to tumor foci cells. Collectively, these data indicate that over-expression of AurA/B is under estrogen control and that the deregulation of Aur kinase protein substrates is implicated in eliciting the alterations observed during oncogenesis.

Introduction

Breast cancer (BC) is the most commonly diagnosed female cancer worldwide (1). It has been well established that both endogenous, and to a lesser extent, exogenous estrogens play a crucial role in both the causation and development of sporadic BC (2), which accounts for 90–95% of all BC cases. (17β-estradiol) E_2-induced and E_2-dependent tumors of the Syrian hamster kidney, first described in 1959 by Kirkman, have become an intensively studied model of estrogen oncogenesis because 100% tumor incidence is achieved with a relatively modest elevation in serum E_2 concentrations (~2.3 ± 0.4 nmol l ml^{-1}) (3).

These E_2-induced tumors arise in the corticomedullary region of the kidney from ectopic uterine-like germinal stem cells (4, 5). These ectopic cells remain dormant unless exposed to a sustained E_2 levels. Moreover, both E_2-induced early tumor foci and primary tumors are highly aneuploid (92–94%) (5, 6). In addition to being aneuploid, these tumors and human ductal BC share many crucial early molecular changes during E_2-induced oncogenesis including: over-expression and amplification of early response nuclear proto-oncogenes *c-myc* (7, 8), *c-fos* (9), and *c-jun* (10), deregulation of cell-cycle entities including cyclins D1 and E1, cdk2 and 4, pRB, and the cdk inhibitor p27, centrosome amplification, and chromosome instability (11, 12). However, it is unclear how these E_2-induced molecular changes lead to chromosomal instability and tumorigenesis.

Aurora A/B (AurA/B) over-expression has been reported in a number of human solid tumors including BC (13). The over-expression of these mitotic kinases is believed to be a crucial event leading to the cascade of centrosome amplification, chromosomal instability, aneuploidy, and eventual tumorigenesis. The Aur kinases are essential for the proper execution of various mitotic events including centrosome duplication, maturation and separation, spindle assembly and stability, chromosome condensation and segregation, and cytokinesis (14–18). AurA/B participate in these various mitotic events through interactions with their many known substrates, at least 15 for AurA and 10 for B (19). It is through the aberrant phosphorylation of these protein substrates that lead to duplication/segregation errors, thus driving breast oncogenesis.

To begin to access whether the over-expression of AurA/B contributes to the deregulation of the centrosome cycle via the hyperphosphorylation of specific protein substrates, we show that both histone H3, a substrate for AurA/B, and targeting protein for Xklp2 (TPX2), a substrate for AurA, are over-expressed in E_2-induced primary tumors of the Syrian hamster kidney. The phosphorylation of histone H3 is required for proper chromosome condensation and cell-cycle progression (20). This role is important in facilitating centrosome amplification via inappropriate or excessive phosphorylation in the presence of AurA/B over-expression. TPX2 interacts with AurA as both an activator and a substrate (21, 22). TPX2 activates AurA through a conformational change that protects AurA from inactivation via PP1 dephosphorylation. Once active, AurA phosphorylates TPX2, which then recruits AurA to the spindle microtubules (22). These studies will allow us to begin to determine whether or not inappropriate and possibly differential phosphorylation of Aur kinase substrates occurs, and what specific substrates may be most important, during the development of E_2-induced tumors.

Materials and Methods

Animals and Treatment. Male Syrian castrated hamsters were treated with 20 mg pellets of 17β-estradiol (E_2), provided by Hormone Pellet Press, Leawood, KS, as previously described (4). For the withdrawal studies, hamsters bearing tumors, treated for 6.0 months with E_2 were divided into four groups ($n = 3$): Group 1 was maintained on E_2; Group 2 had their E_2 withdrawn; Group 3 had one E_2 pellet removed, and additionally implanted with 2 pellets of 20 mg Tamoxifen (Tam) citrate; Group 4 included age-matched cholesterol-treated controls. All animals were killed 10 days after either the removal of the E_2 pellets or Tam treatment. Tumors were individually harvested from groups 1–3 and kidneys from group 4 for subsequent analysis of AurA/B protein expression.

Immunohistochemistry. At different treatment intervals, hamster kidneys were harvested and fixed in 10% paraformaldehyde and embedded in paraffin wax. Five to six micrometer sections were deparaffinized for standard immunohistochemisty with the following primary antibodies: AurA BL656 (Bethyl) at 1:200, AurB ab2254 (Abcam) at 1:50 and TPX2 sc-32863 (Santa Cruz) at 1:40, and appropriate secondary antibodies. Protein expression was evaluated by light microscopy.

Western Blot Analysis. Western blot analysis was performed in untreated, E_2-treated kidneys and tumor samples probed with anti-AurA BL-656 (Bethyl, 1:1000), AurB ab2254 (Abcam, 1:1000), histone H3 sc-10809 (Santa Cruz, 1:500) and TPX2 sc-32863 (Santa Cruz, 1:500), and appropriate secondary antibodies. Protein expression was visualized with ECL reagent and equal loading was confirmed by reprobing with anti-GAPD antibodies.

Statistical Analysis. Statistical evaluation was performed using one way analysis of variance (ANOVA) with Tukey post-hoc tests with significance set at $P < 0.05$. Values represent the mean + SEM.

Results

Western Blot Analysis of AurA/B During E_2-induced Oncogenesis. AurA/B kinase protein expression was determined in kidney tissue lysates after 3, 4, 5, and 6 months of E_2 treatment, and in primary tumors (Fig. 1A). A modest increase in AurA protein expression was detected in early E_2 treatment periods; however, a significant 8.7-fold increase was observed in primary tumors (Fig. 1B) when compared with control cholesterol-treated kidneys. The upper band, observed only in the E_2-induced tumors of the kidney, likely represents the phosphorylated form of AurA, significant changes in AurB expression were not observed after 3–5 months of E_2 treatment. However, a slight increase was seen in protein expression at 6 months followed by a significant 4.6-fold increase in AurB protein expression in E_2-induced primary tumors (Fig. 1B) compared with control cholesterol-treated kidneys.

Regulation of AurA/B Kinase Protein Expression via Estrogens. AurA/B kinase protein expression was determined in tumor samples from groups of 6-month tumor-bearing hamsters that were continuously treated with E_2, hamster that had their E_2 pellets removed, or hamster that had one E_2 pellet removed while undergoing concomitant treatment with Tam for 10 days. The 10 day withdrawal/treatment period was chosen because it takes ~72 h for estrogens to completely clear from the

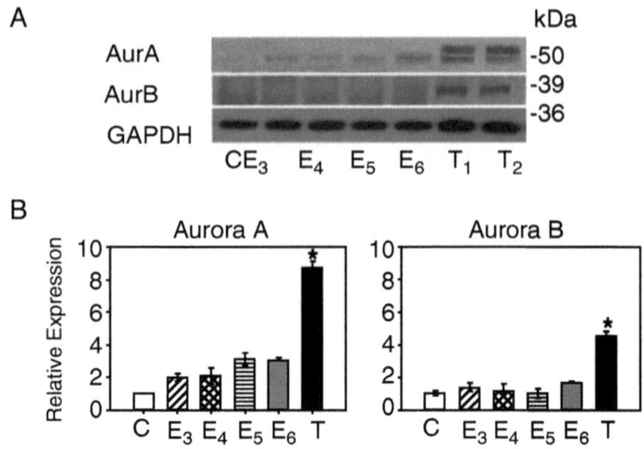

Fig. 1 (**A**) Western blot analysis of AurA/B from untreated controls (C), E_2-treated Syrian hamster kidneys [3.0 months (E_3), 4.0 months (E_4), 5.0 months (E_5), or 6.0 months (E_6)] and E_2-induced kidney tumors (T) ($n = 6$). (**B**) Relative expression of AurA and B. Columns, mean; bars, SE

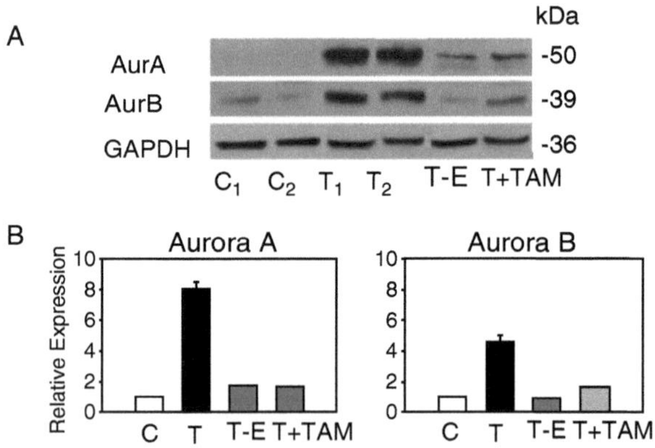

Fig. 2 (**A**) Western blot analysis of AurA/B from untreated controls (C) and tumors from hamsters continuously treated with E_2 (T), 10-day withdrawal of E_2 (T-E) and 10-day concomitant TAM (T + TAM). A significant decline in AurA/B expression was observed after the 10-day E_2-withdrawal period and the 10-day concomitant TAM treatment. (**B**) Relative expression of AurA/B. Columns, mean; bars, SE

serum of treated animals (3). A significant increase in AurA/B expression, 8.0- and 4.7-fold, respectively, was observed in the animals that were continuously treated with E_2 when compared with age-matched cholesterol-treated control kidneys (Fig. 2B).

After a 10-day E_2-withdrawal period, both AurA/B expression was significantly reduced, 78% and 81%, respectively, when compared with tumors receiving sustained E_2 treatment. Similarly, coadministration with Tam resulted in a significant 79% and 64% decline in AurA/B protein expression in these tumors.

Western Blot Analysis of AurA/B Substrates During E_2-induced Oncogenesis. Protein expression of histone H3 and TPX2 was determined in lysates of whole hamster kidney samples after 4.0 months of E_2 treatment and in hamster kidney primary tumors (Fig. 3A). A modest increase in histone H3 expression was observed after 4.0 months of E_2 treatment followed by a significant 3.7-fold increase in primary tumors (Fig. 3B) when compared with cholesterol-treated control kidney samples. A TPX2 doublet was observed in the cholesterol-treated controls and 4.0 months E_2-treated hamsters (Fig. 3A). The primary tumors, however, only exhibited a single TPX2 band and showed a 1.6-fold increase in expression (Fig. 3B) when compared with cholesterol-treated controls.

TPX2 Protein Localization During E_2-Induced Oncogenesis. The precise cellular localization of TPX2 was assessed after various E_2-treatment intervals in kidney sections containing early tumor foci. Sections were examined by H&E staining and immunohistochemistry. TPX2 expression was overwhelmingly confined to cells present in the tumor foci. TPX2 expression increased as the size of the tumor foci increased (Fig. 4A–C). Positively stained cells were not detected in tumor foci sections in the absence of primary antibody (Fig. 4D).

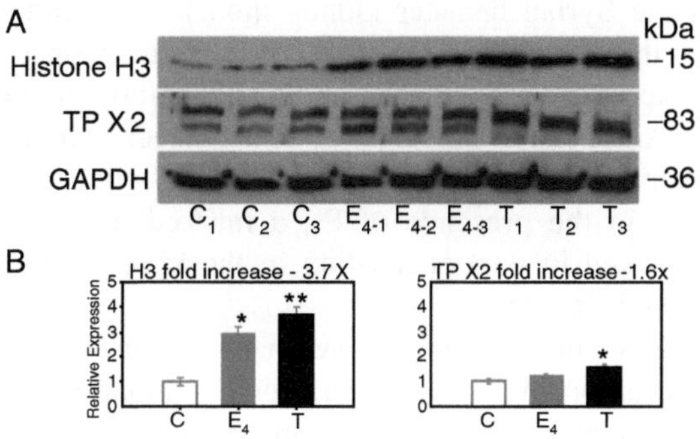

Fig. 3 (**A**) Western blot analysis of Histone H3 and TPX2 from untreated controls (C), 4.0-month E_2-treated (E_4) Syrian hamster kidneys and E_2-induced kidney tumors (T) (n = 3). (**B**) Relative expression of Histone H3 and TPX2. Columns, mean; bars, SE. $^{**}p < 0.001$ vs. control and $^{*}p < 0.05$ vs. control

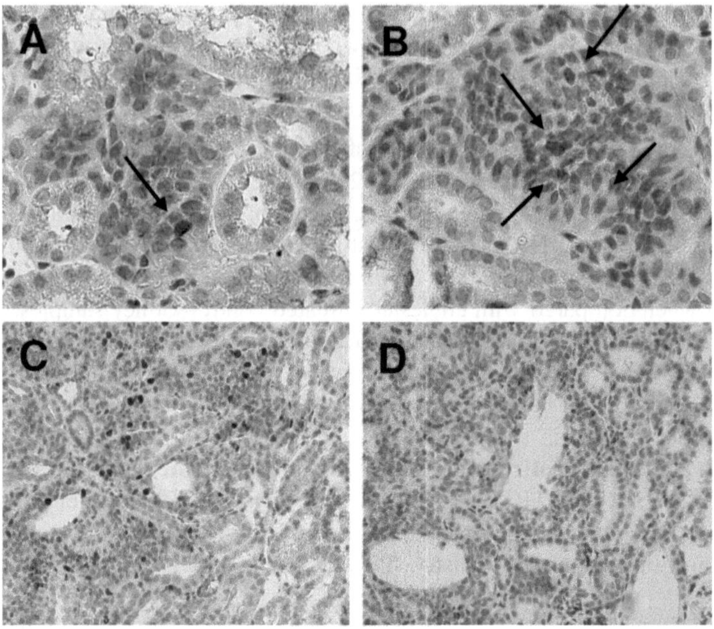

Fig. 4 TPX2 expression in: (**A**) A small tumor focus from a 4.0-month E_2-treated kidney. *Arrow* points to TPX2-positive cell. (**B**) An intermediate tumor focus from a 6.0-month E_2-treated kidney. *Arrows* point to TPX2-positive cells. (**C**) A large tumor focus from a 6.0-month E_2-treated kidney. (**D**) Kidney serial section without primary antibodies. Magnification: ×40

Conclusions

Using a hamster kidney model of E_2-induced oncogenesis, both AurA/B were shown to be over-expressed in early tumor foci in the kidney. The over-expression of Aur kinases in the Syrian hamster kidney model is in tandem with previous results published in the female ACI rat model of E_2-induced oncogenesis reporting that AurA is over-expressed in dysplasias and ductal carcinoma in-situ (23). These results suggest that AurA expression is under either direct or indirect estrogen control. This is now supported by our findings that upon E_2-withdrawal or concomitant treatment with Tam in the presence of E_2, a marked decline in both AurA/B expression was observed in tumors residing in the kidney when compared with tumors maintained on E_2 alone.

Combined, AurA/B phosphorylate more than 20 currently known substrates (19). The phosphorylation of any individual or combination of these substrates might affect the deregulation of the centrosome cycle and downstream molecular changes leading to tumor formation. Our data show that two of these Aurora substrates, histone H3 and Tpx2, are over-expressed in early tumor foci in the kidney. These results, while preliminary, suggest that these two substrates may prove to be important players during early stages of E_2-induced oncogenesis. In addition, Tpx2 overwhelmingly localizes to cells in tumor foci.

In summary, we have shown that AurA/B are over-expressed in the Syrian hamster kidney model of E_2-induced oncogenesis and that this over-expression is under the control of estrogens, either directly or indirectly. We have also shown that two substrates of Aur, histone H3 and Tpx2, are also over-expressed in this model. However, the importance of the later observation is yet to be put into context. Not surprisingly, our results replicate the high frequencies of AurA over-expression reported in the female ACI rat (23) and in human sporadic BC (~94%) (13). Although the precise relationship between the sustained over-expression of Aur and downstream events leading to tumorigenesis are not yet clear, it cannot be ignored that this may represent a common mechanism whereby estrogens elicit tumor development.

Acknowledgments This research was supported in part by NCI/NIH Grant R01 CA102849.

References

1. Hortobagyi GN, de la Garza Salazar J, Pritchard K, et al. (2005) The global breast cancer burden: variations in epidemiology and survival. Clin Breast Cancer, 6, 391–401.
2. Bilimoria MM, Morrow M (1995) The woman at increased risk for breast cancer: evaluation and management strategies. CA Cancer J Clin, 45, 263–78.
3. Li SA, Xue Y, Xie Q, et al. (1994) Serum and tissue levels of estradiol during estrogen-induced renal tumorigenesis in the Syrian hamster. J Steroid Biochem Mol Biol, 48, 283–6.
4. Gonzalez A, Oberley TD, Li JJ (1989) Morphological and immunohistochemical studies of the estrogen-induced Syrian hamster renal tumor: probable cell of origin. Cancer Res, 49, 1020–8.
5. Li JJ, Weroha SJ, Davis MF, et al. (2001) ER and PR in renomedullary interstitial cells during Syrian hamster estrogen-induced tumorigenesis: evidence for receptor-mediated oncogenesis. Endocrinology, 142, 4006–14.
6. Papa D, Li SA, Li JJ (2003) Comparative genomic hybridization of estrogen-induced ectopic uterine-like stem cell neoplasms in the hamster kidney: nonrandom chromosomal alterations. Mol Carcinog, 38, 97–105.
7. Escot C, Theillet C, Lidereau R, et al. (1986) Genetic alteration of the c-myc protooncogene (MYC) in human primary breast carcinomas. Proc Natl Acad Sci USA, 83, 4834–8.
8. Whittaker JL, Walker RA, Varley JM (1986) Differential expression of cellular oncogenes in benign and malignant human breast tissue. Int J Cancer, 38, 651–5.
9. Walker RA, Cowl J (1991) The expression of c-fos protein in human breast. J Pathol, 163, 323–7.
10. Tiniakos DG, Scott LE, Corbett IP, et al. (1994) Studies of c-jun oncogene expression in human breast using a new monoclonal antibody, NCL-DK4. J Pathol, 172, 19–26.
11. Liao DZ, Hou X, Bai S, et al. (2000) Unusual deregulation of cell cycle components in early and frank estrogen-induced renal neoplasias in the Syrian hamster. Carcinogenesis, 21, 2167–73.
12. Li JJ, Hou X, Banerjee SK, et al. (1999) Overexpression and amplification of c-myc in the Syrian hamster kidney during estrogen carcinogenesis: a probable critical role in neoplastic transformation. Cancer Res, 59, 2340–6.

13. Tanaka T, Kimura M, Matsunaga K, et al. Centrosomal kinase AIK1 is overexpressed in invasive ductal carcinoma of the breast. Cancer Res, 59, 2041–4.
14. Carmena M, Earnshaw WC (2003) The cellular geography of aurora kinases. Nat Rev Mol Cell Biol, 4, 842–54.
15. Dutertre S, Descamps S, Prigent C (2002) On the role of aurora-A in centrosome function. Oncogene, 21, 6175–83.
16. Andrews PD, Knatko E, Moore WJ, Swedlow JR (2003) Mitotic mechanics: the auroras come into view. Curr Opin Cell Biol, 15, 672–83.
17. Ducat D, Zheng Y (2004) Aurora kinases in spindle assembly and chromosome segregation. Exp Cell Res, 301, 60–7.
18. Meraldi P, Honda R, Nigg EA (2004) Aurora kinases link chromosome segregation and cell division to cancer susceptibility. Curr Opin Genet Dev, 14, 29–36.
19. Li JJ, Li SA (2006) Mitotic kinases: the key to duplication, segregation, and cytokinesis errors, chromosomal instability, and oncogenesis. Pharmacol Ther, 111, 974–84.
20. Van Hooser A, Goodrich DW, Allis CD, et al. (1998) Histone H3 phosphorylation is required for the initiation, but not maintenance, of mammalian chromosome condensation. J Cell Sci, 111 (Part 23), 3497–506.
21. Eyers PA, Erikson E, Chen, et al. (2003) A novel mechanism for activation of the protein kinase Aurora A. Curr Biol, 13, 691–7.
22. Kufer TA, Sillje HH, Korner R, et al. (2002) Human TPX2 is required for targeting Aurora-A kinase to the spindle. J Cell Biol, 158, 617–23.
23. Li JJ, Weroha SJ, Lingle WL, et al. (2004) Estrogen mediates Aurora-A overexpression, centrosome amplification, chromosomal instability, and breast cancer in female ACI rats. Proc Natl Acad Sci USA, 101, 18123–8.

Global Quantitative Analysis of Protein Phosphorylation Status in Fish Exposed to Microcystin

Karim Mezhoud, Daniele Praseuth, Jean-Christophe Francois, Cecile Bernard, and Marc Edery

Summary The hepatotoxins, microcystins (MCs) are potent inhibitors of protein phosphatases PP1 and PP2A. These nonribosomal peptides are getting more and more attention because of their acute toxicity and potent tumor-promoting activity. These toxins are produced by freshwater cyanobacteria. Herein, we report a toxicological study conducted on aquatic animal models such as the medaka fish. To date, the detailed mechanisms underlying the toxicity of microcystins are unknown. MC-leucine-arginine (MC-LR) is the most toxic and the most commonly encountered variant of MCs in aquatic environment. It has been used for toxicological investigations on the liver of intoxicated medaka. We performed differential proteome analyses of MC-LR-treated and untreated medaka fish to investigate the mechanisms of establishment of early responses to the toxin. The identification of proteins involved in these early responses might constitute candidates of biomarkers of MC-LR exposure. Cytosolic proteins from livers of exposed or nonexposed medaka were resolved by 2D electrophoresis and detected using stains specific for phosphoproteins and for whole protein content. Overall, 15 spots were found to vary significantly on the proteomic 2D maps or on the phosphoproteomic 2D maps. Of these 15 proteins, only two could not be identified by mass spectrometry. Among the other proteins that were identified, phenylalanine hydroxylase and keratin 18 (type I) showed variations in phoshoryl content in agreement with inhibition of PP2A activity after exposure of the fish to MC-LR. The other identified proteins exhibited variations in their expression level. The identified proteins appear to be involved in cytoskeleton assembly, cell signalling, oxidative stress, and apoptosis. The functional implications of responses to MC-LR exposure of these proteins are discussed. The methodology described in this report should be widely used to a number of tissues and organisms, thus helping in the search for biomarkers of MC-LR contamination.

Introduction

Cyanobacteria, photoautotroph organisms, are a frequent component of many freshwater and marine ecosystems. Since many species and strains of these organisms produce toxins (cyanotoxins), blooms can cause health risks to humans or animal.

J.J. Li et al. (eds.), *Hormonal Carcinogenesis V.*

Microcystins (MCs) are a family of hepatotoxic toxins produced by cyanobacteria. About 70 structural variants of MCs have been characterized. MCs have been implicated in the cause of death both of domestic animals that had consumed water containing cyanotoxins and of humans submitted to hemodialysis using water contaminated with MCs. The uptake of MCs into cells is mediated by the bile acid carrier, which is found in liver and, to a lesser extends, in intestinal epithelia. The presence and accumulation of MCs in different fish tissues have been reported by experimental and field studies. Rat or mouse primary hepatocytes have been extensively used to investigate the effects of MCs. The most common and toxic variant of all the 70 MCs is MC-leucine-arginine (MC-LR). This variant induces cytoskeleton damage and oxidative stress. In rat primary hepatocytes, MC-LR increases the mitochondrial permeability transition (MPT), the level of reactive oxygen species (ROS), and initiation of apoptosis. MC-LR induces apoptosis in mouse liver cells after intraperitoneal injection. In this study, we used a rapid approach that takes about a large scale of information such as the protein expression and phosphorylation for all subcellular fractions (cytosol, membranes/organelles, nuclear, and cytoskeleton).

Recently, several proteomic studies have elucidated many proteins as implicated in MC effects. However, most of these studies have focused on expression pattern using in-vitro exposure, but proteomics using in-vivo assays have not been extensively investigated. MCs toxic effects result from the inhibition of serine/threonine protein phosphatases PP1 and PP2A, as demonstrated in vitro and in vivo. Cellular biochemical responses to MCs might be confusing because these are involved in the usual conflict between apoptosis and tumor promotion. Many physiological processes such as protein expression, cell division, and apoptosis are regulated by the delicate balance between the phosphorylated/unphosphorylated status of key control proteins. The effectors of that balance are kinases and phosphatases. In this study, analysis of the phosphoprotein status in liver of medaka fish exposed to MC-LR was performed. Our main goal was to gain deeper insight into the expression regulation and the phosphorylation of medaka liver proteins in response to MC-LR treatment. The implications of identified proteins in the process of MC-LR-induced biochemical alterations are discussed, which will aid in the understanding of some of the molecular mechanisms of MC-LR toxicity, as well as providing the possibility for finding new biomarkers of exposure to the toxin. This study is the first toxicoproteomic approach concerning the effects of a cyanotoxin on an aquatic animal model. The phosphoproteomic approach described in this study is also innovative in the field of toxicology as well as in the area of mechanisms of action involved in toxin activity.

Materials and Methods

Fish. Adult medaka fishes of the inbred cab strain were used in all experiments.

Sodium Dodecyl Sulphate-Polyacrylamide Gel Electrophoresis (SDS-PAGE). Approximately, 20 μg of proteins from different subcellular fractions were separated and loaded in SDS/PAGE [2% (v/v) acrylamide/bisacrylamide].

Gel Staining. First, gels were stained in Pro-Q Diamond phosphoprotein gel stain (Molecular probes, Eugene, OR, USA) and then imaged. Second, gels were stained in SYPRO Ruby stain (Molecular probes) and imaged using Image Master 2D Platinum software, version 5.00 (GE Healthcare).

Data Analysis. The gel images were grouped on to 3 classes (5 gels by class), control, and 30- and 60-min treatment time. The spots were detected simultaneously and matched. Then treated classes were compared with the control class statistical analysis of the spot values within a class that was performed in Image Master 2D Platinum with the Student's *t* test.

Capillary Liquid Chromatography (LC) Electrospray Ionization (ESI) Quadrupole Time-of-Flight. Protein digests were resolved by HPLC and the acquired mass spectral data were queried against NCBI Br with all entries taxonomy protein database using Mascot accessible on line. The acquired mass spectra were converted into Mascot compatible format using the mascot.dll script in Analyst v1.1 software from Applied Biosystems. The identified proteins were then matched in Medaka Expression Data Base for confirmation.

Matrix-Assisted Laser Desorption Time-of-Flight Mass Spectrometry (MALDI-TOF-MS) Analysis. Mass spectra were recorded in positive ionisation mode on a MALDI-TOF-MS (Voyager-DE Pro, Applied Biosystems). Mass/charge (*m/z*) ratios were measured in the reflector/delayed extraction mode with an accelerating voltage of 20 kV, grid voltage of 150 V, 64.5% guide wire voltage, and low mass gate of 600. Protein identification using peptide mass fingerprinting (PMF) was performed by feeding mass data to the Mascot search.

Results

Protein and Phosphoprotein Expression Patterns in Response to MC-LR. Fishes were exposed to a single dose (1 $\mu g\ ml^{-1}$ MC-LR) in water for the experimental time course (30 and 60 min). The accumulation of MCs in different fish tissues have been reported by experimental and field studies. The major toxicity of MCs appears in the liver. Therefore, we pointed our efforts on liver proteomic pattern and its evolution in intoxicated medaka. To both simplify quantitative analysis and ease the detection of phosphorylation, we fractionated the fish liver of homogenate into the cytosolic, membranous, nuclear, and cytoskeletal fractions.

Initial studies were aimed at determining the necessary time to induce phosphorylation modulations in liver cells. As observed on SDS-PAGE gels stained with a phosphoryl-specific reagent, 40 min of treatment were sufficient to obtain an increase of phosphorylation as detected with the ProQ Diamond reagent. When the MC-LR incubation was for longer durations (60 min), the phosphorylation signal started to decrease. Longer exposure times did not reverse the observed trend. Therefore, treatment time was based between 30 and 40 min. We focused our analysis on the cytosolic subcellular fraction that were analysed by 2DE-gels.

Data Analysis Collected For a Set of Five Experiments. Each gel was stained first with the ProQ Diamond dye (phosphorylated proteins) and second with the Sypro Ruby dye (total proteins). These two stains enabled us to perform the quantification of the variation of phosphorylated protein content and total protein content, between control and treated samples. The analysis thus encompassed a total of 30 images (15 images for the Sypro Ruby dye and 15 images for the ProQ-Diamond dye).

In our studies, a typical 2-DE map of cytosolic proteins of medaka hepatocytes was obtained with 200 μg of proteins. On average, 1 143 spots were detected in each gel of which 147 appeared as phosphorylated proteins, which corresponds to 13% Sypro Ruby-stained proteins. Among these, 15 spots were shown to change significantly from control to treated (Table 1).

Protein Regulation. Five phosphoprotein spots of ProQ-Diamond staining were found to have their relative volumes significantly changed ($p > 0,05$). Spots 1, 2, 3 were upregulated after 30-min exposure and spots 1, 2, 4, and 5 were upregulated

Table 1 Protein list of differentially modulated proteins during MC-LR treatment

		Fold-change			Mascot results all taxa	Medaka EST database
id	MW/pI	30 min	60 min	Proteins	NCBI accession 1	NCBI accession 2
1	55/5.2	3.04	2.0	Phenylalanine hydroxylase	AAT39424	BJ737725
2	56/5.0	3.20	3.5	Phenylalanine hydroxylase	AAT39424	BJ737725
3	54/6.5	1.9		Se binding protein1	XP_707845	BJ007475
					AAH5690	BJ007475
4	49/5.7		2.3	Keratin 18 type I	CAA74664	BJ498018
					AAC38007	BJ747804
5	29/5.8		2.3	Spindlin protein B	AAT47135,1	AAT47135.1
6	20/5.8	0.12	0.2	RKIP	CAG08164	BJ747456
				DJ-1	BAD67176	BAD67176
				RKIP	EST	AU170544
7	32/5.9	5.55		RpP0	37779102	
8	22/6.8	0.80		NKEF	AAY25400	BJ714211
				Prx 4	AAH19578	
				RAhpC-TSA family	AV669883	
9	22/6.0	0.80		Nonidentified		
10	22/6.5	0.53		Medaka eye clone NGY45,07f	AU244165	
11	25/6.1	0.4		HGPRT	CAA35648	BJ729370
12	30/5.8	0.53		Nonidentified		
13	27/5.7	0.65		Actin capping protein B subunit	AAA52222	BJ729321
14	47/7.0		2.94	Enolase	AAA70080	BJ722994
15	27/4.0	0.55	0.54	PKC inhibitor	AAB22282	BJ727482
				14-3-3 protein	AAB22943	

after 60 min compared with the control (Table 1). Only spot 6 was shown as down regulated. Moreover, 10 protein spots were found to have their relative volumes in Sypro Ruby dye changed significantly after 30-and 60-min exposure time. Among these, 8 have a down-regulation (i.d., 7, 9, 10, 11, 12, 13, 14, 15) (Table 1).

On the basis of these observations, the proteins corresponding to these spots have been identified by either MALDI-TOF (i.d. 1, 2, 3, 4, 5) and/or LC-ESI-Q-TOF (i.d. 3, 4, 6, 7, 8, 9, 11, 13, 14, 15). *Oryzias Latipes* is not available as repertoired specie in Mascot or Profound peptide fingerprint or MS/MS ions search. We performed our studies in all taxa, eventually restricting at *Chordata* then *Danio rerio* or *Takifugu rupripes*. First, proteins were identified in NCBInr protein or NCBI EST (expressed sequence tags) database at all taxa using mascot or profound as a research engine. In this step, generally the proposed proteins did not belong to the *Oryzias latipes* specie. Second, proteins labeled by accession I were matched to medaka EST database (http://medaka.lab.nig.ac.jp/) or medaka expression pattern database (MEPD, http://ani.embl.de:8080/mepd/) using tblastn program (translating basic local alignment search tool for nucleotide) for the searching the clone corresponding to the protein (NCBI accession I). In this step, the clone that we have to choose (NCBI accession II) contains in its virtual protein at least two sequence peptides sequenced or detected in the first identification (NCBI accession I). The third step consists of a confirmation that the first (NCBI accession I) and the second (NCBI accession II) identifications target the same protein or the same family of proteins.

Phenylalanine hydroxylase (PAH) was identified as phosphorylated protein on two spots (1, 2) shiefted by 0.2 pI and 1 kDa. It was identified by peptide mass fingerprinting (PMF) in mascot and profound search program by homology to *Takifugu rubriques* protein in NCBI database. From medaka EST database, a list of clone was then sorted as a function of their homology percentage. MF015DA clone appears the most interesting. Its virtual sequence covers several of matched peptides in PMF result. This clone was identified on NCBI nucleotide database with BJ737725 accession. Then, the virtual protein of BJ737725 was blasted using protein–protein program (blastp) on NCBInr. PAH appeared as the most probable function of the most homologous proteins retrieved from many species including fishes, birds, and mammals. Only spindlin protein B (i.d. 5) was founded directly at *Oryzias latipes* with 35% of coverage using PMF research. In the other cases, MS/MS ions search was used for confirming the result.

Selenium binding protein (SeBP) and keratin 18 (K18) were identified first by MALDI-TOF at *Danio rerio* (XP_707845) and *Oncorhynchus mykiss* (CAA74664), respectively, and confirmed by MS/MS. SeBP spectra from MS/MS sequenced peptides were matched to NCBInr database then to NCBI EST database. The NCBInr research gave the same protein that MALDI result (SeBP) at the same specie (*Danio rerio*) but with other accession number (AAH5690). The blast result (tblastn) of the sequence protein of AAH5690 in medaka EST database gave the same accession number (BJ007475) that the MALDI result (Table 1). After PMF identification (CAA74664), the medaka clone corresponding to this sequence was found (BJ498018). On the same time, the MS/MS research was found the same protein (K18) with a high score (127/50), but not at the same specie (AAC38007); the medaka clone corresponding was BJ747804. In this case, these researches gave K18 but with different accession number. The sequence alignment of these protein sequences (CAA74664 and AAC38007) gave 78% coverage and 2% for the nucleotide sequences (BJ498018 and BJ747804). It was possible that these clones code for two different parts of k18 sequence. The pI and PM of here virtual protein sequences were calculated (BJ498018: 21 kDa, 4.87 and BJ747804: 30 kDa, 4.99). In fact, the alignment of virtual sequence protein of BJ747804 with keratin sequence (AAC38007) was shown that the BJ747804 clone encodes the C-terminal part of k18 sequence. It was possible in this case that the BJ498018 clone corresponds to N-terminal part. The alignment of the virtual sequence protein of BJ498018 with keratin sequence (AAC38007) indicated that BJ498018 clone encodes effectively for the N-terminal of keratin 18 sequence. The sequenced peptide AMQNLNDR did not match with a virtual sequence protein of BJ747804 clone because it was located in N-terminal of k18 but matched with a virtual sequence protein of BJ498018 clone.

After protein identification, we confirmed protein sequence by blasting result in the Medaka Expression Protein Database (http://ani.embl.de: 8080/mepd/). (a) MALDI-TOF identification; (b) LC-ESI-Q-TOF identification. *i.d.* identity; *can* implicated in cancer or tumoral promotion; *ap* implicated in apoptosis; *mc* implicated in response to MCs

Two cases of MS/MS ions search gave the direct identification at Oryzias latipes. Spot 7 was found as orla C3-1 protein and spot 10 was found after EST database research as protein corresponding to medaka eye clone (AU244165), similar to gamma1-crystallin at African claw.

In the other cases MS/MS ions search gave the identification of protein with other species that Oryzias latipes. The research of their corresponding clone at medaka EST database was done. It was the case for hypoxanthine guanine phosphoribosyl transferase, acting capping protein B and enolase (i.d. 11, 13, 14). In these cases no multiple candidates were suggested and no confusion was observed for the choosing of protein.

In spot 8 identification, the NCBInr database search found two interesting candidates, natural killer enhancing factor (164/46, AAY25400) and peroxiredoxin 4 (61/46, AAH19578). The EST NCBI search gave a list of medaka clones giving the same score (81/65). The virtual protein translation of the clone AV669883 indicates that it belongs to AhpC-TSA family of proteins with a 99% homology with NKEF (AAY25400) and 84% homology with peroxiredoxin-4 (Q9BGI2). AhpC-TSA family contains proteins related to alkyl hydroperoxide reductase (AhpC) and thiol specific antioxidant (TSA). NKEF and peroxiredoxin belong to this family of proteins. On two types of search NKEF was the most probable result. Therefore, all of sequenced peptides of this sample were matched to NKEF but not also to peroxiredoxin. In this case the MS/MS ions search gave two possibilities of proteins that have a low score of homology (66%). In the case of the spot 15, the result gave two possibilities of candidates (PKCi and 14-3-3 domain proteins) that have 98% homology. They were the same sequence. The medaka clone corresponding was BJ727482.

In spot 6 identification, the NCBInr database search found two interesting candidates, Unnamed protein with high score (113/46) at *Tetraodon nigroviridis* (CAG08164) and DJ-1 with lower score 51/46 but at *Oryzias latipes* (BAD67176). The alignment on these proteins shows a very low score (2%), which indicates that it does not belong to the same family of proteins and its functions should be different. The blast of unnamed protein sequence to NCBI protein homology indicates that it could belong to a highly conserved family of phospholipid-binding proteins represented in all three major phylogenetic divisions (eukaryotes, bacteria, and archaea). It was the raf kinase inhibitor protein (RKIP) and its bacterial/archael homolog phosphatidylethanolamine-binding protein (PEBP). The MS/MS ions NCBI EST database search gave directly a list of homolog clones at *Oryzias latipes* with high confidence score (149/60). The blast of the virtual protein of one of these clones (AU170544) showed that it was a protein from RKIP family.

We suggest that there were two proteins in the same spot. The significant down-regulation (0,12) was detected in phospho-specific stain. It was known that RKIP was phosphorylated but not DJ-1. So, we consider that the modulation concerns only RKIP.

Discussion

The protein expression and phosphorylation signatures obtained and identified in this work provide new information contributing to elucidate possible mechanisms of toxicity of the studied cyanotoxin in medaka liver. The identified proteins indicate that the main cause of the changes in the proteome after exposure to MC-LR were proteins involved in cytoskeleton structure maintenance, oxidative stress, and apoptosis.

Globally, the proteins identified in the present study display a down-regulation in the case of antioxidative and inhibitor of kinase proteins. Moreover, some proteins involved in cell structure (K18, acting capping) and in metabolism or synthesis such as phe-OH and enolase were characterized. In fact, in embryologic studies of Mcs effects, it was shown that physiological responses to microcystins required energetic costs, which were compensated to the expense of the energy resources of the yolk for the production of hydrogen peroxide and the antioxidative enzymes, guajacol peroxidase (POD), catalase (CAT), glutathione peroxidase (GPx), and the glutathione restoring enzyme glutathione reductase (GR). These variations could be the basis for inducing apoptosis in the medaka liver.The major proteins identified in this study have clearly relationship with the process of apoptosis, cytoskeleton alteration, and metabolic activity.

Acknowledgments This work was supported by ARC 3270 and ECODYN

References

1. Jacquet C, et al. (2004) Effects of microcystin-LR on development of medaka fish embryos (*Oryzias latipes*).Toxicon 43, 141–147.
2. Schulenberg B, Aggeler R, Beechem, et al. (2003) Analysis of steady-state protein phosphorylation in mitochondria using a novel fluorescent phosphosensor dye. J. Biol. Chem. 278, 27251–27255.
3. Imanishi S, Harada K (2004) Proteomics approach on microcystin binding proteins in mouse liver for investigation of microcystin toxicity. Toxicon 43, 651–659.

Breast Cancer

ERK/MAPK Regulation of the Androgen Responsiveness of Breast Cancer Cells

Diana G. Azzam, Jasmine W.T. Tay, Melissa A. Greeve, Jennet M. Harvey, and Jacqueline M. Bentel

Summary The androgen receptor (AR) is the most widely expressed steroid hormone receptor in human breast cancers and androgens including 5α-dihydrotestosterone are potent inhibitors of breast cancer cell proliferation. The extracellular signal-regulated mitogen activated protein kinase (ERK/MAPK) pathway is hyperactivated in a proportion of breast tumors and can interact with steroid hormone receptor signaling by altering receptor phosphorylation, turnover, ligand, and cofactor interactions. To examine the effects of ERK/MAPK hyperactivity on AR levels, MCF-7 cells were stably transfected with a plasmid encoding a constitutively active MEK1 protein to create MCF-7-ΔMEK1 cells. Treatment of MCF-7-ΔMEK1 with androgens caused a transient increase in AR protein levels, similar to that observed in untransfected MCF-7 cells treated with androgens. Androgens also inhibited the proliferation of MCF-7-ΔMEK1 cells by 50–60% following 8 days of treatment in association with increased accumulation of cells in the G1 phase of the cell cycle. These results indicate that although ERK/MAPK hyperactivation in breast cancer cells is associated with reduced estrogen receptor (ERα) levels and antiestrogen resistance, AR levels are maintained and breast cancer cells remain susceptible to the growth inhibitory effects of androgens.

Introduction

The androgen receptor (AR) is the most widely expressed steroid hormone receptor in breast tumors with 70–90% of primary breast tumors and up to 75% of breast cancer (BC) metastases expressing the receptor (1, 2). The AR is frequently coexpressed with estrogen receptor (ERα) and progesterone receptors (PR) in primary breast tumors; however, in 25% of metastatic deposits it is the sole steroid hormone receptor detected (1, 2). Previous use of androgenic compounds including fluoxymesterone and testosterone propionate as treatments for human breast tumors has demonstrated therapeutic efficacies comparable with other hormonal therapies (2–4). Acting via the AR, androgens are also growth inhibitory to BC cells in vivo in experimental animal models and in vitro in BC cell lines (2, 5, 6).

J.J. Li et al. (eds.), *Hormonal Carcinogenesis V.*

The extracellular signal-regulated mitogen activated protein kinase (ERK/MAPK) signaling pathway transmits signals from the cell surface to the nucleus and plays important roles in critical cellular responses including proliferation, differentiation, and survival (7). A variety of extracellular stimuli including activation of cell surface receptors induce sequential activation of Raf, MEK, and ERK protein kinases, which leads to the phosphorylation of cytoplasmic and nuclear targets. The role of ERK/MAPK signaling in cell cycle progression, in particular the transition of cells from G1 into S phase, is well-characterized as is the elevation of ERK/MAPK activity in malignancy, including breast tumors (7).

Hyperactivity of ERK/MAPK signaling in cancers has been attributed to overexpression of cell surface receptors and/or their ligands, including insulin-like growth factor receptor, the HER family of receptors [epidermal growth factor receptor (EGFR), HER2, HER3, and HER4], and others (8, 9). In addition, activating mutations of upstream activators including Ras and Raf also upregulate ERK/MAPK signaling. Activation of ERK1 (p44) and ERK2 (p42) via phosphorylation of specific tyrosine and threonine residues by MEK1/2 may be detected biochemically or immunohistochemically using antibodies to phosphorylated ERK1/2 (pERK1/2) and has been widely documented in breast tumors.

Elevated expression of MAPK and increased ERK1/2 activity have been reported in human BCs in comparison to benign breast tissues (10) and ERK1/2 hyperactivation has also been correlated with reduced responses to antiestrogen treatment and decreased survival (10, 11). These clinical findings have been supported by laboratory evidence demonstrating downregulation of ERα levels and decreased estrogen and antiestrogen responsiveness of BC cell lines following ERK/MAPK hyperactivation due to overexpression of cell surface receptors (HER2) or transfection of constitutively active upstream activators (MEK1, Raf) (12). As treatment of cells exhibiting elevated ERK/MAPK activity with inhibitors of ERK1/2 signaling has been shown to restore ERα levels and antiestrogen sensitivity, these studies have indicated a potential causal relationship between the reduced responses to antiestrogen therapies of human breast tumors expressing increased ERK/MAPK activation (12–14).

Alterations in the androgen responsiveness of BC cells with hyperactivated ERK/MAPK signaling are uncharacterized. However, overexpression of HER2 in a prostate cancer cell line, LNCaP, has been shown to increase AR transactivation activity via ERK1/2 signaling (15), suggesting divergent cellular effects of ERK1/2 activity on ERα and AR function. ERK1/2 effects on the AR in prostate cancer cells may also involve ERK1/2 mediated phosphorylation of the receptor (16) and the promotion of interactions between the AR and its coregulators (15). Although both ERK1/2 and AR signaling stimulate proliferation of prostate cancer cells, they mediate opposing effects in BC cells where androgens are growth inhibitory (2, 6). Furthermore, the differential expression of AR coregulators in breast and prostate cancer cells, which may account for cell type specific androgen responses, may also lead to contrasting effects of ERK1/2 activation on AR function. To investigate the consequences of ERK1/2 hyperactivation on the

AR and the androgen responsiveness of BC cells, this study has generated MCF-7 cells with elevated ERK1/2 activation following transfection with a constitutively active mutant MEK1 (ΔMEK1).

Results

Previous studies in our laboratory have reported inhibition of the proliferation of MCF-7 cells, which express AR, ER, and PR, when cultured in the presence of 10^{-9}–10^{-7} M DHT (17). DHT treatment causes a transient increase in AR protein levels (Fig. 1A), which is likely to result from similar ligand-induced stabilization of the AR to that previously reported in prostate cancer (PC) cells.

In the present study, the ERK/MAPK pathway was hyperactivated by stable transfection of MCF-7 cells with a plasmid (pMEVHA-MEK-CA (Biomyx Technologies) that encodes a constitutively active MEK1 protein, ΔMEK1 (18). The resultant MCF-7-ΔMEK1 cells exhibited similar monolayer morphology to parental MCF-7 cells and expressed high levels of HA-tagged ΔMEK1 protein (Fig. 2). Western blot analysis indicated that total MEK1 levels were elevated in MCF-7-ΔMEK1 cells in association with increased phosphorylated ERK1/2 protein (pERK1/2) (Fig. 2). With no differences in total ERK1/2 levels in MCF-7 and MCF-7-ΔMEK1 cells, this corresponded to an increased ratio of pERK1/2: ERK1/2 or hyperactivation of the ERK/MAPK pathway. AR levels were similar in MCF-7 and MCF-7-ΔMEK1 cells, indicating that hyperactivation of ERK/MAPK signaling did not alter basal AR levels (Fig. 2).

MCF-7-ΔMEK1 cells proliferated more rapidly than MCF-7 cells, an expected result considering the well-characterized effects of ERK/MAPK signaling on cell

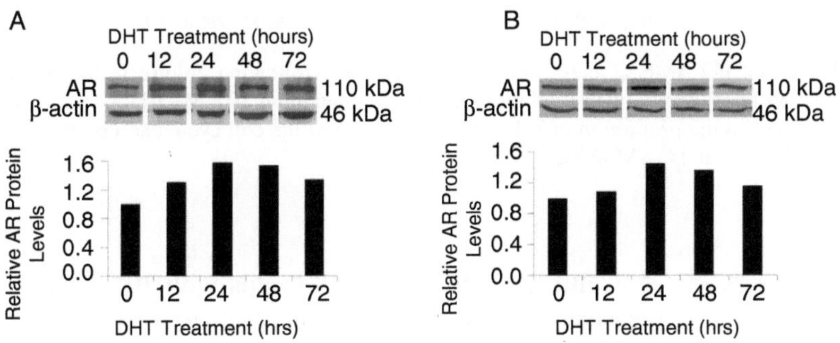

Fig. 1 Treatment of (**A**) MCF-7 and (**B**) MCF-7-ΔMEK1 cells with 10^{-8} M 5α-dihydrotestosterone (DHT) resulted in progressive and transient increases in AR protein levels that peaked at 24–48 h of DHT treatment. AR levels were determined by Western blotting and normalised using β-actin blots

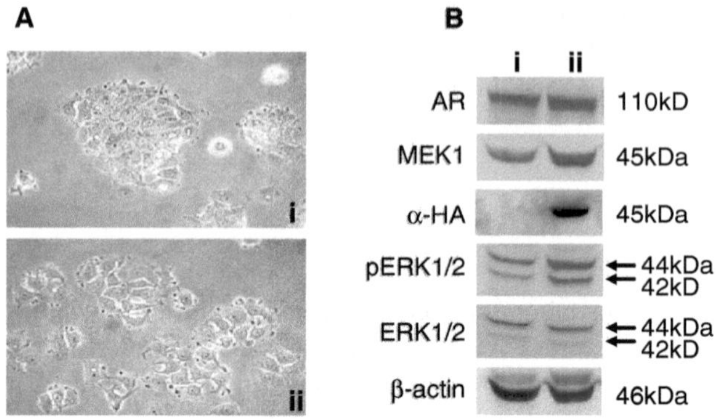

Fig. 2 (**A**) MCF-7 cells were stably transfected with the pMEVHA-MEK-CA plasmid (Biomyx Technologies) to generate the MCF-7-ΔMEK cell line (**i**) MCF-7 and (**ii**) MCF-7-ΔMEK1 cells were morphologically similar when growing in monolayer cultures. (**B**) Western blot of (**i**) MCF-7 and (**ii**) MCF-7-ΔMEK1 cells had similar levels of AR expression levels between the cell lines. MEK1 levels were increased in MCF-7-ΔMEK1 cells, which expressed HA-tagged ΔMEK1, and pERK1/2 levels were elevated in MCF-7-ΔMEK1 cells

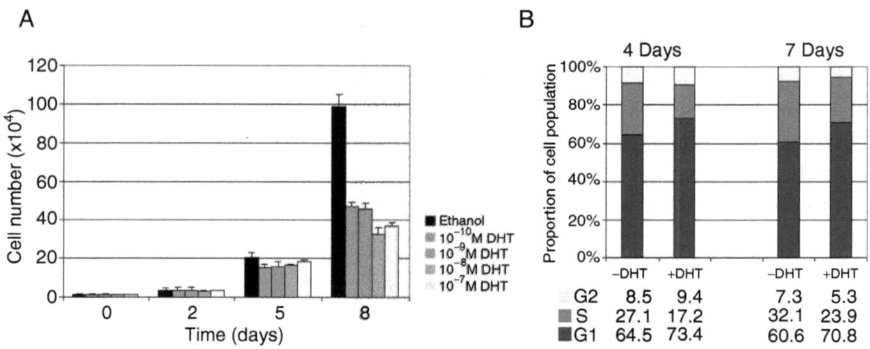

Fig. 3 (**A**) MCF-7-ΔMEK1 cells were growth inhibited following culture in the presence of 10^{-10}–10^{-8} M DHT. (**B**) Growth inhibition in cultures containing 10^{-8} M DHT was accompanied by an increase in the proportion of cells in the G1 phase of the cell cycle and decreased proportions of cells in S and G2/M phases

cycle progression (7). Culture of MCF-7-ΔMEK1 cells with 10^{-8} M DHT was associated with progressively increased AR protein levels to a peak at 1–2 days (Fig. 1B). The findings, which were similar to that detected in parental MCF-7 cells (Fig. 1A) (17), indicated that ligand-induced changes in AR levels, which are important for AR signaling, were not inhibited by ERK/MAPK hyperactivation. Treatment of MCF-7-ΔMEK1 cells with DHT at concentrations of 10^{-10}–10^{-7} M resulted in a marked inhibition (~50–60%) of proliferation by 8 days of culture

(Fig. 3A). Growth inhibition was accompanied by the accumulation of cells in the G1 phase of the cell cycle, with concomitantly decreased proportions of cells in S and G2/M phases (Fig. 3B). The proliferative responses of MCF-7-ΔMEK1 cells to DHT were similar to our previously reported results for untransfected MCF-7 cells (17).

Discussion

The androgen responsiveness of breast tumors is poorly characterised despite the widespread expression of AR in BC cells and the proven efficacy of androgenic drugs in BC treatment (2–4). As hyperactivation of ERK/MAPK signaling in BC cells is well-recognized and is associated with reduced ERα expression and decreased responsiveness to antiestrogen therapies, investigation of the effects of ERK/MAPK hyperactivation on AR expression and function is timely and may indicate therapeutic opportunities for these poor prognosis tumors.

In the present study, ERK/MAPK activation has been elevated by stable transfection of MCF-7 BC cells to overexpress a constitutively active mutant form of MEK1. This approach was chosen as overexpression of a cell surface receptor (i.e., HER2) may lead to hyperactivation of multiple signaling pathways including ERK/MAPK and AKT, consequently obscuring characterization of ERK1/2 effects. Overexpression of ΔMEK1 did not alter total ERK1/2 protein levels, a finding consistent with previous reports using this cDNA (12, 18); however, levels of pERK1/2 were markedly increased, representing elevated MAPK signaling.

AR levels were similar in MCF-7-ΔMEK1 and untransfected MCF-7 cells, suggesting that expression of ΔMEK1 and elevated pERK1/2 do not alter basal AR expression. This contrasts ERK/MAPK effects on ERα levels, which are reduced in MCF-7 cells following ERK/MAPK hyperactivation by transfection of ΔMEK1 or a mutant Raf (Δraf) (12). PR levels in T47Dco BC cells are also decreased by ERK/MAPK activation and this is reported to be due to phosphorylation of serine 294 of the PR, which targets the receptor for degradation by the 26S proteasome (19). The mechanism of ERK/MAPK downregulation of ERα protein levels is presently uncharacterized; however, accelerated protein degradation or transcriptional effects are possible (12).

Ligand-induced activation of the AR in PC cells is accompanied by increased receptor protein levels, an effect attributed to stabilization of the androgen-AR complex. Phosphorylation, nuclear translocation, and dimerization of ligand-bound AR are also involved in mediating cellular responses to androgens and these processes may contribute to the transiently increased levels of AR in androgen-treated cells. Our previous work has documented transient increases in AR protein levels in DHT-treated MCF-7 cells that are similar to results obtained in PC cells (17). In the present study, DHT treatment of MCF-7-ΔMEK1 cells resulted in progressive increases in AR protein levels to a peak at 24 h, followed by a decline to basal levels. Although this aspect of cellular processing of the AR is not markedly altered by ERK/MAPK hyperactivation, additional studies of DHT-induced intracellular trafficking and AR transcriptional activity will be required to determine whether ERK/MAPK hyperactivity induces qualitative or quantitative differences in AR biological activity that have potential implications for the long term administration of androgenic drugs as BC treatments.

The basal proliferation rate of MCF-7-ΔMEK1 cells was increased compared with that of untransfected MCF-7 cultures, a finding that was expected considering the well-characterized effects of ERK signaling on cell cycle progression (7). Activation of ERK1/2 in the cytoplasm results in its nuclear translocation where it phosphorylates and activates nuclear targets including transcription factors (i.e., Elk-1), leading to the downstream induction of cell cycle regulatory genes such as the D-type cyclins (7). Upregulation of cyclin D expression results in increased cyclin D-CDK4/6 complex formation and kinase activity, subsequent phosphorylation and inactivation of Rb and cell cycle progression. These events contribute to the mitogenic effects cell surface receptor activation associated with Ras/MAPK signaling and the increased proliferation of cells with hyperactivated ERK/MAPK.

We have reported previously that DHT treatment of MCF-7 cells reduces levels of a number of cell cycle regulators including cyclins D1 and D3, CDK2 and CDK4, with concomitant decreases in the levels of phosphorylated and total Rb (17). As such, the AR may directly oppose ERK/MAPK effects on the cell cycle. Androgen-induced inhibition of BC cell proliferation has been consistently associated with increased proportions of G1 phase cells in studies using MCF-7-ΔMEK1 (present study), MCF-7 (17), CAMA1 (20), and MDA-MB-453 (21) cultures. Initiating and downstream events leading to androgen-induced growth inhibition have not been characterized and recent reports suggest that androgen effects may be cell type specific. DHT treatment of MCF-7 cells resulted in accumulation of the cell cycle inhibitor p21 on CDK2 complexes (17), while androgen treatment of CAMA1 BC cells and administration of high dose androgens to LNCaP PC cells has been shown to cause accumulation of p27 on CDK2 complexes (21, 22). In all cell lines tested, androgen treatment was associated with decreased CDK2-associated kinase activity and accumulation of cells in the G1 phase of the cell cycle; however mechanisms leading to utilization of the different Cip/Kip inhibitors are unknown.

In the present study, decreased proliferation of MCF-7-ΔMEK1 cells was observed following 8 days of DHT treatment. Of interest was the finding of similar kinetics of proliferation responses of MCF-7-ΔMEK1 and untransfected MCF-7 (17) cells to a range of DHT concentrations (10^{-10}–10^{-7} M). Thus, ERK/MAPK hyperactivation results in distinct effects on the steroid hormone responsiveness of BC cells, conferring diminished sensitivity to estrogenic and antiestrogen ligands while permitting androgen effects that oppose the proproliferative actions of ERK/MAPK signaling. Evidence of the maintenance of AR levels and androgen sensitivity of MCF-7-ΔMEK1 cells and the prevalence of AR expression in early stage and advanced breast tumors therefore supports the further investigation of androgenic treatments for human BCs and in particular the subset of tumors with hyperactivated ERK/MAPK signaling.

References

1. Lea OA, Kvinnsland S, Thorsen T (1989) Improved measurement of androgen receptors in human breast cancer. Cancer Res 49:7162–7167.
2. Birrell SN, Hall RE, Tilley WD (1998) Role of the androgen receptor in human breast cancer. J Mammary Gland Biol Neoplasia 3:95–103.
3. Cooperative Breast Cancer Group CBC (1964) Testosterone propionate therapy in breast cancer. J Amer Med Assoc 188:1069–1072.
4. Ingle JN, Twito DI, Schaid DJ, et al. (1991) Combination hormonal therapy with tamoxifen plus fluoxymesterone versus tamoxifen alone in postmenopausal women with metastatic breast cancer. An updated analysis. Cancer 67:886–891.
5. Dauvois S, Li SM, Martel C, et al. (1989) Inhibitory effect of androgens on DMBA-induced mammary carcinoma in the rat. Breast Cancer Res Treat 14:299–306.
6. Ortmann J, Prifti S, Bohlmann MK, et al. (2002) Testosterone and 5 alpha-dihydrotestosterone inhibit in vitro growth of human breast cancer cell lines. Gynecol Endocrinol 16:113–120.
7. Torii S, Yamamoto T, Tsuchiya Y, et al. (2006) ERK MAP kinase in G1 cell cycle progression and cancer. Cancer Sci 97:697–702.
8. Keen JC, Davidson NE (2003) The biology of breast carcinoma. Cancer 97:825–833.
9. Roskoski R, Jr (2004) The ErbB/HER receptor protein-tyrosine kinases and cancer. Biochem Biophys Res Commun 319:1–11.
10. Sivaraman VS, Wang H, Nuovo GJ, et al. (1997) Hyperexpression of mitogen-activated protein kinase in human breast cancer. J Clin Invest 99:1478–1483.
11. Gee JM, Robertson JF, Ellis IO, et al. (2001) Phosphorylation of ERK1/2 mitogen-activated protein kinase is associated with poor response to anti-hormonal therapy and decreased patient survival in clinical breast cancer. Int J Cancer 95:247–254.
12. Oh AS, Lorant LA, Holloway JN, et al. (2001) Hyperactivation of MAPK induces loss of ERalpha expression in breast cancer cells. Mol Endocrinol 15:1344–1359.
13. Kurokawa H, Lenferink AE, Simpson JF, et al. (2000) Inhibition of HER2/neu (erbB-2) and mitogen-activated protein kinases enhances tamoxifen action against HER2-overexpressing, tamoxifen-resistant breast cancer cells. Cancer Res 60:5887–5894.
14. Shou J, Massarweh S, Osborne CK, et al. (2004) Mechanisms of tamoxifen resistance: increased estrogen receptor-HER2/neu cross-talk in ER/HER2-positive breast cancer. J Natl Cancer Inst 96:926–935.
15. Yeh S, Lin HK, Kang HY, et al. (1999) From HER2/Neu signal cascade to androgen receptor and its coactivators: a novel pathway by induction of androgen target genes through MAP kinase in prostate cancer cells. Proc Natl Acad Sci USA 96:5458–5463.
16. Gioeli D, Ficarro SB, Kwiek JJ, et al. (2002) Androgen receptor phosphorylation. Regulation and identification of the phosphorylation sites. J Biol Chem 277:29304–29314.
17. Greeve MA, Allan RK, Harvey JM, et al. (2004) Inhibition of MCF-7 breast cancer cell proliferation by 5alpha-dihydrotestosterone; a role for p21(Cip1/Waf1). J Mol Endocrinol 32:793–810.
18. Mansour SJ, Matten WT, Hermann AS, et al. (1994) Transformation of mammalian cells by constitutively active MAP kinase kinase. Science 265:966–970.
19. Lange CA, Shen T, Horwitz KB (2000) Phosphorylation of human progesterone receptors at serine-294 by mitogen-activated protein kinase signals their degradation by the 26S proteasome. Proc Natl Acad Sci USA 97:1032–1037.
20. Lapointe J, Labrie C (2001) Role of the cyclin-dependent kinase inhibitor p27(Kip1) in androgen-induced inhibition of CAMA-1 breast cancer cell proliferation. Endocrinology 142:4331–4338.
21. Yeap BB, Krueger RG, Leedman PJ (1999) Differential posttranscriptional regulation of androgen receptor gene expression by androgen in prostate and breast cancer cells. Endocrinology 140:3282–3291.
22. Tsihlias J, Zhang W, Bhattacharya N, et al. (2000) Involvement of p27Kip1 in G1 arrest by high dose 5 alpha-dihydrotestosterone in LNCaP human prostate cancer cells. Oncogene 19:670–679.

Vascular Endothelial Growth Factor Is a Target Gene for Estrogen Receptor and Contributes to Breast Cancer Progression

Martine Perrot Applanat, Helene Buteau-Lozano, Marie Astrid Herve, and Armelle Corpet

Summary Tumor growth requires the development and remodeling of the vascular system, involving paracrine signaling between various growth factors and endothelial receptors. Vascular endothelial growth factor (VEGF) is a key regulator of developmental, physiological and pathological neovascularization, especially involved in tumor growth. Recent studies indicate that 17β-estradiol (E2) modulates VEGF expression in breast cancer cells through transcriptional activation. We have investigated both the molecular mechanisms of E_2-induction of VEGF expression and of VEGF control of breast cancer angiogenesis. In transient transfection assays using the VEGF promoter-luciferase construct, E_2 increased VEGF transcriptional activity in MCF-7 cells and in MDA-MB-231 cotransfected with estrogen receptor (ERα or ERβ). The positive effect was abolished when MCF-7 cells were treated with the pure antiestrogen ICI 182,780 or the agonist/antagonist tamoxifen. We further identified an imperfect estrogen responsive element (ERE1520) in the VEGF promoter, which formed a complex with ERα or ERβ proteins in gel shift assay using MCF-7 or MDA-MB-231 nuclear extracts; the ERE sequence is involved in the transcriptional regulation of VEGF in our experimental conditions. These results demonstrate that in breast cancer (BC) cells *VEGF* is a target gene for ERα or ERβ. To determine the role of VEGF in the progression of human breast carcinoma, we generated stable human breast carcinoma cells (MCF-7) overexpressing VEGF165 (V165 clones). Cells or control vector clones were implanted subcutaneously in athymic mice. Our in vivo findings show that overexpression of VEGF significantly decreased tumor uptake and increased tumor growth and angiogenesis in a murine model of BC.

Introduction

The progressive growth and metastasis of breast cancer (BC) and other tumors are angiogenesis-dependent processes (1). Vascular endothelial growth factor (VEGF) is a crucial angiogenic agent in tumors, including those of endocrine-responsive tissues. VEGF expression is elevated in human breast tumors and negatively influences survival (2–3). Suppression of VEGF inhibits the growth of human tumors implanted in nude mice.

J.J. Li et al. (eds.), *Hormonal Carcinogenesis V.*

Being an endocrine-dependent carcinoma, BC requires hormones, especially estrogen and progesterone for its development (4). In vivo, VEGF expression has been shown to be rapidly induced by 17β–estradiol (E_2) in the uterus, vagina, and pituitary (5–7), and in rodent and primate models of mammary carcinogenesis (8–9). In vitro, most studies have shown that E_2 increases VEGF expression in BC cells (10–12), in addition to hypoxia and growth factors (13). Recent studies have analyzed the molecular mechanisms of VEGF regulation in BC cells.

Estrogens act on cell function through tissue-dependent specific intracellular estrogen receptor (ER), acting as ligand-activated transcription factors. Activated ERs regulate the expression of many genes via direct interaction with an estrogen responsive element (ERE) sequence located in the promoter or via interaction with other transcription factors, such as members of AP-1, STATs, NFkB, and SP families. Selective effects of estrogens may be modulated by subtype ERα or ERβ, the relative expression of each receptor in a target tissue and the promoter context of estrogen-regulated genes.

In this study, we have investigated both the molecular mechanisms of E_2-induction of VEGF expression and the role of VEGF in the control of BC angiogenesis.

Materiels and Methods

Cell Culture. MCF-7 ($ER\alpha^+$) and stable MCF-7 cells overexpressing VEGF165 were maintained in DMEM/F12 supplemented with 10% fetal bovine serum, 100 units ml^{-1} penicillin/streptomycin (PS), and 100 µg μl^{-1} sodium pyruvate. Stable clones were maintained in geneticine supplemented medium.

Transient Transfections and Luc Assay. Different constructions containing VEGF promoter fragments coupled with luciferase were previously described (11). Transient transfection was carried out using specific reporter plasmid and hER expression plasmid (11). Cells were stimulated with either E_2 (1 nM) or vehicle (0.1% ethanol) in phenol-red DMEM/F12 medium with 2% stripped FBS for 24 h at 37 °C. Luciferase activities were normalized to β-galactosidase activity to correct for differences in transfection efficiency, and were further normalized to the empty vector response.

RT-PCR Analysis. RNAs from 70% confluent MCF-7 cells were isolated using Trizol reagent. Reverse transcription of RNA was performed using Superscript II RNase H-reverse transcriptase with random hexamers. Transcript quantification for VEGF was performed using TaqMan® technology (LightCycler 2.0 Roche).

VEGF Immunoassay. VEGF concentrations were measured in cell conditioned culture media using an ELISA kit assay (R&D Systems). Results were expressed as nanogram of VEGF per milligram of cell protein extract.

Western Blot Analysis. Protein samples were loaded onto a 12.5% SDS-polyacrylamide gel for electrophoresis under reducing conditions.

***In-vivo* Tumorigenesis.** Ovariectomized mice were sc implanted with pellets containing E_2 release or placebo pellets 7 days before tumor induction. Subconfluent cells were injected subcutaneously in the flank of female SCID mice. Tumor volume was calculated every 2 days.

Immunohistochemistry. Paraffin or cryostat sections were processed for immunostaining of VEGF and CD31 at room temperature in a moist chamber. VEFG immunostaining was performed as previously described (7). For visualisation of endothelial cells, sections were incubated with CD31 antibodies followed by incubation with biotinylated rabbit anti-goat immunoglobulins. All samples were further incubated with streptavidin-biotin peroxydase, using diaminobenzidine tetrahydrochloride as the chromogen. The quantification of angiogenesis was performed by measuring microvessel density on different sections of the whole tumor. Results were expressed as mean (±SD) of vessel number mm^{-2}.

Results

Molecular Mechanisms of E_2-Induction of VEGF Expression. E_2 (1 nM) increases VEGF expression through transcriptional activation in MCF-7 cells (Fig. 1). This effect involves ERα. Pretreatment of MCF-7 cells with the pure antiestrogen ICI 182,780 or the agonist/antagonist tamoxifen (1 μM) blocks this expression (11). In contrast, no activation was observed in MCF-7 cells cotransfected with ERβ under similar conditions. Interestingly, we show that ERβ can regulate VEGF transcription in MDA-MB231 cells transfected with the receptor (11).

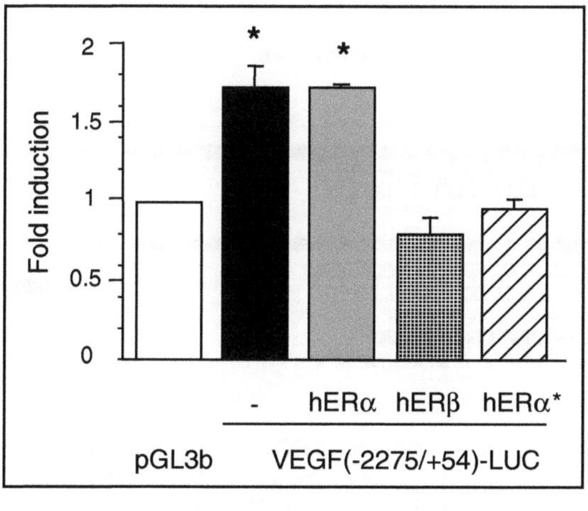

Fig. 1 ERα activates MCF-7 cell VEGF transcription. Cells were transiently transfected with VEGF-LUC construct or pGL3b control vector in the presence of ERα or ERβ expression vectors (1). Cells were treated with 1 nM E_2. Data are expressed as fold LUC activity induction in treated and untreated cells; data represent mean ± SE of 4–8 experiments. $^*p < 0.01$

We have further analyzed the promoter region involved in the regulation of VEGF by E_2 using transfection experiments. VEGF promoter is complex with multiple consensus sequences for different transcription factors including ER, AP1, SP1, and SP3 (14). The VEGF promoter also contains one hypoxia response element (HRE), which binds the transcription factor HIF1α. In our experimental conditions, the upregulation of VEGF expression by E_2 in MCF-7, or in MDA-MB231 cells transfected with either ERα or ERβ, and a plasmid containing the 2.3 kb VEGF promoter linked to the luciferase reporter gene is dose-related. The upregulation depends on a distal region of the VEGF promoter (Fig. 2). An ERE sequence (ERE1520) in the VEGF promoter forms a complex with ERα proteins in a gel shift assay using MCF-7 nuclear extracts (not shown, 11). In addition, we also observed a good correlation of E_2 induction of VEGF expression (mRNA and protein) in BC cells and their estrogenic property using a classical ERE (vitellogenin) transactivation assay (data not shown).

Attempts to identify the sites through which ERs act to induce VEGF expression in vitro and in vivo have yielded different results involving an ERE (11, 15), SP1 and/or SP3 (12) and the hypoxia-responsive element (HRE) (16). These results

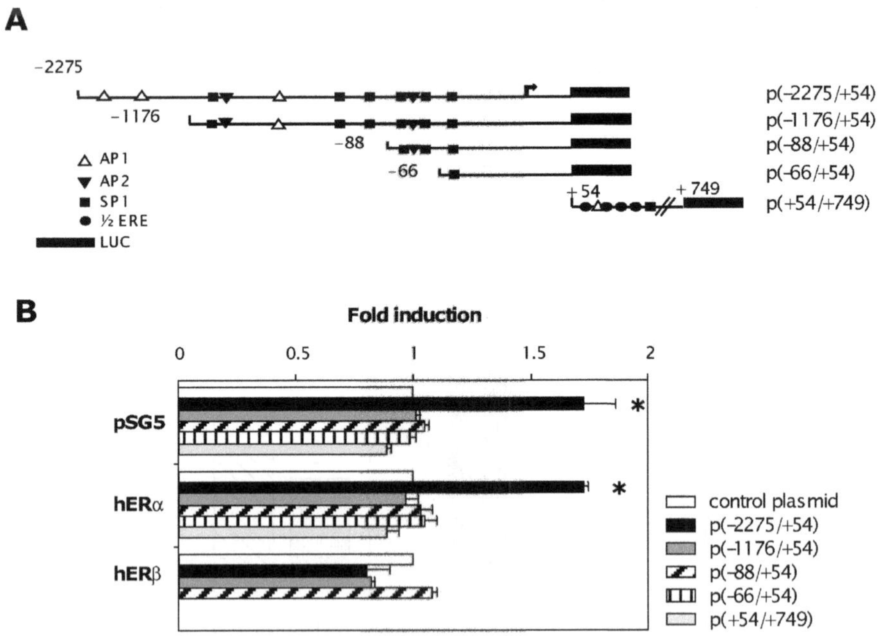

Fig. 2 Transcriptional activity of different VEGF constructs in MCF-7 cells treated with E_2. (**A**) Specific deletions of the 5′ sequence of the human VEGF gene promoter (11). The *arrow* indicates the transcriptional start site of the *VEGF* gene. (**B**) Cells were cotransfected with VEGF(−2275/+54)-LUC plasmid or mutants (−1176/+54, −88/+54, −66/+54, or +54/+749), CMV-GAL plasmid, and with pSG5-hERα, pSG5-hERβ. Cells were grown in the presence of 1 nM E_2 or vehicle alone. Values represent the means of four separate experiments. $^{*}p < 0.01$, E_2-stimulated vs. unstimulated cells

could depend on the nature of cells and the microenvironment, including hypoxia/normoxia, differential expression of other nuclear factors, and their interactions with cis-regulatory elements. Altogether, these results demonstrate that in several human BC cells (MCF-7, ZR-75) the levels of VEGF mRNA and protein are higher in the presence of E_2 (2.0- and 4.0-fold, respectively), when compared with controls.

VEGF Overexpression Contributes to Breast Tumor Growth and Angiogenesis. It has been reported that VEGF produced by carcinoma cells stimulates angiogenesis through a paracrine mechanism in tumor endothelial cells. In the human, VEGF exists as several isoforms, of which VEGF 121 and VEGF 165 predominate (13). We have investigated whether overexpression of VEGF by MCF-7 cells enhances E_2-dependent tumor growth in vivo.

We have generated MCF-7 cells stably transfected with VEGF165 cDNA (referred as V165) or with control vector. Using quantitative real time PCR and oligonucleotide primers of VEGF165 sequence we have first selected two clones of V165 expressing VEGF transcripts at high levels (V165-8 and V165-9), when compared with the level of each specific isoform in parental cells or control clones. Each clone of V165 cells secreted VEGF 23–24 kD isoform in the conditioned medium, using Western blotting. V165-8 and V165-9 secreted high levels of VEGF proteins in the conditioned medium (7–10 ng mg^{-1} cell extract after 24-h of cell culture, using Elisa assay). In contrast, parental MCF-7 or control cells secreted 1 ng mg^{-1} cell extract into the conditioned medium. The biological activity of secreted VEGF has been confirmed using an endothelial cell proliferation/survival assay (Fig. 3). This effect was inhibited in the presence of anti-VEGF antibodies, showing its specificity. The doubling time of selected V165 transfectants and control vector clones was similar in vitro by cell counting (not shown).

The effects of overexpression of VEGF165 on tumor growth were then investigated by using in vivo subcutaneous transplantation into SCID mice of V165

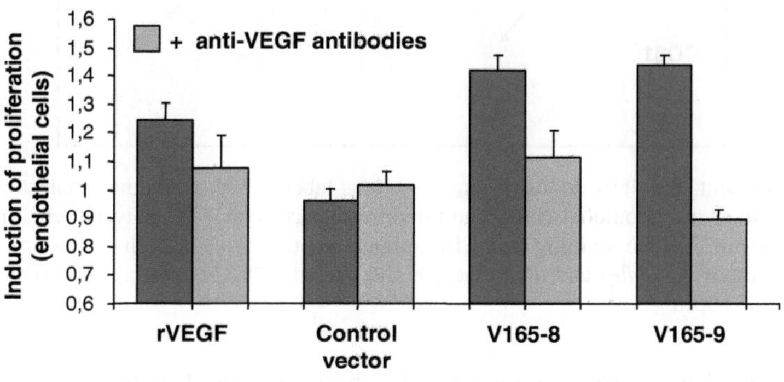

Fig. 3 Endothelial cell proliferation by VEGF 165 produced by selected V165 clones. HUVEC (5×10^3/well) were incubated with the conditioned medium of tumor cells; the proliferation assay was performed using the XTT assay. Data were normalized to control clones. Positive control is rVEGF165 (10 ng ml^{-1})

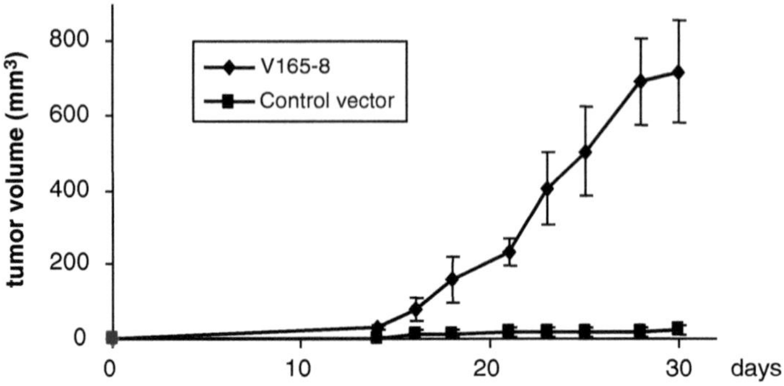

Fig. 4 Tumor growth induced in SCID mice by sc injection of V165 transfectant. After a palpable tumor was observed, tumor diameters were measured every 2 days. Results are expressed as the mean tumor volumes (9 tumors/experimental group)

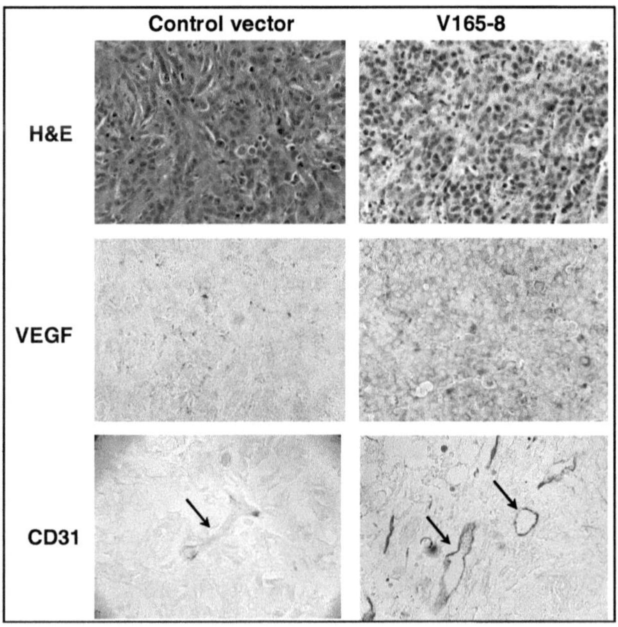

Fig. 5 Representative illustrations of tumor sections labeled with antibodies against VEGF and CD31, a marker of endothelial cells. The tumor vascularization was analyzed when all tumors reached 800 mm^{-3}. H&E staining (*top*). Immunochemistry revealed high expression of VEGF in V165-8 tumors (*middle*) and identified the endothelial cells (*bottom*). *Arrows* indicate blood vessels; bar = 10 μm

clones parallel to control vector cells. As shown in Fig. 4, using E_2-treated mice, VEGF165-8 showed significantly enhanced tumor uptake and growth rate, when compared with vector transfectant ($n = 9$; $p < 0.001$). These tumors strongly expressed VEGF protein (Fig. 5, middle). Immunostaining of sections with antibodies anti-CD1, a specific endothelial cell marker, revealed that VEGF overexpressing tumors were highly vascularized, when compared with control tumors (Fig. 5, bottom). Morphometric analysis showed that vascular density was increased in V165-8 ($10.2\,mm^{-2}$) compared with control tumors ($3.1\,mm^{-2}$ for control vector) (not shown). In addition, these vessels expressed VEGF-R1 and VEGF-R2 (not shown). Our in vivo findings show that overexpression of VEGF165 by E_2-dependent MCF-7 BC cells increases breast tumor growth in E_2-treated ovariectomized mice. In addition, VEGF strongly stimulates neovascularization in MCF-7 tumors. These results agree with other observations reported by Zhang et al. (17) and Guo et al. (18).

Conclusion and Perspectives

Sex steroids, especially E_2, affect VEGF production in BC cells. It has been demonstrated that stimulated VEGF production results in the enhancement of neovascularization in vivo. However, the contribution to mammary carcinogenesis of the E_2-dependent VEGF increase had not yet been clearly established. Our study (using sc injection), as well as other studies (using fat pad or intracranial injection), demonstrate that the enhanced VEGF165 expression increases MCF-7 breast tumor growth through increased neovascularization. Further studies are now in progress to analyze gene expression involved in tumor growth and angiogenesis in our model of breast tumor. Whether ERα or ERβ expression in BC biopsies correlates with VEGF expression and vascular density also remains to be established. This may have implications for the increased incidence of BC after exposure to sex steroids. Improving our understanding of angiogenic regulators in BC and their relation to sex steroids is necessary to optimise treatment options and develop new strategies against BC.

Acknowledgments The authors are grateful to Dr. Jean Plouet for MCF-7 V165 clones and E. Savariau for infography services (Institut Universitaire d'Hématologie, Hôspital Saint Louis, Paris, France). This work was supported by the INSERM, CNRS and the Ligue Contre le Cancer.

References

1. Folkman J (1995) Angiogenesis in cancer, vascular, rheumatoid, and other disease. Nat Med 1, 27–31.
2. Obermair A, Kucera E, Mayerhofer K, et al. (1997) Vascular endothelial growth factor (VEGF) in human breast cancer: correlation with disease-free survival. Int J Cancer 74:455–8.
3. Yoshiji H, Harris SR, Thorgeirsson UP (1997) Vascular endothelial growth factor is essential for initial but not continued in vivo growth of human breast carcinoma cells. Cancer Res 57:3924–8.
4. Osborne CK, Hobbs K, Clarke GM (1985) Effects of estrogens and antiestrogens on growth of human breast cancer cells in athymic nude mice. Cancer Res 45:5846590

5. Cullinan-Bove K, Koos RD (1993). Vascular endothelial growth factor/vascular permeability factor expression on the rat uterus: rapid stimulation by estrogen correlates with estrogen-induced increases in uterine capillary permeability and growth. Endocrinology 133:829–7.
6. Shiffren JL, Tseng J, Zaloudek C, et al. (1996) Ovarian steroid regulation of vascular endothelial growth factor in human endometrium: implications for angiogenesis during the menstrual cycle and in the pathogenesis of endometriosis. J Clin Endocrinol Metab 81:3112–18.
7. Bausero P, Cavaillé F, Meduri G, et al. (1998) Paracrine action of vascular endothelial growth factor in the human endometrium: production and target sites, and hormonal regulation. Angiogenesis 2:167–82.
8. Nakamura J, Savinov A, Lu Q, et al. (1996) Estrogen regulated vascular endothelial growth/permeability factor expression in 7,12-dimethylbenz(a)anthracene-induced rat mammary tumors. Endocrinology 137:5589–96.
9. Dabrosin C, Margetts PJ, Gauldie J (2003) Estradiol increases extracellular levels of vascular endothelial growth factor in vivo in murine mammary cancer. Int J Cancer 107:535–40.
10. Ruohola JK, Valve M, Karkkainen MJ, et al. (1999) Vascular endothelial growth factors are differentially regulated by steroid hormones and antiestrogens in breast cancer cells. Mol Cell Endocrinol 149:29–40.
11. Buteau-Lozano H, Ancelin M, Lardeux B, et al. (2002) Transcriptional regulation of vascular endothelial growth factor by estradiol and tamoxifen in breast cancer cells: a complex interplay between estrogen receptors alpha and beta. Cancer Res 62:4977–84.
12. Stoner M, Wormke M, Saville B, et al. (2004) Estrogen regulation of vascular endothelial growth factor gene expression in ZR-75 breast cancer cells through interaction of estrogen receptor α and SP proteins. Oncogene 23:1052–63.
13. Ferrara N, Davis-Smith T (1997) The biology of vascular endothelial growth factor. Endocr Rev 18:4–25.
14. Tisher E, Mitchell R, Hartman T, et al. (1991) Multiple protein forms are encoded through alternative exon splicing. J Biol Chem 266:11947–54.
15. Mueller MD, Vigne JL, Minchenko A, et al. (2000) Regulation of vascular endothelial growth factor (*VEGF*) gene transcription by estrogen receptor α. and β. Proc Natl Acad Sci USA 97:10972–7.
16. Kazi AA, Jones JM, Koos RD (2005) Chromatin immunoprecipitation analysis of gene expression in the rat uterus *in vivo*: estrogen-induced recruitment of both estrogen receptor alpha and hypoxia-inducible factor 1 to the vascular endothelial growth factor promoter. Mol Endocrinol 19:2006–19.
17. Zhang HT, Scott PA, Morbidelli L, et al. (2000). The 121 amino acid isoform of vascular endothelial growth factor is more strongly tumorigenic than other splice variants in vivo. Br J Cancer 83:63–68.
18. Guo P, Fang Q, Tao HQ, et al. (2003). Overexpression of vascular endothelial growth factor by MCF-7 breast cancer cells promotes estrogen-independent tumor growth in vivo. Cancer Res 63:4684–91.

Identification of Downstream Targets of Estrogen and c-myc in Breast Cancer Cells

Elizabeth A. Musgrove, C. Marcelo Sergio, Luke R. Anderson, Claire K. Inman, Catriona M. McNeil, M. Chehani Alles, Margaret Gardiner-Garden, Christopher J. Ormandy, Alison J. Butt, and Robert L. Sutherland

Summary Estrogen (E) plays a pivotal regulatory role in the control of cell proliferation in the normal breast and breast cancer (BC). To identify genes with likely roles in proliferation control that are regulated by E and its downstream target c-myc, we compared transcript profiles of antiestrogens-arrested cells stimulated to reinitiate cell cycle progression by E treatment or c-myc induction. Approximately 2/3 of the probe sets significantly regulated by E (adjusted $p < 0.01$) increased in expression. Half of the E-regulated probe sets were also regulated by c-myc. Genes involved in cell growth, cell proliferation, and cell survival were over-represented in the E-regulated geneset. Analysis of selected candidates has identified a nucleolar protein whose expression is correlated with c-myc expression in BC cell lines. These data indicate that a significant component of E-induced mitogenesis is mediated by c-myc and that selected c-myc target genes may be surrogate markers of c-myc expression in BC.

Introduction

The female sex steroid, estrogen (E), regulates cell proliferation and cell survival in breast epithelial cells and breast cancer (BC). The early events in E-induced mitogenesis include the transcriptional activation of two proto-oncogenes, c-myc and cyclin D1, the expression of which is rate limiting for progression through the G1 to S phase transition of the cell cycle (1). Inducible expression of either c-myc or cyclin D1 can activate cyclin E/cdk2 and over-ride anti-estrogen-induced growth arrest, mimicking the effects of E treatment (2). This has led to the hypothesis that E induces cell cycle progression through two initially independent pathways, one downstream of c-myc and the other involving cyclin D1. However, comparison of publicly available databases (3, 4) reveals that only 20% of the known E target genes are also known targets of c-myc. Thus, recent work in this laboratory has focused on delineating the pathways downstream of c-myc in BC cells and determining their role both in E-induced mitogenesis and in BC.

Overexpression of c-myc occurs at high frequency in both estrogen receptor $(ER\alpha)^+$ and $ER\alpha^-$ BC (5, 6). The rapid decrease in c-myc expression following

J.J. Li et al. (eds.), *Hormonal Carcinogenesis V.*

anti-estrogen treatment, and the ability of an antisense oligonucleotide-mediated decrease in c-myc expression to recapitulate many of the effects of anti-estrogen treatment (7), imply that c-myc overexpression could modulate sensitivity to endocrine therapies. Support for this concept is provided by data from laboratories, including our own, demonstrating that inducible or constitutive overexpression of c-myc in E-responsive MCF-7 cells confers resistance to tamoxifen and ICI 182780 in vitro [(5, 8, 9), McNeil et al. in preparation].

The evidence for a pivotal role for c-myc in E-induced cell proliferation and the possibility that this might be associated with endocrine resistance raises the question of which downstream targets of c-myc might be required for these effects. Our goal in this work was to identify those acutely E-regulated genes that are also downstream targets of c-myc, using an experimental model designed to enhance sensitivity to the mitogenic effects of E.

Results

Experimental Model and Transcript Profiling. To study the effects of E and c-myc on cell proliferation, we have utilized the well-characterized E-responsive BC cell line MCF-7. A series of clonal MCF-7 cell lines was developed that expressed wild-type c-myc or c-Zip (a deletion mutant lacking the transactivation domain) under the control of the zinc-inducible metallothionein promoter (Fig. 1). Representative clones with E and antiestrogen responses matched to those of the parental MCF-7 cells were chosen for further experiments. Zinc treatment resulted in increased c-myc or c-Zip expression within 3 h, similar to the timing of E induction of c-myc and consistent with our previous data using this zinc-inducible construct (2). The majority of the cell population synchronously reinitiated cell cycle progression following E treatment. Induction of c-myc also led to reinitiation of cell cycle progression to a degree comparable with the effects of E. However, although cells transfected with c-Zip could respond to E treatment, c-Zip induction did not lead to cell cycle progression (Fig. 1).

RNA for transcript profiling was collected 6 h after E or zinc treatment, within the time-frame during which critical E-dependent events necessary for S phase entry occur (10). Transcript profiling was undertaken in triplicate following E treatment (compared with vehicle treatment) and c-myc or c-Zip induction (compared with zinc-treated empty vector cells), using Affymetrix HG-U133 plus 2.0 arrays. Analysis of the microarray data used Bayesian linear modeling methods in the limma package and the Benjamini and Yuketieli adjustment was applied for multiple-hypothesis comparisons (11, 12). Probe sets that were significantly up- or down-regulated following E treatment compared with vehicle-treated cells were identified (adjusted $p < 0.01$, Fig. 1 and Table 1).

These E-regulated probe sets were further divided into those that were significantly regulated following c-myc but not c-Zip induction, designated "E2 and myc," and the remainder, which we designated "E2 not myc." Approximately

Fig. 1 Experimental model for identifying E-regulated c-myc target genes in BC cells. MCF-7 cells inducibly expressing c-myc (wild type or c-Zip), parental cells, and cells bearing the empty vector were pretreated with 10 nM ICI 182,780 for 48 h. Parental cells were then treated with E (100 nM 17β-estradiol) or vehicle (ethanol), and cells transfected with c-myc, c-Zip or the empty vector were treated with zinc (65 μM). (**A**) Quantitation of Western blots 6 h after E treatment or c-myc-induction in representative MCF-7 clones. (**B**) Cells were additionally treated with nocodazole to block E- or c-myc-stimulated cells in G2/M. Cell cycle phase distribution was determined after 36 h E treatment using flow cytometry. (**C**) RNA was prepared from cells harvested 6 h following treatment for transcript profiling. The E-regulated probe sets were compared with those regulated by c-myc but not c-Zip to identify probe sets regulated by both E and c-myc (but not c-Zip), or by E but not c-myc or c-Zip

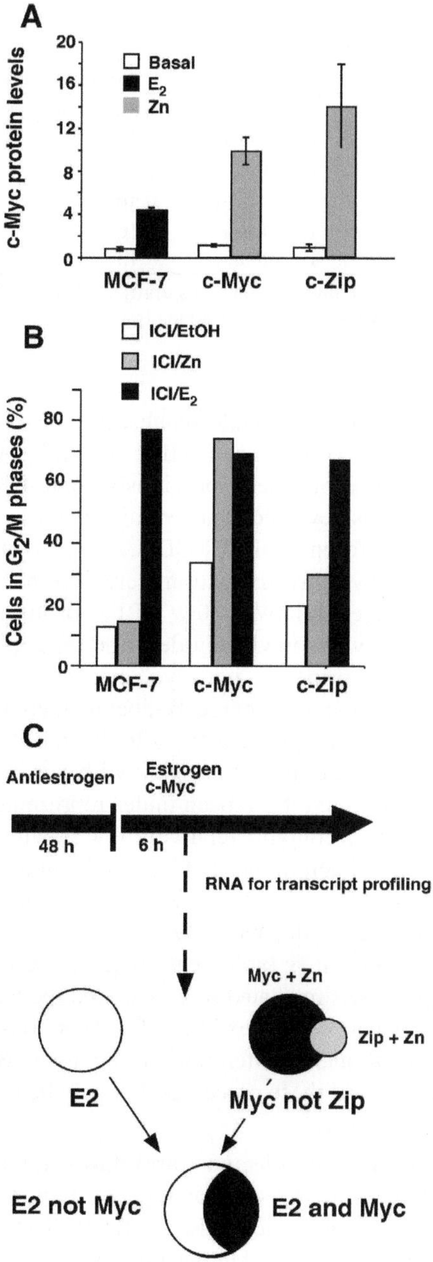

two-thirds of the E-regulated genes were upregulated and in total, half of the probe sets significantly regulated by E in this model system were also significantly regulated by c-myc.

Table 1 Proportion of probe sets identified as E-regulated that are known to be E-responsive or targets of c-myc

		# of probe sets	% in ERGDB[a]	% in *myc* database[b]
Up	E_2	635	188 (29%)	172 (27%)
	E_2 not *myc*	303	118 (39%)	43 (14%)
	E_2 and *myc*	332	70 (21%)	129 (39%)
Down	E_2	303	42 (14%)	37 (12%)
	E_2 not *myc*	166	27 (16%)	22 (13%)
	E_2 and *myc*	137	15 (11%)	15 (11%)

[a]http://defiant.i2r.a-star.edu.sg/projects/Ergdb-v2/index.htm (3)
[b]http://www.myc-cancer-gene.org (4)

Publicly available databases of E-responsive genes (*ERGDB*) (3) and c-myc targets (4) were searched to determine what proportion of the probe sets in each category had been previously identified as either E- or c-myc-regulated. This analysis revealed that overall 29% of probe sets significantly upregulated by E were present in the ERGDB, and 27% of those that were also significantly upregulated by c-myc induction were present in the myc target gene database (Table 1). However, relatively few of the probe sets down-regulated in this experimental model were previously-described E or c-myc targets (12–14%, Table 1). The probe sets from the "E_2 not myc" category that increased in expression had the highest proportion of previously-documented E targets (39%, Table 1). Within the E-upregulated probe sets, 118/188 (63%) of the known E targets were regulated by E but not c-myc, consistent with our initial premise that genes regulated by both E and c-myc have been under-represented in previous studies. Although the latter category contains relatively few previously identified E targets, it likely contains a significant number of bona fide targets since a high proportion of the probe sets in the "E_2 and myc" category that increased in expression were present in the myc target gene database (39%).

To identify biological processes that were significantly over-represented in the probe sets regulated by E, we used the data-mining tool Onto-Express (http://vortex.cs.wayne.edu/index.htm). The functions were essentially all related to either cell growth and proliferation or apoptosis. Bioinformatics analysis to identify specific aspects of these processes that may be regulated by both E and c-myc or by E alone is ongoing.

Analysis of Selected Candidates Arising from Transcript Profiling. Sixteen of the most highly-regulated genes in the transcript profiling experiments were examined using semiquantitative RT-PCR to document the time-course of regulation by E and c-myc. All but one displayed regulation consistent with the transcript profiling experiment. One gene that was regulated by both E and c-myc has undergone more detailed characterization and was designated E- and c-myc-upregulated (EMU4). E treatment or c-myc induction both led to increased EMU4 mRNA abundance within 3 h, and this was followed by increased protein abundance. Chromatin immunoprecipitation experiments demonstrated the recruitment of c-myc to the EMU4

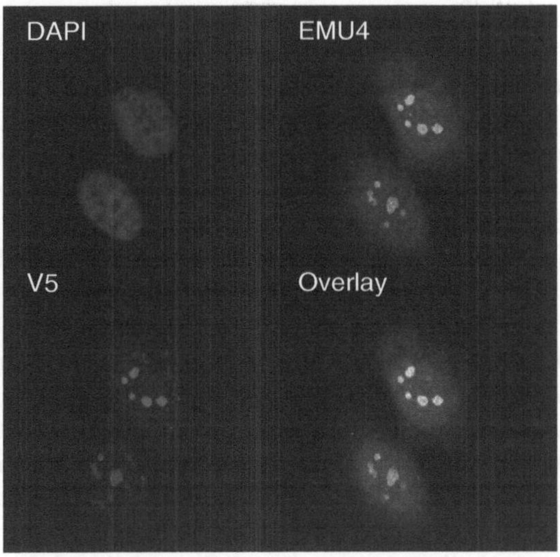

Fig. 2 Localization of EMU4 protein. MCF-7 cells constitutively expressing V5-tagged EMU4 were paraformaldehyde-fixed, permeabilized with Triton X100, and immunostained. The cells were counterstained with the DNA-specific dye DAPI (4,6-diamidino-2-phenylindole) to visualize the nuclei

promoter in response to E treatment (5), suggesting that EMU4 is a direct target of c-myc and that this accounts for its regulation by E. In BC cell lines, EMU4 expression was correlated with c-myc expression (5), (Butt et al. in preparation) and preliminary analysis suggests that expression of these two genes is also correlated in BC specimens.

The EMU4 protein is well conserved in mammals but less well-conserved in flies and worms, with no obvious yeast homologues, and is of unknown function. To gain some insight into possible functions for this protein, we used confocal microscopy to determine the subcellular localization of V5-tagged EMU4 protein. This revealed prominent EMU4 immunoreactivity in the nucleolus, with only faint staining apparent elsewhere in the nucleus (Fig. 2). Since many proteins involved in ribosome biogenesis and cell cycle control are localized to the nucleolus, ongoing experiments are aimed at determining whether EMU4 plays a role in these processes.

Discussion

To better understand the effects of E on BC cells, we have developed an experimental model that is particularly sensitive to E regulation of cell proliferation. Transcript profiling at an early time-point during E stimulation of cell cycle progression has revealed many potential targets for further functional characterization and determination of their putative roles in E action. Although some are known E targets, the majority are not. In particular, by focusing on the overlap between E-regulated targets and those downstream of c-myc, we have been able to substantially extend the list of genes regulated by both E and c-myc in BC cells. Although increasing data from various cell culture models argues for an important role for c-myc in E-stimulated cell cycle progression, the proportion of genes regulated by E that are also regulated by c-myc is unexpectedly high. It will be of particular interest to determine the roles of these genes in E action, for example by determining which are essential for E-induced mitogenesis or cell survival.

One approach to dissecting these pathways is to undertake functional characterization of selected targets. Our characterization of one novel E- and c-myc-target, EMU4, has revealed it to be a direct c-myc target. The importance of c-Myc in determining EMU4 expression is illustrated by the relationship between c-Myc and EMU4 expression (5), suggesting that irrespective of its role in c-myc action, EMU4 may be a surrogate marker of c-myc expression. There have been surprisingly few studies examining c-myc expression and its relationship to clinicopathological parameters including outcome in BC, probably because of technical difficulties in measuring c-myc in clinical specimens. Thus, any validated surrogate markers of c-myc expression or activity that can be readily assessed in pathological specimens may have particular utility in clinical applications.

Although the approach of analyzing individual genes is expected to yield important novel information that enhances our understanding of E action, it is increasingly apparent that complementary studies using a more "systems biology" approach may be required to give an overall picture of the complexity of the processes regulated by this important physiological hormone. Contemporary bioinformatics tools offer a starting point for such analyses and our initial studies using such tools indicate that, as previously documented, E has major effects on cell proliferation and cell death. A more unexpected finding was the close association of E and c-myc in the regulation of several nucleolar proteins with roles in ribosome biogenesis. The role of these target genes in E-induced RNA and protein synthesis is currently under investigation.

Acknowledgments Work in this laboratory is supported by grants from the following organisations: Cancer Institute NSW, Australian Cancer Research Foundation (ACRF), Association for International Cancer Research (AICR), and the National Health and Medical Research Council of Australia (NH&MRC). Elizabeth Musgrove and Alison Butt are Cancer Institute NSW Research Fellows, and Catriona McNeil is a Cancer Institute NSW Scholar.

References

1. Sutherland RL, Prall OW, Watts CK, et al. (1998) Estrogen and progestin regulation of cell cycle progression. J Mammary Gland Biol Neoplasia 3:63–72.
2. Prall OWJ, Rogan EM, Musgrove EA, et al. (1998) c-Myc or cyclin D1 mimics estrogen effects on cyclin E-Cdk2 activation and cell cycle reentry. Mol Cell Biol 18:4499–4508.

3. Tang S, Han H, Bajic VB (2004) ERGDB: Estrogen Responsive Genes Database. Nucleic Acids Res 32:D533–D536.
4. Zeller KI, Jegga AG, Aronow BJ, et al. (2003) An integrated database of genes responsive to the Myc oncogenic transcription factor: identification of direct genomic targets. Genome Biol 4:R69.
5. McNeil CM, Sergio CM, Anderson LA, et al. (2006) c-Myc overexpression and endocrine resistance in breast cancer. J Steroid Biochem Mol Biol doi:10.1016/j.jsbmb.2006.09.028.
6. Jamerson MH, Johnson MD, Dickson RB (2004) Of mice and Myc: c-Myc and mammary tumorigenesis. J Mammary Gland Biol Neoplasia 9:27–37.
7. Carroll JS, Swarbrick A, Musgrove EA, et al. (2002) Mechanisms of growth arrest growth arrest by c-myc antisense oligonucleotides in MCF-7 breast cancer cells: Implications for the antiproliferative effects of antiestrogens. Cancer Res 62:3126–3131.
8. Venditti M, Iwasiow B, Orr FW, et al. (2002) C-myc gene expression alone is sufficient to confer resistance to antiestrogen in human breast cancer cells. Int J Cancer 99:35–42.
9. Mukherjee S, Conrad SE (2005) c-Myc suppresses p21WAF1/CIP1 expression during estrogen signaling and antiestrogen resistance in human breast cancer cells. J Biol Chem 280:17617–17625.
10. Musgrove EA, Wakeling AE, Sutherland RL (1989) Points of action of estrogen antagonists and a calmodulin antagonist within the MCF-7 human breast cancer cell cycle. Cancer Res 49:2398–2404.
11. Smyth GK 2005 Limma: linear models for microarray data. In: Gentleman R, Carey V, Dudoit S, Irizarry R, Huber W, editors. Bioinformatics and Computational Biology Solutions using R and Bioconductor. New York: Springer, pp. 397–420.
12. Benjamini Y, Drai D, Elmer G, et al. (2001) Controlling the false discovery rate in behavior genetics research. Behav Brain Res 125:279–284.

Cathepsin D Overexpressed by Cancer Cells Can Enhance Apoptosis-dependent Chemo-sensitivity Independently of Its Catalytic Activity

Melanie Beaujouin and Emmanuelle Liaudet-Coopman

Summary The aspartic protease cathepsin D (CD) is a key mediator of induced-apoptosis and its proteolytic activity has been generally involved in this event. During apoptosis, CD is translocated to the cytosol. Since CD is one of the lysosomal enzymes that requires a more acidic pH to be proteolytically-active relative to the cysteine lysosomal enzymes such as cathepsin-B and cathepsin-L, it is therefore open to question whether cytosolic CD might be able to cleave substrate(s) implicated in the apoptotic cascade. Here, we have investigated the role of (wild-type) wt CD and its proteolytically inactive counterpart overexpressed by 3Y1-Ad12 cancer cells during chemotherapeutic-induced cytotoxicity and apoptosis, as well as the relevance of CD catalytic function. We demonstrate that wt or mutated catalytically inactive CD strongly enhances chemo-sensitivity and apoptotic response to etoposide. Both wt and mutated inactive CD are translocated to the cytosol, increasing the release of cytochrome c, the activation of caspases-9 and caspases-3 and the induction of a caspase-dependent apoptosis. In addition, pretreatment of cells with the aspartic protease inhibitor, pepstatin A, does not prevent apoptosis. Interestingly, therefore, the stimulatory effect of CD on cell death is independent of its catalytic activity. Overall, our results imply that cytosolic CD stimulates apoptotic pathways by interacting with a member of the apoptotic machinery rather than by cleaving specific substrate(s).

Introduction

Cathepsin D (CD) is a ubiquitous lysosomal aspartic protease extensively reported as being an active player in breast cancer (1, 2). More recently, CD has also been discovered as a key mediator of apoptosis induced by many apoptotic agents such as IFN-γ, Fas/APO, TNFα (3), oxidative stress (4–8), adriamycin, and etoposide (9, 10), as well as staurosporine (11). The role of CD in apoptosis has been linked to the lysosomal release of mature 34 kDa CD into the cytosol leading in turn to the mitochondrial release of cytochrome c (cytc) into the cytosol (4–7, 11, 12), activation of procaspases (5, 11–13), in vitro cleavage of Bid at pH 6.2 (13), or Bax activation independently of Bid cleavage (14). Numerous studies have shown that pepstatin A,

J.J. Li et al. (eds.), *Hormonal Carcinogenesis V.*

an aspartic protease inhibitor, could partially delay apoptosis induced by IFN-γ and Fas/APO (3), staurosporine (11, 14, 15), TNFα (3, 13, 16), serum deprivation (17), oxidative stress (5–8), or even when CD was micro-injected (12). These authors have thus concluded that CD plays a key role in apoptosis mediated via its catalytic activity. However, CD is one of the lysosomal enzymes that require a more acidic pH to be proteolytically active, relative to the cysteine lysosomal enzymes such as cathepsin-B and cathepsin-L. Acidification of the cytosol down to pH values of about 6.7–7 is a well-documented phenomenon in apoptosis (18). In vitro CD can cleave its substrates up to a pH of 5.8, but not above (19). Hence, it is predictable that the proteolytic activity of cytosolic CD would be drastically impaired under adverse pH conditions unfavorable for its catalytic function. It is, therefore, open to question whether CD might be able to cleave cytosolic substrate(s) implicated in the apoptotic cascade. In accordance with this proposal, it was recently described that pepstatin A did not prevent the death of cells treated by etoposide, doxorubicin, anti-CD95, or TNFα (10, 16, 20). Furthermore, pepstatin A did not suppress Bid cleavage or procaspases-9 and procaspases-3 activation by photodynamic therapy in murine hepatoma cells (21). Finally, deficiency in CD activity did not alter cell death induction in CONCL fibroblasts (20).

Results

In the present work, we investigated the role of (wild-type) wt CD and its proteolytically-inactive counterpart overexpressed by cancer cells in chemotherapeutic-induced cytotoxicity and apoptosis, and more precisely the relevance of CD catalytic function. We used the 3Y1-Ad12 cancer cell line, stably transfected with either wt CD or catalytically-inactive ^{D231N}CD or an empty vector, as a tumor model of CD overexpression. CD, ^{D231N}CD, and mock-transfected (control) 3Y1-Ad12 transfected cell lines were first treated with increasing concentrations of the topoisomerase II-inhibitor etoposide for two doubling time (e.g., 4 days) (Fig. 1). Cell viability curves revealed that CD significantly enhanced (fivefold) the chemo-sensitivity of cancer cells exposed to etoposide concentrations ranging from 100 nM to 1 μM, compared with control cells. Interestingly, ^{D231N}CD provided the same decrease in viability as CD (Fig. 1). We, therefore, conclude that CD overexpressed by cancer cells increases their cytotoxic responses to chemotherapeutic agents independently of its catalytic activity.

To dissect out the mechanism involved, we first determined whether chemosensitivity induced by overexpressed CD might be the consequence of an enhancement of apoptosis. Control, CD and ^{D231N}CD cancer cells were first treated with 10 μM etoposide for 24 h and their nuclei were stained with DAPI (Fig. 2a). Cancer cells overexpressing wt CD or ^{D231N}CD exhibited characteristic apoptotic morphological features, such as cell shrinkage, chromatin condensation, and the formation of apoptotic bodies, whereas control cancer cells presented only few apoptotic nuclei (Fig. 2a).

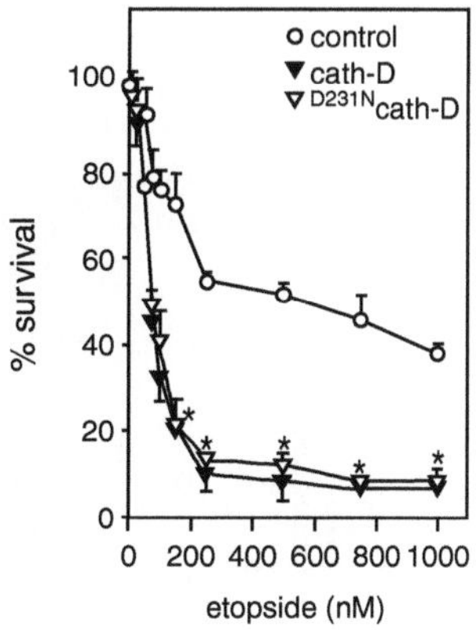

Fig. 1 Overexpression of both catalytically active and inactive CD by 3Y1-Ad12 cancer cells enhances etoposide-induced cytotoxicity. Cells were treated or not with increasing concentrations of etoposide and survival was evaluated by MTT. $*p < 0.0005$ vs. control cells (Student's t test)

Control, CD, and ^{D231N}CD cancer cells were then treated with increasing low (Fig. 2b, left panel) or higher (Fig. 2b, right panel) concentrations of etoposide for 24 h, and apoptotic status was analysed by an ELISA that detects histones associated with fragmented DNA. Both wt CD and ^{D231N}CD enhanced apoptosis in a dose-dependent manner (Fig. 2b). A threefold increase in apoptosis was observed with 1 μM etoposide in CD and ^{D231N}CD cells compared with control cells (Fig. 2b, left panel). Similar results were obtained over a wide range of etoposide concentrations, indicating that CD increased apoptosis for a large scale of etoposide treatment, e.g. from 75 nM to 50 μM (Fig. 2b). We finally examined the DNA fragmentation by TUNEL staining in cell cultures treated with 10 μM etoposide for 24 h and found that the nuclei of the CD and ^{D231N}CD transfected cells were positively stained by TUNEL, whereas almost no TUNEL-positive cells were seen in control cells (Fig. 2c, top panel). Quantification of the TUNEL assay, performed with cells treated with either 1 or 10 μM etoposide, then revealed a significant (10- to 30.0-fold) increase of apoptosis in both CD and ^{D231N}CD cells, compared with control cells (Fig. 2c, bottom panel). Moreover, pretreatment of wt CD cells with 100 μM pepstatin A for 16 h did not inhibit the increase of apoptosis induced by CD (Fig. 2d, left panel). Furthermore, pepstatin A had no effect on apoptosis enhanced by catalytically-inactive ^{D231N}CD (Fig. 2d, left panel). Time-course experiments of CD cells pretreated with pepstatin A confirmed its inefficiency to reverse CD-induced apoptosis (Fig. 2d, right panel). We verified that CD catalytic activity was inhibited by 100 μM pepstatin A in CD cell extracts from untreated or etoposide-treated cells (Fig. 2e).

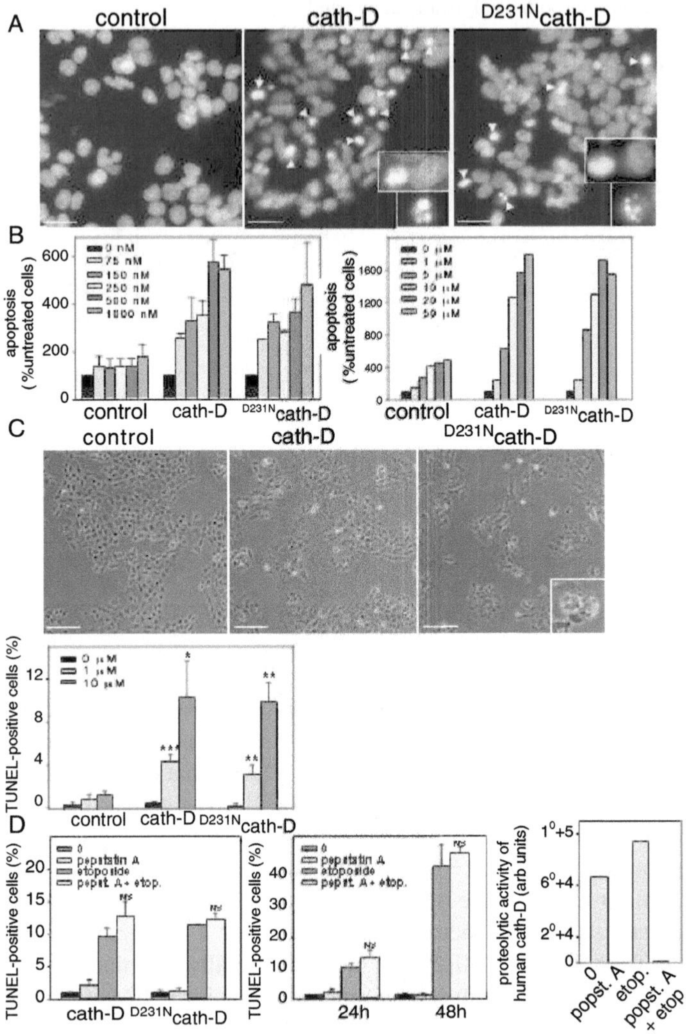

Fig. 2 Catalytically-active and inactive CD amplify etoposide-induced apoptosis. (**A**) Staining of nuclei with DAPI. *Arrows* indicate condensed chromatin and apoptotic bodies. *Insets* show DNA condensation and apoptotic bodies at a higher magnification. *Bars*, 23 µm. (**B**) ELISA apoptosis. Apoptosis induction was quantified on pooled floating and adherent cells using a cell death ELISA. (**C**) TUNEL assay. Representative immunocytochemistry of apoptotic cells (*top panel*). Apoptosis was analysed on adherent cells and floating cells cytospun on glass slides using the TUNEL method. *Inset* shows apoptotic bodies. For quantification of apoptotic cells

Overall, our results demonstrate that CD overexpressed by cancer cells enhances etoposide-induced apoptosis independently of its catalytic activity. CD has been described as being readily translocated to the cytosol in response to various apoptotic stimuli (6, 7, 10, 11, 13, 14). We, therefore, investigated the subcellular distribution of wt and ^{D231N}CD by immunostaining (Fig. 3A) and cell fractionation (Fig. 3B).

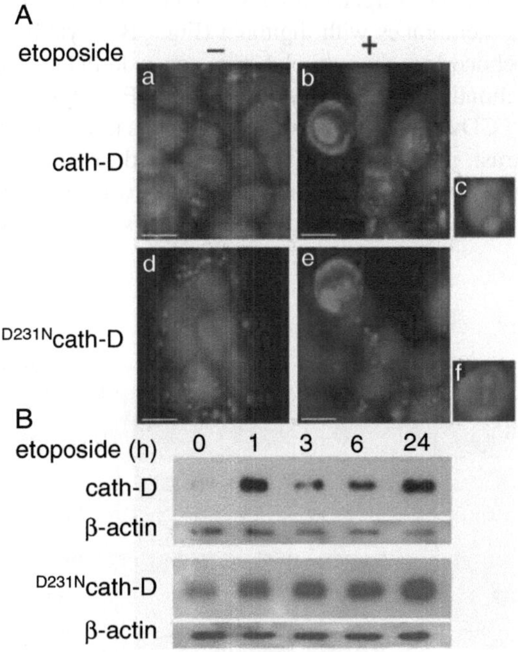

Fig. 3 Catalytically-active and inactive CDs are released early in etoposide-induced apoptosis. (**A**) IF of CD. Cells were either exposed to 10 μM etoposide for 24 h (**b**, **c**, **e**, **f**) or left untreated (**a**, **d**). Cells were incubated first with anti-human CD M1G8 mouse monoclonal antibody followed by a RITC-conjugated goat anti-mouse antibody. *Bars*, 7.4 μm. (**B**) Release of cytosolic CD. Western blotting of wt CD (*top panel*) and ^{D231N}CD (*bottom panel*) in cytosols extracted from cells either exposed to 10 μM etoposide. Cytosol was extracted using 20 μg ml^{-1} and 45 μg ml^{-1} digitonin for CD and ^{D231N}CD cells, respectively, for 10 min as previously described (11). CD immunoblotting was performed using an anti-human CD monoclonal mouse antibody followed by an anti-mouse peroxidase-conjugated antibody

Fig. 2 (continued) (*bottom panel*), 3 fields/experimental condition were chosen by random sampling and the total # of cells and the # of TUNEL positive nuclei were counted from adherent and cytospun floating cells. The results are expressed as a percentage of the estimated total # of cells present in the field. $^{*}p < 0.01$, $^{**}p < 0.025$, and $^{***}p < 0.0025$ vs. control cells (*t* test). *Bars*, 69 μm. (**D**) Effect of pepstatin A on CD-induced apoptosis. Cells pretreated or not with pepstatin A and then treated or not with 10 μM etoposide for 24 h (*left panel*); NS (non significant), $p < 0.1$ and 0.15 vs. etoposide for CD and ^{D231N}CD cells (*t* test). Time-course analysis of apoptosis in the presence/absence of 100 μM pepstatin A was performed with CD cells (*right panel*). $p < 0.25$ vs. etoposide (*t* test). (**E**) Effect of pepstatin A on CD catalytic activity analysed using a fluorogenic substrate

In untreated CD or ^{D231N}CD cells, immunostaining of CD displayed a punctuate distribution using immunofluorescence (IF) microscopy, consistent with a lysosomal location (Fig. 3A). By contrast, in etoposide-treated CD or ^{D231N}CD cells, CD displayed a mixed punctuate/diffuse distribution, suggesting that a portion of it was released from lysosomes into the cytosol (Fig. 3A). Western blot analysis confirmed that the mature (34 kDa) form of CD was indeed present in the cytosolic fractions obtained from wt CD or ^{D231N}CD cells treated for 1 h with etoposide, following permeabilization of the plasma membranes with digitonin (Fig. 3B). Cytc has been shown to be released from mitochondria to the cytosol during etoposide-induced apoptosis (22).

A typical mitochondrial pattern was observed by IF analysis of cyt c in untreated control, CD or ^{D231N}CD cells, as well as in control cells treated with 10 μM etoposide (Fig. 4a). By contrast, in CD or ^{D231N}CD cells exposed to 10 μM etoposide, translocation of cytc from the mitochondria into the cytosol was observed (Fig. 4a). These results indicate that CD over-expressed by cancer cells contributes to mitochondrial

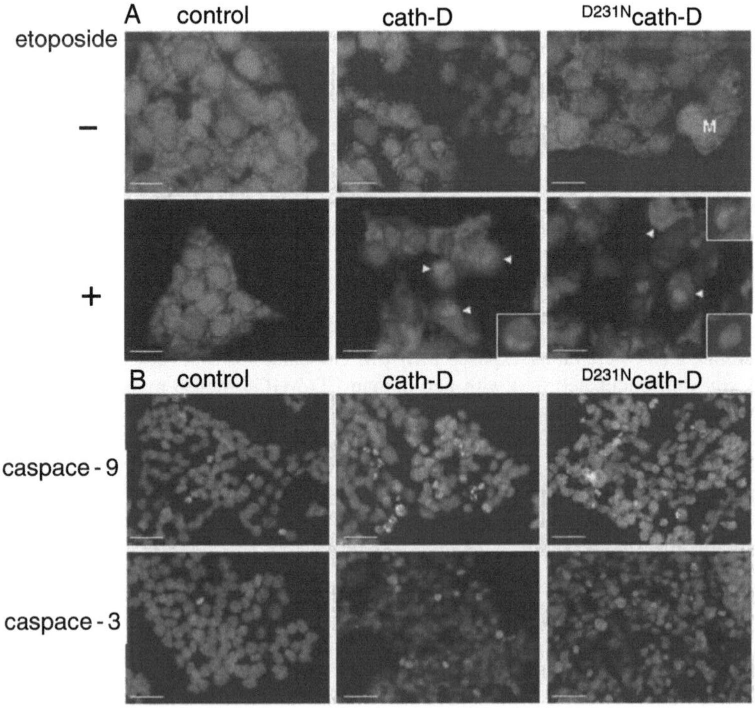

Fig. 4 Catalytically-active and inactive CD induces mitochondrial release of cytc and activation of caspases-9 and caspases-3. (**a**) Cytc analyzed by IF using an anti-rat cytc mouse antibody. *Bars*, 16 μm. (**b**) Activation of caspase-9 and caspases-3 analyzed by IF using an anti-rat cleaved caspase-9 antibody or an anti-rat cleaved caspase-3 antibody followed by a TexasRed-conjugated or FITC-conjugated goat antibody. *Bars*, 43 μm

cytc release in etoposide-treated cells independently of its catalytic function. We then investigated whether CD plays a role in the activation of caspases-9 and caspases-3 since these caspases have been shown to play a pivotal role in etoposide-induced apoptosis (22). As illustrated in Fig. 4b, caspases-9 and caspases-3 were both activated in CD and ^{D231N}CD cells treated with 10 μM etoposide whereas almost

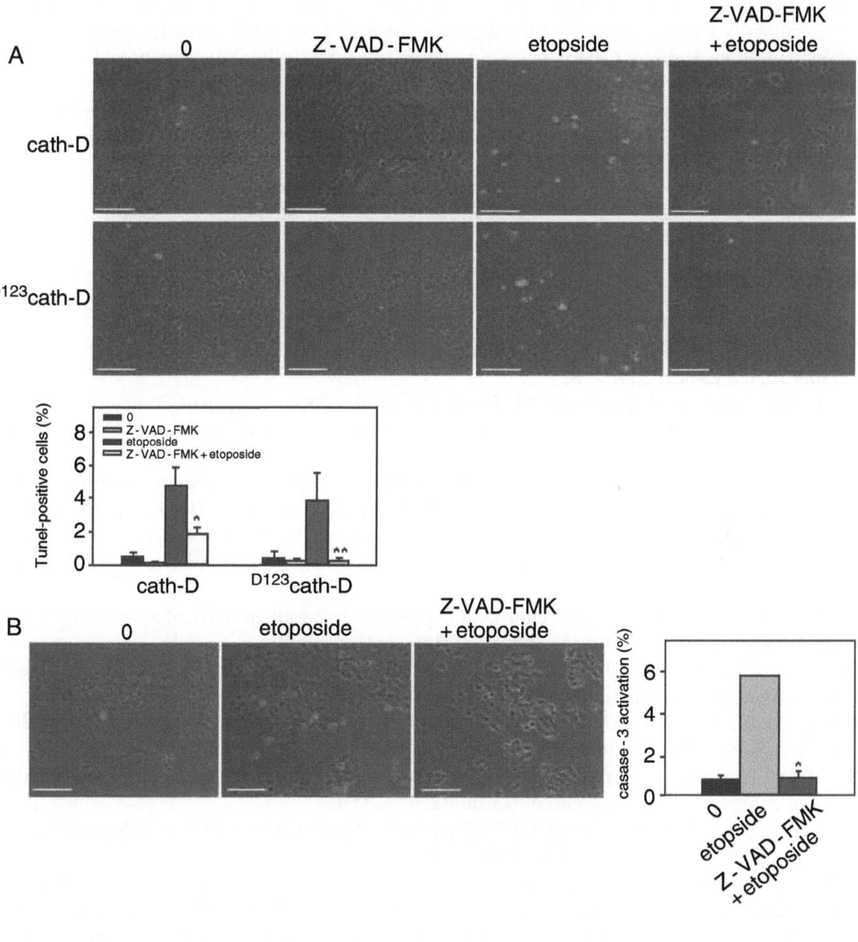

Fig. 5 Catalytically-active and inactive CD increase caspase-dependent apoptosis. (**A**) Effect of Z-VAD-FMK on apoptosis. CD (*top panel*) and ^{D231N}CD (*bottom panel*) cells were first either pretreated for 1 h with 100 μM Z-VAD-FMK or left untreated and later either left untreated or treated with 10 μM etoposide for 24 h. *Bars*, 69 μm $^{*}p < 0.05$ and $^{**}p < 0.025$ vs. etoposide for CD and ^{D231N}CD cells, respectively (*t* test). (**B**) Cells were pretreated or not for 1 h with 100 μM Z-VAD-FMK and were then treated or not with 10 μM etoposide for 24 h. *Bars*, 69 μm $^{*}p < 0.0025$ vs. etoposide treatment (*t* test)

no signal was detected in control cells in similar conditions. To confirm that CD participates in the activation of caspases during etoposide-induced apoptosis, CD or ^{D231N}CD were either left untreated or pretreated with the general caspase inhibitor, Z-VAD-FMK, and then with 10 μM etoposide (Fig. 5a). TUNEL assays revealed that wt or ^{D231N}CD enhanced a caspase-dependent apoptosis since etoposide-induced apoptosis was significantly inhibited by Z-VAD-FMK treatment (Fig. 5a). In parallel experiments, we verified that Z-VAD-FMK efficiently inhibited caspase-3 activation in our cells (Fig. 5b).

Acknowlegments We thank Dr Karin Öllinger (Linköping University, Sweden) for her kind help and valuable discussions concerning the digitonin method, Jean-Yves Cance for the photographs, Nicole Lautredou-Audouy (Centre de Ressources en Imagerie Cellulaire) for confocal microscopy and Nadia Kerdjadj for her secretarial assistance. This work was supported by Institut National de la Santé et de la Recherche Médicale, University of Montpellier I, Association pour la Recherche sur le Cancer (Grant 3344, E Liaudet-Coopman), and Ligue Nationale Contre le Cancer who provided a fellowship for Mélanie Beaujouin.

References

1. Rochefort H (1996) The prognostic value of cathepsin D in breast cancer. A long road to the clinic. Eur J Cancer 32A:7–8.
2. Liaudet-Coopman E, Beaujouin M, Derocq D, et al. (2006) Cathepsin D: newly discovered functions of a long-standing aspartic protease in cancer and apoptosis. Cancer Letters 237:167–179.
3. Deiss LP, Galinka H, Berissi H, et al. (1996) Cathepsin D protease mediates programmed cell death induced by interferon gamma, Fas/APO-1 and TNF alpha. EMBO J 15:3861–3870.
4. Roberg K, Ollinger K (1998) Oxidative stress causes relocation of the lysosomal enzyme cathepsin D with ensuing apoptosis in neonatal rat cardiomyocytes. Am J Pathol 152:1151–1156.
5. Ollinger K (2000) Inhibition of cathepsin D prevents free-radical-induced apoptosis in rat cardiomyocytes. Arch Biochem Biophys 373:346–351.
6. Roberg K (2001) Relocalization of cathepsin D and cytochrome c early in apoptosis revealed by immunoelectron microscopy. Lab Invest 81:149–158.
7. Kagedal K, Johansson U, Öllinger K (2001) The lysosomal protease cathepsin D mediates apoptosis induced by oxidative stress. FASEB J 15:1592–1594.
8. Takuma K, Kiriu M, Mori K, et al. (2003) Roles of cathepsins in reperfusion-induced apoptosis in cultured astrocytes. Neurochem Int 42: 153–159.
9. Wu GS, Saftig P, Peters C, et al. (1998) Potential role of cathepsin D in p53-dependent tumor suppression and chemosensitivity. Oncogene 16:2177–2183.
10. Emert-Sedlak L, Shangary S, Rabinovitz A, et al. (2005) Mol Cancer Ther 4:733–742.
11. Johansson AC, Steen H, Öllinger K, et al. (2003) Cathepsin D mediates cytochrome c release and caspase activation in human fibroblast apoptosis induced by staurosporine. Cell Death Differ 10:1253–1259.
12. Roberg K, Kagedal K, Ollinger K (2002) Microinjection of cathepsin d induces caspase-dependent apoptosis in fibroblasts. Am J Pathol 161:89–96.
13. Heinrich M, Neumeyer J, Jakob M, et al. (2004) Cathepsin D links TNF-induced acid sphingomyelinase to Bid-mediated caspase-9 and -3 activation. Cell Death Differ 11:550–563.
14. Bidère N, Lorenzo HK, Carmona S, et al. (2003) Cathepsin D triggers Bax activation, resulting in selective apoptosis-inducing factor (AIF) relocation in T lymphocytes entering the early commitment phase to apoptosis. J Biol Chem 278: 31401–31411.

15. Terman A, Neuzil J, Kagedal K, et al. (2002) Decreased apoptotic response of inclusion-cell disease fibroblasts: a consequence of lysosomal enzyme missorting? Exp Cell Res 274:9–15.
16. Demoz M, Castino R, Cesaro P, et al. (2002) Endosomal-lysosomal proteolysis mediates death signalling by TNFalpha, not by etoposide, in L929 fibrosarcoma cells: evidence for an active role of cathepsin D. Biol Chem 383: 1237–1248.
17. Shibata M, Kanamori S, Isahara K, et al. (1998) Participation of cathepsins B and D in apoptosis of PC12 cells following serum deprivation. Biochem Biophys Res Commun 251:199–203.
18. Matsuyama S, Llopis J, Deveraux QL, et al. (2000) Changes in intramitochondrial and cytosolic pH: early events that modulate caspase activation during apoptosis. Nat Cell Biol 2:318–325.
19. Capony F, Morisset M, Barrett AJ, et al. (1987) Phosphorylation, glycosylation, and proteolytic activity of the 52-kD estrogen-induced protein secreted by MCF7 cells. J Cell Biol 104:253–262.
20. Tardy C, Tyynela J, Hasilik A, et al. (2003) Stress-induced apoptosis is impaired in cells with a lysosomal targeting defect but is not affected in cells synthesizing a catalytically inactive cathepsin D. Cell Death Differ 10:1090–1100.
21. Reiners JJ, Jr, Caruso JA, Mathieu P, et al. (2002) Release of cytochrome c and activation of pro-caspase-9 following lysosomal photodamage involves bid cleavage. Cell Death Differ 9:934–944.
22. Sawada M, Nakashima S, Banno Y, et al. (2000) Ordering of ceramide formation, caspase activation, and Bax/Bcl-2 expression during etoposide-induced apoptosis in C6 glioma cells. Cell Death Differ 7:761–772.

A Synthetic Peptide Derived from Alpha-fetoprotein Inhibits the Estradiol-induced Proliferation of Mammary Tumor Cells in Culture through the Modulation of p21

Walter D. Sierralta, María J. Epuñan, José M. Reyes, Luis E. Valladares, and Ana M. Pino

Summary A stable cyclized 9-mer peptide (cP) containing the active site of α-alpha fetoprotein (αFP) has been shown to be effective for prevention of estrogen-stimulated tumor cell proliferation in culture or of xenographt growth in immunodeficient mice. cP does not block 17β-estradiol (E_2) binding to its receptors, but rather appears to interfere with intracellular processing of the signal that supports growth. To obtain insight on that mechanism we studied the effect of cP on the proliferation of MCF-7 cells in culture. Proliferation in the presence of 2 μM E_2 is decreased up to 40% upon addition of 2 μg ml^{-1} cP to the medium; the presence of cP did not increase cell death. cP reduced also the proliferation of estrogen-dependent ZR75-1 cells but had no effect on autonomous MDA-MB-231 cells. cP did not modify the number of binding sites for labeled E_2 or affected cell death. We detected increased nuclear $p21^{Cip1}$ immunoreactivity after cP treatment. Our results suggest that cP acts via p21Cip1 to slow the process of MCF-7 cells through the cycle.

Introduction

The most common cancer in women worldwide affects the mammary gland, a tissue composed of numerous highly-sensitive target cells for estradiol and containing the estradiol receptors (ERα and β) (1). As a transcription factor, ERα binds to response elements in the promoter region of target genes (2) or interacts with proteins in other pathways (3) to regulate the transcription of specific genes in estrogen-dependent cells. There is also evidence that estrogen has a proliferative effect on breast cancer (BC) cells via the regulation of several polypeptide growth factors, growth factor receptors, or other signaling molecules (4, 5). 17β-estradiol (E_2) can also activate phosphatidylinositol-3-kinase and mitogen-activated protein kinase pathways directly or indirectly to regulate cell proliferation (6–8). Finally, there is growing evidence that some rapid cell responses to estradiol are mediated via membrane-associated receptors (9). Several studies indicate that estrogen can induce and promote BC and about 50% of primary BCs are $ER\alpha^+$. For these reasons, decreasing estrogen production by aromatase inhibitors or blocking its action with antiestrogens have been the main treatments for $ER\alpha^+$ tumors (10, 11).

α-alpha fetoprotein (αFP) is a protein normally produced during pregnancy by the fetal liver. From epidemiologic studies on pregnancy-associated protection against BC, Jacobson et al. hypothesized that AFP acts in an endocrine manner to extinguish premalignant foci in pregnant women (12) and demonstrated that αFP exert antiestrogenic effects on BC growth in rodent models (13). More recently, they have identified the active site of αFP responsible for this antiestrogenic activity and developed a stable, cyclized 9-mer analog (cP), cyclo-EMTOVNOGQ that has the ability to inhibit the estrogen-stimulated growth of E_2-dependent human BC xenografts growing in SCID mice, hinders the estrogen-stimulated growth of the uterus in immature animals, and is active even administered orally (14, 15).

The aim of this study was to evaluate the antiestrogenic effect of cP on the proliferation of mammary tumor cells in culture. We report on the effect of the αFP-derived cyclic peptide on cell proliferation and $p21^{Cip1}$ expression of MCF7 cells cultured in the presence of E_2.

Results

The proliferation of MCF7 cells in the presence of 2 nM E_2 was lowered following addition of 2 μg ml^{-1} cyclic peptide to the culture (Fig. 1a). The replacement of cP by a scrambled peptide did not affect the E_2-dependent growth. The effect of cP was reversible since its removal allowed a normalized proliferation rate (Fig. 1b). The inhibitory effect of cP on E_2-stimulated cell proliferation was maximal at 2 μg ml^{-1} cP (Fig. 1c).

The proliferation of E_2-insensitive MDA-MB231 cells was not modified by cP, but this compound reduced the proliferation of ER^+ human BC ZR75-1 cells in the presence of 2 nM E_2 (Fig. 2). Neither the association constant nor the number of binding sites for E_2 were modified by cP (not shown).

MCF7 cells in CDSS formed cell clusters of different size after 16 h of seeding depending on the incubation condition; those developed under 2 nM E_2 were larger

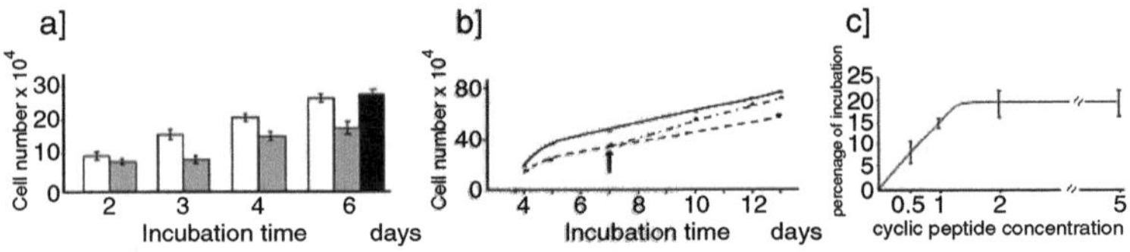

Fig. 1 Effect of cP on E_2-stimulated proliferation of MCF7 cells. **a:** Media: 5% charcoal-treated calf serum (CDSS) + 2 nM E_2 (*open columns*), or + E_2 + 2 μg ml^{-1} cP (*grey columns*), or + E_2 + 2 μg ml^{-1} scrambled peptide (*black column*). Significant values ($p < 0.05$) at 3, 4, and 6 days. **b:** Media: CDSS + 2 nM E_2 (*solid line*) or 2 nM E_2 + 2 μg ml^{-1} cP (*dotted line*), and cP removal at day 7 on the E_2-stimulated cell growth (*dash–dotted line*). **c:** Reduction in proliferation at 96 h of MCF7 cells caused by the simultaneous addition of 2 nM E_2 and increasing amounts of cP. The data represent mean ± SD of 4 experiments

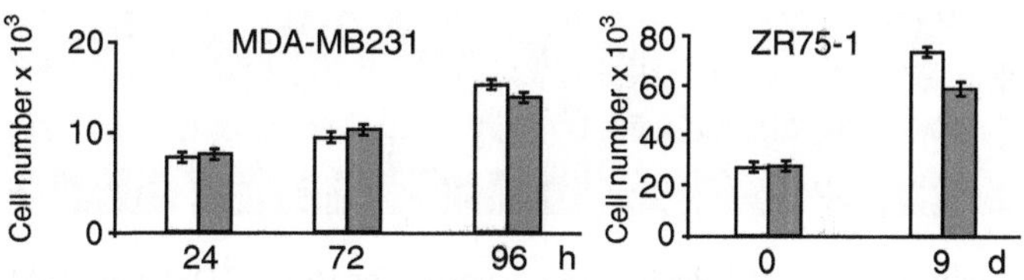

Fig. 2 Effects of cP on other mammary tumor cells in culture. *Left side*: lack of response of MDA-MB-231cells to cP addition. Proliferation in D-Mem/F12 with 5% calf serum in the absence (*open columns*) or presence of 2 μg ml^{-1} cP (*grey columns*). *Right side*: Effects of E_2 and cP on proliferation of ZR75-1 cells in CDSS and 2 nM E_2 (*open columns*) or plus E_2 and 2 μg ml^{-1} cP (*grey columns*). The data represent mean ± SD of four experiments

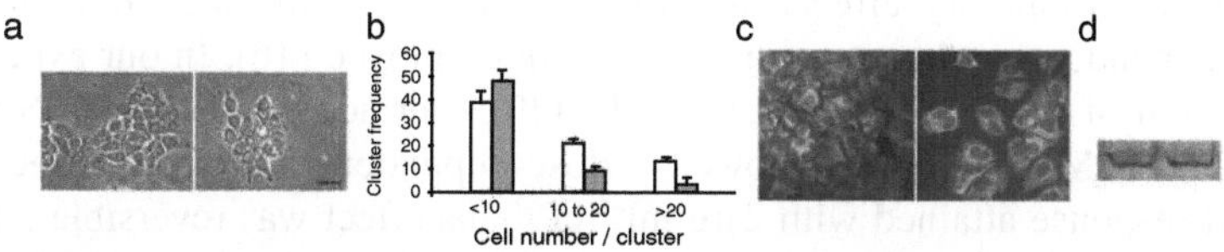

Fig. 3 Clustering of MCF7 cells after 16 h seeding. **a.** MCF7 clustered on coverslips in CDSS + 2 nM E_2 (*left*), or + 2 nM E_2 and 2 μg ml^{-1} cP (*right*). Bar represents 100 μm. **b.** Frequency of cluster size of cells incubated in CDSS medium + E_2 (*white bar*) or E_2 + cP (*grey bar*). **c.** Expression of E-cadherin in MCF-7 cells grown in CDSS + 2 nM E_2 alone (*left side*) or E_2 + 2 μg ml^{-1} cP (*right side*). **d.** Immunoblot of E-cadherin in MCF7 cells grown in CDSS with E_2 alone (*left*) or + cP (*right*). Proliferation analysis demonstrated that the addition of cP slowed-down the cell growth observed under E_2 stimulation.

than those developed under E_2 + cP (Fig. 3a). Figure 3b shows cluster distribution under both experimental conditions. Since the presence of the cP reduced both cell proliferation and initial cluster sizes of E_2-stimulated MCF7 cells, we analyzed by immuno-fluorescence whether cP treatment induced a down-modulation of E-cadherin. As shown in Fig. 3c, E_2-stimulated MCF7 cells exhibit a fluorescence signal for E-cadherin similar to that of the small clusters of MCF7 cells formed in the presence of estradiol and cP. Western blot analysis confirmed that E_2- and E_2 + cP-treated cells contained equivalent amounts of E-cadherin (Fig. 3d).

The subcellular localization of $p21^{Cip1}$ is an important component in the regulation of action, because the inhibitor and the complexes of the cyclin-dependent kinases must interact in the same subcellular compartment. When studied by indirect immunofluorescence at 16 h after seeding, $p21^{Cip1}$ localized within the nucleus only in few estradiol-stimulated. The simultaneous presence of 2 μg ml^{-1} cP and E_2 increased the number of $p21^{Cip1}$ immunostained nuclei (Fig. 4).

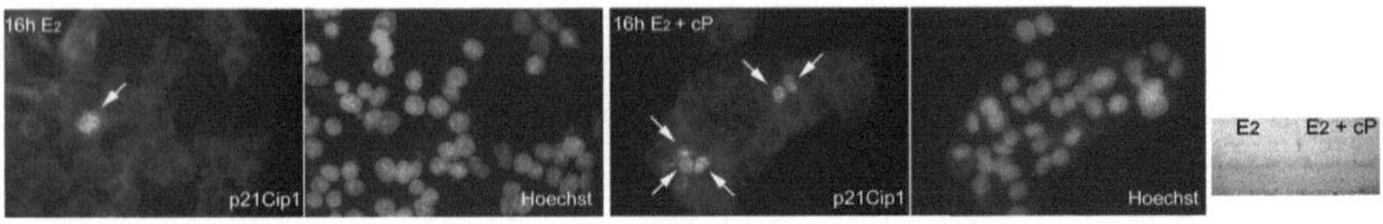

Fig. 4 Expression of p21 in MCF 7cells. Cells expossed for 16 h to CDSS plus 2 nM E_2, or CDSS + 2 nM E_2 and 2 µg ml^{-1} cP. The panels show the expression of p21^{Cip1}, detected by indirect IF with AlexaFluor 488-labeled antibody and the nuclei present in the corresponding fields (Hoechst 33258 DNA staining). The *arrows* point to nuclear p21^{Cip1} positive cells. Original magnification ×240. At the *right* an immunoblot of p21^{Cip1} from whole extracts of the cells 40 g protein is applied

Discussion

The growth stimulatory effects of estrogens in cell culture are often subtle and heavily dependent on the experimental system employed (16). In our experiments, proliferation of MCF7 cells decreased 20–42% after adding cP to CDSS medium containing 2 nM E_2. The cP showed a dose-dependent inhibitory effect with a maximal response attained with 2 µg ml^{-1} cP. This effect was reversible and structure-dependent since removal of the peptide from the medium allowed the cells to recover the proliferation rate they exhibited in the presence of 2 nM E_2, and no inhibitory activity was detected when the cyclic peptide was replaced by one with a scrambled amino acid sequence. Further, cP also inhibited E_2-dependent growth of ZR75-1 cells but had no effect on MDA-MB231, a known E_2-unresponsive cell line. All these results complement recent observations on the effect of cP on the growth of mammary tumor cell xenografts in severely combined immunodeficient mice (14).

The inhibitory effect of the cP on E_2-dependent proliferation does not result from a direct effect on E_2 binding, or on the endogenous generation of this hormone, as values of E_2 dissociation constant and number of binding sites were not affected by the presence of 2 µg ml^{-1} cP during tritium-labeled E_2 binding to MCF7 cells. Earlier experiments have shown no interference of this peptide with E_2 binding to a receptor preparation (17).

We observed differences in the number of cells per cluster formed after 16 h of exposure to cP and/or E_2, suggesting that treatments could affect inter-cell attraction. For this reason, the effect of treatments on E-cadherin was analyzed by immunofluorescence and immunoblotting. cP did not change cellular E-cadherin expression, ruling out an effect of the peptide on cell adhesion.

Comparison of cell cycle progression between E_2- and E_2 + cP treated cells showed no apparent change in the percentage of cells at the different stages of the cell cycle because the effects were overcome by serum addition (not shown). However, there was a significant reduction in the total number of cells, without any indication of increased cell death, suggesting that the reduced proliferation could be a consequence of lengthening of the cell cycle.

The cyclin-dependent kinase inhibitor $p21^{Cip1}$ is considered a pivotal controller of the cell fate. Therefore, modulation of its expression or function is fundamental for regulation of proliferative activity. It has been demonstrated that redistribution of $p21^{Cip1}$ from cyclin D1–Cdk4/6 cytoplasmic complexes to cyclin E–Cdk2 nuclear complexes alterate the cell cycle (18). In cultured tumor mammary cells, $p21^{Cip1}$ is an estrogen-regulated intracellular protein involved in the progression of cells through the cycle and whose expression can be activated by p53-dependent and also independent mechanisms (19). The low levels of $p21^{Cip1}$ in wild type MCF7 cells difficult quantitative measurements of p21 by immunoblotting. For this reason, we used the sensitive but qualitative immunofluorescence analysis and observed changes in the frequency of nuclear $p21^{Cip1}$ immunoreactivity in MCF7 cells exposed to cP in the presence of estradiol. The presence of $p21^{Cip1}$ in the cell nucleus significantly increased after exposure to cP. Nuclear expression of $p21^{Cip1}$ among MCF7 cells was nonhomogeneous, a phenomenon observed in cells exposed to other stimuli (20). The results presented here suggest that the cP could slowering cell cycle progression by promoting nuclear $p21^{Cip1}$ complexing with cyclin E/A-Cdk2, among others. In concordance with our observations, it has been shown that cP interference on estrogen-stimulated growth of human BC xenografts is related to increased p53 phosphorylation (21). We postulate that, cP could transiently remove the E_2-dependent block of $p21^{Cip1}$ expression, reducing the positive effect of E_2 on MCF7 cell proliferation. Alternatively, following cP treatment an increased expression of the $p21^{Cip1}$ gene may result; the newly synthesized $p21^{Cip1}$ could bind cyclin E/A-Cdk2, slowing the progress of MCF7 cells through the cycle. Further studies are required to determine whether other cell cycle regulators are involved and characterize their mechanisms.

Acknowledgments We gratefully acknowledge Drs. HI Jacobson, TT Andersen, and JA Bennett, The Albany Medical College for supplying us with cP and the scrambled peptide. Supported by Fondecyt Chile, Project Nr 1040881.

References

1. Sommer S, Fuqua SA (2001) Oestrogen receptor and breast cancer. Sem Cancer Biol 11:339–352.
2. Tsai M, O'Malley BW (1994) Molecular mechanism of action of steroid/thyroid superfamily members. Ann Rev Biochem 63:451–486.
3. Batistuzzo de Madeiros SR, Krey G, Hihi AK, et al. (1997) Functional interaction between the estrogen receptor and the transcription factor Sp1 regulate the estrogen-dependent transcriptional activity of the vitellogenin A1 promoter. J Biol Chem 272:18250–18260.
4. Bates SE, Davidson NE, Valverius EM, et al. (1998) Expression of transforming growth factor-alpha and its messenger ribonucleic acid in human breast cancer: its regulation by estrogen and its posssible functional significance. Mol Endocrinol 2:543–555.
5. Huynh H, Yang X, Pollak M (1996) Estradiol and antiestrogens regulate a growth inhibitory insulin-like growth factor binding protein 3 autocrine loop in human breast cancer cell. J Biol Chem 271:1016–1021.

6. Kato S, Endoh H, Masuhiro Y, et al. (1995) Activation of the estrogen receptor through phosphorylation by mitogen-activated protein-kinase. Science 270:1491–1494.
7. Simoncini T, Hafezi-Moghadam A, Brazil DP, et al. (2000) Interaction of estrogen receptor with the regulatory subunit of phophatidylinositol-3-OH kinase. Nature 407:538–541.
8. Santen RJ, Song RX, MacPherson R, et al. (2002) The role of mitogen-activated protein (MAP) kinase in breast cancer. J Steroid Biochem Mol Biol 80:239–256.
9. Levin ER (2002) Cellular functions of plasma membrane estrogen receptors. Steroids 67:471–475.
10. Santen (2002) To block estrogen's synthesis or action: that is the question. J Clin Endo Met 87:3007–3012.
11. Gradishar WJ, Jordan VC (1997) Clinical potential of new antie-strogens. J Clin Oncol 15:840–852.
12. Jacobson HI, Thompson WD, Janerich DT (1989) Multiple births and maternal risk of breast cancer. Am J Epidemiol 129:865–873.
13. Jacobson HI, Bennet JA, Mizejewski GJ (1990) Inhibition of estrogen-dependent breast cancer growth by a reaction product of alpha-fetoprotein and estradiol. Cancer Res 50:415–420.
14. DeFreest LA, Mesfin FB, Joseph L, et al. (2004) Synthethic peptide derived from alpha-fetoprotein inhibits growth of human breast cancer: investigation of the pharmacophore and synthesis optimization. J Pept Res 63:409–419.
15. Bennett JA, DeFreest L, Anaka I, et al. (2006) AFPep: an anti-breast cancer peptide that is orally active. Breast Cancer Res Treat 98:133–141.
16. Doisneau-Sixou SF, Sergio CM, Carroll JS, et al. (2003) Estrogen and antiestrogen regulation of cell cycle progression in breast cancer cells. Endocrine-Related Cancer 10:179–186.
17. Bennett JA, Mesfin FB, Andersen TT, et al. (2002) A peptide derived from α-fetoprotein prevents the growth of estrogen-dependent human breast cancers sensitive and resistant to tamoxifen. Proc Natl Acad Sci USA 99:2211–2215.
18. Planas-Silva MD, Weinberg RA (1997) Estrogen-dependent cyclin E-Cdk2 activation through p21 redistribution. Mol Cell Biol 17:4059–4069.
19. Margueron R, Licznar A, Lazennec G, et al. (2003) Oestrogen receptor α increases $p21^{WAF1/CIP1}$ gene expression and the antiproliferative activity of histone acetylase inhibitors in human breast cancer cells. J Endocrinol 179:41–53.
20. Hattinger CM, Jochemsen AG, Tanke HJ, et al. (2002) Induction of p21 mRNA synthesis after short-wavelength UV light visualized in individual cells by RNA FISH. J Histochem Cytochem 50:81–89.
21. Parikh RR, Gildener-Leapman N, Narendran A, et al. (2005) Prevention of *n*-methyl-*N*-nitrosourea-induced breast cancer by alpha-fetoprotein (AFP)-derived eptide, a peptide derived from the active site of AFP. Clin Cancer Res 11:8512–8520.

Hormonal Prevention of Breast Cancer: Significance of Promotional Environment

Rajkumar Lakshmanaswamy, Raphael C. Guzman, and Satyabrata Nandi

Summary Early full-term pregnancy reduces the risk of mammary cancer in humans. Rats and mice also exhibit this phenomenon of parity protection. Short-term treatment with pregnancy levels of estradiol (E_2) is also highly effective in preventing mammary carcinogenesis. Earlier it has been demonstrated that parous rats treated with carcinogen develop latent microscopic mammary tumors that do not progress further to form overt mammary cancers. In the current investigation, we wanted to find out if short-term treatment with pregnancy levels of E_2 also prevents mammary carcinogenesis similar to parity. Rats were injected with *N*-methyl-*N*-nitrosourea at 7 weeks of age and treated with 20 μg, 100 μg, 200 μg, or 30 mg of E_2 in silastic capsules for 3 weeks. 100 μg (17%), 200 μg (17%), and 30 mg (17%) doses of E_2 resulted in levels of E_2 equivalent to pregnancy level and were effective in preventing overt mammary cancer incidence compared with control (100%) or 20 μg (73%) E_2 treatment, which did not result in pregnancy levels of E_2 in the circulation. Although a significant reduction of overt cancers was observed in the pregnancy levels of E_2 treated groups, there was no difference in the incidence of latent microscopic mammary cancers between the E_2 treated and the controls. Proliferation of latent microscopic mammary cancers was examined using immunohistochemistry for cyclin D1 expression. Proliferation in the latent microscopic mammary cancers of the protected groups was significantly lower (~2.0–3.0-fold) than the latent microscopic mammary cancers in the unprotected groups. These findings indicate that mammary cancer development can be blocked by inhibiting or blocking promotion and progression of carcinogen initiated cells.

Introduction

The reproductive history of a woman and her risk for breast cancer (BC) has been long associated. International studies of women conducted by MacMahon et al. (1) suggested that women undergoing a full-term pregnancy before the age of 18, with or without lactation, have about one third the risk of developing BC compared with women who have never undergone a full-term pregnancy. Rats (2) and mice (3) that undergo a full term pregnancy also have a greatly reduced susceptibility to chemical carcinogen induced mammary carcinogenesis compared with nulliparous animals. Hormonal prevention strategies have used short-term exogenous hormonal treatment to mimic the protective effect of pregnancy against BC. Huggins reported that that high levels of estradiol (E_2) and progesterone given for 30 days beginning 15 days after carcinogen administration inhibited mammary carcinogenesis in SD rats treated with the polycyclic hydrocarbon carcinogen 7,12-dimethylbenz(a)anthracene (DMBA) (4). Grubbs demonstrated that treatment of SD rats with high levels of E_2 and progesterone after treatment with *N*-methyl-nitrosourea, a direct acting alkylating agent was as effective as ovariectomy in preventing mammary carcinogenesis (5). The Russo's have reported that short-term treatment with human chorionic gonadotropin (hCG) induces protection against DMBA-induced mammary cancers (6). We have demonstrated that short-term treatment with pregnancy levels of E_2 alone or E_2 + progesterone for 7–21 days is highly effective in conferring protection against methylnitrosourea (MNU)-induced mammary carcinogenesis in female Lewis rats (7, 8).

The mechanisms for the protective effect of early pregnancy or short-term hormone treatment have not been defined. There are two theories to explain the protective effect of early pregnancy and short-term hormone treatment, which mimics the protective effect of pregnancy. The first theory hypothesizes that the protective effect of pregnancy or short-term hormone treatment induces differentiation of the target structures for carcinogenesis, the terminal end buds and terminal ducts resulting in the loss of target cells for carcinogenesis. The second theory explaining the protective effect demonstrates that it is due to persistent changes in the systemic hormonal milieu after pregnancy or short-term hormone treatment resulting in a decreased promotional environment. Further, it has been reported that mammary glands of parous rats treated with carcinogen had a high incidence of latent microscopic mammary cancers, indicating that parous mammary glands have target cells and are highly susceptible to mammary carcinogenesis (9).

Hence, in the current study our objective was to determine whether hormone protected rats also have latent microscopic mammary cancers and if so, to investigate whether the protective hormone treatment results in persistently altered proliferation of the latent microscopic mammary cancers.

Materials and Methods

Animals. Virgin Lewis rats were purchased from Harlan Sprague Dawley (Indianapolis, IN and San Diego, CA). The rats were housed in a temperature-controlled room with 12-h light/12-h dark schedule. They were fed food (Teklad, Madison, WI) and water ad libitum. All the procedures followed University of California Animal Care and Use Committee guidelines.

Carcinogen Treatment. A single ip injection of MNU (Ashe Stevens, Detroit) at a dose of 50 mg kg^{-1} of body weight was given to all the rats at 7 weeks of age. MNU was dissolved in physiological saline that had been adjusted to pH 5.0.

Hormone Treatment. The hormones were packed in individual silastic capsules (Dow Corning; size 0.078 in. i.d. × 0.125 in. o.d., 2 cm in length). All doses of 17β-E_2 (Sigma, St. Louis) were packed in the silastic capsules in a cellulose matrix except for 30 mg E_2 dose that was packed with the hormone alone. All these capsules were primed by soaking over night in media 199 (Gibco) at 37 °C. Silastic capsules were implanted subcutaneously dorsally. All the control rats received empty silastic capsules.

Effect of Short-term E_2-treatment on Overt and Latent Microscopic Mammary Cancers. At 7 weeks of age, the rats were treated with a single intra-peritoneal injection of MNU. When the rats were 9 weeks of age, they were divided into 5 groups, each group consisting of 12 rats and receiving one of the following treatments: (1) control, and E_2 (2) 20 µg, (3) 100 µg, (4) 200 µg, and (5) 30 mg. Each group was treated for 3 weeks, and at the end of the treatment, the silastic capsules were removed from the animals.

Overt Mammary Cancers. Rats were palpated once every week beginning 1 month (mo) after carcinogen exposure for 9 mo to monitor for mammary cancer development. Histopathological examination was performed to confirm the carcinomatous nature of the palpable cancers.

Latent Microscopic Mammary Cancers. After 9 mo of the MNU injection, the rats were terminated. All the mammary glands were removed, fixed in 10% formalin in phosphate buffered saline (PBS), stained with hematoxylin, and later stored in Histoclear (National Diagnostics). Putative latent microscopic mammary lesions were detected under the microscope and microdissected from the glands. Histopathological examination was done to confirm the neoplastic nature of the microscopic mammary lesions.

Cyclin D1 Immunocytochemistry. On confirmation of the neoplastic nature, the mammary cancer sections were analyzed for cyclin D1 expression using standard immunocytochemistry techniques. A minimum of ten latent microscopic mammary cancers/group and at least 5 000 cells per group were counted. The percentage of positively stained cells was determined by dividing the number of percent stained cells by the total number of cells counted and multiplying by 100.

Statistics. The effects of the different hormonal treatments were analyzed by using the χ^2 test for 2 × 2 contingency tables and Student's t test. Values of $p < 0.05$ were considered significant.

Results

The overt mammary cancer incidence data obtained in the present study is in accordance with our earlier findings (7, 8). They demonstrate, once again, that short-term treatment with pregnancy levels of E_2 alone is highly effective in conferring protection against MNU-induced mammary carcinogenesis. Seventy three percent of rats in the 20 µg group, and 17% in 100 µg, 200 µg, and 30 mg groups had mammary cancers compared with 100% in the control group. E_2 treatment also decreases the multiplicity of cancers in each rat (Table 1). The multiplicity in the 20, 100, and 200 µg, and 30 mg groups were lower than the 3.0 ± 0.5 cancers in the control group. There were 1.1, 0.6, 0.3 and 0.2 mammary cancer/rat treated with 20, 100, and 200 µg, and 30 mg of E_2, respectively. The 100 µg, 200 µg, and 30 mg of E_2 treatment resulted in E_2 pregnancy circulating levels, 67.8 ± 13.9, 94.9 ± 17.4, and 143.5 ± 23.9 pg ml^{-1}, respectively. The rats in the 20 µg E_2 treatment group did not result in pregnancy circulation E_2 levels (49.9 ± 12.4 pg ml^{-1}), and therefore, was not protected against the development of mammary cancer.

Nine months after MNU-exposure, mammary glands were dissected and fixed for observation of latent microscopic mammary cancers (Fig. 1). The incidence of latent microscopic mammary cancers in the E_2-treated groups were not significantly different from that of the control group (Table 2) indicating that initiation is not preventable in either the unprotected or protected groups. The controls and E_2-treated rats had 100% incidence of latent mammary cancer (Table 2). Although the number of overt mammary cancers were significantly lower in the different E_2-treated groups than those in the control group, the number of latent microscopic mammary cancers were not different among the treated and control groups.

Sections of the overt and latent cancers were stained for cyclin D1 by immunohistochemistry to compare the proliferation rates of the latent microscopic mammary

Table 1 Effect of short-term E_2 treatment on overt mammary cancer incidence

Treatment	Overt mammary cancer incidence	% of rats with overt mammary cancer	Average # of overt mammary cancers/rat
Control	11/11	100	3.0 ± 0.5
30 mg E_2	2/12	17***	0.2 ± 0.4***
200 µg E_2	2/12	17***	0.3 ± 0.2***
100 µg E_2	2/12	17***	0.6 ± 0.4**
20 µg E_2	8/11	73	1.2 ± 0.4*

*$p < 0.05$, **$p < 0.01$, ***$p < 0.001$ (compared with control)

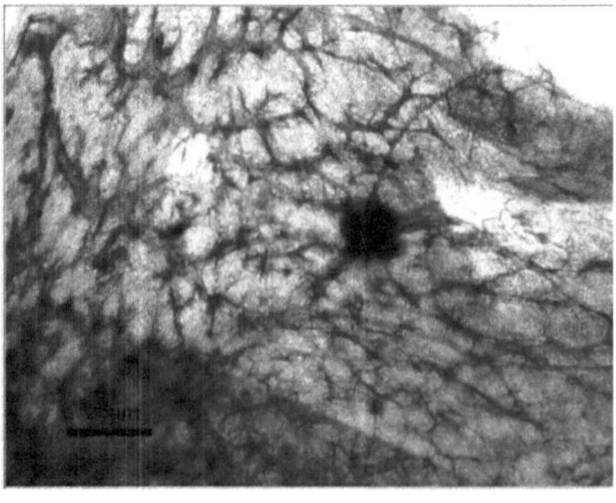

Fig. 1 Latent microscopic mammary cancer

cancers from the protected and unprotected groups. There was a high expression of cyclin D1 in the overt cancers (43.8% ± 5.1%). Cyclin D1 expression was lower in the latent microscopic mammary cancers of protected group compared with the unprotected group (Fig. 2).

Table 2 Effect of short-term E_2 treatment on latent microscopic mammary cancer incidence

Treatment	Overt mammary cancer incidence	% of rats with latent microscopic mammary cancer
Control	11/11	100
30 mg E_2	12/12	100
200 μg E_2	12/12	100
100 μg E_2	12/12	100
20 μg E_2	11/11	100

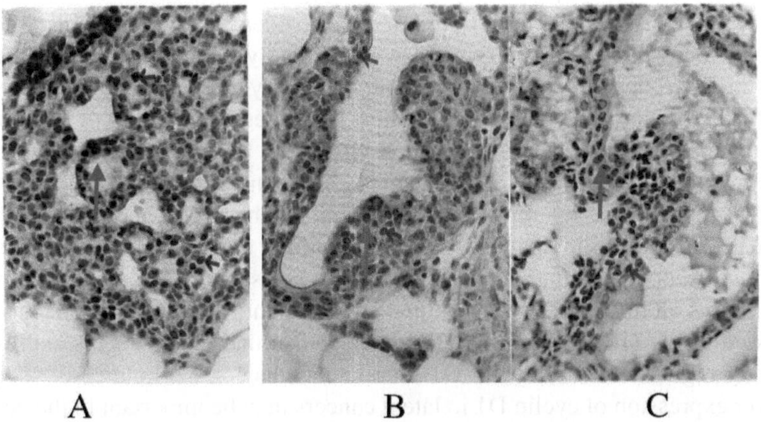

Fig. 2 Cyclin D1 immunohistochemistry of overt and latent microscopic mammary cancers. (**A**) Overt mammary cancer; (**B**) Latent microscopic mammary cancer from the control group; and (**C**) Latent microscopic mammary cancer from the protective E_2-treated group

Table 3 Effect of short-term E_2 treatment on latent microscopic mammary cancer proliferation

Treatment	Latent microscopic mammary cancer % cyclin D1 expression
Control	29.2 ± 4.3
30 mg E_2	12.4 ± 3.1***
200 μg E_2	12.8 ± 2.9***
100 μg E_2	10.4 ± 3.4***
20 μg E_2	24.9 ± 2.7

***$p < 0.001$ (compared with control)

In control latent microscopic mammary cancers, 29.2% ± 4.3% of the cells were positive for cyclin D1, while those from 100 μg, 200 μg, and 30 mg E_2-treated rats had 10.4% ± 3.4%, 12.8% ± 2.9%, and 12.4% ± 3.1% CDI positive cells, respectively. In latent microscopic mammary cancers from the 20 μg E_2-treated rats, 24.9% ± 2.7% were positive for cyclin D1 (Table 3).

Discussion

Full-term pregnancy is effective in preventing the development of BC in rats and humans; likewise, short term E_2 treatment, conferring pregnancy levels of E_2, also provides long term protection against mammary cancer in rats. Even though the incidence of overt carcinoma is drastically lowered in parous rats, the incidence of latent carcinoma is not reduced when compared with nulliparous controls (9). In this study, the effects of different doses of E_2 on the presence of latent and overt cancers were examined to determine the difference between the cancers.

Although short term pregnancy levels of E_2 treatment reduce the overt cancer incidence, like parity short-term E_2 treatment also does not protect against initiation of latent microscopic mammary cancer. This clearly demonstrates that initiation occurs in all groups but latent microscopic mammary cancers in protected animals do not progress into overt carcinoma. Short term pregnancy levels of E_2 treatment may create an insufficient promotional environment for the initiated latent microscopic mammary cancer cells to advance to overt cancers.

Latent microscopic mammary cancer resembles the overt cancer histologically, despite the differences in incidence. The difference in cyclin D1 expression between latent microscopic mammary cancers further supports that E_2 pregnancy circulating levels causes an insufficient promotional environment. Cyclin D1 is normally over expressed in BC (10). Previous studies have shown that overt cancers express significantly higher levels of cyclin D1 than latent cancers. Korkola et al. suggested that over expression of cyclin D1 in latent cancers may be important in the development of mammary cancer (11). High levels of cyclin D1 expression may effectively promote latent cancers to proliferate further into overt cancers.

However, the latent microscopic mammary cancers in the protected groups expressed significantly lower levels of cyclin D1 than the unprotected groups. Low levels of cyclin D1 expression play a role in the resistance toward mammary cancer (11). Thus, the latent microscopic mammary cancers in the protected groups are resistant to overt cancer development. The levels of cyclin D1 expression in E_2 treated rats may reflect that the systemic environment is not optimal for promotion. This suggests that pregnancy levels of E_2 inhibit the promotional environment for latent microscopic mammary cancers to proliferate into overt cancers, and it decreases proliferation of latent microscopic mammary cancers in the protected groups. This study demonstrates that like in parous rats the target cells for carcinogenesis are not lost during the short-term hormone treatment, but they result in decreased promotional environment for mammary carcinogenesis. Thus, these studies indicate that the mammary epithelial cells of parous rats are highly susceptible to carcinogen-initiation and are not totally refractory to the carcinogenic process.

Our studies have shown that latent microscopic mammary cancers in MNU-exposed short-term E_2-treated rats have significantly reduced immunohistochemically detectable cyclin D1 positive cells compared with those in nulliparous rats. Analogous phenomena have been observed between Japanese men in Japan with low incidence of prostate cancer and Caucasians in USA with high incidence (12). Men from both countries have a similar incidence of latent prostate cancer. The lower incidence of overt prostate cancer in Japanese men is also likely to be due an inadequate promotional hormonal environment. Overall, the short term pregnancy levels of E_2 reduce the incidence of overt cancers by preventing the promotion of the latent microscopic mammary cancers to overt mammary cancers.

References

1. MacMahon B, Cole P Lin TM, et al. (1970) Age at first birth and breast cancer risk. Bull WHO 43:209–221.
2. Moon RC (1969) Relationship between previous reproductive history and chemically induced mammary cancer in rats. Int J Cancer 4:312–317.
3. Marchant J (1955) Influence of pregnancy and lactation on the incidence of mammary carcinoma induced with methylcholanthrene in female mice of the IF strain. J Pathol Bacteriol 70:415–418.
4. Huggins C, Moon RC, Morii S (1962) Extinction of experimental mammary cancer. I. Estradiol-17beta and progesterone. Proc Natl Acad Sci USA 48:379–386.
5. Grubbs CJ Juliana MM, Whitaker LM (1988) Short-term hormone treatment as a chemopreventive method against mammary cancer initiation in rats. Anticancer Res 8:113–117.
6. Russo IH, Koszalka M, Russo J (1990) Effect of human chorionic gonadotropin on mammary gland differentiation and carcinogenesis. Carcinogenesis 11:1849–1855.
7. Guzman RC, Yang J, Rajkumar L, et al. (1999) Hormonal prevention of breast cancer: mimicking the protective effect of pregnancy. Proc of the Natl Acad Sci USA 96:2520–2525.
8. Rajkumar L, Guzman RC, Yang J, et al. (2001) Short-term exposure to pregnancy levels of estrogen prevents mammary carcinogenesis, Proc Natl Acad Sci USA 98:11755–11759.
9. Yang J, Yoshizawa K, Nandi S, et al. (1999) Protective effects of pregnancy and lactation against *N*-methyl-*N*-nitrosourea-induced mammary carcinomas in female Lewis rats. Carcinogenesis 20:623–628.
10. McMahon C, Suthigphongchai T, DiRenso J, et al. (1999) P/CAF associates with cyclin D1 and potentiates it s activation of the estrogen receptor. Proc Natl Acad Sci USA 96:5382–5387.
11. Korkola J, Wood G, Archer M (1999) Cyclin D1 expression during rat mammary tumor development and its potential role in the resistance of the Copenhagen rat. Breast Cancer Res 1:88–94.
12. Yatani R, Shiraishi T, Nakakuki K, et al. (1998) Trends in frequency of latent prostate carcinoma in Japan from 1965–1979 to 1982–1986. J Natl Cancer Inst 80:683–687.

Hormones of Pregnancy, α-Feto Protein, and Reduction of Breast Cancer Risk

Herbert I. Jacobson, Nicole Lemanski, Amithi Narendran, Anu Agarwal, James A. Bennett, and Thomas T. Andersen

Cancer is not a growth of anarchic cells; laws control the survival of the cancer cell and these it must obey. (Charles Brenton Huggins, MD, 1979)

Summary Parity profoundly reduces breast cancer (BC) risk later in life. It has been reasoned that hormones (either estradiol E_2 or estriol E_3), progesterone (P) or human chorionic gonadotropin (hCG) in the serum of pregnant women might lead to that reduction in risk. These agents have been shown to reduce BC incidence in nonpregnant rats. We investigated the hypothesis that exogenously added E_2, E_3, P, or hCG are not the proximal effectors of risk reduction, but that they elicit α-fetoprotein (αFP) from the nonpregnant liver, and that αFP is the proximal agent by which reduction of BC risk is obtained. Methylnitrosourea (MNU)-exposed animals were treated with saline, E_3, E_2 + P, E_3 + P, hCG, or were allowed to experience pregnancy, and AFP levels were measured in the serum and subsequent tumor incidence was recorded. Human HepG2 liver cells in culture were treated with E_3, E_2 + P, P, or hCG and elicited AFP was measured in the media. The HepG2 culture media containing elicited AFP was assessed for its ability to inhibit proliferation of T47D cells when applied to these human BC cells in culture, and to inhibit the estrogen-induced phosphorylation of the estrogen receptor in T47D cells. For each condition in the prevention studies, hormone treatment reduced the incidence of BC to an extent similar to that reported by the original studies. In each condition, αFP levels in serum were elevated over that in control animals. In culture, treatment of human liver cells with E_3, E_2 + P, or hCG, but not P alone, led to increased levels of AFP in the media. Media containing hCG-elicited AFP inhibited the estrogen-stimulated proliferation of T47D cells in culture, and inhibited phosphorylation of the estrogen receptor, whereas, estrogens and hCG did not inhibit the growth of these tumor cells in culture. In conclusion, since the hormones of pregnancy elicit αFP from the liver, and αFP but not the hormones of pregnancy has direct antitumor properties, it is concluded that αFP is the proximal agent through which reduction in BC incidence is realized from the experience of pregnancy.

Introduction

Early parity profoundly affects women's breast cancer (BC) risk, reducing it to about half that of nulliparous women (1). What is the mechanism? Can it be developed as therapy? Researchers seeking modalities for reduction of BC risk have reasoned that some component of the serum during pregnancy must be an effector of the reduction in risk; posited effectors are the steroidal estrogens (E_2 or E_3) (2, 3), P (2, 3), and human chorionic gonadotropin (hCG) (4). Additionally, α-fetoprotein (αFP) has been posited to be the risk-reducing agent (5).

The carcinogens MNU and dimethylbenzanthracene (DMBA) are potent generators of BC in SD rats. Carcinogen-exposed female rats that are subsequently mated and bear litters ultimately generate half as many tumors as those that remain nulliparous (6). This striking parallel to the effect of pregnancy on human BC risk (1) has suggested studies of cancer yield to be performed in carcinogen-exposed rats that have been administered pregnancy-associated hormones as surrogates for pregnancy. In such studies, Russo's laboratory (4) administered hCG, Lemon (7) used E_3, and Grubbs (2), Nandi et al. (3) and others have employed combinations of steroid hormones (progesterone (P) + either E_2 or E_3). Interestingly, all of these challenges produced similar results, a reduction of BC incidence among carcinogen-exposed rats. We are aware of no hypotheses that would explain the same result from all of these various surrogates. There is need to account for this apparent redundancy of mechanisms available to the mammary gland for protection against malignant transformation.

We have studied the role of αFP in the parity-risk phenomenon, and have noted as well that this protein is an effective inhibitor of human BC xenograft growth (8). We identified the active site of the 69,000kD αFP molecule, synthesized a 9-mer cyclic peptide analog, αFPep, and have shown that it is effective for the prevention of BC in MNU-treated rats (9), as well as for the inhibition of BC xenograft growth. We hypothesize that E_3, E_2 + P, E_3 + P, and hCG may be effective cancer preventive agents by virtue of their ability to elicit αFP secretion from the nonpregnant rat liver, rather than by having a direct effect on the mammary gland, and that αFP is the proximal agent by which the reduction in BC risk is obtained. To test that hypothesis, we have repeated the hormone treatments that were reported to be effective at reducing rat BC, in the same experimental configuration used by the previous researchers, and have measured serum αFP levels in these carcinogen-exposed, hormone-treated animals. Further, as a surrogate for human liver, we examined the ability of these hormones to elicit αFP from human liver cells in culture. We report here that these hormone treatments do, in fact, lead to the production of αFP and that hormone-elicited αFP directly inhibits growth of BC cells in culture.

Materials and Methods

Carcinogen Exposure. Groups of 30 female SD rats received 50mg kg^{-1} body weight of MNU in sterile physiological saline at 50 days of age as a single intrajugular vein injection. Treatment with hormones was initiated within 10–21 days after MNU exposure and for each experiment continued for the durations used by the earlier workers. For tumor detection, animals were palpated twice weekly, continuing for 123 days.

Hormone Treatments. These followed the doses and schedules specified in the publications being replicated, and therefore differ from one another in modest ways. In every case, we used MNU, though some of the published investigations used DMBA. **E_2 + P**. Following Grubbs et al. (2), 10 days after carcinogen administration, rats received 20 µg E_2 + 4 mg P dissolved in sesame oil, daily, by subcutaneous injection (0.2 ml), for 40 days. **E_3 + P**. Following Rajkumar et al. (3), 13 days after MNU, and under isofluorane anesthesia, rats received two individual sc silastic capsules, one packed with 30 mg E_3 and another packed with 30 mg of P. These were left in place for 21 days to mimic pregnancy. **E_3**. Similarly, 13 days following MNU administration, rats received a sc silastic capsule packed with 30 mg E_3 that remained in place for 21 days to mimic pregnancy. **hCG**. Similar to Russo et al. (4), 21 days after MNU administration, rats received 100 IU hCG in deionized water, ip daily, for 60 days. **Pregnancy**. Following Grubbs et al. (11), 10 days following MNU, female rats were introduced to males. Females stayed with males for 7 days, then were removed and separated into individual cages. After 21–23 days, 19 females bore litters and were allowed to breast feed for 15 days. Females that did not become pregnant were excluded from the study. **Controls**. The no-treatment group was exposed to MNU without subsequent treatment, and experienced the maximal number of tumors over the course of the ensuing four months.

Blood Samples. Tail vein blood was drawn from animals in each of the 6 groups of rats at four time points: prior to first dose of treatment (for pregnancy, on the fourth and final day females were housed with males), at the midpoint of the treatment regimens, on the last day of the treatment regimen, and lastly, 14 days later. Blood was allowed to clot at room temperature, centrifuged, and sera stored at −20°C. No animal had more than one blood draw.

Rat αFP Measurements. Detection and semiquantitation of rat αFP in sera was by Western blot employing a goat antibody raised against recombinant rat AFP and a standard curve based on AFP from rat amniotic fluid taken at pregnancy day 15. Measurement of elicited human αFP: HepG2 cells (10^5 cells per well) were maintained and grown as a monolayer in αMEM (Invitrogen Corporation, Carlsbad, CA) supplemented with 5% serum (40% bovine calf serum, 60% FBS), penicillin G (100 units ml^{-1}), and streptomycin (100 µg ml^{-1}) in 24-well culture plates in triplicates. Cells were changed to serum-free medium when confluent and media were changed every three days. Hormones, E_3, E_3 + P, E_2 + P, or P were dissolved in 95% ethanol, diluted in buffer and brought to 10^{-8} M in the cell culture dish. hCG was dissolved in sterile deionized water and used at a final concentration of 10^{-9} M. Control cells were grown in serum-free medium alone. Hormones were administered daily for 21 days. Media containing secreted αFP were stored at −20°C and were assayed for human AFP on the Beckman-Coulter Access Immunoassay System (Beckman Coulter Inc, Fullerton, CA).

Effect of HepG2 Supernatant on T47D Proliferation. HepG2 cells were grown as above in 25 cm^3 flasks while supplemented with 10^{-9} M hCG daily. Supernatants were collected every third day, centrifuged, concentrated and retained for T47D proliferation experiments. T47D Cells. T47D cells (1.2×10^5 per well) were plated in estrogen-depleted medium and treated with 100 µL HepG2 supernatant the day following plating. After incubation for 1 h, 10 µl of 10^{-8} M E_2 (10^{-10} M final concentration) was added. Cells were treated for 7 days with media changes every other day and counted on day 8 using a hemacytometer.

Results

Five groups of carcinogen-exposed rats were treated with different pregnancy-hormone regimens, while a sixth group received no additional treatments. In each case, the reduction of BC incidence in hormone-treated groups is similar to that reported by earlier workers. The decreased incidence is statistically significant for each treatment (compared with control), and multiplicity and tumor volume decreased by each of these treatments.

Rat AFP levels were quantitated using Western blot analysis. On the basis of eight replicate studies, treatment with hCG elevated serum αFP concentration 2.4-fold over untreated rat levels, E_2 + P leads to 1.9-fold elevation, E_3 alone produced 2.9-fold elevation, and E_3 + P produced a 3.2-fold rise in αFP levels. Treatment with P alone did not elevate αFP levels persistently (Fig. 1).

To evaluate the effect of hormones on human liver cells, HepG2 cells were incubated with either E_3, E_2 + P, hCG, or P alone, and human αFP released into the cell culture medium was quantified. Figure 2 indicates that treatment with E_3, E_2 + P or hCG led to elevated levels of human αFP above control levels in the media, whereas treatment with P alone did not stimulate αFP production. Control levels (no hormone added to HepG2 cells) ranged from 0.2 to 0.4 µg ml^{-1}.

To assess whether elicited human αFP could inhibit BC cell growth in culture, 10×-concentrated supernatants from hCG-treated HepG2 cells were added to cultured T47D human BC cells. E_2 strongly stimulated growth while αFP-containing

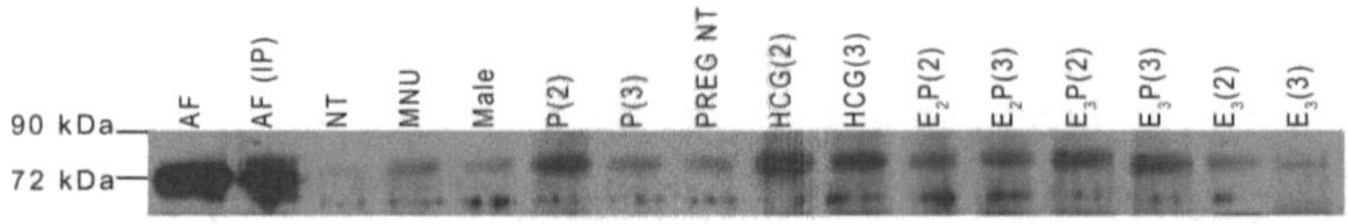

Fig. 1 Measurement of αFP in the sera of hormone-treated animals. Western blot analysis for αFP in rat sera obtained from bleeds at mid-treatment (bleed 2), at conclusion of treatment (bleed 3), and 14 days later (not shown). Controls are rat amniotic fluid (AF), sera from untreated rats (NT) given MNU only, or from male rats. The blot is representative of eight replicate experiments

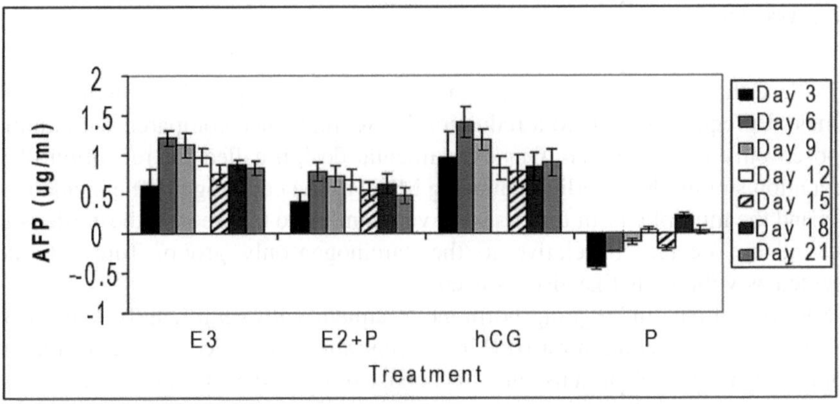

Fig. 2 Production of αFP by cultured human HepG2 cells exposed to hormone treatment. αFP levels produced in supernatants of confluent HepG2 cells (10^5 per well) cultured with different hormones were quantified by ELISA and is shown as the difference between the treatment and NT groups

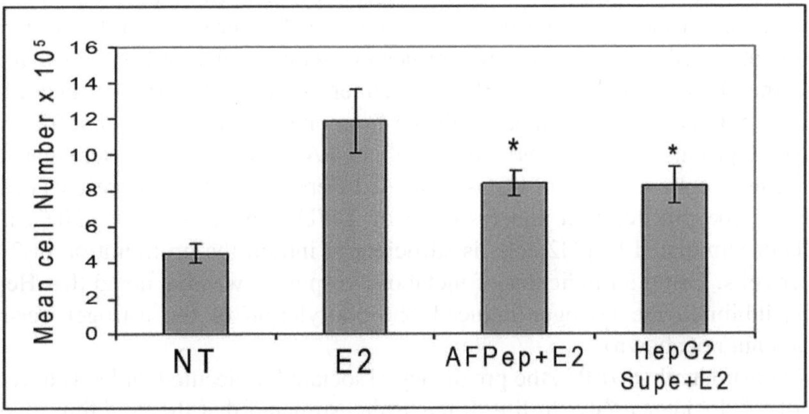

Fig. 3 Inhibition of T47D cell proliferation by hCG-treated HepG2 supernatant. T47D cells (1.2 × 10^5) treated with AFPep (10^{-6} M) or hCG (10^9 M)-stimulated HepG2 supernatant (×10) followed by E_2 (10^{-10} M) stimulation per 7 days. Cells were harvested and counted. *P values (Student's t test) were 0.006 for inhibition by HepG2 supernatant and 0.01 for αFPep ($n = 5$)

supernatant significantly inhibited it. Similar inhibition of E_2-stimulated T47D cells growth was produced by a peptide analog of the active site of αFP (Fig. 3). In contrast hCG did not directly inhibit T47D cells (data not shown).

Discussion

It is clear that treatment of nonpregnant, carcinogen-exposed rats with the hormones of pregnancy leads to a reduction in BC incidence compared with animals exposed only to carcinogen. Our experimental design called for repetition of the prevention studies done earlier, but using MNU as the carcinogen. The incidence of BCs and the multiplicity of tumors were very similar to that seen in the earlier studies, and are decreased relative to the carcinogen-only group. Tumor burden, assessed as volume, is likewise reduced.

Sera of animals undergoing hormone treatment with significantly reduced BC risk were found to contain rat αFP in concentrations significantly higher than the basal levels in control animals, and this persisted for 14 days following cessation of hormone treatment. A significant control observation was provided by rats treated with P, which fails to provide protection against BC (6). When treated with progesterone (only), they failed to produce sustained high concentrations of αFP. If αFP is the agent of pregnancy responsible for reduction of BC risk, the data suggest that sustained exposure to that protein is required. It may be questioned whether the cancer prevention experiments in rats constitute a rational extension of the phenomenon occurring in human BC. Since there were no experimental means to elicit αFP from human liver, we undertook a more practical alternative, specifically a challenge of cultured HepG2 human liver cancer cells with these same hormones. With challenges of E_3, E_2 + P, or hCG, the level of human αFP appearing in culture medium was sustained at significantly elevated levels over the basal level secreted by untreated cells. Again, a significant negative control observation was that progesterone alone did not elicit αFP. Using either assay (sera of rats or media from cultured human cells), treatment with the hormones of pregnancy is sufficient to elicit αFP production. To assess whether elicited AFP is sufficient to attenuate cancer, we used culture media of hCG-stimulated HepG2 (no steroids present) cells to influence the proliferation and response of T47D human BC cells. The culture media of stimulated HepG2 cells is sufficient to inhibit the proliferation of T47D cancer cells. Using an indicator of metabolic response, we also noted that HepG2 media inhibited the estrogen-induced phosphorylation of the estrogen receptor alpha (data not shown).

We postulate that αFP is the pregnancy-associated molecule that leads to reduction of BC incidence later in life. It has been estimated that the doubling time for human BCs is about 4 months. Making the rough approximation that growth rate (net mitosis less apoptosis) is constant throughout the duration of an eventual tumor, it can be estimated that to grow from a single cell to a minimally detectable cell mass (approximately 10^9 cells) would require 30 doublings, or 10 years. Exposure to the AFP of pregnancy may decrease the mitotic rate relative to apoptosis, leading to tumor extinction, or at least to decreased growth, so that fewer tumors are ever detected. Bennett et al. (8, 10) have reported on experiments showing human AFP can inhibit the growth of human BC xenografts. It may be possible to capitalize on these observations (i.e., the concept that cancer could be present but undetected for 10 years) in that administration of a safe and effective agent during that window may decrease tumor incidence in humans. An analog of the active site of αFP, termed αFPep, may be such an agent (9). Molecules that mimic E_3, E_2, P, or hCG would seem less desirable as preventive agents. If αFP could indeed mimic the chemo preventive effect of pregnancy and were utilized, then the incidence of BC in the USA alone would be reduced from 200,000 new cases per year to under 100,000 new cases, possibly below 40,000.

Investigators have given hormones to carcinogen-exposed rats and achieved reduction of BC risk, suggesting that postcarcinogen hormone treatment to reduce BC risk functions by the same mechanism as does pregnancy. This experiment mimics the viewpoint that prolonged exposure to environmental carcinogens occurs prior to pregnancy in humans. However, several investigators have administered the hormone prior to the carcinogen (7), and achieved reduction of BC appearance, which they attribute to hormone-produced changes in the susceptibility of the mammary gland to mutagenic insult. An alternative viewpoint, however, would be that hormones administered prior to the carcinogen produce αFP which, as we have shown, persists sufficiently so as to be present days later, at the time the carcinogen is administered, thus achieving the same result as if the hormones were delivered after the carcinogen.

In summary, it has been clearly shown that BC risk is reduced as a consequence of parity. Definitive studies have shown that the hormones of pregnancy may act independently as surrogates to reduce BC risk in nonpregnant rats. The studies reported herein demonstrate that these treatments also concomitantly elevate αFP levels in nonpregnant adult rat serum. Since this study also shows that αFP inhibits BC growth directly and the hormones of pregnancy do not, it is, therefore, logical to develop αFP analogs for future chemoprevention of BC.

Acknowledgments This work was supported by DOD grant W81XWH-04-1-0486.

References

1. Kelsey JL, Gammon MD, John EM (1993) Reproductive factors and breast cancer. Epidemiol Rev 15: 36–47.
2. Grubbs CJ, Peckham JC, McDonough KD (1983) Effect of ovarian hormones on the induction of 1-methyl-1-nitrosourea-induced mammary cancer. Carcinogenesis 4: 495–497.
3. Rajkumar L, Guzman RC, Yang J, et al. (2004) Prevention of mammary carcinogenesis by short-term estrogen and progestin treatments. Breast Cancer Res 6: R31–R37.
4. Russo IH, Koszalka M, Russo J (1990) Human chorionic gonadotropin and rat mammary cancer prevention. J Natl Cancer Inst 82: 1286–1289.
5. Jacobson HI, Janerich DT (1989) Pregnancy-altered breast cancer risk: mediated by AFP? In: Mizejewski GJ and Jacobson HI, editors. Biological activities of alpha-fetoprotein. Boca Raton, FL: CRC Press, 1989, pp. 93–100.
6. Grubbs CJ, Hill DL, McDonough KC, et al. (1983) *N*-nitroso-*N*-methylurea-induced mammary carcinogenesis: effect of pregnancy on preneoplastic cells. J Natl Cancer Inst 71: 625–628.

7. Lemon HM (1975) Estriol prevention of mammary carcinoma induced by 7,12-dimethylbenzanthracene and procarbazine. Cancer Res 35: 1341–1353.
8. Bennett JA, Zhu S, Pagano-Mirarchi A, et al. (1998) Alpha-fetoprotein derived from a human hepatoma prevents growth of estrogen-dependent human breast cancer xenografts. Clin Cancer Res 4: 2877–2884.
9. Parikh RR, Gildener-Leapman N, Narendran A, et al. (2005) Prevention of N-methyl-N-nitrosourea-induced breast cancer by alpha-fetoprotein (AFP)-derived peptide, a peptide derived from the active site of AFP. Clin Cancer Res 11: 8512–8520.
10. Jacobson HI, Bennett JA, Mizejewski GJ (1990) Inhibition of estrogen-dependent breast cancer growth by a reaction product of alpha-fetoprotein and estradiol. Cancer Res 50: 415–420.
11. Guzman RC, Yang J, Rajkumar L, et al. (1999) Hormonal prevention of breast cancer: mimicking the protective effect of pregnancy. Proc Natl Acad Sci USA 96: 2520–2525.

Role of Estrogens and Their Receptors in Adhesion and Invasiveness of Breast Cancer Cells

Marie Maynadier, Philippe Nirdé, Jean-Marie Ramirez, Anne Marie Cathiard, Nadine Platet, Monique Chambon, and Marcel Garcia

Summary Estrogen receptors (ERs) are overexpressed in human breast cancers (BCs) and associated with differentiated tumors and with a more favorable prognosis. Paradoxically, ERs mediate the mitogenic action of estrogens in human BC cells and the efficacy of antiestrogens in adjuvant therapy of primary tumors. The exact mechanism underlying the ER protection against cancer progression to metastasis remains to be investigated. Herein, we show that ERs decrease invasiveness of BC cells. Detailed studies revealed that the unliganded and the E_2-activated ERs decrease cancer cell invasion in vitro through two distinct mechanisms. In the presence of ligand, ERα inhibits invasion through a mechanism requiring the functional ERα domains involved in the transcriptional activation of target genes. Moreover, using different approaches, we found that cell–cell contacts were markedly increased by 17β-estradiol (E_2) treatment and decreased by the pure antiestrogen, $ICI_{182,780}$. This cell–cell adhesion was associated with an increase of the major intercellular junctions, desmosomes. Conversely, in the absence of ligand, ERα also inhibits invasion through a distinct mechanism involving protein–protein interaction with the region of the first zinc finger of ERα. The relationship of these data with clinical studies and their potential therapeutic consequences will be discussed.

Introduction

Breast cancer (BC) is the most frequent malignancy diagnosed in women in western countries, and estrogen receptor (ER) is overexpressed in about two thirds of breast tumors. Estrogens are the major sex steroid hormones involved in the development of normal mammary gland and in the etiology of human BC. The promoter role of estrogens in BC has been evidenced by epidemiological studies indicating an increased incidence in women with prolonged exposure to estrogens and a drastic decrease of incidence in women having nonfunctional ovaries (1). Most of the estrogen effects are mediated by two specific receptors ERα and ERβ that have a structure characteristic of members of the large nuclear receptor superfamily (2, 3). These receptors are subdivided into several functional domains (Fig. 1).

J.J. Li et al. (eds.), *Hormonal Carcinogenesis V.*

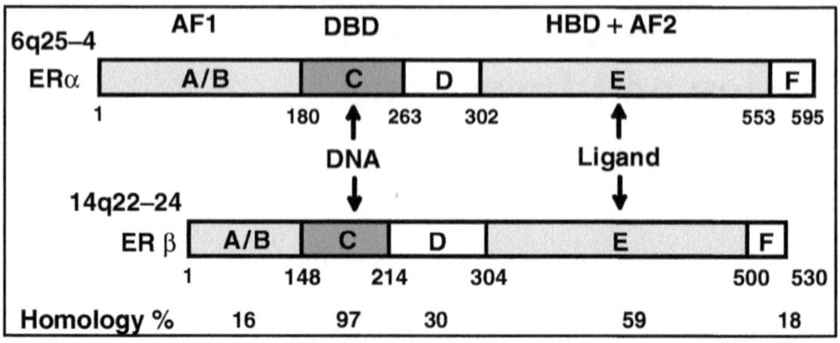

Fig. 1 Structures and functions of estrogen receptors

Functional domains have been defined as a DNA binding domain (DBD, C) containing two zinc fingers involved in DNA binding and receptor dimerization, a ligand binding domain (LBD, E) important for ligand binding, receptor dimerization, and interaction with transcriptional coactivators and corepressors. The ligand binding domain contains a ligand activable transactivation function AF2 whereas the N-terminal A/B region contains a ligand independent transactivating function AF1. The hinge domain (D) contains the nuclear localization signal of the receptor and contributes to the flexibility of the two moieties. The F domain participates to the transactivation capacity.

In malignant mammary cells, the effects of estrogens are more complex since estrogens affect directly the growth, motility, and invasiveness of cancer cells but, could also influence some biological responses of the host such as neoangiogenesis and immune response. The presence of ERα is associated with more differentiated and less invasive tumors and, clinical studies have demonstrated the favorable prognostic value of ER in primary breast tumors (4). The routine practice of ERα assay in primary tumors has been established to predict the efficacy of antiestrogens, widely used as first-line adjuvant therapy (5). The second receptor, ERβ, has likewise been detected in primary BCs by immunohistochemistry but its own clinical relevance in prognosis and tumor progression remains to be established (6).

In this report, we focus on the effects of estrogens and ERα on the invasiveness of BC cells. Estrogens are classically known to favor growth of cancer cells, and estrogen antagonists are clinically important in the treatment of hormone-dependent BCs. However, several reports have demonstrated that ER may protect against cancer cell invasion of basement membranes, an important step of the metastatic process required for cancer dissemination.

Results

Estrogens Inhibit Invasion via ERE-regulated Genes. The effects of estrogens on cell invasiveness have been studied in vitro using a two-chamber culture model and Matrigel, a reconstituted basement membrane. The initial studies indicated that

17β-estradiol (E_2) significantly reduces invasiveness that is reversed by antiestrogens (7, 8). This conclusion was noted in several $ER\alpha^+$ cancer cell lines established from breast or ovary, and in different $ER\alpha^-$ cancer cells constitutively expressing ERα after stable transfection. These in vitro data were confirmed in nude mice, since the formation of experimental lung metastases from metastatic $ER\alpha^-$ MDA-MB-231 BC cells was inhibited by E_2 after ERα expression by transfection (9, 10). The mechanism by which E_2 inhibits invasion was studied using an invasion assay based on the transient expression of ERα in the $ER\alpha^-$ MDA-MB-231 cell line (11). E_2 treatment decreased by twofold the invasiveness in ERα-transfectant (Fig. 2). The inhibitory effect of E_2 is reversed by both types of antiestrogens, OH-tamoxifen (4-OH-tamoxifen, the active metabolite of tamoxifen), and ICI $_{164,384}$, a pure antiestrogen. By contrast, E_2 or antiestrogen treatments did not significantly affect invasion of control ER^- cells. Moreover, the analysis of different ERα deletion mutants strongly suggested that some estrogen-regulated genes negatively control

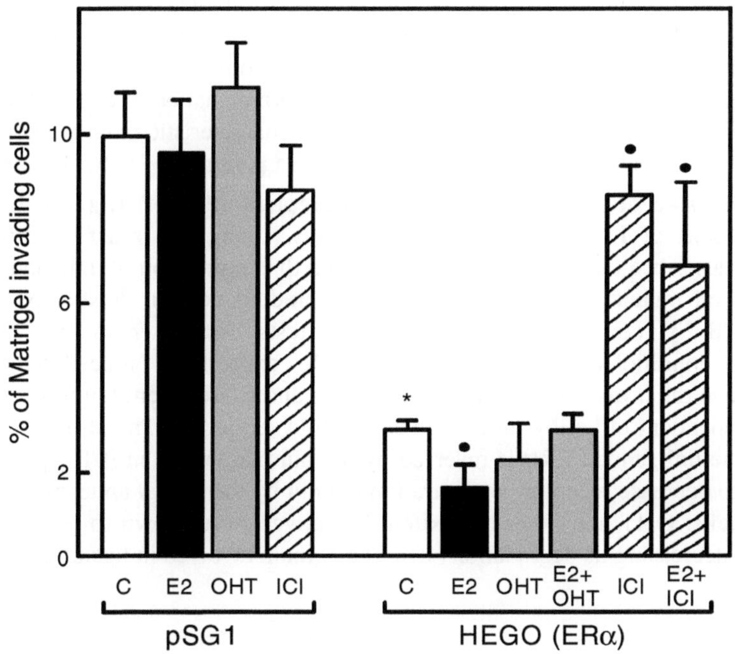

* $p < 0.01$ vs pSG1 control. • $p<0.05$ vs HEGO control. Reproduced from (11) with permission.

Fig. 2 Effect of ERα transient transfection and estradiol treatment on cell invasion. ERα-negative MDA-MB-231 cells were transiently cotransfected with ERα-expressing vector (HEGO) or control vector (pSG1), and a luciferase expressing vector used as a marker of transfected cells. The percentage of cells invading Matrigel was estimated in the presence of 20 nM estradiol (E_2), 100 nM 4-OH-tamoxifen (OHT), 100 nM $ICI_{164,384}$ (ICI), or ethanol alone (C)

invasion since the integrity of the hormone-binding domain, the DNA-binding domain and activating function 2 (AF2) of ERα was required (11). In contrast, the N-terminal domain containing the AF1 function is not involved since a deletion ΔAB mutant was as efficient as the wild-type receptor. As possible candidates among estrogen-regulated proteins, those that increase cell–cell adhesion, such as E-cadherin, or that decrease matrix degradation, such as α1-antichymotrypsin, should be considered (7, 8).

Recently, we studied the effects of estrogens on the adhesiveness of BC cells. Using different approaches, we found that cell–cell contacts were markedly increased by E_2 treatment and decreased by the pure antiestrogen, $ICI_{182,780}$ (Table 1). E_2-treated cells formed aggregates in standard culture conditions or on a collagen layer and were more resistant to trypsin treatment. In suspension culture conditions, only estrogen-treated cells constituted large spherical multilayered cell structures. Using electron microscope analysis, the increase in cell–cell adhesion was associated with the formation of the major intercellular junctions, desmosomes. E_2 increased both the number and the shape of desmosomes with enlarged dense plates in MCF7 BC cells but also in normal mammary cells isolated from reduction mammoplasties. The estrogen-induced proteins involved in desmosome production are currently investigated. Overall, these data indicated that the upregulation of cell–cell adhesion by estrogen found in normal mammary cells is a characteristic maintained in $ER\alpha^+$ BC cells, and could participate to the low invasive potential of $ER\alpha^+$ tumors.

Estrogen Receptors Inhibit Invasion Independently of Ligand Binding: Involvement of Protein–Protein Interactions. Using the transfection/invasion method described earlier, we also demonstrated that expression of unliganded ERα and several mutants deleted in the hormone-binding domain drastically reduced MDA-MB-231 cell invasiveness in Matrigel tests. As shown in Fig. 2, in E_2-deprived conditions, transient wild type ERα expression induced a threefold decrease in the invasiveness of transfected cells and E_2 treatment reinforced the ligand-independent effect by an additional twofold reduction. The strong inhibition due to the unliganded ERα is reversed by the pure antiestrogen $ICI_{164,384}$, which is known to decrease receptor concentration but not by the partial antiestrogen, OHT. In BC cells, OHT and other tamoxifen derivatives were shown to upregulate the ERα by decreasing its degradation (12). The domain involved in ligand-independent

Table 1 Effects of E_2 and $ICI_{182,780}$ on cell–cell interactions[a]

Effects on cell–cell interactions and treatment	Control	E_2	$E_2 + ICI_{182,780}$
Trypsin resistance	–	++	+
Cell aggregation in collagen gel	–	++	–
Spheroid structures in liquid culture	–	+	–
Number of desmosomes	+	++	+
Enlarged dense plates	±	+++	±

[a] Steroid-stripped MCF7 cells were treated for 5 days with 1 nM E_2, E_2 + 1 μM $ICI_{182,780}$, or treated with vehicle alone (control), and the cell–cell interactions were studied in the indicated cultures conditions

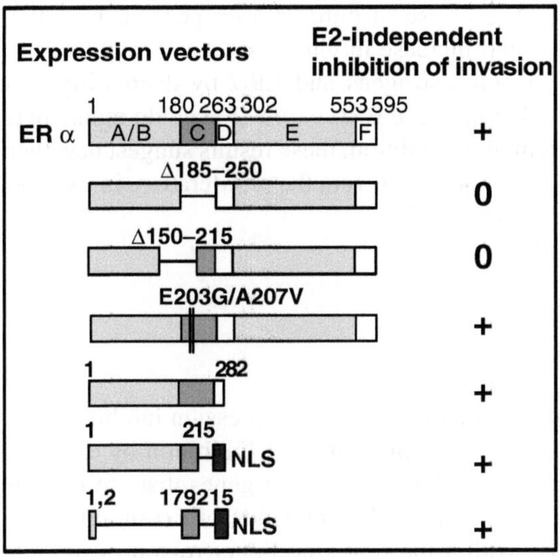

Fig. 3 E_2-independent inhibition of cancer cell invasion requires the DNA binding domain first zinc finger. Effects of ERα mutants on MDA-MB-231 cell invasiveness were determined in transfection/invasion assay in Matrigel. A nuclear localization signal (NLS) was added to maintain nuclear targeting of mutants deleted of the endogenous signals. Reproduced from (11) with permission

inhibition of invasion has been further characterized by progressive deletions in the ERα sequence.

As shown in Fig. 3, the first zinc finger of the DNA-binding domain (i.e., amino acids 179–215) is responsible for the antiinvasive activity. This activity is independent of the two key aminoacids that are essential for ERE binding and the estrogen specificity of the responses (13). Among the different nuclear receptors, invasion was specifically decreased by the expression of ERα (threefold) and to a lesser extent by ERβ (twofold), but was not affected by thyroid hormone receptor α1, vitamin D receptor, retinoid acid receptor α, or glucocorticoid and androgen receptors. On the basis of these data, it was proposed that unliganded ER decreases invasiveness via interaction of the first zinc finger region with an unknown nuclear factor.

Moreover, immunocytochemical studies of $ER\alpha^+$ BC cell lines (MCF7, ZR75.1, T47D) indicated that in hormone-deprived conditions, ERα expression was inversely correlated with cell motility (11). Migrating cells had lower ERα levels than nonmigrating cells. Finally, treatments such as phorbol ester or pure antiestrogen, known to decrease ERα levels in MCF7 BC cells, significantly increased in vitro invasiveness (11, 14). Taken together, these in vitro data indicate a protective role of ERα against the invasiveness of BC cells.

These data obtained on cell cultures cannot be extrapolated to the in vivo situation, where the endocrine and paracrine effects of estrogens may have major consequences on the invasiveness of cancer cells. However, their possible implications in the monitoring of BC should be discussed. Particularly, adverse effects should be anticipated in the clinical use of pure antiestrogens such as $ICI_{164,384}$ or $ICI_{182,780}$, since these drugs could increase in vitro cancer cell invasiveness by inhibiting the protective effect of both estrogens and ERα by decreasing its content. Selected adjuvant therapies that preserve ERα expression could maintain differentiation and a low invasive potential. In addition, these results suggest new therapeutic strategies based on ERα re-expression to prevent the proliferation, invasiveness, and metastatic potency of $ER\alpha^-$ BC cells (15).

Discussion

Our data demonstrated that E_2 and ERα expression inhibit cancer cell invasiveness in BC cells through different mechanisms. Inhibition by estrogens is dependent to transcriptional activation of specific target genes that are probably involved in an increase of cell–cell adhesion. The unliganded ERα inhibits invasion via protein–protein interactions within the first zinc finger region of the receptor. These data could explain the protective role of ERα against tumor invasion and metastasis previously found in these cell lines. ERα expression has been associated with a low invasiveness and a low motility in culture tests (14, 15). Moreover, when $ER\alpha^+$ cells were implanted in nude mice, tumors are poorly metastatic when compared with those developed from $ER\alpha^-$ BC cell lines (8).

Some clinical data support that estrogens prevent invasion. In mammary carcinogenesis, even though the mitogenic effect of estrogens is well demonstrated, the presence of ERα is associated with more differentiated and less invasive tumors and a more favorable prognosis. Moreover, there is some clinical evidences indicating that estrogens and their receptors protect against invasion. Epidemiological studies have evaluated the BC risk in women using hormone replacement therapy where 80% were taking preparations containing estrogen alone (ERT) (16, 17). Among the women using ERT, the risk of BC slightly increased, but the tumors under E_2 treatment were confined to localized disease with more favorable prognosis. Tumors in ERT-users were less invasive into axillary lymph nodes and distant sites. Other studies of tamoxifen therapy of primary BC suggest that tamoxifen increased the spreading of $ER\alpha^+$ primary tumor cells to contralateral sites (18). All together, these clinical data are in agreement with an antiinvasive effect of estrogens.

In conclusion, we present evidence that estrogens inhibit invasiveness of BC cells via a classical activation of ERE-regulated genes leading to an increase in cell–cell adhesion. In the absence of ligand, the ER could also prevent invasion through interaction with an unknown protein. Nonclassical mechanisms of action, in which the ER may bind to other transcription factors instead of DNA or to the proteins involved in pathways such as motility and invasion, require further investigation. The identification of the factors that inhibit the invasiveness of $ER\alpha^+$ cells would be a useful step in the development of new therapeutic targets to cure the most aggressive $ER\alpha^-$ tumors.

Acknowledgments We thank Michel Gleizes for excellent technical assistance and Jean-Yves Cance for artwork. This work was supported by the "Institut National de la Santé et de la Recherche Médicale," the "Association pour la Recherche sur le Cancer, and the "Ligue Contre le Cancer, Comité de l'Hérault" (fellowship for MM).

References

1. Pike MC, Krailo MD, Henderson BE, et al. (1983) Hormonal risk factors, breast tissue age and the age-incidence of breast cancer. Nature 303:767–770.
2. Mangelsdorf DJ, Thummel C, Beato M, et al. (1996) The nuclear receptor superfamily: the second decade. Cell 86:835–839.
3. Gustafsson JA (1999) Estrogen receptor ß – a new dimension in estrogen mechanism of action. J Endocrinol 163:379–383.
4. McGuire WL (1978) Hormone receptor: their role in predicting prognosis and response to endocrine therapy. Semi Oncol 5:2428–2433.
5. Fisher B, Jeong JH, Dignam J, et al. (2001) Findings from recent National Surgical Adjuvant Breast and Bowel Project adjuvant studies in stage I breast cancer. J Natl Cancer Inst Monogr 30:62–66.
6. Fuqua SA, Schiff R, Parra I, et al. (2003) Estrogen receptor beta protein in human breast cancer: correlation with clinical tumor parameters. Cancer Res 63:2434–2439.
7. Rochefort H, Platet N, Hayashido Y, et al. (1998) Estrogen receptor mediated inhibition of cancer cell invasion and motility: an overview. J Steroid Biochem Molec Biol 65:163–168.
8. Garcia M, Rochefort H (2000) Estrogen receptor targeted therapies of breast cancer. Cur Opin Oncol Endocr Metab Invest Drugs 2:60–67.
9. Garcia, M, Derocq D, Freiss G, et al. (1992) Activation of estrogen receptor into a receptor-negative breast cancer cell line decreases the metastatic and invasive potential of the cells. Proc Natl Acad Sci USA 89:1538–11542.
10. Long BJ, Rose DP (1996) Invasive capacity and regulation of urokinase-type plasminogen activator in estrogen receptor (ER)-negative MDA-MB-231 human breast cancer cells, and a transfectant (S30) stably expressing ER. Cancer Lett 99:209–215.
11. Platet N, Cunat S, Chalbos D, et al. (2000) Unliganded and liganded estrogen receptors protect against cancer invasion via different mechanisms. Mol Endocrinol 14:999–1009.
12. Laios I, Journe F, Laurent G, et al. (2003) Mechanisms governing the accumulation of estrogen receptor alpha in MCF-7 breast cancer cells treated with hydroxytamoxifen and related antiestrogens. J Steroid Biochem Mol Biol 87:207–221.
13. Green S, Kumar V, Theulaz I, et al. (1988) The N-terminal DNA-binding 'zinc finger' of the oestrogen and glucocorticoid receptors determines target gene specificity. EMBO J 7:3037–44.
14. Platet N, Prevostel C, Derocq D, et al. (1998) Breast cancer cell invasiveness: correlation with protein kinase C activity and differential regulation by phorbol ester in estrogen receptor-positive and -negative cells. Int J Cancer 75:750–756.
15. Thompson EW, Paik S, Brünner N, et al. (1992) Association of increased basement membrane-invasiveness with absence of estrogen receptor and expression of vimentin in human breast cancer cell lines. J Cell Physiol 150:534–544.
16. Marsden J, Backs NPM (1996) Hormone replacement therapy and breast cancer. Endocr Relat Cancer 3:81–97.
17. Collaborative group on hormonal factors in breast cancer (1997) Lancet 350:1047–1059.
18. Li CI, Malone KE, Weiss NS, et al. (2001) Tamoxifen therapy for primary breast cancer and risk of contralateral breast cancer. J Natl Cancer Inst 13:1008–1013.

Proteomic Analysis of Autocrine/Paracrine Effects of Human Growth Hormone in Human Mammary Carcinoma Cells

Cécile M. Vouyovitch, Laurent Vidal, Sahra Borges, Mireille Raccurt, Cécile Arnould, Jean Chiesa, Peter E. Lobie, Joël Lachuer, and Hichem-Claude Mertani

Summary Human growth hormone (hGH) is expressed by mammary epithelial cells and associated with proliferative disorders of the human breast. Our goal is to characterize the paracrine effects of hGH on morphological and functional changes of mammary carcinoma cells using MCF7 cells stably transfected with the hGH gene (MCFhGH). To identify the molecular actors involved in autocrine hGH-induced cell proliferation, we have used a protein chip technology using a commercial antibody microarray. The results enabled us to qualitatively characterize MCF-hGH cell's proteome from a panel of 500 proteins. Statistical analysis of variations in protein levels between the two cell lines did not highlight any significant differences. Thus, we concluded that variations in MCF-hGH proteome are more likely to reside in the activation status rather than drastic variations in the expression level of the 500 spotted proteins. To test this hypothesis, we confronted the protein chip result to the study of the regulation of the transcriptional factor Pax (Paired-box)-5 whose expression was not found to be altered on the protein chip. Surprisingly, we found that autocrine production of hGH in MCF7 cells was associated with a strong nuclear accumulation of Pax5 in a JAK2-dependent manner associated with an increase in Pax5-DNA binding activity. Our work indicates that subtle changes mediated by Pax5 are responsible for autocrine hGH-induced cell proliferation.

Introduction

The human growth hormone (hGH) is secreted episodically by pituitary somatotroph and exerts pleiotropic effects on postnatal growth, cellular differentiation, and metabolism. Extra pituitary synthesis of hGH has been highlighted in many cell types including human mammary epithelial cells and plays an important role in growth of the mammary gland (1). We have shown that hGH is upregulated in breast cancer with the highest level found in metastatic carcinoma cells and that de novo stromal expression of hGH is correlated to advanced stages of breast cancer (2). The oncogenic potential of hGH in mammary cells was later demonstrated (3, 4) and shown to be dependent on the activation of a transcriptional platform including

J.J. Li et al. (eds.), *Hormonal Carcinogenesis V.*

cell cycle activators (c-myc, cyclin D1) and antiapoptotic factors (Bcl-2) (5, 6). The aim of the present study was to characterize the proteomic changes induced by autocrine hGH in mammary carcinoma cells that could explain its role in tumor progression.

Results

We demonstrated that hGH protein was synthesized and secreted as expected by MCF-hGH cells (Fig. 1). MCF-MUT cells transcribe the *hGH* gene with the start codon mutated to a stop codon, which resulted in absence of the protein. Inhibition of JAK2 tyrosine kinase activity using AG490 did not modify hGH production but impeded its secretion by half (Fig. 1). The hGH protein was also detected at the correct molecular weight of 22 kDa, was not affected by AG490 (Fig. 2), and prominently localized in the perinuclear and Golgi area (Fig. 3).

Strong JAK2 protein phosphorylation was detected in MCF-hGH cells and was almost totally inhibited by AG490 (Fig. 4). Phospo-JAK2 exhibited a prominent plasma membrane ruffles and cytoplasmic localization (Fig. 5).

The number of live MCF-hGH cells was significantly higher than MCF-MUT cells at 24, 48, and 72 h culture in SF and in 10% FBS (Fig. 6a). hGH production by MCF7 cells counteracted the inhibition of cell proliferation induced by FBS deprivation. Addition of 10% FBS markedly increased the proliferation of MCF-hGH (Fig. 6b). The proliferation of MCF-MUT cells was also increased by 10% FBS; however, autocrine hGH and FBS synergistically activated cell proliferation (Fig. 6b). AG490 exerted inhibitory effects on MCF-hGH cell proliferation (Fig. 6c). This reduction in cell number induced by AG490 indicates its cytotoxic effect,

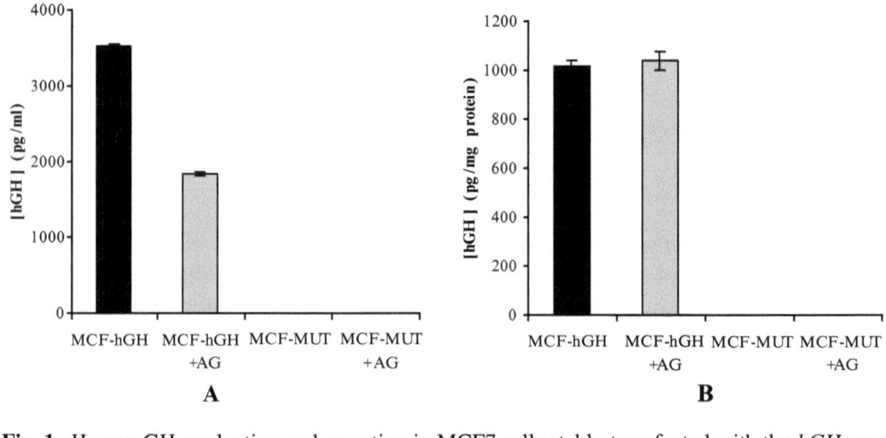

Fig. 1 Human GH production and secretion in MCF7 cells stably transfected with the *hGH* gene (MCF-hGH). MCF-MUT cells were used as negative control. hGH was measured by Elisa in serum free (SF) media with or without Jak2 inhibitor AG490 (24 h). The results represent: (**A**) the concentration of hGH synthesis and (**B**) the secretion into the medium

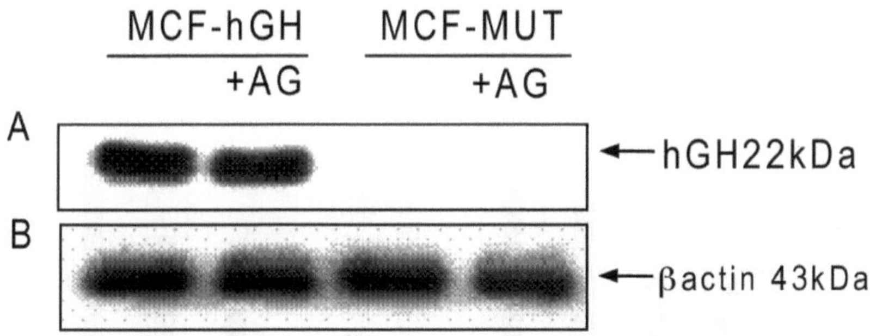

Fig. 2 (**A**) Western blot analysis of hGH expression in MCF-hGH and MCF-MUT cells grown in SF media or supplemented with AG490 (25 μg ml^{-1}, 24 h) (+AG). (**B**) β actin detection was used as loading control

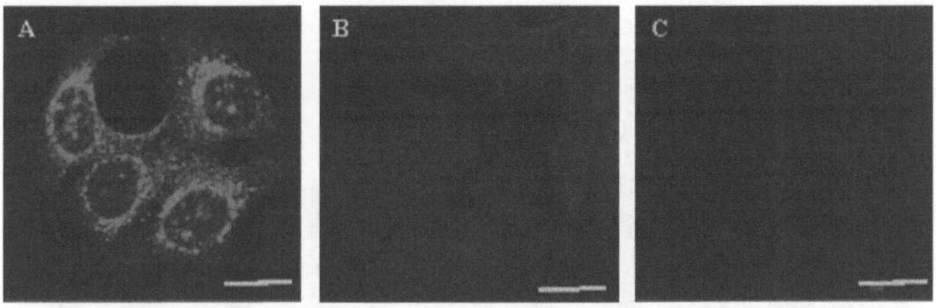

Fig. 3 Confocal detection of hGH expression in (**A**) MCF-hGH cells and (**B**) MCF-MUT cells. Lack of hGH in MCF-hGH cells when the hGH antibody was preincubated with excess recombinant hGH (**c**). Bar = 10 μm

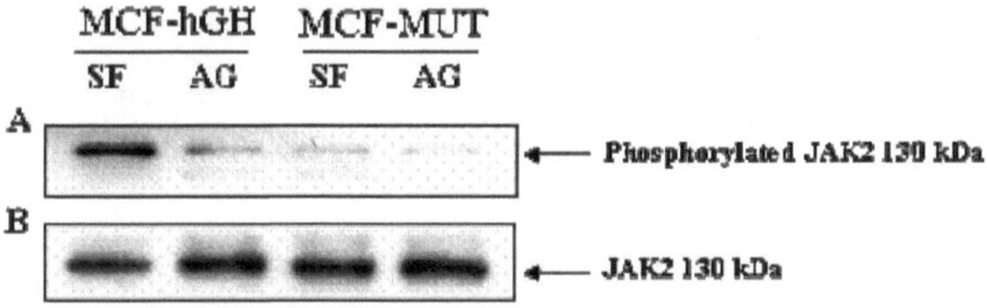

Fig. 4 Western blot analysis of the effects of autocrine hGH on the level of (**A**) phosphor-JAK2 protein in MCF-hGH and (**B**) MCF-MUT cells grown in SF +- AG490 (25 μg ml^{-1}, 12 h). JAK2 loading control

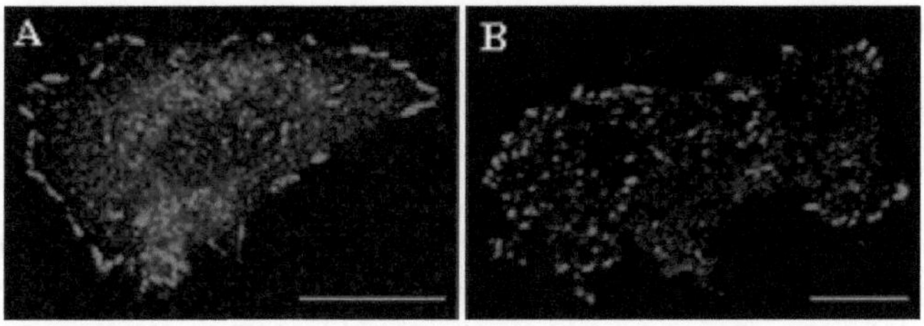

Fig. 5 Level and localization of (**A**) phospho-JAK2 in MCF-hGH and (**B**) MCF-MUT cells. Bar = 10 μm

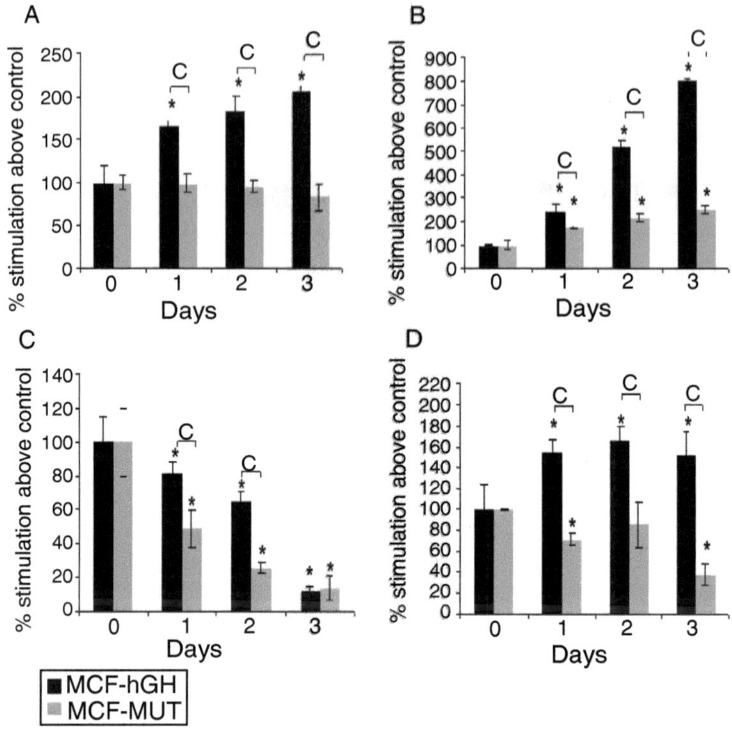

Fig. 6 Proliferation rate of MCF-hGH and MCF-MUT cells expressed as percentage of cell number at the beginning of the experiment (Day 0). Live cells were quantified using the MTS technique in: (**a**) SF medium or in (**c**) SF medium supplemented with 25 µg ml^{-1} AG490, (**b**) in medium supplemented with 10% FBS, or (**d**) in medium supplemented with 10% FBS and 25 µg ml^{-1} AG490

particularly in MCF-MUT cells, leading to 80% cell death at 72 h (Fig. 6c). Autocrine production of hGH by MCF-hGH cells delayed the cytotoxic effect of AG490 on the first day of treatment but did not prevent the drastic reduction in cell number at days 2 and 3. When MCF-hGH cells were grown in a medium supplemented with 10% FBS, AG490 could not completely block their proliferation (Fig. 6d), indicating that FBS stimulation of MCF-hGH cells enable cell renewal and counterbalance the cytotoxic effects of AG490. However, FBS stimulation is partially efficient in counteracting the cytotoxic effects of AG490 in absence of autocrine hGH (Fig. 6d).

We used a commercial antibody microarray (Clontech®) to identify and compare protein expression pattern of autocrine hGH in MCF7 cells ± AG490. Proteins from MCF-hGH and MCF-MUT cells were labeled with cyanine 3 and cyanine 5 fluorochromes and incubated on the chip spotted with 500 antibodies. Fluorescence intensity of the green and red spots that indicates relative level of protein expression was scanned (Axon GenePix 4000B) and converted using the Ab microarray analysis workbook (Microsoft Excel® 97/98, provided by the manufacturer) into an internally normalized ratio (INR) for each antigen on the array (Fig. 7). This value represents the abundance of a given antigen in one sample relative to that of another sample. As recommended by the manufacturer, it was necessary to calculate the mean INR of every spot and then multiply it by 1.3 to obtain the upper threshold value and by 0.77 to obtain the lower threshold value.

Image analysis of the fluorescent spots showed that approximately 80% of the 500 antibodies printed on the slide reacted with cellular antigens extracted from MCF7 cells with variable intensity, but all spots displayed a yellow fluorescence that indicates equal amount of proteins between the two cell lines (Fig. 8). Besides, the statistical microarray analysis show that none of the calculated INR were ≤ 0.77 or ≥ 1.33 indicating no significant differences between protein expression of MCF-hGH and MCF-MUT cells. Thus, quantitative variations in MCF-hGH proteome are not detectable using this Ab microarray.

The stimulation of Pax5 transcriptional activity by autocrine hGH was first demonstrate using the TranSignal Protein/DNA Arrays (Panomics ®). Using this system, we demonstrate that MCF-hGH cells higher DNA binding activity for Pax5 (Fig. 9). Protein expression level of Pax5 was not affected by autocrine production of hGH as shown on the protein chip (Fig. 8) and subsequent Western blot experiment on whole cell protein extract demonstrated equal levels of Pax5 in both MCF-hGH and MCF-MUT cells (data not show). We next studied the localization of Pax5 by confocal microscopy and found a prominent nuclear localization with a higher intensity in MCF-hGH cells (Fig. 10).

Pax5 was identified as a protein with the expected molecular weight of 50 kDa. Identical level of Pax5 protein was observed in the cytoplasmic fractions of each

A

Name	Localization	Median F635-B635	Median F634-B634	Ratio 1= (ACy5/BCy3)	Ratio 2= BCy5/ACy3	Ratio1/Ratio2	INR
Pax5	B4a5/b5	3683	4953	0,743589744	0,580482456	1,27	1,13

B

Lower threshhold	INR average	Upper threshold
0,780746623 ← x0.77	1,013956653	x1.33 → 1,318143649

Fig. 7 Statistical analysis of the fluorescent signals obtained on the protein chip. (**A**) Calculation of Pax5 INR carried out with the Excel® software «Analysis Workbook» from raw data corresponding to the fluorescence intensity. Ratio 1 and 2 are calculated from median values of the intensity of fluorescence (F635, F532) for Pax5, measured at 635 nm (Cy3) and at 532 nm (Cy5) from which the background noise was subtracted (B635, B532). The INR corresponds to the √ (ratio1/ratio2). (**B**) Threshold values of INR, indicating significant differences in protein expression, were calculated from the mean of the 500 INR

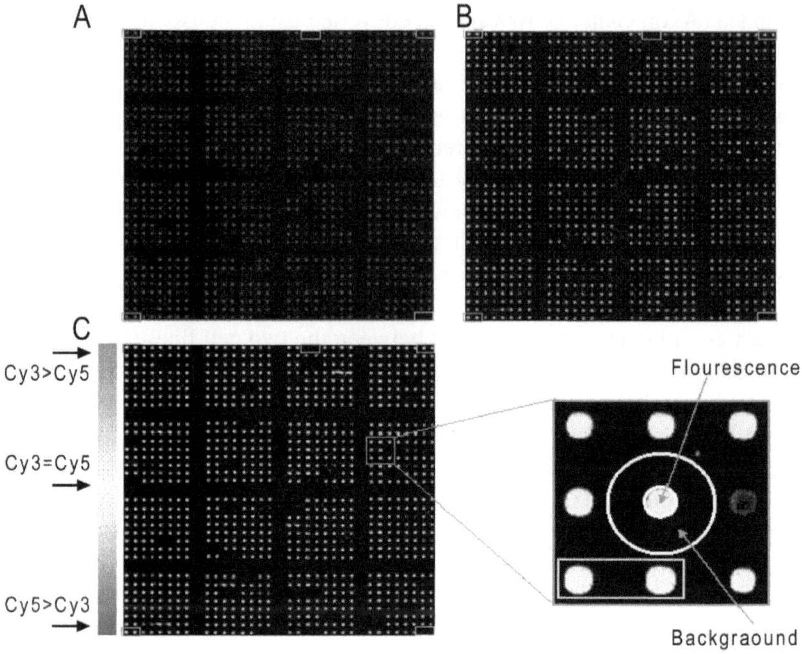

Fig. 8 Protein expression pattern observed in MCF-hGH and MCF-MUT cells by use of Ab Microarray (Clontech®). (**A**) Raw image obtained with wavelength scan of Cy5. (**B**) Raw image obtained with wavelength scan of Cy3. (**C**) Superimposed raw images. The two spots framed on the enlargement on the right correspond to Pax5 protein detection. The slide contains unlabeled BSA as negative control (□)and BSA prelabeled with Cy3 and Cy5 (□)as orientation markers

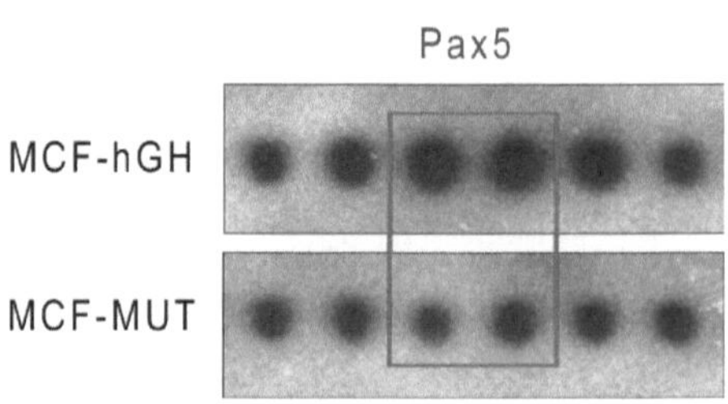

Fig. 9 Autocrine effects of hGH on transcription factor activity using TranSignal Protein/DNA Array (Panomics ®). The frame corresponds to Pax5 protein activity detected in MCF-hGH vs. MCF-MUT cells

cell type, while autocrine hGH induced a strong nuclear accumulation of Pax5 (Fig. 11). Inhibition of JAK2 activity induces a loss in nuclear accumulation of Pax5 in MCF-hGH cells without altering basal level of nuclear Pax5 in MCF-MUT cells (Fig. 12).

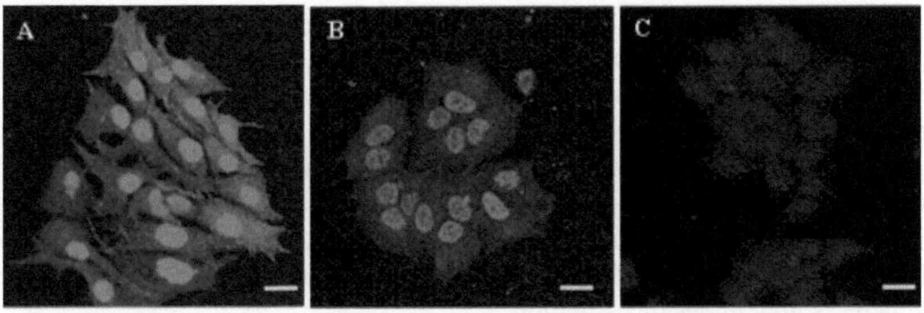

Fig. 10 Confocal laser scanning microscopic analysis of the effect of autocrine hGH on the level of (**A**) Pax5 protein in MCF-hGH and (**B**) MCF-MUT cells. Cells were grown in SF media one night, fixed, permeabilized before proceeding to IF detection, (**C**) control was performed in the absence of primary antibody on MCF-hGH cells. Bar = 10 μm

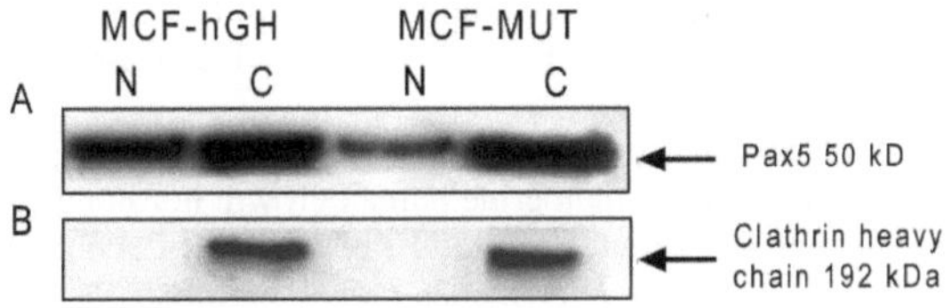

Fig. 11 (**A**) PAX5 protein detection in nuclear (N) and cytoplasmic fractions (C) in cells grown in SF. (**B**) The membrane was stripped and blotted for clathrin detection to demonstrate equivalent loading and quality of cell fractions

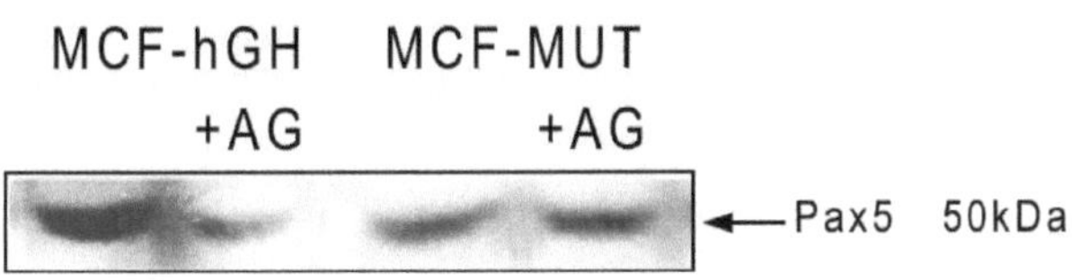

Fig. 12 Western blot analysis of AG490 treatment (+AG, 25 μg ml^{-1}, 12 h) on Pax5 levels in the nuclear fraction of mammary carcinoma cells MCF-hGH and MCF-MUT grown in SF media

Discussion

We have used a proteomic approach to identify molecular targets of autocrine hGH in human mammary carcinoma cells. The synthesis and release of hGH has been detected in MCF-hGH cells at the concentration of 1 000 pg ml^{-1} in cells and 3 500 pg ml^{-1} in the medium. We analyzed the effects of AG490, a JAK2 tyrosine kinase inhibitor on the amount of hGH produced and released. Our results show that hGH production is not altered by AG490; however, it does reduce hGH secretion by 50%. Thus, AG490 partly disturbs the release of autocrine hGH. We next show that autocrine hGH exerts proliferative effects on MCF7 cells and is protective against the cytotoxicity induced by sustained presence of AG490, and we sought to elucidate the underlying molecular mechanisms. Surprisingly, statistical analysis of the Ab microarray results did not show significant differences between the quantity of proteins expressed in MCF-hGH and MCF-MUT cells. The absence of a range of antibody concentration on the chip indicates that it might not be sensitive enough to detect subtle differences in protein levels. This unexpected result prompted us to test the hypothesis that proteome modifications are more likely to reside in the activation status of signaling proteins. Accordingly, we demonstrate that DNA binding activity of Pax5 transcription factor is stimulated by autocrine hGH while no variation in Pax5 protein expression is found on the protein chip or by Western blot. We have found that nuclear accumulation of Pax5 in MCF-hGH cells is also JAK2 dependent. The role of Pax5 in the hyperproliferative state induced by autocrine hGH is now under investigation.

References

1. Mukhina S, Liu D, Guo K, et al. (2006) Autocrine growth hormone prevents lactogenic differentiation of mouse mammary epithelial cells. Endocrinology 147:1819–1829.
2. Raccurt M, Lobie PE, Moudilou E, et al. (2002) High stromal and epithelial human *GH* gene expression is associated with proliferative disorders of the mammary gland. J Endocrinol 175:307–318.
3. Zhu T, Starling-Emerald B, Zhang X, et al. (2005) Oncogenic transformation of human mammary epithelial cells by autocrine human growth hormone. Cancer Res 65:317–324.
4. Mukhina S, Mertani HC, Guo K, et al. (2004) Phenotypic conversion of human mammary carcinoma cells by autocrine human growth hormone. Proc Natl Acad Sci USA 101:15166–15171.
5. Mertani HC, Zhu T, Goh EL, et al. (2001) Autocrine human growth hormone (hGH) regulation of human mammary carcinoma cell gene expression. Identification of CHOP as a mediator of hGH-stimulated human mammary carcinoma cell survival. J Biol Chem 276:21464–21475.
6. Zhang X, Zhu T, Chen Y, et al. (2003) Human growth hormone-regulated HOXA1 is a human mammary epithelial oncogene. J Biol Chem 278:7580–7590.

Endometrial Cancer

Expression Analysis of Human Endometrial Adenocarcinoma in an Inbred Rat Model

Sandra Karlsson and Karin Klinga-Levan

Summary Endometrial cancer (EC) is the most abundant female gynaecologic malignancy, ranking fourth in incidence among invasive tumors in women. Hormone-related (estrogen-dependent) EC is the prevalent subtype and accounts for approximately 75% of these cancers. Females of the BDII inbred rat strain are extremely prone to endometrial adenocarcinoma, (EAC) and approximately 90% of virgin females spontaneously develop EAC during their life span. Thus, these rats serve as a useful model for the genetic analysis of this malignancy. In the present work, gene expression profiling, by means of cDNA microarrays, was performed on cDNA from endometrial tumor cell lines and from cell lines derived from nonmalignant lesions/normal tissues of the endometrium without specific findings (WSF). We identified numerous genes differentially expressed between endometrial cell lines and WSFs employing clustering analysis and statistical inference analysis. Many of the genes identified are located within or close to the chromosomal regions earlier identified to be associated with EAC susceptibility and development. Several of the genes identified are involved in pathways commonly altered in carcinogenesis, such as the TGF-pathway.

Introduction

Endometrial cancer (EC) develops from the endometrium of the uterus, a tissue that is the target for the steroid hormones produced by the ovaries. It is the most common pelvic malignancy diagnosed in women in the western countries. There are two major categories of endometrial tumors: Type I tumors are estrogen-dependent endometrioid carcinomas (~80%), which are frequently preceded by complex atypical hyperplasias. Endometrial adenocarcinoma (EAC) originates from the glandular cells of the surface epithelium/endometrium and is the most common type I tumor (75%). There are clear indications that genetic predisposition contributes to human EAC pathogenesis, probably involving both high- and low-penetrance genes (1).

The genetic heterogeneity present in the human population and the influence of environmental factors is a major problem when analyzing complex patterns of genetic interactions involved in human carcinogenesis. One approach to reduce the

J.J. Li et al. (eds.), *Hormonal Carcinogenesis V.*

variation in the genetic and the environmental background is to apply an inbred animal model system, thus the identification of genes involved in the progression of different experimental tumors will be significantly facilitated.

In the BDII/Han strain, the incidence of spontaneously developed EACs is very high. More than 90% of EC during their life span among which the majority of the tumors are EAC. The endometrial carcinogenesis in female BDII/Han rats is hormone-dependent and thus represents an excellent model for spontaneous hormonal carcinogenesis in human. The BDII rat model, first described in 1987, is genetically well characterized and has become a recognized model for understanding the genetic elements involved in the endometrial tumorigenesis (2–4).

The present investigation aims to identify differences in the expression pattern between endometrial tumors and normal or premalignant tissues from the endometrium by means of cDNA microarrays. In this manner, we may be able to identify changes in gene expression occurring in the early stages of transformation and hence facilitating further investigations for elucidating the underlying genetic mechanisms of EAC development. Earlier work within this rat model system by means of genome wide screens with microsatellite markers has revealed several regions associated with EAC susceptibility (RNO1q35–36, RNO11q23, RNO17p11–q11, and RNO20p12). From these studies, we concluded that the onset of tumors depend on presence of several susceptibility genes with minor and cooperating effects and by the contribution of genetic components derived from nonsusceptible strains (5, 6). Previous studies also included studies of chromosomal aberrations that occurred in tumors by means of comparative genome hybridization (CGH) in the cross progenies. Characteristic chromosomal changes, such as amplifications and deletions, were scanned for in the tumors that appeared in the progeny by CGH. From these studies, it could be concluded that certain chromosomal regions were recurrently engaged in changes such as increases in copy number (e.g., hyperploidy and amplifications) or decreases (e.g., hypoploidy and deletions) (7–9). The combination of analysis on both the genomic and transcriptional level provides us with a more complete picture of the mechanisms underlying EAC etiology and hence facilitates further investigations on the gene function and the phenotype level.

Animal Crosses and Tumor Material

Crosses between BDII females and two nonsusceptible Brown Norway (BN) or Sprague–Dawley-curly3 (SPRDcu) 3 males were made to produce an F1 progeny. The F1 progeny was subsequently backcrossed to BDII females to produce an N1 generation. Female progeny with suspected tumors were euthanized and tumors were surgically removed and subsequently subjected to pathological characterization. Tumor tissues were collected for DNA extraction and cell culture establishment.

The tumors that developed in the N1 progeny were pathologically classified either as EAC or as other uterine tumors (Table 1). In some cases, no cancer cells were detected in the removed cell mass when pathologically analyzed. These tumors were referred to as “without specific findings” (WSF).

Table 1 Tumor classification

Animal group	BDIIxBN BC (N1)	BDIIxSPRDcu3 BC (N1)	Total	No of cell cultures
Total no. of animals	116	110	226	
Animals with tumors diagnosed as EAC	26	29	55	38
Animals with tumors diagnosed as WSF	60	50	110	42
Animals with tumors diagnosed as other uterine tumor	10	13	23	19

In Vitro Cell Culturing. Cell cultures established from endometrial tumors and other endometrial cell types were cultured in Dulbecco's modified Eagle medium supplemented with 10% heat-inactivated fetal bovine serum (WSF cell lines was cultured in medium containing 20% serum), 100 IU/100 μg ml^{-1} penicillin/streptomycin, L-glutamine, MEM amino acids, MEM nonessential amino acids, MEM vitamins solution. The cells were grown at 37°C in an atmosphere of 95% humidity and 5% CO_2. The cells were harvested by trypzination at 80–90% confluence (~1×10^6 cells).

RNA Extraction. Total RNA samples were automatically extracted from the different cell lines in culture with a KingFisher Instrument (Thermo Electron Corporation) according to the manufacturer's protocol (MagAttract Tissue Mini M48 Kit, Qiagen). RNA was quantified using a NanoDrop (NanoDrop technologies). RNA quality was assessed using a 2100 Bioanalyser instrument (Agilent Technologies) and/or run on a 1.5% agarose gel.

Reverse Transcription and Labeling. cDNA was synthesized and labeled with the green fluorescent dye, Cy3 (Amersham) from 5 μg of each total RNA from the tumor samples, using chipshot reverse transcriptase (pronto! PLUS direct labeling kit v1.2, Corning). Universal rat reference RNA (Stratagene) was used as a common RNA reference in all hybridizations. For each hybridization, 5 μg of reference RNA was used for cDNA synthesis and labeling with the red fluorescent dye, Cy5 (Amersham). Labeled cDNA was purified and subsequently quantified using the NanoDrop. Then 60–80 pmol Cy3-labeled cDNA and 45–60 pmol Cy5-labeled cDNA were pooled and dried using a DNA120 speedvac system (Thermo Electron Corporation).

Design of the Microarray Study. For this study, we have employed the two-channel cDNA microarray format. The 18 K (6 000 clones in triplicates) rat 70mer oligonucleotide arrays used were printed at the Swegene DNA microarray resource center in Lund, Sweden, by a BioRobotics MicroGrid 2 arrayer (Cambridge, UK) together with a split pin system. Each probe in the probe set (rat 70mer oligonucleotide set, ver 1.0, OPERON) was printed in triplicates at random positions on the arrays.

Microarray Hybridization. Prior to hybridizations, the slides were subjected to presoaking, dehydration, and prehybridization, according to the manufacturer's protocol (pronto! PLUS direct labeling v1.2, Corning). Pooled and dried samples were cohybridized to cDNA microarrays in Corning hybridization chambers in Pronto Hybridization buffer (Corning Life Sciences). Hybridization was performed at 42°C for 16–20 h in a volume of 40 μl after which the slides were washed and dried prior to scanning in an Agilent Microarray scanner (Agilent technologies).

Statistical Analysis. After image acquisition and analysis, the microarray data were normalized and the corresponding arrays were grouped according to annotations (EAC and WSF) and a *t* test was performed to calculate the probability that there is a significant difference for reporter *x* between two annotation groups. The significance level was set to 0.05. For exploratory data analysis, hierarchical clustering of both arrays and genes was performed. Euclidean distance was used for the array distance metric and the Pearson correlation coefficient was used for the gene distance metric. Average linkage was used as the linking distance.

Results and Discussion

The objective of this study was to define a pattern of gene expression that would serve as a signature discriminating between early and late stages of EAC. With this approach we may be able to identify gene expression aberrations essential to EAC carcinogenesis.

Design of the Microarray Studies. We have used a universal rat reference RNA constituting RNA from a mixture of 14 cell lines from different tissues as an internal control for all the microarray experiments. This common reference design allows us to compare the expression profiles of the different groups of endometrial cell lines against each other but, in addition, it also allows the comparison of the normal, premalignant, and malignant cell lines against the rat reference. The most desirable and natural choice of a reference would be to use the animal's normal matching tissue. This approach may, however, be intricate for several reasons: the expression pattern may vary substantially over time (there are for example major cyclic changes in gene expression in the ovaries and the uterus during different phases of the menstrual cycle), sufficient quantities of normal endometrium are technically hard to retrieve and the normal or adjacent tissue might actually contain cancerous cells. Moreover, by using universal rat reference as a standardizing reference for all hybridizations, a large number of genes will be expressed on the array and thus hybridization signals at the majority of the spots will be obtained.

Hierarchical Clustering and Statistical Analysis of the Expression Data. Previous work within the BDII rat model system has been focused on finding chromosomal regions associated with EAC susceptibility and development by associations with microsatellites, CGH, and allelic imbalance studies. These efforts have resulted in the identification of several chromosomal regions with abnormalities/changes such as deletions and amplifications. This information allows us, in addition to the global expression analysis, to put specific emphasis to those regions and obtain a more complete picture of the changes that have occurred.

We believe that the tissues classified as WSFs might be premalignant hyperplasias or normal endometrium tissue, which in either case provides us with a valuable material when trying to elucidate the EAC susceptibility and development/progression from normal to early to late stage. As expected, the hierarchical clustering of the microarray data showed a clear segregation of the expression pattern of the endometrial tumors and WSFs (Fig. 1). However, we observed one WSF cell line (NUT118) clustering together with the EACs, and therefore seem to share similar expression profile with the tumors. This implies that the WSFs cell lines might be premalignant hyperplasias and in different phases of differentiation, which also is supported by the differences in morphology and physiology of the cells observed when cultured in vitro.

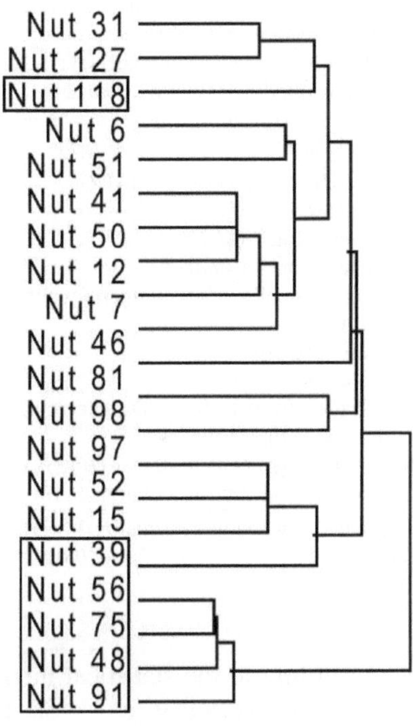

Fig. 1 Dendrogram of the arrays from the hierarchical cluster analysis. The WSF cell lines are marked with a *red rectangle*. NUT denotes rat uterine tumor that was developed in the back cross progeny

Table 2 Genes significantly differentially expressed between EACs and WSFs

Gene Name	Chr	Direction of change in tumors vs. WSF	*p*-value
Tgfβ pathway associated genes			
TGFβ1-induced transcript 1 (*Tgfb1i1*)	**1q36**	⇓	0.003**
TGFβ 3 (*Tgfb3*)	6q31	⇓	0.001**
Cyclin-dependent kinase inhibitor 2B (p15) (*Cdkn2b*)	5q32	⇑	0.0005***
Latent TGFβ binding protein (*Ltbp2*)	6q31	⇓	0.002**
Inhibin β a chain precursor (*Inhba*)	**17q12.1**	⇓	0.0005***

(continued)

Table 2 (continued)

Gene Name	Chr	Direction of change in tumors vs. WSF	*p*-value
Smad1	19q11	⇑	0.03*
Cyclin-dependent kinase inhibitor 1A (*Cdkn1a*, p21)	**20p12**	⇓	0.01*
Collagen. type V. alpha 1 (*Col5a1*)	3p12	⇓	0.03*
Collagen. Alpha 1 (I) chain (fragments) (*Col1a1*)	**10 (±)**	⇓	0.0005***
Procollagen. Type II. alpha 1 (*Col1a2*)	No info	⇓	0.001**
Fibronectin 1 (*Fn1*)	9q33	⇓	0.01*
Connective tissue growth factor (*Ctgf*)	1p12	⇓	0.002**
Collagen alpha 1 (III) chain (*Ca13_rat*)	No info	⇓	0.007**
Wnt pathway associated genes			
Wnt 1 inducible signaling pathway protein 2 (*Wisp2*)	3q42	⇓	0.002**
Wingless-related MMTV integration site 4 (*Wnt4*)	**5q36** (–)	⇓	0.007**
Dickkopf homolog 3 (Xenopus laevis) (*Dkk3*)	**1q33**	⇓	0.01*
C-terminal binding protein 2 (*Ctbp2*)	No info	⇑	0.002**
Beta-catenin (*Catnb*)	No info	⇑	0.05*
Casein kinase 1 gamma 1 (*Csnk1g1*)	8q24	⇑	0.04*
Mismatch repair system (MMR)			
DNA ligase I (*Lig1*)	1q12	⇑	0.02*
MutS homolog 2 (*Msh2*)	**6q12** (+)	⇑	0.03*
Menage a trois 1 (*Mnat1*)	6q24	⇓	0.005**
Proliferating cell nuclear antigen (cyclin) (*PCNA*)	3q36	⇓	0.003**
Apoptosis			
Caspase 2 (*Casp2*)	**4q23** (+)	⇑	0.0005***
Caspase 7 (*Casp7*)	1q55	⇑	0.0005***
BCL2/adenovirus E1B 19 kDa-interacting protein 3-like (*Bnip3l*)	**15p12** (–)	⇓	0.0005***
Inhibitor of apoptosis protein 1 (*Birc3*)	8q11	⇑	0.02*
Baculoviral IAP repeat-containg 4 (*Birc4*)	Xq11	⇑	0.01*

The data (prin-tip loess normalized) were analyzed by performing a *t* test (the significant level was set to 0.05, $*p < 0.05$, $**p < 0.01$, $***p < 0.001$, and $****p < 0.0001$) and hierarchical clustering

According to the statistical inference analysis, ~800 genes were differentially expressed between EAC and WSF. Table 2 represents a list of genes associated with pathways known to be related to carcinogenesis and that were identified to be differentially expressed between EACs and WSFs.

Genes included in the table were primarily selected by visually examining the expression profiles from the cluster analysis. The gene names, chromosome location (Chr), and *p*-values are listed in the table. Chromosome locations identified in CGH analysis are bold where [(+) denotes a gain of the specific region and (–) a loss] and chromosome regions identified as susceptibility regions are denoted red.

The results imply that several pathways related to carcinogenesis are distorted in the majority of the endometrial tumors. Several of these genes are located within or close to the chromosome regions earlier identified to be associated with EAC susceptibility and development. However, these results need to be verified by additional traditional expression analysis methods and further experimentation on the functional level of carcinogenesis. Presently, we are in the process of verifying the results from the microarray data analysis and we are also collecting and analyzing more microarray data from EAC tumors, normal/premalignant lesions and other uterine tumors. The results from these studies facilitate further experimentation on the phenomic level of the specific genes and cellular pathways identified to be disrupted.

Acknowledgments This work was supported by the Swedish National Research School in Bioinformatics and Genomics and by a financial support from the Nilsson-Ehle Foundation. We also thank the Swegene DNA Microarray Resource Center for their excellent services and support.

References

1. Gruber SB, Thompson WD (1996) A population-based study of endometrial cancer and familial risk in younger women. Cancer and steroid hormone study group. Cancer Epidemiol Biomarkers Preview, 5(6): p. 411–417.
2. Deerberg F, Kaspareit J (1987) Endometrial Carcinoma in BDII/Han rats: model of a spontaneous hormone-dependent tumor. J Natl Cancer Inst, 78: p. 1245–1251.
3. Kaspareit-Rittinghausen J, Deerberg F, Rapp K (1987) Mortality and incidence of spontaneous neoplasms in BDII/Han rats. Z Versuchstierkd, 30(5–6): p. 209–216.
4. Vollmer G (2003) Endometrial cancer: experimental models useful for studies on molecular aspects of endometrial cancer and carcinogenesis. Endocr Relat Cancer, 10(1): p. 23–42.
5. Roshani L, et al. (2005) Genetic analysis of susceptibility to endometrial adenocarcinoma in the BDII rat model. Cancer Genet Cytogenet, 158(2): p. 137–41.
6. Roshani L, et al. (2001) Genetic identification of multiple susceptibility genes involved in the development of endometrial carcinoma in a rat model. Int J Cancer, 94(6): p. 795–799.
7. Helou K, et al. (2000) Comparative genome hybridzation (CGH) analysis in rat uterine endometrial carcinoma. Rat Genome, 6: p. 78.
8. Helou K, et al. (2001) Analysis of genetic changes in rat endometrial carcinomas by means of comparative genomic hybridization. Cancer Genet Cytogenet, 127(2): p. 118–127.
9. Hamta A, et al. (2005) Cytogenetic aberrations in spontaneous endometrial adenocarcinomas in the BDII rat model as revealed by chromosome banding and comparative genome hybridization. Cancer Genet Cytogenet, 159: p. 123–128.

Amplification Studies of MET and Cdk6 in a Rat Endometrial Tumor Model and Their Correlation to Human Type I Endometrial Carcinoma Tumors

Emma Samuelson, Carola Nordlander, Göran Levan, and Afrouz Behboudi

Summary Cancer is known to be a genetic disease that is both polygenic and heterogeneous, in most cases involving changes in several genes in a stepwise fashion. The spectrum of individual genes involved in the initiation and progression of cancer is greatly influenced by genetic factors unique to each patient. A study of complex diseases such as cancer is complicated by the genetic heterogeneous background and environmental factors in the human population. Endometrial cancer (EC) is ranked fourth among invasive tumors in women. In Sweden, approximately 1300 women (27/100,000 women) are diagnosed annually. To be able to study the genetic alterations in cancer, the use of an animal model is very convenient. Females of the BDII strain are genetically predisposed to EC and 90% of female BDII rats develop EC during their lifetime. Thus, BDII rats have been used to model human EC with respect to the genetics of susceptibility and of tumor development. A set of rat EC tumors was analyzed using conventional cytogenetics and comparative genome hybridization (CGH). Chromosomal aberrations, i.e., gains, were found on rat chromosome 4 (RNO4). Using FISH analysis, we concluded that the *Met* oncogene and *Cdk6* (cyclin-dependent kinase 6) were amplified in this set of EC tumors. The data from this investigation were used to analyze a set of human endometrial tumors for amplification of *Cdk6* and *Met*. Our preliminary data are indicative for a good correlation between our findings in the BDII rat model for EAC and the situation in human EC. These data provide strong support for the use of animal model systems for better understanding and scrutinizing of human complex disease of cancer.

Introduction

Cancer may occur in any tissue of the human body and can be divided into at least 100 subtypes. Cancer is classified according to organ location and tissue and cell type where it started. Most tumors develop from a single cell with the acquired ability to uncontrolled growth often within a tissue. Tumorigenesis and tumor progression are processes involving accumulation of genetic changes in a stepwise fashion. The onset of this process is not clear; however, it may depend on genetic predisposition as well as additional somatic mutations due to environmental or

J.J. Li et al. (eds.), *Hormonal Carcinogenesis V.*

stochastic factors. When the tumor cells acquired the ability to invade and colonize neighboring and distant tissues in the body, metastases develop. The tumor cells are genotypically and phenotypically heterogeneous, and they are genomically unstable. The most common types of genomic instability are microsatellite (MIN) and chromosomal instabilities (CIN). MIN tumors often contain a diploid karyotype, but harbor severe nucleotide sequence instability, while CIN tumors develop an aneuploid karyotype. CIN tumors contain gross chromosomal alterations such as loss of whole or parts of chromosomes, translocations, and amplifications. An amplification results from the multiplication of a DNA segment into several copies. There are several different mechanisms leading to amplification and often a specific gene is amplified on the bases of its properties, i.e., oncogenes. Amplification is cytogenetically visible as elongated chromosomes containing homogeneously stained regions (HSR) and/or extra chromosomal material known as double minutes (DM). Amplifications may lead to over expression of the gene(s) involved. Recent advances in cytogenetic and molecular genetics has shown that different types of changes, such as amplifications, might be used to improve clinical classification and to establish novel biological classification to individualize treatments for patients.

Analysis of human cancer genetics is often complicated as two pathologically identical tumors are likely to differ in the genotype and origin of the tumors. This complexity of cytogenetic and molecular level springs from the inherent genetic heterogeneity of the human population. It is also caused by the influences from environmental differences, life style, and diet of each individual.

The rat has been used as a model for complex human diseases for decades. In cancer research, the rat has been used to model fibrosarcoma (1), breast cancer (2), and endometrial cancer (EC) (3). The BDII rat is genetically prone to develop endometrial tumors and 90% of the BDII females develop EC during their life span spontaneously. The main types of tumors are endometrial adenocarcinoma (EAC). This strain is, therefore, used in this project to elucidate the genetic alterations occurred during EAC tumorigenesis and also to analyze pathways involved in initiation and progression of EAC tumors.

EC, also known as carcinoma of the endometrium or corpus cancer, is ranked forth among invasive cancers in women and is the most frequently diagnosed malignancy in the female reproductive tract (4). However, the underlying molecular genetic events involved in EAC initiation and progression is poorly understood.

In the present project, genetic crosses between the susceptible BDII females and rats from the resistant SPRD or Brown Norway (BN) strains have been performed, followed by subsequent intercrosses and back crosses, and a population of rats with various degrees of susceptibility to spontaneously develop EAC was developed. Tumors from a number of intercross and backcross animals were subjected to genome-wide screening analysis (CGH), which revealed several common genetic changes in these tumors (5). One such chromosomal aberration was recurrent gains of rat chromosome 4 (RNO4), predominantly involving the cytogenetic bands RNO4q12 to q22. To further analyze the nature and extent of the RNO4 gains in EAC tumors, and to narrow down the target region, we analyzed tumors by fluorescent in situ hybridization (FISH) experiments. RNO4-specific chromosome paint and gene-specific probes were used to pinpoint the target oncogene(s) involved in the amplification events. The methodology for these studies has been described earlier (6, 7).

Result and Discussion

Twenty-two EAC tumors, one peritoneal mesothelioma (PM), and one endometrial squamous cell carcinoma (ESCC) were analyzed in this study. All 24 tumors were subjected to karyotyping and the analysis showed that the tumor chromosome number varied from near diploid to near tetraploid ($2n = 42–85$). There was a considerable variation regarding the chromosome number as shown in Table 1. In some

Table 1 RNO4-specific chromosome paint analysis in rat EAC, peritoneal mesothelioma (PM), and endometrial squamous cell carcinoma (ESCC)

Cross BDII crossed with	Tumor type	Tumor name	Modal Chr. No	No. of expected RNO4	No. of positive RNO4 fragment	Translocation or HSR chromosome	No translocation
SPRD[3]	EAC	NUT6	62	2.9	4	0	4
SPRD[3]	EAC	NUT11	75	3.6	6	0	6
SPRD[3]	EAC	NUT55	44	2.1	3	0	3
BN[3]	EAC	NUT52	42	2.0	4	0	4
BN[3]	EAC	NUT82	55	2.6	4	0	4
SPRD[3]	EAC	NUT84	42	2.0	2	0	2
BN[3]	EAC	NUT128	64	3.0	4	0	4
BN[3]	EAC	NUT97	52	2.5	4	0	4
BN[3]	EAC	NUT98	48	2.3	4	0	4
BN[3]	EAC	NUT14	42	2.0	4	0	4
BN[3]	EAC	NUT42	44	2.1	4	0	4
BN[3]	EAC	NUT46	60	2,3	3	0	3
SPRD[3]	EAC	NUT47	48	2.3	3	0	3
SPRD[3]	EAC	NUT62	60	2.8	3	0	3
BN[3]	EAC	NUT76	57	2.7	3	0	3
BN[3]	EAC	NUT4	62	3.0	4	1	3
BN[3]	EAC	NUT7	80	3.8	4	1	3
		Subpop.	85	4.0	5	2	3
BN[2]	MPM	RUT29	44	2.1	4	2	2
BN[3]	EAC	NUT12	58	2.8	5	2	3
BN[2]	ESCC	RUT5	75	3.5	4	2	2
BN[1]	EAC	RUT7	43	2.0	4	2	2
SPRD[3]	EAC	NUT50	71	3.4	5	3	2
BN[3]	EAC	NUT100	64	3.0	5	3	2
SPRD[1]	EAC	RUT13	62	3.0	5	4	1

Subpop, a common subpopulation in the NUT7 tumor
[1] F1 (first intercross progeny)
[2] F2 (second intercross progeny)
[3] BC (back cross progeny)

cases, the chromosome number in different metaphases of the same tumor varied. Therefore, the average values are presented in Table 1.

RNO4-specific Chromosome Paint. Tumors were further analyzed using RNO4-specific chromosome paint making it possible to identify all RNO4 and RNO4-derived elements in the metaphases. The results indicated that RNO4 gains were common among EAC tumors. In all of the tumors 1–4 normal copies of RNO4 were present and in some cases RNO4-derivatives were detected as translocations and/or other chromosomal aberrations. In 75% of cases (in 18 tumors), the observed number of RNO4 elements exceeded the expected number of chromosome 4 (Table 1). Several tumors (42%) displayed RNO4 gains as rearranged chromosomes of various sizes, sometimes with an extra arm (i.e., p-arm). Seven tumors (NUT7, NUT4, RUT7, RUT13, RUT29, RUT5, and NUT12) showed RNO4 amplifications as homogenously stained regions (HSR).

Comparative Genomics. Gene content and genome size are highly conserved in mammals (8). Thereby, comparative genome analysis could potentially make animal models more useful as tools in cancer research. Within a conserved chromosomal segment, the gene order between two closely related species is usually identical. This provides platforms allowing us to exchange information from one mammalian system to another. It is practically possible to use information about gene order and locations in one mammal to predict location and gene order in another. It is thereby possible to transfer and apply findings from an animal cancer model to human systems.

FISH Analysis. Metaphase spreads from 8 tumors with translocations or HSR marker chromosomes were further analyzed using FISH analysis. Previous studies showed two separate amplified regions on RNO4 in a number of tumors, centering round *Cdk6* at RNO4q13 and *Met* at RNO4q21 (6). In this study, additional genes were selected on the basis of their biological function or physical location on chromosome 4, spanning the region RNO4q12 to 4q21 (Fig. 1). In total, 13 genes, with an average distance of 1.56 Mb, were used to investigate the structure of the two amplified regions and the region in between. Gene specific PAC or BAC probes were used to analyze chromosome metaphase spreads from six EAC, one PM, and one ESCC (Table 2).

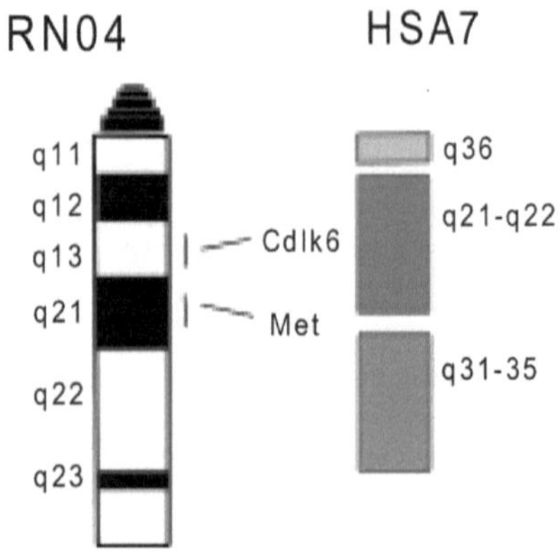

Fig. 1 The proximal part of rat chromosome 4, showing *Cdk6* at position 4q13 and *Met* at 4q21, and homologous segments in human chromosome 7

Table 2 FISH analysis of the region RNO4q12–q21 with 13 gene specific PAC/BAC clones on metaphase chromosomes from six EAC, one PM, and one ESCC

Cytogenetic position	4q12.1	4q13	4q13	4q13	4q13	4q13	4q13	4q21	4q21	4q21	4q21	4q21.1	4q21.1
Mb on RNO4	17	26.7	26.94	27.3	27.36	28.94	29.65	32.88	33.42	40.20	43.1	43.3	43.6
Genes / Tumors	Hgf	Cyp51	Mterf	Loc362320	Cdk6	Bet1	Ppp1r9a	Asns	Rpa3	Ppp1r3a	Cav1	Met	Wnt2
NUT7	+	3+	3+	2+	3+	2+	Nt	+	2+	+	+	2+	3+
NUT11	+	2+	2+	2+	3+	2+	2+	2+	2+	+	+	4+	3+
RUT5	+	–	3+	3+	3+	3+	+	–	+	+	3+	3+	–
RUT13	+	3+	3+	3+	3+	+	+	+	Nt	3+	3+	3+	3+
RUT29	–	2+	2+	Nt	2+	2+	+	+	+	Nt	3+	3+	3+
NUT4	+	+	+	+	3+	+	Nt	+	+	Nt	+	4+	3+
NUT12	+	+	+	+	+	+	Nt	+	+	Nt	+	4+	+
RUT7	+	–	Nt	Nt	–	Nt	Nt	–	Nt	Nt	+	3+	+

Nt, not tested; **–**, no signal; **+,** normal, 2–6 copies; **2+,** 7–15 copies; **3+,** 16–30 copies; **4+,** over 30 copies

Two independent amplification regions were detected in the tumors. The distal amplified region was centered at *Cdk6* (cyclin-dependent kinase 6) and was detected in six tumors. In three tumors (NUT7, NUT11, and RUT29), the amplified region was of an approximate size of 3 Mb, starting at *Cyp51*, located 1.3 Mb proximal to *Cdk6*, to *Bet1*, located 1.5 Mb distal to *Cdk6*. One tumor (NUT11) displayed an amplified region of approximately 7 Mb, starting at *Cyp51*, (1.3 Mb proximal to *Cdk6*) to *Rpa3*, 6 Mb distal to *Cdk6*. In the sixth tumor (NUT4), *Cdk6* was exclusively amplified. The observed pattern of amplification suggested that *Cdk6* most likely is the main target for gene amplification in the proximal region of RNO4.

Cdk6, a cyclin-dependent kinase, is an important regulator of cell cycle progression. The active phosphorylated form of *Cdk6* is shown to actively regulate the activity of the well known tumor suppressor protein Rb. Human *CDK6* maps to HSA7q21–22 and has been implicated in human prostate cancer (9), human gliomas (10), and nodal metastasis (11).

In the distal part of RNO4, at RNO4q21, the *Met* oncogene was found to be the target for amplification in all tumors. In two cases, NUT12 and RUT7, high levels of amplification were exclusively detected for *Met*. In the remainder six cases the amplified region was including one or more of the neighboring genes (Table 2).

Our data suggest that *Met* to be the most likely target for the observed amplification events. There are numerous reports suggesting involvement of *Met* in tumor progression in a rat tumor model (6), as well as in human tumors; glioblastomas and gastric tumors (12–14).

Comparative Mapping Analysis. Presently, we are in the process of transferring our data (6, 7, 15) into the human system. CGH analysis of human endometrial carcinomas (EC) showed recurrent gains at HSA7q11–q21 and HSA7q21–32 (data not shown). HSA7q21–q22 is homologous to RNO4q12–q21 and HSA7q31 is homologous to RNO4q21, where *Cdk6* and *Met*, respectively, are located. Accordingly, we want to investigate the status of the candidate target genes for amplification that we found in the BDII rat model for EAC in the human EC samples. Our preliminary results are suggestive of a good correlation between our findings in the BDII rat model and in the human tumor samples. This data provides stronger support for use of animal model systems for better understanding and scrutinizing of human complex diseases such as cancer.

Acknowledgments The RNO4 paint was kindly provided from Professor Malcolm. Ferguson-Smith. The Swedish Cancer Foundation and the Nilsson-Ehle Foundation supported this project.

References

1. Sjöling Å, Walentinsson A, Helou K, et al. (2000) A genome wide study of genetic alterations in DMBA induced rat sarcomas. Rat Genome 6:75.
2. Laes J-F, Quan X, Ravoet M, et al. (2001) Analysis of candidate genes included in the mammary cancer susceptibility 1 (Mcs1) region. Mamm Genome 12:199–206.
3. Helou K, Walentinsson A, Beckmann B, et al. (2001) Analysis of genetic changes in rat endometrial carcinomas by means of comparative genome hybridization. Cancer Genet Cytogenet 127:118–127.
4. Cavanagh D, Fiorica JV, Hoffman MS, et al. (1999) Adenocarcinoma of the endometrium: an institutional review. Cancer Control; JMCC 6:354–360.
5. Hamta A, Adamovic T, Helou K, et al. (2005) Cytogenetic aberrations in spontaneous endometrial adenocarcinomas in the BDII rat model as revealed by chromosome banding and comparative genome hybridization. Cancer Genet Cytogenet 159:123–128.
6. Walentinsson A, Helou K, Wallenius V, et al. (2001) Independent amplification of two gene clusters on chromosome 4 in rat endometrial cancer: identification and molecular characterization. Cancer Res 61:8263–8273.
7. Adamovic T, Trossö F, Roshani L, et al. (2005) Oncogene amplification in the proximal part of chromosome 6 in rat endometrial adenocarcinomas as revealed by combined BAC/PAC FISH analysis, chromosome painting, zoo-FISH and allelotyping analysis. Genes Chromosom Cancer 44:139–153.
8. Andersson L, Archibald A, Ashburner M, et al. (1996) Comparative genome organization of vertebrates. The First International Workshop on Comparative Genome Organization. Mamm Genome 7:717–734.
9. Lim J, Mansukhani M, Weinstein I (2005) Proc Natl Acad Sci USA 14:5156–5161.
10. Costello JF, Plass C, Arap W, et al. (1997) Cyclin-dependent kinase 6 (CDK6) amplification in human gliomas identified using two-dimensional separation of genomic DNA. Cancer Res 57:1250–1254.

11. Hirai Y, Utsugi K, Takeshima N, et al. (2004) Putative gene loci associated with carcinogenesis and metastasis of endocervial adenocarcinomas of uterus determined by conventional and array-based CGH. Am J Obstet Gynecol 191:1173–1182.
12. Wullich B, Muller HW, Fischer U, et al. (1993) Amplified met gene linked to double minutes in human glioblastoma. Eur J Cancer 29A:1991–1995.
13. Wullich B, Sattler HP, Fischer U, et al. (1994) Two independent amplification events on chromosome 7 in glioma: amplification of the epidermal growth factor receptor gene and amplification of the oncogene MET. Anticancer Res 14:577–579.
14. Nakajima M, Sawada H, Yamada Y, et al. (1999) The prognostic significance of amplification and overexpression of c-met and c-erb B-2 in human gastric carcinomas. Cancer 85:1894–1902.
15. Karlsson A, Helou K, Walentinsson A, et al. (2001) Amplification of Mycn, Ddx1, Rrm2, and Odc1 in rat uterine endometrial carcinomas. Gene Chromosome Cancer 31:345–356.

Recurrent Chromosome 10 Aberrations and *Tp53* Mutations in Rat Endometrial Adenocarcinomas

Carola Nordlander, Emma Samuelson, Karin Klinga-Levan, and Afrouz Behboudi

Summary Human genetic heterogeneity and differences in the environment and life style make analysis of complex diseases such as cancer difficult. By using inbred animal strains, the genetic variability can be minimized and the environmental factors can be reasonably controlled. Endometrial adenocarcinoma (EAC) is the most common gynecologic malignancy, ranking fourth in incidence among tumors in women. The inbred BDII rat strain is genetically prone to spontaneously develop hormone-related EAC, and can be used as a tool to investigate and characterize genetic changes in this tumor type. In the present project, BDII females were crossed to males from two nonsusceptible rat strains and F1, F2, and backcross progeny were produced. Genetic and molecular genetic analysis of tumors showed that rat chromosome 10 (RNO10) was frequently involved in genetic changes. Our data indicate that often there was loss of chromosomal material in the proximal to middle part of the chromosome followed by gains in distal RNO10. This suggested that there is a tumor suppressor gene(s) in the proximal to middle part of RNO10 and an oncogene(s) in the distal part of the chromosome with potential significance in EAC development. The *Tp53* gene, located at band RNO10q24–q25, was a strong candidate target for the observed aberrations affecting the middle part of the chromosome. However, our *Tp53* gene mutation analyses suggested that a second gene situated very close to *Tp53* might be the main target for the observed pattern of genetic changes.

Introduction

Endometrial cancer (EC), cancer of the uterine corpus, is the most common type of female genital cancer in the western world. Two general tumor subtypes of EC have been described: Type I, E-related and Type II, non-E-related. The E-related (type I) is generally associated with endometrial hyperplasia and comprises ~75% of all ECs. These tumors are associated with risk factors such as use of estrogen replacement therapy or obesity. The EC risk decreases with each pregnancy. The nonestrogen-related (type II) is of nonendometrioid histologic type, i.e., serous carcinoma (1, 2).

J.J. Li et al. (eds.), *Hormonal Carcinogenesis V.*

Detection of specific genetic changes leading to this complex disease is difficult because of genetic heterogeneity and differences in the environment and life style in human populations. Therefore, the variability in genetic background of human tumors complicates the distinction between significant and insignificant changes during tumor development. Using inbred experimental model systems for this analysis, the genetic variability background is considerably minimized, and the environmental factors can be reasonably controlled. Thus, it should be easier to track, detect, and characterize the important changes involved in the development of the disease. Results from animal system studies can readily be transferred to the human condition by comparative genomics.

The inbred BDII rat is a suitable experimental animal model for genetic studies of endometrial adenocarcinoma (EAC). BDII females are highly prone to develop spontaneous EC, mainly of the EAC subtype (3, 4). Herein, genetic crosses between the susceptible BDII females and the resistant SPRD-Cu3 or Brown Norway (BN) rat strains were performed, followed by subsequent inter and back crosses. A population of rats with various degrees of susceptibility was produced. The resulting rats were genome-wide genotyped, and subsequently the genotype data were related to cancer phenotypes. Several loci harboring genetic determinants of susceptibility to EAC development were identified (5, 6).

Tumor cytogenetic and comparative genome hybridization (CGH) analysis revealed a nonrandom pattern of cytogenetic changes affecting several chromosomes. Characteristic findings included amplifications in RNO4 and RNO6, deletion in the proximal area combined with gain in the distal area of RNO10 and RNO15 (7, 8). Analysis of allelic imbalance (AI) in the proximal area of RNO10 showed three frequently deleted regions in F1 and F2 progeny (9). Later, we extended the material considerably by including a set of backcrossed (N1) tumors. Furthermore, we included the distal area of the chromosome, in which gains were implicated from the CGH data (10).

To determine whether gross chromosomal aberrations such as amplifications, deletions, or translocations affected RNO10 in the tumors, we applied a RNO10-whole chromosome paint and dual-color gene-specific fluorescence in situ hybridization (FISH) with PAC DNA for the *Tp53* gene located in the middle part of the chromosome and the *Thra* gene located in the distal part of the chromosome were used as probes (11).

The AI results combined with the results from whole RNO10-specific paint and dual-color FISH provide us with a more comprehensive picture of RNO10 aberrations. Our goal was to define the affected regions precisely and to find candidate genes targeted by the RNO10 aberrations in EAC development.

One common deletion region was located in the central part of the chromosome. The *Tp53* gene (located at band 10q24–q25) was singled out as a candidate gene to be affected by aberrations in this region. To investigate whether the *Tp53* was in fact the target, we performed sequencing for gene mutations and found mutations in a fraction of the tumors (11). However, there were several tumors with detected AI at RNO10q24–q25 and no *Tp53* mutation. Therefore, the analysis suggested that *Tp53* might not be the only molecular target in the region.

Results and Discussion

Allelic Imbalance (AI)/Loss off Heterozygosity (LOH). In the present study, we used a marker panel with polymorphic markers in the backcrosses BDIIx(BDIIxBN) and BDIIx(BDIIxSPRD) to determine AI occurrence in our tumor material. For each region we attempted to define the smallest region of overlap (SRO) with AI. Four different segments of recurrent AI were identified along the chromosome, two in the proximal region, one in the middle region, and one in the distal region (10). On the basis of the cytogenetic and CGH data, we know that the observed AI in regions 1–3 is due to chromosomal loss, whereas AI in region 4 is due to chromosomal gain (10). Thus, we would expect the target genes to be tumor suppressor genes (TSGs) in the three proximal regions and an oncogene in the distal region 4. Using the sequence databases as well as comparative maps among rat, mouse, and human, it is possible to pin point a number of candidate gene(s) for each respective region, and to determine the homologous regions in human and mouse. Even with these approaches, it is difficult to find a "hot candidate gene" for each region, since there are many unknown genes and also the function of many known genes is still unknown. In our comparative genomic analysis, we suggested Axin1 (axis inhibitor 1) and Tsc2 (tuberous sclerosis 2) as best TSG candidates for the first region of common allelic loss; both of these genes are found in LOH regions in several types of human cancer. Irf1 (interferon regulatory factor 1) is a good candidate for the second region of AI. The third LOH region corresponds to a well-known TSG that is commonly lost in many types of human tumors, including endometrioid-type adenocarcinomas. This region harbors the *Tp53* gene and also other well-known cancer-related genes such as Hic1 (hypermetylated in cancer 1) and Ovca 1 and 2 (ovarian cancer 1 and 2). In region 4, the *Grb2* (growth factor receptor bound protein 2) and the *Erbb2* (avian erythroblastosis viral, *v-erb-B2*, oncogene homologue 2) genes are the best protooncogene candidates, both with recognized tumor development involvement.

Chromosome Painting and Dual-color Fluorescence *in-situ* Hybridization (FISH). The chromosome painting that was performed in 27 EAC tumor cell lines and 2 non-EAC tumor cell lines (MPM, malignant peritoneal mesothelioma; ESCC, endometrial squamous cell carcinoma) revealed that RNO10 was often involved in different chromosomal aberrations such as translocations and/or deletions in these tumors. In most cases the rearranged chromosomes appeared to have been derived by a break somewhere in the middle part. To make a more detailed analysis of the aberrations, we performed dual-color FISH with two gene-specific probes. We used two PAC clones containing *Tp53* (located at RNO10q24–q25) and *Thra* (located at RNO10q32.1) as probes representing the proximal/middle part and distal part of the chromosome, respectively (11). By counting the number of the signals from each of these genes in the dual-color FISH, it was shown that in five EAC tumors one copy of the *Tp53* gene was missing, in three EAC tumors there was one extra copy of the *Thra* gene and in eight EAC tumors there was both reduction in *Tp53* and gain in *Thra*. In conclusion, the analysis showed that breakage between the *Tp53* and *Thra* loci were very common in the EAC tumors. The *Tp53* gene was reduced in 13 of the 27 EAC tumors and 11 of the 27 EAC tumors displayed extra copies *Thra*. (Table 1) (11).

Our preliminary data suggest that the breakpoint most likely is not the same among different tumors. Thus, the driving force behind the breaks probably is that there is powerful selection against some gene(s) in the proximal to middle part and/or powerful selection for some gene(s) in the distal part of the chromosome. *Tp53* or another TSG(s) located close to *Tp53* are the best candidates for the loss events in the proximal/middle region, whereas the oncogene *Erbb2*, which is located 0.4 Mb proximal to *Thra* may be target for the amplification event in distal RNO10.

Table 1 Molecular cytogenetic analysis of RNO10 aberrations in 27 EAC and 2 non-EAC tumors

Tumor designation	Tumor type	Chr. Nr	Found RN010 Chr	Copies of *Tp53*		Copies of *Thra*	
				Found	Dev[a]	Found	Dev[a]
NUT39	EAC	39	2	2		2	
NUT16	EAC	42	2	2		2	
NUT81	EAC	42	2	2		2	
NUT52	EAC	42	3	2		2	
NUT42	EAC	44	2	2		2	
NUT55	EAC	44	2	2		2	
NUT47	EAC	48	2	2		2	
NUT97	EAC	52	2	2		2	
NUT51	EAC	71	4	3		3	
RUT7	EAC	43	2	1	−1	2	
RUT6	EAC	58	4	2	−1	4	1
NUT6	EAC	60	4	2	−1	4	1
RUT30	EAC	61	4	2	−1	4	1
NUT100	EAC	67	4	2	−1	4	1
NUT7	EAC	46	5	2		3	1
NUT12	EAC	59	3	2	−1	3	
RUT13	EAC	62	3	2	−1	3	
NUT128	EAC	64	3	2	−1	3	
NUT50	EAC	66	4	2	−1	4	1
NUT31	EAC	67	3	2	−1	3	
RUT2	EAC	40	3	2		2	
NUT84	EAC	42	3	1	−1	3	1
RUT12	EAC	51	3	2		2	
RUT25	EAC	48	4	2		3	1
NUT4	EAC	62	5	3		4	1
NUT99	EAC	67	5	2	−1	4	1
NUT127	EAC	70	5	2	−1	5	2
RUT29	MPM	43	2	2		2	
RUT5	ESCC	77	4	4		4	

[a]Deviation from expected number of copies based on the tumor cell ploidy

Mutation in the *Tp53* Tumor Suppressor Gene. The *TP53* is a well-known TSG and has been called "the guardian of the genome". Its central role is to eliminate genomic damage by induction of cell cycle arrest or apoptosis. The inactivation of *TP53* often starts with a first-hit by mutation followed by a second-hit leading to loss of the other functional allele. Ninety percent of the reported mutations are in the core DNA-binding domain of the gene, which corresponds to exons 5–10 in the human *TP53* gene. Since *Tp53* was an obvious candidate to be the target TSG in the proximal/middle RNO10 region, we screened the tumors for *Tp53* mutations. The core DNA-binding domain on genomic DNA from the tumor cell cultures (27 EAC tumors and two non-EAC tumors) as well as from normal liver DNA of the three parental inbred strains (BDII, BN, and SPRD) was sequenced and analyzed. We found *Tp53* mutations in 18 out of 27 EAC tumors, truncated or highly abnormal products were found in four tumors and single amino acid substitutions were found in 14 (Table 2) (11).

Conclusion

From the DNA sequence data, it appeared that only one *Tp53* allele (the mutated one) was present in the majority of the tumors, in spite of the fact that the FISH data had shown that in most tumors, more than one copy of *Tp53* was present (Table 2). Thus, we concluded that there was an abnormal function of Tp53 in the majority of these tumors. However, our combined *Tp53* mutation and AI data suggested that *Tp53* mutations might represent late events in the tumor development, and that *Tp53* probably is not the only target for mechanisms generating AI/LOH in the proximal region of RNO10 in these tumors. Because there were six informative tumors with clear AI at RNO10q24–q25 that exhibited no *Tp53* mutations (Table 2). Thus, the occurrence of AI, at least in these tumors, is targeting another TSG(s) in the region.

Taken together, the occurrence of recurrent chromosome breaks and losses close to *Tp53* suggest that RNO10q24 may harbor another tumor suppressor gene (hypothetical tumors suppressor gene, *Htsg*), which plays an important role in EAC development. High frequency of LOH at 17p13.3 (HSA17p13.3) accompanied by low frequency of *Tp53* mutation has been reported to be common in a variety of human malignancies including breast, lung, and hepatocellular carcinomas as well as neuronal tumors (12–17). It could be a possibility that the candidate tumor suppressor region implicated by the present study in rat EAC tumors may be the same that is discussed in these human cancer studies. Further analysis of this locus might serve as an important foundation for future efforts toward the identification of the putative tumor suppressor gene(s) at 17p13.3 in human tumors.

Acknowledgments The RNO10 paint probe was generously provided by Dr. Malcolm Ferguson-Smith. This project was supported by grants from the Swedish Cancer Society, the Nilsson-Ehle Foundation and the IngaBritt and Arne Lundberg Research Foundation.

Table 2 Data on *Tp53* mutations in 27 EAC and two non-EAC tumor cell cultures

Design	Tumor type	Chr. nr	*Tp53* data Copies (FISH)	Mutation	Exon	AA	AI data (Reg 3) AA No.	Allele retained	I/U
RUT7	EAC	43	1	GCG-CCG	5	Ala-Pro	136	**BN**	**I**
RUT12	EAC	51	2	DEL 1041	8	Truncated	287	HET	I
RUT25	EAC	48	2	CGG-CAG	6/7	Arg-Gln	247	**BN**	**I**
RUT30	EAC	61	2	DEL1040	8	Truncated	286	–	U
NUT6	EAC	60	2	CAC-CGC	5	His-Arg	166	BDII	U
NUT16	EAC	42	2	CGC-CAC	5	Arg-His	173	HET	I
NUT31	EAC	67	2	CGT-CAT	5	Arg-His	156	BDII	U
NUT50	EAC	66	2	—				**BDII**	**I**
NUT51	EAC	71	3	CGC-CCC	6/7	Arg-Pro	246	–	U
NUT52	EAC	42	2	GTG-ATG	5	Val-Met	171	**BDII**	**I**
NUT81	EAC	42	2	GAA-TAA	9	Glu-stop	324	BDII	U
NUT97	EAC	52	2	CGC-CAC	5	Arg-His	173	BDII	U
NUT99	EAC	67	2	DEL1016-40	8	Truncated	278	BDII	U
NUT100	EAC	67	2	—				**BN**	**I**
NUT127	EAC	70	2	CGT-CAT	5	Arg-His	156	**BN**	**I**
NUT128	EAC	64	2	—				BDII	U
RUT2	EAC	40	2	—				**BDII**	**I**
RUT6	EAC	58	2	CGT-TGT	8	Arg-Cys	271	SP	U
RUT13	EAC	62	2	CCT-TCT	8	Pro-Ser	276	–	U
NUT4	EAC	62	3	—				HET	I
NUT7	EAC	46	2	AGA-GGA	8	Arg-Gly	278	BDII	U
NUT12	EAC	59	2	AAG-AGG	5	Lys-Arg	130	**SP**	**I**
NUT39	EAC	39	2	—				**SP**	**I**
NUT42	EAC	44	2	—				**BDII**	**I**
NUT47	EAC	48	2	—				HET	I
NUT55	EAC	44	2	—				BDII	U
NUT84	EAC	42	1	GTG-ATG	6/7	Val-Met	214	BDII	U
RUT29	MPM	43	2	CGT-TGT	8	Arg-Cys	271	BDII	U
RUT5	ESCC	77	4	TGC-TTC	6/7	Cys-Phe	240	**BDII**	**I**

[a] **AI data** (from Nordlander et al. 2005): I, U – informative or uninformative in RNO10 region 3, respectively, determined by genotyping liver DNA with polymorphic microsatellite markers, bold marked are tumors with AI in reg. 3, grey marked are tumors with AI without *Tp53* mutation

References

1. Boyd J (1996) Molecular biology in the clinicopathologic assessment of endometrial carcinoma subtypes. Gynecologic Oncology 61:163–165.
2. Lax SF (2004) Molecular genetic pathways in various types of endometrial carcinoma: from a phenotypical to a molecular-based classification. Virchows Archiv 444(3):213–223.
3. Kaspareit-Rittinghausen J, Deerberg F, Rapp K (1987) Mortality and incidence of spontaneous neoplasms in BDII/Han rats. Zeitschrift für Versuchstierkunde 30:209–216.
4. Deerberg F, Kaspareit J (1987) Endometrial carcinoma in BDII/Han rats: model of a spontaneous hormone-dependent tumor. J Natl Cancer Inst 78:1245–1251.

5. Roshani L, Wedekind D, Szpirer J, et al. (2001) Genetic identification of multiple susceptibility genes involved in the development of endometrial carcinoma in a rat model. Int J Cancer 94:795–799.
6. Roshani L, Mallon P, Sjostrand E, et al. (2005) Genetic analysis of susceptibility to endometrial adenocarcinoma in the BDII rat model. Cancer Genet Cytogenet 158:137–141.
7. Helou K, Walentinsson A, Beckmann B, et al. (2001) Analysis of genetic changes in rat endometrial carcinomas by means of comparative genome hybridization. Cancer Genet Cytogenet 127:118–127.
8. Hamta A, Adamovic T, Helou K, et al. (2005) Cytogenetic aberrations in spontaneous endometrial adenocarcinomas in the BDII rat model as revealed by chromosome banding and comparative genome hybridization. Cancer Genet Cytogenet 159:123–128.
9. Behboudi A, Levan G, Hedrich HJ, et al. (2001) High-density marker loss of heterozygosity analysis of rat chromosome 10 in endometrial adenocarcinoma. Gene Chromosome Cancer 32:330–341.
10. Nordlander C, Behboudi A, Levan G, et al. (2005) Allelic imbalance on chromosome 10 in rat endometrial adenocarcinomas. Cancer Genet Cytogenet 156:158–166.
11. Nordlander C, Karlsson S, Karlsson Å, et al. (2007) Analysis of chromosome 10 aberrations in rat endometrial cancer – evidence for a tumor suppressor locus distal to Tp53. Int J Cancer 120:1472–1481.
12. Coles C, Thompson AM, Elder PA, et al. (1990) Evidence implicating at least two genes on chromosome 17p in breast carcinogenesis. Lancet 336:761–763.
13. Jung HL, Wang KC, Kim SK, et al. (2004) Loss of heterozygosity analysis of chromosome 17p13.1–13.3 and its correlation with clinical outcome in medulloblastomas. J Neuro Oncol 67:41–46.
14. Konishi H, Sugiyama M, Mizuno K, et al. (2003) Detailed characterization of a homozygously deleted region corresponding to a candidate tumor suppressor locus at distal 17p13.3 in human lung cancer. Oncogene 22:1892–1905.
15. Roncuzzi L, Brognara I, Baiocchi D, et al. (2005) Loss of heterozygosity at 17p13.3-ter, distal to TP53, correlates with negative hormonal phenotype in sporadic breast cancer. Oncol Rep 14:471–474.
16. Sarkar C, Chattopadhyay P, Ralte AM, et al. (2003) Loss of heterozygosity of a locus in the chromosomal region 17p13.3 is associated with increased cell proliferation in astrocytic tumors. Cancer Genet Cytogenet 144:156–164.
17. Zhao X, He M, Wan D, et al. (2003) The minimum LOH region defined on chromosome 17p13.3 in human hepatocellular carcinoma with gene content analysis. Cancer lett 190:221–232.

Prostate Cancer

Specific Properties of a C-terminal Truncated Androgen Receptor Detected in Hormone Refractory Prostate Cancer

Gaëlle Lapouge, Gemma Marcias, Eva Erdmann, Pascal Kessler, Marion Cruchant, Sebastian Serra, Jean-Pierre Bergerat, and Jocelyn Céraline

Summary Mutations in the human androgen receptor (AR) gene that lead to C-terminus truncated AR variants are frequently detected in prostate cancer (PC). These AR variants lack both the ligand-binding domain (LBD) and the AF-2 region. The aim of this study was to delineate the alternative mechanisms that lead to the activation of such AR variants as they are unresponsive to hormone stimulation, and to outline consequences of the loss of the LBD/AF-2 region on their functional properties. By using an MMTV-luciferase reporter construct and LY294002, UO126, or ZD1839, inhibitor of PI3K, MEK1/2, and EGFR signaling pathway respectively, we demonstrated that phosphorylation was required for full transcriptional activities of one these AR variants, the Q640X mutant AR. Western-blot analyses confirmed that these inhibitors affect the phosphorylation status of this AR variant. Furthermore, studies of the intranuclear colocalization of the Q640X AR with cofactors, such as CBP, GRIP-1, and c-Jun, reveal that the transcriptional complex that forms around the mutant AR is different to that formed around the wild type AR. We demonstrated that CBP and c-Jun are highly recruited by the mutant AR, and this leads to an unexpected activation of AP-1, NFAT, and NFκB transcriptional activities. Similar enhanced activities of these transcription factors were not observed with the wild type AR. The importance of the LBD/AF-2 for the regulation of AR transcriptional activities, the impact of the presence of such AR variants on PC cells proliferation and survival, and on progression to androgen independence are discussed.

Introduction

Androgen receptor (AR) mutations are recurrent events during the progression of prostate cancer (PC) to a hormone-refractory status. Selected AR mutations confer new functional properties to the AR, and favor PC cell growth and survival in an androgen-depleted environment. Mutations in the ligand-binding domain (LBD) and the activation function-2 (AF-2) of the AR have been shown to affect receptor activation. Indeed, the C-terminal end (CTE), including the LBD and AF-2 regions, is essential for AR regulation notably being the target of direct or indirect phosphorylation

J.J. Li et al. (eds.), *Hormonal Carcinogenesis V.*

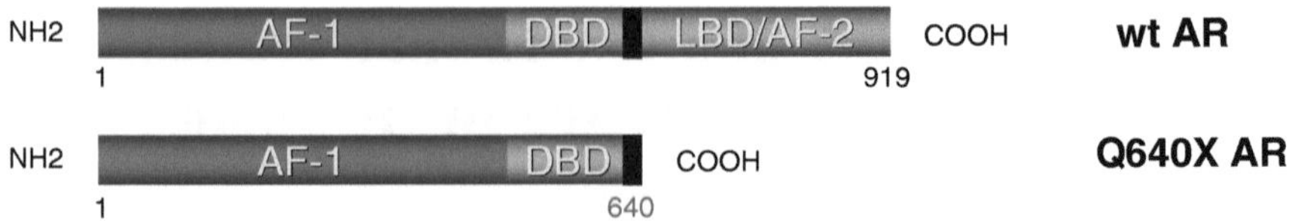

Fig. 1 Schema of the C-terminal truncated Q640X AR variant compared with the wild-type (wt) AR. AR is constituted by three major domains: the N-terminal domain containing the transactivation function AF-1, the central DNA-binding domain (DBD), which also contains a nuclear localization signal, and the C-terminus domain containing the LBD and the second transactivation function AF-2

after ligand-binding or by interaction with other signal transduction pathways (1, 2). The control of the AR transcriptional activities is also upon the control of numerous cofactors. The AR requires interaction with cofactors to activate (coactivators) or to inhibit (corepressor) target genes (3). The cAMP response element binding protein (CREB)-binding protein (CBP)/p300, is one of these coactivators that interplay with the AR at the target gene promoter to facilitate DNA occupancy, chromatin remodeling, or the recruitment of general transcription factors associated with the RNA polymerase II (3). Also, CBP being a partner for other transcription factors allows crosstalk of the AR with other signaling pathways. Also, the glucocorticoid receptor interacting protein-1 (GRIP-1), coactivator of the AR, induces the appropriate folding of the AR or facilitates AR N/C-terminal interactions to direct target gene expression. Similarly, the proto-oncoprotein c-Jun acts as an AR coactivator by enhancing AR N/C-terminal interactions (3, 4). In summary, the CTE is the target of many regulation pathways and the loss of these parts of AR should be significant.

In clinical studies, we detected several nonsense mutations in *AR* gene of localized and metastatic PC patients that lead to C-terminal truncated AR proteins. One of these mutations, the Q640X AR variant, has been previously described as a ligand-independent and constitutive transcriptional factor (Fig. 1) (5). In this study, we demonstrate that the Q640X AR requires interactions with other signal transduction pathways to be fully active. Moreover, we show that the lack of CTE leads to a differential recruitment of cofactors and then could affect remarkably target gene involved in cell proliferation, survival, and differentiation.

Results

Activation of the Q640X AR Variant Needs Phosphorylation. The AR is a phosphoprotein in which the serine 16, 81, 94, 256, 308, 424, 650, and 791 are phosphorylated (6). The phosphorylation of the serine 650 is activated by the epidermal growth factor receptor (EGFR) pathway (1), although serine 213 and 791 are regulated in vitro by PI3K pathway (7, 8). Moreover, the phosphorylation of serine 514 requires the MEK1/2 pathway and regulates the AR-induced cytotoxicity

(9). To investigate the role of these pathways on the activation of the Q640X AR variant, we evaluated the impact of their inhibition on the transcriptional properties of the Q640X AR variant by using MMTV-luciferase reporter assays. The LY294002, UO126, and ZD1839, are inhibitors of PI3K, MEK1/2, and EGFR signaling pathway, respectively. As previously described, the Q640X AR exhibited higher and constitutive transcriptional activities from MMTV when compared with the wt AR (Fig. 2) (5).

As expected and illustrated in Fig. 2a, the three inhibitors affect the DHT-activated transcriptional activities of wt AR. Interestingly, the Q640X AR transcriptional activities from MMTV promoter were also decreased in the presence of each kinase inhibitor (Fig. 2b). These results suggest that despite the lack of the LBD/AF-2 region, the Q640X AR variant remains sensitive to activating phosphorylation of the signal transduction of the EGFR, PI3K, or MEK1/2 signaling pathways.

Specific Recruitment of Cofactors in Presence of the Q640X AR. The AR is regulated by the N/C interaction and by cofactor interactions. Many of these interactions are located in the LBD/AF-2 region (10–12). The Q640X AR, with the lack of the LBD/AF-2, probably loses several interaction sites important for the regulation of its transcriptional activities. To investigate the impact of this mutation on cofactor recruitment, we analyzed the nuclear distribution of three cofactors, CBP, GRIP-1, and c-Jun using confocal microscopy. As expected, colocalization between the wt AR and these cofactors was observed in the presence of DHT (Fig. 3a). Interestingly, CBP and c-Jun were recruited in the presence of the Q640X AR in the presence or absence of DHT. Conversely, lack of colocalization and GRIP-1 recruitment were observed in the presence of the Q640X AR (Fig. 3b). These

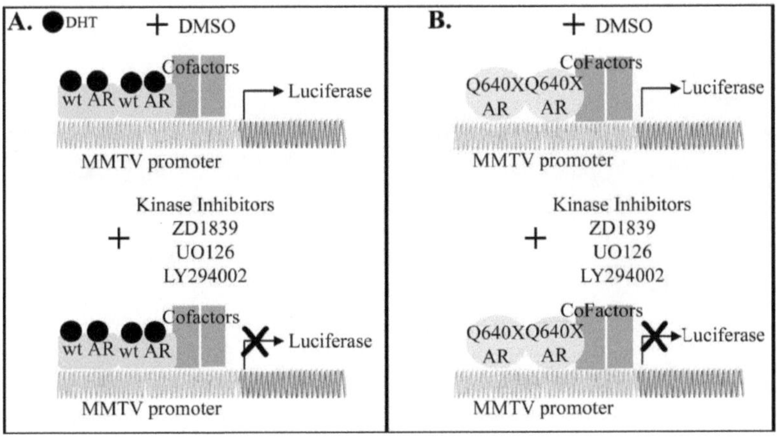

Fig. 2 Transcriptional activities of the wt and the Q640X mutant ARs from MMTV promoter in CV-1 cells after treatment with kinase inhibitors. In these assays, the CV-1 cells were cotransfected with the MMTV-luc construct and the wt AR (**A**) or the Q640X AR variant (**B**). After 24 h, cells were treated 10 μM of ZD1839, 10 μM of LY294002, 1 μM of UO126, or DMSO as vehicle, and luciferase activities were measured 24 h later

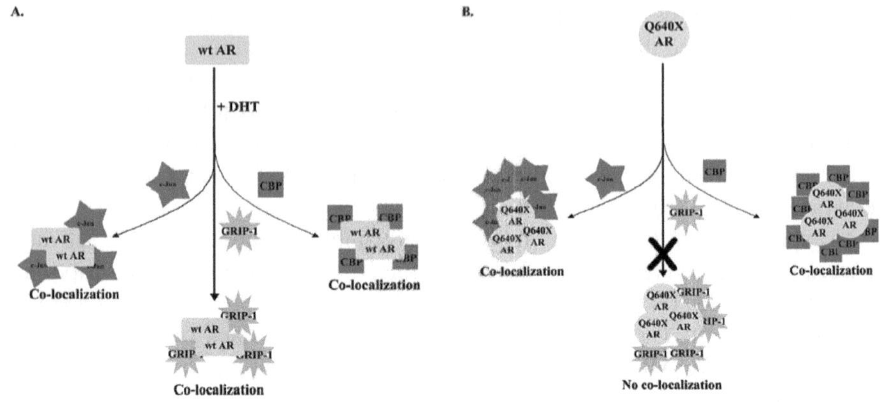

Fig. 3 Schema of cofactors recruitment by the wt (**A**) or the Q640X mutant (**B**) ARs. The wt or the Q640X mutant ARs recruits differently c-Jun, GRIP-1, or CBP

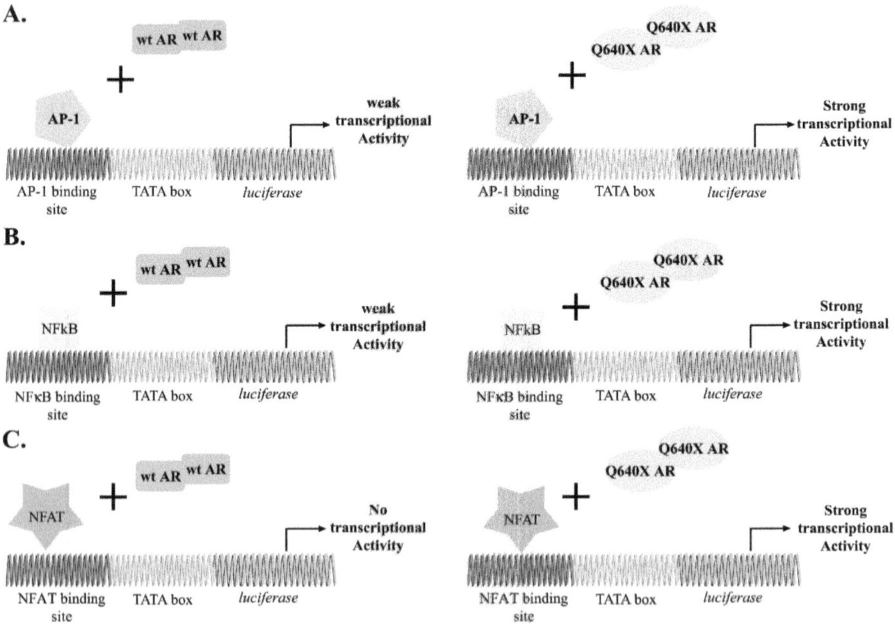

Fig. 4 Impact of the wt and the Q640X mutant ARs on activities of transcription factors involved in cell proliferation and differentiation. LNCaP cells were cotransfected with the wt or the Q640X mutant AR and with a luciferase gene reporter construct, and were treated with 100 nM of DHT. In these construct, the luciferase gene is controlled by a minimal promoter linked to (A) AP-1, (B) NFκB, or (C) NFAT binding sites and luciferase activities were measured 24 h later

results indicate that in the absence of the LBD/AF-2 region, the Q640X AR could recruit differentially cofactors, exhibiting specific transcriptional activities.

Unexpected Activation of AP-1, NFAT, and NFκB Transcriptional Activities in the Presence of the Q640X AR. The AR is a transcriptional factor that activates AR specific target genes and regulates cell proliferation and cell survival. To explore the impact of Q640X mutant AR on these two cellular processes, the transcriptional activities of AP-1, nuclear factor κB (NFκB), and nuclear factor of activated T cells (NFAT) were analyzed in the presence of the mutant AR using luciferase gene reporter assays. AP-1, dimeric complex of Jun and Fos proteins, induced by growth factors, cytokines and oncoprotein, is implicated in the proliferation, survival, and transformation of cells (13). NFκB is an important transcription factor involved in immune inflammatory responses, cell growth, differentiation, and apoptosis (14). NFAT is a transcriptional factor that regulates cytokines and cell-surface receptor gene in immune cells but NFAT plays a key role in nonimmune cells by regulating gene expression in diverse cellular events such as cell differentiation (15, 16). In the PC cell line, LNCaP, AP-1 transcriptional activities were dramatically increased in the presence of the Q640X AR (Fig. 4a), compared with the wt AR.

Similarly NFκB and NFAT activities were strongly enhanced in the presence of the Q640X AR, while only a weak NFκB was observed in presence of the wt AR. Moreover, the wt AR does not stimulate NFAT activities (Fig. 4b, c). These results suggest that the Q640X AR expression has an impact on the transcription factors involved in cell proliferation and differentiation.

Discussion

In this study, we investigated how a C-terminal truncated AR could be regulated, and how this mutant AR affected PC cell proliferation. In that respect, we studied the influence of several kinase pathways on Q640X AR activity and phosphorylation. Our data suggest that, similarly to the wt AR, the Q640X AR without CTE still requires these kinase pathways and specifically the EGFR, PI3K, and MEK1/2 signaling pathways for full transcriptional activities. Nevertheless, the Q640X AR regulation is different than that of the wt AR as illustrated by cofactors recruitment. Indeed, the impaired recruitment of GRIP-1 and the strong recruitment of CBP and c-Jun in the dots formed by the Q640X AR may modify the receptor transcriptional activities. Interestingly, this study shows that the Q640X AR mediates transcriptional activities of AP-1, NFκB, and NFAT linking somehow the Q640X mutant AR to cell proliferation. Moreover, we wonder whether the different cofactors recruitment by the Q640X AR may be linked to the unexpected enhanced activities of these transcription factors.

In conclusion, the study of this C-terminal truncated Q640X AR variant demonstrates the importance of the LBD/AF-2 for the regulation of AR transcriptional activities, as well as the potential role of such mutant ARs on PC progression during hormone therapy.

Acknowledgments This work was supported by the associations ATGC, ARECHO, Ligue Contre le Cancer, ARC, ARTP, and the Hôpitaux Universitaires de Strasbourg.

References

1. Gioeli D, Ficarro SB, Kwiek JJ, et al. (2002) Androgen receptor phosphorylation. Regulation and identification of the phosphorylation sites. J Biol Chem. 277, 29304–29314.
2. Culig Z, Steiner H, Bartsch G, et al. (2005) Mechanisms of endocrine therapy-responsive and -unresponsive prostate tumours. Endocr Relat Cancer. 12, 229–244.
3. Heinlein CA, Chang C (2002) Androgen receptor (AR) coregulators: an overview. Endocr Rev. 23, 175–200.
4. Bubulya A, Chen SY, Fisher CJ, et al. (2001) c-Jun potentiates the functional interaction between the amino and carboxyl termini of the androgen receptor. J Biol Chem. 276, 44704–44711.
5. Ceraline J, Cruchant MD, Erdmann E, et al. (2004) Constitutive activation of the androgen receptor by a point mutation in the hinge region: a new mechanism for androgen-independent growth in prostate cancer. Int J Cancer. 108, 152–157.
6. Taneja SS, Ha S, Swenson NK, et al. (2005) Cell-specific regulation of androgen receptor phosphorylation in vivo. J Biol Chem. 280, 40916–40924.
7. Wen Y, Hu MC, Makino K, et al. (2000) HER-2/neu promotes androgen-independent survival and growth of prostate cancer cells through the Akt pathway. Cancer Res. 60(24), 6841–6845.
8. Lin HK, Yeh S, Kang HY, et al. (2001) Akt suppresses androgen-induced apoptosis by phosphorylating and inhibiting androgen receptor. Proc Natl Acad Sci USA. 98(13), 7200–7205.
9. LaFevre-Bernt MA, Ellerby LM (2003) Kennedy's disease. Phosphorylation of the polyglutamine-expanded form of androgen receptor regulates its cleavage by caspase-3 and enhances cell death. J Biol Chem. 278(37), 34918–34924.
10. He B, Kemppainen JA, Voegel JJ, et al. (1999) Activation function 2 in the human androgen receptor ligand binding domain mediates interdomain communication with the NH(2)-terminal domain. J Biol Chem. 274, 37219–37225.
11. Ikonen T, Palvimo JJ, Janne OA (1997) Interaction between the amino- and carboxyl-terminal regions of the rat androgen receptor modulates transcriptional activity and is influenced by nuclear receptor coactivators. J Biol Chem. 272, 29821–29828.
12. Berrevoets CA, Doesburg P, Steketee K, et al. (1998) Functional interactions of the AF-2 activation domain core region of the human androgen receptor with the amino-terminal domain and with the transcriptional coactivator TIF2 (transcriptional intermediary factor2). Mol Endocrinol. 12, 1172–1183.
13. Eferl R, Wagner EF (2003) AP-1: a double-edged sword in tumorigenesis Nat Rev Cancer. 3(11), 859–868.
14. Ghosh S, Karin M (2002) Missing Pieces in the NF-κB Puzzle. Cell. 109 (Suppl): S81–S96.
15. Rao A, Luo C, Hogan PG (1997) Transcription factors of the NFAT family: regulation and function. Annu Rev Immunol. 15, 707–747.
16. Hogan PG, Chen L, Nardone J, et al. (2003) Transcriptional regulation by calcium, calcineurin, and NFAT. Genes Dev. 17(18), 2205–2232.

The Role of the Transcriptional Coactivator p300 in Prostate Cancer Progression

Hannelore V. Heemers, Jose D. Debes, and Donald J. Tindall

Summary Although the factors contributing to the progression of prostate cancer (PC) remain incompletely understood, androgens have long been recognized to play a central role in this process. Upon entering PC cells, androgens bind to a cognate nuclear receptor, the androgen receptor (AR). The activated AR translocates to the nucleus, binds as a dimer to androgen response elements (AREs) in the promoter of target genes, where it recruits the coactivator proteins necessary for the formation of a productive transcriptional complex, an event crucial for PC cell viability. For many decades, the androgen dependency of PCs has been exploited therapeutically by androgen ablation strategies. Although initially successful, these forms of therapy almost inevitably fail eventually, and an androgen depletion independent (ADI) disease emerges, for which currently no cure is available. Studies from our laboratory and others demonstrate that despite low circulating levels of functional androgens, the AR is critical for the proliferation and survival of ADI PC cells. Recent data indicate that alterations in the expression and/or activity of AR coactivator proteins occur during PC progression that can foster ADI activation of the AR. Here, we have investigated the role of the coactivator p300 in AR transcriptional activity and progression of PC.

Introduction

Activation of the androgen receptor (AR) in androgen depletion independent (ADI) prostate cancer (PC) cells has been attributed to mechanisms of AR hypersensitivity (AR amplification and/or mutations), promiscuous activation of the AR (by adrenal androgens, nonandrogenic steroids, and even anti-androgens), and outlaw AR pathways (AR activated by growth factors and cytokines, thereby bypassing the need for androgens) (1, 2). Recently, the importance of the involvement of AR coactivator proteins in ADI AR activation is increasingly being recognized. Under physiological conditions, coactivators are necessary for the formation of a productive transcriptional AR complex by facilitating DNA occupancy, chromatin remodeling, recruitment of general transcription factors associated with the RNA polymerase II holocomplex, as well as ensuring appropriate folding of the AR, AR protein stability, and/or

J.J. Li et al. (eds.), *Hormonal Carcinogenesis V.*

proper AR subcellular distribution (3). In the progression of PC a subset of these coactivators has been shown to be overexpressed, and this overexpression has been demonstrated to substantially contribute to the ADI mechanisms of AR activation described above. Therefore, AR coactivators have been suggested as valuable targets for therapeutic intervention (4).

The transcriptional coactivator p300 has been shown to regulate gene transcription through several distinct mechanisms. In addition to its potential to act as a direct bridge between DNA-bound transcription factors and the basal transcriptional machinery and to serve as a scaffold interacting with and assembling a number of other transcriptional regulators, p300 possesses histone acetyl transferase (HAT) activity by which it renders the chromatin environment more easily accessible for the transcriptional machinery. Apart from histones, p300 has also been shown to acetylate other proteins such as transcription factors and coregulators, resulting in modulation of transcription through altered protein/protein interactions, protein/DNA interactions, nuclear retention, or protein half life of certain proteins (5). In terms of its involvement in AR signaling in PC cells, a role for p300 has previously been described in androgen-dependent activation of the AR. Full ligand-induced transcriptional activity of the AR, recruitment of coactivators to the AR transcriptional complex, and prostate cancer cell growth was shown to depend on its direct acetylation by p300 HAT activity (6, 7). Here, we describe the importance of p300 in ligand-independent ADI AR activity and PC progression.

Results

p300 Expression is Increased in PC Cells and Correlates with PC Cell Proliferation and Aggressive Tumor Features. To investigate whether p300 may be involved in the development and/or progression of PC, we assessed the expression of p300 in tissue samples of 95 patients with biopsy-proven PC by performing immunohistochemistry. Expression of p300 was evaluated by digital image analysis as well as visual grade. These studies demonstrated overall p300 expression to be elevated in neoplastic tissues when compared with benign adjacent tissues (Fig. 1). Moreover, our analysis revealed a positive correlation of p300 expression on PC biopsy with expression of MIB-I, an in situ marker of cell proliferation ($p = 0.009$). In line with these findings, siRNA-mediated silencing of p300 expression in the PC cell line LNCaP led to a decrease in cell proliferation in vitro. Review of the clinical information of the patient population studied as well as biopsy and prostatectomy findings allowed for a correlative analysis of p300 expression and several clinico-pathological parameters. This analysis demonstrated that high levels of p300 on biopsy predict larger tumor volumes ($p < 0.001$), extraprostatic extension of disease ($p = 0.003$), seminal vesicle involvement at prostatectomy ($p = 0.002$) as well as PC progression after surgery ($p = 0.01$). In terms of genotypic and cellular changes, higher levels of p300 correlated positively with nondiploid DNA content and p300 expression tended to be positively correlated with high Gleason scores, which

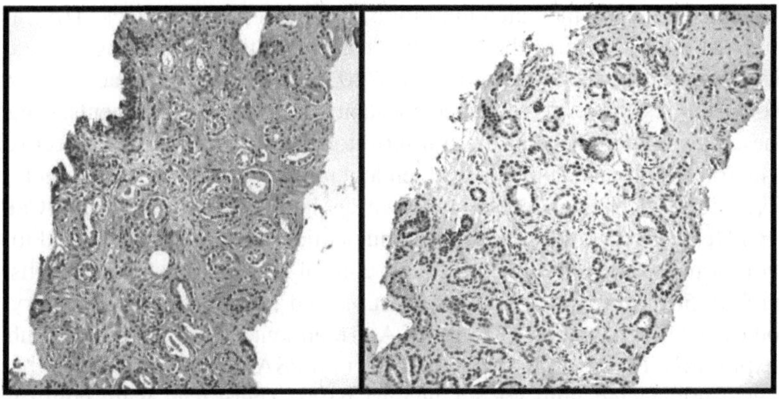

Fig. 1 High p300 expression in PC tissues. Formalin-fixed, paraffin-embedded PCa tissues from needle biopsies were stained with H&E (*left panel*) or p300 immunohistochemistry was performed as described (*right panel*) (8)

correspond to more undifferentiated tumors (8). Taken together, these data clearly point toward an association of p300 expression in PC with cell proliferation and more aggressive disease.

p300 is Required for ADI Activation of the AR. Having established that p300 expression is elevated in PC and correlates with aggressive tumor features, we were interested in exploring the impact of increased p300 expression on some key features of PC progression. To this end, we first investigated whether p300 is important for ADI activation of the AR by nonandrogenic ligands such as interleukin-6 (IL-6), a cytokine of particular interest to prostate cancer disease. Both IL-6 and its receptor have been shown to be expressed in PC cells. Several independent studies have revealed that plasma levels of IL-6 are increased in patients with ADI PC disease. Moreover, high IL-6 serum levels correlate with poor prognosis. Importantly, IL-6 has been shown to regulate AR transcriptional activity in the absence of androgens. Mechanistically, IL-6 exerts its effect on PC cells and AR activity through activation of the JAK/STAT, MAPK, and PI3K signaling pathways (9).

To determine whether p300 is involved in ADI transactivation of the AR by IL-6, LNCaP cells were transfected with an AR-dependent prostate specific antigen (PSA) promoter reporter construct and stimulated with 50 ng ml^{-1} IL-6. In agreement with previous reports, we found increased reporter activity following IL-6 treatment. Blocking the MAPK pathway with the inhibitor PD98059 reduced the IL-6-mediated activation of the AR. Interestingly, transfection of p300 into cells reversed this inhibition, suggesting that p300 is a final target for the MAPK pathway during IL-6 stimulation. To further assess the role of p300 in this model, we used adenoviral E1A, which sequesters p300, thereby inhibiting its HAT activity. Transfection of E1A into LNCaP cells inhibited IL-6-mediated activation of the AR. To assess if the E1A-mediated repression is due to its specific interaction with p300, cells were cotransfected with increasing amounts of p300. Overexpression of p300 reversed the E1A-mediated inhibition. A mutant p300 that lacks HAT activity (p300-HAT) did not reverse the E1A-mediated inhibition, indicating that the HAT activity of p300 is necessary in order for p300 to functionally interact with the AR. AR protein expression levels remained constant during these experiments, suggesting that these events take place at the transcriptional level. To assess a direct role for p300 in transcriptional activation of endogenously expressed AR target genes, small interfering RNAs were used to downregulate p300 expression. LNCaP cells were transfected with either siRNA oligonucleotides specifically designed to target p300 or nonspecific oligonuleotides as control. After transfection, cells were treated with 50 ng ml^{-1} IL-6. Cells were subjected to immunocytochemistry using antibodies specific to p300, AR, and PSA. Treatment of cells with IL-6 enhanced PSA expression. However, IL-6 had no effect on PSA expression in cells that were previously transfected with siRNA against p300. These data were confirmed by Western blot to detect PSA protein expression. Again, AR protein levels remained constant during these studies, including during disruption of p300. Our data indicate that p300 is directly involved in transactivation of the AR by IL-6 under ADI conditions. It should be noted, however, that p 300 transfection did not further increase activation of the AR by IL-6 and did not induce AR activation when transfected in the absence of IL-6, indicating that p300 is necessary but not sufficient to induce AR activation (10).

p300 Modulates Nuclear Morphology in PC. Morphologic changes in the structure of cells, primarily the nuclei, are characteristic features of most cancers. Significantly, quantifiable nuclear features of PC cells have been shown to correlate with disease progression. Because we had demonstrated that p300 expression correlates with PC progression following prostatectomy, we addressed whether p300 plays a role in nuclear morphologic changes in PC cells. To this end, we examined tissue samples from our patient population with biopsy-proven PC, and tested whether p300 expression correlated with nuclear alterations measured and quantified by digital image analysis (DIA). Interestingly, we found significant associations between p300 expression and several nuclear alterations in PC tissues, such as nuclear area, minimum diameter, DNA mass, and DNA index. To test whether these alterations were a direct effect of p300 on PC cell structure, we transfected a p300 expression construct or empty vector into the LNCaP-C4-2 PC cell line, a well-characterized model for ADI PC. Following transfection, cells were stained with Feulgen dye and subjected to DIA to assess nuclear alterations. Remarkably, we observed increases in many of these same features, such as nuclear area, perimeter, and minimum/maximum diameter, in cells transfected with p300. Transfection of LNCaP-C4-2 cells with an expression construct encoding the p300 homologue CBP or a vector harboring a noncoding sequence of similar size did not induce changes in nuclear features, indicating these effects to be specific for p300. Upon further analysis, we determined that the nuclear alterations associated with and induced by p300, including nuclear perimeter, area, and minimum/maximum diameter, each independently correlated with a more aggressive phenotype as judged by Ki-67 expression and extraprostatic extension of the tumor at the time of surgery (11).

These results, therefore, provide compelling evidence that the changes in nuclear morphology induced by p300 in PC cells are of clinical importance and correlate with aggressive PC. These observations further support our hypothesis that p300 plays an important role in the ADI progression of PC.

Alterations in nuclear structure frequently involve deregulation of nuclear matrix proteins, such as lamin, PC-1, and NMP149, which show differential expression in cancer tissues. Therefore, we examined whether an increase in p300 expression could affect expression of factors involved in nuclear structure formation. To this end, LNCaP-C4-2 cells were transfected with an expression construct for p300 or empty vector. Expression of several nuclear structure-related factors, including lamin A/C, lamina-associated peptide 2, laminins, and tubulins, 48 h after transfection, was evaluated. Real time PCR and semiquantitative PCR measurements revealed increases in the expression of lamin A/C. Elevated lamin A/C expression was confirmed by Western blotting on nuclear extracts from LNCaP-C4-2 (11). In summary, these data are the first to show direct regulation of quantifiable nuclear features by a transcriptional coactivator, through a mechanism involving regulation of nuclear lamin A/C levels.

Discussion

Taken together, our data point toward a central role for p300 expression as a determinant for the development of aggressive PC features. Noteworthy, comparison of the findings obtained from the clinical material available from 95 patients with biopsy-proven PC that were treated with radical retropubic prostatectomy without neoadjuvant therapy and studies on p300 overexpression in PC cell lines in culture reveals remarkable similarities (Fig. 2). Indeed, both naturally occurring and

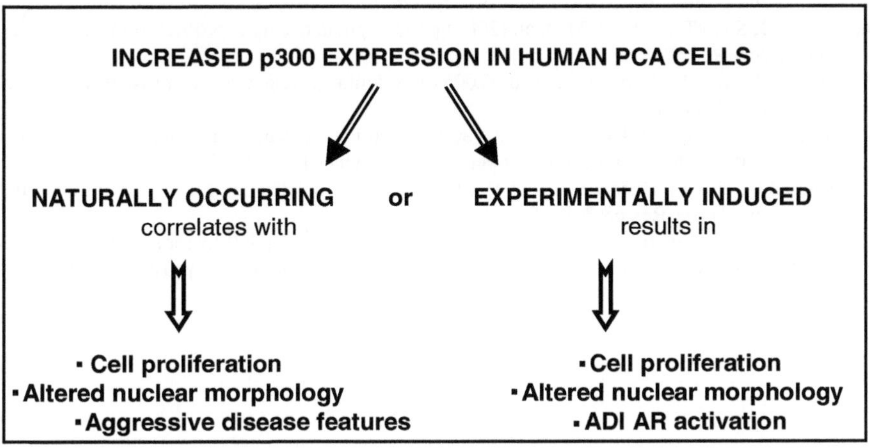

Fig. 2 p300 expression in PC cell lines in culture

experimentally induced increases in p300 expression in PC cells correlates with enhanced cell proliferation and modulation of nuclear morphology. Moreover, consistent with a role for p300 in AR activation in the ADI state of the disease, p300 was shown to be necessary for IL-6 induced activation of AR activity under androgen deprived conditions.

Recently, the neuropeptide bombesin was shown to be able to enhance the HAT activity of p300 protein while leaving p300 expression levels unaltered (12). In view of these findings and the observation that p300 HAT activity is critical for ADI AR activity (10), it will be important that the explore whether growth factors and cytokines such as IL-6 that are frequently overexpressed in ADI can have an additional affect on p300 activity. Finally, as p300 as well as other coactivators are emerging as potential therapeutic targets, efforts should be directed toward identifying the mechanisms underlying and the factors driving p300 overexpression.

References

1. Grossmann ME, Huang H, Tindall DJ (2001) Androgen receptor signaling in androgen-refractory prostate cancer. J Natl Cancer Inst 93:1687–97.
2. Debes JD, Tindall DJ (2004) Mechanisms of androgen-refractory prostate cancer. N Engl J Med 351:1488–90.
3. Heinlein CA, Chang C (2002) Androgen receptor (AR) coregulators: an overview. Endocr Rev 23:175–200.
4. Heemers HV, Tindall DJ (2005) Androgen receptor coregulatory proteins as potential therapeutic targets in the treatment of prostate cancer. Curr Cancer Ther Rev 1:175–86.
5. Kalkhoven E (2004) CBP and p300: HATs for different occasions. Biochem Pharmacol 68:1145–55.
6. Fu M, Wang C, Reutens AT, et al. (2000) p300 and p300/cAMP-response element-binding protein-associated factor acetylate the androgen receptor at sites governing hormone-dependent transactivation. J Biol Chem 275:20853–60.
7. Fu M, Rao M, Wang C, et al. (2003) Acetylation of androgen receptor enhances coactivator binding and promotes prostate cancer cell growth. Mol Cell Biol 23:8563–75.
8. Debes JD, Sebo TJ, Lohse CM, et al. (2003) p300 in prostate cancer proliferation and progression. Cancer Res 63:7638–40.
9. Culig Z, Steiner H, Bartsch G, et al. (2005) Interleukin-6 regulation of prostate cancer cell growth. J Cell Biochem 95:497–505.
10. Debes JD, Schmidt LJ, Huang H, et al. (2002) p300 mediates androgen-independent transactivation of the androgen receptor by interleukin 6. Cancer Res 62:5632–6.
11. Debes JD, Sebo TJ, Heemers HV, et al. (2005) p300 modulates nuclear morphology in prostate cancer. Cancer Res 65:708–12.
12. Gong J, Zhu J, Goodman OB, Jr, et al. (2006) Activation of p300 histone acetyltransferase activity and acetylation of the androgen receptor by bombesin in prostate cancer cells. Oncogene. 25:2011–21.

Characterization of Androgen Regulation of ZEB-1 and PSA in 22RV1 Prostate Cancer Cells

Bynthia M. Anose, Lisa LaGoo, and Jamie Schwendinger

Summary Two-thirds of patients who present with metastatic prostate cancer (PC) are dead within 5 years of diagnosis. The comparable survival rate for patients with localized disease is 100%, which clearly stresses the need for pursuing and developing bioassays that allow prediction of which localized cases are most likely to metastasize. The commonly assayed prostate specific antigen (PSA), while touted as a transformation biomarker, has recently proven to be problematic in the area of false positive diagnoses. It remains, however, a hallmark gene for studying androgen regulation as its expression is reliably stimulated by androgens such as dihydrotestosterone (DHT). Herein, we have elucidated the effects of flutamide (a defined anti-androgen) and DHT on the expression of PSA and Zinc finger E-box Binding factor (ZEB-1). Additionally, we assayed the androgenic capabilities of two DHT derivatives on expression of *PSA*. Our previous research had identified *ZEB-1* as a putative biomarker for the onset of metastasis in prostate cancer. The expression of this gene is regulated by androgen and decreases sharply at metastasis. In the current study, the effects of 1 and 10 nM flutamide, in combination with 1 and 10 nM DHT, on expression of *ZEB-1* and *PSA*, were investigated in 22Rv1, an androgen-responsive human PC cell line. Also in this cell line, the effects of testosterone propionate and dehydroisoandrosterone were studied. Our research confirmed the feasibility of considering *ZEB-1* a metastatic PCa biomarker, using the highly sensitive technique of real-time polymerase chain reaction (RT-PCR). Interestingly, it also revealed the danger of using flutamide as a therapeutic antagonist, as we demonstrate herein its alarming capability to behave as an agonist.

Introduction

Prostate cancer (PC) is the most common form of cancer diagnosed in USA men. According to an April 2006 report of the American Cancer Society, the five-year survival rate for a patient with localized PC is 100%; however, if the tumor metastasizes, the survival rate drops significantly to 34% (1). Because of this severe increase in mortality with the onset of metastasis, the urgent need for researchers to identify a highly specific biomarker to predict metastasis in PC patients is clear.

J.J. Li et al. (eds.), *Hormonal Carcinogenesis V.*

Prostate specific antigen (PSA) is a protein secreted by the prostate epithelium and is relatively easy to assay clinically. Expression of the gene is remarkably induced by androgen (2). PSA protein has long been touted as a prostate cancer transformation biomarker because its concentration in the bloodstream has been shown to increase with the onset of metastasis. However, PSA expression can be stimulated by a variety of other prostate problems, including benign prostatic hyperplasia and prostate infection (3). Because of this dilemma of yielding false-positives, it is desirable to locate a more specific biomarker for the onset of human metastatic PC. Prior work in our laboratory identified human Zinc finger E-box Binding factor (ZEB)-1 as a promising metastatic biomarker (4); *ZEB-1* expression decreases sharply at the onset of metastasis in human PC.

To confirm the feasibility of using ZEB-1 as a metastatic biomarker, its regulation must be characterized. We have previously shown that the expression of *ZEB-1* is induced by the androgen dihydrotestosterone (DHT) (4), but no one has yet shown a decrease in its expression with anti-androgen treatment. Current PC treatment involves the administration of flutamide (5–7), an anti-androgen that inhibits the expression of AR-regulated genes by binding the androgen receptor (AR). The expectation is that flutamide will decrease the expression of AR-regulated genes involved in the proliferation of tumors (8).

Despite the questionable nature of PSA expression as an indicator of PC metastasis, the gene remains quite useful in androgen research, since its expression is reliably and positively regulated by androgens. In fact, the regulation of this gene by androgen derivatives has yet to be characterized and may provide insights into the PSA phenomenon of false positives. The research presented herein investigates the effects of 1 and 10 nM DHT and flutamide on ZEB-1 and PSA expression in 22Rv1, an androgen-responsive human prostatic carcinoma cell line. Also in this cell line, the effects of testosterone propionate and dehydroisoandrosterone were studied, to determine the effects of these testosterone derivatives on *PSA* expression.

Cell Culture and Reverse Transcription. Human PC cell line 22Rv1 (American Type Culture Collection, ATCC) was obtained and maintained in RPMI 1640 medium (ATCC). The cells were treated for 24 h with 0, 8, or 9 nM DHT; 0, 8, or 9 nM flutamide; or 8 nM DHT + 8 nM flutamide (Sigma-Aldrich). Total RNA was harvested using the RNeasy© Plus Mini Kit (Qiagen). Reverse transcription was performed according to manufacturer's protocol using ThermoScript™ (Invitrogen) to create cDNA.

Quantitative Real-Time PCR. cDNA from the DHT-, flutamide-, and DHT + flutamide-treated cells was amplified in parallel reactions, either with primers for PSA or with primers for ZEB-1. Quantitative RT-PCR was carried out in 96-well plates on the *iCycler* PCR machine from BioRad. Each well contained 0.334 μg cDNA, 2 μl 20 mM ZEB-1 or PSA primers, 12.5 μl SYBR® Green *iQ* PCR Master Mix (BioRad), and 7.5 μl sterile water. The reactions were thermocycled as follows:

Results

22Rv1 is an androgen-responsive human PC cell line. Upon treatment with the potent androgen DHT, both PSA and ZEB-1 expression levels in 22Rv1 cells increased greatly. To determine relative changes in gene expression by RT-PCR,

Cycle 1:	(1×)	Step 1:	95.0°C	for 03:00
Cycle 2:	(40×)	Step 1:	95.0°C	for 00:10
		Step 2:	55.0°C	for 00:45
Cycle 3:	(1×)	Step 1:	95.0°C	for 01:00
Cycle 4:	(1×)	Step 1:	55.0°C	for 01:00

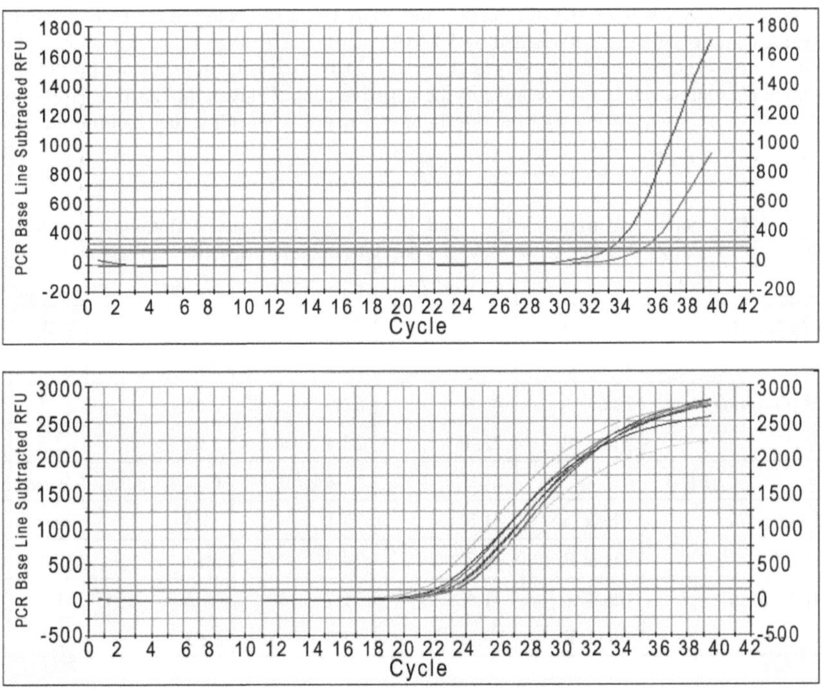

Fig. 1 The point at which the *curves* cross the *dotted line* is known as the threshold-crossing point. For the negative control, the average C_T value is 34. The average C_T value for PSA expression in DHT-treated cells is 23. Therefore, ΔC_T for PSA expression in DHT-treated cells equals 11. The equation $\Delta Expression = 2^{\Delta CT}$ calculates the overall change in gene expression and reveals that DHT-treated cells produce PSA mRNA in amounts roughly 2 000 times greater than untreated 22Rv1 cells

the average threshold-crossing (C_T) points of PSA or ZEB-1 cDNA from treated cells were compared with the corresponding C_T values of untreated cells (ethanol vehicle alone, negative control). The difference between these two values, known as ΔC_T, served to elucidate the relationship between mRNA expression in treated and untreated cells, using the equation: $\Delta Expression = 2^{\Delta CT}$. Illustrative graphs

Table 1 22Rv1 cells treated with 1 and 10 nM DHT for 24 h[a]

PSA		C_T *well 1*	C_T *well 2*	C_T *well 3*	*Average* C_T
1 nM DHT	Flask A	21.9	21.4	21.6	21.6
	Flask B	22.9	22.7	22.8	22.8
10 nM DHT	Flask A	24.1	23.9	24.7	24.2
ZEB-1		C_T *well 1*	C_T *well 2*	C_T *well 3*	*Average* C_T
1 nM DHT	Flask A	29.8	30	30.3	30.0
	Flask B	33.9	31.1	31.5	32.2
10 nM DHT	Flask A	33	33.2	32.7	33.0

[a] The harvested RNA was reverse transcribed into cDNA and subjected to RT-PCR. Changes in PSA expression levels were determined by RT-PCR, using the average C_T scores of treated cells compared with negative control C_T scores ($C_T = 34$). ZEB-1 expression levels are provided only as average C_T scores, rather than relative to the negative control, given that ZEB-1 expression in the negative control was so low that the threshold was not crossed (C_T values indeterminable)

Table 2 22Rv1 cells were treated with 1 and 10 nM concentrations of flutamide for 24 h, and changes in PSA and ZEB-1 expression levels were determined as described in legend to Table 1

PSA		C_T *well 1*	C_T *well 2*	C_T *well 3*	*Average* C_T
1 nM flutamide	Flask A	21.7	21.4	22.4	21.8
	Flask B	20.6	20.7	20.9	20.7
10 nM flutamide	Flask A	22.6	22.9	22.6	22.5
	Flask B	22.3	22.4	22.1	22.3
ZEB-1		C_T *well 1*	C_T *well 2*	C_T *well 3*	*Average* C_T
1 nM flutamide	Flask A	31.5	33.1	31.5	32.0
	Flask B	30.7	29.7	30	30.1
10 nM flutamide	Flask A	31.4	31.7	31.5	31.9
	Flask B	32.3	32.3	32.3	32.3

Table 3 22Rv1 cells were treated with a combination of 10 nM DHT + 10 nM flutamide, and changes in *PSA* and *ZEB-1* expression levels were determined as described in legend to Table 1

DHT + Flutamide	C_T *well 1*	C_T *well 2*	C_T *well 3*	*Average* C_T
PSA	22.5	21.9	21.8	22.1
ZEB-1	30.8	31.1	31.8	31.2

showing curves for PSA expression in ethanol- and DHT-treated cells are shown in Fig. 1.

In this study, two concentrations of DHT (1 and 10 nM) were tested. DHT concentration to maximally induce PSA expression was 1 nM (Table 1). This same cell line was also treated with 1 and 10 nM concentrations of flutamide, an anti-androgen. Surprisingly, flutamide exhibited similar stimulation of PSA and ZEB-1 levels in the 22Rv1 cells (Table 2).

In addition, a 10 nM combination of both flutamide + DHT increased PSA and ZEB-1 expression levels to an even greater extent than DHT treatment alone, yet to a lesser extent than treatment with flutamide alone (Table 3).

To investigate the phenomenon of false-positive diagnoses of advanced PC due to elevated *PSA* expression levels, we hypothesized that PSA may be stimulated in vivo by metabolic breakdown products of potent androgens. We studied the ability of several heretofore uninvestigated DHT derivatives to stimulate *PSA* expression. Although the positive control, testosterone, induces a marked increase in PSA mRNA levels, treatment with either dehydroisoandrosterone or testosterone propionate failed to cause an increase in PSA expression, indicating that neither of these derivatives take part in the reoccurrence of prostate cancer after initial androgen ablation treatments (data not shown).

Conclusions

These results illustrate that ZEB-1 remains a feasible alternative to PSA as a transformation biomarker due to its responsiveness to androgens and its previously demonstrated decrease in expression upon PC metastasis. However, this study calls into question the suitability of flutamide as an androgen ablation treatment for reproductive cancers. We show flutamide treatment not only fails to block androgen stimulation of target genes, but synergistically enhances androgen stimulation in some cases. Additionally, we negated the possibility that two DHT derivatives, dehydroisoandrosterone and testosterone propionate, play a role in secondary PC invasions. Future work is required to corroborate the evidence provided herein that flutamide stimulates androgen-regulated target gene expression, and to confirm whether flutamide does indeed increase expression of human *ZEB-1*.

References

1. What are the key statistics about prostate cancer? (2006) American Cancer Society, Inc. Accessed 12 May, 2006. Available at http://www.cancer.org/docroot/CRI/content/CRI_2_4_1X_What_are_the_key_statistics_for_prostate_cancer_36.asp?sitearea
2. Hartel A, Didier A, Pfaffl MW, et al. (2003) Characterisation of gene expression patterns in 22RV1 cells for determination of environmental androgenic/antiandrogenic compounds. J Steroid Biochem Mol Biol 84:231–238.

3. Barry MJ (2001) Prostate specific antigen testing for early diagnosis of prostate cancer. New Engl J Medicine 344:1373–1377.
4. Anose BM, Landry MM, and Sanders MM (2004) Investigating the Role and Regulation of ZEB-1 by Quantitative Real-Time PCR. 12th International Congress of Endocrinology, pp. 1473–1478.
5. Ornstein DK, Rao GS, Johnson B, et al. (1996) Combined finasteride and flutamide therapy in men with advanced prostate cancer. Urology 48:901–905.
6. Stone NN, Clejan SJ (1991) Response of prostate volume, prostate-specific antigen, and testosterone to flutamide in men with benign prostatic hyperplasia. J Andrology 12:376–380.
7. Airhart RA, Barnett TF, Sullivan JW, et al. (1978) Flutamide therapy for carcinoma of the prostate. S Med J 71:798–801.
8. Taplin ME, Bubley GJ, Shuster TD, et al. (1995) Mutation of the androgen-receptor gene in metastatic androgen-independent prostate cancer. New Engl J Med 332:1393–1398.

Mcl-1 is Regulated by IL-6 and Mediates the Survival Activity of the Cytokine in a Model of Late Stage Prostate Carcinoma

Ilaria Teresa Cavarretta, Hannes Neuwirt, Mohamed H. Zaki, Hannes Steiner, Alfred Hobisch, Jeffrey A. Nemeth, and Zoran Culig

Summary The proinflammatory cytokine interleukin-6 (IL-6) has been considered a positive growth factor in late stage prostate cancer (PC) cells and a potential target for therapeutic interference. We studied the effects of inhibition of IL-6 in LNCaP-IL6$^+$ cells, a model system for advanced PC, which produce IL-6. By using the chimeric anti-IL-6 antibody, CNTO 328, we showed that the autocrine IL-6 loop is responsible for decreased sensitivity of LNCaP-IL-6$^+$ cells to die by apoptosis. Dysregulation of Bcl-2 family members could be implicated in the acquisition of resistance to apoptosis in malignant cell lines. Myeloid cell leukemia 1 (Mcl-1) is an antiapoptotic member of this family that is overexpressed in the IL-6 selected cells compared with control. Specific knock-down of *Mcl-1* gene expression by siRNA yielded an increase in apoptosis of LNCaP-IL-6$^+$ cells. Interestingly, inactivation of IL-6 autocrine loop was not able to increase apoptosis levels in the absence of Mcl-1, thus suggesting this molecule as a mediator of the survival action of IL-6. Finally, using selective kinase inhibitors we provide evidence for the involvement of p38 and ERK1/2 mitogen-activated protein kinases pathways in the IL-6-mediated regulation of Mcl-1. In conclusion, these data suggest that endogenous IL-6 acts as an antiapoptotic factor in LNCaP-IL-6$^+$ cells and that Mcl-1 is critical for its survival activity. CNTO 328, in our experimental conditions, is able to render LNCaP-IL-6$^+$ cells more sensitive to apoptosis. These data support the concept of anti-IL-6 therapy in human PC.

Introduction

Prostate cancer (PC) could be cured only if it is detected as an organ-confined tumor and the treatments include radical prostatectomy or radiation therapy. Other patients receive endocrine therapy that is aimed to diminish the levels of circulating androgen or block activation of the androgen receptor (AR). However, prostate growth and proliferation are regulated also by other steroid and peptide hormones. Interleukin-6 (IL-6), a cytokine mostly involved in modulation of immuno- and inflammatory responses, is implicated in the development and progression of several tumors including those of the prostate. In vivo, IL-6 levels are higher in sera

J.J. Li et al. (eds.), *Hormonal Carcinogenesis V.*

and prostate tissues from patients with therapy-resistant metastatic disease and correlate with a bad prognosis (1). IL-6 binds to its specific receptor, IL-6 receptor α (IL-6Rα), an 80 kDa glycosylated transmembrane protein, belonging to the cytokine receptor superfamily. The complex IL-6/IL-6Rα associates with the protein gp130, a transmembrane common cytokine receptor, thereby inducing its dimerization followed by activation of multiple signaling pathways, in particular those of Janus kinases (JAK)/signal transducers and activators of transcription (STAT) factors and mitogen-activated protein kinase (MAPK), in a cell-specific manner. ERK1/2 MAPK have been frequently associated with cellular proliferation. Recently, it became evident that they also have a role in promoting survival in different cellular models (2–4). Phospho-ERK1/2 are overexpressed in LNCaP-IL-6^+ PC cells (5). Another MAPK, p38MAPK, is activated by cytokines and growth factors (IL-1, TNFβ, EGF, TGFβ) as well as from stress events (6). p38 MAPK has been shown to have a differential expression/activation during the initiation and progression of prostate cancer.

In the attempt to obtain an in vitro model that represents late stage prostate tumor cells (i.e., exposed to high cytokine levels), a new cell line was derived from parental LNCaP cells maintained in the presence of IL-6 (7). Continuous exposure to the cytokine was responsible for selection of LNCaP-IL-6^+ cells that acquired growth advantage compared with control LNCaP-IL-6^- cells passaged at the same time in the absence of IL-6. Differently from parental and control cells, LNCaP-IL-6^+ express and secrete the cytokine and do not undergo growth inhibition when exposed to exogenous IL-6. The higher rate of proliferation of the LNCaP-IL-6^+ cells can be explained by alterations in expression of cyclin-dependent kinases, p27, and retinoblastoma, as well as by differences in JAK/STAT3 and MAPKs expression and phosphorylation. On the basis of experimental results with prostate cancer cell lines and xenografts, most studies support the view that IL-6 is a target for novel therapies in prostate cancer (8).

Deregulation of Bcl-2 family members is often implicated in acquisition of resistance to apoptosis in malignant cell lines (9). We verified whether the levels of some crucial proapoptotic and antiapoptotic proteins are altered in LNCaP-IL-6^+ cells compared with the control cells. In particular, we focused on Mcl-1 (myeloid cell leukemia-1). Its role in the regulation of cell death has been mostly investigated in myeloma cells in which it mediates the prosurvival action of IL-6 (10). Altered expression of Mcl-1 in specimens from PC compared with those from normal, hyperplastic, or prostate intraepithelial neoplasia (PIN) (11–13) suggests an involvement of this gene also in PC.

In this study, we investigated sensitivity of LNCaP-IL-6^+ cells to apoptosis compared with IL-6-negative control cells. By applying the siRNA technology and using the chimeric anti-IL-6 antibody CNTO 328, we highlight the importance of Mcl-1 in mediating survival activity of IL-6 in prostate cancer. We also explored the signaling pathways involved in the regulation of this protein. In summary, we demonstrated that endogenous IL-6 acts as a survival factor in cells representing advanced PC and that Mcl-1 is an important mediator of its activity. The p38 and ERK1/2 MAPKs are specifically involved in this process. CNTO 328 is able to render LNCaP-IL-6⁺ cells more sensitive to apoptosis. Taken together, these results support the concept of anti-IL-6 therapy in human prostate cancer.

Results

Neutralization of IL-6 Autocrine Loop by CNTO 328 Inhibits Cell Viability and Increases Apoptosis of LNCaP-IL-6⁺ Cells. We asked whether the autocrine IL-6 loop is responsible for decreased sensitivity of LNCaP-IL-6⁺ to apoptosis. To this aim, LNCaP-IL-6⁺ cells were treated with the chimeric anti-IL-6 antibody CNTO 328, at a concentration of $10\,\mu g\ ml^{-1}$, in the absence of serum. After 4 and 6 days of treatment, apoptosis and cell viability, respectively, were evaluated. Neutralization of IL-6 activity by CNTO 328 decreases cell viability and enhances apoptosis of LNCaP-IL-6⁺ by about 30% compared with levels measured in the presence of control cells treated with nonimmune hIgG (Fig. 1). These data show that endogenous IL-6 protects LNCaP-IL-6⁺ cells from apoptosis and that CNTO 328, in these experimental conditions, is able to antagonize the cytokine's effect.

Levels of Bcl-2 Family Members are Altered in the IL-6 Positive Cell Line. Bcl-2 family members are frequently implicated in the acquisition of resistance to apoptosis in malignant cell lines. We examined possible differences between control and LNCa-P-IL-6⁺ cells in terms of expression of pro- and antiapoptotic members of this family. In particular, Bax, Bcl-2, and Mcl-1 proteins were quantified by Western blot analyses. We found that, differently from control cells, LNCaP-IL-6⁺ cells lack the proapoptotic Bax protein and present with significantly higher levels of the antiapoptotic Bcl-2 and Mcl-1 proteins. We, therefore, investigated whether deregulation of Bcl-2 proteins is linked to the production of IL-6. To clarify this issue, we quantified Bax, Bcl-2, and Mcl-1 protein levels after neutralization of the endogenous cytokine. Results from Western blot analyses show that treatment with

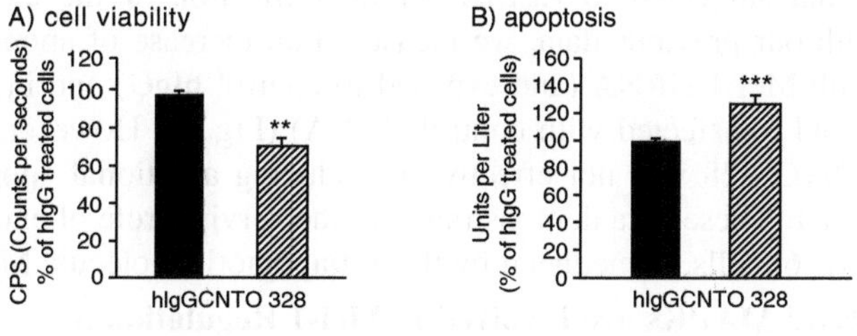

Fig. 1 Effect of the anti-IL-6 antibody on cell viability and apoptosis. Cells were exposed to serum-free medium containing $10\,\mu g\ ml^{-1}$ of the anti-IL-6 antibody (CNTO 328) or equal amounts of hIgG (as a control). Treatment was repeated every two days. (**A**) ATP assay. Mean values ± SE from at least three independent experiments are shown (**$p < 0.01$. Mann-Whitney-U Test). (**B**) Cleaved CK 18 ELISA. Data represent the average of four independent experiments performed in quadriplicates ± SE (***$p < 0.001$. Mann-Whitney-U-Test)

the anti-IL-6 antibody does not restore Bax protein expression, whereas it yielded reduced Bcl-2 and Mcl-1 levels in LNCaP-IL-6$^+$ cells. As expected, CNTO328 did not induce any change in expression of these proteins in the IL-6$^-$ cells. In conclusion, these results show an increased Bcl-2/Bax ratio in LNCaP-IL-6$^+$ cells (compared with control cells) that are strongly in favor of survival and a specific response of these cells to CNTO 328.

Increased Mcl-1 Protein Levels in LNCaP-IL-6$^+$ Cells are Responsible for Resistance to Apoptosis: To get more insight into the mechanisms responsible for the reduced cell viability of LNCaP-IL-6$^+$ cells, we focused our following studies on the Mcl-1 protein because of its association with IL-6 mediated-survival activity in myeloma cells. We investigated whether Mcl-1 accounts for the decreased sensitivity of LNCaP-IL-6$^+$ cells to apoptosis. To this end, Mcl-1 expression was inhibited by using specific siRNA and levels of cleaved CK18 were measured thereafter, as an index of apoptosis. Efficacy of siRNA in decreasing Mcl-1 expression was verified by comparing Mcl-1 protein levels in cells transfected with the specific siRNA to those measured in cells transfected with scrambled siRNA as a control (Fig. 2). Specific siRNA is highly effective in decreasing Mcl-1 levels after 1, 2, or 3 days of siRNA exposure, whereas Mcl-1 levels in cells transfected with the control siRNA did not change at any time. Thus, cells were transfected with 10 nM Mcl-1- or control siRNA, and cleaved CK18 was measured 2 days later. We found that apoptosis increased by about 60% in LNCaP-IL-6$^+$ cells after suppression of Mcl-1. We have performed a similar experiment in DU-145 cells that also express and secrete IL-6. In DU-145 cells, apoptosis increases by 60% after down-regulation of Mcl-1. Taken together, these data show that Mcl-1 is critical for survival of IL-6 positive advanced PC cell lines.

CNTO 328 is not Effective in Increasing Apoptosis in the Absence of Mcl-1. To obtain further support for our hypothesis that the IL-6 survival effect is mediated by Mcl-1, we investigated whether reduction of its expression affects the extent of apoptosis induced by CNTO 328. For this purpose, LNCaP-IL-6$^+$ cells were treated with 10 μg ml^{-1} CNTO 328 or hIgG (+ CNTO 328 vehicle) in serum-free medium one day after transfection of 10 nM Mcl-1 or control siRNA. Medium was collected after 2 days and subjected to ELISA for quantification of the cleaved CK18. Consistent with our previous data, we measured an increase of apoptosis in cells transfected with Mcl-1 siRNA (and exposed to control hIgG) and in cells treated with CNTO (and transfected with control siRNA) (Fig. 3). However, inactivation of IL-6 by CNTO 328 was not effective in inducing additional apoptosis in the absence of Mcl-1. These data demonstrate that the survival role of endogenous IL-6, in LNCaP-IL-6$^+$ cells, is mediated by the antiapoptotic molecule Mcl-1.

p38 and ERK1/2 MAPKs are Involved in Mcl-1 Regulation in IL-6$^+$ PC Cells. We sought to uncover signaling pathways involved in the regulation of Mcl-1 protein. Data from the literature have demonstrated regulation of Mcl-1 by ERK1/2 MAPKs in different cellular models (14–16). We previously showed that LNCaP-IL-6$^+$ cells overexpress the activated form of these kinases (i.e., phospho ERK1/2) (17). Their overexpression is linked to the endogenous production of IL-6 as evi-

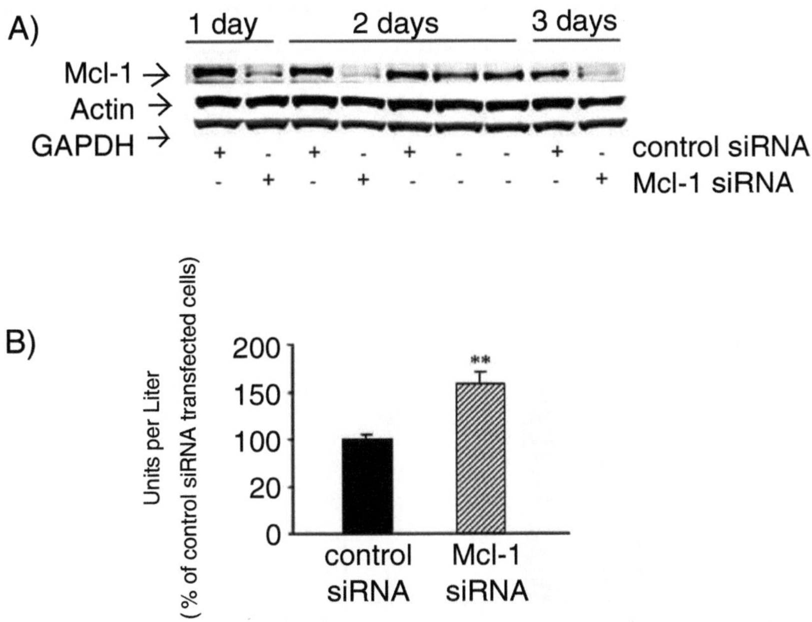

Fig. 2 Implications of down-regulation of Mcl-1 on apoptosis in LNCaP-IL-6+ cells. (**a**) Western Blot Analyses: Representative blots of Mcl-1, Actin, and GAPDH in LNCaP-IL-6+ after transfection, by Lipofectamine, of Mcl-1- or control siRNA. (**b**) Cleaved CK 18 ELISA: Apoptosis in LNCaP-IL-6+ cells after transfection of either Mcl-1- or control siRNA. Data represent the mean of three independent experiments ± SE (**$p < 0.01$. MWU-Test)

denced in experiments in which treatment of LNCaP-IL-6+ cells with CNTO 328 decreased levels of phospho-ERK1/2 by 30%. We investigated whether blockade of this MAPK has a consequence on the levels of Mcl-1 protein. For this reason, LNCaP-IL-6- and LNCaP-IL-6+ cells were exposed for 2 days to the specific MAPK inhibitor PD 98059, at concentrations of 25 and 50 μM. We measured decreased Mcl-1 levels in LNCaP-IL6+ but not in the control cells after PD98059 treatment. This suggests that the IL-6-caused upregulation of Mcl-1 is mediated, at least partially, by ERK1/2 MAPKs. To extend our findings, the levels of Mcl-1 and phosphor-ERK1/2 were compared in PC3, DU-145, and LNCaP-IL-6+ cells. Interestingly, we found very different levels of both Mcl-1 and activated ERK1/2 in the three cell lines and, consistent with our hypothesis, the levels of the antiapoptotic protein were higher in cells with higher expression of phospho MAPKs.

We finally investigated another MAPK, p38, which is known to be activated by different cytokines. Also in this case, we found that blockade of p38 activity by the specific inhibitor SB202190 was responsible for a dose-dependent inhibition of Mcl-1 protein levels in LNCaP-IL-6+ cells but not in the counterpart, LNCaP-IL-6−.

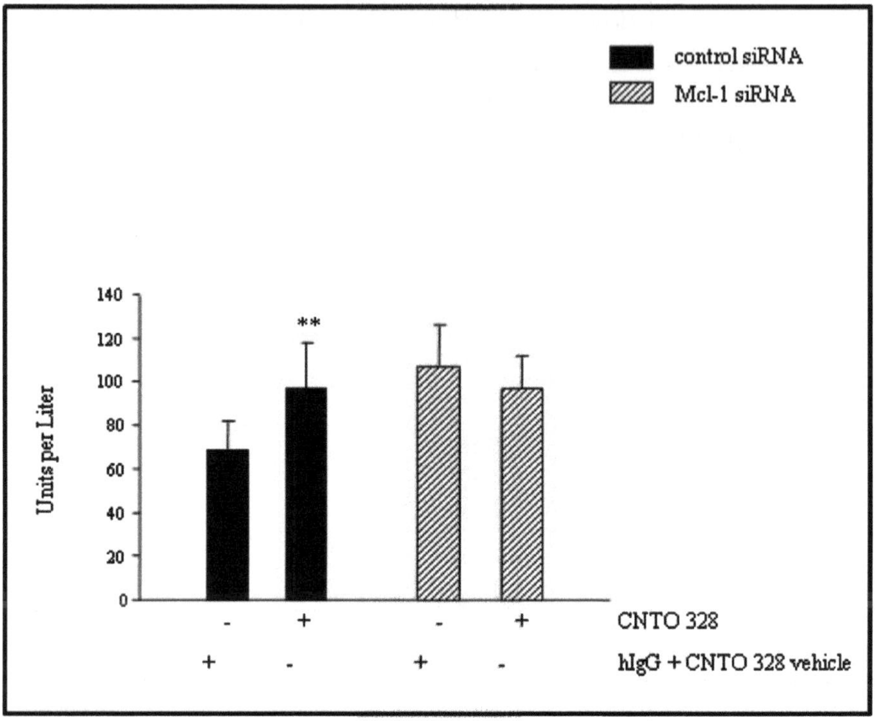

Fig. 3 Effect of the anti-IL-6 antibody CNTO 328 following down-regulation of Mcl-1. Cleaved CK 18 ELISA: After transfection of Mcl-1- or control siRNA, LNCaP-IL-6⁺ cells were exposed for 2 days to either CNTO 328 or hIgG antibody. Afterward, apoptosis rate was determined by M30 apoptosense ELISA kit. Data represent the average of three independent experiments ± SE. (*$p < 0.05$. Mann-Whitney-U-Test)

Further studies to elucidate the possible contribution of this signaling pathway to the survival of LNCaP-IL-6^+ cells are in progress.

Discussion

The survival role of IL-6 was investigated in LNCaP-IL-6^+ cells that represent a model of advanced stage of prostate carcinoma. We previously showed that LNCaP-IL-6^+ cells respond with a lesser extent than the controls (LNCaP-IL-6^-) to induction of apoptosis by IL-6 and the calcium ionophore A23187. The autocrine IL-6 loop established in these IL-6 selected cells is at least partly responsible for acquisition of resistance to apoptosis. Its neutralization by the chimeric anti-IL-6 antibody CNTO 328 decreased their viability and increased apoptosis. These results are consistent with recent data obtained in other IL-6^+ PC cells, PC3, and DU-145, in which neutralization of endogenous IL-6 activity increases their sensitivity to induction of apoptosis by cytotoxic drugs (18–21). In preclinical studies, neutralization of IL-6 by either murine or chimeric form of CNTO 328 induced apoptosis and regression of PC3 and inhibition of growth of LuCaP 35 prostate tumors xenografted in nude mice, respectively (18). Besides, in the same animal model, the anti-IL-6 antibody prevented conversion to an androgen-independent phenotype.

Consistent with the less sensitivity to apoptosis displayed by LNCaP-IL-6$^+$ cells (compared with controls), the ratio between the antiapoptotic and proapoptotic Bcl-2 family members was found increased (in the same cells). In particular, Mcl-1 was shown to have an active role in protection from apoptosis as evidenced in experiments in which siRNA technology was applied. Despite the known crucial role of Mcl-1 in protecting cells from apoptosis, in the last decade only a few studies had explored Mcl-1 involvement in prostate carcinogenesis (11–13, 22).

Between the IL-6 activated signaling pathways, STAT3 was shown to be involved in Mcl-1 regulation in different cancer cells (23–27). However, we can exclude an involvement of the JAK/STAT3 pathway in the regulation of Mcl-1 in LNCaP-IL-6$^+$ cells since these cells lack the phosphorylated form of STAT3 (17). On the contrary, they express higher levels of phospho-ERK1/2 MAPKs compared with control cells. We demonstrated that these kinases are involved in the regulation of the antiapoptotic Mcl-1 protein as shown in experiments in which the ERK1/2 specific inhibitor was utilized. Interestingly, the same inhibitor was recently shown to increase docetaxel-induced apoptosis of androgen-independent C-81 LNCaP PC cells (28). Also the p38 MAPK contributed to the basal levels of Mcl-1, specifically in LNCaP-IL-6$^+$ and not in the control cells highlighting a complex and articulated regulation of Mcl-1 in IL-6$^+$ cell lines.

Finally, the inability of CNTO 328 to cause additional apoptosis in cells transfected with siRNA against Mcl-1 provides further support for the hypothesis that the IL-6 survival effect is mediated by Mcl-1. Taken together, results of our group and other researchers imply that CNTO 328 has a potential in PC novel experimental therapies.

References

1. Twillie DA, Eisenberger MA, Carducci MA, et al. (1995) Interleukin-6: a candidate mediator of human prostate cancer morbidity. Urology 45: 542–9.
2. Yoon S, Seger R (2006) The extracellular signal-regulated kinase: multiple substrates regulate diverse cellular functions. Growth Factors 24(1): 21–44.
3. Rubinfeld H, Seger R (2005) The ERK cascade: a prototype of MAPK signaling. Mol Biotechnol 31(2): 151–74.
4. Sridhar SS, Hedley D, and Siu LL (2005) Raf kinase as a target for anticancer therapeutics. Mol Cancer Ther 4(4): 677–85.
5. Shimada K, Nakamura M, Ishida E, et al. (2002) Contributions of mitogen-activated protein kinase and nuclear factor kappa B to *N*-(4-hydroxyphenyl)retinamide-induced apoptosis in prostate cancer cells. Mol Carcinog 35(3): 127–37.

6. Ricote M, Garcia-Tunon I, Bethencourt F, et al. (2006) The p38 transduction pathway in prostatic neoplasia. J Pathol 208(3): 401–7.
7. Hobisch A, Ramoner R, Fuchs D, et al. (2001) Prostate cancer cells (LNCaP) generated after long-term interleukin 6 (IL-6) treatment express IL-6 and acquire an IL-6 partially resistant phenotype. Clin Cancer Res 7(9): 2941–8.
8. Culig Z, Steiner H, Bartsch G, et al. (2005) Interleukin-6 regulation of prostate cancer cell growth. J Cell Biochem 95: 497–505.
9. Fadeel B, Orrenius S (2005) Apoptosis: a basic biological phenomenon with wide-ranging implications in human disease. J Intern Med 258(6): 479–517.
10. Jourdan M, Veyrune JL, Vos JD, et al. (2003) A major role for Mcl-1 antiapoptotic protein in the IL-6-induced survival of human myeloma cells. Oncogene 22(19): 2950–9.
11. Royuela M, Arenas MI, Bethencourt FR, et al. (2001) Immunoexpressions of p21, Rb, mcl-1 and bad gene products in normal, hyperplastic and carcinomatous human prostates. Eur Cytokine Netw 12(4): 654–63.
12. Krajewska M, Krajewski S, Epstein JI, et al. (1996) Immunohistochemical analysis of bcl-2, bax, bcl-X, and mcl-1 expression in prostate cancers. Am J Pathol 148(5): 1567–76.
13. Krajewski S, Bodrug S, Krajewska M, et al. (1995) Immunohistochemical analysis of Mcl-1 protein in human tissues. Differential regulation of Mcl-1 and Bcl-2 protein production suggests a unique role for Mcl-1 in control of programmed cell death in vivo. Am J Pathol 146(6): 1309–19.
14. Derouet M, Thomas L, Cross A, et al. (2004) Granulocyte macrophage colony-stimulating factor signaling and proteasome inhibition delay neutrophil apoptosis by increasing the stability of Mcl-1. J Biol Chem 279(26): 26915–21.
15. Boucher MJ, Morisset J, Vachon PH, et al. (2000) MEK/ERK signaling pathway regulates the expression of Bcl-2, Bcl-X(L), and Mcl-1 and promotes survival of human pancreatic cancer cells. J Cell Biochem 79(3): 355–69.
16. Selzer E, Thallinger C, Hoeller C, et al. (2002) Betulinic acid-induced Mcl-1 expression in human melanoma – mode of action and functional significance. Mol Med 8(12): 877–84.
17. Steiner H, Godoy-Tundidor S, Rogatsch H, et al. (2003) Accelerated in vivo growth of prostate tumors that up-regulate interleukin-6 is associated with reduced retinoblastoma protein expression and activation of the mitogen-activated protein kinase pathway. Am J Pathol 162(2): 655–63.
18. Borsellino N, Belldegrun A, Bonavida B (1995) Endogenous interleukin 6 is a resistance factor for cis-diamminedichloroplatinum and etoposide-mediated cytotoxicity of human prostate carcinoma cell lines. Cancer Res 55(20): 4633–9.
19. Borsellino N, Bonavida B, Ciliberto G, et al. (1999) Blocking signaling through the Gp130 receptor chain by interleukin-6 and oncostatin M inhibits PC-3 cell growth and sensitizes the tumor cells to etoposide and cisplatin-mediated cytotoxicity. Cancer 85(1): 134–44.
20. Pu YS, Hour TC, Chuang SE, et al. (2004) Interleukin-6 is responsible for drug resistance and anti-apoptotic effects in prostatic cancer cells. Prostate 60(2): 120–9.
21. Smith PC, Hobisch A, Lin DL, et al. (2001) Interleukin-6 and prostate cancer progression. Cytokine Growth Factor Rev 12(1): 33–40.
22. Royuela M, Arenas MI, Bethencourt FR, et al. (2002) Regulation of proliferation/apoptosis equilibrium by mitogen-activated protein kinases in normal, hyperplastic, and carcinomatous human prostate. Hum Pathol 33(3): 299–306.
23. Raje N, Kumar S, Hideshima T, et al. (2005) Seliciclib (CYC202 or R-roscovitine), a small-molecule cyclin-dependent kinase inhibitor, mediates activity via down-regulation of Mcl-1 in multiple myeloma. Blood 106(3): 1042–7.
24. Puthier D, Derenne S, Barille S, et al. (1999) Mcl-1 and Bcl-xL are co-regulated by IL-6 in human myeloma cells. Br J Haematol, 107(2): 392–5.
25. Lee SO, Lou W, Johnson CS, et al. (2004) Interleukin-6 protects LNCaP cells from apoptosis induced by androgen deprivation through the Stat3 pathway. Prostate 60(3): 178–86.
26. Nam S, Buettner R, Turkson J, et al. (2005) Indirubin derivatives inhibit Stat3 signaling and induce apoptosis in human cancer cells. Proc Natl Acad Sci USA 102(17): 5998–6003.

27. Niu G, Bowman T, Huang M, et al. (2002) Roles of activated Src and Stat3 signaling in melanoma tumor cell growth. Oncogene 21(46): 7001–10.
28. Zelivianski S, Spellman M, Kellerman M, et al. (2003) ERK inhibitor PD98059 enhances docetaxel-induced apoptosis of androgen-independent human prostate cancer cells. Int J Cancer 107(3): 478–85.

Sex Hormone-binding Globulin Influences Gene Expression of LNCaP and MCF-7 Cells in Response to Androgen and Estrogen Treatment

Scott M. Kahn, Yu-Hua Li, Daniel J. Hryb, Atif M. Nakhla, Nicholas A. Romas, Janice Cheong, and William Rosner

Summary Sex hormone-binding globulin (SHBG), a plasma protein that binds androgens and estrogens, also participates in the initial steps of a membrane-based steroid signaling pathway in human prostate and breast. We have recently shown that SHBG is expressed at the mRNA and protein levels in the prostate and breast. In this study, we addressed whether locally expressed SHBG: (1) Functions to regulate activation of membrane-based steroid signaling and (2) influences activation of the androgen (AR) and estrogen (ER) receptors. Using microarray analysis, we identified specific genes that are influenced by SHBG expression in LNCaP and MCF-7 cells in a manner consistent with each of these properties. These findings suggest that locally expressed SHBG can play a functional role in the steroid responsiveness of prostate and breast cells through multiple signaling pathways and that perturbations in local SHBG expression could contribute to prostate and breast cancer.

Introduction

The importance of androgens and estrogen in the development and progression of prostate cancer (PC) and breast cancer (BC) is extensively documented (1, 2). A number of approaches to the treatment of patients with prostate or BC are based upon their responsiveness to, respectively, androgens and estrogens. However, prostate and breast tumors often progress to steroid "independent" states, ultimately rendering these therapies ineffective. Although changes in nuclear androgen receptor (AR) and estrogen receptor (ER) signaling are likely to be involved in the progression of these tumors, we have been interested in a possible role for sex hormone-binding globulin (SHBG), a protein that mediates an alternative, membrane-based steroid signaling pathway (3, 4).

The initial step of androgen and estrogen signaling through SHBG requires binding of unliganded SHBG to its specific membrane receptor, R_{SHBG}. This primes, but does not activate downstream signaling. Subsequent binding of an appropriate androgen or estrogen to the SHBG-R_{SHBG} complex is required for its activation; this results in the generation of cAMP and protein kinase A activation. The early steps of this

J.J. Li et al. (eds.), *Hormonal Carcinogenesis V.*

pathway cascade rapidly and independently of the AR and ER. Downstream effects reported in prostate and breast cell lines include PSA induction (5), inhibition of progesterone receptor expression (6), increased apoptosis (7), and regulation of cell growth (8, 9); however, the overall biology of this pathway is not well understood.

In humans, SHBG is expressed most abundantly by the liver, and secreted into plasma, where it binds certain androgens and estrogens. It is also expressed in the testis for intratesticular utilization, and in the prostate and breast (3, 4, 10), classic target tissues for androgens and estrogens. Interestingly, in the prostate, SHBG is most abundant in luminal epithelial cells, whereas stromal cells from prostate explants possess the greatest SHBG binding ability, i.e., contain R_{SHBG}. The human LNCaP (PC) and MCF-7 (BC) cell lines express SHBG mRNA and immunoreactive SHBG protein, and can be engineered to overexpress and secrete elevated amounts of SHBG into their surrounding growth medium (data not shown). Both cell lines also display membrane binding of SHBG and can activate the R_{SHBG} signaling pathway (10–12).

We have incorporated these findings into a model that posits that endogenously expressed SHBG in the prostate and breast serves to regulate the SHBG-R_{SHBG} signaling pathway in an autocrine/paracrine manner. Furthermore, intracellular SHBG could, through its binding properties, affect free intracellular steroid concentrations, and thus modulate AR and ER activity. If this model is correct, locally expressed SHBG, through its effects on androgen and estrogen signaling, could be involved in normal breast and prostate development and function. Because the *SHBG* gene is located within 35 kb of the *p53* gene on chromosome 17p13.1 (13), a known hotspot for genomic alterations in prostate and breast tumors, perturbations in local SHBG expression in cancer cells could alter their normal response to androgens and estrogen, and thus contribute to tumor progression.

To test this hypothesis, we asked whether locally expressed SHBG, in human prostate and breast cells, exerts steroid-induced membrane effects through R_{SHBG} and whether it acts intracellularly to bind steroids and thus modulate activation of the AR and ER. SHBG was inducibly overexpressed in LNCaP cells, and constitutively overexpressed in MCF-7 cells with the objective of amplifying R_{SHBG} signaling and dampening AR and ER activation in response to dihydrotestosterone (DHT) or 17β-estradiol (E_2) treatment. Microarray analysis was performed to identify specific genes whose expression was consistent with their regulation by SHBG in two different ways: (1) through the SHBG-R_{SHBG} pathway or (2) through modulation of AR or ER-mediated induction or repression. Our initial results provide support for a role of locally expressed SHBG in the regulation of gene expression by both of these mechanisms.

Methods

Generation of Cell Lines. The inducible L5S2 clonal cell line [which expresses SHBG in response to Ponasterone A (PonA)] is indirectly derived from LNCaP cells, through an intermediate cell line, L5. L5 was generated by stably transfecting

LNCaP cells with the plasmid, pVgRXR (Invitrogen, Carlsbad, CA). pVgRXR encodes a hybrid transactivator that is activated by PonA. This transactivator recognizes and directs transcription from a promoter within a second plasmid, pINDhygro (Invitrogen). L5S2 was generated by stably transfecting L5 cells with a pINDhygro construct that contains the full length human SHBG cDNA coding sequence cloned directly downstream of the inducible promoter. The L5V4 vector control cell line was generated by stably transfecting L5 cells with the empty vector, pINDhygro. The MCF-7SHBG-myc23-2 cell line was created by stably transfecting MCF-7 cells with a recombinant plasmid generated via the Xi-Clone system (Gene Therapy Systems, San Diego, CA). This plasmid encoded a full length SHBG protein containing an immunogenic myc peptide at its carboxyl terminus, under the transcriptional control of a constitutively active CMV promoter. SHBG protein expression is greatly elevated in PonA-treated L5S2 cells, MCF-7SHBG-myc23-2 cells, and their conditioned media, as measured by Western blot, immunostaining, and ELISA assays (data not shown).

Cell Culture, Treatment Conditions, and Microarray Analysis. The experimental design compared the respective effects of DHT and E_2 on LNCaP and MCF-7 cells that do or do not overexpress SHBG. In the inducible LNCaP system, we controlled for treatment with the inducing agent, PonA. L5V4 vector control cells and inducible L5S2 cells were each seeded into two groups of multiple six-well plates in RPMI-1640 medium (Mediatech, Herndon, VA) supplemented with 1 mM sodium pyruvate (Mediatech), 100 units ml^{-1} Penicillin–Streptomycin (Invitrogen), and 10% charcoal stripped fetal calf serum (Gemini Bio-Products, Woodland, CA) for 24 h. One group was then treated with the inducing agent, PonA (10 μM) (In Vitrogen), and the other with an equal volume of carrier ethanol, for 24 h. Triplicate wells of PonA-treated cells were treated for an additional 24 h with carrier or 10 nM DHT, giving six treatment conditions:

1. L5V4 negative treatment control (carrier treated)
2. L5V4 PonA-treated only
3. L5V4 PonA-treated, DHT-treated 24 h
4. L5S2 negative treatment control (carrier treated)
5. L5S2 PonA-treated only
6. L5S2 PonA-treated, DHT-treated 24 h

MCF-7 and MCF-7SHBG-myc23-2 cells were seeded into multiple six-well plates in DMEM (Invitrogen) supplemented with 1× nonessential amino acid solution (Mediatech), 1 mM sodium pyruvate, 100 units ml^{-1} of Penicillin–Streptomycin, and 10% charcoal stripped fetal calf serum, and incubated for 48 h. Each cell line was then treated in triplicate for 24 h with 10 nM E_2 or with the carrier ethanol. This gave four additional treatment conditions

7. MCF-7 SHBG-myc23-2 negative treatment control (carrier treated)
8. MCF-7 SHBG-myc23-2 E_2-treated 24 h
9. MCF-7 negative treatment control (carrier treated)
10. MCF-7 E_2-treated 24 h

Total RNA was isolated with Trizol (Invitrogen) followed by a Qiagen clean up procedure (Qiagen, Valencia, CA). RNA integrity was assessed using an Agilent 2100 Bioanalyzer and RNA 6000 Nano Lab Chip LabChips (Agilent, Palo Alto, CA). RNA samples showed a 260/280 ratio between 1.8 and 2.0 and 28S:18S ratio of 1.5 and higher. In this pilot study, only one sample of each triplicate RNA preparation was used for microarray analysis. First-strand cDNAs were synthesized from 5 µg of each RNA sample using a T7-Oligo(dT) promoter primer and SuperScript II. After RNase H-mediated second-stranded cDNA synthesis, double-stranded cDNAs were purified using a GeneChip sample clean-up module. Biotinylated complementary RNAs (cRNAs) were generated by in vitro transcription using T7 RNA Polymerase and a biotinylated nucleotide analog/ribonucleotide mix. Biotinylated cRNAs were cleaned up, fragmented, and hybridized to Affymetrix Human Genome U133 Plus 2.0 Array chips, representing 54 675 transcripts (Affymetrix, Santa Clara, CA), at 45°C for 16 h with constant rotation at 60 rpm. Chips were processed using an Affymetrix fluidics station and scanned on an Affymetrix scanner 3000 with workstation. Images were processed with GeneChip Operating Software (GCOS) and raw data were analyzed with GeneSpring 7.2 software (Silicon Genetics, Redwood City, CA) to identify differentially expressed genes between conditions. Data were normalized to the 50th percentile of measurements taken from the chip to reduce chip-wide variations in intensity. Each gene was normalized to the average measurement of the gene throughout the experiment to enable comparison of relative changes in gene expression levels between different conditions. Data filtration was performed based on flags present or marginal. For this report, we focused on the top 100 genes that showed the greatest fold induction or suppression in DHT- and PonA-treated L5S2 cells compared with similarly treated L5V4 vector control cells or to PonA-treated L5S2 cells, in E_2-treated MCF-7 SHBG-myc23-2 cells compared with E_2-treated MCF-7 cells, in DHT and PonA-treated L5V4 control cells compared with PonA-treated L5V4 cells, and in E_2-treated MCF-7 cells compared with mock-treated MCF-7 cells.

Results

Induced L5S2 and MCF-7 SHBG-myc23-2 Cells Overexpress SHBG. SHBG mRNA expression was the first parameter to be evaluated. PonA treated L5S2 cells displayed an 1 800-fold increase in SHBG mRNA expression compared with untreated vector control L5V4 cells. PonA exerted a comparatively modest ninefold increase in SHBG mRNA expression in L5V4 cells. Unexpectedly and significantly, uninduced L5S2 cells had relatively high SHBG mRNA expression, a 56-fold increase compared with L5V4 cells not treated with PonA. This could adversely impact our data analysis. MCF-7SHBG-myc23-2 cells displayed a 1 017-fold increase in SHBG mRNA expression vs. MCF-7 cells. Twenty-four hours E_2-treated MCF-7SHBG-myc23-2 cells displayed a 1 032-fold increase vs. MCF-7 cells. E2 had no significant effect on SHBG expression.

Identification of Candidate R_{SHBG} Responsive Genes in LNCaP and MCF-7 Cells. To identify downstream candidates of SHBG-R_{SHBG} signaling in LNCaP cells, we identified genes (1) induced or repressed in PonA stimulated (SHBG overexpressing) L5S2 cells treated with DHT; (2) unaffected by PonA in DHT-treated L5V4 cells; (3) equally expressed in untreated L5S2 and L5V4 cells; (4) unaffected by PonA alone in L5S2 cells. PonA and DHT treatment of L5S2 cells induced 16 genes by ≥ 9.0-fold, and 12 genes were repressed by ≥ 95% [Table 1, undocumented gene (*U* gene)].

Candidate downstream targets of the SHBG-R_{SHBG} pathway in MCF-7 cells include genes induced or repressed by E_2 in MCF-7SHBG-myc23-2 cells and unaffected by E_2 in MCF-7 cells. Three genes were induced ≥ 16.0-fold by E_2 in MCF-7SHBG-myc23-2 cells, and 13 were repressed by ≥ 94% (Table 2).

SHBG Suppression of DHT Responsive Genes in LNCaP Cells and E_2-responsive Genes in MCF-7 Cells. To test whether intracellular SHBG acts as an androgen buffer, inhibiting AR activation, we identified DHT responsive genes in L5V4 cells unresponsive to PonA, and which showed reduced DHT response in PonA stimulated L5S2 cells. Candidate SHBG affected AR-responsive genes were: (1) induced/repressed by DHT in control L5V4 cells in the presence of PonA; (2) significantly less responsive to DHT in PonA induced L5S2 cells that overexpress SHBG; (3) similarly expressed in untreated L5V4 and L5S2 cells; (4) similarly expressed in PonA-treated L5V4 cells L5S2 cells. Six genes displayed ≥ 95%

Table 1 Candidate SHBG-R_{SHBG} responsive genes in LNCaP cells

Induced genes			
BC035072	*U* gene	AW058459	*U* gene
AC005620	*U* gene	AF100640	Metastasis related protein (MB2)
AW197196		*U* gene	BC034050
AB038041	Phosphodiesterase 11A2	AW074853	*U* gene
BM873997	*U* gene	NM_130759	Immunity associated protein 1 (IMAP1)
AK091277	*U* gene	NM_001208	Basic transcription factor3 like1 (BTF3L1)
AA650281	*U* gene	W52934	Weakly similar to serine-threonine kinase
AI079134	*U* gene	AW276240	Weakly similar to p80
Repressed genes			
AI693378	*U* gene	AW182934	*U* gene
BE15799	*U* gene	BE858373	*U* gene
AW298171		*U* gene	AI683261
AB014532	*U* gene	AA576947	*U* gene
AI829603	*U* gene	AL832339	*U* gene
NM_002774	Kallikrein 6 (neurosin)		
BC030808	Similar to endosome-associated FYVE-domain protein		

Table 2 Candidate SHBG-R_{SHBG} responsive genes in MCF-7 cells

Induced genes			
X16323	Hepatocyte growth factor	BI092935	Zinc finger protein 42
BE740743	Thyroid stimulating hormone receptor		
Repressed genes			
NM_015858	Growth regulating protein	NM_002886	RAP 2B
NM_002286	Lymphocyte activating precursor 3	BC029855	Similar to KR18
NM_001036	Ryanodine receptor 3	AL021786	Internal membrane protein
BG434272	Prothymosin alpha	AV725364	GPCR family C group 5
AF130116	*U* gene	AL359626	U gene
W52934	Similar to serine threonine kinase	AB029025	U gene AK021928
AK021928	Rab GTPase activating protein, noncatalytic subunit		

Table 3 Androgen responsive genes suppressed by SHBG in LNCaP cells

Genes less responsive to androgen-induction in PonA treated L5S2 cells			
AJ431619	*U* gene	AL162077	*U* gene
NM_021105	Phospholipid scramblase 1	AA404269	*U* gene
AL049337	*U* gene	AK057458	*U* gene
Genes less responsive to androgen-repression in PonA treated L5S2 cells			
AW772079	*U* gene	AC005620	*U* gene
AL117625	*U* gene	AA975530	*U* gene
D17262	*U* gene	AF055011	*U* gene
AI982754	Clusterin	AF118068	*U* gene
AA733172	*U* gene	AL832211	*U* gene
BC037529	*U* gene	BE644917	Nuclear receptor subfamily 1, group I,#3
AW058580	*U* gene	AL574912	Similar to serine protease

Table 4 Estrogen responsive genes suppressed by SHBG in MCF-7 Cells

Genes less responsive to E_2-induction in MCF-7SHBG-myc23-2 cells			
NM_004473	Forkhead box E1 thyroid transcription factor 2	NM_006898	Homeobox D3
NM_012404	Related to pp32	M88107	Formyl peptide receptor
AL024493	*U* gene	AU158247	*U* gene
NM_004473	Forkhead box E1 thyroid transcription factor 2	NM_006898	Homeobox D3
NM_012404	Related to pp32	M88107	Formyl peptide receptor
Genes less responsive to estrogen-repression in MCF-7SHBG-myc23-2 cells			
NM_020361	Carboxypeptidase beta precursor	AW016250	*U* gene
NM_002443	Microseminoprotein, β- (MSMB)	AI580966	*U* gene
U55185	Oral cancer candidate gene	NM_002172	Interferon alpha14
AI822082	Similar to synaptic glycoprotein SC2	AK023546	Phospholipase C ε 2
NM_001704	Brain specific angiogenesis inhibitor 3	NM_004932	Cadherin 6 (k-cadherin)
NM_005408	Small inducible cytokine family A member 13	NM_080475	Serpin B11

inhibition of induction by DHT in PonA induced L5S2 cells, and 14 genes displayed ≥ 97% lower repression (Table 3).

Similarly, we found E_2-induction or repression of specific genes in MCF-7 cells to be dampened in MCF-7SHBG-myc23-2 that constitutively overexpress SHBG. Ten genes had ≥ 95% suppression of induction by E_2 in MCF-7SHBG-myc23-2 cells, and 12 genes had ≥ 96% dampened repression (Table 4).

Conclusions

We obtained initial evidence that endogenously expressed SHBG can affect the expression of genes in LNCaP and MCF-7 cells. The effects of SHBG appear to be mediated through at least two different mechanisms, R_{SHBG} signaling and modulation of the AR and ER genomic response. It is not yet clear from our initial screen whether specific pathways are activated by SHBG expression, since the genes identified are involved in a broad spectrum of cellular functions. This initial data analysis and experimental approach requires careful follow up, and such work is currently in progress. Confirmation of these results would implicate SHBG in the steroid responsiveness of LNCaP and MCF-7 cells, while the identities of genes that are regulated by SHBG would provide important clues as to its functions and biologic effects. From a hormonal carcinogenesis standpoint, we can speculate that deletions targeting the SHBG/p53 locus could: (1) dampen the SHBG-R_{SHBG} signaling pathway; and (2) reduce the intracellular binding of certain steroid hormones thereby increasing AR and ER activity in prostate and breast tumors. These speculations await further elucidation of SHBG-mediated steroid signaling functions in the prostate and breast.

Acknowledgments These studies were supported by awards from the AHEPA Cancer Research Foundation, Starr Foundation, and St. Luke's Roosevelt-Associate Trustees.

References

1. Jensen EV (1996) Steroid hormones, receptors, and antagonists Ann N Y Acad Sci 784:1–17
2. Ko Y, Balk S (2004) Targeting steroid hormone receptor pathways in the treatment of hormone dependent cancers. Curr Pharm Biotech (5):459–470.
3. Kahn SM, Hryb DJ, Nakhla AM, et al. (2002) SHBG is synthesized in target cells. J Endocrinol 175:113–120.
4. Kahn SM, Hryb DJ, Nakhla AM, et al. (2005) Immunohistochemical and in situ detection of SHBG expression in breast and prostate cancer: implications for hormone regulation. Hormonal Carcinogenesis, Vol. 4. Jonathan Li, Sara Li, Antonio Llombart-Bosch, editors. Springer, Berlin Heidelberg New York 508–514.
5. Nakhla AM, Romas NA, Rosner W (1997) Estradiol activates the prostate AR and PSA secretion through the intermediacy of SHBG. J Biol Chem 272:6838–6841.
6. Fazzari A, Catalano MG, Comba A, et al. (2001) The control of progesterone receptor expression in MCF-7 breast cancer cells: effects of estradiol and sex hormone-binding globulin (SHBG). Mol Cell Endocrinol 172(1–2):31–36.

7. Catalano MG, Frairia R, Boccuzzi G, Fortunati N (2005) Sex hormone-binding globulin antagonizes the anti-apoptotic effect of estradiol in breast cancer cells. Mol Cell Endocrinol 230(1–2):31–37.
8. Fortunati N, Becchis M, Catalano MG, et al. (1999) SHBG, its membrane receptor, and breast cancer: a new approach to the modulation of estradiol action in neoplastic cells. J Ster Biochem Mol Biol 69(1–6):473–479.
9. Nakhla AM, Rosner W (1996) Stimulation of prostate cancer growth by androgens and estrogens through the intermediacy of SHBG. Endocrinology 137:4126–4129.
10. Hryb DJ, Nakhla AM, Kahn, SM, et al. (2002) SHBG in the human prostate is locally synthesized and may act as an autocrine/paracrine effector. J Biol Chem 277:26618–26622.
11. Nakhla AM, Khan MS, Rosner W (1990) Biologically active steroids activate receptor-bound human sex hormone-binding globulin to cause LNCaP cells to accumulate adenosine 3´,5´-monophosphate. J Clin Endo Metab 71:398–404.
12. Porto C, Musto N, Bardin C, Gunsalus G (1992) Binding of an extracellular steroid-binding globulin to membranes and soluble receptors from human breast cancer cells (MCF-7 cells). Endocrinology 130:2931–2936.
13. April, 2003 UCSC Human Genome Project Working Draft- http://genome.ucsc.edu/cgibin/hgTracks?positionX16351&hgsid=22396239

Inactivation of *ID-1* Gene Induces Sensitivity of Prostate Cancer Cells to Chemotherapeutic Drugs

Yong-Chuan Wong, Xiao-Meng Zhang, Ming-Tak Ling, and Xiang-Hong Wang

Summary Resistance to anticancer drugs is one of the major reasons of treatment failure for androgen-independent prostate cancer (PC). Increase in expression of Id-1 has been reported in several types of advanced cancer including PC. It has been suggested that overexpression of Id-1 may provide an advantage for cancer cell survival and thus inactivation of Id-1 may be able to increase the susceptibility of cancer cells to apoptosis. In this study, using small RNA interfering (siRNA) technology, we inactivated the *Id-1* gene in two androgen-independent PC cell lines, DU145 and PC3, and investigated whether down-regulation of Id-1 could lead to increased sensitivity of these PC cells to a commonly used anticancer drug, taxol (Tx). Our results showed that inactivation of Id-1 by siId-1 resulted in decrease in both colony forming ability and cell viability in prostate cancer cells after Tx treatment. Furthermore, the siId-1 induced sensitization to Tx was associated with activation of apoptotic pathway. In addition, c-Jun N-terminal kinase (JNK), one of the common pathways responsible for Tx-induced apoptosis, was also activated in the si-Id-1 transfected cells. Inhibition of JNK activity by a specific inhibitor, SP600125, blocked the siId-1-induced sensitivity to Tx. These results indicate that increased Id-1 expression in PC cells may play a protective role against apoptosis, and down-regulation of Id-1 may be a potential target to increase sensitivity of Tx-induced apoptosis in PC cells.

Introduction

Prostate cancer (PC) is one of the most common cancers in Western countries and is the second highest killer cancer in the US (1). Although the figure in Asia including China is much lower, but with the growing affluence and changing lifestyle in the last several decades, the incidence of PC has been increasing steadily (2). Despite extensive effort worldwide, relatively little progress in term of prolongation of life for advanced PC sufferers has been made. The major stumbling block is the tendency of PC to transform to androgen reflexive after an initial period of remission after androgen ablation therapy. Once PC reemerges from its "dormancy" and become androgen reflexive, there is to date no effective way of controlling the

J.J. Li et al. (eds.), *Hormonal Carcinogenesis V.*

growth of the cancer and the patients often die within a period of several months. Therefore, new therapeutic strategies are urgently needed to improve the survival of this group of patients.

Using in vitro system, in the last several years, we have established that Id-1 (inhibitor of differentiation or DNA binding) stimulates cancer cell proliferation by down-regulation of p16/Rb (4), a well known tumor suppressive pathway on one hand while promoting cell proliferation pathway through activation of MAPK and NF-κB pathways (5, 6). Over expression of Id-1 also protects prostate cancer cells from TNF-alpha induced apoptosis (6). We further demonstrate that over-expression of Id-1 up-regulates the expression of EGF-R and stimulates LNCaP cells to secrete PSA, which are closely related to the transformation of PC cells from androgen-dependent to androgen-independent phenotype (3). More recently, we have shown further that over expression Id-1 could stimulate prostate cancer cells to secrete VEGF thus promotes angiogenesis (7). Out latest data show that in addition to angiogenesis, Id-1 over expression promotes development of multidrug resistance of PCs through activation of JNK pathway, one of the common pathways responsible for taxol (Tx)-induced apoptosis (8). Inactivation of Id-1, can thus increase the sensitivity of PC cells to common therapeutic drug such as Tx, may be a new strategy for treatment of androgen independent PC.

Experimental Design and Results

Down-Regulation of Id-1 by siRNA in PC3 and DU145 Cells. PC3 and DU145 are well-known androgen-independent PC cell lines, which express high level of Id-1 protein constitutively. We first designed siRNA sequences targeted to Id-1 and transfected through a retroviral vector into these cells and selected two stable clones each from each of the two cell lines. The results showed that si-Id-1s were effective in down-regulating the Id-1 levels of both PC3 and DU145 cells. These stably transfected lines were used for the following experiments.

Effect of Id-1 Inactivation on Sensitivity of Cells to Taxol (Tx) Treatment. The current study was designed to check the effect of down regulation of Id-1 on sensitivity of PC cells to Tx treatment using the colony forming assay and MTT assay. The results showed that down-regulation of Id-1 by siId-1 was associated with decreased colony forming ability of both PC3 and DU145 cells in a dose-dependent manner to Tx treatment compared with the parental lines and vector controls (Fig. 1). The effect was not restricted only to Tx. Similar results were observed for several other commonly used chemodrugs (i.e., cisplatin, carboplatin, epirubicin, and vincristine), though with slightly different sensitivities.

Effect of Id-1 Inactivation on Tx-Induced Apoptosis. In the current experiment, we examined whether the increase in sensitivity of PC cells to Tx treatment was related to increased sensitivity to apoptosis. We performed DAPI and DNA laddering assays on the siId-1 clones of both PC3 and DU145 cells. The results showed that

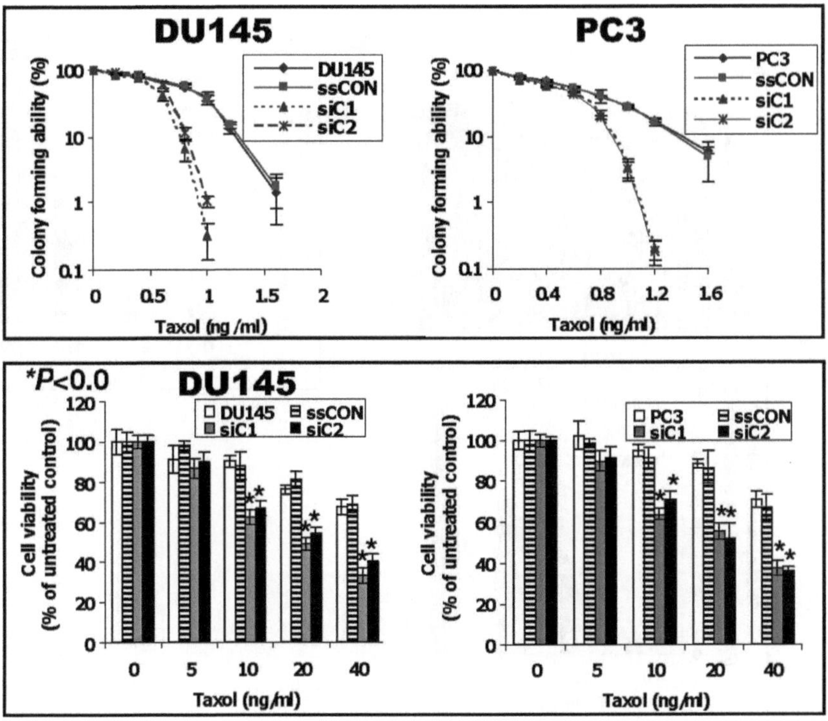

Fig. 1 Effect of si-Id-1 on cellular sensitivity to Tx. (**A**) Colony forming ability of si-Id-1 transfectants and vector controls after exposure to Tx. Results presented as the mean ± SD from three experiments. Note that colony forming ability of si-Id-1 transfectants (*dotted lines*) is lower compared with that of vector control (*solid lines*). (**B**) Cell viability of si-Id-1 transfectants and the vector controls after exposure to Tx as examined by MTT assay. Results represented the OD ratio between treated and untreated cells. Note that cell viability of si-Id-1 transfectants (*solid columns*) is lower than the vector controls (*open columns*). ($^*P < 0.05$ compared with the vector control, analysed by two-tailed student *t* test)

all cell lines showed increased apoptotic rate in a dose-dependent manner regardless of Id-1 levels (Fig. 2). However, the apoptotic rate was significantly higher in siId-1 clones than the parental and vector controls in both cell lines. DNA laddering assay was able to confirm this results and that DNA fragmentation was much more evident in the siId-1 clones compared with the vector controls. The results suggest that siId-1-induced sensitivity to taxol may be mediated through activation of the apoptotic pathway.

Effect of Taxol on Cell Cycle Distribution. We analysed whether taxol-induced apoptosis was related to G2/M arrest by flow cytometry after 24 and 48 h treatment of taxol. The results showed that there was no significant effect on G2/M phase cells in either the siId-1 clones or the vector controls compared with the untreated cells. However, there was an increase in sub-G1 phase cells in siId-1 clones which further supported the results from apoptosis studies that decreased Id-1 was associated with

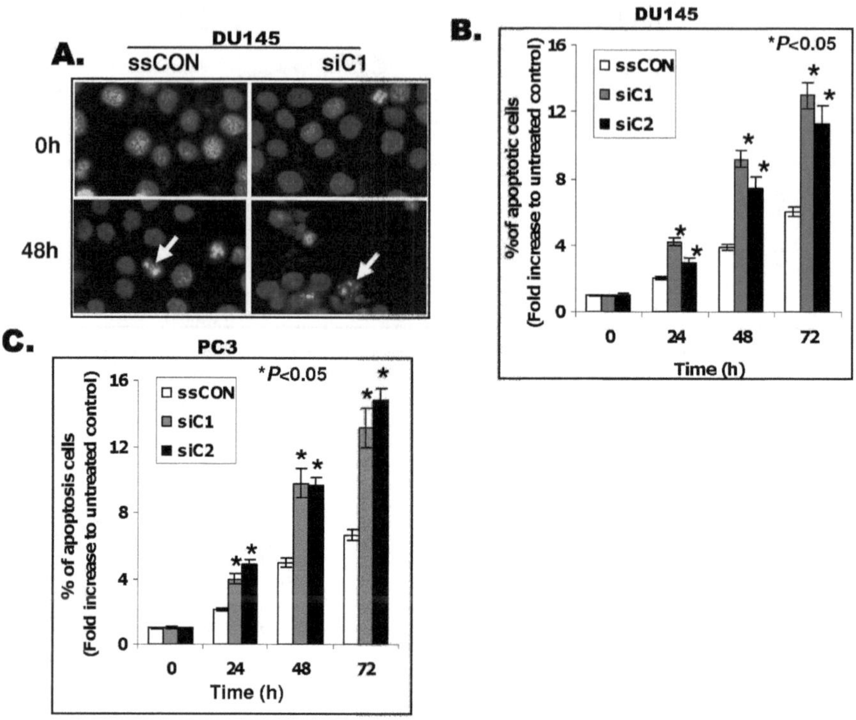

Fig. 2 Effect of Id-1 inactivation on taxol-induced apoptosis. (**A**) Representative photos of apoptotic cells by DAPI staining. Cells that showed nuclear fragmentation were considered to be undergoing apoptosis. (**B**, **C**). Summary of apoptosis rate in the si-Id-1 transfectants and the vector controls after exposed to taxol. Results represent the mean ± SD of three independent experiments. Note that the apoptosis rates in si-Id-1 transfectants (*solid columns*) are higher than the vector control (*open columns*) after exposure to taxol in both cell lines ($^*P < 0.05$ compared with the vector control)

increased sensitivity to taxol-induced apoptosis (Fig. 3). The results suggest that taxol-induced apoptosis is independent of G2/M arrest in PC3 and DU145 cells.

Effect of Id-1 Inactivation on the Expression of Apoptosis-related Proteins. To substantiate the results as observed earlier, we examined at the molecular level the apoptotic related proteins, such as Bcl-2, Bcl-xL, Bax, Caspase-3, and PARP in siId-1 clones compared with the vector controls. The results showed that Bcl-2 level was much lower in siId-1 clones compared with the vector controls indicating that down-regulation of Id-1 was associated with decreased Bcl-2 (Fig. 4). After taxol treatment, the Bcl-2 level remained low compared with vector control. However, the expression of Bax in DU145 was barely detectable probably because of a mutated *Bax* gene in DU145 cells. The Bax level of PC3 cells was slightly increased. The decrease in Bcl-2 in siId-1 clones was correlated with higher levels of cleaved Caspase-3 and PARP at later time points especially in response to taxol. These results indicate that decreased Id-1 expression is associated with increased

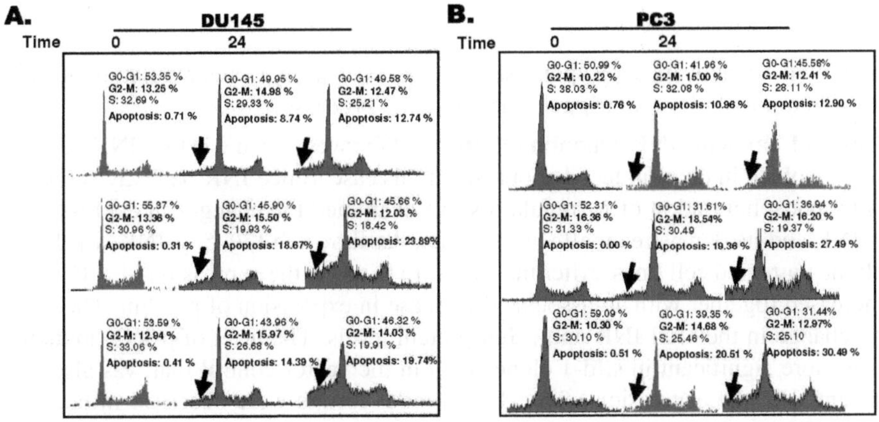

Fig. 3 Effect of taxol on cell cycle distribution. Flow cytometric analysis on the si-Id-1 transfectants and the vector controls of DU145 and PC3 cells after exposure to taxol (5 ng ml^{-1}) for 24 and 48 h, respectively. Note that there is no notable G2/M arrest after taxol treatment but apparent sub-G1 phase is observed, especially in the si-Id-1 transfectants

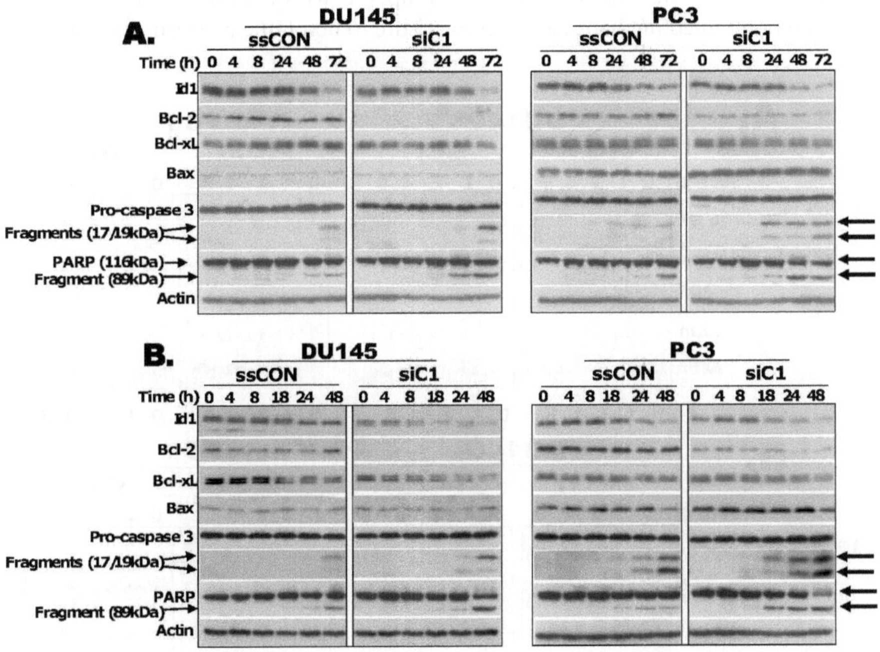

Fig. 4 Effect of Id-1 inactivation on the expression of apoptosis-related proteins after exposure to taxol. Apoptosis-related protein expression in the si-Id-1 transfectants and the vector controls examined by Western blotting after exposure to 5 ng ml^{-1} (**a**) and 10 ng ml^{-1} (**b**) taxol. Note that the expression of Id-1 and Bcl-2 is much lower in si-Id1 transfectants than the vector control before taxol treatment. After exposure to taxol, the expression of Id-1 and Bcl-2 remains low in the si-Id-1 transfectants compared with the vector control that is associated with increased expression of cleaved Caspase 3 and PARP

sensitivity to taxol-induced apoptosis pathway activation. Down regulation of Bcl-2 may be a possibility that triggers the apoptotic processes.

Effect of Taxol on JNK Signaling Pathway. We investigated whether JNK pathway was involved in taxol-induced apoptosis in this case. Since JNK pathway activation is by phosphorylation of its regulators, we examined the changes in expression of p-JNK and its downstream effector p-c-Jun. As shown in Fig. 5, p-JNK was low in all the untreated cell lines. After the exposure to taxol, the expression of p-JNK was increased together with an associated increase in expression of p-c-Jun. There was no change in the total JNK and c-Jun protein levels. The level of p-JNK induction was more significant in siId-1 clones than in the vector control that was also consistent with the expression of p-c-Jun. Similar results were observed in both cell lines. The results suggest that activation of JNK signaling pathway may play a role in the increased apoptosis.

Effect of JNK Inhibitor on Taxol-induced Apoptosis in siId-1 Clones. To confirm the role of JNK activation in siId-1 induced sensitivity to taxol, we investigated whether specific inhibitor, SP600125, could inhibit the protection of siId-1 cells from apoptosis. As shown in Fig. 6, treatment with SP60015 inhibited the expression of p-c-Jun in the two siId-1 clones compared with the untreated control. The total c-Jun remained unchanged. After exposure to taxol, the cells treated with taxol

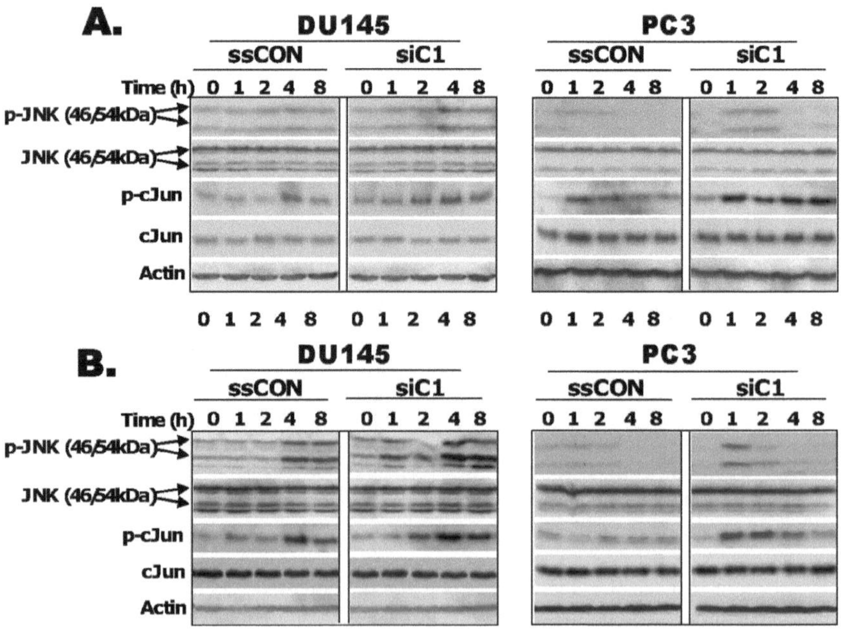

Fig. 5 Effect of taxol on JNK pathway. Expression of p-JNK and p-c-Jun as examined by Western blotting in the si-Id-1 transfectants and the vector controls after exposed to 5 ng ml^{-1} (**a**) and 10 ng ml^{-1} (**b**) taxol. Note that p-JNK is low in all untreated cell lines and the expression of p-JNK and p-c-Jun is increased after treated with taxol, especially in the si-Id-1 transfectants

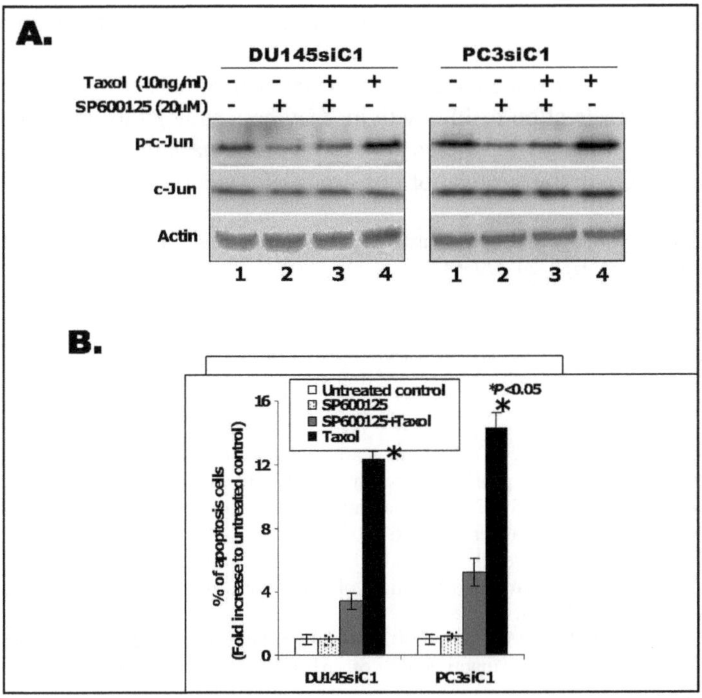

Fig. 6 Effect of JNK inhibitor SP600125 on taxol-induced apoptosis. (**a**) Western blotting analysis of DU145siC1 and PC3siC1 cells after exposed to SP600125 and taxol. Note that the expression of p-JNK and p-c-Jun is decreased after treatment with SP600125 (20µM) and taxol (10ng ml^{-1}) compared with the cells treated with taxol alone. (**b**) Apoptosis rate of DU145siC1 and PC3siC1 cells after exposed to SP600125 and taxol. Note that the apoptosis rate of DU145siC1 and PC3siC1 is decreased after exposure to SP600125 and taxol compared with the cells treated with taxol alone ($*P < 0.05$)

alone showed increased p-c-Jun while the SP600125 treated cells showed much lower p-c-Jun expression indicating a successful suppression of JNK signaling. In contrast treatment with both SP600125 and taxol or taxol alone led to much higher apoptotic rate. But the apoptotic rate in cells treated with both SP600125 and taxol was higher than treatment with taxol alone. This suggests that inactivation of JNK in siId-1 cells leads to protection against taxol-induced apoptosis.

Conclusion

Taken together, our studies show that Id-1 is not only crucial in the progression of advanced prostate cancer from androgen-dependent to androgen-independent phenotype and tumor angiogenesis, but also plays a pivotal role in the development of multidrug resistance. Furthermore, our results suggest that Id-1 over-expression provides prostate cancer cells with survival advantage against commonly used chemodrugs including taxol. Down-regulation of Id-1 activity by siId-1 promotes chemosensitivity of prostate cancer cells to taxol treatment through activation of JNK pathway. Inactivation of Id-1 activity may be a strategy to improve the treatment efficiency of taxol for patients with advanced metastatic prostate cancer, thus improve and prolong the survival of this group of patients.

Acknowledgments Supported by research grants from the RGC: HKU7314/01M, HKU7490/03M, HKU7470/04M, HKU7478/03M.

References

1. Penson DF, Albertsen PC (2000) Lessons learnt about early prostate cancer from large scale databases: population-based pearls of wisdom. Surg Oncol 11:3–11.
2. Hsing AW, Tsao L, Devesa SS (2000). International trends and patterns of prostate cancer incidence and mortality. Int J Cancer 85:60–67.
3. Ling MT, Wang X, Lee DT, et al. (2004). Id-1 expression induces androgen-independent prostate cancer cell growth through activation of epidermal growth factor receptor (EGF-R). Carcinogenesis 25:517–525.
4. Ouyang XS, Wang X, Ling MT, et al. (2002). Id-1 stimulates serum independent prostate cancer cell proliferation through inactivation of p16(INK4a)/pRB pathway. Carcinogenesis 23:721–725.
5. Ling MT, Wang X, Ouyang XS, et al. (2002). Activation of MAPK signaling pathway is essential for Id-1 induced serum independent prostate cancer cell growth. Oncogene 21:8498–505.
6. Ling MT, Wang X, Ouyang XS, et al. (2003). Id-1 expression promotes cell survival through activation of NF-kappaB signalling pathway in prostate cancer cells. Oncogene 22:4498–508.
7. Ling MT, Lau TC, Zhou C, et al. (2005). Overexpression of Id-1 in prostate cancer cells promotes angiogenesis, through the activation of vascular endothelial growth factor (VEGF). Carcinogenesis 26(10):1668–1676.
8. Zhang XM, Wang XH, Ling MT, et al. (2006) Inactivation of Id-1 in prostate cancer cells: A potential therapeutic target in inducing chemosensitization to taxol through activation of JNK pathway. Int J Cancer 118:2072–2081.

Membrane Initiated Signaling by 1,25α-dihydroxyvitamin D_3 in LNCaP Prostate Cancer Cells

Dennis Larsson, Malin Hagberg, Nahren Malek, Charlotte Kjellberg, Edina Senneberg, Neda Tahmasebifar, and Viktoria Johansson

Summary Prostate cancer (PC) is one of the most common cancers among men, and vitamin D and its metabolites are candidates for prevention and therapy of this disease. The vitamin D metabolites, 1, 25-dihydroxyvitamin D_3 (1,25D) and 25-hydroxyvitamin D_3, decreases cellular proliferation and invasiveness, and stimulates differentiation of PC cells. However, the underlying mechanisms are not fully clarified, and there is evidence that some of these effects of the vitamin D system are mediated by specific membrane-associated receptors/binding proteins in addition to its nuclear receptor, suggesting multiple regulatory pathways. The aim of the present study was to examine the role of membrane initiated pathways mediating effects of 1,25D on cell invasiveness in LNCaP cells. Treatment with 1,25D evoked a dose-dependent activation of the JNK/SAPK MAPK signaling pathways within 10 min, demonstrating membrane initiated signaling of 1,25D in LNCaP cells. Furthermore, treatment with 1,25D decreased LNCaP cell invasiveness by approximately 20% after 48 h. Using an inhibitor (SP600125) for the JNK/SAPK MAPK signaling pathway in combination with 1,25D on LNCaP cells, the inhibitory action of 1,25D on invasiveness was eliminated. In conclusion, 1,25D decrease invasiveness of LNCaP cells by interaction with a putative membrane associated receptor, which activate membrane, initiated signaling via the JNK/SAPK MAPK signaling pathway.

Introduction

The active components of the vitamin D endocrine system, 1,25D and 24,25-dihydroxyvitamin D_3 (24,25D), regulates cell growth and differentiation in various tissues (1–4). This control is carried out through regulation of transcription and activity of key components governing cell growth, differentiation, and apoptosis (1, 3). Two signaling pathways for 1,25D (3, 5) and one for 24,25D (6) have been identified. 1,25D regulates cells via the classic nuclear/cytosolic vitamin D receptor (nVDR), and a recently described membrane associated protein, 1,25D-membrane associated rapid response steroid binding protein (1,25-MARRS). Several independent studies on chondrocytes, enterocytes, and osteoblasts have indicated

J.J. Li et al. (eds.), *Hormonal Carcinogenesis V.*

presence of 24,25D-receptors/binding proteins associated to the cell membrane and recently, a putative new receptor protein for 24,25D have been isolated from enterocytes and was identified as catalase (24,25-MARRS) (6). 24,25-MARRS was found to be associated to the plasma membrane as well as intracellular membranes, binds 24,25D with high affinity and specificity (6), and has been postulated to be involved in regulating intestinal calcium transport (7, 8). Studies with conformation restricted analogs have confirmed existence of two receptor populations for 1,25D. 6-trans restricted analogs of 1,25D interact largely with nVDR and 6-cis restricted analogs of 1,25D interact with 1,25-MARRS or nVDR associated to the plasma membrane, inducing distinctly different cellular effects (3). Thus, the recently identified 1,25-MARRS and 24,25-MARRS might constitute the key components for the origin of 1,25D and 24,25D initiated signaling at the cell membrane, and may solve a 60-year-old enigma in steroid research.

Two different hypotheses have been proposed for the mechanisms underlying steroid hormone regulation of cell differentiation, proliferation, and apoptosis (3):

- Steroid hormones may act through interaction of their cytosolic/nuclear receptors with specific DNA sequences, thereby regulating the expression of genes governing cell differentiation, proliferation, and apoptosis (Fig. 1).
- Steroid hormones may act through receptors coupled to cytosolic- or membrane-associated signal-transducing proteins, and trigger activation of signal cascades upon steroid binding to the receptor complex. This leads to induction of genes regulating cell growth, proliferation and apoptosis (Fig. 1).

Androgens and 1,25D play important roles in the development and growth of prostate cells (3–5). The androgen receptor (AR) and nVDR are expressed in normal prostate and in most CaP specimens that have been evaluated (9). Androgens promote cell growth and androgen ablation causes apoptosis of prostate epithelial cells (10), whereas 1,25D inhibits cell growth, induces differentiation and apoptosis (4). 25-hydroxyvitamin D_3 (25D) decrease CaP proliferation at high concentrations and is an equally potent inhibitor as 1,25D. Chen and colleagues (12) have suggested that 25D is primarily converted to 1,25D in CaP cells, and that 1,25D exert the growth inhibitory effects via the on CaP cells. The role of 24,25D in regulating CaP cell biology has not been evaluated.

One of the signaling cascades initiated at the cell membrane that are activated by androgens, as well as other autocrine and paracrine growth factors, is mitogen-activated protein kinases (MAPK) (13, 14). The MAPK signaling pathway is intimately involved in cellular proliferation, differentiation, and apoptosis (15, 16). It consists of three distinct layers of protein kinases, including a MEK kinase (MEKK), which activates MAPK kinase (MEK), which in turn activates MAPK. Three MAPK pathways have been identified in mammalian cells, leading to the activation of the MAPK extra cellular signal regulated kinases (ERK), the c-Jun N-terminal kinases (JNK), and p38. Once activated, the MAPK translocates to the nucleus where they phosphorylate and activate a number of target proteins (Fig. 1) (15, 16).

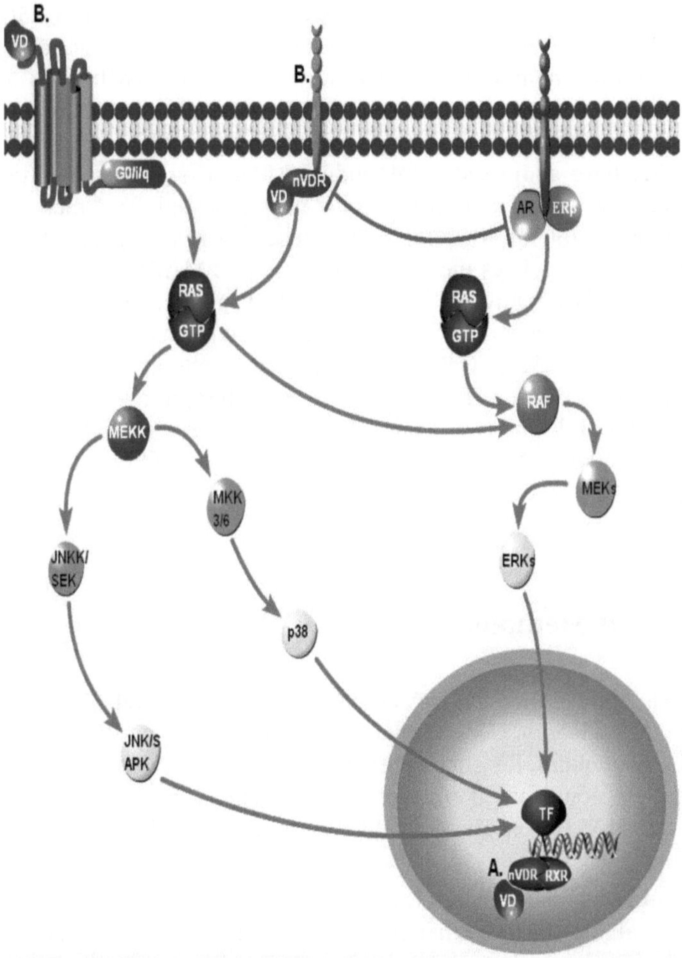

Fig. 1 Proposed mechanism of vitamin D activity in prostate cells. **A.** 1,25D binds to the cytosolic/nuclear nVDR causing nVDR to heterodimerize with the retinoid X receptor (RXR). The heterodimer binds to specific vit D response elements and induces gene transcription. **B.** 1,25D binds to a membrane-associated (nVDR) or an integral/associated membrane receptor (1,25-MARRS) causing activation of MAPK cascades, ERK 1/2, SAPK/JNK, or p38. Activation of ERK1/2 induces gene transcription of factors important for cell survival, and activation of SAPK/JNK and p38 induces transcription of factors that cause cell cycle arrest and apoptosis

In LNCaP cells, presence of androgen and estradiol triggers concurrent association of the AR and estradiol receptor α with Src, leading to the activation of the Src/Raf/Erk MAPK-pathway. As a result, cells enter the S-phase and proliferation increases (13). This observation is not unique for androgens and estrogens in

LNCaP cells. Steroid hormones, such as progesterone (13, 17) and 1,25D (18–20), share the ability to activate MAPK signaling cascades with androgens and estrogens. This indicates that membrane initiated steroid signaling (MISS) is a common phenomenon among steroid hormones and comprises an alternative pathway for steroids to regulate gene expression and subsequently cell biology.

Few studies have addressed 1,25D-induced MAPK-signaling in prostate cells, and at present, only long-term effects have been assessed. Treatment with 1,25D increases transcription of MEKK1 and induces caspase-dependent cleavage of MEK, and thus, up-regulates the JNK/SAPK and p38 MAPK pathways and down-regulates the ERK 1/2 pathway. The overall effect in prostate cells is an induction of the apoptotic pathways (Fig. 1). So far, no attempts have been made to study MISS of MAPK by 1,25D in CaP cells. Androgens increase proliferation of prostate cells via ERK 1/2 activation. Thus, MAPK activation represents a "node" or "intersection point" for vit D and androgens. The aim of the present study was to examine the role of the JNK/SAPK MAPK pathway in mediating effects of 1,25D on cell invasiveness in LNCaP cells via a rapid membrane initiated pathway.

Materials and Methods

Cell Culture. Human prostate cancer (PC) cells from the cell line LNCaP clone FCG (ECACC, Salisbury, UK) were grown as a monolayer in RPMI 1640 medium (Invitrogen, Scotland UK) supplemented with 1% MEM-vitamins, 2% MEM-amino acids, 1% nonessential amino acids, 1% penicillin, 10% fetal bovine serum (FBS) and 1,2% 1M Hepes. The cell cultures were maintained at 37°C in a humidified atmosphere of 5% CO_2.

Western Blot Analysis. Cell lysates were prepared from LNCaP cells treated for 10min with vehicle (0.01% ethanol, final concentration) or 1,25D (10^{-11}, 10^{-10}, 10^{-9}, 10^{-8} or 10^{-7} M) and were resolved on 12% SDS-polyacrylamide gels. Gels were blotted to PVDF membranes (Biorad) and the membranes were incubated with 1:2 500 dilutions of rabbit polyclonal antibodies to phosphorylated p-54 and p-56 JNK/SAPK MAPK (Cell Signalling Technology, Beverly, MA), followed by incubations with 1:5 000 dilutions of alkaline phophatase conjugated antirabbit IgG (Sigma). The JNK/SAPK MAPK bands (p-p54 and p-p46) were visualized using the nitro blue tetra zolium/5 bromo-4chloro-3-indolyl-phosphate reagent (Sigma).

Invasion Assays. LNCaP cells were starved in serum-free media 48h before the experiment. Multiscreen-MIC filter wells (pore size of 8μM, Millipore) were coated with SERVA Biomatrix EHS (30μl, Serva Electrophoresis, Germany) and air dried at room temperature overnight. Supplemented RPMI 1640 media (140μl) were placed in the receiver wells and the filter wells were placed on top. LNCaP cells with a density of 100,000 cells per filter well were placed in the Multiscreen-MIC plates in supplemented RPMI 1640 media. The cells were treated with 1,25D in three different concentrations, 10^{-12}, 10^{-10}, 10^{-8} M, or ethanol (0.01%, vehicle control), and with or without additional treatment with the JNK-inhibitor SP600125

(20 μM, Sigma). The plates were incubated for 48 h at 37°C and 5% CO_2. The media in the receiver wells were transferred to a Costar 96-well plate. The filter wells were incubated in PBS-5 mM EDTA (1251) at 37°C and 5% CO_2 for 30 min. Then the plates were placed in a shaker for 20 min to dislodge cells from the membrane underside. The filter wells were removed, and the PBS-5 mM EDTA was transferred to the media in the Costar 96-well plate. Every well were incubated with Calcein-AM (0.25 μM, Sigma). Fluorescence's were measured by a fluorometer (Fluo Star galaxy) after 30 min incubation.

Results and Discussion

LNCaP cells incubated with 1,25D showed a dose-dependent increase in the activation of both p-p54 and p-p46 JNK/SAPK MAPK compared with control treated cells (Fig. 2). The activation was observed within 10 min and suggests that 1,25D interacts with a putative receptor protein associated to the cell membrane. Thus, in the prostate, as in other target organs such as the intestine, cartilage, and bone, 1,25D regulate cell physiology by two different signaling pathways, MISS and nuclear initiated signaling (NISS) (3–5). Similar to responses observed in intestine, cartilage, and bone, MISS signaling by 1,25D involves downstream activation of MAPK signaling cascades (3–5). The nature of the membrane receptor for 1,25D is still under debate, and dependent on the physiological parameters observed and on the organ and cell systems tested, either the classical nVDR or 1,25-MARRS are reported to cause MISS (3–5).

1,25D inhibited LNCaP invasion at all concentrations tested (10^{-12}, 10^{-10}, 10^{-8} M) compared with vehicle treated cells (Fig. 3). Presence of the JNK/SAPK MAPK inhibitor SP600125 abolished the inhibitory effects of 1,25D, and LNCaP cells treated with a combination of 1,25D and SP600125 demonstrated an increased invasiveness. Thus, the data presented in this study suggest that 1,25D mediate its inhibitory actions on LNCaP cell invasiveness by the JNK/SAPK MAPK signaling pathway.

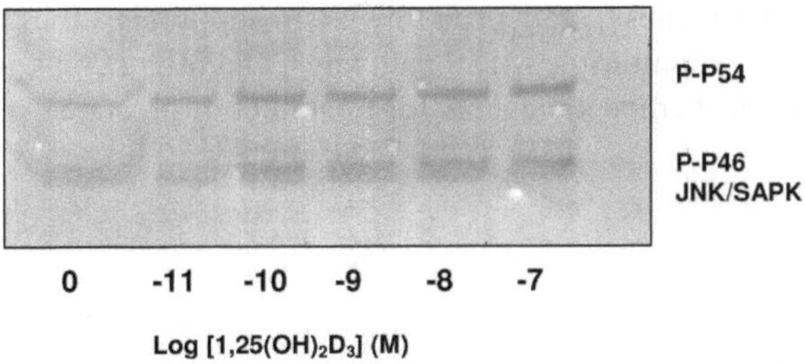

Fig. 2 Effects of 1,25D on SAPK/JNK MAPK activation. LNCaP cells were treated with 1,25D (10^{-11}, 10^{-10}, 10^{-9}, 10^{-8} and 10^{-7} M) or vehicle (ethanol, 0.01% final concentration) for 10 min. The cytosol fraction was collected by centrifugation and activated SAPK/JNK (p-p54 and p-p46) was detected by Western blot

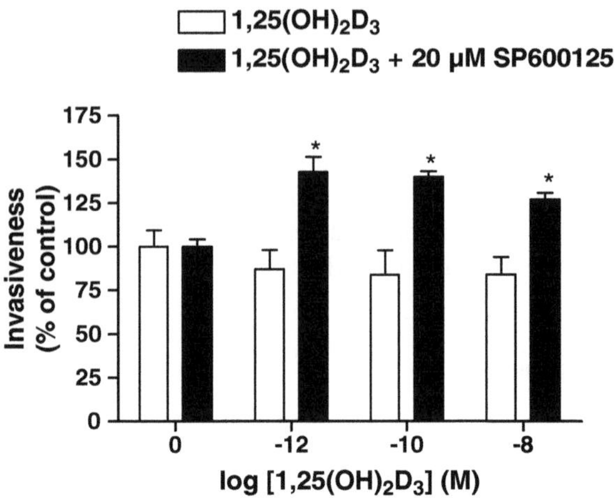

Fig. 3 Effects of 1,25D on invasiveness of LNCaP cells. Cells were incubated with 1,25D (final concentrations of 0, 10^{-12}, 10^{-10}, and 10^{-8} M) or a combination of 1,25D and the JNK/SAPK MAPK inhibitor SP600125 (20 μM) for 48 h. LNCaP cell invasiveness was inhibited by 1,25D, at all concentrations tested. Addition of SP600125 abolished the inhibitory action evoked by 1,25D on LNCaP cell invasiveness, suggesting that the inhibitory response by 1,25D on LNCaP cell invasiveness is mediated by the JNK/SAPK MAPK signaling pathway

Recent publications by Bonaccorsi et al. (20) and Marchiani et al. (21) demonstrate that a nonhypercalcemic analog of vitamin D, BXL-628, inhibit proliferation and invasiveness of two PC cell lines, DU145 and PC3. In line with the results presented in the present study, the reports by Bonaccorsi et al. (20) and Marchiani et al. (21) suggests that the effects of BXL-628 are of MISS origin. The nature of the putative membrane receptor for 1,25D in prostate cells is still unknown, but existence of a putative membrane receptor is supported by the observations that activation of the membrane associated VDR in LNCaP cells enhance JNK/SAPK MAPK signaling, whereas activation of the membrane associated VDR in PC3 and DU145 suppress keratinocyte growth factor receptor (KGFR) autophosphorylation, which decrease downstream signaling via the PI3K/AKT signaling pathway. Thus, current experimental data suggests that 1,25D regulates prostate cancer cell proliferation, invasiveness, and probably also apoptosis (JNK/SAPK MAPK signaling pathway) via multiple MISS pathways, and might constitute new targets for designing better therapeutic agents for treatment of prostate cancer.

References

1. Holick MF (2006) Vitamin D: its role in cancer prevention and treatment. Prog Bioph Mol Biol 92:49–59.
2. Haussler MR, Whitfield GK, Haussler CA, et al. (1998) The nuclear vitamin D receptor: biological and molecular regulatory properties revealed. J Bone Miner Res 13:325–349.

3. Norman AW, Mizwicki MT, Norman DP (2004) Steroid-hormone rapid actions, membrane receptors and a conformational ensemble model. Nat Rev Drug Discov 3:27–41.
4. Boyan BD, Bonewald LF, Sylvia VL, et al. (2002) Evidence for distinct membrane receptors for 1 alpha,25-$(OH)_2D_3$ and 24R,25-$(OH)_2D_3$ in osteoblasts. Steroids 67:235–246.
5. Nemere I, Farach-Carson MC, Rohe B, et al. (2004) Ribozyme knockdown functionally links a 1,25$(OH)_2D_3$ membrane binding protein (1,25D_3-MARRS) and phosphate uptake in intestinal cells. PNAS 101(19):7392–7397.
6. Larsson D, Anderson D, Smith NM, et al. (2006) 24,25-dihydroxyvitamin D_3 binds to catalase. J Cell Biochem 97(6):1259–1266.
7. Nemere I, Yazzie-Atkinson D, Johns DO, et al. (2002) Biochemical characterization and purification of a binding protein for 24,25-dihydroxyvitamin D_3 from chick intestine. J Endocrinol 172:211–219.
8. Larsson D, Nemere I (2001) 24,25-Dihydroxyvitamin D_3 and intestinal calcium transport: from receptors to physiological response. Recent Res Dev Endocrinol 2:183–198.
9. Chen TC, Holick MF (2003) Vitamin D and prostate cancer prevention and treatment. Trends Endocrinol Metab 14(9):423–430.
10. Trapman J, Brinkman AO (1996) The androgen receptor in prostate cancer. Path Res Pract 192:752–760.
11. Chen TC, Schwartz GG, Burnstein KL, et al. (2000) The in vitro evaluation of 25-hydroxy-vitamin D_3 and 19-nor-1alpha,25-dihydroxyvitamin D_2 as therapeutic agents for prostate cancer. Clin Cancer Res 6:901–908.
12. Migliaccio A, Castoria G, Di Domenico M, et al. (2000) Steroid-induced androgen receptor-oestradiol receptor beta-Src complex triggers prostate cancer cell proliferation. EMBO J 19(20):5406–5417.
13. Peterziel H, Mink S, Schonert A, et al. (1999) Rapid signalling by androgen receptor in prostate cancer cells. Oncogene 18:6322–6329.
14. Schaeffer HJ, Weber MJ (1999) Mitogen-activated protein kinases: specific messages from ubiquitous messengers. Mol Cell Biol 19:2435–2444.
15. Pearson G, Robinson F, Beers Gibson T, et al. (2001) Mitogen-activated protein (MAP) kinase pathways: regulation and physiological functions. Endocr Rev 22:153–183.
16. Ballare C, Uhrig M, Bechtold T, et al. (2003) Two domains of the progesterone receptor interact with the estrogen receptor and are required for progesterone activation of the c-Src/Erk pathway in mammalian cells. Mol Cell Biol 23:1994–2008.
17. Bhatia M, Kirkland JB, Meckling-Gill KA (1995) Monocytic differentiation of acute promyelocytic leukemia cells in response to 1,25-dihydroxyvitamin D_3 is independent of nuclear receptor binding. J Biol Chem 270:15962–15965.
18. Song X, Bishop JE, Okamura WH, et al. (1998) Stimulation of phosphorylation of mitogen-activated protein kinase by 1alpha,25-dihydroxyvitamin D_3 in promyelocytic NB4 leukemia cells: a structure-function study. Endocrinology 139:457–465.
19. Schwartz Z, Ehland H, Sylvia VL, et al. (2002) 1alpha,25-dihydroxyvitamin D_3 and 24R,25-dihydroxyvitamin D_3 modulate growth plate chondrocyte physiology via protein kinase C-dependent phosphorylation of extracellular signal-regulated kinase 1/2 mitogen-activated protein kinase. Endocrinology 143:2775–2786.
20. Bonaccorsi L, Marchiani S, Ferruzzi P, et al. (2006) Non-genomic effects of the androgen receptor and Vitamin D agonist are involved in suppressing invasive phenotype of prostate cancer cells. Steroids 71:304–309.
21. Marchiano S, Bonaccorsi L, Ferruzzi P, et al. (2006) The vitamin D analogue BXL-628 inhibits growth factor-stimulated proliferation and invasion of DU145 prostate cancer cells. J Cancer Res Clin Oncol 132:408.

Gonadotrophin Releasing Hormone-Based Vaccine, an Effective Candidate for Prostate Cancer and Other Hormone-sensitive Neoplasms

Jesús A. Junco, Roberto Basalto, Franklin Fuentes, Eddy Bover, Osvaldo Reyes, Eulogio Pimentel, Lesvia Calzada, Maria D. Castro, Niurka Arteaga, Yovisleidis López, Héctor Hernández, Ricardo Bringas, Hilda Garay, Peter Peschke, José Bertot, and Gerardo Guillén

Summary Prostate growth, development, functions, and neoplastic transformation is androgen dependent. Estrogens have similar effects in the ovary and breast. Previous studies using gonadotrophin releasing hormone (GnRH/LHRH) vaccines have shown the usefulness of immunization against this hormone in prostate (PC) and breast cancer (BC). We have synthesized a peptide mutated at position 6 and attached to the 830–844 tetanic toxoid (TT) helper T cell sequence in the same synthesis process. After repeated pig immunizations, we have demonstrated a vaccine that significantly decreased testes size ($p < 0.001$), prostate ($p < 0.01$), seminal vesicles ($p < 0.01$), and testosterone (T) castration [0.05 nM ml^{-1} ($p < 0.01$)]. Similar results were obtained in adult male and female healthy dogs and *Macaca fascicularis* models. These data indicate that this GnRHm1-TT vaccine is safe and able to induce significant tumor growth inhibition in the Dunning R3327-H rat androgen responsive prostate tumor model. In these rats, the immunization induced high anti-GnRH titers concomitant with T castration reduction ($p < 0.01$) in 90% of the animals tested. In addition, 70% of the responders exhibited tumor growth inhibition ($p = 0.02$) and a survival rate approximately three times longer that those of untreated rats. These data indicate that GnRHm1-TT vaccine may be a potential candidate in the treatment of PC, BC, and other hormone-dependent cancers.

Introduction

A number of studies had shown that gonadotrophin releasing hormone (GnRH) or its receptors are expressed in some reproductive organs, including the prostate and mammary glands, and others tumors derived from these glands. Additionally, it has been demonstrated that GnRH analogs and antagonists exert direct inhibitory effects on the proliferation of human prostate cancer (PC). An alternative approach to the use of GnRH analogs is the immunization with the native or mimetic GnRH for the production of anti-GnRH antibodies that may neutralize the biological activity of this molecule, resulting in similar castration effects caused by GnRH drugs.

J.J. Li et al. (eds.), *Hormonal Carcinogenesis V.*

The candidate vaccine reported herein (GnRHm1-TT) is a synthetic immunogen consisting of a GnRH mimetic decapeptide adhered to a 15 amino acid tetanic toxoid (TT). In healthy animals, this vaccine was highly immunogenic, indicating high anti-GnRH antibodies titers, and resulted in T reduction and significant reduction in prostate and testicles weight in pigs, dogs, and rats. Our data regarding the effect of this vaccine in the Dunning R3327-H rat tumor model indicates that it may be a potential candidate for the treatment of PC and other hormone-dependent neoplasms.

Results

Design of LHRH Synthetic Peptide Immunogens. A diverse array of chimeric LHRH/GnRH molecules has been published (2, 3). They include LHRH peptide immunogens carrying different helper T cell epitopes. The construction of a GnRHm1-TT synthetic consisted of linking a GnRH molecule modified at position 6 in which the glycin was substituted by proline and joined through a couple of glycin residues to a T helper (Th) epitope corresponding to the 830–844 region of the tetanic toxoid. This peptide fragment has been reported to be a promiscuous Th epitope that stimulates a broad range of histocompatibility backgrounds. As a result of the chemical synthesis, we obtained a highly purified 27 amino acid peptide formulated as water in oil (W/O) vaccine candidate (Fig. 1).

Studies in Prepuberal Pigs. After immunizing pigs twice with GnRHm1-TT in oil adjuvant, significant decreases ($p < 0.001$) in the size of testes, prostate, seminal vesicles, epididimus, and bulbourethral glands ($p < 0.05$) were observed (Fig. 2). In the same animals, the T levels fell to 0.05 nM ml^{-1} ($p < 0.01$)

Studies in Adult Male and Female Healthy Dogs. Similar studies were performed in healthy male and female dogs (Fig. 3). Macroscopic and microscopic examination of the testes and ovaries showed degenerative changes.

Studies in Healthy and Tumor-implanted Copenhagen Adult Male Rats. After immunization of Copenhagen adult male rats with 750 µg of GnRHm1-TT, we observed high anti-GnRH antibody titers, which correlate with T castration levels and a signification size reduction of testicles, prostate, and seminal glands ($p < 0.01$) (Fig. 4).

Natural GnRH - **QHWSYGLRPG**

GnRHm1-TT - **QHWSYPLRPG** *GG* Q*YIKANSKFIGITEL*

Fig. 1 Primary structure of the native LHRH/GnRH molecule and the GnRHm1-TT peptide. The GnRHm1 decapeptide is synthesized with the epitope 830–844 of TT and covalently linked to the carboxy terminal GnRHm1. High levels of anti GnRH/LHRH antibodies were raised against the native molecule but not against the carrier

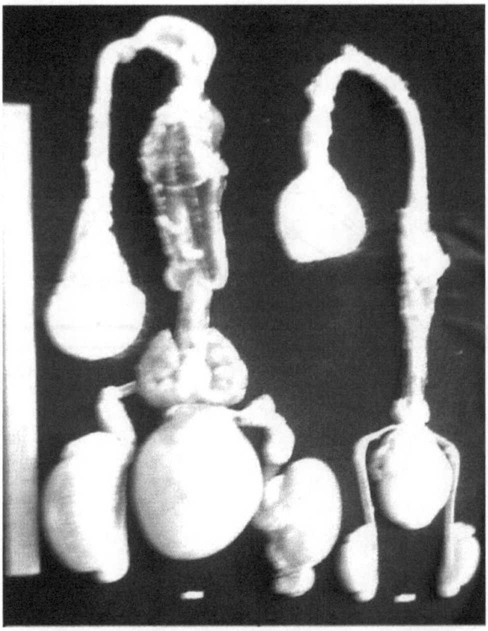

Fig. 2 Effect of GnRHm1-TT vaccination on the size of the prostate, testicles, and genitourinary glands of prepuberal pigs: (**a**) Placebo; (**b**) Immunized

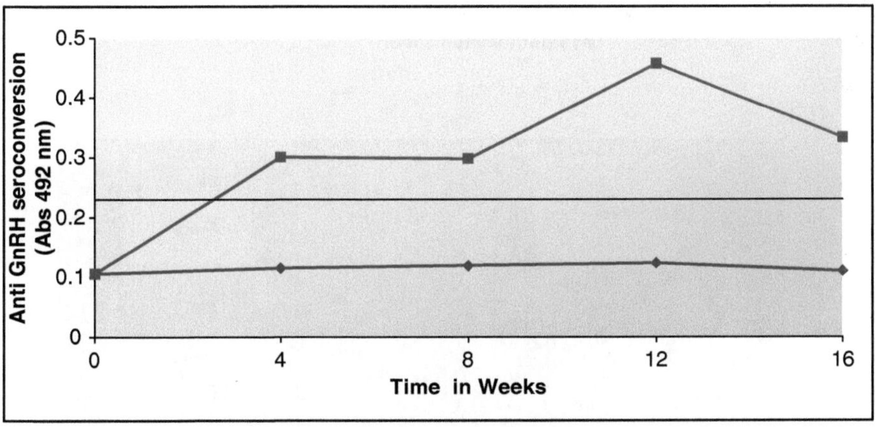

Fig. 3 Anti-GnRH seroconversion of adult male and female Beagle dogs after two immunizations with the GnRHm1-TT peptide emulsified in oil adjuvant

Therapeutic GnRHm1-TT Immunization-induced Tumor Growth Retardation in Preimplanted Copenhagen Rats. GnRHm1-TT immunized Copenhagen rats were classified as: good, moderate, and not responders. Under these categories,

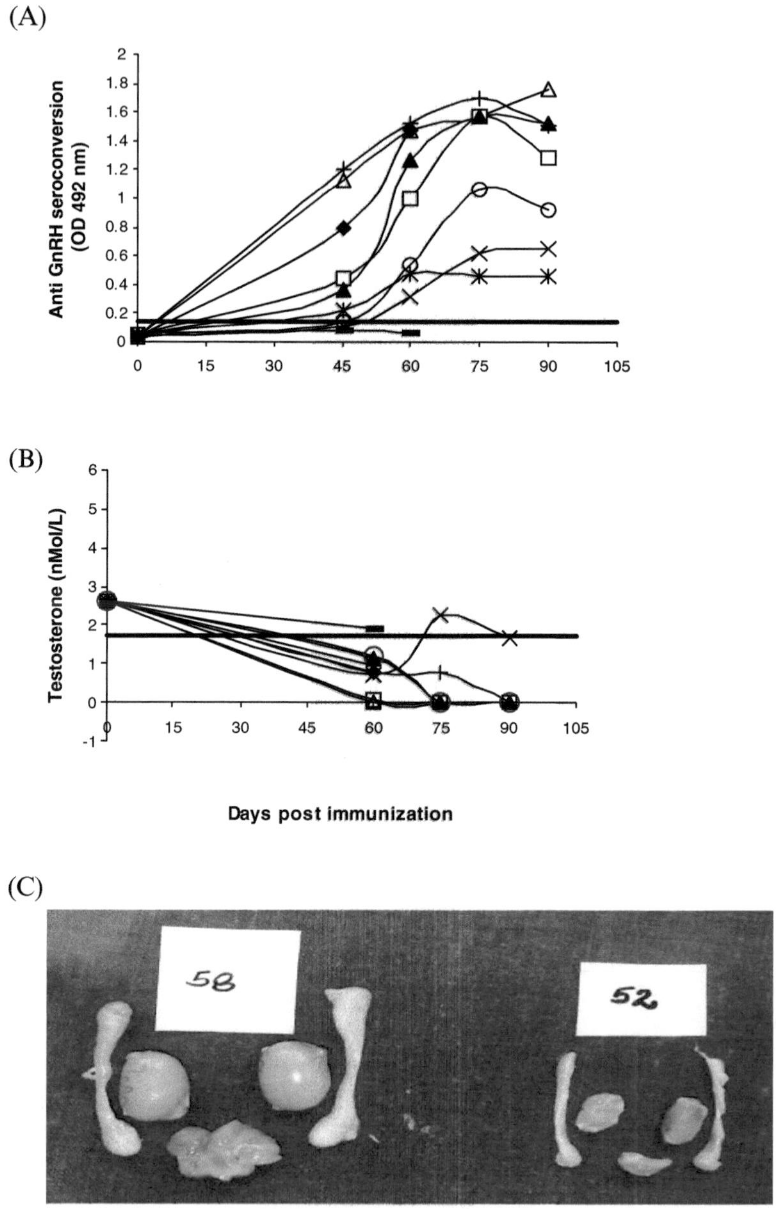

Fig. 4 Effect of GnRHm1-TT immunization in Copenhagen rats. (**A**) Indirect ELISA determination of anti-GnRH antibodies after six immunizations. (**B**) Individual T levels. (**C**) Macroscopic aspect of prostate and testes of intact (*left*) and immunized Copenhagen rats (*right*)

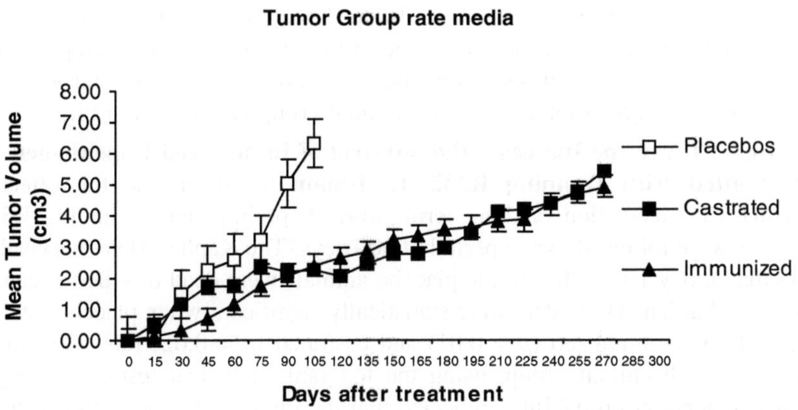

Fig. 5 Tumor growth in placebo, immunized, and castrated animals at day 105 of the study. By day 105, 100% of the placebo animals were dead or killed because of their tumor volume. Immunized and castrated animals showed a slow growing rate of Dunning R3327-H prostate tumor with respect to the controls

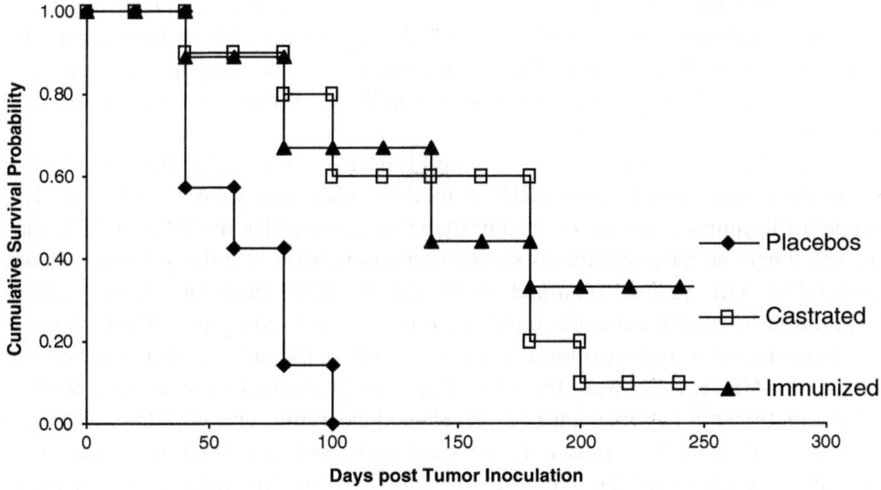

Fig. 6 Kaplan-Meier representation of the cumulative survival probability of the different experimental groups

75% were considered responders. In addition, three rats were included in the first category and three in the second one. One of the responder rats developed a large tumor. We observed tumor growth inhibition in 6 out of 10 rats in the control castrated group. The tumor growth delay for each animal group is depicted in Fig. 5.

GnRHm1-TT Vaccine Increases the Survival of Immunized Copenhagen Rats Preimplanted with Dunning R3327-H Tumor. To determine the effects of GnRHm1-TT vaccination on long term survival, preimplanted Dunning R3327-tumor rats were followed over a period of 300 days. The Kaplan-Meier survival plot shows that at day 105, 100% of the placebo animals were dead or killed because of their tumor burden. These data were statistically significant when placebo rats were compared with immunized ($p = 0.01$) and castrated ($p = 0.02$) rats. The survival analysis of each animal group, using the log rank statistical test comparing the cumulative survival probability, indicated that the immunized and castrated groups survived approximately two times more than the placebo group (Fig. 6).

Discussion

In the studies presented herein, a synthetic peptide, GnRHm1-TT, was tested as an immunotherapeutic vaccine for treatment of androgen responsive PC in pigs, dogs, and rats. The ability of GnRHm1-TT to induce anti-GnRH seroconversion and T reduction support the benefits of a single step GnRH-carrier molecule instead of the conjugation process used in previous studies (4). In addition, GnRHm1-TT has shown to be effective in generating anti-LHRH antibody levels that overcome the cut off barrier (>0.134 D0 at 1/50 dilution). These data are comparable to the candidate vaccine UBITh®, in which four different T helper cell sequences were used simultaneously for immunization (1).

The rat immunization with the vaccine GnRHm1-TT revealed that (88.8%) of the animals neutralized anti-GnRH antibodies when analyzed by ELISA. The number of immunoresponders is higher than that reported for the VAXTRATE vaccine (5), where an 84% of animals showed seroconversion. Similarly, these data are comparable with results obtained using the D17DT candidate from Apthon Corporation in preclinical experiments and in a clinical setting (6). When comparing seroconversion and castration, we observed that the animals that showed the highest anti-GnRH titers were the same that quickly reached the castration levels.

The relationship between antibody levels and castration was ~0.3 OD, when the serum was diluted 1/50. These data are similar to those described by Talwar et al. (1), and supported by antibody quality as isotype and avidity, responsible for many of the antibody functions (7, 8). Those parameters are under investigation.

The hormone deprivation action of GnRHm1-TT adjuved in Montanide ISA 51 to inhibit T synthesis has different uses and applications. The most prominent is related to the immunotherapy of androgen-sensitive PC and other hormone sensitive-diseases as BC. At the same time, these candidates can be used for reversible treatment of benign prostate hyperplasia (BPH) (9), pets contraception (10), and in the swine industry to avoid the boar taint caused by androgens (11). Therefore, the most important end point for GnRH immunotherapy is T levels.

In summary, we have demonstrated that the vaccine originated from a modified GnRHm1 molecule and the single epitope 830–844 of TT induced similar immunity responses and T reduction in healthy animals, when compared with the D17DT vaccine from Aphton. At the same time, a single 27 amino acid peptide that served as active principle of the vaccine avoided the possible carrier-induced suppression that results from the use of the whole DT molecule, as well as the conjugation process needed for the D17DT preparation. Similarly, the results described here are comparable to those described using UBITh® immunogen in which 4 GnRH peptides were bound indiscriminately to four different T Helper sequences (2). Therefore, the single molecule GnRHm1-TT vaccine adjured in Montanide ISA 51 appear to cause an important tumor inhibition as demonstrated in the in Copenhagen rat model.

References

1. Hougten RA, De Graw ST, Bray MK, et al. (1986) Simultaneous multiple peptide Synthesis: The rapid preparation of large numbers of discrete peptides for biological, immunological and methodologicals studies. Biotecniques; 4:522–526.
2. Finstad CL, Wang CY, Kowalsky J, et al. (2004) Synthetic luteinizing hormone releasing hormone (LHRH) vaccine for effective androgen deprivation and its application to prostate cancer immunotherapy. Vaccine; 22:1300–1313.
3. Basulto R, Milanés C, Rojas A, et al. (2003) Effects of GnRH immunization in the testicular structure and function of adult dogs. Biotecnología Aplicada; 20:20–24.
4. Subash S, Chaujan VS, Arunan K, et al. (1993) Synthetic Gonadotrophin-releasing hormone (GnRH) vaccines incorporating GnRH and synthetic T helper epitopes. Vaccine; 11:1145–1150.
5. Hoskinson RM, Rigby RDG, Mattner PE, et al. (1990) Vaxstrate: An anti-reproductive vaccine for cattle. Australian Journal of Biotechnology; 4:166–170.
6. Simms MS, Schofield DP, Jacobs E, et al. (2000) Anti GnRH antibodies can induce castrate levels of testosterone in patients with advanced prostate cancer. British Journal of Cancer; 83:443–446.
7. Fraomme B, Eftekhari P, van Regenmontel M, et al. (2003) A novel retro-inverso gonadotropin-releasing hormone (GnRH) immunogen elicits antibodies that neutralize the activity of native GnRH. Endocrinology; 144:73262–73269.
8. Paul. WE (1993) Fundamental Immunology, 3rd ed. Chapter 32 pp. 1143–1178. Reven Press, Ltd., 1185, New York.
9. Fuerst J, Fiebiger E, Jungwirth A, et al. (1997) Effect of active immunization against luteinizing hormone releasing hormone on the androgen sensitive Dunning R3327-PAP and androgen-independent Dunning R3327-AT2.1 prostate cancer sublines. Prostate; 32:77–84.
10. Ladd A, Tsong YY, Prabhu G, et al. (1998) Effects of long term immunization against LHRH and androgen treatment on gonadal function. Journal of Reproductive Immunology; 15:85.
11. Oonk HB, Turkstra JA, Schaaper WM, et al. (1998) New GnRH-like peptide construct to optimize efficient immunocastration of male pigs by immunoneutralization of GnRH. Vaccine; 36:64–67.

Liver Cancer

Pigment Epithelium-derived Factor is an Angiogenesis and Lipid Regulator that Activates Peroxisome Proliferator-activated Receptor α

Chuhan Chung, Jennifer A. Doll, Veronica M. Stellmach, John Gonzales, Sailesh Surapureddi, Mona Cornwell, Janardan K. Reddy, and Susan E. Crawford

Summary Pigment epithelium-derived factor (PEDF) is an endogenous antiangiogenic protein that also possesses antitumor activity. The mechanisms by which PEDF exerts its actions remains poorly understood. We sought to understand the role of PEDF in hepatocellular carcinoma (HCC), a hypervascular malignancy that has been shown to upregulate enzymes involved in fatty acid synthesis. PEDF expression occurs in two HCC cell lines and is oxygen dependent. Migration studies confirm PEDF's role as an endogenous inhibitor of angiogenesis in HCC cells. Loss of PEDF in an animal model leads to hepatocyte lipid accumulation, proliferation, and cellular atypia. To investigate potential interactions with transcription factors that are involved in fatty acid metabolism and cellular proliferation, we examined PEDF's interaction with PPARα in vitro and its functional activity through transactivation assays. We show that PEDF binds to PPARα but minimally to PPARγ. In the presence of the ligand, ciprofibrate, PEDF binding to PPARα decreases whereas the presence of troglitazone does not alter PEDF interactions with PPARγ. Transfection of the *PEDF* gene in the presence of the PPARα/RXR heterodimer demonstrates transcriptional activation of PPARα by PEDF. These data show that PEDF regulates lipid metabolism through activation of the transcription factor PPARα.

Introduction

Hepatocellular carcinoma (HCC) is a common malignancy throughout the world with a rising incidence in the USA. A notable characteristic is its hypervascularity, which is now the target of a variety of antiangiogenic compounds. Another prominent feature of HCC is the presence of fat and the induction of lipogenic enzymes (1). The relationship between fatty acid metabolism and cellular proliferation is poorly understood in both neoplastic and nonneoplastic conditions. Herein, we provide data documenting pigment epithelium-derived factor's (PEDF's) role as a lipid regulator in the liver through activation of the nuclear transcription factor PPARα.

First isolated in 1989 from the conditioned medium (CM) of retinal pigmented epithelial cells, PEDF is a 50 kDa multifunctional protein that is abundant within

J.J. Li et al. (eds.), *Hormonal Carcinogenesis V.*

most compartments of the eye and is also present in high concentration in other organs such as the heart, placenta, prostate, and liver (1, 2). Its function initially appeared to be neurotrophic and neuroprotective with the ability to transform retinoblastoma cells into a nonreplicating neuronal phenotype (1) and to promote the survival of multiple cell types within the nervous system including cerebella granule cells, retinal neurons, spinal motor neurons, and Schwann cells (1). The observation that PEDF could induce differentiation of retinoblastoma cells into a more mature phenotype in vitro and its presence within the human retina at 7.4 weeks of gestation suggests that it may play a developmental role in the eye and in other neural structures (1). A recent experimental report indicated that PEDF acts as a survival factor for neuronal stem cells adding complexity to its functional effects on neuronal differentiation (1).

The anti-angiogenic activity of PEDF was first reported by the laboratory of Noel Bouck (1). Isolated from the conditioned media of a retinoblastoma cell line, PEDF inhibited endothelial cell migration more effectively than other anti-angiogenic proteins such as angiostatin, thrombospondin-1, and endostatin (3–8). PEDF inhibited the effects of multiple proangiogenic factors in vitro including vascular endothelial growth factor (VEGF). In avascular regions of the eye such as the cornea and vitreous humor, PEDF plays the major role in maintaining vascular quiescence with neutralization of PEDF permitting neovascularization. Its regulation appears oxygen dependent with lower PEDF expression associated with hypoxia and increased PEDF expression occurring in states of hyperoxia (8). Since the discovery of PEDF's anti-angiogenic activity within the eye, we have pursued our investigation of this protein in the pathological angiogenesis that accompanies prostate cancer and hepatocellular carcinoma.

PPARα is a transcription factor that upon interaction with its heterodimer, retinoid X receptor (RXR), binds to PPAR-response elements (PPRE) within target genes. PPARα plays a crucial role in intracellular lipid metabolism regulating a host of genes that are involved in peroxisomal and mitochondrial oxidation of long chain fatty acids (1, 2). These include enzymes responsible for esterifying fatty acids, transport of fatty acids into the mitochondria, and β-oxidation. The use of PPARα ligands in rodent models also induces cellular proliferation and HCC. In short, PPARα is a key regulator of fatty acid metabolism and cellular proliferation. The data presented below postulates that PEDF acts as a transcriptional activator of PPARα, and provides a mechanism to link fat metabolism with cellular proliferation.

Results

PEDF Expression from HCC is Oxygen-dependent and Inhibits Endothelial Cell Migration. PEDF expression was confirmed in the CM of two HCC cell lines. We showed that PEDF expression decreased in conditions that mimic hypoxia and exposure to 1% hypoxia (Fig. 1a). The decrease in PEDF expression in response to hypoxic stress was completely blocked in the presence of anti-VEGF antibody

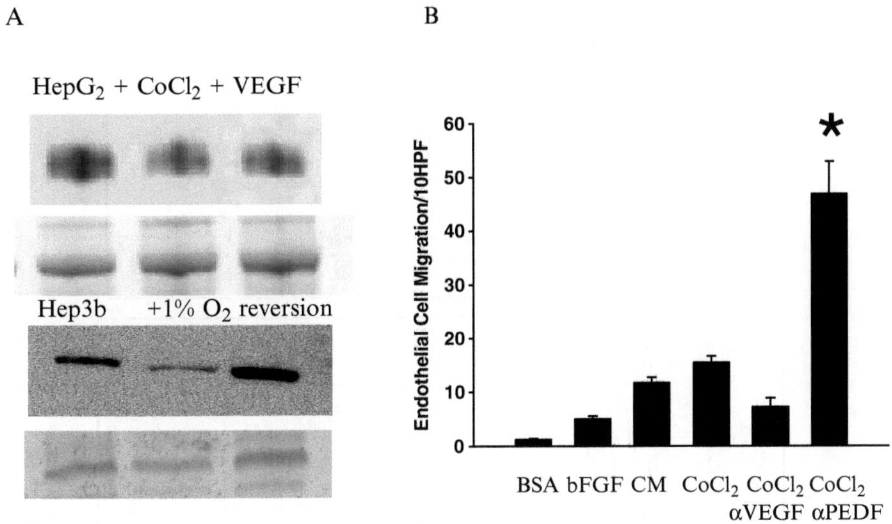

Fig. 1 (**A**) CM from two HCC cell lines secretes PEDF. Hypoxia mimetics ($CoCl_2$ 100 μM or VEGF 200 pM, top) and 1% O_2 (lower) lead to decreased PEDF expression. Exposure to normoxia after 24 h of hypoxia restores PEDF expression to baseline levels. (**B**) Endothelial cell migration assay using CM from HCC cell lines (Hep3b shown) demonstrates that depletion of PEDF from the CM results in a marked increase in endothelial cell migration. BSA (bovine serum albumin) served as the negative control; bFGF (basic fibroblast growth factor) served as the positive control. Endothelial cell migration was averaged by counting 10 (*$p < 0.05$)

(data not shown). We further showed that PEDF in the conditioned medium inhibits angiogenesis in human HCC cell lines in an endothelial cell migration assay. The addition of neutralizing antibodies to PEDF in the CM significantly increased endothelial cell migration in both HCC cell lines compared with CM alone and CM from $CoCl_2$-exposed conditions, (Fig. 1b) ($p < 0.05$).

Livers from PEDF Null Animals Display Increased Cell Proliferation and Fat. A PEDF null-animal was created as previously described (1). Liver tissue from 14 PEDF null-animals and age-matched controls (ages 6–16 months) was obtained and processed for immunohistochemistry. Histological examination revealed fatty change in all PEDF null animals compared with none for the controls (Fig. 2a). Staining for proliferating cell nuclear antigen (PCNA) revealed significantly elevated cellular proliferation in PEDF null animals compared to wild type (Fig. 2b).

PEDF Binds PPARα in a Ligand-independent Manner. Seeking potential nuclear functions for PEDF, we asked whether PEDF binds proteins known to regulate lipid metabolism and hepatocyte proliferation. Binding studies were performed with [^{35}S]-methionine-labeled PEDF. PEDF was cloned into pcDNA3.1 (Invitrogen), and used to synthesize [^{35}S]-methionine-labeled PEDF using TNT quick coupled transcription/translation system (Promega). Glutathione-*S* transferase (GST) alone served as a control for nonspecific binding, or GST-fusion proteins of

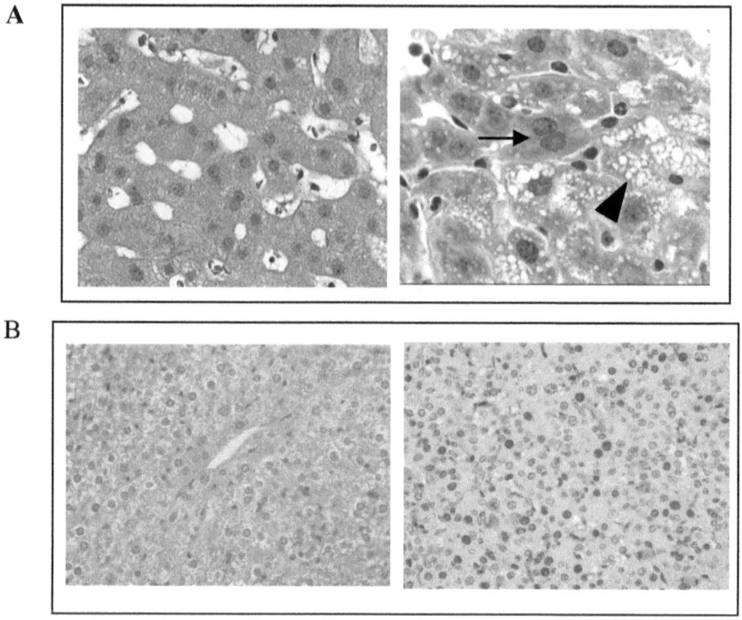

Fig. 2 (**A**) Steatosis and increased hepatocyte proliferation in the PEDF null animal liver (*right*) compared with age-matched control (left). A binucleated hepatocyte is seen in the center of the field (*arrow*); steatosis, primarily micro-vesicular, is present within hepatocytes (*arrowhead*). H&E stain, ×100. (**B**) Marked increase in PCNA staining in PEDF null animal (*right*) compared with control liver (*left*) ×40

the indicated nuclear receptors and coregulators were immobilized on GSH-Sepharose beads and incubated with aliquots of [^{35}S]-methionine-labeled PEDF in binding buffer. Aliquots of [^{35}S]-methionine-labeled PEDF in binding buffer were incubated with GST-fusion proteins of the indicated nuclear receptors immobilized on GSH-Sepharose beads or with Glutathione-*S* transferase (GST) alone, which served as a control for nonspecific binding. PEDF bound strongly to the nuclear receptor PPARα. This binding was most pronounced in the absence of the ligand, ciprofibrate (Fig. 3). Minimal binding of PEDF was noted to PPARγ and RXR in the absence and presence of troglitazone and retinoic acid, respectively.

Transactivation Assays Demonstrates PEDF Activation of PPARα. To determine the functional effect of PEDF on the PPARα/RXR heterodimer, we used a PPRE transactivation assay in the presence and absence of SLED, the *PEDF* gene. Plasmids containing the PPRE cloned upstream of the firefly luciferase reporter (tk-PPRE-Luc) were cotransfected with the control *Renilla* luciferase construct (Promega, WI) along with the PPARα and RXR constructs into HEK 293 cells. Twenty-four hours after transfection, cells were treated with the PPARα agonist, WY-14643 20 μM and incubated for an additional 24 h. Luciferase activity was measured in triplicate using the Dual-Luciferase Reporter Assay system (Promega). The presence of PEDF significantly increased PPARα/RXR activity from 1.56 ±

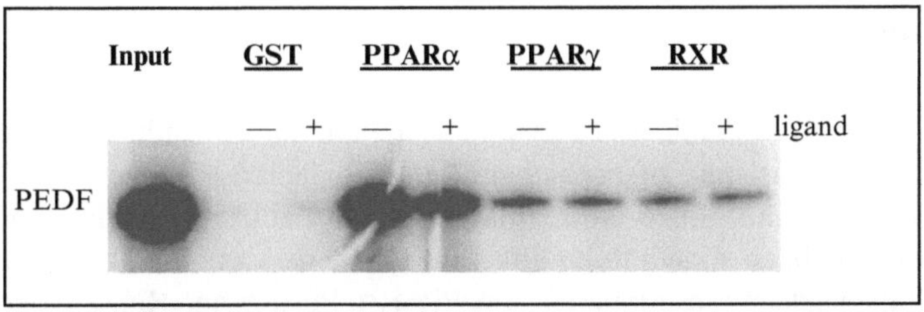

Fig. 3 GST pull-down assay demonstrates that PEDF binds PPARα. GST alone served as a control for nonspecific binding, and GST-fusion proteins of PPARα, PPARγ, and RXR were immobilized on GSH-Sepharose beads and incubated with [^{35}S]-methionine-labeled PEDF in binding buffer in the presence and absence of the appropriate ligands (ciprofibrate for PPARα, troglitazone for PPARγ, retinoic acid for RXR). In the presence of ciprofibrate, PEDF binding to PPARα decreases

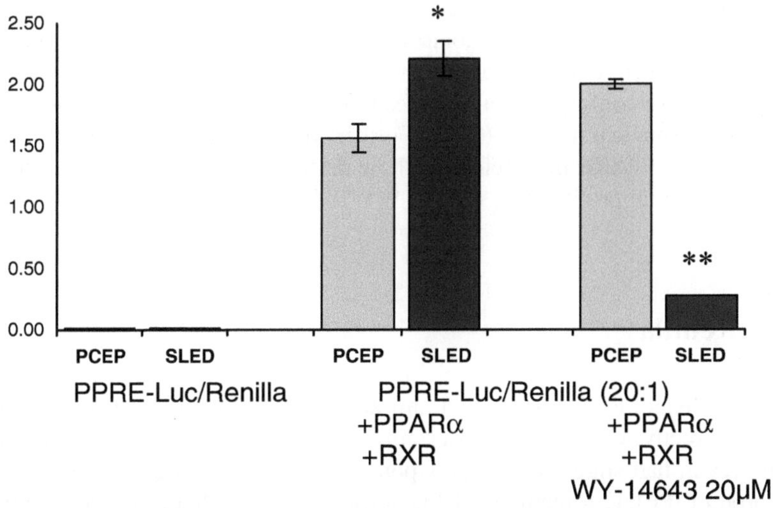

Fig. 4 Transactivation assay demonstrates that PEDF (SLED) activates the PPARα/RXR heterodimer compared with the control vector (PCEP). The presence of SLED increased luciferase activity by 42% (*p < 0.05) (n = 6). In the presence of the PPARα agonist, WY-14643, SLED activating function was significantly suppressed (**p < 0.01) (n = 3) compared with the PCEP treated group. PPRE-Luc represents the PPAR response element cloned upstream of the Firefly luciferase construct

0.12 to 2.21 ± 0.14 ($p < 0.05$), a 42% increase (Fig. 4). Treatment of the transfected cells treated with WY-14643 demonstrated activity similar to the SLED-treated animals. In the WY-14643-treated cells transfected with SLED, activity significantly decreased to 0.27 ± 0.01 ($p < 0.01$). The reasons for this decrease are unclear but may represent an allosteric conformational change that occurs upon binding of PPARα to its ligand. PEDF binding to PPARα was similarly reduced in the GST pull down assay in the presence of ligand.

Discussion

In this study, we show that loss of PEDF in an animal model results in a precancerous liver phenotype consisting of hepatocyte steatosis, increased cellular proliferation, and hepatocyte atypia. Since the regulation of fatty acid metabolism and cellular proliferation has been shown to be linked in HCC, we examined the interaction of PEDF with the nuclear transcription factors PPARα, PPARγ, and RXR. In our findings, PEDF interacted strongly with PPARα but minimally with PPARγ and RXR. In the presence of ligand, PEDF interactions with PPARα decreased suggesting an allosteric change to the PEDF binding site.

Transactivation assays confirmed that the *PEDF* gene, *SLED*, significantly increased the activity of the PPARα/RXR heterodimer more than that of the carrier plasmid PCEP. This activity was comparable to the activity found after the addition of the potent PPARα agonist, WY-14643. In the presence of both agonist and SLED, PPARα/RXR heterodimer activity decreased significantly. The mechanism by which the presence of a known PPARα agonist and PEDF diminishes the responsiveness of the PPARα/RXR heterodimer to activators is unknown. However, high affinity ligands for PPARα induce conformational changes to its secondary structure consistent with its function as a nuclear receptor (9–14). Further work delineating the precise binding sites of PEDF to PPARα may determine how these interactions alter the responsiveness to peroxisomal proliferators such as WY-14643.

Firefly/Renilla

To our knowledge, PEDF is the first anti-angiogenic protein that has been shown to interact directly with known transcription factors. Previous experimental experience, however, had suggested PEDF's potential function as a transcriptional regulatory protein. PEDF, a member of the *serpin* family without protease inhibitory function, has diverse functions including neuronal survival, angiogenesis, cell-cycle regulation, and tumor growth inhibition. Cristafalo showed that recombinant PEDF inhibits the ability of rapidly dividing fibroblasts to undergo DNA synthesis while addition of neutralizing antibodies vs. PEDF was permissive for S-phase entry (2). Previously, our group noted that PEDF's antitumor properties were multiple with PEDF promoting the survival of cells known to increase its own production, while increasing apoptosis of tumor cells (2). Our current results suggest that PEDF likely binds other transcription factors that are involved in control of the cell cycle since PPARα is considered a stimulator of cellular proliferation. The structure of PEDF lends support to the proposed novel function as a regulator of transcriptional proteins. PEDF contains a putative nuclear localization signal (NLS) that is conserved across multiple species (2). Future experiments involving mutagenesis of its NLS may discern PEDF's functional interactions.

Acknowledgments CMX-mPPARα and tk-PPRE-Luc constructs were a generous gift from Dr. David Mangelsdorf (UT-Southwestern). RXR and *Renilla*-luciferase constructs were a generous gift from Dr. James Boyer (Yale University). This work was supported by the American Liver Foundation Liver Scholar Award, Hemophilia of Georgia, NIH Digestive Diseases Research Core Center DK34989 (C.C); NCI 64329 (S.C.)

References

1. Yahagi N, Shimano H, Hasegawa K, et al. (2005) Co-ordinate activation of lipogenic enzymes in hepatocellular carcinoma. Eur J Cancer;41:1316–22.
2. Tombran-Tink J, Barnstable CJ (2003) PEDF: a multifaceted neurotrophic factor. Nat Rev Neuro;4:628–36.
3. Chader GJ (2001) PEDF: raising both hopes and questions in controlling angiogenesis. Proc Natl Acad Sci USA;98:2122–44.
4. Steele FR, Chader GJ, Johnson LV, et al. (1993) Pigment epithelium-derived factor: neurotrophic activity and identification as a member of the serine protease inhibitor gene family. Proc Natl Acad Sci USA;15:1526–30.
5. Becerra SP (1997) Structure-function studies on PEDF. A noninhibitory serpin with neurotrophic activity. Adv Exp Med Biol;425:223–37.
6. Karakousis PC, John SK, Behling KC, et al. (2001) Localization of pigment epithelium derived factor (PEDF) in developing and adult human ocular tissues. Mol Vis;7:153–63.
7. Ramirez-Castillejo C, Sanchez-Sanchez F, Andreu-Agullo C, et al. (2006) Pigment epithelium-derived factor is a niche signal for neural stem cell renewal. Nat Neuro;9:331–39.
8. Dawson DW, Volpert OV, Gillis P, et al. (1999) Pigment epithelium-derived factor: a potent inhibitor of angiogenesis. Science;285:245–48.
9. Aoyama, T, Peters, J, et al. (1998) Altered constitutive expression of fatty acid-metabolizing enzymes in mice lacking the peroxisome proliferator-activated receptor alpha (PPARα). J Biol Chem;273:5678–84.
10. Reddy JK (2004) Peroxisome proliferators and peroxisome proliferator-activated receptor alpha: biotic and xenobiotic sensing. Am J Pathol;164:2305–21.
11. Doll JA, Stellmach VM, Bouck NP, et al. (2003) Pigment epithelium-derived factor regulates the vasculature and mass of the prostate and pancreas. Nat Med 2003;9:774–80.
12. Hostetler, H, Petrescu, AD, et al. (2005) Peroxisome proliferator-activated receptor alpha interacts with high affinity and is conformationally responsive to endogenous ligands. J Biol Chem;280:18667–82.
13. Pignolo, RJ, Francis MK, et al. (2003) Putative role for EPC-1/PEDF in the G0 growth arrest of human diploid fibroblasts. J Cell Physiol;195:12–20.
14. Crawford SE, Stellmach VM, et al. (2001) epithelium-derived factor (PEDF) in neuroblastoma: a multifunctional mediator of Schwann cell antitumor activity. J Cell Sci;114:4421–8.

Hormone Metabolism

Estrogenic Activity of the Equine Estrogen Metabolite, 4-Methoxyequilenin

Minsun Chang, Cassia R. Overk, Irida Kastrati, Kuan-wei Peng, Ping Yao, Zhi-Hui Qin, Pavel Petukhov, Judy L. Bolton, and Gregory R.J. Thatcher

Summary Oxidative metabolism of estrogens has been associated with genotoxicity. *O*-methylation of catechol estrogens is considered as a protective mechanism. 4-Methoxyequilenin (4-MeOEN) is the *O*-methylated product of 4-hydroxyequilenin (4-OHEN). 4-OHEN, the major catechol metabolite of the equine estrogens present in the most widely prescribed hormone replacement therapeutics, causes DNA damage via quinone formation. In this study, estrogen receptor (ERα) binding of 4-MeOEN was compared with estradiol (E_2) and equilenin derivatives including 4-BrEN using computer modeling, estrogen response element (ERE)-luciferase induction in MCF-7 cells, and alkaline phosphatase (AP) induction in Ishikawa cells. 4-MeOEN induced AP and luciferase with nanomolar potency and displayed a similar profile of activity to E_2. Molecular modeling indicated that MeOEN could be a ligand for ERα despite no binding being observed in the ERα competitive binding assay. Methylation of 4-OHEN may not represent a detoxification pathway, since 4-MeOEN is a full estrogen agonist with nanomolar potency.

Introduction

Prolonged exposure to estrogens has been associated with an increased risk of breast (BC) and endometrial cancers through early menarche, late menopause, or hormone replacement therapy (HRT) (1–3). In animal models, estrogens have been proven to induce mammary, pituitary, cervical, and uterine tumors (4). Although the molecular mechanisms underlying estrogen-associated carcinogenesis are not completely established, the ERα is seen as a prime mediator in mechanisms of hormonal carcinogenesis. Estrogenic (agonist) activity at the ERα leads to cellular proliferation commonly seen as providing increased opportunity for accumulation of the genetic damage that can initiate carcinogenesis and for proliferation of preneoplastic and neoplastic cells (5).

Estrogens are oxidatively metabolized to form catechols, which are subject to conjugative metabolism including formation of methoxy-estrogens by catechol-*O*-methyltransferase (COMT) (6). Oxidative metabolism of estrogens generates genotoxic, electrophilic, and redox active intermediates, whereas conjugative

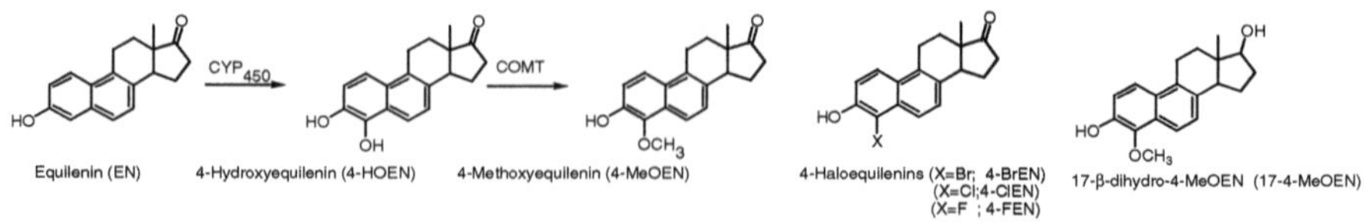

Scheme 1 Oxidative metabolism pathway from EN to 4-MeOEN, and structures of EN derivatives used in this study

metabolism is considered as a detoxification pathway (7). In contrast to endogenous estrogens, the equine estrogens are predominantly hydroxylated at the 4-position, yielding catechols that are more reactive and cytotoxic than their human estrogen homologs. The catechol, 4-hydroxyequilenin (4-OHEN), is the major phase I oxidative metabolite of the equine estrogens, equilenin and equilin, which constitute 50% of the estrogens in the most widely prescribed HRT formulations (8). 4-OHEN autoxidizes to an *o*-quinone (Scheme 1), while conjugative metabolism of 4-OHEN gives the stable methyl ether metabolite, 4-methoxyequilenin (4-MeOEN), which in addition to 17β-dihydro-4-methoxy-equilenin (17-4-MeOEN) was identified in BC cells (Scheme 1) (9, 10). 4-OHEN has also been reported to induce estrogenic effects in cell cultures (11), although the chemical instability of 4-OHEN hinders study of its hormonal actions. A question to be asked is whether catechol *O*-methylation represents a detoxification pathway, or these metabolites are themselves proliferative or mitotic agents able to elicit responses via ER or other relevant receptors. The potential capacity of equine estrogen metabolites to function as hormonal carcinogens, added to their known ability to function as chemical carcinogens, indicates that metabolic bioactivation contributes to carcinogenesis.

4-MeOEN has shown to be a proliferative agent in estrogen-sensitive BC and uterine cancer cells, to activate gene expression via the ERE, but to be an extremely weak ligand for both ERα and ERβ as determined by radioligand competitive binding assays. These results are relevant to the carcinogenic potential of equine estrogen bioactivation and hint at the complexity of ER mediated mechanisms of estrogenic action.

Methods

Materials. All chemicals and reagents were purchased from Sigma (St. Louis, MO) or Fisher Scientific (Itasca, IL) unless stated otherwise. EN derivatives were synthesized as described elsewhere (9, 12, 13).

Computational Methods. All molecular modeling studies were performed on an SGI computer with the SYBYL 7.2 software packages. The coordinates for the ERα ligand binding domain (LBD) were extracted from the co-crystal structure data of the complex between ERα LBD and E_2 (PDB code: 1ERE). The active site was designated to consist of the amino acid residues within a radius of 6.5 Å from the original ligand E_2. Glu353, Arg394, and His524 were set as a core subpocket. The following FlexX-Pharm settings were used to restrict the binding of the ligands to the E_2 binding site: Glu353 is an optional hydrogen bond acceptor.

Estrogenic Compounds. Equilenin and its 4-substituted analogs (4-FEN, 4-ClEN, 4-BrEN,) used as ligands because of availability of ER binding affinity data. After ligand docking was performed by the FlexX and Flex-Pharm modules in SYBYL, the 30 best poses were selected for each ligand and saved for analysis by CScore. We used a combination of several functions and the criterion of consensus ≥3 to select the best pose for each ligand. The binding modes of the docked ligands were viewed and found consistent with those expected and observed for estrogens. The selection of the binding poses was based on FlexX score. Scoring and ranking of the poses using just the CScore scoring function performed poorly. Once other empirical functions were incorporated such as Lipo Score, Clash Score, and Max Volume overlap score, the performance was improved. Thus, a combination of these functions was used for further docking experiments. The two most meaningful poses for each ligand were cominimized with the receptor LBD using the Tripos Force Field and Gasteiger-Huckel charges. The minimized pose was reevaluated with CScore functions.

Cell Culture. All cell culture reagents were purchased from Invitrogen (Carlsbad, CA) unless stated otherwise. MCF-7 human breast cancer cells were obtained from American Type Culture Collection (Manassas, VA) and the clonal variant MCF-7 WS8 cells were a kind gift from Dr. V.C. Jordan, Fox Chase Cancer Center, Philadelphia, PA, and maintained in RPMI 1640 containing 10% fetal bovine serum (FBS; Atlanta Laboratory, Atlanta, GA), 2 mM L-glutamine, 0.1 mM nonessential amino acids, and 6 μg ml^{-1} bovine insulin (Sigma). Estrogen-free media were prepared by supplementing 3× dextran-coated charcoal-treated FBS to phenol-red free RPMI 1640 media while other components remained the same. The Ishikawa cell line was provided by Dr. R.B. Hochberg (Yale University, New Haven, CT) and maintained in Dulbecco's Modified Eagle medium (DMEM/F12) containing 1% sodium pyruvate, 1% nonessential amino acids (NEAA), 1% glutamax-1, 0.05% insulin, and 10% heat-inactivated FBS. A day prior to treating the cells, the medium was replaced with phenol red-free DMEM/F12 medium containing charcoal/dextran-stripped FBS and supplements.

Transient Transfection and Luciferase Assays. Cells were cultured in estrogen-free media for 4 days before transfection. The cells were transfected with 2 μg of the pERE-luciferae plasmid, which contains three copies of the *Xenopus laevis* vitellogenin A2 ERE upstream of fire fly luciferase (a gift from Dr. V.C. Jordan) (14). To normalize transfection efficiency, pRL-TK plasmid (1 μg, Promega) was cotransfected. Cells (5×10^6) in serum-free media were transfected by electroporation in a 0.4-cm cuvette (Bio-Rad Laboratories) at a voltage of 0.320 kV and a high capacitance of 950 μF in a GenePulser X-cell (Bio-Rad Laboratories). The cells were resuspended in estrogen-free media, transferred to 12-well plates immediately after electroporation, and incubated overnight. The cells were treated with the appropriate compounds for 24 h. The luciferase activities in cell lysates were measured using Dual Luciferase Assay system (Promega) with a FLUOstar OPTIMA (BMG LABTECH, Durham, NC). Data are reported as relative luciferase activity, which is the fire fly luciferase reading divided by the *Renilla* luciferase reading.

Induction of Alkaline Phosphatase in Ishikawa Cells. The procedure of Pisha et al. was used as described previously (15). Briefly, Ishikawa cells (5×10^4 cells ml^{-1}) were plated and incubated overnight with estrogen-free media in 96-well plates, test compounds or vehicle were added and cells were incubated for a further 4 days before assay of alkaline phosphatase (AP) activity.

Results and Discussion

Docking Analysis. E_2 is anchored in the hydrophobic core of ERα LBD by hydrogen bonding interactions. The phenolic hydroxyl of the A-ring makes hydrogen bonding contacts to the carboxylate of Glu 353, guanindinium group of Arg 394, and a water molecule. The 17β-OH makes hydrogen bonding contacts with His 524. Recognizing the importance of the first hydrogen bond cluster, we imposed the FlexPharm constrain. As a result, the poses closely mimicked E_2 binding, striving for hydrogen bonding interactions with Glu353 and Arg 394. All ligands including the test compounds 4-OHEN and 4-MeOEN reached for this interaction.

A major distinction between E_2 and the EN series is the B-ring; EN has an aromatic B-ring (naphthalene core), imposing more rigidity and planarity that further extends to the C-ring. Besides the entropic gain from the rigidity of the B-ring, and the favorable hydrophobic contacts, there is the possibility for an edge-to-face interaction between the AB-rings and Phe 401 residue. The angle between the plane defined by the AB-rings and Phe401 phenyl ring is 97° and the distance is less than 3 Å, indicative of such a stabilizing electrostatic interaction. There is a strong hydrophobic component driving binding to ERα, because of the hydrophobic lining of the ERα binding site (Fig. 1a), thus supporting the decision to include functions such as Lipo Score, Clash Score, and Max Volume overlap score in the overall scoring.

The flexibility potential map (Fig. 1b), derived from the crystal structure, shows that the portion of the ligand-binding site in proximity to Glu 353 and Arg 394 is relatively rigid. This would not automatically translate into rigidity of the physiological receptor, but there is some indication of the higher ordering in this portion of ERα LBD and clearly there is limited space for the bulkier 4-methoxy group. Yet, from the FlexX docking result where the ligand backbone is relaxed, but the protein rigid binding of 4-MeOEN falls in between the scores of 4-ClEN and 4-BrEN, translating into a nanomolar ligand for ERα based upon experimental ERα ligand-binding affinities (ligand ranking for ERα affinity: 4-FEN > 4-OHEN > 4-ClEN > 4-MeOEN > 4-BrEN). After relaxation of the protein in cominimization using the Tripos force field, the binding cavity relaxed to accommodate the methoxy substituent for 4-MeOEN (Fig. 1a) improving the docking score of 4-MeOEN to

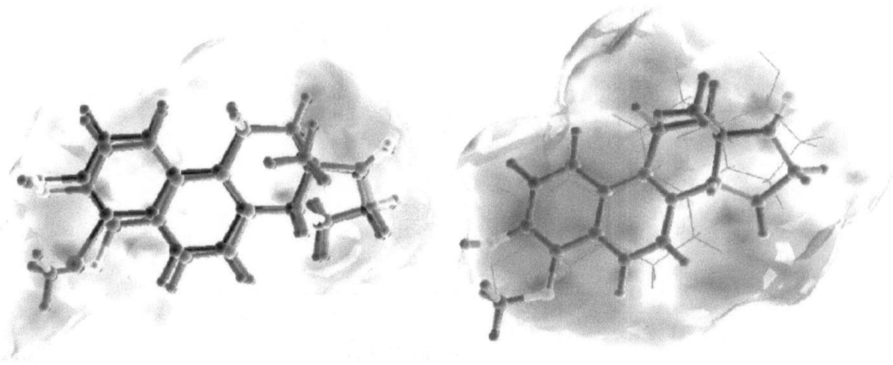

Fig. 1 (*Left*) 4-BrEN and 4-OMeEN docked and cominimized with the LBD of ERα using the Tripos force field; the interior surface of the LBD binding site is *shaded* showing distortion to accommodate the 4-MeO group. (*Right*) 4-OMeEN and E_2 (*stick structure*) docked in the LBD of ERα using the structure of E_2 cocrystal without relaxation of protein structure. The interior of the binding site is *shaded*

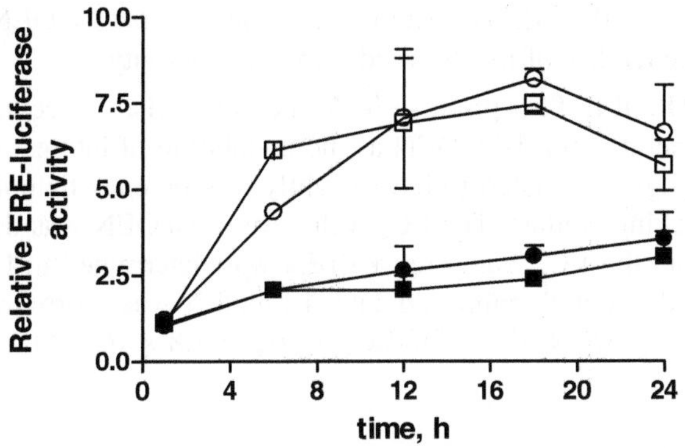

Fig. 2 Time-course of induction of ERE-luciferase in MCF-7 cells by 4-MeOEN and E_2. *Open square* E_2 (1 nM); *open circle* 4-MeOEN (100 nM); *closed square* E_2 (0.02 nM); *closed circle* 4-MeOEN (5 nM)

better than that of 4-OHEN. The modification of the protein in this cominimization perturbs helix 3 and helix 6, but has no effect on helix 12 (Fig. 1a). This perturbed structure obtained in silico is unlikely to have physical meaning, since presumably the overall stability of ERα would be compromised, but it does emphasize that there is some flexibility in the ER LBD to accommodate structures such as 4-MeOEN.

Activation of an ERE-luciferase Reporter in MCF-7 Cells. To compare the estrogenic activity of 4MeOEN to that of E_2, the time course of ERE-dependent luciferase induction was examined at 1, 6, 12, 18, and 24 h in MCF-7 cells (Fig. 2). No significant differences were seen between the time-course for induction by

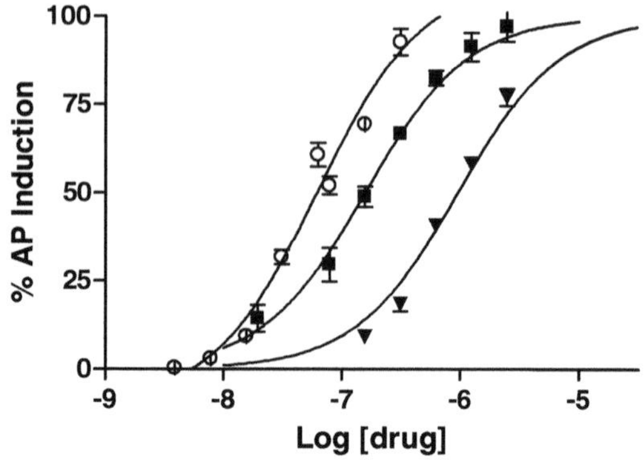

Fig. 3 AP induction in Ishikawa cells. Concentration-response curves for 17-4-MeOEN (*filled triangles*); 4-MeOEN (*filled squares*); 4-BrEN (*open circles*)

4-MeOEN (100 nM) and E_2 (1 nM), or between 4-MeOEN (1 nM) and E_2 (10 pM). Full concentration–response curves were obtained for 4-MeOEN and E_2 in the MCF-7 WS8 clonal variant. The EC_{50} values obtained for 4-MeOEN and E_2 were 5×10^{-9} M and 2×10^{-11} M. The nanomolar potency of 4-MeOEN quantitatively confirmed the generality of its observed estrogenic activity.

Induction of Alkaline Phosphatase in Ishikawa Cells. Concentration–response curves were generated for 4-MeOEN as the metabolite of interest and 4-BrEN, as a synthetic analogue predicted to have slightly less potency than 4-MeOEN from the in-silico docking studies. The EC_{50} value for 4-MeOEN was determined to be 6.2×10^{-8} M and the EC_{50} value for 4-BrEN was determined to be 1.6×10^{-7} M (Fig. 3) in accordance with published data. 4-MeOEN was approximately 2.5 times more potent than 4-BrEN; E_2 had an EC_{50} value of 1.8×10^{-10} M.

Conclusion

The purpose of this study was to evaluate 4-MeOEN as an estrogenic compound using molecular modeling and in vitro assays. 4-MeOEN is a product of a pathway anticipated to detoxify the catechol 4-OHEN. Computational studies predicted that 4-MeOEN would be a ligand for ERα, but no binding to ERα was measurable in the standard competitive radioligand displacement assay preformed on full length recombinant hERα. However, 4-MeOEN demonstrated nanomolar potency in the Ishikawa AP induction assay and the ERE-luciferase assay in breast cancer cells. These results show that 4-MeOEN is not an inactive metabolite, but rather an estrogenic compound of nanomolar potency and a potential hormonal carcinogen. The interactions of equine estrogen metabolites with potential hormonal and chemical carcinogenic activity warrants further study.

References

1. Colditz GA (1998) Relationship between estrogen levels, use of hormone replacement therapy, and breast cancer. J. Natl. Cancer Inst., 90: 814–823.
2. Service, RF (1998) New role for estrogen in cancer? Science, 279: 1631–1633.
3. Grady D, Gebretsadik T, Kerlikowske, et al. (1995) Hormone replacement therapy and endometrial cancer risk: a meta-analysis. Obstet. Gynecol., 85: 304–313.
4. Liehr JG (1990) Genotoxic effects of estrogens. Mutation Res., 238: 269–276.
5. Santen RJ (2002) To block estrogen's synthesis or action: that is the question. J. Clin. Endocrinol. Metab., 87: 3007–3012.
6. Zhu BT, Conney AH (1998) Functional role of estrogen metabolism in target cells: review and perspectives. Carcinogenesis, 19: 1–27.
7. Zhu BT, Liehr JG (1996) Inhibition of catechol O-methyltransferase-catalyzed O-methylation of 2- and 4-hydroxyestradiol by quercetin. Possible role in estradiol-induced tumorigenesis. J. Biol. Chem., 271: 1357–1363.
8. Zhang F, Chen Y, Pisha E, et al. (1999) The major metabolite of equilin, 4-hydroxyequilin autoxidizes to an o-quinone which isomerizes to the potent cytotoxin 4-hydroxyequilenin-o-quinone. Chem. Res. Toxicol., 12: 204–213.
9. Spink DC, Zhang F, Hussain MM, et al. (2001) Metabolism of equilenin in MCF-7 and MDA-MB-231 human breast cancer cells. Chem. Res. Toxicol., 14: 572–581.
10. Yao J, Li Y, Chang M, et al. (2003) Catechol estrogen 4-hydroxyequilenin is a substrate and an inhibitor of catechol-O-methyltransferase. Chem. Res. Toxicol., 16: 668–675.
11. Liu X, Yao J, Pisha E, et al. (2002) Oxidative DNA damage induced by equine estrogen metabolites: role of estrogen receptor alpha. Chem. Res. Toxicol., 15: 512–519.
12. Rao PN, Somawardhana CW (1987) Synthesis of 2-methoxy and 4-methoxy equine estrogens. Steroids, 49: 419–432.
13. Liu X, Zhang F, Liu H, et al. (2003) Effect of halogenated substituents on the metabolism and estrogenic effects of the equine estrogen, equilenin. Chem. Res. Toxicol., 16: 741–749.
14. Catherino WH, Jordan VC (1995) Increasing the number of tandem estrogen response elements increases the estrogenic activity of a tamoxifen analogue. Cancer Lett., 92: 39–47.
15. Pisha E, Pezzuto JM (1997) Cell-based assay for the determination of estrogenic and anti-estrogenic activities. Methods Cell Sci., 19: 37–43.

Dynamics of Oxidative Damage at Early Stages of Estrogen-dependant Carcinogenesis

Jarek Kobiela, Tomasz Stefaniak, Jacek Krajewski, Beata Kalinska-Blach, Dorota Zuawa-Janicka, Andrzej Lachinski, Daniel Gackowski, Ryszard Olinski, Jerzy Nowak, Narcyz Knap, Barbara Lipinska, Zbigniew Sledzinski, and Michal Wozniak

Summary The objective of this study was to assess the dynamics of oxidative damage to cellular macromolecules such as proteins, lipids, and DNA under conditions of oxidative stress triggering early stages of estrogen-dependent carcinogenesis. A rodent model of carcinogenesis was used. Syrian hamsters were sacrificed after 1, 3, 5 h and 1 month from the initial implantation of 17β-estradiol (E_2). Matching control groups were used. Kidneys as target organs for E_2-mediated oxidative stress were excised and homogenized for biochemical assays. Subcellular fractions were isolated. Carbonyl groups (as a marker of protein oxidation) and lipid hydroxyperoxides were assessed. DNA was isolated and 8-oxodGuo was assessed. Electron paramagnetic resonance spectroscopy was used to confirm the results for lipid peroxidation.

Exposition to E_2 in rodent model leads to a damage of macromolecules of the cell, including proteins and DNA, but not lipids. Proteins appear to be primary target of the damage but are shortly followed by DNA. It has previously been speculated that protein peroxides can increase DNA modifications. This time sequence was observed in our study. Nevertheless, direct relation between protein and DNA damage still remains unsolved.

Introduction

Breast cancer (BC) is the most commonly diagnosed cancer among women worldwide (1). The role of estrogens in the development of estrogen-dependent human BC has been previously underlined but has not been precisely settled so far (2). Although not conclusive, ample biochemical evidence exists to suggest that estradiol (E_2) and/or estrogen metabolites play a role in breast cancer initiation and progression. Supposedly, estrogen receptor (ERα) interactions and/or catechol estrogens (CE) formation involves an oxidative stress-mediated pathway (2). Under the condition of oxidative stress various biomolecules are damaged. Those include proteins and lipids and most interestingly DNA. Oxidative damage to DNA is of particular importance since, contrary to proteins and lipids, entirely new molecules cannot be synthesized to replace the damaged ones. Although background levels of

J.J. Li et al. (eds.), *Hormonal Carcinogenesis V.*

oxidative damaged DNA exist, oxidative stress can lead to damage increase, which has been described in various pathological conditions, such as carcinogenesis, chronic inflammation/infection, ageing, neurodegenerative, and cardiovascular disease (3). In the case of estrogens, the DNA damage is being potentiated by proliferation stimulation, and that is why even quantitatively insignificant DNA damage may result in tumorigenesis (4).

The objective of this study was to assess the dynamics of oxidative damage to cellular macromolecules such as proteins and lipids under conditions of oxidative stress triggering early stages of estrogen-dependent carcinogenesis. Furthermore, the time sequence obtained would be used for possible explanation of interactions between the oxidized macromolecules i.e., proteins, leading to DNA oxidation.

Material and Methods

Chemicals. All reagents were of the highest grade commercially available and were used without further purification. 17β-E_2 was purchased from Sigma. Hormone pellets and suspensions were prepared instantly before administration to the animals as described before (5).

Animals. Male Syrian hamsters (60) aged 4 weeks, weighing ~80 g each were used in this study. After arrival, the animals were acclimatized for 1 week, and divided into four treatment groups: (A) 15 hamsters were implanted subcutaneously with 25 mg of E_2, (B) 30 hamsters were injected i.p. with 75 mg of E_2 (1 ml of 0.5% carboxymethylcellulose saline solution). (C) 15 animals were injected i.p. with the sole vehicle. All the animals were kept in the animal facility at room temperature, standard humidity, and 12 h day/night circadian cycle, fed with standard chow and were given water ad libitum. All the procedures concerning animals were approved by Local Ethical Committee (LEC) and performed according to the instructions authorized by LEC.

Experimental Design. The experiments were performed in two series: S (short) and U (ultra short). Animals of S series (25 mg of E_2 and controls) were sacrificed after 1 month from the initial implantation of E_2. The same number of animals of U series (75 mg of E_2 and controls) were sacrificed after 1, 3, and 5 h of the experiment. The kidneys as target organs for E_2-mediated oxidative stress were excised for further biochemical assays. The excised organs were immediately placed on ice.

Preparation of Homogenates and Isolation of Subcellular Fractions. Kidneys from animals of each treatment group were combined and a 20% homogenate was prepared to obtain the microsomal fraction.

Carbonyl Group Assessment. This was performed as described by Oliver et al. (6).

Lipid Hydroxyperoxide Assessment. Lipid hydroperoxides were determined according to Thomas and Poznansky (7).

DNA Isolation and 8-Oxodguo Assessment. DNA was isolated using the method as described earlier (8). Determination of 8-oxodGuo by means of HPLC/EC technique as described previously (9).

Electron Paramagnetic Resonance Spectroscopy and Spin Labeling Techniques. Characteristics of microsomal membrane polarity can be examined by 5-doxyl-stearic acid (5-DSA) by the order parameter S (10). This value varies from zero for totally anisotropic motion of the label molecule to one – when its motion is ideally anisotropic.

Statistical Analysis. Each assessment was performed three times and as the data were not significantly different, an average of the three measurements was calculated. The analysis was performed using Student's *t* tests with Bonferroni's correction.

Results

The weight of the testes was decided for a marker of estrogenization efficiency in series S and proved to be at least 12-fold lower in each of the animals treated with the estrogen when compared with the control. Assessment of lipid hydroperoxides and EPR spin labeling with 5-doxylstearic acid as lipid peroxidation induced perturbances of ordering of membrane phospholipids did not reveal statistically significant changes at any point of the experiment comparing to the control (Figs 1, 2).

Protein oxidation presented significant changes when compared with the controls. Levels of carbonyl groups increased as soon as after 1 h exposition to the hormone and progressed further after 3 h to decrease substantially after 5 h. However, the increase remained almost twofold after 1 month of the exposition (Fig. 3).

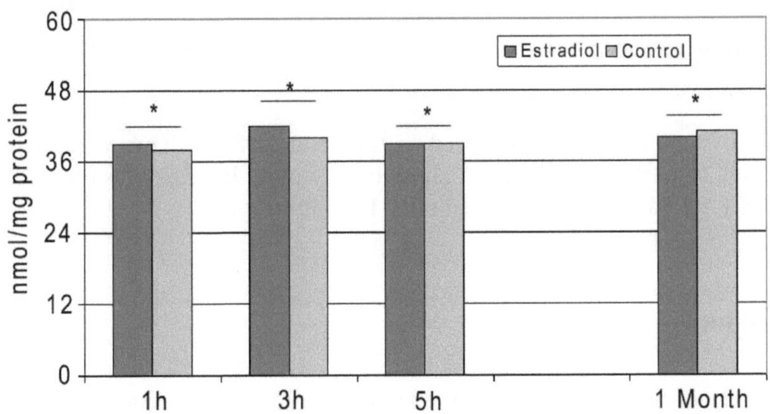

Fig. 1 Level of lipid hydroperoxides as markers of lipid peroxidation in hamster kidney homogenates

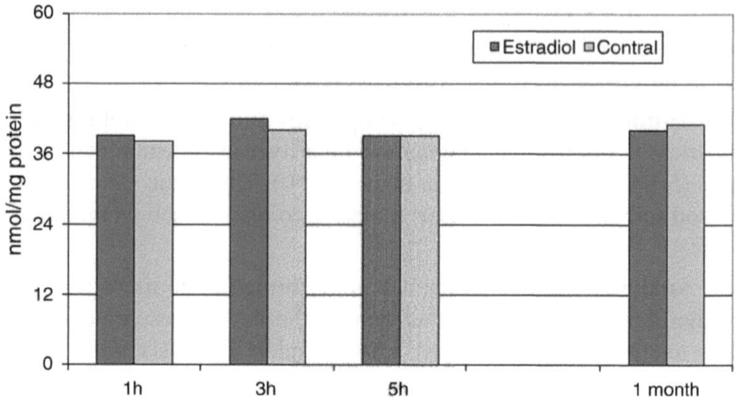

Fig. 2 Order parameter *S* in microsomal fraction of hamster kidney homogenates

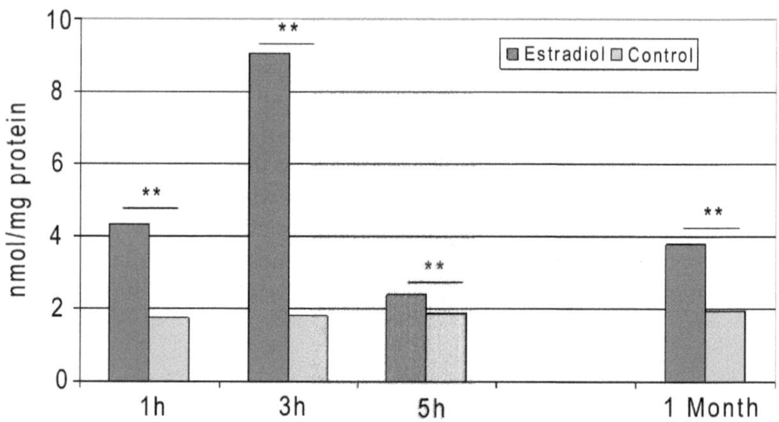

Fig. 3 Level of carbonyl groups as markers of protein oxidation in hamster kidney homogenates

DNA oxidative damage measured as the level of 8-oxodGuo was unchanged after 1 and 3 h of the experiment. A significant twofold increase was observed after 5 h and a 1.5-fold increase remained after 1 month (Fig. 4).

Discussion

In our study, during the acute exposition to E_2, there was a remarkable time-dependant limitation of both protein and DNA oxidative damage. These could be due counteraction activity of antioxidants such as GSH present in the cells, which

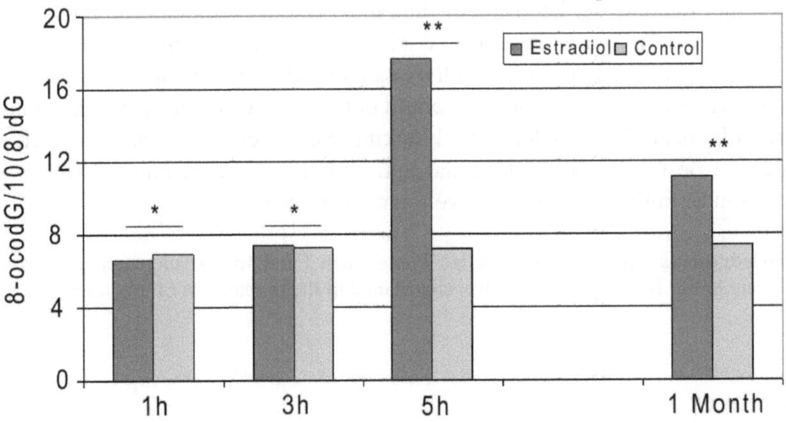

Fig. 4 8-oxodGuo levels as markers of DNA oxidation in hamster kidney homogenates

would be exploited to a certain point, but the severity of the oxidative stress crossed the protective capacity of intracellular or easily accessible extracellular antioxidants. Moreover there are several antioxidative defense and repair enzymes, just to mention glutathione-*S*-transferase, catalase, superoxide dysmutase (11, 12) or catechol *O*-methyl transferase (2), whose was reported to be increased (because of activation or induction) under the condition of oxidative stress and in several neoplasms.

When considered in terms of cell survival, DNA would be the ultimate target of oxidative stress and that implies two possible pathways depending on the severity of the damage and energetic status of the cell. Those two pathways would be cell death via apoptosis/necrosis or replication of the changes that would lead to carcinogensis/mutagensis. It should be remembered that the DNA damage also includes mitochondrial DNA (mtDNA).

In our study, the activity of DNA repair enzymes seemed effective enough to ensure the cell survival; however, it was insufficient to prevent any DNA damage under condition of protein oxidation. It should be added that 8-oxodGuo is not the only possible DNA alteration, as there are several different base modifications as well as strand breaks, apuric sites, and others, which are more or less susceptible to repairing mechanisms (13–15). The lifelong, free-radical derived accumulation of alterations in human tissue during aging is implicated both in degenerative and proliferative diseases. And thus, it may well be the major cause of these diseases, including hormonal carcinogenesis (16).

To conclude, exposition to E_2 in rodent model mimicking human breast carcinoma leads to formation of CE, which via redox cycling damage the macromolecules of the cell, including proteins and DNA, but not lipids. Proteins appear to be the primary target of the damage but are shortly followed by DNA. It could also be concluded that the protective and repairing DNA mechanisms, though initially active, are no longer efficient as the exposition to the carcinogen prolongates.

On the other hand, it has been established that protein peroxides can increase DNA modifications (17). This time sequence was observed in our study. Nevertheless, the direct relation between protein and DNA damage still remains unsolved. In practice, better understanding of this process could contribute to better molecular modeling of antioxidants aiming at selectively blocking the protein peroxidation. Such molecules could be used in the additional antioxidative therapy, parallel to hormonal substitution or contraception to prevent cancer induction.

Acknowledgments Authors acknowledge Pomeranian Fund for Development in Surgery for granting an Award N7/3/2007, which has contributed to the preparation of this manuscript.

References

1. American Cancer Society (2001) Cancer Facts and Figs. American Cancer Society, Atlanta, Georgia.
2. Mobley JA, Brueggemeier RW (2002) Increasing the DNA damage threshold in breast cancer cells. Toxicol Appl Pharmacol 180: 219–226.
3. Cooke MS, Olinski R, Evans MD (2006) Does measurement of oxidative damage to DNA have clinical significance? Clin Chim Acta 365: 30–49.
4. Malins DC, Holmes EH, Polisssar NL, et al. (1993) The etiology of breast cancer. Characteristic alterations in hydroxyl radical-induced DNA base lesions during oncogenesis with potential for evaluating incidence risk. Cancer 17: 3036–3043.
5. Roy D, Liehr JG (1999) Estrogen, DNA damage and mutations. Mutat Res 424:107–115.
6. Oliver CN, Ahn BW, Moerman EJ, et al. (1987) Age-related changes in oxidized proteins. J Biol Chem 262: 5488–5491.
7. Thomas PD, Poznansky MJ (1990) A modified tetramethylbenzidine method for measuring lipid hydroperoxides. Anal Biochem 188: 228–232.
8. Gackowski D, Kruszewski M, Bartlomiejczyk T, et al. (2002) The level of 8-oxo-7,8-dihydro-2'-deoxyguanosine is positively correlated with the size of the labile iron pool in human lymphocytes. J Biol Inorg Chem 7: 548–550.
9. Foksinski M, Bialkowski K, Skiba M, et al. (1999) Evaluation of 8-oxodeoxyguanosine, typical oxidative DNA damage, in lymphocytes of ozone-treated arteriosclerotic patients. Mutat Res 438: 23–27.
10. Schreier S, Polnaszek CF, Smith IC (1978) Spin labels in membranes. Problems in practice. Biochim Biophys Acta 515: 395–436.
11. Li Y, Huang TT, Carlson EJ, et al. (1995) Dilated cardiomyopathy and neonatal lethality in mutant mice lacking manganese superoxide dismutase, Nat Genet 11: 376–381.
12. Lebovitz RM, Zhang H, Vogel H, et al. (1996) Neurodegeneration, myocardial injury, and perinatal death in mitochondrial superoxide dismutase-deficient mice. Proc Natl Acad Sci USA 93: 9782–9787.
13. Chen Y, Shen L, Zhang F, et al. (1998) The equine estrogen metabolite 4-hydroxyequilenin causes DNA single-strand breaks and oxidation of DNA bases in vitro. Chem Res Toxicol 11:1105–1111.
14. Zhang F, Chen Y, Pisha E, et al. (1999) The major metabolite of equilin, 4-hydroxyequilin, autoxidizes to an *o*-quinone which isomerizes to the potent cytotoxin 4-hydroxyequilenin-*o*-quinone. Chem Res Toxicol 12: 204–213.
15. Shen L, Pisha E, Huang Z, et al. (1997) Bioreductive activation of catechol estrogen-ortho-quinones: Aromatization of the B ring in 4-hydroxyequilenin markedly alters quinoid formation and reactivity. Carcinogenesis 18: 1093–1101.

16. Friedberg EC, Walker GC, Siede W (1995) DNA Repair and Mutagenesis, American Society for Microbiology, Washington, DC.
17. Gebicki S, Gebicki JM (1999) Crosslinking of DNA and proteins induced by protein hydroperoxides. Biochem J 338: 629–636.

Gene Expression of 17β-Estradiol-metabolizing Isozymes: Comparison of Normal Human Mammary Gland to Normal Human Liver and to Cultured Human Breast Adenocarcinoma Cells

Leane Lehmann and Jörg Wagner

Summary Metabolic activation of 17β-estradiol (E_2) to catechols and quinones together with lack of deactivation constitute risk factors in human breast carcinogenesis. E_2-catchols are generated by cytochrome P450-dependent monooxygenases (CYPs). Deactivation of E_2, E_2-catechols, and E_2-quinones is mediated by UDP-glucuronosyltransferase (UGT), sulfotransferase (SULT), catechol-*O*-methyltransferase (COMT), glutathione-*S*-transferase (GST), and NADPH-quinone-oxidoreductase (QR) isozymes, respectively. The aim of the present study was to quantify mRNA levels of E_2-metabolizing isozymes expressed in MCF-7 cells cultured in the presence/absence of steroids by reverse transcription/competitive PCR in relation to the housekeeping gene hypoxanthine-guanine phosphoribosyltransferase and compare them with expression levels in normal human mammary gland (MG) and liver tissue. CYP1A1, 1B1, SULT1A1, 1A2, membrane-bound and soluble COMT, GSTT1, QR1, and UGT2B7 were detected in both tissues and MCF-7 cells; however, most enzymes were expressed at least tenfold higher in liver. Yet, CYP1B1 was expressed as high in breast as in liver and UGTs were not detected in MCF-7 cells cultured with steroids. MCF-7 cells cultured steroid-free additionally expressed CYP1A2 as well as UGT1A4, 1A8, and 1A9. Normal human liver but not MG expressed CYP1A2, 3A4, UGT1A1, 1A3, 1A4, 1A9, and SULT2A1. UGT1A8 was only detected in MCF7 cells but was not found in human liver. Thus, our study provides a comprehensive overview of expression levels of E_2-metabolizing enzymes in a popular in vitro model and in human tissues, which will contribute to the interpretation of in vitro studies concerning the activation/deactivation of E_2.

Introduction

Estrogen exposure is associated with an increased risk of breast cancer (BC) (1). Besides the induction of proliferation by estrogens such as E_2 (Fig. 1) that could favor tumor progression (2), 17β-estradiol (E_2) can be activated to genotoxic metabolites (3). Metabolic hydroxylation of E_2 by cytochrome P450-dependent monooxygenases (CYPs) generates the catechol estrogens 2-OH- and 4-OH-E_2 (4),

J.J. Li et al. (eds.), *Hormonal Carcinogenesis V.*

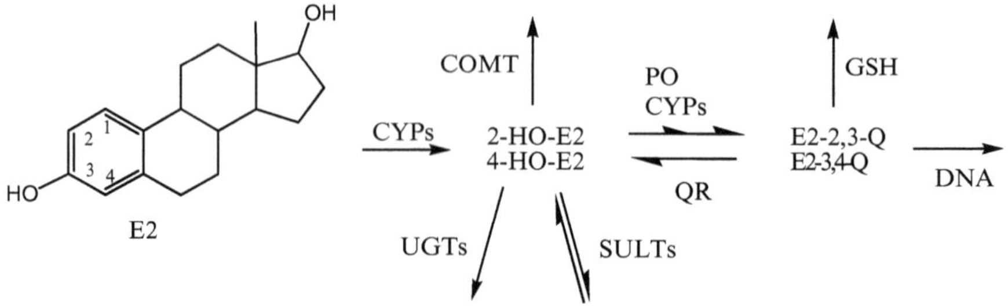

Fig. 1 Simplified overview of E_2 metabolism in the human MG; *PO* peroxidase

which can be further oxidized to the quinones E_2-2,3-quinone (E_2-2,3-Q) and E_2-3,4-quinone (E_2-2,3-Q). In breast tissue, the formation of 2-HO-E_2 is catalyzed by CYP1A1 isozyme (5), whereas the formation of 4-HO-E_2 is catalyzed by CYP1B1 isozyme. Further oxidation, which is catalyzed either by CYPs, peroxidase or non-enzymatically, generates the respective E_2 quinones. E_2 quinones may react with DNA to form adducts (3).

There are several metabolic pathways that lead to the inactivation of E_2 and E_2 metabolites (Fig. 1): E_2 and E_2 catechols constitute substrates for nearly all of the UDP-glucuronosyltransferase (UGT)1A isozymes as well as for UGT2B7, with UGT2B7 exhibiting the highest specific activity for 4-HO-E_2 > E_2 (6). However, only UGT1A3, 1A4, 1A8, and 2B7 have been detected in total mRNA of human normal MG tissue (7). Additionally, UGT1A8/9 were identified immunohistochemically in breast epithelium (8). UGT1A1 has been detected in various BC cell lines (9), however, not in total mRNA from normal MG (7). Furthermore, E_2 and E_2 catechols are substrates for sulfotransferase (SULT) 1A1, 1A3, 1E1, 2A1, all of which are reported to be substantially expressed in malignant and nonmalignant breast tissue (10). Yet, the affinity of SULT1E1 for E_2 (11) is at least 100 times higher than for any other SULT (12, 13), indicating a major role for SULT1E1 in the conjugation of physiologic concentrations of E_2 and catechol estrogens. Another common pathway of inactivation of E_2 catechols in the human MG occurs by *O*-methylation catalyzed by the ubiquitous catechol-*O*-methyltransferase (COMT), which preferably methylates 2-HO-E_2 (14). E_2 quinones can be reduced to the catechols by the NADPH-quinone-oxidoreductase (QR) (15) or are conjugated with glutathione by glutathione-*S*-transferases (GSTs) M1 and P1 (16, 17).

MCF-7 cells originate from a human MG adenocarcinoma; therefore, they represent a widespread and frequently used test system for the investigation of carcinogenesis in the breast. The relevance of in vitro studies is often questioned because of restricted transferability to the in vivo situation. To meet the requirements for an appropriate test system, key enzymes should be defined and characterized and compared with the in vivo situation. The aim of this study was to compare cDNAs from human tissue with that from differently cultured MCF-7 cells and thus clarify differences and analogies between them focusing on the gene expression of E_2-metabolizing enzymes.

Results

The present study provides an overview of the expression of enzymes involved in E_2 metabolism in commercially available mRNA of human liver and MG tissue as well as in MCF-7 BUS cells that where grown in complete medium (DMEM + 5% fetal calf serum (FCS) and in steroid-deprived medium (DMEM – phenol red, + 5% charcoal/dextran (CD)-treated FCS). To determine the relative gene expression of various enzymes involved in E_2 metabolism (Table 1), a competitive polymerase chain reaction (PCR) method using internal DNA competitors was applied using the housekeeping gene hypoxanthine-guanine phosphoribosyl-transferase (HPRT) for standardization.

To give an overview of the characteristic expression profile of each cDNA tested, isozymes of each enzyme family were summed up and the relative proportion of each enzyme family to all E_2-metabolizing enzymes was calculated (Fig. 2).

In human MG tissue and MCF-7 cells deprived of steroids, the single enzyme families contributed in the same order to the expression profile of E_2-metabolizing enzymes (Fig. 2): COMT (53%/92%) > CYPs (44%/7.4%) > GSTs (2.0%/7.2%) > QR1 (1.3%/1.5%) > SULTs (0.37%/1.5%) > UGTs (0.008%/0.005%, Fig. 2). MCF-7 cells cultured in the presence of complete medium expressed less COMT, resulting in the following succession (Fig. 2): CYPs (63%) > COMT (27%) > GSTs (0.56%) > QR1 (0.27%) > SULTs (0.057%). UGTs were below the detection limit in those MCF-7 cells.

In nearly all cDNAs originating from MG, the same isozymes of each enzyme family contributed most to the expression profile (Fig. 3); differences between human MG tissue and cultured cells were only observed with respect to GSTs.

Like in human MG tissue, normal human liver tissue expressed COMT (48%) > CYPs (30%) > GSTs (11%). In contrast to cDNAs with MG origin, the next frequent E_2-metabolizing enzyme family in liver was UGTs (5.2%) > SULTs (4.2) > QR-1 (1.0%, Fig. 3). Furthermore, the main CYP, SULT, and GST isozymes were different from those in breast-derived cells and tissue (Figs. 3 and 4).

Also the amounts of mRNA from different enzyme families in relation to the housekeeping gene *HPRT* were different in breast-derived cells and tissue and could be divided in to three classes (Fig. 3): (1) those with more than tenfold higher expression than HPRT, i.e. COMT and CYPs, (2) those between 0.1 and 10-fold

Table 1 Enzyme families and isozymes involved in the metabolism of E_2 and E_2 metabolites[a]

Enzyme family	Isozyme	Reference
CYP	1A1, 1A2, 1B1, 3A4	(5)
UGT	1A1, 1A3, 1A4, 1A8, 1A9, 2B7	(6–8)
SULT	1A1, 1A2, 2A1, 2E1	(10, 18)
COMT	s + mb	(14)
GST	M1, P1, T1	(16, 17)
QR	1	(15, 19)

[a] *s* soluble; *mb* membrane-bound

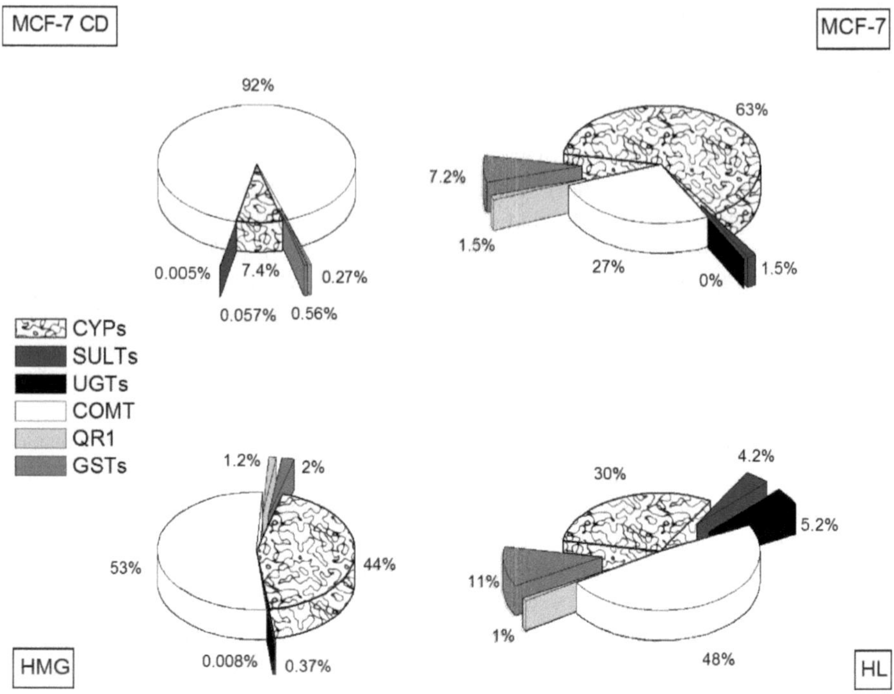

Fig. 2 Relative mRNA levels of E_2-metabolizing enzymes in MCF-7 cells cultured with steroid-depleted medium (MCF-7 CD) or complete FCS (MCF-7) and in normal human MG (HMG) and liver (HL) tissue

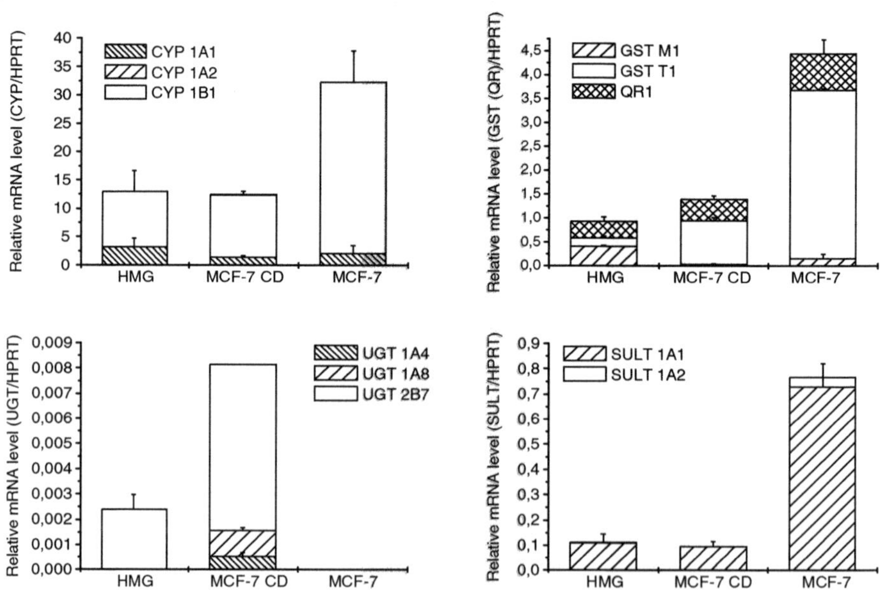

Fig. 3 Relative mRNA levels of CYP, UGT, SULT, GST, and QR isozymes known to metabolize E_2 or its metabolites in normal human mammary gland (HMG) tissue and in MCF-7 cells cultured with medium containing steroid-depleted FCS (CD-FCS) or full FCS. Data represent mean + SD of three independent determinations

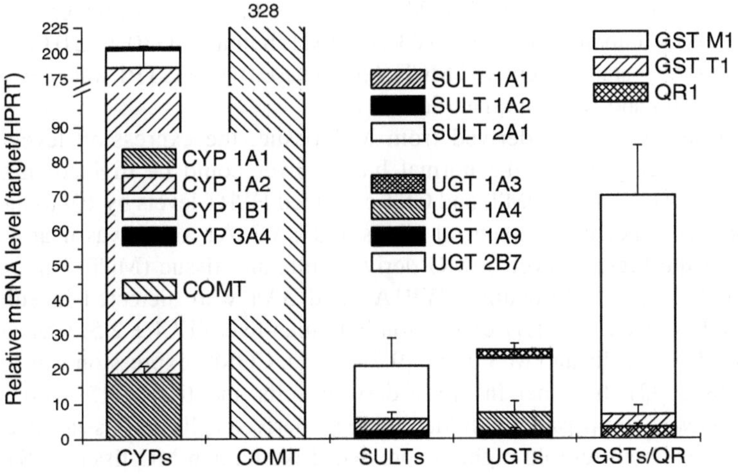

Fig. 4 Relative mRNA levels of CYP, COMT, UGT, SULT, GST, and QR isozymes known to metabolize E_2 or its metabolites in normal human liver tissue. Data represent mean ± SD of three independent determinations

expression of HPRT (i.e., QR-1, GSTs, and SULTs), and (3) those with less than 0.1-fold expression of HPRT (i.e., UGTs and CYP1A2, Fig. 3). SULT1E1 was detected in the cDNA of normal human MG tissue but below the quantification threshold (ten copies per reaction) and was therefore not included in Fig. 3.

In MG-derived mRNA, the COMT to HPRT relations were highest in steroid-deprived MCF-7 cells (160 ± 35), followed by MCF-7 cells grown in complete medium (35 ± 18), and normal human MG tissue (15 ± 3, data not shown). In all three cDNAs of breast origin, the relative amount of CYP1B1 mRNA was higher than that of CYP1A1. CYP1B1 expression was highest in MCF-7 cells grown in complete medium (30 ± 5) > steroid-deprived MCF-7 cells (11.5 ± 0.7) > MG tissue (9.7 ± 3.7). CYP1A2 was only detected in MCF-7 cells cultured with steroid-deprived medium and in a much lower amount than the other CYP isozymes (0.018 ± 0.003). In human MG tissue, only UGT2B7 (0.0024 ± 0.0006) was found. In steroid-deprived MCF-7 cells, thrice as much UGT2B7 (0.0073 ± 0.0006) and additionally UGT1A4 (0.00055 ± 0.00015) and 1A8 (0.0011 ± 0.0001) were detected (Fig. 3).

SULT1A1 was found in every kind of breast cDNA: MCF-7 cells grown in steroid-deprived medium had the highest relative amount of SULT1A1 (0.73 ± 0.09), followed by human MG tissue (0.11 ± 0.03) and MCF-7 cells cultured in steroid-deprived medium (0.09 ± 0.02). SULT1A2 expression was highest in MCF-7 cells grown in complete medium (0.039 ± 0.001) and relative SULT1A2 levels were much lower in human MG tissue (0.00143 ± 0.00003) and in steroid-deprived MCF-7 cells (0.0002 ± 0.0001).

QR-1, GSTT1, and GSTM1 were found in all three breast-derived cDNAs in varying amounts. QR1: MCF-7 (0.75 ± 0.29) > human MG (0.35 ± 0.08) ≈steroid-deprived

MCF-7 cells (0.47 ± 0.07); GSTT1: MCF-7 (3.53 ± 0.02) > steroid-deprived MCF-7 (0.94 ± 0.08) > human MG (0.17 ± 0.03); GSTM1: human MG (0.41 ± 0.03) > MCF-7 (0.15 ± 0.09) > steroid-deprived MCF-7 (0.038 ± 0.018). GSTP1 was only found in the human MG and was not quantified.

Likewise the cDNAs derived from MG tissue, the expression levels of E_2-metabolizing enzymes in the normal human liver could be divided in to three classes (Fig. 4): CYPs, COMT, and GST exceeded HPRT levels by 10- to more than 100-fold. As expected, the sum of CYPs and CYP1A1 alone was nearly tenfold more than the highest level in MG derived cells and tissue (MCF-7 in complete medium, Fig. 3). Additionally, CYP1A2 and 3A4 were detected in substantial amounts. However, CYP1B1 expression in human liver (17.4 ± 2.5) was comparable with MCF-7 cells and MG tissue (9.7 – 30.2). COMT expression was higher in liver (328 ± 82) than that in breast-derived cells and tissue (15–160). GSTT1 expression was comparable with that in breast-derived cells and tissue. In contrast, GSTM1 was about tenfold higher expressed. Like human MG tissue, GSTP1 was detected in human liver tissue but was not quantified.

Expression levels comparable to HPRT were observed for SULTS, UGTs, and QR. SULTs were expressed at least 20-fold higher than in breast-derived cells and tissues, the main isozyme was SULT2A1 (20 ± 8) > SULT1A1 (5.6 ± 2) > SULT1A2 (2.0 ± 0.2). In addition, SULT1E1 was detected in human liver tissue (not shown). The most striking difference between liver and MG was the UGT expression: the sum of UGT cDNA from human liver was 2 000-fold higher than in the maximum breast-derived cDNA from steroid-deprived cultured MCF-7 cells and more than 10 000-fold higher than in human MG tissue. UGT2B7 was expressed most, followed by UGT1A4, UGT1A3, and UGT1A9 (Fig. 4). UGT1A1 was detected qualitatively (not shown).

In conclusion, striking differences were observed between the various cDNAs derived from MG tissue and cultured MCF-7 cells. Normal human MG tissue had lower expression levels of E2-metabolizing enzymes than the MCF-7 cells, with the exception of UGT expression, which was completely absent in MCF-7 cultured with complete medium. Also the expression pattern of MG tissue and MCF-7 differed particularly in the UGT isozyme expression profile. Comparing the MCF-7 cells cultured with and without steroids, it can be concluded that MCF-7 cultured in the absence of steroids possessed a wider spectrum of isozymes, whereas the total expression level of enzymes were higher in MCF-7 cultured in complete medium. Compared with liver tissue, the mammary cells and tissue-derived cDNAs had a lower total expression of enzymes and a smaller number of different isozymes.

Discussion

In the present study, commercially available mRNA of the MG and liver tissue were derived from a single healthy donor each. The origin in only one donor represents a severe limitation due to high inter-individual as well as intra-individual variations in the expression of enzymes involved in biotransformation (i.e., actual hormone levels of the breast tissue donor are unknown). However, nearly all CYP, COMT, SULT, GST, and QR isozymes reported previously to be expressed in normal human MG tissue (5, 7, 10, 15, 16, 20) were detected; yet some (i.e., CYP1A2, SULT1A2, 1E1) at very low levels. In contrast, less UGT isozymes (only UGT2B7) were detected in the human MG mRNA than expected (UGT1A3, 1A4, and 1A8).

In the human liver cDNA, all expected isozymes were detected. Since only isozymes were chosen that have been reported to be involved in the activation/inactivation of E_2 and metabolites, the expression profile particularly of the human liver tissue is certainly incomplete and their % cannot be compared with published expression profiles of human liver. Yet, the present study focuses on the expression of the key enzymes involved in E_2 metabolism. Despite the abundance of studies concerning the expression of biotransformation enzymes, most focus on certain enzyme families, not specializing on the isozymes, involved in E_2 activation/inactivation. Because of the complex E_2 metabolism, a conclusive overview of the key players is important for the planning and interpretation of studies on E_2 activation. In particular, in vitro genotoxicity studies might gain significance by knowledge about the high influence of culture conditions on the expression of E_2-metabolizing enzymes.

Acknowledgment We thank A. Soto for providing the MCF-7 BUS cell line.

References

1. Cavalieri E, Chakravarti D, Guttenplan J, et al. (2006) Catechol estrogen quinones as initiators of breast and other human cancers: Implications for biomarkers of susceptibility and cancer prevention. Biochim Biophys Acta 1766(1): 63–78.
2. Henderson BE, Feigelson HS (2000) Hormonal carcinogenesis. Carcinogenesis 21: 427–433.
3. Cavalieri E, Chakravarti D, Guttenplan J, et al. (2006) Catechol estrogen quinones as initiators of breast and other human cancers: Implications for biomarkers of susceptibility and cancer prevention. Biochim Biophys Acta 1766: 63–78.
4. Spink DC, Spink BC, Cao JQ, et al. (1998) Differential expression of CYP1A1 and CYP1B1 in human breast epithelial cells and breast tumor cells. Carcinogenesis 19: 291–298.
5. Badawi AF, Cavalieri EL, Rogan EG (2001) Role of human cytochrome P450 1A1, 1A2, 1B1, and 3A4 in the 2-, 4-, and 16alpha-hydroxylation of 17beta-estradiol. Metabolism 50: 1001–1003.
6. Turgeon D, Carrier JS, Levesque E, et al. (2001) Relative enzymatic activity, protein stability, and tissue distribution of human steroid-metabolizing UGT2B subfamily members. Endocrinology 142: 778–787.
7. Chouinard S, Tessier M, Vernouillet G, et al. (2006) Inactivation of the pure antiestrogen fulvestrant and other synthetic estrogen molecules by UDP-glucuronosyltransferase 1A enzymes expressed in breast tissue. Mol Pharmacol 69: 908–920.
8. Thibaudeau J, Lepine J, Tojcic J, et al. (2006) Characterization of common UGT1A8, UGT1A9, and UGT2B7 variants with different capacities to inactivate mutagenic 4-hydroxylated metabolites of estradiol and estrone. Cancer Res 66: 125–133.

9. Guillemette C, Millikan RC, Newman B, et al. (2000) Genetic polymorphisms in uridine diphospho-glucuronosyltransferase 1A1 and association with breast cancer among African Americans. Cancer Res 60: 950–956.
10. Aust S, Obrist P, Klimpfinger M, et al. (2005) Altered expression of the hormone- and xenobiotic-metabolizing sulfotransferase enzymes 1A2 and 1C1 in malignant breast tissue. Int J Oncol 26: 1079–1085.
11. Nishiyama T, Ogura K, Nakano H, et al. (2002) Sulfation of environmental estrogens by cytosolic human sulfotransferases. Drug Metab Pharmacokinet 17: 221–228.
12. Harris RM, Waring RH, Kirk CJ, et al. (2000) Sulfation of "estrogenic" alkylphenols and 17beta-estradiol by human platelet phenol sulfotransferases. J Biol Chem 275: 159–166.
13. Wang LQ, James MO (2005) Sulfotransferase 2A1 forms estradiol-17-sulfate and celecoxib switches the dominant product from estradiol-3-sulfate to estradiol-17-sulfate. J Steroid Biochem Mol Biol 96: 367–374.
14. Ball P, Knuppen R (1980) Catecholoestrogens (2-and 4-hydroxyoestrogens): chemistry, biogenesis, metabolism, occurrence and physiological significance. Acta Endocrinol Suppl (Copenh) 232: 1–127.
15. Danson S, Ward TH, Butler J, et al. (2004) DT-diaphorase: a target for new anticancer drugs. Cancer Treat Rev 30: 437–449.
16. Albin N, Massaad L, Toussaint C, et al. (1993) Main drug-metabolizing enzyme systems in human breast tumors and peritumoral tissues. Cancer Res 53: 3541–3546.
17. Dawling S, Hachey DL, Roodi N, et al. (2004) In vitro model of mammary estrogen metabolism: structural and kinetic differences between catechol estrogens 2- and 4-hydroxyestradiol. Chem Res Toxicol 17: 1258–1264.
18. Aust S, Jaeger W, Klimpfinger M, et al. (2005) Biotransformation of melatonin in human breast cancer cell lines: role of sulfotransferase 1A1. J Pineal Res 39: 276–282.
19. Montano MM, Bianco NR, Deng H, et al. (2005) Estrogen receptor regulation of quinone reductase in breast cancer: implications for estrogen-induced breast tumor growth and therapeutic uses of tamoxifen. Front Biosci 10: 1440–1461.
20. Tenhunen J, Heikkila P, Alanko A, et al. (1999) Soluble and membrane-bound catechol-*O*-methyltransferase in normal and malignant mammary gland. Cancer Lett 144: 75–84.

Phytoestrogens Modulate the Expression of 17α-Estradiol Metabolizing Enzymes in Cultured MCF-7 Cells

Jörg Wagner, Ling Jiang, and Leane Lehmann

Summary The activation of 17β-estradiol (E_2) to 2-hydroxyestradiol (2-HO-E_2), the more genotoxic 4-hydroxyestradiol (4-HO-E_2), and the oxidation to the respective quinones constitutes a risk factor in hormonal carcinogenesis. 2-HO-E_2 is formed by cytochrome P450 CYP1A1, and 4-HO-E_2 is formed by CYP1B1. Both are detoxified by catechol-O-methyltransferase (COMT), whereas their quinones are inactivated by NADPH-quinone-oxidoreductase (QR). Since the soy isoflavones genistein (GEN) and daidzein (DAI) are widely consumed due to their putative protective function in breast carcinogenesis, we examined the influence of E_2, GEN, and DAI on CYP1A1/1B1, COMT, and QR expression in MCF-7 cells by reverse transcription/competitive PCR. CYP1A1 and COMT enzyme activity were determined using ethoxyresorufin and quercetin as substrates. Furthermore, estrogen receptor (ER)-regulated cell proliferation was determined by E-screen. E_2, GEN, and DAI inhibited the expression of CYP1A1, COMT, and QR. The maximum effect (reduction by 40–80%, depending on the gene product and compound) was obtained at 100 pM E_2, 1 μM GEN, and 10 μM DAI, which also induced the most pronounced cell proliferation in the E-screen. In contrast, expression of CYP1B1 was only slightly affected. CYP1A1 and COMT mRNA levels correlated with enzyme activities. The ER antagonist ICI 182,780 reversed the E_2- and isoflavone-mediated effects. Thus, GEN and DAI at estrogen-active concentrations stimulate the formation of the more E_2 genotoxic metabolites and inhibit the detoxification of catechol and quinone estrogens in estrogen-responsive tumor cells.

Introduction

Exposure to estrogens is commonly known to be associated with an increased risk of breast cancer (BC). Besides the induction of proliferation by estrogens such as 17β-estradiol (E_2, Fig. 1), which could favor tumor progression, E_2 can be activated to genotoxic metabolites (1). It has been hypothesized that homeostasis of tissue concentrations of E_2 and reactive metabolites may be influenced significantly by the balance between the activity of local estrogen-producing enzymes and those involved in conjugation and inactivation (2). In the mammary gland, hydroxylation

J.J. Li et al. (eds.), *Hormonal Carcinogenesis V.*

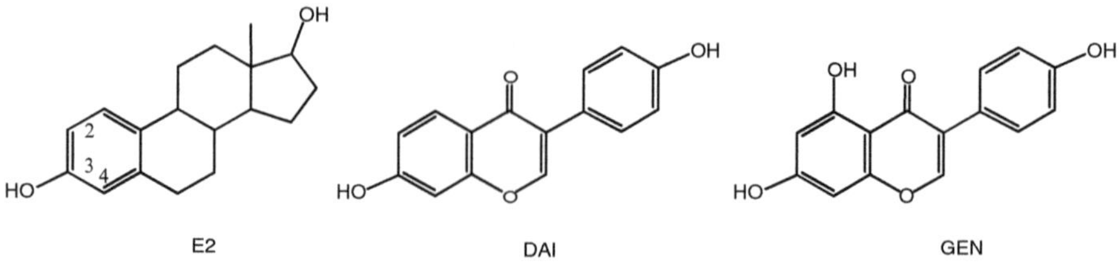

Fig. 1 Chemical structures of E_2, DAI, and GEN

of E_2 to 2- and 4-hydroxyestradiol (4-HO-E_2) is catalyzed predominately by CYP1A1 and CYP1B1, respectively (3, 4). The catechol estrogens can be further oxidized to the E_2-2,3-quinone (E_2-2,3-Q) and E_2-3,4-Q. E_2-2,3-Q yields stable adducts that might be subject to DNA repair processes, whereas DNA-adducts from E_2-3,4-Q with adenine quickly depurinate, and thus generate a highly promutagenic DNA lesion. In general, E_2 as well as catechol estrogens are inactivated by conjugating reactions, such as glucuronidation, sulfonation, and *O*-methylation. The inactivation of E_2 catechols in the human mammary gland is catalyzed by the ubiquitous catechol-*O*-methyltransferase (COMT), which preferably methylates 2-hydroxyestradiol (2-HO-E_2) (5), which is therefore inactivated more efficiently than 4-HO-$E_2$2. If conjugation of catechol estrogens via *O*-methylation, sulfonation, or glucuronidation becomes deficient in the mammary gland, competitive catalytic oxidation of 4-HO-E_2 to E_2-3,4-Q can occur. Protection at the quinone level takes place by conjugation with glutathione, catalyzed by glutathione-*S*-transferase (GST) isozymes (6, 7). A second inactivating process for estrogen quinones is their reduction to catechol estrogens by NADPH-quinone oxidoreductase (QR) type 1 (8). If these two inactivating processes are not effective, estrogen quinones may contribute to the initiation of cells (1). Because of the lower BC incidence of Asian women consuming a soy-based diet, the search for food constituents providing chemoprevention against BC soon focused on soy isoflavones, especially genistein (GEN, Fig. 1) and daidzein (DAI, Fig. 1) (9). Despite an abundance of studies concerning the effects of GEN, DAI in vivo and in vitro, the mechanisms of the putative chemoprevention by isoflavones remain elusive and the safety of high-dosed food supplements containing GEN and DAI still needs to be clarified (9–11).

Human breast tumors exhibit higher tissue concentration of 4-HO-E_2, an abnormal ratio of 4- to 2-HO-E_2 and less *O*-methylated products than normal breast tissue (12). Furthermore, CYP1B1 is higher expressed and COMT and QR1 are less expressed in breast carcinoma tissue compared with normal female breast tissue (13). Either interindividual genetic differences in the metabolic pathways of E_2, i.e., by genetic polymorphisms of E_2 metabolizing enzymes (14) or the direct interaction of endogenous and exogenous compounds with E_2-metabolizing enzymes (15) may cause this difference. Also, the gene expression of key proteins in E_2 metabolism might be modulated by exogenous factors, i.e., diet. Therefore, the aim of the present study was to investigate the influence of E_2, GEN, und DAI on the levels of

CYP1A1, CYP1B1, COMT, and QR mRNA in cultured MCF-7 cells by reverse transcription/competitive quantitative polymerase chain reaction using internal DNA standards using the ER-independent housekeeping gene hypoxanthine-guanine phosphoribosyltransferase (HPRT) for standardization. To assess the physiological relevance of changes of mRNA levels, ethoxyresorufin-*O*-deethylase (EROD) activity, which can be predominately attributed to CYP1A1 in MCF-7 cells, was determined after treatment with the estrogens for 48 h by means of the conversion of ethoxyresorufin to the fluorescent dye resorufin (16). COMT activity was determined by HPLC analysis of the methylation products of the model substrate quercetin. As reference for the stimulation of the estrogen receptor (ER), estrogen-induced cell proliferation was determined by the E-screen (17, 18) using flow cytometric quantification of cell nuclei.

Results

The extent of steroid depletion in the culture medium and the sensitivity of the MCF-7 BUS cells to stimulation of the ER was assessed by E-screen: only slight proliferation is expected for the cells of the solvent control and maximum stimulation of proliferation is expected at 100 pM E_2 (18). Therefore, each RNA expression experiment was accompanied by an E-screen using the same cell suspension and incubation medium as for the main experiments. RNA was isolated from every main experiment and RNAs from experiments with satisfying results of the E-screen (Fig. 2) were analyzed by RT-PCR.

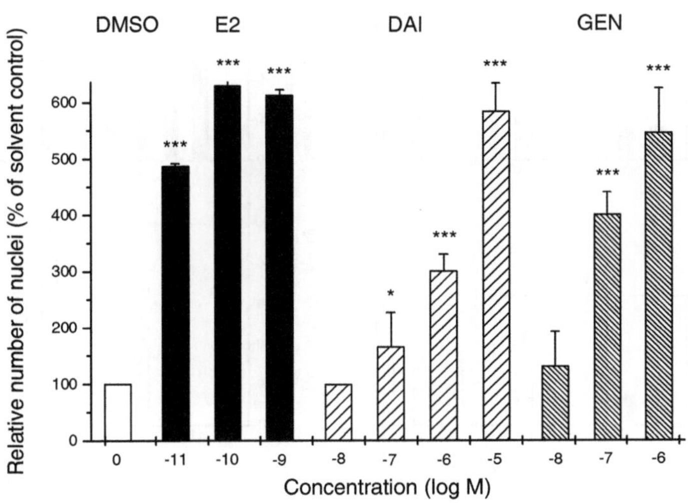

Fig. 2 Relative number of nuclei after treatment of MCF-7 cells with E2, DAI, and GEN for 6 days. 100%, number of cells treated with solvent (0.1% DMSO) only. Data are mean ± SD of three determinations. $^*p \leq 0.05$; $^{***}p \leq 0.001$ (Student's t test)

After treatment of MCF-7 BUS cells with 100 pM E_2, 10 μM DAI, and 1 μM GEN for 24 and 48 h, respectively, the relative mRNA levels of CYP1A1 as well as EROD activities were reduced significantly compared with cells kept in steroid-depleted medium (Fig. 3). The reduction of mRNA levels amounted to 26.5 ± 5.9% (100 pM E_2), 22.0 ± 3.6% (1 nM E_2), 50.6 ± 27.8% (10 μM DAI), and 44.0 ± 17.2% (1 μM GEN) of that of the solvent control. In concordance with the mRNA data, EROD activities were reduced to 26.5 ± 5.9% (100 pM E_2), 22.0 ± 3.6% (1 nM E_2), 50.6 ± 27.8% (10 μM DAI), and 44.0 ± 17.2% (1 μM GEN) of that of the solvent control. Treatment with tetrachloro[p]dibenzodioxine (TCDD) at 10 nM increased CYP1A1 mRNA levels about 100-fold and EROD activities more than 16-fold. In contrast to CYP1A1 mRNA, relative mRNA levels of CYP1B1 were not affected significantly after treatment of MCF-7 BUS cells with 10–1 000 pM E_2, 100 nM and 10 μM DAI, and 10 nM and 1 μM GEN for 24 h (data not shown).

After treatment of MCF-7 BUS cells with various concentrations of E_2, DAI, and GEN for 24 and 48 h, respectively, the relative mRNA levels of COMT were significantly reduced compared with cells kept in steroid-depleted medium (Fig. 3).

The reduction of mRNA levels amounted to 50.0 ± 11.3%, 26.7 ± 13.3%, and 19.9 ± 1.4% (10, 100, and 1 000 pM E_2), 51.0 ± 5.9% and 33.3 ± 13.6% (1 and 10 μM DAI), and 24.2 ± 33.4% (1 μM GEN) of that of the solvent control. In comparison with mRNA levels, the reduction of COMT activities was less pronounced

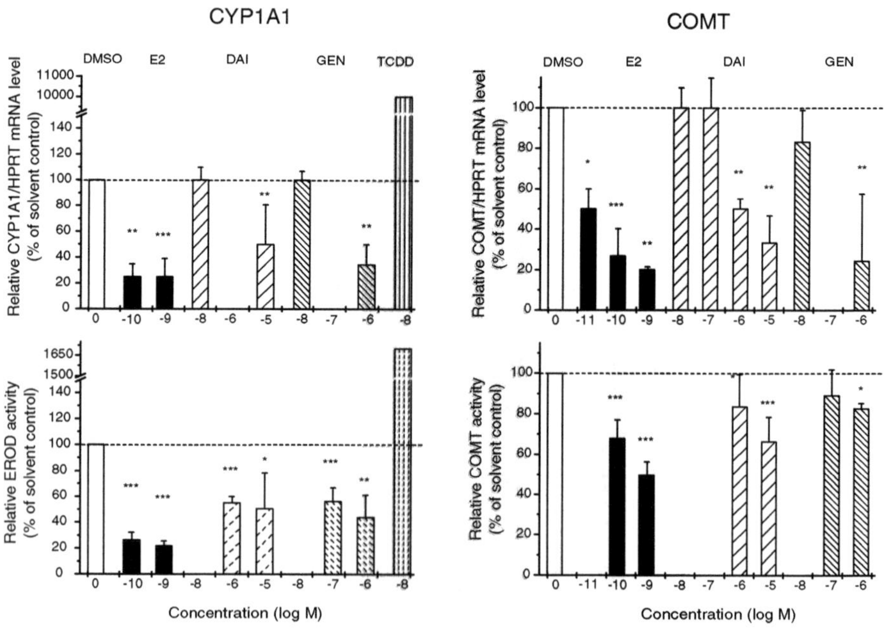

Fig. 3 Relative mRNA levels (*upper panel*) and enzyme activities (*lower panel*) of CYP1A1 (*left panel*) and COMT (*right panel*) after treatment of MCF-7 BUS cells with E_2, GEN, and DAI for 24 h (mRNA) or 48 h (activity). Data are mean ± SD of three independent experiments. $*p \le 0.05$; $**p \le 0.01$; $***p \le 0.001$ (Student's *t* test)

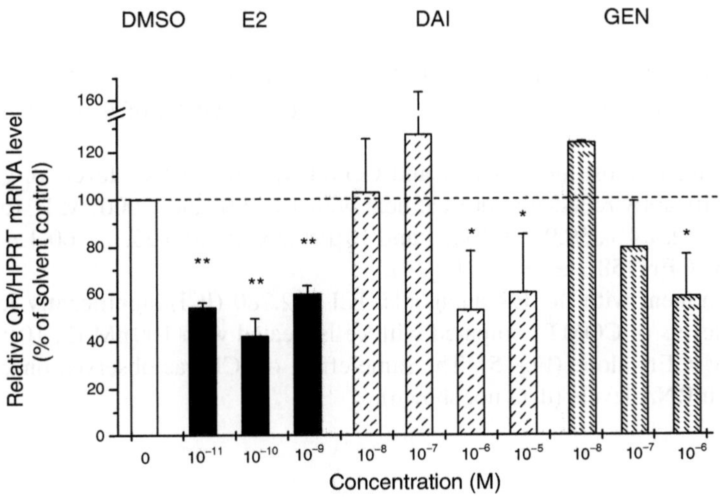

Fig. 4 Relative mRNA levels of QR after treatment of MCF-7 BUS cells with E_2, GEN, and DAI for 24 h. Data are mean ± standard deviation of three independent experiments. $^{*}p \leq 0.05$; $^{**}p \leq 0.01$; $^{***}p \leq 0.001$ (Student's *t* test)

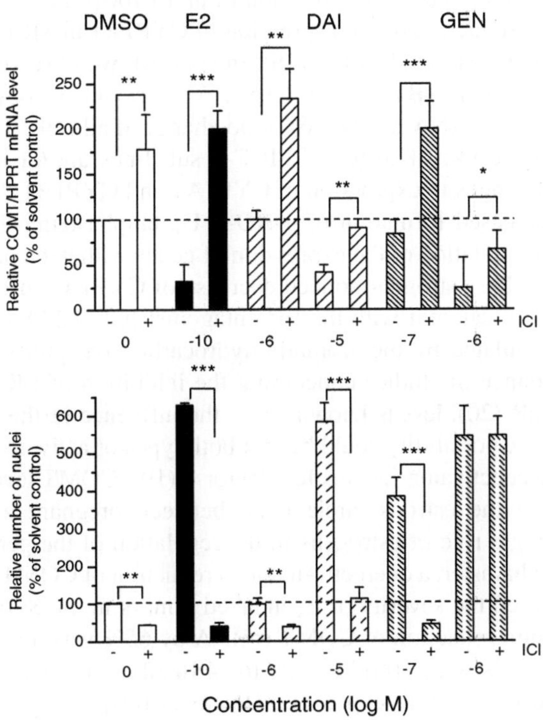

Fig. 5 Relative mRNA levels of COMT (*upper panel*) and relative number of cells (*lower panel*) after treatment of MCF-7 BUS cells with E_2, GEN, and DAI in the absence (–) and presence (+) of 10 nM ICI for 24 h. 100% (*line*), number of cells treated with solvent (0.1% DMSO) only. Data are mean ± SD of three determinations. $^{*}p \leq 0.05$; $^{**}p \leq 0.01$; $^{***}p \leq 0.001$ (Student's *t* test)

with 67.7 ± 9.2 and 49.6 ± 6.7% (100 and 1000 pM E_2), 83.7 ± 15.8% and 66.4 ± 12.1% (1 and 10 μM DAI), and 82.5 ± 2.8% (1 μM GEN) of that of the solvent control.

Likewise to those of CYP1A1 and COMT, relative mRNA levels of QR were also significantly reduced after treatment with 10, 100, and 1 000 pM E_2 (to 54.3 ± 2.0%, 40.0 ± 7.4%, 59.9 ± 3.7%), 1 and 10 μM DAI (52.9 ± 25.1%, 60.4 ± 24.6%), and 1 μM GEN (58.3 ± 18.1%, Fig. 4).

Cotreatment with the ER antagonist ICI 182,780 (ICI) significantly increased mRNA levels of COMT compared with cells treated with 100 pM E_2, 10 μM DAI, and 1 μM GEN alone (Fig. 5). The same effect of ICI was observed on CYP1A1 and QR mRNA levels (data not shown).

Discussion

The present study demonstrates a reduction of the expression of CYP1A1 and COMT in MCF-7 cells on the mRNA as well as on the enzyme activity level by E_2, GEN, and DAI. The effective concentration of up to 100 pM E_2 was in the physiological range. The reduction of the expression of CYP1A1 in MCF-7 cells exposed to 1 nM E_2 has been described earlier (16). In contrast, we observed only an insignificant increase in CYP1B1 mRNA levels, whereas others observed either an increase of CYP1B1 mRNA levels (16) or no change at all (19). These differences might be due to the use of different MCF-7 sub lines and/or different culture conditions. The changes in expression of CYP1A1 and CYP1B1 described by (16) resulted in an increased formation of 4-HO-E E_2, emphasizing the physiological relevance of the modulation of the expression of genes with gene products involved in E_2 metabolism. The estrogen-induced decrease of CYP1A1 mRNA levels could be prevented by cotreatment with the ER antagonist ICI 182,780. The expression of CYP1A1 is regulated by the aromatic hydrocarbon receptor (AhR). Although there is an abundance of studies concerning the inhibition of ER-dependent gene expression by AhR (20), less is known about the influence of the activated ER on AhR signaling. One possibility could be that both types of activated receptors compete for the same coactivator, i.e., nuclear factor-1 (19). COMT levels are known to vary according to the estrous circle (21), between pregnant and nonpregnant women, suggesting a role of estrogens in the regulation of the gene expression of COMT. In Syrian hamster, a quercetin-induced reduction of COMT activity by only about 30% increased the severity of E_2-induced tumors in the Syrian hamster kidney (22). Maximum reduction of COMT mRNA by 65% was observed after treatment of MCF-7 cells with 100 nM E E_2 for 48 h (23). In contrast, we observed maximum reduction of COMT expression already at 100 pM E E_2 and no difference between cells treated for 24 and 48 h (data not shown). Since the MCF-7 cell line is inhomogeneous, the choice of the MCF-7 sub line as well as the culture conditions and the kind of serum influence the outcome of the experiments (17, 24). Also the expression of QR has been reported to be increased by antiestrogens and decreased by E_2 (25). Besides the ER, the binding of transcription factor Nrf to antioxidant responsive elements might be involved in the regulation of the gene expression of QR (26). After consumption of soy-derived food supplements, plasma concentrations of total (free + conjugated) GEN and DAI in the low micromolar range can be achieved (27) and despite the main proportion of the isoflavones is conjugated (i.e., ineffective) hormonal responses such as variations in the estrous cycle can be observed in women consuming soy or isoflavone supplements (28). On the basis of these observations, women chronically consuming high doses of isoflavones might undergo changes in the metabolism of E_2. However, the present study is limited to a tumor cell line. Further studies using normal mammary gland tissue will yield further insight in the influence of soy isoflavones on the metabolism of E_2. In conclusion, the present study indicates that estrogen-active concentrations of GEN und DAI stimulate the formation of the more genotoxic catechol metabolite of E_2 but inhibit the detoxification of both catechol metabolites in estrogen-responsive tumor cells.

Acknowledgments The authors thank M. Lehmann and Eric Lerch for their help in the laboratory and A. Soto for providing the MCF-7 BUS cell line. Supported by DFG, Grant Me574/21–1.

References

1. Cavalieri E, Chakravarti D, Guttenplan J, et al. (2006) Catechol estrogen quinones as initiators of breast and other human cancers: implications for biomarkers of susceptibility and cancer prevention. Biochim Biophys Acta 1766: 63–78.
2. Guillemette C, Belanger A, Lepine J (2004) Metabolic inactivation of estrogens in breast tissue by UDP-glucuronosyltransferase enzymes: an overview. Breast Cancer Res 6: 246–254.
3. Badawi AF, Cavalieri EL, Rogan EG (2001) Role of human cytochrome P450 1A1, 1A2, 1B1, and 3A4 in the 2-, 4-, and 16alpha-hydroxylation of 17beta-estradiol. Metabolism 50: 1001–1003.
4. Zhu BT, Conney AH (1998) Functional role of estrogen metabolism in target cells: review and perspectives. Carcinogenesis 19: 1–27.
5. Ball P, Knuppen R (1980) Catecholoestrogens (2-and 4-hydroxyoestrogens): chemistry, biogenesis, metabolism, occurrence and physiological significance. Acta Endocrinol Suppl (Copenh) 232: 1–127.
6. Albin N, Massaad L, Toussaint C, et al. (1993) Main drug-metabolizing enzyme systems in human breast tumors and peritumoral tissues. Cancer Res 53: 3541–3546.
7. Dawling S, Hachey DL, Roodi N, et al. (2004) In vitro model of mammary estrogen metabolism: structural and kinetic differences between catechol estrogens 2- and 4-hydroxyestradiol. Chem Res Toxicol 17: 1258–1264.
8. Danson S, Ward TH, Butler J, et al. (2004) DT-diaphorase: a target for new anticancer drugs. Cancer Treat Rev 30: 437–449.
9. Adlercreutz H, Heinonen SM, Penalvo-Garcia J (2004) Phytoestrogens, cancer and coronary heart disease. Biofactors 22: 229–236.
10. Cornwell T, Cohick W, Raskin I (2004) Dietary phytoestrogens and health. Phytochemistry 65: 995–1016.
11. Limer JL, Speirs V (2004) Phyto-oestrogens and breast cancer chemoprevention. Breast Cancer Res 6: 119–127.

12. Rogan EG, Badawi AF, Devanesan PD, et al. (2003) Relative imbalances in estrogen metabolism and conjugation in breast tissue of women with carcinoma: potential biomarkers of susceptibility to cancer. Carcinogenesis 24: 697–702.
13. Singh S, Chakravarti D, Edney JA, et al. (2005) Relative imbalances in the expression of estrogen-metabolizing enzymes in the breast tissue of women with breast carcinoma. Oncol Rep 14: 1091–1096.
14. Mitrunen K, Hirvonen A (2003) Molecular epidemiology of sporadic breast cancer. The role of polymorphic genes involved in oestrogen biosynthesis and metabolism. Mutat Res 544: 9–41.
15. Zhu BT, Liehr JG (1996) Inhibition of catechol O-methyltransferase-catalyzed O-methylation of 2- and 4-hydroxyestradiol by quercetin. Possible role in estradiol-induced tumorigenesis. J Biol Chem 271: 1357–1363.
16. Spink DC, Katz BH, Hussain MM, et al. (2003) Estrogen regulates Ah responsiveness in MCF-7 breast cancer cells. Carcinogenesis 24: 1941–1950.
17. Payne J, Jones C, Lakhani S, et al. (2000) Improving the reproducibility of the MCF-7 cell proliferation assay for the detection of xenoestrogens. Sci Total Environ 248: 51–62.
18. Soto AM, Sonnenschein C, Chung KL, et al. (1995) The E-SCREEN assay as a tool to identify estrogens: an update on estrogenic environmental pollutants. Environ Health Perspect 103 (Suppl 7): 113–122.
19. Ricci MS, Toscano DG, Mattingly CJ, et al. (1999) Estrogen receptor reduces CYP1A1 induction in cultured human endometrial cells. J Biol Chem 274: 3430–3438.
20. Safe S, Wormke M, Samudio I (2000) Mechanisms of inhibitory aryl hydrocarbon receptor-estrogen receptor crosstalk in human breast cancer cells. J Mammary Gland Biol Neoplasia 5: 295–306.
21. Parvez S, Ismahan G, Raza-Bukhari A, et al. (1978) Activity of catechol-*o*-methyltransferase in brain regions and adrenal gland during the oestrus cycle. J Neural Transm 42: 305–312.
22. Zhu BT, Liehr JG (1994) Quercetin increases the severity of estradiol-induced tumorigenesis in hamster kidney. Toxicol Appl Pharmacol 125: 149–158.
23. Xie T, Ho SL, Ramsden D (1999) Characterization and implications of estrogenic down-regulation of human catechol-*O*-methyltransferase gene transcription. Mol Pharmacol 56: 31–38.
24. Rasmussen TH, Nielsen JB (2002) Critical parameters in the MCF-7 cell proliferation bioassay (E-Screen). Biomarkers 7: 322–336.
25. Montano MM, Bianco NR, Deng H, et al. (2005) Estrogen receptor regulation of quinone reductase in breast cancer: implications for estrogen-induced breast tumor growth and therapeutic uses of tamoxifen. Front Biosci 10: 1440–1461.
26. Lee JM, Anderson PC, Padgitt JK, et al. (2003) Nrf2, not the estrogen receptor, mediates catechol estrogen-induced activation of the antioxidant responsive element. Biochim Biophys Acta 1629: 92–101.
27. Setchell KD, Faughnan MS, Avades T, et al. (2003) Comparing the pharmacokinetics of daidzein and genistein with the use of 13C-labeled tracers in premenopausal women. Am J Clin Nutr 77: 411–419.
28. Kurzer MS (2002) Hormonal effects of soy in premenopausal women and men. J Nutr 132: 570S–573S.

Index

C

D

E

F

I

N

P

R

MIX
Papier aus verantwortungsvollen Quellen
Paper from responsible sources
FSC® C105338

If you have any concerns about our products,
you can contact us on
ProductSafety@springernature.com

In case Publisher is established outside the EU,
the EU authorized representative is:
Springer Nature Customer Service Center GmbH
Europaplatz 3, 69115 Heidelberg, Germany

Printed by Libri Plureos GmbH
in Hamburg, Germany